D1604429

Agrobacterium

Agrobacterium: From Biology to Biotechnology

Edited by

Tzvi Tzfira

Department of Molecular, Cellular and Developmental Biology
University of Michigan
Ann Arbor, MI, USA

and

Vitaly Citovsky

Department of Biochemistry and Cell Biology
State University of New York
Stony Brook, NY, USA

Editors
Tzvi Tzfira
Department of Molecular,
Cellular and Developmental Biology
University of Michigan
830 N. University Ave.
4071D Natural Science Building
Ann Arbor, MI 48109-1048
USA
ttzfira@umich.edu

Vitaly Citovsky
Department of Biochemistry and
Cell Biology
State University of New York
Stony Brook, NY 11792-5215
USA
vitaly.citovsky@stonybrook.edu

ISBN: 978-0-387-72289-4 e-ISBN: 978-0-387-72290-0

Library of Congress Control Number: 2007942364

Printed on acid-free paper.

9 8 7 6 5 4 3 2 1

springer.com

We dedicate this book to our families and Alla Z.

Contributing Authors

Roni Aloni
Department of Plant Sciences
Tel Aviv University
Tel Aviv 69978
Israel

Krishnamohan Atmakuri
Department of Microbiology and Molecular Genetics
University of Texas Medical School at Houston
6431 Fannin street
Houston, TX 77030
USA

Lois M. Banta
Department of Biology
Williams College
Williamstown, MA 01267
USA

Andrew N. Binns
Department of Biology and Plant Sciences Institute
University of Pennsylvania
Philadelphia, PA 19104
USA

Monica T. Britton
Department of Plant Sciences
University of California
One Shields Ave
Davis, CA 95616
USA

Thomas Burr
Department of Plant Pathology
Cornell University
Geneva, NY 14456
USA

Peter J. Christie
Department of Microbiology and Molecular Genetics
University of Texas Medical School at Houston
6431 Fannin street
Houston, TX 77030
USA

Vitaly Citovsky
Department of Biochemistry and Cell Biology
State University of New York
Stony Brook, NY 11794
USA

Abhaya M. Dandekar
Department of Plant Sciences
University of California
One Shields Ave
Davis, CA 95616
USA

Sylvie De Buck
Department of Plant Systems Biology
Flanders Interuniversity Institute for Biotechnology
Ghent University, Technologiepark 927
B-9052 Gent
Belgium

Ann Depicker
Department of Plant Systems Biology
Flanders Interuniversity Institute for Biotechnology (VIB)
Ghent University, Technologiepark 927
B-9052 Gent
Belgium

Michael Elbaum
Department of Materials and Interface
Weizmann Institute of Science
Rehovot 76100
Israel

Matthew A. Escobar
Department of Biological Sciences
California State University San Marcos
333 S. Twin Oaks Valley Road
San Marcos, CA 92096
USA

Clay Fuqua
Department of Biology
Indiana University
Bloomington, IN 47405
USA

Stanton B. Gelvin
Department of Biological Sciences
Purdue University
West Lafayette, IN 47907
USA

Barry S. Goldman
Monsanto Company
800 North Lindbergh Boulevard
St. Louis, MO 63167
USA

Brad W. Goodner
Hiram College
Department of Biology
Hiram, OH 44234
USA

G. Paul H. van Heusden
Department of Molecular and Developmental Genetics
Institute of Biology
Leiden University
Wassenaarseweg 64,
2333 AL Leiden
The Netherlands

Barbara Hohn
FMI for Biomedical Research
Maulbeerstrasse 66
CH-4058 Basel
Switzerland

Paul J.J. Hooykaas
Department of Molecular and Developmental Genetics
Institute of Biology
Leiden University
Wassenaarseweg 64,
2333 AL Leiden
The Netherlands

Benoît Lacroix
Department of Biochemistry and Cell Biology
State University of New York
Stony Brook, NY 11794
USA

Yi-Han Lin
Center for Fundamental and Applied Molecular Evolution
Departments of Chemistry and Biology
Emory University
Atlanta, GA 30322
USA

David G. Lynn
Center for Fundamental and Applied Molecular Evolution
Departments of Chemistry and Biology
Emory University
Atlanta, GA 30322
USA

Kathrine H. Madsen
Danish Institute of Agricultural Sciences
Research Centre Flakkebjerg
DK-4200 Slagelse
Denmark

Maywa Montenegro
Department of Biology
Williams College
Williamstown, MA 01267
USA

Eugene W. Nester
Department of Microbiology
University of Washington
Box 357242
Seattle, WA 98195-7242
USA

Carol Nottenburg
Cougar Patent Law
Seattle, WA
USA

Léon Otten
Institut de Biologie Moléculaire des Plantes
Rue du Général Zimmer 12
67084 Strasbourg
France

Walt Ream
Department of Microbiology
Oregon State University
Corvallis, OR 97331
USA

Carolina Roa Rodríguez
Australian National University
Canberra, ACT
Australia

Peter Sandøe
Royal Veterinary and Agricultural University
Rolighedsvej 25
DK-1958 Frederiksberg C
Denmark

João C. Setubal
Virginia Bioinformatics Institute and Department of Computer Science
Virginia Polytechnic Institute and State University Bioinformatics 1, Box 0477
Blacksburg, VA 24060-0477

Steven C. Slater
Arizona State University
The Biodesign Institute and Department of Applied Biological Sciences
7001 E. Williams Field Road
Mesa, AZ 85212
USA

Jalal Soltani
Department of Molecular and Developmental Genetics
Institute of Biology
Leiden University
Wassenaarseweg 64,
2333 AL Leiden
The Netherlands

Ernö Szegedi
Research Institute for Viticulture and Enology
P.O. Box 25, 6000
Kecskemét
Hungary

Nobukazu Tanaka
Center for Gene Science
Hiroshima University
Kagamiyama 1-4-2
Higashi-Hiroshim
Hiroshima 739-8527
Japan

Tzvi Tzfira
Department of Molecular, Cellular and Developmental Biology
The University of Michigan
Ann Arbor, MI 4810
USA

Cornelia I. Ullrich
Institute of Botany,
Darmstadt University of Technology
Schnittspahnstr. 3
64287 Darmstadt
Germany

Catharine E. White
Department of Microbiology
Cornell University
Ithaca, NY 14853
USA

Stephen C. Winans
Department of Microbiology
Cornell University
Ithaca, NY 14853
USA

Pieter Windels
Pioneer Hi-Bred International
Avenue des Arts 44
B-1040 Brussels
Belgium

Derek W. Wood
Department of Biology
Seattle Pacific University
3307 3rd Ave W., suite 205
Seattle, WA 98119
USA

John M. Young
Landcare Research
Private Bag 92170
Auckland
New Zealand

Alicja Ziemienowicz
Department of Molecular Genetics
Faculty of Biochemistry, Biophysics and Biotechnology
Jagiellonian University
Gronostajowa 7
30-387 Krakow
Poland

Table of Contents

Chapter 2 A brief history of research on *Agrobacterium tumefaciens*: 1900-1980s

Chapter 3 *Agrobacterium* and plant biotechnology

Chapter 5 *Agrobacterium*—taxonomy of plant-pathogenic *Rhizobium* species

Chapter 4 The *Agrobacterium tumefaciens* C58 genome

Chapter 10 Intracellular transport of *Agrobacterium* T-DNA

Chapter 11 Mechanisms of T-DNA integration

Chapter 12 *Agrobacterium tumefaciens*-mediated transformation: patterns of T-DNA integration into the host genome

Chapter 13 Function of host proteins in the *Agrobacterium*-mediated plant transformation process

Chapter 14 The oncogenes of *Agrobacterium tumefaciens* and *Agrobacterium rhizogenes*

Chapter 15 Biology of crown gall tumors

Chapter 16 The cell-cell communication system of *Agrobacterium tumefaciens*

Chapter 17 Horizontal gene transfer

Chapter 18 *Agrobacterium*-mediated transformation of non-plant organisms

Chapter 19 The bioethics and biosafety of gene transfer

Chapter 20 ***Agrobacterium*****-mediated gene transfer: a lawyer's perspective**

Preface

The bacterial origin of crown gall tumors was recognized a hundred years ago; 70 years later, stable integration of bacterial DNA in the crown gall cells was discovered, positioning *Agrobacterium* as the only cellular organism on Earth that is naturally capable of transferring genetic material between the kingdoms of life, from prokaryotes to eukaryotes. Since then, *Agrobacterium* has faithfully served plant biologists in a uniquely dual role: as a primary tool for genetic engineering, for both industrial and research applications, and as an extremely useful experimental system for studies on a wide range of basic biological processes, such as cell-cell recognition, cell-to-cell transport, nuclear import, assembly and disassembly of protein-DNA complexes, DNA recombination, and regulation of gene expression, within plant cells. These studies have uncovered a wealth of information on the process of *Agrobacterium*-mediated genetic transformation and on the bacterial and host cell factors involved in the infection. Furthermore, *Agrobacterium* has been shown to genetically transform, under laboratory conditions, numerous non-plant species, from fungi to human cells, indicating the truly basic nature of the transformation process. It is therefore not surprising that *Agrobacterium* and its ability to produce genetically modified organisms has also become the focus of numerous ethical and legal debates. These aspects of *Agrobacterium* research—its history, application, basic biology, and sociology, are reviewed in the present book. We begin the book with a description of the crown gall disease that initially attracted scientists' attention to this microorganism, followed by a historical essay on highlights of the first 70 years of *Agrobacterium* research. The book continues with a description of how *Agrobacterium* is used as a tool in plant biotechnology. The next two chapters describe our knowledge of the *Agrobacterium* genome gained with the advent of genomics approaches and place *Agrobacterium* in the taxonomic context of related bacterial species. The main portion of the book, which comprises 11 chapters, provides a detailed review of virtually all molecular aspects of the genetic transformation process, including chemistry, biochemistry and molecular biology of host recognition and attachment, production of the transferable DNA molecule (T-DNA) and secretion of this DNA—together with bacterial protein effectors—into the host cell, transport of the invading bacterial DNA-protein complex (T-complex) through the host-cell cytoplasm into the nucleus and targeting to the host chromatin, and mechanisms and patterns of T-DNA integration into the host genome. Special attention is paid to a description of the host factors involved in the transformation process, and the biology of the crown gall disease and bacterial oncogenes that cause these neoplastic growths. The next two chapters focus on interactions of *Agrobacterium* with non-plant species, from communication with its sister agrobacteria to fungi, algae, and mammalian cells, and on horizontal gene transfer from *Agrobacterium* to plants. The final two chapters of the book discuss ethical and safety issues associated with the use of *Agrobacterium* for interspecies gene transfer and look at the legal issues surrounding patents that involve *Agrobacterium.* The result is a comprehensive book which we hope the readers will find useful as a reference source for all major—biological, ethical, and legal—aspects of the *Agrobacterium*-mediated genetic transformation of plant and non-plant organisms.

Tzvi Tzfira

July 2007, Ann Arbor

Vitaly Citovsky

July 2007, New York

Color Figures

Acknowledgments

We would like to express our sincere gratitude to all of the authors for their outstanding contributions, and to the staff of Springer Life Sciences for their help and patience during the book's production. Special thanks are reserved for Vardit Zeevi, who was highly instrumental in preparing the book for print.

Chapter 1

AGROBACTERIUM: A DISEASE-CAUSING BACTERIUM

Léon Otten[1], Thomas Burr[2] and Ernö Szegedi[3]

[1]Institut de Biologie Moléculaire des Plantes, Rue du Général Zimmer 12, 67084 Strasbourg, France; [2]Department of Plant Pathology, Cornell University, Geneva, NY 14456, USA; [3]Research Institute for Viticulture and Enology, P.O. Box 25, 6000 Kecskemét, Hungary

Abstract. The common use of *Agrobacterium* as a gene vector for plants has somewhat obscured the fact that this bacterium remains an important plant pathogen. Pathogenic strains of the genus *Agrobacterium* cause unorganized tissue growth called crown gall or profuse abnormal root development called hairy root. *Agrobacterium tumefaciens* induces galls on roots and crowns of several fruit and forest trees and ornamental plants. *A. vitis* is responsible for the crown gall disease of grapevine, while *A. rhizogenes* induces abnormal rooting on its hosts. Plants tissues that become diseased undergo physiological changes resulting in weak growth, low yields or even death of the entire plant. Tumors originate from dividing plant cells, e. g. from cambium. Thus the cambial region becomes unable to differentiate into efficient phloem and xylem elements leading to deficient nutrient transport. Symptoms may appear on roots, crowns and aerial parts of attacked plants (*Figure 1-1*). Tumors are usually comprised of unorganized tissue, but sometimes they differentiate into roots or shoots. This depends on the host plant, the position on the infected plant or the inducing bacterium (*Figure 1-2*). As indicated by several reviews, crown gall has been a worldwide problem in agriculture for over hundred years (Moore and Cooksey, 1981; Burr et al., 1998; Burr and Otten, 1999; Escobar and Dandekar, 2003).

1 INTRODUCTION

1.1 Strain classification

Early taxonomy distinguished agrobacteria on the basis of their pathogenic properties. Thus strains causing crown gall were classified as *A. tumefaciens*, those inducing cane gall on raspberry (*Rubus idaeus*) were described as *A. rubi* and hairy root-inducing isolates were allocated to *A. rhizogenes*. Non-pathogenic strains were called *A. radiobacter* (Allen and Holding, 1974). Later, strains were identified on the basis of their biochemical and physiological properties which led to the definition of three biotypes (Kerr and Panagopoulos, 1977; Süle, 1978). Species- and biotype-based taxonomies do not coincide (Kersters and De Ley, 1984). Biotype 3 strains were isolated almost exclusively from grapevine (*Vitis vinifera*) and allocated to *A. vitis* (Ophel and Kerr, 1990). Similarly, several isolates from weeping fig (*Ficus benjamina*) form a distinct group and were classified as *A. larrymoorei* (Bouzar and Jones, 2001). For a recent review on *Agrobacterium* taxonomy, see Chapter 5 in this book.

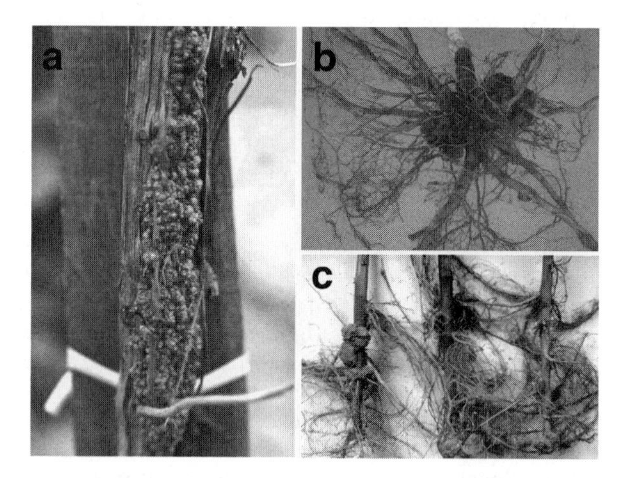

Figure 1-1. Natural crown galls on different hosts. a, grapevine (*Vitis vinifera*, cv. 'Ezerfürtü'); b, raspberry; c, apple. (photo a was kindly provided by Jozsef Mikulas, Kecskemét, Hungary, photo b by Thomas Burr, Cornell University, Geneva and photo c by Christine Blaser, University of Guelph, Laboratory Services, Pest Diagnostic Clinic.)

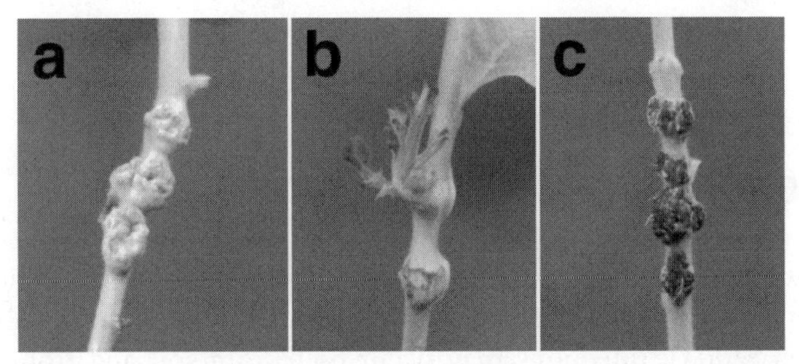

Figure 1-2. Experimental infections of *Nicotiana tabacum* cv. *Samsun* with different *Agrobacterium vitis* strains, showing differences in crown gall morphology. a, strain AB4, undifferentiated tumors; b, strain AT1, shooty tumors (teratomata); c, Tm4, necrotic tumors. (Photos by Ernö Szegedi).

1.2 The infection process

During the infection process a segment of the Ti (tumor-inducing) plasmid, called T(transferred)-DNA, is exported from *Agrobacterium* to the plant cell nucleus where it is integrated into the chromosomal DNA and expressed. Hairy root is caused in a similar way by a root-inducing or Ri plasmid. The T-DNA transfer and integration processes involve a large number of bacterial and host factors, and finally results in genetically transformed plant cells. Details of this unique natural example of inter-kingdom DNA transfer have been reviewed (Zhu et al., 2000; Zupan et al., 2000; Gelvin, 2003; Tzfira et al., 2004 and other chapters in this book). During the infection process agrobacteria suppress plant defense mechanisms via the chromosomally encoded degradation of hydrogen peroxide (Xu and Pan, 2000) and by Ti plasmid-related functions (Veena et al., 2003). Transformation of plant cells results in elevated hormone (auxin and cytokinin) production and sensitivity. Both trigger abnormal proliferation leading to tumorous growth or abnormal rooting (Petersen et al., 1989; Gaudin et al., 1994; Costacurta and Vanderleyden, 1995; see also chapter 15). Tumors and hairy roots produce and secrete specific amino acid and sugar derivatives, called opines. These opines serve as selective nutrients for the inducing bacterium and promote conjugal transfer of their Ti/Ri plasmids. Their central role in the disease has been summarized in the 'opine concept' (Guyon et al., 1980; Petit et al., 1983; Dessaux et al., 1998; see also Chapter 14). Although opines are known as highly specific nutrients for agrobacteria, they can also be used by some other microbes like fluorescent pseudomonads, coryneform bacteria and even by fungal species belonging to the *Cylindrocarpon* and *Fusarium* genera (Rossignol and Dion, 1985; Tremblay et al., 1987a; Tremblay et al., 1987b; Beauchamp et al., 1990; Canfield and Moore, 1991).

2 *AGROBACTERIUM* HOST RANGE

Crown gall has been found to occur on approximately 40 economically important plants (De Cleene, 1979; Burr et al., 1998; Escobar and Dandekar, 2003). Infections may occur both from soil and from infected propagating material. During the 20th century crown gall has become a major bacterial disease both in nurseries and in plantations although in some cases, like cherry trees, no harmful effects have been demonstrated (Garrett, 1987). In other cases, like grapevine, reduction of yield and growth vigor might reach 40% even under the moderate climatic conditions of California (Schroth et al., 1988). Damage due to crown gall is generally more serious in cold climates frequently causing loss of the infected plants (Burr et al., 1998). During the recent decades dissemination of crown gall disease has highly increased due to the intensive exchange and marketing of latently infected propagating material (Burr et al., 1998; Pionnat et al., 1999). Little is known on the effect of hairy root disease on affected plants. Some authors have presented evidence that secondary root formation induced by *A. rhizogenes* can have beneficial effects on infected plants. *A. rhizogenes*-inoculated almond and olive trees showed better growth rate, higher yield and better drought resistance than non-inoculated ones (Strobel and Nachmias, 1985; Strobel et al., 1988). Therefore natural or artificial infection using this 'pathogen' has considerable potential in agriculture, especially in arid regions. This approach could be improved by stable introduction of the *A. rhizogenes* root-inducing genes into plants by genetic engineering (Rinallo et al., 1999).

The host range of *Agrobacterium* is determined by bacterial and plant factors. The former include bacterial virulence genes and T-DNA oncogenes, the latter plant genes required for transformation and tumor formation (see other chapters in this volume). Besides these genetic factors tissue type and physiological status of the plant may also influence efficient transformation and tumor formation. For example, monocottyledons are known as non-hosts of *Agrobacterium*, but meristematic cells of several monocotyledons have been successfully transformed under laboratory conditions (reviewed in Smith and Hood, 1995). The genetic diversity of the pathogen and its potential hosts results in rather different host range patterns (Anderson and Moore, 1979; Thomashow et al., 1980; Knauf et al., 1982).

An early comprehensive review on the host range of crown gall reported 643 susceptible hosts (approximately 60% of the tested species) belonging to 93 families, mainly gymnosperms and dicotyledons (De Cleene and De Ley, 1976). On the other hand, only 3% of the tested monocotyledons (257 species belonging to 27 families) were found to be susceptible to infection with *A. tumefaciens* B6 on the basis of tumor formation (De Cleene, 1985). Further studies carried out with a large set of different agrobacteria showed that they can infect a much wider range of monocotyledons than thought previously (Conner and Dommisse, 1992). The host range of hairy root is also wide and shows several overlaps with the host range of *A. tumefaciens* (De Cleene and De Ley, 1981; Tepfer, 1990; Porter, 1991).

The host range of *A. tumefaciens* and *A. rhizogenes* in the above reports was determined by monitoring the appearance of crown gall or hairy root symptoms. However, several studies have shown that *Agrobacterium* can transform a significantly wider range of plants without causing symptoms. The first evidence came from opine analysis of calli formed at the inoculated wounds both in monocotyledonous (Hooykaas-Van Slogteren et al., 1984) and dicotyledonous (Facciotti et al., 1985; Szegedi et al., 1989) non-host plants. In other experiments using the 'agroinfection' method, plant virus genes cloned into T-DNAs could be transferred into maize and wheat, both non-hosts (Grimsley et al., 1988; Hayes et al., 1988; Boulton et al., 1989). Soon after these reports several economically important monocotyledons, e.g. rice (Raineri et al., 1990; Hiei et al., 1994), maize (Ishida et al., 1996) (Ishida et al., 1996) and wheat (Cheng et al., 1997) were successfully transformed with *Agrobacterium*-based vectors. Subsequently, genetic transformation of yeast, fungi and human cells with *Agrobacterium* was also reported (see Chapter 18).

3 DIVERSITY OF NATURAL ISOLATES

3.1 Strain diversity

In order to detect, identify and eradicate *Agrobacterium* we require more knowledge about the natural diversity of this plant pathogen and its main pathogenic determinant, the Ti/Ri plasmid. A large number of *Agrobacterium* strains have been isolated. They were obtained from all over the world and from widely different host plants including weeping fig (Bouzar et al., 1995; Vaudequin-Dransart et al., 1995; Zoina et al., 2001; Raio et al.,

2004), roses (Marti et al., 1999; Pionnat et al., 1999), poplar (Nesme et al., 1987; Michel et al., 1990; Nesme et al., 1992), chrysanthemum (Bazzi and Rosciglione, 1982; Bush and Pueppke, 1991a; Ogawa et al., 2000), Lippia (Unger et al., 1985) several fruit trees (Albiach and Lopez, 1992; Pulawska et al., 1998; Ridé et al., 2000; Moore et al., 2001; Peluso et al., 2003) and grapevines (Panagopoulos et al., 1978; Burr and Katz, 1983; Paulus et al., 1989; Ridé et al., 2000). Isolates from a given host species usually show high biochemical and genetic diversity. For example, rose isolates belonged to biotype 1 or 2 with a nearly equal occurrence of succinamopine and nopaline Ti plasmids. They could be further subclassified into several chromosomal and Ti plasmid groups using PCR-RFLP analysis (Pionnat et al., 1999). Chromosomal and Ti plasmid diversity was also observed among poplar isolates (Nesme et al., 1987; Nesme et al., 1992). Crown galls on weeping fig (*Ficus benjamina*) were caused by *A. tumefaciens* biotype 1 isolates and by a new species named *A. larrymoorei* (Bouzar et al., 1995; Zoina et al., 2001; Raio et al., 2004). On grapevines crown gall is caused by *A. vitis*. The diversity of *A. vitis* is illustrated by the occurrence of octopine/cucumopine, nopaline and vitopine strains (Paulus et al., 1989; Burr et al., 1998; Ridé et al., 2000). Among the *A. tumefaciens* strains that occasionally occur on grapevine two types are predominant. Some isolates have an octopine/cucumopine type pTi which is characteristic for *A. vitis*. The others are similar to the *A. tumefaciens* nopaline strains (Szegedi et al., 2005). Although members of the genus *Agrobacterium* are primarily known as plant pathogens they have also been found in human clinical samples. *Agrobacterium* infections in humans are frequently associated with the use of plastic catheters or with immunocompromising diseases like HIV (Edmont et al., 1993; Hulse et al., 1993; Manfredi et al., 1999; Landron et al., 2002). None of these strains carried a Ti/Ri plasmid.

3.2 pTi and pRi plasmid diversity

3.2.1 Opine classification

The Ti/Ri plasmids used in genetic engineering are derived from only a few natural plasmids. Thus, it is not unusual to encounter descriptions of 'the *Agrobacterium* Ti plasmid' that refer in fact to one particular Ti plasmid, and ignore the existence of numerous other types of Ti/Ri plasmids. Few of these plasmids have been entirely sequenced: pTiC58 (Goodner et al., 2001; Wood et al., 2001), pTi15955 (AF242881), pTi-SAKURA

(Suzuki et al., 2000), pRi1724 (Moriguchi et al., 2001). Traditionally, opines have been used for classifying both *Agrobacterium* strains and their Ti/Ri plasmids. Sofar, octopine, nopaline, succinamopine, agropine, agropine/mannopine, mannopine, chrysopine/succinamopine, chrysopine/nopaline, cucumopine/mikimopine (Petit et al., 1983; Dessaux et al., 1998; Pionnat et al., 1999; Moriguchi et al., 2001), octopine/cucumopine and vitopine (Szegedi et al., 1988) strains and plasmids have been identified; a null-type category has been proposed for cases in which no opine could be detected. The possible relations between plasmid type and host range have been little studied sofar. Vitopine pTis occur exclusively in grapevine isolates (*A. vitis*) indicating an ecological adaptation of this group. Vitopine is a condensation product of glutamine and pyruvate (Chilton et al., 2001). Glutamine occurs at an extremely high concentration (2-4 mM) in grapevine xylem sap, accounting for up to 75-85% of the total free amino acid content (Prima-Putra and Botton, 1998). Its conjugation with pyruvate to vitopine and the specific uptake and degradation of vitopine provide *A. vitis* with large supplies of an abundant and metabolically important compound allowing it to outcompete other grapevine-associated microorganisms. Further details on the occurrence and classification of different pTi types in *Agrobacterium* spp. and the role of opines have recently been reviewed (Dessaux et al., 1998).

The opine classification is not perfect. First, some Ti/Ri plasmids induce tumors that do not contain any known opine type. Some (pTiAT181, pTiEU6 and pTiT10/73) were initially classified as defective nopaline-type Ti plasmids because of restriction pattern similarities to the nopaline plasmid pTiT37 (Guyon et al., 1980). Later they were found to induce the synthesis of leucinopine (Chang et al., 1983) and succinamopine leading to a re-classification as succinamopine plasmids (Chilton et al., 1984). Others, like pTiBo542 and pTiAT1, were first classified as null-type plasmids, then found to induce agropine synthesis (Guyon et al., 1980) and called agropine-type plasmids, and later reclassified as succinamopine plasmids (Hood et al., 1986). Another example is the *Lippia canescens* strain AB2/73. The opines induced by AB2/73 (Unger et al., 1985) are still unknown and AB2/73 is therefore part of the null-type group. This group is clearly artificial.

Secondly, opine synthesis and/or utilization genes occupy only relatively small parts of the large Ti/Ri plasmids. Even if Ti/Ri plasmids carry similar opine genes, their remaining sequences can be be quite different. A classification exclusively based on opines would lead to artificial groups based on only partially related plasmid structures.

Thirdly, one plasmid generally induces several types of opines and different plasmids can specify different combinations of opines, for example, agropine and mannopine (pRiA4), agropine, mannopine and octopine (pTi15955), or octopine and cucumopine (pTiTm4). The choice of one opine rather than another to define a plasmid group has so far been largely arbitrary. In summary, opine genes only seem to be of limited value for pTi/pRi classification.

3.2.2 Incompatibility

Plasmids can also be classified according to their incompatibility properties (Couturier et al., 1988). Plasmid incompatibility is defined as the failure of two co-resident plasmids to be stably inherited in the absence of external selection. In the case of the Ti/Ri plasmids, four incompatibility groups have been defined. The octopine and nopaline Ti plasmids from *A. tumefaciens* have been grouped together in the IncRh1 group (Hooykaas et al., 1980). The octopine/cucumopine Ti plasmid pTiB10/7 and the nopaline Ti plasmid pTiAT66 from *A. vitis* also belong to IncRh1 (Szegedi et al., 1996). The succinamopine Ti plasmid pTiBo542 belongs to IncRh2 (Hood et al., 1986), the agropine pRi plasmids to IncRh3 (White and Nester, 1980) and the vitopine Ti plasmid pTiS4 to IncRh4 (Szegedi et al., 1996). In most of the large plasmids of the *Rhizobiaceae* incompatibility properties are encoded by *repABC* genes. Incompatible plasmids have similar *repC* regions although exceptions have been found (Cevallos et al., 2002). As in the case of the opine genes, the *rep* genes may be of limited interest for a natural classification system, since similar *rep* sequences can be associated with large stretches of different DNA sequences. Given these difficulties, it would seem logical to base plasmid comparison and classification on full plasmid sequences. Early restriction enzyme analysis of purified Ti/Ri plasmids and hybridization of plasmid probes to restriction digests of total or plasmid DNA (Drummond and Chilton, 1978; White and Nester, 1980; Thomashow et al., 1981; Knauf et al., 1983; Huffman et al., 1984) showed a great variability of Ti/Ri plasmid structures. Subsequently, Ti/Ri plasmids were shown to be evolutionary chimaeras resulting from extensive horizontal gene transfer, gene loss and insertion sequence activity (Otten et al., 1992; Otten and De Ruffray, 1994; see also chapter 17). The mosaic nature of the Ti/Ri plasmids makes it impossible to calculate evolutionary distances from global DNA homology values. As an example of this problem, consider the *A. vitis* pTiAB4 and pTiTm4 plasmids. About 75% of pTiAB4 is colinear with pTiTm4 and practically identical, the remainder is entirely different, and partly resembles pTiC58. Similarly, pTiTm4 and pTi2608 are identical, except for a specific 50 kb fragment (Fournier et al., 1994). Clearly, overall homology values based on this type of differences fail to give correct estimations about phylogenetic relationships. The patchwork nature of the Ti/Ri plasmids is also evident at a smaller scale. This will be illustrated by a detailed comparison between T-DNAs.

3.3 T-DNA diversity

T-DNAs are composite structures derived from related units (*Figure 1-3*). The smallest units are individual genes. The socalled *plast* genes (like *lso, b, c', d, e, 3', 4', 5, 6a, 6b, 7, rolB, rolC, ORF13, ORF14, ORF18* and other genes) are weakly related and have functions in tumor and root initiation that still remain to be defined (Levesque et al., 1988; Otten and Schmidt, 1998; Otten et al., 1999). They have been combined in different ways in various T-DNA structures. Some units are much larger. One is encountered in several T-DNAs and has the gene order *acs-5-iaaH-iaaM-ipt-6a-6b*. Derivatives of this structure can be separated into three groups: those with an *ocs* gene (commonly called octopine T-DNAs), those with a *nos* gene (nopaline T-DNAs) and those without an *ocs* or *nos* gene at their right border (succinamopine T-DNAs). Octopine T-DNAs are quite diverse: some (like pTiTm4 and pTiAB3) have a partial *6a* deletion, one has an IS*869* element between *6b* and *ocs* and an IS*870* element in *iaaM* (pTiCG474), one an IS*866* insertion in *iaaH* (pTiTm4 and other related plasmids), others have an *iaa-ipt-6a* deletion (pTiAB3, pTiAg57 and pTiNW233, each of which with further changes). T-DNAs with a partial *acs* deletion and an additional gene (gene *7*) are found on pTiAch5, the pTiA66 T-DNA is similar but has an IS*66* insertion in *iaaH*. The TL-DNAs of the succinamopine plasmids pTiChry5 and pTiBo542 also contain the *acs*-to-*6b* fragment, the full sequence is still unknown. In pTiBo542 an IS*1312* element is found between gene *5* and *7*. Octopine and succinamopine Ti plasmids carry a second T-DNA. They can be transferred independently of the other T-DNA and are divided in TR-DNAs (pTiAch5 and pTiA66) and TB-DNAs (the remainder, see below). Although octopine T-DNAs are basically similar, the plasmids on which they reside can be quite different.

This also applies to nopaline T-DNAs. The best known is from pTiC58; the central *acs*-to-*6b* fragment carries a *nos* gene at the right border and a large 11 kb T-DNA extension at its left with genes *a-b-c-c'-d-e-f* (*a*-to-*f* region, Willmitzer et al., 1983; Otten et al., 1999) Genes *a* and *b* are related to *acs* and gene *5*. An *a-b-c* fragment is also found at the left end of the pTiTm4 TB-DNA (see below and Otten et al., 1999). Thus, the *acs-5* fragment is found in three different contexts (left part of *acs*-to-*6b* fragment, left part of pTiC58 T-DNA and left part of pTiTm4 TB-DNA) and may constitute an ancestral functional unit. The pTiT37 T-DNA is similar to that of pTiC58 but a spontaneous variant carries an IS*136* element between the *6a* and *6b* genes (Vanderleyden et al., 1986). In pTiAB4 and pTi82.139, the *acs*-to-*6b* fragment is modified by replacement of the *6a-6b* fragment with an unrelated *6b-3'* gene fragment (Drevet et al., 1994; Otten and De Ruffray, 1994). The *3'* gene is also found on the pTiAch5 and pTiChry5 TR-DNA (see below). Whereas pTi82.139 carries the *a*-to-*f* region, it is lacking in the strongly related pTiAB4. Other nopaline T-DNA variations have been reported (Wabiko et al., 1991).

The TR-DNA of octopine and succinamopine plasmids has the common structure *4'-3'-mas2'-mas1'-ags* and is situated either close to the TL-DNA (pTiAch5) or further away (pTiChry5 and pTiBo542). In pTiChry5 a gene

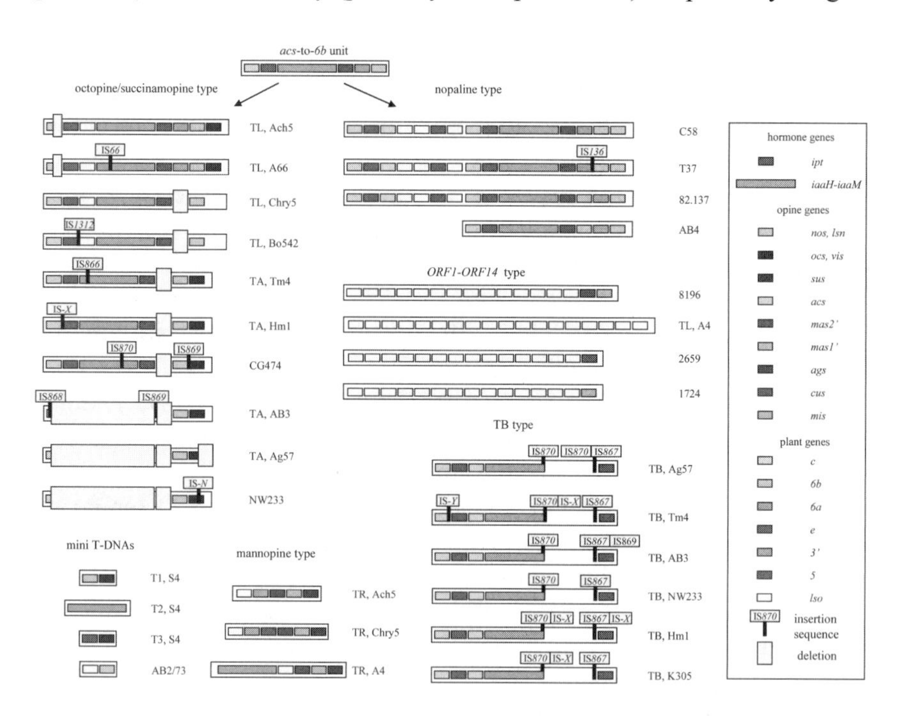

Figure 1-3. Schematic maps of different T-DNA structures. Genes discussed in the text are marked in color. During T-DNA evolution, genes and gene groups were combined in different ways. Maps are not to scale.

with homology to gene *e* is found between *3'* and *mas2'* (Otten, unpublished). A *mas2'-mas1'-ags* fragment is also found at the right end of the pRiA4 TR-DNA (Bouchez and Tourneur, 1991) where it is linked to an *iaaH-iaaM-rolBTR* fragment, a *mas2'-mas1'* fragment (without *ags*) is found in the unique T-DNA of pRi8196 (Hansen et al., 1991).

The T-DNAs or TL-DNAs of the Ri plasmids carry a large conserved fragment with gene order ORF1-ORF14. This fragment is linked to various specific fragments at the right end: ORF15-18 in pRiA4 (Slightom et al., 1986), *mas2'-mas1'* in pRi8196, *cus* in pRi2659 and the *cus*-related *mis* in pRi1724 (Suzuki et al., 2001). A second T-DNA (TB-DNA) is also found on o/c Ti plasmids like pTiTm4. The left part with the *a-b-c* fragment is followed by an *iaaH-iaaM* fragment and a *cus* gene (Otten et al., 1999). Six TB-DNA variants have been identified (Otten and van Nuenen, 1993).

The pTiAB2/73 plasmid carries a minimal T-DNA with one oncogene, *lso*, and one opine gene, *lsn* (Otten and Schmidt, 1998). This T-DNA is also found on the vitopine-type *A. vitis* pTiS4 plasmid (Otten, unpublished data). Interestingly, pTiS4 carries three other small T-DNAs: T1 (*6b-vis*), T2 (*iaaH-iaaM*) and T3 (*ipt-vis'*) (Canaday et al., 1992).

In view of such a large T-DNA diversity, we expect that more T-DNA structures remain to be discovered. Comparative T-DNA studies allow a number of interesting conclusions. (i) Isolates with identical or nearly identical T-DNA structures are found in widely different locations and on different host plants. This indicates a recent clonal origin for these structures, followed by rapid expansion. (ii) The mosaic T-DNA structures were probably created by horizontal pTi/Ri transfer followed by recombination with resident plasmids. (iii) The known T-DNAs do not form a gradual evolutionary series. Thus, there seem to be no transition forms between the small T-DNAs and the larger units. It is likely that many of the intermediates were lost. (iv) Due to the lack of intermediates, it is probably impossible to reconstruct the evolutionary history of the many present-day T-DNAs. (v) The large range of structures probably reflects adaptation to different selection factors like plasmid stability, metabolic coherence, hosts, soils, climates and competing organisms.

3.4 Other ecologically significant plasmids

Apart from the Ti/Ri plasmids, agrobacteria contain other plasmids of ecological interest. The agrocin plasmids, like pAgK84 and pAgK434 from *A. radiobacter* strain K84 (also mentioned as *A. rhizogenes* K84) encode the synthesis of antibiotics that can kill certain virulent agrobacterial strains. K84 can be used for biocontrol (see below). The sequence of pAgK84 has been determined (NC006277). Its occurrence in natural isolates remains to be studied (for details see part 4). Tartrate utilization plasmids (pTr) of *A. vitis* provide another example. They promote growth on grapevine, a species with high tartrate levels (Kado, 1998; Salomone et al., 1998). Strain AB3 carries two related tartrate utilization (tar) regions (TAR-I and TAR-II) on pTrAB3 and pTiAB3 respectively (Otten et al., 1995). The tar region consists of the *ttuA-E* genes, *orfX*, *orfY* and a second *ttuC* copy. AB4 carries another tar region variant on pTrAB4, TAR-III (Salomone et al., 1996). The tar regions are thus found in different plasmid contexts and are also present in other bacteria, like *Pseudomonas putida* (Tipton and Beecher, 1994). Two tumor-inducing plasmids, pTiAB3 and pTiAT6, also carry tar genes. pTiAB3 and pTiAT6 belong to the IncRh1/2 group, pTrS4 to IncRh2, pTrAT6, pTrAB4 and pTrRr4 to IncRh4, while pTrTm4, pTrAB3 and pTrNW221 belong to other, as yet unidentified Inc groups (Szegedi and Otten, 1998). Thus, the pTr plasmids are a heterogeneous group and it is likely that their tar regions were acquired by horizontal gene transfer to different plasmids. The different incompatibility properties of pTr plasmids provide ecological flexibility since this allows them to coexist with pTi plasmids from various incompatibility groups. It would be interesting to know what other functions tartrate plasmids carry.

In the case of *A. rhizogenes* agropine strains, mannopine, mannopinic acid and agropinic acid catabolism is encoded by a non-oncogenic plasmid that can form a cointegrate with the pRi plasmid (Petit et al., 1983). In strain C58, the large cryptic pAtC58 plasmid plays a role in virulence, although the mechanism is not clear (Nair et al., 2003). The antagonistic strain F2/5 (see also section 4.6) contains conjugative tartrate and octopine utilization plasmids. Neither of these is associated with biocontrol on grapevine (Szegedi et al., 1999).

4 SOURCES OF INFECTION AND CONTROL OF CROWN GALL DISEASE

Agrobacterium usually infects fruit trees from soil and ground water of galled orchards (Moore and Cooksey, 1981), while in the case of grapevine the systemically infected propagating material is the main source of infection. Therefore the traditional chemical control methods cannot be used to prevent crown gall disease of crop plants. In spite of these difficulties there are some protocols to reduce the occurrence of epidemic crown gall. These include detection of the pathogen in soils, the use of pathogen-free propagation material, biological control, and traditional or molecular breeding for new resistant varieties. In spite of several promising results the genetic diversity of the pathogen could limit the efficiency of these methods. In this paragraph we give a brief overview of possibilities that can be envisaged or applied to reduce economic losses caused by crown gall.

4.1 Diagnostic methods

Early protocols for detection of agrobacteria in propagation material included isolation of bacteria from plant samples on selective media, their identification by physiological and biochemical tests, and finally determination of pathogenicity on test plants (Moore et al., 2001). Since a virulence assay for *Agrobacterium* typically takes 3-4 weeks, several other methods have been developed, including serological tests (Bazzi et al., 1987; Ophel et al., 1988; Bishop et al., 1989) and DNA hybridization (Burr et al., 1990). The introduction of the polymerase chain reaction (PCR) in plant pathology (Louws et al., 1999) opened up new possibilities for rapid detection and identification of *Agrobacterium* in agriculturally important plants. First studies were started in the early 90s (Dong et al., 1992; Schulz et al., 1993). Primers designed for the amplification of pathogenic strains have been based on specific chromosomal (Ponsonnet and Nesme, 1994; Eastwell et al., 1995; Szegedi and Bottka, 2002), or Ti plasmid sequences including the *vir*-region (Ponsonnet and Nesme, 1994; Haas et al., 1995; Sawada et al., 1995) or T-DNA (Dong et al., 1992; Schulz et al., 1993; Haas et al., 1995; Kauffmann et al., 1996; Pulawska and Sobiczewski, 2005). In order to increase the sensitivity of detection PCR methods were combined with serological techniques ('immunocapture', Kauffmann et al., 1996). To avoid false positives and to increase the specificity of reaction (semi-)nested PCR can be used, this involves the use of additional primer(s) to yield a new specific fragment (Pulawska and Sobiczewski, 2005). The high genetic diversity of agrobacteria, even on a single host, may require the use of multiplex PCR techniques. Analysis of restriction fragment length polymorphism of PCR-amplified fragments (PCR-RFLP) and random amplified polymorphic DNA (RAPD) techniques are additional helpful tools in genomic typing of agrobacteria (Ponsonnet and Nesme, 1994; Irelan and Meredith, 1996; Otten et al., 1996; Momol et al., 1998). Since these protocols allow not only the detection of pathogenic strains but also their precise identification they can be efficiently used to trace back the origin of infection in plant materials (Pionnat et al., 1999; Llop et al., 2003).

The pathogen can be detected directly from total DNA prepared from plant samples, or following isolation of bacterial colonies. Although the former protocol is fast and simple, the uneven distribution and low concentration of the pathogen in the host, and the presence of polymerase inhibitors in plant samples may limit its efficiency. Therefore it is advisable to prepare DNA from plant samples with multistep organic extraction or DNA purification columns (Eastwell et al., 1995; Cubero et al., 1999; Llop et al., 1999). Interestingly, it was more efficient to force an SDS-based lysis solution through grapevine canes and to analyze this material by PCR than to force water through canes and isolate colonies (Eastwell et al., 1995). This is in agreement with other observations (Bazzi et al., 1987) showing only 2-12% recovery of bacterial cells previously introduced into canes. The most likely explanation of this phenomenon is that bacteria are firmly attached to host cell walls (Pu and Goodman, 1993a; Cotado-Sampayo et al., 2001) or trapped in the complex xylem structure. In other experiments (Stover et al., 1997a) it was determined that if cuttings were frozen at –20°C and then incubated for 48 hr at 28°C recovery of *A. vitis* colonies with the xylem flushing method increased several-fold. Interestingly, no increase was obtained if canes were assayed immediately after freezing. The freezing/incubation treatment may provide a useful step for improving the sensitivity of indexing methods. Although *Agrobacterium* can be detected by PCR amplification of plant material (Eastwell et al., 1995; Cubero et al., 1999; Szegedi, 2003) this method is limited to plant tissues which are poor in PCR polymerase inhibitors and contain significant numbers of agrobacteria, like fresh tumors or heavily infected plant tissues. To exclude false negatives caused by polymerase inhibitors PCR controls can be included in the samples (Cubero et al., 2002), or one may first isolate bacterial cultures that are then analyzed by PCR (Szegedi and Bottka, 2002).

4.2 Soil as a potential source of infection

Members of *Agrobacterium* spp. are known as soil-borne plant pathogens, therefore soils are generally considered as sources for infection. Although this is true for fruit trees, grapevine crown gall is usually due to infected propagating material. Agrobacteria may occur in fallow as well as in virgin soils. In the USA, savanna and prairie soils and roots of endemic plants contained predominantly non-pathogenic *A. radiobacter* biotype

2 bacteria (Bouzar and Moore, 1987). On the other hand, fallow soils in Algeria were populated with nearly equal ratios of *Agrobacterium* biotype 1 and 2 isolates with a relatively high proportion of pathogenic biotype 1 bacteria (Bouzar et al., 1993). The number of pathogenic bacteria showed significant seasonal fluctuation. During the vegetation period the numbers of pathogenic bacteria were high but drastically decreased during the fall and winter indicating an essential role of the weed rhizosphere in the regulation of *A. tumefaciens* populations in nursery soils (Krimi et al., 2002). In France, soil contained almost exclusively non-virulent biotype 1 strains. In soil samples collected from fruit tree and raspberry nurseries no biotype 3 (*A. vitis*) was found (Mougel et al., 2001). In other cases the presence of pathogenic *A. tumefaciens* closely correlated with the presence of diseased plants (Pulawska and Sobiczewski, 2005).

Studies in vineyards in the USA have shown that *A. vitis* can only be isolated from the rhizosphere of diseased plants (Burr and Katz, 1983; Burr and Katz, 1984; Burr et al., 1987). Pu and Goodman (Pu and Goodman, 1993b) found about 9000 *Agrobacterium* colonies per gram of vineyard soil, although their biotype was not determined. Further experiments were carried out in Germany where 128 soil samples from 19 vineyards were analysed. These samples contained exclusively *A. tumefaciens*, no *A. vitis* was detected (Jäger et al., 1990) confirming that in the case of grapevine not the soil but the infected plant material is primarily responsible for spreading of the disease. It should be noted however that *A. vitis* survives in root pieces for years and may therefore initiate new infections from soils in which formerly infected plants were grown (Burr et al., 1995). This may be very important in grapevine nurseries where the turnover of plant material is high. To prevent long-term survival of *A. vitis* in vineyards, grape roots should be removed as much as possible when plantations are cut down and the area should be left fallow or non-host (e.g. monocotyledonous) crops that do not favour the persistence of *A. vitis* should be grown on it (Bishop et al., 1988). A survey for non-host annual crops has already been published (Novák et al., 1998) but similar investigations with regional fallow plant and weed populations should also be of interest.

Various studies have demonstrated significant transport of opines from tumors into healthy plant tissues and into the rhizosphere (Messens et al., 1985; Szegedi et al., 1988; Savka et al., 1996), thus providing opines for free-living agrobacteria. Indeed, opine-producing plants selectively promote growth of opine-utilizing agrobacteria associated with the root system of opine-producing *Lotus corniculatus* plants (Guyon et al., 1993; Dessaux et al., 1998). Additional experiments confirmed that opine-producing plants affect the composition of the bacterial population in their rhizosphere as well as in the surrounding soil (Oger et al., 1997; Savka and Farrand, 1997; Oger et al., 2000; Mansouri et al., 2002). This promotes long-term persistence of Ti plasmid-containing (pathogenic) agrobacteria. The essential role of plants in influencing microbial populations of soil has also been established for other plant-microbe associations (Wieland et al., 2001). Therefore crop rotation was proposed in order to eliminate pathogens from soils (Oger et al., 2000).

Although infected propagating material is the primary source for spreading *A. vitis* in grapevine, infection of *Agrobacterium*-free grapevines from soil has also been demonstrated (Pu and Goodman, 1993b). *Agrobacterium* infection from soil is most probably enhanced by nematodes as documented for cotton (Zutra, 1982), raspberry (Vrain and Copeman, 1987) and grapevine (Süle et al., 1995). Nematode-resistant plum rootstocks did not become infected with crown gall, while symptoms appeared on the roots of sensitive rootstocks (Rubio-Cabetas et al., 2001) indicating that agrobacteria can enter the roots through nematode wounds. Thus the nematode population of soils may contribute to epidemic crown gall. Recently, it was shown that nematodes enhance transformation by *Agrobacterium in vitro* (Karimi et al., 2000). Whether nematodes are able to carry the bacteria from one plant to another is still unknown.

In view of these data it is advisable to establish conditions that do not favor the long-term persistence of pathogenic agrobacteria in soils, i.e. by growing non-hosts plants and by eliminating nematodes as far as possible. Additionally, the use of efficient methods to detect the presence of agrobacteria in soil prior to planting (Mougel et al., 2001; Pulawska and Sobiczewski, 2005) is also essential.

4.3 Propagating material as a source of infection

Pathogenic agrobacteria invade the whole plant from roots and tumors via the vessels; healthy parts of the plant can become latently infected without appearance of symptoms. Systemic spreading of *Agrobacterium* has been described in several host plants, e.g. in chrysanthemum (Miller, 1975; Jones and Raju, 1988), rose (Martí et al., 1999; Pionnat et al., 1999), weeping fig (Zoina et al., 2001), and perhaps most extensively in grapevine (Lehoczky, 1968, 1971; Burr and Katz, 1984; Süle, 1986; Tarbah and Goodman, 1987; Thies et al., 1991). Since in the case of grapevine the infected propagating material is the primary source in the spreading of the pathogen it is of basic importance to select or produce pathogen-free stocks to reduce economic loss caused by crown gall. Some details of this field have previously been reviewed (Burr et al., 1998; Burr and Otten, 1999).

4.4 Selection for pathogen-free plant material: The grapevine story

The systemic presence of *Agrobacterium* has been demonstrated in several plant species causing dissemination of the pathogen with propagating material. Since the most intensive studies have been carried out on grapevine we would like to discuss the grapevine case, being confident that the results may also be of interest for other plants.

For reliable detection of *Agrobacterium* in grapevine propagating material indexing should be carried out at the different steps of propagation and include the testing of stock plantations, wooden canes used for rooting or graftings, as well as rooted material prior to planting. The presence of *Agrobacterium* in stock plantations used to produce propagating material can be simply monitored by the analysis of bleeding sap. Grape bleeding sap is rich in nutrients so it provides optimal growth conditions for several microbes including *Agrobacterium* spp. The presence of *Agrobacterium* in bleeding sap was first described in Hungary (Lehoczky, 1968) and later confirmed by other laboratories (Burr and Katz, 1983; Mohammadi and Fatehi-Paykani, 1999) demonstrating the systemic nature of the crown gall bacterium in grapes. Using bleeding sap analysis Pu and Goodman (Pu and Goodman, 1993b) found that 53% of originally pathogen-free grapes planted in infested soil contained *Agrobacterium* two years after plantation. On the other hand, Stover et al. (Stover et al., 1997a) found agrobacteria in only 5,3% of infected plants in greenhouse experiments showing their slow and uneven distribution in the host plant. In other experiments *A. vitis* was detected in one of six healthy, and two of ten galled Riesling and Cabernet Sauvignon plants confirming the uneven distribution in grapevine plants (Szegedi and Bottka, 2002). Thus this method can be used to show the appearance of agrobacteria in plantations if sufficient plant numbers are tested, but for the detection in individual plants the bleeding sap test is less reliable.

Under natural conditions wounded plants form calli that cannot always be clearly distinguished from crown galls by visual inspection alone. For example, phloem wounding may stop auxin transport leading to large auxin-induced callus formation at the wound site. During grapevine propagation wounds are made at the base of rootstocks, and also at disbudding and grafting sites. The extent of callus formation at these wounds depends on the plant genotype as well as on environmental and physiological conditions, such calli frequently cannot be distinguished from crown gall. For rapid examination of these tissues opine analysis can be useful (Ophel et al., 1988; Szegedi, 2003). It was recently observed that the position of wound relative to growing shoots affected the development of crown gall (Creasap et al., 2005). Such differences were hypothesized to be related to growth induction of cells sensitive to infection following transport of IAA to the wounds since exogenous application of IAA stimulated crown gall development at otherwise recalcitrant wounds.

For vegetative propagation of grapevines one-year-old wooden canes are most widely used, these are planted directly or used for graftings. Thus it is of great importance that they are free of pathogens and pests. For indexing dormant canes bacteria can be recovered from homogenized cane pieces (Stover et al., 1997a), or by forcing sterile water through the xylem using vacuum (Bazzi et al., 1987; Ophel et al., 1988; Burr et al., 1989), pressure chambers (Tarbah and Goodman, 1986; Goodman et al., 1987), or centrifugal force (Burr et al., 1988). The recovery of agrobacteria from canes was usually less than 10% (see above and Bazzi et al., 1987). Seasonal variation of bacterial cell numbers in grape plants (Pu and Goodman, 1993b; Bauer et al., 1994) is an additional limiting factor for efficient detection. These seasonal changes are probably determined by the availability of nutrients in the host plants (Pu and Goodman, 1993b). Few *Agrobacterium* spp. were usually recovered when dormant canes were used directly for isolation of bacteria, although a relatively high rate was obtained with Californian samples collected in March when bleeding had started (Goodman et al., 1987). In another experiment bleeding sap analysis carried out in April showed 32% infection of a young plantation. During fall and winter, agrobacteria were detected in only 2 and 0% of the same plants respectively. Next spring 25% of bleeding sap samples were again positive. To overcome difficulties caused by the low recovery of agrobacteria dormant cuttings were rooted or callused under sterile, moist conditions and the freshly formed young roots and calli were used for analysis (Lehoczky, 1971). Inducing growth of new young tissues on canes prior to isolation of agrobacteria increased their detectability with about one order of magnitude (Burr and Katz, 1984; Burr et al., 1989).

Although crown gall symptoms have rarely been observed on grapevine roots, several authors reported the presence of *Agrobacterium* in roots (Lehoczky, 1971; Süle, 1986; Burr et al., 1987; Thies et al., 1991). Later it was suggested that the root system of grapevines provides optimal conditions for growth and survival of agrobacteria and constitutes a reservoir for systemic infection of whole plants. Bacteria overwinter in the roots but at spring when bleeding starts they invade the aerial parts of plants by xylem transport where they can serve as an initial source for crown gall induction (Lehoczky, 1978). It was subsequently confirmed that two years after vines were removed agrobacteria still persisted in root debris remaining in the soil (Burr et al., 1995). Interestingly, roots of feral grapes (*Vitis riparia*) contained only non-pathogenic *A. vitis* that atypically did not utilize tartrate (Burr et al., 1999). Results of grafting analysis including root, rootstock, and scion parts of one-year-old dormant graftings showed that approximately 90% of the *A. vitis* cells were concentrated in roots (Szegedi and Dula, 2005). These data are in agreement with the earlier observations by Lehoczky (Lehoczky, 1978).

In summary it can be concluded that indexing wooden plant material like grapevine canes for *Agrobacterium* has several limiting factors like attachment of agrobacteria to plant cell walls, uneven distribution of pathogens in the host plant, complexity of vessel structures and seasonal changes of pathogen cell numbers in plants. Reliable detection requires expensive and time-consuming multistep tests that limit application for large-scale use. Indexing should mainly be used for small-scale selection of pathogen-free stock material that can then be propagated under proper quarantine conditions.

4.5 Production of *Agrobacterium*-free plant material

Dipping dormant grape cuttings in a water bath at 50-52°C for 30-45 minutes eliminates most *A. vitis* cells from canes. Treatment at higher temperatures may be harmful to the survival of dormant grape buds, depending on variety and whether the cuttings were given a post-treatment storage period (Wample, 1993). Hot water treatment to eliminate *A. vitis* from grapevine propagating material has been tested in the USA (Burr et al., 1989; Burr et al., 1996), Australia (Ophel et al., 1990), Italy (Bazzi et al., 1991) and Iran (Mahmoodzadeh et al., 2003). Although such heat treatment kills *A. vitis in vitro*, a few percent of the pathogen cells survive *in vivo* when canes are treated. This limits the application of this simple method. The reason for the difference between *in vivo* and *in vitro* treatments is unknown.

Shoot-tip and apical meristem cultures can also be used to obtain *Agrobacterium*-free plants. Although *in vitro* propagation of crop plants is rather time-consuming and needs laboratory equipment, it is very reliable.

Cultures are usually started from shoot tips or apical meristems since they are free of systemic agrobacteria. Using this method large numbers of *Agrobacterium*-free *Vitis vinifera* L. Pinot Chardonnay plants were produced from approximately 2 cm long shoot tips (Burr et al., 1988). Similarly, Thies and coworkers (Thies et al., 1991; Thies and Graves, 1992) efficiently regenerated sterile *Vitis rotundifolia* Michx. (muscadine grapes) from 0,2-0,4 mm long apical meristems. Plantations established from shoot-tip propagated grapes were still free from crown gall symptoms seven years later, even under cold climatic conditions (Burr et al., 1998).

As an alternative to meristem cultures, green internodes can also be used for propagation: new shoots growing out from older wooden parts of grape trunks become systemically infected only after the old (wooden) and new (green) xylem elements have fused after the lignification of the new shoots. Thus, in New York State, internodal fragments of Chenin blanc, Pinot Chardonnay and Riesling grapes did not contain detectable *Agrobacterium* until August (Burr et al., 1998). Indeed, crown galls have not been found on new green shoots, although they are sensitive to infection and are wounded during cultivation. Thus, not only shoot-tips and apical meristems but also young green shoots can be used as initial sources to produce *Agrobacterium*-free plants. Since single-node softwood grape cuttings can be routinely rooted under greenhouse conditions (Thomas and Schiefelbein, 2001, 2004) this method may become a simple, efficient and economical protocol for mass production of *Agrobacterium*-free plants.

4.6 Biological control

One of the strategies to combat plant diseases is the use of non-pathogenic antagonistic organisms, like viruses, bacteria, fungi or insects. An efficient biocontrol agent should not only be antagonistic, but also able to survive stably in the target plant and/or in its environment. Sofar, most of the antagonistic microorganisms used against pathogenic agrobacteria belong to the genus *Agrobacterium*. The first efficient microbe that showed an inhibitory effect on pathogenic agrobacteria on peach, *Agrobacterium radiobacter* K84, was isolated in Australia (Kerr, 1972). After the first trials K84 was rapidly tested on several additional crops (Moore and Warren, 1979). Strain K84 contains three plasmids. The 47 kb pAgK84 encodes the production of agrocin 84 and provides agrocin resistance to the producing strain K84 (Slota and Farrand, 1982). Agrocin 84 is an adenine nucleotide analogue that resembles agrocinopine and inhibits DNA synthesis. The second plasmid (173 kb) carries functions for nopaline and agrocinopine uptake and catabolism. This IncRh1 plasmid has significant homology with the nopaline Ti plasmid pTiC58, but lacks virulence and oncogenic functions (Clare et al., 1990). The third, less well characterised large plasmid is 300-400 kb in size and encodes the production of a second antibiotic, agrocin 434 (Donner et al., 1993; McClure et al., 1998). Besides these antibiotics, K84 produces also a siderophore called ALS84 that inhibits the growth of several agrobacteria at low-iron conditions (Penyalver et al., 2001). Nopaline and succinamopine Ti plasmids encode sensitivity to agrocin 84 (Van Larebeke et al., 1975; Watson et al., 1975; Chilton et al., 1984), this sensitivity is associated with the Ti plasmid-encoded uptake and catabolism of certain agrocinopines (reviewed in Dessaux et al., 1998). Dipping seeds or roots into a K84 suspension prior to planting efficiently prevents crown gall formation in the field on the roots of roses (Kerr, 1980; Farkas and Haas, 1985; Jones et al., 1991) and on several stone fruits like peach (Kerr, 1972), almond (Jones and Kerr, 1989) and cherry (Moore, 1977). A further advantage of K84 is its excellent survival in the plant rhizosphere (Macrae et al., 1988; Stockwell et al., 1993; Penyalver and Lopez, 1999). Besides these benefits of K84 there are some factors that limit its application. First, only a limited range of *A. tumefaciens* strains, those having nopaline or succinamopine Ti plasmids, are sensitive to agrocin 84. *A. tumefaciens* carrying octopine-type pTis and all types of *A. vitis* irrespective of their pTi plasmids are resistant (Engler et al., 1975; Kerr and Roberts, 1976; Kerr and Panagopoulos, 1977; Panagopoulos et al., 1978; Burr and Katz, 1983; Knauf et al., 1983; Bien et al., 1990). Second, some pathogenic agrocin 84 sensitive agrobacteria readily mutate to an agrocin-resistant phenotype (Süle and Kado, 1980). Third, pAgK84 may be transferred from K84 into virulent agrobacteria by conjugal transfer (Panagopoulos et al., 1979) and thereby introduce agrocin production and resistance into such strains. To overcome this limitation a transfer-minus (Tra$^-$) mutant, called K1026 was established from the wild type K84 strain (Jones et al., 1988). This ecologically safe K1026 showed similar biocontrol activity as the wild type *A. radiobacter* K84 strain (Jones and Kerr, 1989; Jones et al., 1991). Fourth, *A. radiobacter* K84 may acquire a Ti plasmid from oncogenic agrobacteria which could result in virulent strains that produce agrocin and are resistant to it (Vicedo et al., 1996).

Since *A. radiobacter* K84 is inefficient against *A. vitis*, several experiments were carried out to isolate antagonistic strains able to prevent grapevine crown gall disease. The pathogenic biotype 2 *A. tumefaciens* J73 strain isolated in South-Africa inhibited the growth of several *A. vitis* strains *in vitro*, and showed some activity on grapevine *in planta* (Thomson, 1986; Webster et al., 1986). The non-virulent biotype 1 *A. radiobacter* HLB-2 strain isolated in China was also antagonistic to several *A. vitis* strains (Pu and Goodman, 1993c). The non-pathogenic *A. vitis* F2/5 from South-Africa inhibited the growth of several pathogenic *A. vitis* strains *in vitro* and was able to prevent crown gall formation on grapevine in greenhouse experiments. It is interesting to note that agrocin-minus mutants of F2/5 were still able to inhibit *A. vitis*-induced tumor growth, but only on grapevine. It is still unclear which infection step is blocked in this specific tripartite system (antagonistic *A. vitis* F2/5-pathogenic *A. vitis*-grapevine, Burr and Reid, 1994; Burr et al., 1997). *A. vitis* F2/5 has two chromosomally located *luxR*-like (quorum-sensing regulated) genes, *aviR* and *avhR* that are associated with a hypersensitive response on tobacco and a necrotic reaction on grapevine (Zheng et al., 2003; Hao et al., 2005). Recent studies indicate that the agrocin-independent biocontrol activity of F2/5 is caused by necrotic reactions induced on grape cambium (Creasap et al., 2005) but a final proof using necrosis-negative mutants is still lacking. Until now this strain proved to be the most efficient one in controlling gall formation on grape. Several additional attempts have been made to use non-pathogenic *A. vitis* isolates in biological control since *A. vitis* presumably can permanently colonize grapevines (Bazzi et al., 1999; Burr et al., 1999; Szegedi et al., 1999; Wang et al., 2003).

Members of other bacterial genera have also been tested for biological control of crown gall. Bell et al. (1995) tested approximately 850 diverse bacterial isolates from grapevine in Canada as possible *A. vitis* antagonists *in vitro*. Although some of them (e.g. *Enterobacter agglomerans*, *Rahnella aquatilis* and *Pseudomonas* spp.) inhibited the growth of *A. vitis*, they showed variable and usually insufficient efficiency on grapevine plants. Similar results were obtained in Russia and Israel with *Pseudomonas* strains (Khmel et al., 1998). Opine-utilizing fluorescent *Pseudomonas* spp. have been isolated from several crown gall tumors, and nursery or orchard soil samples (Rossignol and Dion, 1985; Tremblay et al., 1987b; Canfield and Moore, 1991). Some of these isolates inhibited growth of *A. tumefaciens* (Canfield and Moore, 1991). These observations indicate that pseudomonads may down-regulate agrobacterial growth in nature. Hypersensitive response reactions induced by *Pseudomonas* spp. also inhibit crown gall induction (Robinette and Matthysse, 1990). It remains to be investigated whether treatment of plants with selected pseudomonads can reduce the occurrence of crown gall disease under natural conditions.

For safe and more efficient practical use several wild type antagonistic strains have been genetically modified. The Tra⁻ derivative of K84 has been mentioned above. The IncRh1 Ti plasmid of *A. tumefaciens* J73 was cured by introducing an IncRh1 *ori* clone, resulting in an avirulent antagonistic strain (Webster and Thomson, 1988). *Rhizobium leguminosarum* bv. *trifolii* strain T24 produces an antibacterial protein called trifolitoxin (TFX) which inhibits the growth of several agrobacteria. Introducing genes for TFX production and resistance into the well characterized antagonistic strains *A. radiobacter* K84 and *A. vitis* F2/5 can extend their biocontrol activity to a wider range of virulent strains and host plants (Herlache and Triplett, 2002). Apart from antagonistic effects other, more indirect effects might be used for biological control. Transfer of Ti plasmids from pathogenic to non-pathogenic agrobacteria is induced by conjugative opines produced in crown galls (Dessaux et al., 1998), as well as by *N*-acylhomoserine lactones (AHLs) produced by bacteria (Farrand et al., 2002; von Bodman et al., 2003; see also Chapter 14). AHL degradation by lactonase producing bacteria may prevent pathogen spreading or even tumor formation (Carlier et al., 2003; Molina et al., 2003).

4.7 Selection and breeding for crown gall-resistant crops

The most efficient solution to prevent crown gall would be the use of *Agrobacterium*-resistant crop plants, both from the environmental and economical point of view. Although *Agrobacterium* spp. have an extremely wide host range considerable differences in susceptibility exist among species and cultivars. A review on the host range of crown gall (De Cleene and De Ley, 1976) provides several examples. More recently, additional data have been obtained on the susceptibility of various crop plants to *Agrobacterium* spp. The aim of these studies was partly to reduce crown gall-induced damage, partly to develop *Agrobacterium*-mediated transformation protocols for recalcitrant plant species. Crown gall resistance can be defined as the capacity of a plant to maintain normal growth in the presence of a given *Agrobacterium* strain. Different cultivars within a crop species usually form tumors of different size. A plant genotype with small but numerous tumors at the inoculation site cannot be considered as 'resistant', since the transformed plant tissue (usually cambium) will be impaired in its normal functions.

In order to select resistant rootstocks, genotypes of several fruit trees have been tested including apple (Stover and Walsh, 1998), walnut (McKenna and Epstein, 2003), *Prunus* spp. (Pierronnet and Salesses, 1996;

Bliss et al., 1999), chrysanthemum and roses. Chrysanthemum cultivars tested with *A. tumefaciens* Chry5 and B6 strains formed four susceptibility groups: some cvs. were susceptible to both strains (1), others to Chry5 (2) or to B6 (3) only, and the remaining ones (4) did not respond to inoculation with either strain. These differences were due to differences in T-DNA transfer and/or integration (Bush and Pueppke, 1991b). Although the rose cultivars were all susceptible, they exhibited significant differences in the frequency of tumor formation and gall size (Reynders-Aloisi et al., 1998). Resistance to crown gall has also been well established for grapevine (*Vitis* spp.). *V. amurensis* clones inherited resistance as a single Mendelian trait (Szegedi et al., 1984; Szegedi and Kozma, 1984). The tested *Vitis* genotypes exhibited three phenotypically different 'resistance' responses to *Agrobacterium* infection. One group formed opine-negative wound calli. In the second group, wound sites contained opines, showing that transformation had taken place without tumor growth and in the third group, necrotic reactions occurred (Szegedi et al., 1989). These data suggest that *Vitis-Agrobacterium* interactions depend on both the host and the pathogen type. Further studies were carried out in South-Africa, Hungary, USA, Switzerland, Germany and more recently in Iran, mainly to select resistant rootstock genotypes and to get an insight into the nature of the resistance (Ferreira and van Zyl, 1986; Goodman et al., 1993; Heil, 1993; Süle et al., 1994; Stover et al., 1997b; Ehemann, 1998; Mahmoodzadeh et al., 2004). These studies revealed additional promising rootstock hybrids, of which *V. riparia* cv. Gloire de Montpellier, 3309 C and Paulsen were the most resistant. The genetic and physiological bases of crown gall resistance in these *Vitis* genotypes are still unknown. In field experiments the use of *V. riparia* cv. Gloire de Montpellier significantly reduced the frequency of crown gall disease (Süle and Burr, 1998).

4.8 Introduction of crown gall resistance by genetic engineering

In the near future, crown gall resistance might be achieved by genetic engineering targeting virulence proteins and tumor functions that contribute to T-DNA transfer and crown gall formation, respectively. Targeting host genes required for the nuclear transport and integration of T-DNA into the plant chromosome may also be envisaged.

4.8.1 Targeting T-DNA transfer and integration

Transport of the T-strand from *Agrobacterium* to the plant cell and integration in host DNA require two key proteins, VirD2 and VirE2 (see Chapter 10). Expression of a mutated *virE2* gene lacking the coding sequence for the ssDNA binding domain in transgenic tobacco significantly reduced tumor formation upon inoculation with *Agrobacterium* (Citovsky et al., 1994). This construct and additional truncated *virE2* genes derived from various strains were used to transform grapevines and several of the selected transgenic lines showed reduced susceptibility to crown gall (Xue et al., 1999; Holden et al., 2003). The mutated VirE2 protein probably competes with functional VirE2 molecules in the plant cytoplasm.

The VirE1 protein has been shown to bind to VirE2, probably to avoid VirE2 self-aggregation, and can prevent binding of VirE2 to the T-strand (Zhao et al., 2001). By blocking binding of VirE2 to the T-strand in the plant cell, VirE1 may interfere with its transport and integration into the plant nucleus. Expression of the *virE1* gene in tobacco resulted in a significant degree of resistance to *A. vitis* octopine strains (Szegedi et al., 2001).

4.8.2 Inhibition of oncogene expression

Transgenes often inhibit the activity of homologous endogenous genes by a phenomenon known as 'silencing'. A 0.7 kbp partial sense *iaaM* fragment from *A. tumefaciens* nopaline strain pTiPO22 prevented tumor formation in tobacco and aspen (Ebinuma et al., 1997). More recently, regions from the major oncogenes *iaaM* and *ipt* were cloned as inverted repeats on the same T-DNA. These constructs produced self-complementary RNAs in tomato and *Arabidopsis thaliana* that triggered silencing of homologous wild type oncogene sequences, thus resulting in a high level of crown gall resistance (Escobar et al., 2001). A transgenic tomato line expressing the self-complementary oncogenes showed resistance to 34 pathogenic *Agrobacterium* strains from each of the three biotypes (Escobar et al., 2003). For efficient silencing of the *iaaM* oncogene the presence of a translation start site is essential (Lee et al., 2003). Oncogene silencing has already been used to produce crown gall resistant walnut (Escobar et al., 2002) and apple rootstock (Viss et al., 2003).

4.8.3 Manipulating plant genes for crown gall resistance

Gelvin and coworkers (Nam et al., 1999) tested about 3000 T-DNA tagged *Arabidopsis* mutants for susceptibility to *Agrobacterium*. About

0.7% of them were resistant indicating that several plant genes are required for *Agrobacterium*-mediated transformation. To date 126 mutants have been selected from approximately 16,500 mutants (Zhu et al., 2003). These studies may lead to candidate plant genes that can be targeted by antisense or microRNA methods to develop new crown gall-resistant crop plants (Gelvin, 2003; see also Chapter 13). Certain plant proteins contribute to the transformation process through interaction with VirD2, VirE2 or VirB2 (Deng et al., 1998; Tzfira et al., 2001; Tzfira and Citovsky, 2002; Hwang and Gelvin, 2004). The gene coding for the VirE2-interacting protein VIP1 has been cloned from an *Arabidopsis* cDNA library and transformation of tobacco plants with an antisense *VIP1* gene construct resulted in resistance to crown gall tumor formation. *Agrobacterium* resistance correlated with a reduced level of *VIP1* transcription and protein synthesis (Tzfira et al., 2001). These observations suggest that crown gall resistance can also be achieved by inhibiting the synthesis of host proteins interacting with bacterial virulence proteins that are essential for transformation.

5 ACKNOWLEDGEMENTS

Due to space limitations, we were unfortunately unable to discuss several interesting contributions to *Agrobacterium* phytopathology. We ask our colleagues to accept our apologies for this. We thank Jozsef Mikulas and Christine Blaser for providing figures. This work was supported by grants NRI Competitive Grants Program/USDA award 2002-35319-12582 and USDA Viticulture Consortium to T. J. Burr and OTKA grant No. T049438 to E. Szegedi.

6 REFERENCES

Albiach MR, Lopez MM (1992) Plasmid heterogeneity in Spanish isolates of *Agrobacterium tumefaciens* from thirteen different hosts. Appl Environm Microbiol 58: 2683-2687

Allen ON, Holding AJ (1974) Genus II. *Agrobacterium. In* RE Buchanan, NE Gibbons, eds, Bergey's Manual of Determinative Bacteriology, 8th Edition. Williams and Wilkins Co., Baltimore, pp 264-267

Anderson AR, Moore LW (1979) Host specificity in the genus *Agrobacterium.* Phytopathology 69: 320-323

Bauer C, Schulz TF, Lorenz D, Eichhorn KW, Plapp R (1994) Population dynamics of *Agrobacterium vitis* in two grapevine varieties during the vegetation period. Vitis 33: 25-20

Bazzi C, Alexandrova M, Stefani E, Anaclerio F, Burr TJ (1999) Biological control of *Agrobacterium vitis* using non-tumorigenic agrobacteria. Vitis 38: 31-35

Bazzi C, Piazza C, Burr TJ (1987) Detection of *Agrobacterium tumefaciens* in grapevine cuttings. EPPO Bulletin 17: 105-112

Bazzi C, Rosciglione B (1982) *Agrobacterium tumefaciens* biotype 3, causal agent of crown gall on *Chrysanthemum* in Italy. Phytopath Z 103: 280-284

Bazzi C, Stefani E, Gozzi R, Burr TJ, Moore CL, Anaclerio F (1991) Hot-water treatment of dormant grape cuttings: Its effects on *Agrobacterium tumefaciens* and on grafting and growth of wine. Vitis 30: 177-187

Beauchamp CJ, Chilton WS, Dion P, Antoun H (1990) Fungal catabolism of crown gall opines. Appl Environ Microbiol 56: 150-155

Bell CR, Dickie GA, Chan JW (1995) Variable response of bacteria isolated from grapevine xylem to control grape crown gall disease *in planta*. Am J Enol Vitic 46: 499-508

Bien E, Lorenz D, Eichhorn K, Plapp R (1990) Isolation and characterization of *Agrobacterium tumefaciens* from the German vineregion Rheinpfalz. J Plant Dis Prot 97: 313-322

Bishop AL, Burr TJ, Mittak VL, Katz BH (1989) A monoclonal antibody specific to *Agrobacterium tumefaciens* biovar 3 and its utilization for indexing grapevine propagation material. Phytopathology 79: 995-998

Bishop AL, Katz BH, Burr TJ (1988) Infection of grapevine by soilborne *Agrobacterium tumefaciens* biovar 3 and population dynamics in host and nonhost rhizospheres. Phytopathology 78: 945-948

Bliss FA, Almehdi AA, Dandekar AM, Schuerman PL, Bellaloui N (1999) Crown gall resistance in accessions of 20 *Prunus* species. Hortscience 34: 326-330

Bouchez D, Tourneur J (1991) Organization of the agropine synthesis region of the T-DNA of the Ri plasmid from *Agrobacterium rhizogenes*. Plasmid 25: 27-39

Boulton MI, Buchholz WG, Marks MS, Markham PG, Davies JW (1989) Specificity of *Agrobacterium*-mediated delivery of maize streak virus DNA to members of the *Gramineae*. Plant Mol Biol 12: 31-40

Bouzar H, Chilton WS, Nesme X, Dessaux Y, Vaudequin V, Petit A, Jones JB, Hodge NC (1995) A new *Agrobacterium* strain isolated from aerial tumors on *Ficus benjamina* L. Appl Environ Microbiol 61: 65-73

Bouzar H, Jones JB (2001) *Agrobacterium larrymoorei* sp. nov., a pathogen isolated from aerial tumours of *Ficus benjamina*. Int J Syst Evol Microbiol 51: 1023-1026

Bouzar H, Moore LW (1987) Isolation of different *Agrobacterium* biovars from a natural oak savanna and tallgrass prairie. Appl Environm Microbiol 53: 717-721

Bouzar H, Quadah D, Krimi Z, Jones JB, Trovato M, Petit A, Dessaux Y (1993) Correlative association between resident plasmids and the host chromosome in a diverse *Agrobacterium* soil population. Appl Environm Microbiol 59: 1310-1317

Burr TJ, Bazzi C, Süle S, Otten L (1998) Crown gall of grape: biology of *Agrobacterium vitis* and the development of disease control strategies. Plant Disease 82: 1288-1297

Burr TJ, Katz BH (1983) Isolation of *Agrobacterium tumefaciens* biovar 3 from grapevine galls and sap, and from vineyard soil. Phytopathology 73: 163-165

Burr TJ, Katz BH (1984) Grapevine cuttings as potential sites of survival and means of dissemination of *Agrobacterium tumefaciens*. Plant Disease 68: 976-978

Burr TJ, Katz BH, Bishop AL (1987) Populations of *Agrobacterium* in vineyard and non vineyard soils and grape roots in vineyards and nurseries. Plant Disease 71: 617-620

Burr TJ, Katz BH, Bishop AL, Meyers CA, Mittak VL (1988) Effect of shoot age and tip culture propagation of grapes on systemic infestations by *Agrobacterium tumefaciens* biovar 3. Am J Enol Vitic 39: 67-70

Burr TJ, Norelli JL, Katz BH, Bishop AL (1990) Use of Ti plasmid DNA probes for determining tumorigenicity of *Agrobacterium* strains. Appl Environm Microbiol 56: 1782-1785

Burr TJ, Ophel K, Katz BH, Kerr A (1989) Effect of hot water treatment on systemic *Agrobacterium tumefaciens* biovar 3 in dormant grape cuttings. Plant Disease 73: 242-245

Burr TJ, Otten L (1999) Crown gall of grape: biology and disease management. Annu Rev Phytopathol 37: 53-80

Burr TJ, Reid CL (1994) Biological control of grape crown gall with non-tumorigenic *Agrobacterium vitis* strain F2/5. Am J Enol Vitic 45: 213-219

Burr TJ, Reid CL, Adams CE, Momol EA (1999) Characterization of *Agrobacterium vitis* strains isolated from feral *Vitis riparia*. Plant Disease 83: 102-107

Burr TJ, Reid CL, Splittstoesser DF, Yoshimura M (1996) Effect of heat treatment on grape bud mortality and survival of *Agrobacterium vitis in vitro* and in dormant grapevine cuttings. Am J Enol Vitic 47: 119-123

Burr TJ, Reid CL, Tagliati E, Bazzi C, Süle S (1997) Biological control of grape crown gall by strain F2/5 is not associated with agrocin production or competition for attachment sites on grape cells. Phytopathology 87: 706-711

Burr TJ, Reid CL, Yoshimura M, Momol EA, Bazzi C (1995) Survival and tumorigenicity of *Agrobacterium vitis* in living and decaying grape roots and canes in soil. Plant Disease 79: 677-682

Bush AL, Pueppke SG (1991a) Characterization of an unusual new *Agrobacterium tumefaciens* strain from *Chrysanthemum morifolium* Ram. Appl Environm Microbiol 57: 2468-2472

Bush AL, Pueppke SG (1991b) Cultivar-strain specificity between *Chrysanthemum morifolium* and *Agrobacterium tumefaciens*. Physiol Mol Plant Pathol 39: 309-323

Canaday J, Gérard JC, Crouzet P, Otten L (1992) Organization and functional analysis of three T-DNAs from the vitopine Ti plasmid pTiS4. Mol Gen Genet 235: 292-303

Canfield ML, Moore LW (1991) Isolation and characterization of opine-utilizing strains of *Agrobacterium tumefaciens* and fluorescent strains of *Pseudomonas* spp. from rootstocks of *Malus*. Phytopathology 81: 440-443

Carlier A, Uroz S, Smadja B, Fray R, Latour X, Dessaux Y, Faure D (2003) The Ti plasmid of *Agrobacterium tumefaciens* harbors an *attM*-paralogous gene, *aiiB*, also encoding N-Acyl homoserine lactonase activity. Appl Environ Microbiol 69: 4989-4993

Cevallos MA, Porta H, Izquierdo J, Tun-Garrido C, Garcia-de-los-Santos A, Davila G, Brom S (2002) *Rhizobium etli* CFN42 contains at least three plasmids of the *repABC* family: a structural and evolutionary analysis. Plasmid 48: 104-116

Chang CC, Chen CM, Adams BR, Trost BM (1983) Leucinopine, a characteristic compound of some crown-gall tumors. Proc Natl Acad Sci USA 80: 3573-3576

Cheng M, Fry JE, Pang S, Zhou H, Hironaka CM, Duncan DR, Conner TW, Wan Y (1997) Genetic transformation of wheat mediated by *Agrobacterium tumefaciens*. Plant Physiol 115: 971-980

Chilton WS, Petit A, Chilton MD, Dessaux Y (2001) Structure and characterization of the crown gall opines heliopine, vitopine and rideopine. Phytochemistry 58: 137-142

Chilton WS, Tempé J, Matzke M, Chilton MD (1984) Succinamopine: a new crown gall opine. J Bacteriol 157: 357-362

Citovsky V, Warnick D, Zambryski PC (1994) Nuclear import of *Agrobacterium* VirD2 and VirE2 proteins in maize and tobacco. Proc Natl Acad Sci USA 91: 3210-3214

Clare BG, Kerr A, Jones DA (1990) Characteristics of the nopaline catabolic plasmid in *Agrobacterium* strains K84 and K1026 used for biological control of crown gall disease. Plasmid 23: 126-137

Conner AJ, Dommisse EM (1992) Monocotyledonous plants as hosts for *Agrobacterium*. Int J Plant Sci 153: 550-555

Costacurta A, Vanderleyden J (1995) Synthesis of phytohormones by plant-associated bacteria. Crit Rev Microbiol 21: 1-18

Cotado-Sampayo M, Segura A, Wuest J, Strasser RJ, Barja F (2001) Interaction of *Agrobacterium vitis* with grapevine rootstocks. Archs Sci Geneve 54: 223-231

Couturier M, Bex F, Bergquist PL, Maas WK (1988) Identification and classification of bacterial plasmids. Microbiol Rev 52: 375-395

Creasap JE, Reid CL, Goffinet MC, Aloni R, Ullrich C, Burr TJ (2005) Effect of wound position, auxin and *Agrobacterium vitis* strain F2/5 on wound healing and crown gall in grapevine. Phytopathology 95: 362-367

Cubero J, Martinez MC, Llop P, Lopez MM (1999) A simple and efficient PCR method for the detection of *Agrobacterium tumefaciens* in plant tumours. J Appl Microbiol 86: 591-602

Cubero J, van der Wolf J, van Beckhoven J, Lopez MM (2002) An internal control for the diagnosis of crown gall by PCR. J Microbiol Methods 51: 387-392

De Cleene M (1979) Crown gall: economic importance and control. Zbl Bakt II Abt 134: 551-554

De Cleene M (1985) Susceptibility of monocotyledons to *Agrobacterium tumefaciens*. Phytopath Z 113: 81-89

De Cleene M, De Ley J (1976) The host range of crown gall. Bot Rev 42: 389-466

De Cleene M, De Ley J (1981) The host range of infectious hairy root. Bot Rev 47: 147-194

Deng W, Chen L, Wood DW, Metcalfe T, Liang X, Gordon MP, Comai L, Nester EW (1998) *Agrobacterium* VirD2 protein interacts with plant host cyclophilins. Proc Natl Acad Sci USA 95: 7040-7045

Dessaux Y, Petit A, Farrand SK, Murphy PJ (1998) Opines and opine-like molecules involved in Plant-*Rhizobiaceae* interactions. *In* HP Spaink, A Kondorosi, PJJ Hooykaas, eds, The *Rhizobiaceae*: Molecular Biology of Model Plant-Associated Bacteria. Kluwer Academic Publisher, Dordrecht-Boston-London, pp 173-197

Dong LC, Sun CW, Thies KL, Luthe DS, Graves CH (1992) Use of polymerase chain reaction to detect pathogenic strains of *Agrobacterium*. Phytopathology 82: 434-439

Donner SC, Jones DA, McClure NC, Rosewarne GM, Tate ME, Kerr A, Fajardo NN, Clare BG (1993) Agrocin 434, a new plasmid-encoded agrocin from the biocontrol *Agrobacterium* strains K84 and K1026, which inhibits biovar 2 agrobacteria. Physiol Mol Plant Pathol 42: 185-194

Drevet C, Brasileiro AC, Jouanin L (1994) Oncogene arrangement in a shooty strain of *Agrobacterium tumefaciens*. Plant Mol Biol 25: 83-90

Drummond MH, Chilton MD (1978) Tumor-inducing (Ti) plasmids of *Agrobacterium* share extensive regions of DNA homology. J Bacteriol 136: 1178-1183

Eastwell KC, Willis LG, Cavileer TD (1995) A rapid and sensitive method to detect *Agrobacterium vitis* in grapevine cuttings using the polymerase chain reaction. Plant Disease 79: 822-827

Ebinuma H, Matsunaga E, Yamada K, Yamakado M (1997) Transformation of hybrid aspen for resistance to crown gall disease. *In* USDA Forest Service Gen Tech Rep RM-GTR-297, pp 161-164

Edmont MB, Riddler SA, Baxter CM, Wicklund BM, Pasculle AW (1993) *Agrobacterium radiobacter*: a recently recognized opportunistic pathogen. Clin Infect Dis 16: 388-391

Ehemann A (1998) Untersuchung von Interaktionen im Wirt-Parasit System *Vitis/Agrobacterium*. Dissertation. University Hohenheim, Stuttgart

Engler G, Holsters M, Van Montagu M, Schell J, Hernalsteens JP, Schilperoort RA (1975) Agrocin 84 sensitivity: a plasmid determined property in *Agrobacterium tumefaciens*. Mol Gen Genet 138: 345-349

Escobar MA, Civerolo EL, Polito VS, Pinney KA, Dandekar AM (2003) Characterization of oncogene-silenced transgenic plants: implications for *Agrobacterium* biology and post-transcriptional gene silencing. Mol Plant Pathol 4: 57-65

Escobar MA, Civerolo EL, Summerfelt KR, Dandekar AM (2001) RNAi-mediated oncogene silencing confers resistance to crown gall tumorigenesis. Proc Natl Acad Sci USA 98: 13437-13442

Escobar MA, Dandekar AM (2003) *Agrobacterium tumefaciens* as an agent of disease. Trends Plant Sci 8: 380-386

Escobar MA, Leslie CA, McGranahan GH, Dandekar AM (2002) Silencing crown gall disease in walnut (*Juglans regia* L.). Plant Sci 163: 591-597

Facciotti D, O'Neal JK, Lee S, Shewmaker CK (1985) Light-inducible expression of a chimeric gene in soybean tissue transformed with *Agrobacterium*. Bio/Technology 3: 241-246

Farkas E, Haas JH (1985) Biological control of crown gall in rose nursery stock. Phytoparasitica 13: 121-127

Farrand SK, Qin Y, Oger P (2002) Quorum-sensing system of *Agrobacterium* plasmids: analysis and utility. Methods Enzymol 358: 452-484

Ferreira JHS, van Zyl FGH (1986) Susceptibility of grapevine rootstocks to strains of *Agrobacterium tumefaciens* biovar 3. South Afr J Enol Vitic 7: 101-104

Fournier P, de Ruffray P, Otten L (1994) Natural instability of *Agrobacterium vitis* Ti plasmid due to unusual duplication of a 2.3-kb DNA fragment. Mol Plant-Microbe Interact 7: 164-172

Garrett CME (1987) The effect of crown gall on growth of cherry trees. Plant Pathol 36: 339-345

Gaudin V, Vrain T, Jouanin L (1994) Bacterial genes modifying hormonal balances in plants. Plant Physiol Biochem 32: 11-29

Gelvin SB (2003) *Agrobacterium*-mediated plant transformation: the biology behind the "gene-jockeying" tool. Microbiol Mol Biol Rev 67: 16-37

Goodman RN, Butrov D, Tarbah F (1987) The occurrence of *Agrobacterium tumefaciens* in grapevine-propagating material and a simplified indexing system. Am J Enol Vitic 38: 189-193

Goodman RN, Grimm R, Frank M (1993) The influence of grape rootstocks on the crown gall infection process and on tumor development. Am J Enol Vitic 44: 22-26

Goodner B, Hinkle G, Gattung S, Miller N, Blanchard M, Qurollo B, Goldman BS, Cao Y, Askenazi M, Halling H, Mullin L, Houmiel K, Gordon J, Vaudin M, Iartchouk O, Epp A, Liu F, Wollam C, Allinger M, Doughty D, Scott C, Lappas C, Markelz B, Flanagan C, Crowell C, Gurson J, Lomo C, Sear C, Strub G, Cielo C, Slater S (2001) Genome sequence of the plant pathogen and biotechnology agent *Agrobacterium tumefaciens* C58. Science 294: 2323-2328

Grimsley NH, Ramos C, Hein T, Hohn B (1988) Meristematic tissues of maize plants are most susceptible to *Agrobacterium* with maize streak virus. Bio/Technology 6: 185-189

Guyon P, Chilton MD, Petit A, Tempé J (1980) Agropine in "null-type" tumors: evidence for the generality of the opine concept. Proc Natl Acad Sci USA 77: 2693-2697

Guyon P, Petit A, Tempé J, Dessaux Y (1993) Transformed plants producing opines specifically promote growth of opine-degrading agrobacteria. Mol Plant-Microbe Interact 6: 92-98

Haas JH, Moore LW, Ream W, Manulis S (1995) Universal PCR primers for detection of phytopathogenic *Agrobacterium* strains. Appl Environ Microbiol 61: 2879-2884

Hansen G, Larribe M, Vaubert D, Tempé J, Biermann BJ, Montoya AL, Chilton MD, Brevet J (1991) *Agrobacterium rhizogenes* pRi8196 T-DNA: mapping and DNA sequence of functions involved in mannopine synthesis and hairy root differentiation. Proc Natl Acad Sci USA 88: 7763-7767

Hao G, Zhang H, Zheng D, Burr TJ (2005) *luxR* homolog *avhR* in *Agrobacterium vitis* affects the development of a grape-specific necrosis and a tobacco hypersensitive response. J Bacteriol 187: 185-192

Hayes RJ, MacDonald H, Coutts RHA, Buck KW (1988) Agroinfection of *Triticum aestivum* with cloned DNA of wheat dwarf virus. J Gen Virol 69: 891-896

Heil M (1993) Untersuchungen zur Resistenz von *Vitis* gegen *Agrobacterium tumefaciens*. Dissertation. University Hohenheim, Stuttgart

Herlache TC, Triplett EW (2002) Expression of a crown gall biological control phenotype in an avirulent strain of *Agrobacterium vitis* by addition of the trifolitoxin production and resistance genes. BMC Biotechnol 2: 2

Hiei Y, Ohta S, Komari T, Kumashiro T (1994) Efficient transformation of rice (*Oryza sativa* L.) mediated by *Agrobacterium* and sequence analysis of the boundaries of the T-DNA. Plant J 6: 271-282

Holden M, Krastanova S, Xue B, Pang S, Sekiya M, Momol EA, Gonzalves D, Burr TJ (2003) Genetic engineering of grape for resistance to crown gall. Acta Hort 603: 481-484

Hood EE, Helmer GL, Fraley RT, Chilton MD (1986) The hypervirulence of *Agrobacterium tumefaciens* A281 is encoded in a region of pTiBo542 outside of T-DNA. J Bacteriol 168: 1291-1301

Hooykaas PJ, den Dulk-Ras H, Ooms G, Schilperoort RA (1980) Interactions between octopine and nopaline plasmids in *Agrobacterium tumefaciens*. J Bacteriol 143: 1295-1306

Hooykaas-Van Slogteren GMS, Hooykaas PJJ, Schilperoort RA (1984) Expression of Ti plasmid genes in monocotyledonous plants infected with *Agrobacterium tumefaciens*. Nature 311: 763-764

Huffman GA, White FF, Gordon MP, Nester EW (1984) Hairy-root-inducing plasmid: physical map and homology to tumor-inducing plasmids. J Bacteriol 157: 269-276

Hulse M, Johnson S, Ferrieri P (1993) *Agrobacterium* infections in humans: experience at one hospital and review. Clin Infect Dis 16: 112-117

Hwang HH, Gelvin SB (2004) Plant proteins that interact with VirB2, the *Agrobacterium tumefaciens* pilin protein, mediate plant transformation. Plant Cell 16: 3148-3167

Irelan NA, Meredith CP (1996) Genetic analysis of *Agrobacterium tumefaciens* and *A. vitis* using randomly amplified polymorphic DNA. Amer J Enol Vitic 47: 145-151

Ishida Y, Saito H, Ohta S, Hiei Y, Komari T, Kumashiro T (1996) High efficiency transformation of maize (*Zea mays* L.) mediated by *Agrobacterium tumefaciens*. Nat Biotechnol 14: 745-750

Jäger J, Lorenz D, Plapp R, Eichhorn KW (1990) Untersuchungen zum latenten Vorkommen von *Agrobacterium tumefaciens* Biovar 3 in der Weinrebe (*Vitis vinifera* L.). Die Weinwissenschaft 45: 14-20

Jones DA, Kerr A (1989) *Agrobacterium radiobacter* strain K1026, a genetically engineered derivative of strain K84, for biological control of crown gall. Plant Disease 73: 15-18

Jones DA, Ryder MH, Clare BG, Farrand SK, Kerr A (1988) Construction of a Tra- deletion mutant of pAgK84 to safeguard the biological control of crown gall. Mol Gen Genet 212: 207-214

Jones DA, Ryder MH, Clare BG, Farrand SK, Kerr A (1991) Biological control of crown gall using *Agrobacterium* strains K84 and K1026. *In* H Komada, K Kiritani, J Bay-Petersen, eds, The Biological Control of Plant Diseases, FTC Book Series no 42, Vol 42. Food and Fertilizer Technology Center for the Asian and Pacific Region, Taipei, Taiwan, pp 161-170

Jones JB, Raju BC (1988) Systemic movement of *A. tumefaciens* in symptomless stem tissue of *Chrysanthemum morifolium.* Plant Disease 72: 51-54

Kado CI (1998) Origin and evolution of plasmids. Antonie Van Leeuwenhoek 73: 117-126

Karimi M, van Montagu M, Gheysen G (2000) Nematodes as vectors to introduce *Agrobacterium* into plant roots. Mol Plant Pathol 1: 383-387

Kauffmann M, Kassemeyer HH, Otten L (1996) Isolation of *Agrobacterium vitis* from grapevine propagating material by means of PCR after immunocapture cultivation. Vitis 35: 151-153

Kerr A (1972) Biological control of crown gall: seed inoculation. J Appl Bacteriol 35: 493-497

Kerr A (1980) Biological control of crown gall through production of agrocin 84. Plant Disease 64: 25-30

Kerr A, Panagopoulos CG (1977) Biotypes of *Agrobacterium radiobacter* var. *tumefaciens* and their biological control. Phytopath Z 90: 172-179

Kerr A, Roberts WP (1976) *Agrobacterium*: correlations between and transfer of pathogenicity, octopine and nopaline metabolism and bacteriocin 84 sensitivity. Physiol Plant Pathol 9: 205-211

Kersters K, De Ley J (1984) Genus III. *Agrobacterium* Conn. *In* NR Krieg, JG Holt, eds, Bergey's Manual of Systematic Bacteriology, Vol 1, Vol 1. Williams and Wilkins Co., Baltimore-London, pp 244-254

Khmel IA, Sorokina TA, Lemanova LB, Lipasova VA, Metlitsky OZ, Burdeynaya TV, Chernin LS (1998) Biological control of crown gall in grapevine and raspberry by two *Pseudomonas* spp. with a wide spectrum of antagonistic activity. Biocontr Sci Technol 8: 45-57

Knauf VC, Panagopoulos CG, Nester EW (1982) Genetic factors controlling the host range of *Agrobacterium tumefaciens*. Phytopathology 72: 1545-1549

Knauf VC, Panagopoulos CG, Nester EW (1983) Comparison of Ti plasmids from three different biotypes of *Agrobacterium tumefaciens* isolated from grapevines. J Bacteriol 153: 1535-1542

Krimi Z, Petit A, Mougel C, Dessaux Y, Nesme X (2002) Seasonal fluctuations and long-term persistence of pathogenic populations of *Agrobacterium* spp. in soils. Appl Environ Microbiol 68: 3358-3365

Landron C, Le Moal G, Roblot F, Grignon B, Bonnin A, Becq-Giraudon B (2002) Central venous catheter-related infection due to *Agrobacterium radiobacter*: a report of 2 cases. Scand J Infect Dis 34: 693-694

Lee H, Humann JL, Pitrak JS, Cuperus JT, Parks TD, Whistler CA, Mok MC, Ream LW (2003) Translation start sequences affect the efficiency of silencing of *Agrobacterium tumefaciens* T-DNA oncogenes. Plant Physiol 133: 966-977

Lehoczky J (1968) Spread of *Agrobacterium tumefaciens* in the vessels of the grapevine, after natural infection. Phytopath Z 63: 239-246

Lehoczky J (1971) Further evidences concerning the systemic spreading of *Agrobacterium tumefaciens* in the vascular system of grapevines. Vitis 10: 215-221

Lehoczky J (1978) Root-system of the grapevine as a reservoir of *Agrobacterium tumefaciens* cells. *In* Proc 4th Internat Conf Plant Path Bact, Angers, France, pp 239-243

Levesque H, Delepelaire P, Rouzé P, Slightom J, Tepfer D (1988) Common evolutionary origin of the central portions of the Ri TL-DNA of *Agrobacterium rhizogenes* and the Ti T-DNAs of *Agrobacterium tumefaciens*. Plant Mol Biol 11: 731-744

Llop P, Caruso P, Cubero J, Morente C, Lopez MM (1999) A simple extraction procedure for efficient routine detection of pathogenic bacteria in plant material by polymerase chain reaction. J Microbiol Methods 37: 23-31

Llop P, Lastra B, Marsal H, Murillo J, Lopez MM (2003) Tracking *Agrobacterium* strains by a RAPD system to identify single colonies from plant tumors. Eur J Plant Pathol 109: 381-389

Louws F, Rademaker J, de Bruijn F (1999) The three ds of PCR-based genomic analysis of phytobacteria: diversity, detection, and disease diagnosis. Annu Rev Phytopathol 37: 81-125

Macrae S, Thomson JA, van Staden J (1988) Colonization of tomato plants by two agrocin-producing strains of *Agrobacterium tumefaciens*. Appl Environm Microbiol 54: 3133-3137

Mahmoodzadeh H, Nazemieh A, Majidi I, Paygami I, Khalighi A (2003) Effects of thermotherapy treatments on systemic *Agrobacterium vitis* in dormant grape cuttings. J Phytopathol 151: 481-484

Mahmoodzadeh H, Nazemieh A, Majidi I, Paygami I, Khalighi A (2004) Evaluation of crown gall resistance in *Vitis vinifera* and hybrids of *Vitis* spp. Vitis 43: 75-79

Manfredi R, Nanetti A, Ferri M, Mastroianni A, Coronado OV, Chiodo F (1999) Emerging gram-negative pathogens in the immunocompromised host: *Agrobacterium radiobacter* septicemia during HIV disease. New Microbiol 22: 375-382

Mansouri H, Petit A, Oger P, Dessaux Y (2002) Engineered rhizosphere: the trophic bias generated by opine-producing plants is independent of the opine type, the soil origin, and the plant species. Appl Environ Microbiol 68: 2562-2566

Marti R, Cubero J, Daza A, Piquer J, Salcedo CI, Morente C, Lopez MM (1999) Evidence of migration and endophytic presence of *Agrobacterium tumefaciens* in rose plants. Eur J Plant Pathol 105: 39-50

McClure NC, Ahmadi AR, Clare BG (1998) Construction of a range of derivatives of the biological control strain *Agrobacterium rhizogenes* K84: a study of factors involved in biological control of crown gall disease. Appl Environ Microbiol 64: 3977-3982

McKenna JR, Epstein L (2003) Susceptibility of *Juglans* species and interspecific hybrids to *Agrobacterium tumefaciens*. Hortscience 38: 435-439

Messens E, Lenaerts A, van Montagu M, Hedges RW (1985) Genetic basis for opine secretion from crown gall tumor cells. Mol Gen Genet 199: 344-348

Michel MF, Brasileiro ACM, Depierreux C, Otten L, Delmotte F, Jouanin L (1990) Identification of different *Agrobacterium* strains isolated from the same forest nursery. Appl Environm Microbiol 56: 3537-3545

Miller HN (1975) Leaf, stem, crown, and root galls induced in *Chrysanthemum* by *Agrobacterium tumefaciens.* Phytopathology 65: 805-811

Mohammadi M, Fatehi-Paykani R (1999) Phenotypical characterization of Iranian isolates of *Agrobacterium vitis*, the causal agent of crown gall disease of grapevine. Vitis 38: 115-121

Molina L, Constantinescu F, Michel L, Reimmann C, Duffy B, Défago G (2003) Degradation of pathogen quorum sensing molecules by soil bacteria: a preventive and curative biological control mechanism. FEMS Microbiol Ecol 45: 71-81

Momol EA, Burr TJ, Reid CL, Momol MT, Otten L (1998) Genetic diversity of *Agrobacterium vitis* as determined by DNA fingerprints of the 5' end of the 23S rRNA gene and Random Amplified Polymorphic DNA. J Appl Microbiol 85: 685-692

Moore LW (1977) Prevention of crown gall on *Prunus* roots by bacterial antagonists. Phytopathology 67: 139-144

Moore LW, Bouzar H, Burr TJ (2001) *Agrobacterium. In* NW Schaad, JB Jones, W Chun, eds, Laboratory Guide for Identification of Plant Pathogenic Bacteria. American Phytopathological Society Press, St. Paul, Minnesota, pp 17-33

Moore LW, Cooksey DA (1981) Biology of *Agrobacterium* tumefaciens: plant interactions. Internat Rev Cytol suppl 13: 15-46

Moore LW, Warren G (1979) *Agrobacterium radiobacter* strain K84 and biological control of crown gall. Annu Rev Phytopathol 17: 163-179

Moriguchi K, Maeda Y, Satou M, Hardayani NS, Kataoka M, Tanaka N, Yoshida K (2001) The complete nucleotide sequence of a plant root-inducing (Ri) plasmid indicates its chimeric structure and evolutionary relationship between tumor-inducing (Ti) and symbiotic (Sym) plasmids in *Rhizobiaceae*. J Mol Biol 307: 771-784

Mougel C, Cournoyer B, Nesme X (2001) Novel tellurite-amended media and specific chromosomal and Ti plasmid probes for direct analysis of soil populations of *Agrobacterium* biovars 1 and 2. Appl Environ Microbiol 67: 65-74

Nair GR, Liu Z, Binns AN (2003) Re-examining the role of the cryptic plasmid pAtC58 in the virulence of *Agrobacterium tumefaciens* strain C58. Plant Physiol 133: 989-999

Nam J, Mysore KS, Zheng C, Knue MK, Matthysse AG, Gelvin SB (1999) Identification of T-DNA tagged *Arabidopsis* mutants that are resistant to transformation by *Agrobacterium*. Mol Gen Genet 261: 429-438

Nesme X, Michel MF, Digat B (1987) Population heterogeneity of *Agrobacterium tumefaciens* in galls of *Populus* L. from a single nursery. Appl Environ Microbiol 53: 655-659

Nesme X, Ponsonnet C, Picard C, Normand P (1992) Chromosomal and pTi genotypes of *Agrobacterium* strains isolated from *Populus* tumors in two nurseries. FEMS Microbiol Ecol 101: 189-196

Novak C, Hevesi M, Keck M, Szegedi E (1998) Susceptibility of vegetable crops to *Agrobacterium vitis* Ophel and Kerr. Acta Phytopathol Entomol Hung 33: 43-47

Ogawa Y, Ishikawa K, Mii M (2000) Highly tumorigenic *Agrobacterium tumefaciens* strain from crown gall tumors of chrysanthemum. Arch Microbiol 173: 311-315

Oger P, Mansouri H, Dessaux Y (2000) Effect of crop rotation and soil cover on alteration of the soil microflora generated by the culture of transgenic plants producing opines. Mol Ecol 9: 881-890

Oger P, Petit A, Dessaux Y (1997) Genetically engineered plants producing opines alter their biological environment. Nat Biotechnol 15: 369-372

Ophel K, Burr TJ, Magarey PA, Kerr A (1988) Detection of *Agrobacterium tumefaciens* biovar 3 in South Australian grapevine propagation material. Australasian Plant Pathol 17: 61-66

Ophel K, Kerr A (1990) *Agrobacterium vitis* sp. nov. for strains of *Agrobacterium* biovar 3 from grapevines. Internat J Syst Bacteriol 40: 236-241

Ophel K, Nicholas PR, Magarey PA, Bass AW (1990) Hot water treatment of dormant grape cuttings reduces crown gall incidence in a field nursery. Am J Enol Vitic 41: 325-329

Otten L, Canaday J, Gérard JC, Fournier P, Crouzet P, Paulus F (1992) Evolution of agrobacteria and their Ti plasmids-a review. Mol Plant-Microbe Interact 5: 279-287

Otten L, Crouzet P, Salomone JY, De Ruffray P, Szegedi E (1995) *Agrobacterium vitis* strain AB3 harbors two independent tartrate utilization systems, one of which is encoded by the Ti plasmid. Mol Plant-Microbe Interact 8: 138-146

Otten L, De Ruffray P (1994) *Agrobacterium vitis* nopaline Ti plasmid pTiAB4: relationship to other Ti plasmids and T-DNA structure. Mol Gen Genet 245: 493-505

Otten L, De Ruffray P, Momol EA, Momol MT, Burr TJ (1996) Phylogenetic relationships between *Agrobacterium vitis* isolates and their Ti plasmids. Mol Plant-Microbe Interact 9: 782-786

Otten L, Salomone JY, Helfer A, Schmidt J, Hammann P, De Ruffray P (1999) Sequence and functional analysis of the left-hand part of the T-region from the nopaline-type Ti plasmid, pTiC58. Plant Mol Biol 41: 765-776

Otten L, Schmidt J (1998) A T-DNA from the *Agrobacterium tumefaciens* limited-host-range strain AB2/73 contains a single oncogene. Mol Plant-Microbe Interact 11: 335-342

Otten L, van Nuenen M (1993) Natural instability of octopine/cucumopine Ti plasmids of clonal origin. Microb Releases 2: 91-96

Panagopoulos CG, Psallidas PG, Alivizatos AS (1978) Studies on biotype 3 of *Agrobacterium radiobacter* var. *tumefaciens*. *In* Proc 4th Internat Conf Plant Path Bact, Angers, France, pp 221-228

Panagopoulos CG, Psallidas PG, Alivizatos AS (1979) Evidence of a breakdown in the effectiveness of biological control of crown gall. *In* B Schippers, W Gams, eds, Soil-Borne Plant Pathogens. Academic Press, London, pp 569-578

Paulus F, Huss B, Bonnard G, Ridé M, Szegedi E, Tempé J, Petit A, Otten L (1989) Molecular systematics of biotype III Ti plasmids of *Agrobacterium tumefaciens*. Mol Plant-Microbe Interact 2: 64-74

Peluso R, Raio A, Morra F, Zoina A (2003) Physiological, biochemical and molecular analyses of an Italian collection of *Agrobacterium tumefaciens* strains. Eur J Plant Pathol 109: 291-300

Penyalver R, Lopez MM (1999) Cocolonization of the rhizosphere by pathogenic agrobacterium strains and nonpathogenic strains K84 and K1026, used for crown gall biocontrol. Appl Environ Microbiol 65: 1936-1940

Penyalver R, Oger P, Lopez MM, Farrand SK (2001) Iron-binding compounds from *Agrobacterium* spp.: biological control strain *Agrobacterium rhizogenes* K84 produces a hydroxamate siderophore. Appl Environ Microbiol 67: 654-664

Petersen SG, Stummann BM, Olesen P, Henningsen KW (1989) Structure and function of root-inducing (Ri) plasmids and their relation to tumor-inducing (Ti) plasmids. Physiol Plant 77: 427-435

Petit A, David C, Dahl GA, Ellis JG, Guyon P, Casse-Delbart F, Tempé J (1983) Further extension of the opine concept: plasmids in *Agrobacterium rhizogenes* cooperate for opine degradation. Mol Gen Genet 190: 204-214

Pierronnet A, Salesses G (1996) Behaviour of *Prunus* cultivars and hybrids towards *Agrobacterium tumefaciens* estimated from hardwood cuttings. Agronomie 16: 247-256

Pionnat S, Keller H, Hericher D, Bettachini A, Dessaux Y, Nesme X, Poncet C (1999) Ti plasmids from *Agrobacterium* characterize rootstock clones that initiated a spread of crown gall disease in Mediterranean countries. Appl Environ Microbiol 65: 4197-4206

Ponsonnet C, Nesme X (1994) Identification of *Agrobacterium* strains by PCR-RFLP analysis of pTi and chromosomal regions. Arch Microbiol 161: 300-309

Porter JR (1991) Host range and implications of plant infection by *Agrobacterium rhizogenes*. Crit Rev Plant Sci 10: 387-421

Prima-Putra D, Botton B (1998) Organic and inorganic compounds of xylem exudates from five woody plants at the stage of bud breaking. J Plant Physiol 153: 670-676

Pu XA, Goodman RN (1993a) Attachment of agrobacteria to grape cells. Appl Environ Microbiol 59: 2572-2577

Pu XA, Goodman RN (1993b) Effects of fumigation and biological control on infection of indexed crown gall free grape plants. Am J Enol Vitic 44: 241-248

Pu XA, Goodman RN (1993c) Tumor formation by *Agrobacterium tumefaciens* is suppressed by *Agrobacterium radiobacter* HLB-2 on grape plants. Am J Enol Vitic 44: 249-254

Pulawska J, Malinowski T, Sobiczewski P (1998) Diversity of plasmids of *Agrobacterium tumefaciens* isolated from fruit trees in Poland. J Phytopathol 146: 465-468

Pulawska J, Sobiczewski P (2005) Development of a semi-nested PCR based method for sensitive detection of tumorigenic *Agrobacterium* in soil. J Appl Microbiol 98: 710-721

Raineri DM, Bottino P, Gordon MP, Nester EW (1990) *Agrobacterium*-mediated transformation of rice (*Oryza sativa* L.). Bio/Technology 8: 33-38

Raio A, Peluso R, Nesme X, Zoina A (2004) Chromosomal and plasmid diversity of *Agrobacterium* strains isolated from *Ficus benjamina* tumors. Eur J Plant Pathol 110: 163-174

Reynders-Aloisi S, Pelloli G, Bettachini A, Poncet C (1998) Tolerance to crown gall differs among genotypes of rose rootstocks. Hortscience 33: 296-297

Ridé M, Ridé S, Petit A, Bollet C, Dessaux Y, Gardan L (2000) Characterization of plasmid-borne and chromosome-encoded traits of *Agrobacterium* biovar 1, 2, and 3 strains from France. Appl Environ Microbiol 66: 1818-1825

Rinallo C, Mittempergher L, Frugis G, Mariotti D (1999) Clonal propagation in the genus *Ulmus*: improvement of rooting ability by *Agrobacterium rhizogenes* T-DNA genes. J Hortic Sci Biotechnol 74: 502-506

Robinette D, Matthysse AG (1990) Inhibition by *Agrobacterium tumefaciens* and Pseudomonas savastanoi of development of the hypersensitive response elicited by *Pseudomonas syringae* pv. *phaseolicola*. J Bacteriol 172: 5742-5749

Rossignol G, Dion P (1985) Octopine, nopaline and octopinic acid utilization in *Pseudomonas*. Can J Microbiol 31: 68-74

Rubio-Cabetas MJ, Minot JC, Voisin M, Esmenjaud D (2001) Interaction of root-knot nematodes (RKN) and the bacterium *Agrobacterium tumefaciens* in roots of *Prunus cerasifera*: evidence of the protective effect of the Ma RKN resistance genes against expression of crown gall symptoms. Eur J Plant Pathol 107: 433-441

Salomone JY, Crouzet P, De Ruffray P, Otten L (1996) Characterization and distribution of tartrate utilization genes in the grapevine pathogen *Agrobacterium vitis*. Mol Plant-Microbe Interact 9: 401-408

Salomone JY, Szegedi E, Cobanov P, Otten L (1998) Tartrate utilization genes promote growth of *Agrobacterium* spp. on grapevine. Mol Plant-Microbe Interact 11: 836-838

Savka MA, Black RC, Binns AN, Farrand SK (1996) Translocation and exudation of tumor metabolites in crown galled plants. Mol Plant-Microbe Interact 9: 310-313

Savka MA, Farrand SK (1997) Modification of rhizobacterial populations by engineering bacterium utilization of a novel plant-produced resource. Nat Biotechnol 15: 363-368

Sawada H, Ieki H, Matsuda I (1995) PCR detection of Ti and Ri plasmids from phytopathogenic *Agrobacterium* strains. Appl Environ Microbiol 61: 828-831

Schroth MN, McCain AH, Foott JH, Huisman OC (1988) Reduction in yield and vigor of grapevine caused by crown gall disease. Plant Disease 72: 241-246

Schulz TF, Lorenz D, Eichhorn KW, Otten L (1993) Amplification of different marker sequences for identification of *Agrobacterium vitis* strains. Vitis 32: 179-182

Slightom JL, Durand-Tardif M, Jouanin L, Tepfer D (1986) Nucleotide sequence analysis of TL-DNA of *Agrobacterium rhizogenes* agropine type plasmid. Identification of open reading frames. J Biol Chem 261: 108-121

Slota JE, Farrand SK (1982) Genetic isolation and physical characterization of pAgK84, the plasmid responsible for agrocin 84 production. Plasmid 8: 175-186

Smith RH, Hood EE (1995) *Agrobacterium tumefaciens* transformation of monocotyledons. Crop Sci 35: 301-309

Stockwell VO, Moore LW, Loper JE (1993) Fate of *Agrobacterium radiobacter* K84 in the environment. Appl Environ Microbiol 59: 2112-2120

Stover E, Walsh C (1998) Crown gall in apple rootstocks: inoculation above and below soil and relationship to root mass proliferation. Hortscience 33: 92-95

Stover EW, Swartz HJ, Burr TJ (1997a) Endophytic *Agrobacterium* in crown gall-resistant and susceptible *Vitis* genotypes. Vitis 36: 21-26

Stover EW, Swartz HJ, Burr TJ (1997b) Crown gall formation in a diverse collection of *Vitis* genotypes inoculated with *Agrobacterium vitis*. Am J Enol Vitic 48: 26-32

Strobel GA, Nachmias A (1985) *Agrobacterium rhizogenes* promotes the initial growth of bare root stock almond. J Gen Microbiol 131: 1245-1249

Strobel GA, Nachmias A, Hess WM (1988) Improvements in the growth and yield of olive trees by transformation with the Ri plasmid of *Agrobacterium rhizogenes*. Can J Bot 66: 2581-2585

Süle S (1978) Biotypes of *Agrobacterium tumefaciens* in Hungary. J Appl Bacteriol 44: 207-213

Süle S (1986) Survival of *Agrobacterium tumefaciens* in *Berlandieri* x *Riparia* grapevine rootstock. Acta Phytopathol Entomol Hung 21: 203-206

Süle S, Burr TJ (1998) The effect of resistance of rootstocks to crown gall (*Agrobacterium* spp.) on the susceptibility of scions in grapevine cultivars. Plant Pathol 47: 84-88

Süle S, Kado CI (1980) Agrocin resistance in virulent derivatives of *Agrobacterium tumefaciens* harboring the pTi plasmid. Physiol Plant Pathol 17: 347-356

Süle S, Lehoczky J, Jenser G, Nagy P, Burr TJ (1995) Infection of grapevine roots by *Agrobacterium vitis* and *Meloidogyne hapla*. J Phytopathol 143: 169-171

Süle S, Mozsar J, Burr TJ (1994) Crown gall resistance of *Vitis* spp. and grapevine rootstocks. Phytopathology 84: 607-611

Suzuki K, Hattori Y, Uraji M, Ohta N, Iwata K, Murata K, Kato A, Yoshida K (2000) Complete nucleotide sequence of a plant tumor-inducing Ti plasmid. Gene 242: 331-336

Suzuki K, Tanaka N, Kamada H, Yamashita I (2001) Mikimopine synthase (*mis*) gene on pRi1724. Gene 263: 49-58

Szegedi E (2003) Opines in naturally infected grapevine crown gall tumors. Vitis 42: 39-41

Szegedi E, Bottka S (2002) Detection of *Agrobacterium vitis* by polymerase chain reaction in grapevine bleeding sap after isolation on a semiselective medium. Vitis 41: 37-42

Szegedi E, Bottka S, Mikulas J, Otten L, Süle S (2005) Characterization of *Agrobacterium tumefaciens* strains isolated from grapevine. Vitis 44: 49-54

Szegedi E, Czakó M, Otten L (1996) Further evidence that the vitopine-type pTi's of *Agrobacterium vitis* represent a novel group of Ti plasmids. Mol Plant-Microbe Interact 9: 139-143

Szegedi E, Czakó M, Otten L, Koncz C (1988) Opines in crown gall tumors induced by biotype 3 isolates of *Agrobacterium tumefaciens.* Physiol Mol Plant Pathol 32: 237-247

Szegedi E, Dula T (2005) Detection of *Agrobacterium* infection in grapevine graftings (in Hungarian with English abstract). Növényvédelem (in press)

Szegedi E, Korbuly J, Koleda I (1984) Crown gall resistance in East-Asian *Vitis* species and in their *V. vinifera* hybrids. Vitis 23: 21-26

Szegedi E, Korbuly J, Otten L (1989) Types of resistance of grapevine varieties to isolates of *Agrobacterium tumefaciens* biotype 3. Physiol Mol Plant Pathol 35: 35-43

Szegedi E, Kozma P (1984) Studies on the inheritance of resistance to crown gall disease of grapevine. Vitis 23: 121-126

Szegedi E, Oberschall A, Bottka S, Oláh R, Tinland B (2001) Transformation of tobacco plants with *virE1* gene derived from *Agrobacterium tumefaciens* pTiA6 and its effect on crown gall tumor formation. Int J Hortic Sci 7: 54-57

Szegedi E, Otten L (1998) Incompatibility properties of tartrate utilization plasmids derived from *Agrobacterium vitis* strains. Plasmid 39: 35-40

Szegedi E, Süle S, Burr TJ (1999) *Agrobacterium vitis* strain F2/5 contains tartrate and octopine utilization plasmids which do not encode functions for tumor inhibition on grapevine. J Phytopath 17: 665-669

Tarbah FA, Goodman RN (1986) Rapid detection of *Agrobacterium tumefaciens* in grapevine propagating material and the basis for an efficient indexing system. Plant Disease 70: 566-568

Tarbah FA, Goodman RN (1987) Systemic spread of *Agrobacterium tumefaciens* biovar 3 in the vascular system of grapes. Phytopathology 77: 915-920

Tepfer D (1990) Genetic transformation using *Agrobacterium rhizogenes*. Physiol Plant 79: 140-146

Thies KL, Graves CH (1992) Meristem micropropagation protocols for *Vitis rotundifolia* Michx. Hortscience 27: 447-449

Thies KL, Griffin DE, Graves CH, Hedgewood CP (1991) Characterization of *Agrobacterium* isolates from muscadine grape. Plant Disease 75: 634-637

Thomas P, Schiefelbein JW (2001) Combined *in vitro* and *in vivo* propagation for rapid multiplication of grapevine cv. *Arka Neelamani*. Hortscience 36: 1107-1110

Thomas P, Schiefelbein JW (2004) Roles of leaf in regulation of root and shoot growth from a single node softwood cuttings of grape (*Vitis vinifera*). Ann Appl Biol 144: 27-23

Thomashow MF, Knauf VC, Nester EW (1981) Relationship between the limited and wide host range octopine-type Ti plasmids of *Agrobacterium tumefaciens*. J Bacteriol 146: 484-493

Thomashow MF, Panagopoulos CG, Gordon MP, Nester EW (1980) Host range of *Agrobacterium tumefaciens* is determined by the Ti plasmid. Nature 283: 794-796

Thomson J (1986) The potential for biological control of crown gall disease on grapevines. Trends Biotechnol 4: 219-224

Tipton PA, Beecher BS (1994) Tartrate dehydrogenase, a new member of the family of metal-dependent decarboxylating R-hydroxyacid dehydrogenases. Arch Biochem Biophys 313: 15-21

Tremblay G, Gagliardo R, Chilton WS, Dion P (1987a) Diversity among opine-utilizing bacteria: identification of coryneform isolates. Appl Environ Microbiol 53: 1519-1524

Tremblay G, Lambert R, Lebeuf H, Dion P (1987b) Isolation of bacteria from soil and crown-gall tumors on the basis of their capacity for opine utilization. Phytoprotection 68: 35-42

Tzfira T, Citovsky V (2002) Partners-in-infection: host proteins involved in the transformation of plant cells by *Agrobacterium*. Trends Cell Biol 12: 121-129

Tzfira T, Li J, Lacroix B, Citovsky V (2004) *Agrobacterium* T-DNA integration: molecules and models. Trends Genet 20: 375-383

Tzfira T, Vaidya M, Citovsky V (2001) VIP1, an *Arabidopsis* protein that interacts with *Agrobacterium* VirE2, is involved in VirE2 nuclear import and *Agrobacterium* infectivity. EMBO J 20: 3596-3607

Unger L, Ziegler SF, Huffman GA, Knauf VC, Peet R, Moore LW, Gordon MP, Nester EW (1985) New class of limited-host-range *Agrobacterium* mega-tumor-inducing plasmids lacking homology to the transferred DNA of a wide-host-range, tumor-inducing plasmid. J Bacteriol 164: 723-730

Van Larebeke N, Genetello C, Schell J, Schilperoort RA, Hermans AK, Van Montagu M, Hernalsteens JP (1975) Acquisition of tumour-inducing ability by non-oncogenic agrobacteria as a result of plasmid transfer. Nature 255: 742-743

Vanderleyden J, Desair J, De Meirsman C, Michiels K, Van Gool AP, Chilton M-D, Jen GC (1986) Nucleotide sequence of an insertion sequence (IS) element identified in the T-DNA region of a spontaneous variant of the Ti-plasmid pTiT37. Nucleic Acids Res 14: 6699-6709

Vaudequin-Dransart V, Petit A, Poncet C, Ponsonnet C, Nesme X, Jones JB, Bouzar H, Chilton WS, Dessaux Y (1995) Novel Ti plasmids in *Agrobacterium* strains isolated from fig tree and chrysanthemum tumors and their opine-like molecules. Mol Plant-Microbe Interact 8: 311-321

Veena, Jiang H, Doerge RW, Gelvin SB (2003) Transfer of T-DNA and Vir proteins to plant cells by *Agrobacterium tumefaciens* induces expression of host genes involved in mediating transformation and suppresses host defense gene expression. Plant J 35: 219-236

Vicedo B, Lopez MJ, Asins MJ, Lopez MM (1996) Spontaneous transfer of the Ti plasmid of *Agrobacterium tumefaciens* and the nopaline catabolism plasmid of *A. radiobacter* strain K84 in crown gall tissue. Phytopathology 86: 528-534

Viss WJ, Pitrak J, Humann J, Cook M, Driver J, Ream W (2003) Crown-gall-resistant transgenic apple trees that silence *Agrobacterium tumefaciens* oncogenes. Mol Breed 12: 283-295

von Bodman SB, Bauer WD, Coplin DL (2003) Quorum sensing in plant-pathogenic bacteria. Annu Rev Phytopathol 41: 455-482

Vrain TC, Copeman RJ (1987) Interactions between *Agrobacterium tumefaciens* and *Pratylenchus penetrans* in the roots of two red raspberry cultivars. Can J Plant Pathol 9: 236-240

Wabiko H, Kagaya M, Sano H (1991) Polymorphism of Nopaline-type T-DNAs from *Agrobacterium tumefaciens*. Plasmid 25: 3-15

Wample RL (1993) Influence of pre- and post-treatment storage on budbreak of hot water treated Cabernet Sauvignon. Am J Enol Vitic 44: 153-158

Wang HM, Wang HX, Ng TB, Li JY (2003) Purification and characterization of an antibacterial compound produced by *Agrobacterium vitis* strain E26 with activity against *A. tumefaciens*. Plant Pathol 52: 134-139

Watson B, Currier TC, Gordon MP, Chilton MD, Nester EW (1975) Plasmid required for virulence of *Agrobacterium tumefaciens*. J Bacteriol 123: 255-264

Webster J, Dos Santos M, Thomson JA (1986) Agrocin-producing *Agrobacterium tumefaciens* strain active against grapevine isolates. Appl Environ Microbiol 52: 217-219

Webster J, Thomson J (1988) Genetic analysis of an *Agrobacterium tumefaciens* strain producing an agrocin active against biotype 3 pathogens. Mol Gen Genet 214: 142-147

White FF, Nester EW (1980) Relationship of plasmids responsible for hairy root and crown gall tumorigenicity. J Bacteriol 144: 710-720

Wieland G, Neumann R, Backhaus H (2001) Variation of microbial communities in soil, rhizosphere, and rhizoplane in response to crop species, soil type, and crop development. Appl Environ Microbiol 67: 5849-5854

Willmitzer L, Dhaese P, Schreier PH, Schmalenbach W, Van Montagu M, Schell J (1983) Size, location and polarity of T-DNA-encoded transcripts in nopaline crown gall tumors; common transcripts in octopine and nopaline tumors. Cell 32: 1045-1056

Wood DW, Setubal JC, Kaul R, Monks DE, Kitajima JP, Okura VK, Zhou Y, Chen L, Wood GE, Almeida Jr. NF, Woo L, Chen Y, Paulsen IT, Eisen JA, Karp PD, Bovee Sr. D, Chapman P, Clendenning J, Deatherage G, Gillet W, Grant C, Kutyavin T, Levy R, Li MJ, McClelland E, Palmieri P, Raymond C, Rouse R, Saenphimmachak C, Wu Z, Romero P, Gordon D, Zhang S, Yoo H, Tao Y, Biddle P, Jung M, Krespan W, Perry M, Gordon-Kamm B, Liao L, Kim S, Hendrick C, Zhao ZY, Dolan M, Chumley F, Tingey SV, Tomb JF, Gordon MP, Olson MV, Nester EW (2001) The genome of the natural genetic engineer *Agrobacterium tumefaciens* C58. Science 294: 2317-2323

Xu X, Pan SQ (2000) An *Agrobacterium* catalase is a virulence factor involved in tumorigenesis. Mol Microbiol 35: 407-414

Xue B, Ling KS, Reid CL, Krastanova S, Sekiya M, Momol EA, Süle S, Mozsar J, Gonsalves D, Burr TJ (1999) Transformation of five grapevine rootstocks with plant virus genes and a *virE2* gene from *Agrobacterium tumefaciens*. In Vitro Cell Dev Biol-Plant 35: 226-231

Zhao Z, Sagulenko E, Ding Z, Christie PJ (2001) Activities of *virE1* and the VirE1 secretion chaperone in export of the multifunctional VirE2 effector via an *Agrobacterium* Type IV secretion pathway. J Bacteriol 183: 3855-3865

Zheng D, Zhang H, Carle S, Hao G, Holden MR, Burr TJ (2003) A *luxR* homolog, *aviR*, in *Agrobacterium vitis* is associated with induction of necrosis on grape and a hypersensitive response on tobacco. Mol Plant-Microbe Interact 16: 650-658

Zhu J, Oger PM, Schrammeijer B, Hooykaas PJ, Farrand SK, Winans SC (2000) The bases of crown gall tumorigenesis. J Bacteriol 182: 3885-3895

Zhu Y, Nam J, Humara JM, Mysore K, Lee LY, Cao H, Valentine L, Li J, Kaiser A, Kopecky A, Hwang HH, Bhattacharjee S, Rao P, Tzfira T, Rajagopal J, Yi HC, Yadav VBS, Crane Y, Lin K, Larcher Y, Gelvin M, Knue M, Zhao X, Davis S, Kim SI, Kumar CTR, Choi YJ, Hallan V, Chattopadhyay S, Sui X,

Ziemienowitz A, Matthysse AG, Citovsky V, Hohn B, Gelvin SB (2003) Identification of *Arabidopsis rat* mutants. Plant Physiol 132: 494-505

Zoina A, Raio A, Peluso R, Spasiano A (2001) Characterization of agrobacteria from weeping fig (*Ficus benjamina*). Plant Pathol 50: 620-627

Zupan J, Muth TR, Draper O, Zambryski PC (2000) The transfer of DNA from *Agrobacterium tumefaciens* into plants: a feast of fundamental insights. Plant J 23: 11-28

Zutra D (1982) Crown gall bacteria (*Agrobacterium radiobacter* var. *tumefaciens*) on cotton roots in Israel. Plant Disease 66: 1200-1201

Chapter 2

A BRIEF HISTORY OF RESEARCH ON *AGROBACTERIUM TUMEFACIENS*: 1900-1980s

Andrew N. Binns

Department of Biology and Plant Sciences Institute, University of Pennsylvania, PA 19104, USA

Abstract. The study of tumorigenesis on plants as a result of their infection by *Agrobacterium tumefaciens* has resulted in enormous advances in our understanding of interspecies genetic transfer. This chapter seeks to trace the earlier studies (from the early 1900s up to mid 1980s) that were involved in defining the biology, genetics and molecular biology of this system. The analysis of these studies will be carried out with the objective of understanding how *Agrobacterium* has become not only a model system in bacterial pathogenesis but also a key player in both basic plant molecular genetics and agricultural biotechnology.

1 INTRODUCTION

Nearly every person picking up this book already knows that virulent strains of *Agrobacterium tumefaciens* and *Agrobacterium rhizogenes* have the capacity to transfer DNA from their Ti or Ri plasmid into plant cells, incorporate this bacterial DNA into the plant chromosomes where its expression results in the formation of 'crown gall tumors' or 'hairy roots', respectively. The various steps in this process will be examined in detail in other chapters of this book. Having been asked to prepare a contribution concerning a 'historical view' of *Agrobacterium* research I found myself asking what that really means. My answer is to consider the current status of *Agrobacterium* research and then look back at the literature for the historical roots of such studies. What experiments led us in the directions we now find ourselves pursuing? Are there other new directions that the earlier studies might suggest?

Fundamentally, *Agrobacterium* research is now carried out on two quite distinct fronts: first, as a model bacterial pathogen and second as a gene vector for modern plant biology and agricultural biotechnology. In terms of *Agrobacterium* as pathogen, important insights have been provided not only to the plant pathologists but those studying bacterial pathogens, generally. Examples of this include recognition of the host by the pathogen, the mechanisms of DNA and protein virulence factor transfer from pathogen to host and the ultimate selective advantage conferred upon the pathogen by the transformation process including issues related to quorum sensing and biofilm formation. Meanwhile, an entirely distinct direction of research is that which develops and utilizes *Agrobacterium* as a means by which to create transgenic plants (and fungi!) for studies in virtually all areas of modern plant biology and agricultural biotechnology. Though these two major streams of research have distinctly different goals and outcomes, they evolved from the same very modest beginnings and are obviously linked together by a common biology. Moreover, they serve as an important reminder of how basic research can lead to ideas and technologies never envisioned by the original students of the system. The objective here is to examine some of the critical studies that revealed the 'common biology' and yet moved the field in these distinct directions.

Obviously, space constraints will force a rather brief consideration of these issues. Should the readers desire a more completely developed view of the earlier studies on *Agrobacterium*, I encourage them to look at a book published twenty five years ago, *The Molecular Biology of Plant Tumors,* edited by Gunter Kahl and Jeff Schell (1982). This contains a series of chapters reviewing much of the work done in the 1960s and '70s to elucidate the mechanism of transformation and tumorigenesis, as well as a remarkable (and controversial) chapter by Armin Braun (1982) on the early history of crown gall research.

2 *AGROBACTERIUM*—THE PATHOGEN

2.1 Early studies

As described in the first chapter of this volume, the "crown gall" disease of higher plants was a particular problem in orchards and vineyards, though a wide variety of plants were known to develop distinct 'galls'. The earliest work identifying bacteria as the cause of these galls, in contrast to the then known limited galls produced as a result of insect or nematode infection, was published by Cavara (1987) who isolated 'white bacteria' that would give rise to galls when inoculated on plants. A much more thorough (and apparently independent – see Braun, 1982) characterization of the causal agent of the crown gall disease was published by Smith and Townsend (1907) in which many of the characteristics of the inciting bacterium (named then as *Bacterium tumefaciens*) were described including its rod shape, size, polar flagella and inability to grow well at 37°C ('blood temperature'). The debate over the nomenclature of *Agrobacterium* species still exist (*Box 2-1* and Chapter 5), and for simplicity, I will refer to *Agrobacterium tumefaciens* as the causal agent of crown gall tumors and *Agrobacterium rhizogenes* as the causal agent of the hairy root disease throughout the course of this chapter.

The nomenclature of *Agrobacterium* species (and genus) has changed several times over the past 100 years. Virulent strains have been called *Bacterium tumefaciens*, *Phytomonas tumefaciens*, *Agrobacterium tumefaciens*, *Agrobacterium rhizogenes*, and *Rhizobium rhizogenes* whereas non virulent strains have been called *Bacterium radiobacter* and *Agrobacterium radiobacter*.

Young et al. (2001) proposed that the *Agrobacterium* genus, a member of the Rhizobiaciea, be renamed as a member of the closely related *Rhizobium* genus and also proposed renaming the species.

Farrand et al. (2003) argue that sufficient differences exist between *Rhizobium* and *Agrobacterium* such that the genus names should not be changed.

Box 2-1. Agrobacterium nomenclature.

Through the next thirty years studies on the crown gall disease described the responses of many plants to various different field isolates, generally concurring with the observations of Smith and Townsend. Of particular interest amongst these early papers were the descriptions by Smith (1916) and later Levin and Levine (Levin and Levine, 1918; Levine, 1919) of 'teratomas' – spontaneously shoot forming tumors – that could be isolated on certain plants by certain bacterial isolates (see below). Nevertheless, despite a good deal of speculation about the relationship of crown gall tumors of plants to neoplasias of animals, no particular insights into the mechanism whereby *A. tumefaciens* might be inducing tumors were developed.

The prospects for progress improved as physiological and genetic tools in both plant biology and bacteriology were developed. During the period of time from 1923-1941 the studies by the Riker lab at UW-Madison did a great deal to the set the stage for future studies, particularly by Braun, which ultimately established *Agrobacterium*-mediated tumorigenesis as a fundamental biological system. Amongst these studies, Riker found that (i) bacteria could be added to wounds after as long as 4-5 days and still yield tumors (Riker, 1923b); (ii) elevated temperature (32°C) appeared to abolish the capacity of the bacterium to induce tumors, even though neither plant nor bacterial growth were significantly effected (Riker, 1926); and (iii) an 'attenuated' strain (A6-6, derived from wild type strain A6) was isolated that, induced significantly smaller tumors which had lower levels of auxin than wild type tumors, could grow to full size if inoculated below a virulent tumor, and, when inoculated on decapitated tomato plants, produced shoot forming tumors (Hendrickson et al., 1934; Locke et al., 1938)). In fact, Locke et al. (1938) showed that the attenuated culture stimulates bud development distal to the tumor (we now know these results from the cytokinin produced by such tumors, see below). This attenuated strain is the first mutant strain of *Agrobacterium* (to which I could find reference) that affected tumorigenesis and it was subsequently used in both physiological and molecular genetic studies (Braun and Laskaris, 1942; Binns et al., 1982 and see below).

2.2 *Agrobacterium* 'transforms' plant cells

The lack of progress in understanding the mechanism of crown gall tumorigenesis was reversed through a series of ground-breaking studies by Armin C. Braun at the Rockefeller Institute (which later became the Rockefeller University) from the 1940's through the 1970s. The first set of these established a quite surprising fact: the continuous proliferation of crown gall tumors did not require the continued presence of the inciting bacterium. One early clue of this had been the difficulty with which a variety of workers had in isolating *Agrobacterium* from primary tumors (for details see Braun 1982). Additionally, Smith et al. (1912) had previously reported that 'secondary' tumors could arise at some distance from the primary inoculation site in at least some plants. Braun and Phillip White (also at the Rockefeller Institute) collaborated, using both plant tissue culture techniques being developed by White and grafting and tumor induction techniques being used by Braun, to show that the secondary tumors of sunflower could grow continuously in culture on a defined medium that did not support the growth of non-transformed plant tissues (White and Braun, 1941; Braun and White, 1943). Importantly, no bacteria could be isolated from these cultured tumors. When small fragments (~20-40 mg) of such cultured tissues were grafted back onto a healthy host, bacteria-free tumors would develop as if from an inoculation. Later studies also demonstrated that bacteria-free primary tumors could be isolated and these, too, could grow continuously in culture conditions that did not support growth of normal tissues (Braun, 1943, 1951a). Together the results from these experiments demonstrated that crown gall tumors do not require the presence of the bacteria to be active neoplasias.

These studies spurred Braun to further define the nature of the event and the roles played by both the plant and the bacterium in the process. One of the studies I find most insightful is that of Braun and Laskaris (1942) examining the attenuated strain A6-6 (A66) isolated by Riker as described above. They confirmed several of the observations by Riker: inoculation of intact plants by the A66 strain resulted in small, slow growing tumors whereas inoculations just under sites of decapitation resulted in virtually no tumor formation. Additionally, shoot forming tumors were also observed. A major difference however, was that the application of two different synthetic auxins (naphthalene acetic acid and indole butyric acid) to the decapitation site just above sites inoculated with strain A66 resulted in the formation of large, tumorous growths, whereas auxin application, by itself, had only a small growth effect on mock inoculations. Moreover, the auxin-stimulated A66 tumors were capable of forming transplantable tumors, that is, when grafted onto a healthy – intact – host plants they continued to grow and divide. These results led Braun and Laskaris (1942) to two quite remarkable insights. First, they proposed that there are two phases in the process of tumor formation: 'inception' followed by 'stimulation' to continued multiplication by growth substances. The attenuated culture appeared capable of the former but was deficient in the latter unless auxin was provided, either from endogenous auxin of the intact plant or via exogenous supplement to the decapitated stem above the infection site. We now know, of course, that the "inception" phase represents the transfer of the T-DNA into plant cells and its integration into chromosomes while the "stimulation" phase is the result of production of plant growth substances via enzymes encoded on the T-DNA. Second, because auxin application, alone, to the tomato stem did not result in continuous growth and cell division, Braun and Laskaris (1942) reasoned that two (or more) growth substances must be involved in tumor growth. We now know that two plant hormones, the auxins and cytokinins, are indeed required for continuous cell proliferation by non-transformed cells but not by tumors induced by wild type strains (Skoog and Miller, 1957; Braun, 1958) and these are produced by crown gall tumor cells (see below). Finally, molecular genetic analysis of strain A66 demonstrated that it carries an insertion element in one of the two genes of the T-DNA that encodes enzymes required for auxin biosynthesis (Binns et al., 1982) by the transformed cells.

2.3 The "Tumor Inducing Principle" (TIP)

The results described above indicate that the continuous presence of the bacteria is not required for tumorous growth and that the 'inception' and 'growth stimulation' phases are distinct. They did not, however, address the mechanism of tumor inception. How and when does this occur? What activities of the plant and bacteria are required? Braun (1943) utilized a temperature regime (originally developed by Kunkel (1941) to eliminate viral infections) whereby periwinkle (*Vinca rosea*) plants inoculated with virulent agrobacteria were incubated at 46°C for 5 days at various times after inoculation and then returned to 25°C. The high temperature is lethal to the bacteria but not the plant. He discovered that as long as the infected plants were held at 25°C for 36-48 hrs after inoculation prior to a 5 day heat treatment, tumors would develop but they would be free of bacteria, confirming that the bacteria are not needed for tumor proliferation. However, if inoculated plants were held for times less than 30 hrs at 25°C prior to heat treatment, few or no tumors would arise. As noted above, Riker (1923b) had shown that after ~ 5-7 days wound sites become much less responsive, and ultimately non-responsive, to the bacteria. Taken together these results strongly suggested that the continuous proliferating state of crown gall tumor cells was a result of a 'transformation' of the plant cells during a very short period of time after infection at a wound site.

Subsequently, Braun (1947, 1952) and Braun and Mandle (1948) took advantage of Riker's observation that 32°C treatment disrupts the transformation process in a series of temperature shift experiments. Two fundamental findings were reported. First, a period of wound healing – at either 25 or 32°C – must precede the actual transformation process but the latter can only occur at 25°C. Second, the transformation can occur in as little as 10-12hrs as long as the plant had 30-96 hrs of time to respond to the wound at an inoculation site. As the period of wound healing increased so to did the magnitude of the tumor response, even when the inoculated plants were transferred to 32°C after only 24 hrs at 25°C. The conclusions from these studies were that the wound healing process was necessary in order to provide an environment in which the bacteria could produce what was termed an "active principle" (Braun, 1947) or "tumor inducing (transforming) principle" (Braun and Mandle, 1948). This principle then acts on the plant cells in a fairly short period of time, resulting in their transformation to the tumorous state. The role of the wound in transformation by *Agrobacterium* continues to intrigue current students in the field (e.g. Brencic et al., 2005; McCullen and Binns, 2006). Clearly, phenolics necessary for the induction of the virulence genes of the Ti plasmid (see other chapters in this volume) are produced in high quantities at wound sites (Baron and Zambryski, 1995). Are other aspects of wound healing, for example, cell division, also influential in optimizing the transformation process? And if so, how? These are questions still to be answered.

What is the tumor inducing principle (TIP)? The fundamental possibilities were outlined by Braun (1947): The TIP "...may fall into one of the following categories: (i) a metabolic product of the crown gall bacterium; (ii) a normal host constituent that is converted by the bacterium into a tumor-inducing substrate; (iii) a chemical fraction of the bacterial cell that is capable of initiating, as in the case of the transforming substance (desoxyribonucleic [sic] acid) of the pneumococci, a specific alteration in the host cell with a resultant consistent, and in this case abnormal, development of these cells; (iv) a virus or other agent which is present in association with the crown gall organism." Nearly thirty years would pass before the molecular basis of the tumor inducing principle was elucidated.

2.4 Identification of T-DNA from the Ti plasmid as the "TIP"

While Braun's studies paved the way for analysis of the crown gall problem, the solution to the identification of the TIP ultimately required technologies not available until the mid-late 1970s. Braun spent a significant portion of the latter stages in his career supporting an 'epigenetic' model of tumor initiation in which the bacterium was envisioned as inducing an autocatalytic pattern of gene regulation that was involved in promoting cell division (see Braun, 1981). This hypothesis stemmed from two major factors. First, molecular studies of the time attempting to use isolated nucleic acid fractions from virulent strains to transform plant cells were unsuccessful or not repeatable. Second, in some cases cultured plant cells, after exposure to prolonged culture or in response to specific inducing conditions, could exhibit hormone independent growth (termed 'habituation') and this phenotype was readily reversible as a result of plant regeneration (e.g. see Meins and N., 1978). Third, Braun had discovered that at least in some instances transformation of the plant cells by *Agrobacterium* left them totipotent: normal, fertile plants could be regularly obtained from single cell clones of certain tumors (Braun, 1959; Braun and Wood, 1976 and see below).

The epigenetic model, however, was seriously challenged with the identification of an unusual amino acid derivative – lysopine – in crown gall tissues (Lioret, 1957) but not in non-tumorous tissues. This work was followed up by the pivotal studies emerging from Morel and colleagues demonstrating not only the existence of a variety of novel amino acid derivatives (termed, generically, opines) in crown gall tumors but also the strain-specificity of their occurrence (Goldmann et al., 1968; Petit et al., 1970). For example, strains A6 and B6 yielded tumors that contained octopine (a condensation product of arginine and pyruvate) whereas strains T37 and IIBV7 contained nopaline (condensation product of arginine and α-ketoglutarate). These opines, as well as numerous others that have been described since, are specific for crown gall tumors: non-transformed plant cells do not make them. Note that while there were, during the 1960s and '70s, several reports of non-transformed cells producing opines, virtually every report of this has been shown to have some flaw (for review see Tempe and Goldmann, 1982). The strain specificity of opines produced by tumors led Goldmann et al., (1968) to conclude: "Cette observation est en faveur de l'hypothèse du transfert d'une information spécifique permanente de la bactèrie dans la cellule vègètale, au cours de la transformation tumorale". The most obvious example of permanent, specific information from the bacteria was, of course, DNA.

Adding to the intrigue of the opines being produced by crown gall tumors was the finding that if the bacteria causes the synthesis of a particular opine it is able to utilize that opine as a carbon and nitrogen source

(Lejeune and Jubier, 1967; Petit et al., 1970). These results were widely considered as indicating that genes within the specific strains were responsible for both the tumorous phenotype, the type of opine synthesized by that tumor and the catabolism of that opine by the bacteria. In particular, Petit and Touneur (1972) carried out repeated platings in medium with octopine as sole nitrogen source and observed occasional small colonies that had lost both virulence and the capacity to degrade octopine. They suggested that such a genetic linkage between these two phenotypes could be in the form of a plasmid ("episome"). This was consistent with both the report by Hamilton and Fall (1971) that growth of a virulent *A. tumefaciens* strain (C58) at 35°C lead to a consistent loss of virulence. The possibility of a plasmid controlling virulence was also consistent with the report by Kerr (1969) of the transfer of virulence from a virulent bacterium to an avirulent one if they were co-resident in a tumor induced by the virulent strain.

The experiments described above showing that strain-specific opines are present in tumors suggested that a genetic transfer might be occurring. Consistent with this were the observations reported by the Lippincotts strongly suggesting that bacterial attachment to plant cells is critical for plant transformation by *Agrobacterium*. In these studies, mixtures of virulent and avirulent *Agrobacterium* strains were tested for their capacity to elicit tumor formation (Lippincott and Lippincott, 1969): while avirulent *Agrobacterium* strains would interfere with tumor formation by virulent strains, unrelated, or distantly related, bacterial species would not. Subsequent experiments demonstrated that incubation of a virulent *Agrobacterium* strain with plant cell wall fractions – which would contain such binding sites – resulted in substantially reduced tumor formation (Lippincott et al., 1977; Lippincott and Lippincott, 1978). Together, these results suggested that *A. tumefaciens* binding to specific sites on plant cell walls is required for tumorigenesis. Later studies using a direct binding assay demonstrated that, indeed, there are a saturable number of *Agrobacterium* binding sites on plant cells (Neff and Binns, 1985). While these sites have not been identified, genomic approaches now present significant new opportunities for their characterization (e.g. Zhu et al., 2003).

Thus, by the early 1970s the evidence strongly indicated that (i) plant cells were transformed, somehow, by virulent strains of *Agrobacterium* through the activity of TIP, (ii) the type of opine produced by the tumors and utilized by the bacteria were specified by the bacteria and genetically linked; and (iii) an intimate association of bacteria and plant cell was important for transformation. This suggested that some type of genetic element might be the TIP and that DNA might be moved into the plant cells. One candidate was a prophage, PS8, found in many *Agrobacterium* strains. While a variety of studies claimed that this, or other DNA from *Agrobacterium*, was found in sterile crown gall tumors all of them were, for one reason or another, not able to supply convincing evidence. An excellent review by Drilica and Kado (1975) goes over the various experimental and technical issues surrounding these experiments.

The pivotal experiments underlying the discovery of the TIP came with the discovery of large plasmids in virulent strains for *Agrobacterium tumefaciens* by Schell and Van Montagu and colleagues. They reasoned that a potentially important (for virulence) but cryptic prophage might be present as a plasmid or reside on a plasmid (Zaenen et al., 1974). A systematic search for plasmids in various virulent and avirulent *Agrobacterium* strains was carried out utilizing a variety of gradient centrifugation protocols developed in the late 1960s and early 1970's for the study of plasmids from *E. coli*. The results were startling: the presence of one or more large plasmids was completely correlated with virulence – all 11 virulent strains tested carried such plasmid whereas all 8 avirulent strains did not. The presence of various phages, prophages or defective phage-like particles, on the other hand, was not correlated with virulence. Zaenen et al. (1974) proposed the following hypothesis: "The tumor-inducing principle (Braun, 1947) in crown-gall inducing *Agrobacterium* strains is carried by one or several large plasmids of various lengths". Intense testing of this hypothesis commenced immediately. Van Larebeke et al. (1974) took advantage of the observations of Hamilton and Fall (1971, see above) to show that strains made avirulent by heat curing lacked the large plasmid seen in the virulent strain. These investigators termed this plasmid the 'tumor-inducing plasmid". Finally, Van Larebeke et al. (1975) and Watson et al. (1975) both demonstrated that the conversion of a non-virulent *Agrobacterium* strain to a virulent one, via the method of virulence transfer discovered by Kerr (1969, see above), was accompanied by the acquisition of the large plasmid. Opine production and utilization specificity accompanied the plasmid transfer as well (Bomhoff et al., 1976) . Shortly after the identification of the Ti plasmid as critical for *A. tumefaciens* virulence, a similar "root-inducing" (Ri) plasmid was demonstrated in *A. rhizogenes* (Moore et al., 1979; White and Nester, 1980).

Thus, by 1975 it had become clear that genetic determinants for virulence, opine production and opine utilization are carried on the tumor inducing (Ti) plasmids. The production by the tumors of opines not found in non-transformed tissues strongly suggested that there is a transfer of genetic information from the bacterium into the plant. Yet the possibility that a factor or factors encoded on the plasmid could induce epigenetic changes resulting in tumor proliferation and expression of cryptic plant genes involved in opine production could not be excluded. The hunt was on for evidence of the Ti plasmid in bacteria-free crown gall tumor lines. Studies from the Nester lab (Chilton et al., 1974) and the Schilperoort lab (Dons, 1975) found no evidence of the Ti plasmid in such lines, even under conditions that allowed the detection of one copy of the Ti plasmid per tobacco tumor cell. This did not, however, eliminate the possibility that only part of the Ti plasmid was transferred into the plant cell. The ultimate solution to this problem came with the utilization of the, then, recently discovered restriction enzymes to digest the Ti plasmid into numerous smaller fragments. Chilton et al. (1977) isolated such fragments and used them, individually, in solution hybridization studies with DNA from tobacco crown gall tumor lines or non-transformed tobacco tissue culture lines. The results were unequivocal: *Sma*I fragment 3c of the Ti plasmid from strain B6 hybridized to DNA in the B6 induced, octopine producing tobacco tumor line, E9, that had been grown, bacteria-free, in culture for several years; moreover, it did not hybridize with DNA from non-transformed cell lines (Chilton et al., 1977). These authors stated: "Our results suggest that the tumor inducing principle first proposed by Braun (1947) is indeed DNA as many investigators have long suspected".

2.5 The T-DNA of the Ti plasmid: Structure, function and transfer

The work described above demonstrated the critical importance of the Ti plasmid in tumorigenesis and the fact that a portion of it is delivered to, and maintained in, transformed plant cells. Three critical questions were subsequently addressed: What, exactly, is the T-DNA and where does it reside? How does the T-DNA result in tumorous growth? What functions encoded by the Ti plasmid – and the chromosome – are necessary for T-DNA transfer and how do they carry out this function? Genetic, molecular and biochemical approaches to these problems ultimately have answered the first two of these questions and made major inroads on the third. (Several of the chapters in this volume will review these questions in more detail than space here can provide.). In relation to the first question, Chilton et al. (1978) and DePicker et al. (1978) simultaneously reported the observation of 'common' DNA in the T-DNA of octopine and nopaline type tumors with the latter study also demonstrating that an insertion (of RP4) in this region resulted in a loss of virulence. Southern blot analysis was soon applied to the T-DNA problem and it quickly became apparent that the T-DNA in crown gall tumor lines contained both this common DNA as well as non-conserved DNA that we now know encodes the synthesis of opine biosynthetic enzymes (Lemmers et al., 1980; Thomashow et al., 1980a). The T-DNA was found integrated into the nuclear genome (Willmitzer et al., 1980; Chilton et al., 1980). Southern blot analysis, and cloning and sequence analysis of fragments of integrated T-DNA, as well as the Ti plasmid, revealed that the boundaries of the T-DNA were marked by 23 bp direct repeats in all Ti plasmids and that the T-DNA could be inserted in both repeated and unique host DNA, that is, approximately randomly (Lemmers et al., 1980; Thomashow et al., 1980b; Yadav et al., 1980; Zambryski et al., 1980; Yadav et al., 1982; Barker et al., 1983; Wang et al., 1984). This latter observation proved crucial in the ultimate development of *Agrobacterium* as an agent of insertional mutagenesis that has proven so powerful in modern plant molecular genetics (see Alonso and Stepanova, 2003 for review).

Genetic and molecular analysis of the Ti plasmid revealed two basic sets of mutations in relation to virulence. First, insertions at various sites within the common region of the T-DNA affected tumor growth and morphology (Garfinkel et al., 1981; Ooms et al., 1981; Binns et al., 1982; Leemans et al., 1982; Binns, 1983). Depending on the site of insertion, these mutant T-DNAs cause either root or shoot forming tumors in tobacco, and this immediately led to the proposal that mutations leading to shoot forming tumors cause a deficiency in auxin accumulation whereas those mutations leading to root forming tumors cause a deficiency in cytokinin accumulation (e.g. Akiyoshi et al., 1984; van Onckelen et al., 1984). Mutations at both of these loci rendered the strain avirulent (Hille et al., 1983; Ream et al., 1983). Molecular analysis revealed the presence of polyadenylated transcripts from these loci as well as from the loci encoding opine synthesis (Willmitzer et al., 1982; Willmitzer et al., 1983). Subsequent biochemical and sequence analysis demonstrated that, in fact, a two step auxin biosynthesis pathway is encoded on the T-DNA (Inzé et al., 1984; Schröder et al., 1984; Thomashow et al., 1984; Thomashow et al., 1986; van Onckelen et al., 1986) as is a one step synthesis of cytokinin (Akiyoshi et al., 1984; Barry et al., 1984; Buchmann et al., 1985). Other genes in the common DNA were discovered that appear to have functions that modify or indirectly affect hormone production and/or response.

The second critical set of genes on the Ti plasmid were those shown to be outside of the T-DNA but required for its transfer into plant cells, first examined in the Schilperoort and Nester labs (Garfinkel and Nester, 1980; Ooms et al., 1980; Klee et al., 1983). Importantly, work from the Schilperoort lab showed that these genes in the "virulence" (*vir*) region of the Ti plasmid work in *trans* to the T-DNA (Hoekma et al., 1983), a fact that has become very useful in the development of *Agrobacterium* as a vector (see below). Chromosomal genes were also discovered to be important in the processes leading to T-DNA transfer (Garfinkel and Nester, 1980) including some necessary for attachment of the bacterium to the plant cell (Douglas et al., 1982). Finally, the landmark works of Stachel and Nester (Stachel and Nester, 1986), and Zambryski and colleagues (Stachel et al., 1985; Stachel et al., 1986; Stachel and Zambryski, 1986; Stachel et al., 1987) identifying the VirA/VirG two component regulatory system that controls *vir* gene expression as well as the discovery of a single stranded DNA intermediate (the T-strand) set the stage for work on the virulence region for the next two decades. As will be described in considerable detail in other chapters of this volume, these experiments as well as those from many other labs, defined the *vir*-region genes that are required for host recognition, the gene products required for production of the T-strand substrate (as well as other transported virulence effectors) and the mechanism of T-strand and protein transfer into the plant cell. Perhaps the most striking aspect of studies on the virulence region is that they have moved *A. tumefaciens* into the position of a model for pathogenic bacteria in general. For example, the VirA/VirG two-component system that regulates virulence gene expression is now recognized as the best developed such system that responds to multiple host-derived signals. Even more impressive has been the characterization of the Type IV secretion system (VirB complex) that mediates transfer of virulence factors to eukaryotic host cells. This clearly serves as one of the model Type IV secretion systems used by bacteria in both pathogenesis and interbacterial conjugation. Both of these are topics of other chapters in this volume.

3 *A. TUMEFACIENS* AS THE VECTOR OF CHOICE FOR PLANT GENETIC ENGINEERING

As the biology of *A. tumefaciens* mediated transformation was coming to be understood, so to were the underlying features of the system that have allowed it to become the vector of choice for gene transfer experiments in current plant biology and agriculture. As described above, the study of *Agrobacterium,* the pathogen, led us to understand that *A. tumefaciens* moves DNA into plant cells and converts them into a population of dividing cells that are dedicated to the production of opines, a source of nitrogen and carbon that can be used by the inciting bacterium, but not by the plant. This is an exquisite system that could be called 'nature's first plant genetic engineer'. The major question confronting students of *A. tumefaciens* in the late 1970's and early '80s was whether this capacity could be exploited to direct the transfer of a specific gene, or genes, selected by an investigator, into a plant and specifically into the germ line of a plant. The convergence of ever-more sophisticated plant cell culture protocols combined with the molecular genetics of the *Agrobacterium* system allowed this field to progress rapidly.

3.1 Setting the stage—the analysis of crown gall teratomas

Not surprisingly, the origins of such work trace back to studies by Armin Braun in the late 1940s and early '50s. He had become interested in a class of crown gall tumors observed by Smith (1916) and Levin and Levine (Levin and Levine, 1918; Levine, 1919) that spontaneously formed a chaotic assemblage of differentiated tissues (leaves, shoot-like structures, etc) and were termed teratomas. A great deal of the early debate centered on the question of whether these differentiated structures were tumor cells that had differentiated or were non-transformed cells that differentiated abnormally under the influence of the tumor. Evidence for the former was that inoculations of, for example, decapitated stems would lead first to an unorganized tumor which would subsequently form differentiated tissues (Levin and Levine, 1918). On the other hand, inoculations near axillary buds would affect their development, particularly after decapitation. Braun's earliest work on this topic (Braun, 1948) demonstrated that some strains, e.g. T37, induced teratomas on *Kalanchoe daigremontiana* whereas other strains, e.g. B2, B6,induced typical unorganized tumors. (The earlier studies reporting on teratoma formation did not always specify strains used in their inoculations, though Smith (1916) reported using a "hop strain").

Braun (1948) took advantage of the interesting developmental pattern exhibited by *Kalanchoe* leaves – they form new plantlets at their margins via vegetative reproduction – in an attempt to understand whether the differentiated tissues of the T37 induced teratomas were normal or tumorous. Leaves from the teratomas exhibiting varying degrees of normalcy were cultured on White's basic medium and plantlets originating from them characterized. Growths from these leaf margins ranged from tumor-like growths to abnormal plantlets with no root system, to very occasional plants with a complete root system. Even though the original teratomic leaves arose from the tumor, Braun noted that these results could be "...explained on the basis of a mixture of tumor and normal cells in the same structure". Similar results were obtained when T37 induced teratomas of tobacco were studied (Braun, 1951b), though the extent to which tissue culture and grafting studies could be utilized were much greater than with *Kalanchoe*. For example, the original teratomas could be cultured over a period of years and still maintain the teratogenic phenotype, but never made root-derived structures. Procedures in which shoot buds from the teratomas were 'scions' in grafts to normal tobacco plants as 'stock' yielded normal appearing shoots (Braun, 1951b). Tissues from these shoots, when returned to culture, reverted to the teratomatous phenotype. Intriguingly, these shoots were fertile and the progeny were completely non-tumorigenic – they formed roots and appeared to have 'recovered' from the effects of the TIP (Braun, 1951b). These data appeared to support the hypothesis that the TIP might be some type of cytoplasmic self-duplicating entity that could be 'diluted' away as a result of forcing rapid growth of the teratoma derived buds. However, it was not until 8 years later, with the advent of single cell cloning procedures, that Braun (1959) could conclude that the capacity of teratoma tissues to generate highly differentiated, organized tissues "...is a reflection of the inherent potentialities of pluripotent tumor cells and not the result...of a mixture of normal and tumor cells". Moreover, grafted shoots derived from the single-cell cloned teratoma lines were, as in the earlier study, fertile and progeny from them were normal in every respect (Braun, 1959).

3.2 Fate of the T-DNA in plants regenerated from *A. tumefaciens*-transformed cells

The results described above indicated that transformation by *A. tumefaciens* does not necessarily result in permanent changes in the plant genome that keep it from being a completely normal cell. As the role of the Ti plasmid and T-DNA in tumorigenesis was elucidated, one important question became: what is the status of the T-DNA in the grafted teratoma shoots and in the progeny? Braun and collaborators used a series of biological, biochemical and, ultimately, molecular assays on teratomas induced by strain T37 to address the issue. They found that the grafted teratoma shoots contained nopaline (Braun and Wood, 1976). Moreover, all the specialized tissues and/or cells of these grafted shoots had the capacity to revert to the tumorous state when returned to culture (Braun and Wood, 1976; Turgeon et al., 1976; Wood et al., 1978; Binns et al., 1981). The oncogenic properties of the T-DNA had, somehow, been suppressed, allowing for the differentiation of specialized tissues but, intriguingly, these shoots could still express the T-DNA encoded enzyme necessary for opine synthesis (Wood et al., 1978). This was very good news for those proposing that the *A. tumefaciens* T-DNA could be a useful system for plant genetic engineering: here was foreign DNA in a normal shoot synthesizing a functional, foreign genome encoded enzyme (nopaline synthase).

The major problem, however, was the fact that although these T37 transformed, grafted shoots were fertile, the progeny, as originally described by Braun, were completely normal: they made roots, did not synthesize nopaline and tissues from them did not grow in culture as tumors (Braun and Wood, 1976; Binns et al., 1981). Moreover, Southern blot analysis of the progeny tissues indicated they did not contain the T-DNA (Yang et al., 1980). Was the genome being actively scanned for foreign DNA, which was somehow removed during meiosis? Or, were the cells that received the oncogenic DNA after meiotic segregation incapable of becoming functional germ cells? Evidence for the latter came from a series of experiments in which the more usual unorganized tumors induced by octopine strains such as B6 were treated with conditions that normally induce shoot formation in non-transformed tissues. Occasional regenerants were observed, and these had generally lost the T-DNA, though in one particular case the opine synthesizing portion of the T-DNA was retained, and, more importantly, it was passed on to its progeny (De Greve et al., 1982). This strongly suggested that if the T-DNA were made incapable of causing tumors, for example by mutagenizing the common or oncogenic DNA, then transmission to progeny would likely occur. This was predicated on the notion that transformation by the non-oncogenic strain could still occur – a good bet given Braun's early work distinguishing inception vs growth (see above).

Of course, the big problem with such a strategy was in the identification of cells transformed by non-oncogenic strains: without the tumorous phenotype the transformed cells would not be at a selective advantage and therefore be difficult to find. However, opine synthesis could be observed in plant tissues transformed strains carrying Ti plasmids with insertion mutations in the common DNA that rendered the strain avirulent – that is no tumors were produced (Barton et al., 1983; Hille et al., 1983; Ream et al., 1983). Barton et al (1983) screened for nopaline production in tissues from tobacco stem segments transformed with strain T37 that carried an engineered insertion in the cytokinin locus of its T-DNA. Such tissues were cultured, single cell cloned, and the nopaline positive clones were treated with standard tobacco shoot regeneration protocols. The resultant shoots were capable of forming roots, contained the full length T-DNA, synthesized nopaline and, importantly, transmitted the full length T-DNA to the progeny (Barton et al., 1983).

3.3 Construction of selectable markers provides the capacity to easily identify transformed cells carrying non-oncogenic T-DNA

Clearly, *Agrobacterium* could be used to regularly generate transgenic plants. But the opine screening protocol was tedious and labor intensive. The next steps setting the stage for current use of *Agrobacterium* as a vector were (i) the development of selectable markers that could be used to identify transformed plant cells without affecting the regenerative potential of the host and (ii) the removal of all the oncogenes without affecting T-DNA transfer. Studies on the expression of the T-DNA in tumors revealed that the opine synthesis genes were highly expressed (Willmitzer et al., 1982; Willmitzer et al., 1983). This suggested that their promoters could be used to drive expression of, for example, antibiotic resistance genes that would protect transformed plant cells from the normally toxic effects of molecules such as kanamycin. Such studies were accomplished nearly simultaneously in the labs of Van Montagu and Schell in Ghent and Köln (Herrera-Estrella et al., 1983), Fraley and colleagues in St. Louis (Fraley et al., 1983) and Bevan and colleagues in Cambridge (Bevan et al., 1983). Once available these were immediately transferred into the T-DNA of appropriately disarmed Ti plasmids (lacking both the auxin and cytokinin biosynthesis genes) and used to select for transformed cells that could be regenerated into fertile transgenic plants that would transmit the engineered DNA to their progeny (Zambryski et al., 1983; Horsch et al., 1984). The development and refinement of these strategies is detailed in Chapter 3.

4 CONCLUSIONS

I would argue that this brief examination of the 'history of crown gall research' has supported lessons that actually have been taught over and over again. First, an intriguing biological problem is important to study even when the ideas foremost in the thoughts of the investigator(s) may ultimately prove incorrect. Smith, for example, was convinced that studies on crown gall would reveal that bacteria were a cause, generally, of cancers in animals as well as plants. Braun was equally certain, based on the studies demonstrating the reversal of the tumorous phenotype, that crown gall specifically, and many cancers generally, were the result of epigenetic changes induced by the causal agent. Though these hypotheses were disproven, the science that generated them was extremely solid and provoked other, equally solid science that ultimately unraveled the story as we now know it. The second major lesson is that key advances in crown gall research have been (and continue to be) driven by technological advances in other arenas of science. In the case of crown gall these include: the development of sterile technique and various other microbiological methodologies used in the elucidation of *Agrobacterium* as the causal agent of crown gall; the advance of plant tissue culture techniques in studies demonstrating bacteria-free crown galls grow autonomously and hence are transformed as well as those studies related to the regeneration of transgenic plants; the biochemistry of amino acid and metabolite analysis used to unravel the opine issue; and the methodologies of plasmid characterization, restriction enzymes, transposon mutagenesis and sequence analysis so critical in the understanding of the Ti plasmid and its role in tumorigenesis. Of course the role of advances in technology as drivers of science is obvious, but it certainly is useful and interesting to see the advances at work as the best minds in the field sought to unravel the incredible biological activity of *Agrobacterium tumefaciens* and develop it into a tool that is so critical to modern plant biology and agricultural biotechnology.

5 ACKNOWLEDGEMENTS

I dedicate this chapter to the memory of Dr. Armin C. Braun who, amongst all other qualities, treasured the power of science, particularly the truth that can be uncovered by good data carefully considered. Thanks to Dr. Arlene Wise for reading an earlier version of this manuscript. I gratefully acknowledge the support of National Science Foundation and the National Institutes of Health for work in my laboratory.

6 REFERENCES

Akiyoshi DE, Klee H, Amasino RM, Nester EW, Gordon MP (1984) T-DNA of *Agrobacterium tumefaciens* encodes an enzyme of cytokinin biosynthesis. Proc Natl Acad Sci USA 81: 5994-5998

Alonso JM, Stepanova AN (2003) T-DNA mutagenesis in *Arabidopsis*. Methods Mol Biol 236: 177-188

Barker R, Idler KB, Thompson DV, Kemp JD (1983) Nucleotide sequence of the T-DNA region from the *Agrobacterium tumefaciens* octopine Ti plasmid pTi 15955. Plant Mol Biol 2: 335-350

Baron C, Zambryski PC (1995) The plant response in pathogenesis, symbiosis and wounding: variations on a common theme? Annu Rev Genet 29: 107-129

Barry GF, Rogers SG, Fraley RT, Brand L (1984) Identification of cloned cytokinin biosynthesis gene. Proc Natl Acad Sci USA 81: 4776-4780

Barton KA, Binns AN, Matzke AJM, Chilton MD (1983) Regeneration of intact tobacco plants containing full length copies of genetically engineered T-DNA, and transmission of T-DNA to R1 progeny. Cell 32: 1033-1043

Bevan MW, Flavell RB, Chilton M-D (1983) A chimeric antibiotic resistance gene as a selectable marker for plant transformation. Nature 304: 184-187

Binns AN (1983) Host and T-DNA encoded determinants of cytokinin autonomy in tobacco cells transformed by *Agrobacterium tumefaciens*. Planta 158: 272-279

Binns AN, Sciaky D, Wood HN (1982) Variation in hormone autonomy and regeneration potential of cells transformed by strain A66 of *Agrobacterium tumefaciens*. Cell 31: 605-612

Binns AN, Wood HN, Braun AC (1981) Suppression of the tumourous state in tobacco crown gall teratomas: A clonal analysis. Differentiation 19: 97-102

Bomhoff G, Klapwijk PM, Kester HCM, Schilperoort RA, Hernalsteens JP, Schell J (1976) Octopine and nopaline synthesis and breakdown genetically controlled by a plasmid of *Agrobacterium tumefaciens*. Mol Gen Genet 145: 177-181

Braun AC (1943) Studies on tumor inception in the crown-gall disease. Am J Bot 30: 674-677

Braun AC (1947) Thermal studies on tumor inception in the crown gall disease. Am J Bot 34: 234-240

Braun AC (1948) Studies on the origin and development of plant teratomas incited by the crown-gall bacterium. Am J Bot 35: 511-519
Braun AC (1951a) Cellular autonomy in crown gall. Phytopathology 41: 963-966
Braun AC (1951b) Recovery of crown gall tumor cells. Cancer Res 11: 839-844
Braun AC (1952) Conditioning of the host cell as a factor in the transformation process in crown gall. Growth 16: 65-74
Braun AC (1958) A physiological basis for the autonomous growth of the crown gall tumor cell. Proc Natl Acad Sci USA 44: 344-349
Braun AC (1959) A demonstration of the recovery of the crown-gall tumor cell with the use of complex tumors of single cell origin. Proc Natl Acad Sci USA 45: 932-938
Braun AC (1981) An epigenetic model for the origin of cancer. Q Rev Biol 56: 33-60
Braun AC (1982) A history of the crown gall problem. *In* G Kahl, J Schell, eds, Molecular Biology of Plant Tumors. Academic Press, New York, pp 155-210
Braun AC, Laskaris T (1942) Tumor formation by attenuated crown-gall bacteria in the presence of growth promoting substances. Proc Natl Acad Sci USA 28: 468-477
Braun AC, Mandle RJ (1948) Studies on the inactivation of the tumor inducing principle in crown gall. Growth 12: 255-269
Braun AC, White PR (1943) Bacteriological sterility of tissues derived from secondary crown-gall tumors. Phytopathology 33: 85-100
Braun AC, Wood HN (1976) Suppression of the neoplastic state with the acquistion of specialized functions in cells, tissues and organs of crown-gall teratomas of tobacco. Proc Natl Acad Sci USA 73: 496-500
Brencic A, Angert ER, Winans SC (2005) Unwounded plants elicit *Agrobacterium vir* gene induction and T-DNA transfer: transformed plant cells produce opines yet are tumor free. Mol Microbiol 57: 1522-1531
Buchmann I, Marner F-J, Schröder G, Waffenschmidt S, Schröder J (1985) Tumor genes in plants: T-DNA encoded cytokinin biosynthesis. EMBO J 4: 853-859
Cavara F (1897) Intorno alla eziologia di alcune malattie di plant cultivate. Stn Sper Agrar Ital Modena 30: 483-487
Chilton M-D, Drummond MH, Merio DJ, Sciaky D, Montoya AL, Gordon MP, Nester EW (1977) Stable incorporation of plasmid DNA into higher plant cells: the molecular basis of crown gall tumorigenesis. Cell 11: 263-271
Chilton M-D, Drummond MH, Merlo DJ, Sciaky D (1978) Highly conserved DNA of Ti plasmid overlaps T-DNA maintained in plant tumors. Nature 275: 147-149
Chilton M-D, Farrand SK, Eden FC, Currier TC, Bendich AJ, Gordon MP, Nester EW (1974) Is there foreign DNA in crown gall tumor DNA? *In* R Markham,

DR Davies, D Hopwood, RW Horne, eds, Modification of the Information Content of Plant Cells. Elsevier, New York, p 297

Chilton M-D, Saiki RK, Yadav N, Gordon MP, Quétier F (1980) T-DNA from *Agrobacterium* Ti plasmid is in the nuclear fraction of crown gall tumor cells. Proc Natl Acad Sci USA 77: 2693-2697

De Greve H, Leemans J, Hernalsteens JP, Thia-Toong L, De Beukeleer M, Willmitzer L, Otten L, Van Montagu M, Schell J (1982) Regeneration of normal and fertile plants that express octopine synthase from tobacco galls after deletion of tumor-controlling functions. Nature 300: 752-757

Depicker A, Van Montagu M, Schell J (1978) Homologous DNA sequences in different Ti-plasmids are essential for oncogenicity. Nature 275: 150-153

Dons JJM (1975) Crown gall – a plant tumor. Investigation on the nuclear content and on the presnce of *Agrobacterium tumefaciens* DNA and phage PS8 DNA in crown gall tumor cells. Ph. D. dissertation. Leiden, The Netherlands

Douglas CJ, Halperin W, Nester EW (1982) *Agrobacterium tumefaciens* mutants affected in attachment to plant cell. J Bacteriol 152: 1265-1275

Drlica KA, Kado CI (1975) Crown gall tumors: are bacterial nucleic acids involved? Bacteriol Rev 39: 186-196

Farrand SK, Van Berkum PB, Oger P (2003) *Agrobacterium* is a definable genus of the family *Rhizobiaceae*. Int J Syst Evol Microbiol 53: 1681-1687

Fraley RT, Rogers SG, Horsch RB, Sanders PR, Flick JS, Adams SP, Bittner ML, Brand LA, Fink CL, Fry JS, Galluppi GR, Goldberg SB, Hoffmann NL, Woo SC (1983) Expression of bacterial genes in plant cells. Proc Natl Acad Sci USA 80: 4803-4807

Garfinkel DJ, Nester EW (1980) *Agrobacterium tumefaciens* mutants affected in crown gall tumorigenesis and octopine catabolism. J Bacteriol 144: 732-743

Garfinkel DJ, Simpson RB, Ream LW, White FF, Gordon MP, Nester EW (1981) Genetic analysis of crown gall: fine structure map of the T-DNA by site-directed mutagenesis. Cell 27: 143-153

Goldmann A, Tempé J, Morel G (1968) Quelques particularités de diverses souches d'*Agrobacterium tumefaciens*. CR Seances Soc Biol Ses Fil 162: 630-631

Hamilton RC, Fall MZ (1971) The loss of tumor initiating ability in *Agrobacterium tumefaciens* by incubation at high temperature. Experientia 27: 229-230

Hendrickson AA, Baldwin IL, Riker AJ (1934) Studies on certain physiological characters of *Phytomonas tumefaciens*, *Phytomonas rhizogenes* and *Phytomonas radiobacter*. J Bacteriol 28: 597-618

Herrera-Estrella L, Depicker A, Van Montagu M, Schell J (1983) Expression of chimaeric genes transferred into plant cells using a Ti plasmid derived vector. Nature 303: 209-213

Hille J, Wullems GJ, Schilperoort RA (1983) Non-oncogenic T-region mutants of *Agrobacterium tumefaciens* do transfer T-DNA into plant cells. Plant Mol Biol 2: 155-164

Hoekma A, Hirsch PR, Hooykaas PJJ, Schilperoort RA (1983) A binary vector strategy based on separation of the vir and T-region of the *Agroabcterium tumefaciens* Ti plasmid. Nature 303: 179-180

Horsch RB, Fraley RT, Rogers SG, Sanders PR, Lloyd A, Hoffmann N (1984) Inheritance of functional foreign genes in plants. Science 223: 496-497

Inzé D, Follin A, Van Lijsebettens M, Simoens C, Genetello C, et al. (1984) Genetic analysis of the individual T-DNA genes of Agrobacterium tumefaciens: further evidence that two genes are involved in indole-3-acetic acid sysnthesis. Mol Gen Genet 194: 265-274

Kahl G, Schell J (1982) Molecular Biology of Plant Tumors. Academic Press, New York

Kerr A (1969) Transfer of virulence between isolates of *Agrobacterium*. Nature 223: 1175-1176

Klee HJ, White FF, Iyer VN, Gordon MP, Nester EW (1983) Mutational analysis of the virulence region of an *Agrobacterium tumefaciens* Ti plasmid. J Bacteriol 153: 878-883

Kunkel LO (1941) Heat cure of aster yellows in periwinkle. Am J Bot 28: 761-769

Leemans J, Deblaere R, Willmitzer L, De Greeve H, Hernalsteens JP, Van Montagu M, Schell J (1982) Genetic identification of functions of T_L-DNA transcripts in octopine crown galls. EMBO J 1: 147-152

Lejeune B, Jubier MF (1967) Etude de la dégradation de la lysopine par *Agrobacterium tumefaciens*. C R Hebd Seances Acad Sci Ser D 264: 1803-1805

Lemmers M, De Beuckeleer M, Holsters M, Zambryski P, Depicker A, Hernalsteens JP, Van Montagu M, Schell J (1980) Internal organization, boundaries and integration of Ti-plasmid DNA in nopaline crown gall tumours. J Mol Biol 144: 353-376

Levin I, Levine M (1918) Malignancy of the crown gall and its analogy to animal cancer. Proc Soc Exp Biol and Med 16: 21-22

Levine M (1919) Studies on plant cancers - I. The mechanism of the formation of the leafy crown gall. Bull Torr Bot Soc 46: 447-452

Lioret C (1957) Les acides aminés libres des tissus de crown-gall. Mise en évidence d'un acide aminé particulier à ces tissus. CR Hebd Seances Acad Sci 244: 2171-2174

Lippincott BB, Lippincott JA (1969) Bacterial attachment to a specific wound site as an essential stage in tumor initiation by *Agrobacterium tumefaciens*. J Bacteriol 97: 620-628

Lippincott BB, Whatley MH, Lippincott JA (1977) Tumor induction by *Agrbabcterium tumefaciens* involves attachment of the bacterium to a site on the host plant cell wall. Plant Physiol 59: 388-390

Lippincott JA, Lippincott BB (1978) Cell walls of crown-gall tumors and embryonic tissues lack *Agrobacterium* adherence sites. Science 109: 1075-1078

Locke SB, Riker AJ, Duggar BM (1938) Growth substance and the development of crown gall. J Agr Res 57: 21-39

McCullen CA, Binns AN (2006) *Agrobacterium tumefaciens* and plant cell interactions and activities required for interkingdom macromolecular transfer. Annu Rev Cell Dev Biol 22: 101-127

Meins FJ, Binns AN (1978) Epigenetic clonal variation in the requirement of plant cells for cytokinins. *In* S Subtelny, IM Sussex, eds, The Clonal Basis of Development. Academic Press, New York, pp 185-201

Moore L, Warren G, Strobel G (1979) Involvement of a plasmid in the hairy root disease of plants caused by *Agrobacterium rhizogenes*. Plasmid 2: 617-626

Neff NT, Binns AN (1985) *Agrobacterium tumefaciens* interaction with suspension-cultured tomato cells. Plant Physiol 77: 35-42

Ooms G, Hooykaas PJJ, Noleman G, Schilperoort RA (1981) Crown gall tumors of abnormal morphology induced by *Agrobacterium tumefaciens* carrying mutated octopine Ti plasmids: analysis of T-DNA functions. Gene 14: 33-50

Ooms G, Klapwijk PM, Poulis JA, Schilperoort RA (1980) Characterization of Tn904 insertions in octopine Ti plasmid mutants of *Agrobacterium tumefaciens*. J Bacteriol 144: 82-91

Petit A, Delhaye S, Tempé J, Morel G (1970) Recherches sur les guanidines des tissus de crown-gall. Mise en evidence d'une relation biochimique spécifique entre les souches d'*Agrobacterium tumefaciens* et les tumeurs qu'elles induisent. Physiol Veg 8: 205-213

Petit A, Tourneur J (1972) Perte de virulence associée à la perte d'une activité enzymatique chez *Agrobacterium tumefaciens*. C R Acad Sci Paris Life Sciences 275: 137-139

Ream LW, Gordon MP, Nester EW (1983) Multiple mutations in the T region of the *Agrobacterium tumefaciens* tumor-inducing plasmid. Proc Natl Acad Sci USA 80: 1660-1664

Riker AJ (1923a) Some morphological resoponses of the host tissue to the crown gall organism. J Agric Res 26: 425-435

Riker AJ (1923b) Some relations of the crown gall organism to its host tissue. J Agric Res 25: 119-132

Riker AJ (1926) Studies on the influence of some environmental factors on the development of crown gall. J Agric Res 32: 83-96

Schröder G, Waffenschmidt S, Weiler EW, Schröder J (1984) The T-region of Ti plasmids codes for an enzyme synthesizing indole-3-acetic acid. Eur J Biochem 138: 387-391

Skoog F, Miller CO (1957) Chemical regulation of growth and organ formation in plant tissues cultured in vitro. Symp Soc Exp Biol 11: 118-131

Smith EF (1916) Crown gall studies showing changes in plant structures due to a changed stimula. Jour Agric Res 6: 179-182 (plus plates)

Smith EF, Brown NA, Townsend CO (1912) The structure and development of crown gall: A plant cancer. US Dept Agric Bur Plant Ind Bull 255: 1-61

Smith EF, Townsend CO (1907) A plant tumor of bacterial origin. Science 25: 671-673

Stachel SE, Messens E, Van Montagu M, Zambryski P (1985) Identification of the signal molecules produced by wounded plant cell that activate T-DNA transfer in *Agrobacterium tumefaciens*. Nature 318: 624-629

Stachel SE, Nester EW (1986) The genetic and transcriptional organization of the *vir* region of the A6 Ti plasmid of *Agrobacterium tumefaciens*. EMBO J 5: 1445-1454

Stachel SE, Timmerman B, Zambryski P (1986) Generation of single-stranded T-DNA molecules during the initial stages of T-DNA transfer for *Agrobacterium tumefaciens* to plant cells. Nature 322: 706-712

Stachel SE, Timmerman B, Zambryski P (1987) Activation of *Agrobacterium tumefaciens vir* gene expression generates multiple single-stranded T-strand molecules from the pTiA6 T-region: requirement for 5' *virD* gene products. EMBO J 6: 857-863

Stachel SE, Zambryski PC (1986) *virA* and *virG* control the plant-induced activation of the T-DNA transfer process of *A. tumefaciens*. Cell 46: 325-333

Tempé J, Goldmann A (1982) Occurence and biosynthesis of opines. *In* G Kahl, JS Schell, eds, Molecular Biology of Plant Tumors. Academic Press, New York, pp 427-449

Thomashow LS, Reeves S, Thomashow MF (1984) Crown gall oncogenesis: evidence that a T-DNA gene from the *Agrobacterium* Ti plasmid pTiA6 encodes an enzyme that catalyzes synthesis of indole-3-acetic acid. Proc Natl Acad Sci USA 81: 5071-5075

Thomashow MF, Hughly S, Buchholz WG, Thomashow LS (1986) Molecular basis for the auxin-independent phenotype of crown gall tumor tissue. Science 231: 616-618

Thomashow MF, Nutter R, Montoya AL, Gordon MP, Nester EW (1980a) Integration and organization of Ti plasmid sequences in crown gall tumors. Cell 19: 729-739

Thomashow MF, Nutter R, Postle K, Chilton M-D, Blattner FR, Powell A, Gordon MP, Nester EW (1980b) Recombination between higher plant DNA and the Ti plasmid of *Agrobacterium tumefaciens*. Proc Natl Acad Sci USA 77: 6448-6452

Turgeon R, Wood HN, Braun AC (1976) Studies on the recovery of crown gall tumor cells. Proc Natl Acad Sci USA 73: 3562-3564

Van Larebeke N, Engler G, Holsters M, Van den Elsacker S, Zaenen I, Schilperoort RA, Schell J (1974) Large plasmid in *Agrobacterium tumefaciens* essential for crown gall-inducing ability. Nature 252: 169-170

Van Larebeke N, Genetello C, Schell J, Schilperoort RA, Hermans AK, Van Montagu M, Hernalsteens JP (1975) Acquisition of tumour-inducing ability by non-oncogenic agrobacteria as a result of plasmid transfer. Nature 255: 742-743

Van Onckelen H, Prinsen E, Inzé D, Rudelsheim P, Van Lijsebettens M, Follin A, Schell J, Van Montagu M, De Greef J (1986) *Agrobacterium* T-DNA gene *1* codes for tryptophan 2-monooxygenase activity in tobacco crown gall cells. FEBS Lett 198: 357-360

Van Onckelen H, Rudelsheim P, Hermans R, Horemans S, Messens E, Hernalsteens J-P, Van Montagu M, De Greef J (1984) Kinetics of endogenous cytokinin, IAA and ABA levels in relation to the growth and morphology of tobacco crown gall tissue. Plant Cell Physiol 25: 1017-1025

Wang K, Herrera-Estrella L, Van Montagu M, Zambryski P (1984) Right 25 bp terminus sequence of the nopaline T-DNA is essential for and determines direction of DNA transfer from *Agrobacterium* to the plant genome. Cell 38: 455-462

Watson B, Currier TC, Gordon MP, Chilton M-D, Nester EW (1975) Plasmid required for virulence of *Agrobacterium tumefaciens*. J Bacteriol 123: 255-264

White FF, Nester EW (1980) Hairy root: plasmid encodes virulence traits in *Agrobacterium rhizogenes*. J Bacteriol 141: 1134-1141

White PR, Braun AC (1941) Crown gall production by bacteria free tumor tissues. Science 94: 239-241

Willmitzer L, De Beuckeleer M, Lemmers M, Van Montagu M, Schell J (1980) DNA from Ti plasmid present in nucleus and absent from plastids of crown gall plant cells. Nature 287: 359-361

Willmitzer L, Dhaese P, Schreier PH, Schmalenbach W, Van Montagu M, Schell J (1983) Size, location and polarity of T-DNA-encoded transcripts in nopaline crown gall tumos-common transcripts in octopine and nopaline tumors. Cell 32: 1045-1056

Willmitzer L, Simons G, Schell J (1982) The T_L-DNA in octopine crown-gall tumors codes for seven well-defined polyadenylated transcripts. EMBO J 1: 139-146

Wood HN, Binns AN, Braun AC (1978) Differential expression of oncogenicity and nopaline synthesis in intact leaves derived from crown gall teratomas of tobacco. Differentiation 11: 175-180

Yadav NS, Postle K, Saiki RK, Thomashow MF, Chilton M-D (1980) T-DNA of a crown gall teratoma is covalently joined to host plant DNA. Nature 287: 458-461

Yadav NS, Vanderleyden J, Bennet DR, Barnes WM, Chilton M-D (1982) Short direct repeats flank th T-DNA on a nopaline Ti plasmid. Proc Natl Acad Sci USA 79: 6322-6326

Yang F, Montoya AL, Merlo DJ, Drummond MH, Chilton M-D, Nester EW, Gordon MP (1980) Foreign DNA sequences in crown gall teratomas and their fate during the loss of the tumorous traits. Mol Gen Genet 177: 707-714

Young JM, Kuykendall LD, Martinez-Romero E, Kerr A, Sawada H (2001) A revision of *Rhizobium* Frank 1889, with an emended description of the genus, and the inclusion of all species of *Agrobacterium* Conn 1942 and *Allorhizobium undicola* de Lajudie et al. 1998 as new combinations: *Rhizobium radiobacter*, *R. rhizogenes*, *R. rubi*, *R. undicola* and *R. vitis*. Int J Syst Evol Microbiol 51: 89-103

Zaenen I, Van Larebeke N, Tenchy H, Van Montagu M, Schell J (1974) Supercoiled circular DNA in crown gall inducing *Agrobacterium* strains. J Mol Biol 86: 109-127

Zambryski P, Holsters M, Kruger K, Depicker A, Schell J, Van Montagu M, Goodman HM (1980) Tumor DNA structure in plant cells transformed by *A. tumefaciens*. Science 209: 1385-1391

Zambryski PC, Joos H, Genetello C, Leemans J, Van Montagu M, Schell J (1983) Ti plasmid vector for the introduction of DNA into plant cells without alteration of their normal regeneration capacity. EMBO J 2: 2143-2150

Zhu Y, Nam J, Humara JM, Mysore KS, Lee LY, Cao H, Valentine L, Li J, Kaiser AD, Kopecky AL, Hwang HH, Bhattacharjee S, Rao PK, Tzfira T, Rajagopal J, Yi H, Veena, Yadav BS, Crane YM, Lin K, Larcher Y, Gelvin MJ, Knue M, Ramos C, Zhao X, Davis SJ, Kim SI, Ranjith-Kumar CT, Choi YJ, Hallan VK, Chattopadhyay S, Sui X, Ziemienowicz A, Matthysse AG, Citovsky V, Hohn B, Gelvin SB (2003) Identification of Arabidopsis *rat* mutants. Plant Physiol 132: 494-505

Chapter 3

AGROBACTERIUM AND PLANT BIOTECHNOLOGY

Lois M. Banta and Maywa Montenegro

Department of Biology, Williams College, Williamstown, MA 01267, USA

Abstract. *Agrobacterium*-mediated transformation has revolutionized agriculture as well as basic research in plant molecular biology, by enabling the genetic modification of a wide variety of plant species. Advances in binary vector design and selection strategies, coupled with improvements in regeneration technology and gene delivery mechanisms, have dramatically extended the range of organisms, including grains, that can be transformed. Recent innovations have focused on methods to stack multiple transgenes, to eliminate vector backbone sequences, and to target transgene insertion to specific sites within the host genome. Public unease with the presence of foreign DNA sequences in crop plants has driven the development of completely marker-free transformation technology and molecular strategies for transgene containment. Among the many useful compounds produced in genetically modified plants are biodegradable plastics, primary and secondary metabolites with pharmaceutical properties, and edible vaccines. Crop plant productivity may be improved by introducing genes that enhance soil nutrient utilization or resistance to viral, bacterial, or fungal diseases. Other transgenes have been shown to confer increased tolerance to many of the environmental constraints, including drought, extreme temperature, high salinity, and heavy metal soil contamination, faced by resource-poor farmers attempting to cultivate marginally arable land. Early applications of plant biotechnology focused primarily on traits that benefit farmers in industrialized regions of the world, but recent surveys document the degree to which this pattern is changing in favor of modified crops that contribute to enhanced ecological and human health. Documented decreases in the use of pesticides attributable to genetically engineered plants are harbingers of the health and environmental benefits that can be expected from transgenic crop plants designed to decrease reliance on harmful agrochemicals. As one thread in a network that also includes integrated pest and soil fertility management, a reduced emphasis on monoculture, and traditional crop breeding, plant genetic modification has the potential to help those who currently suffer from inadequate access to a full complement of nutrients. The development of "golden rice" illustrates the possibility to imbue a plant with enhanced nutritional value, but also the challenges posed by intellectual property considerations and the need to introduce novel traits into locally adapted varieties. Implementation of plant genetic modification within a framework of sustainable agricultural development will require increased attention to potential ecological impacts and technology-transcending socioeconomic ramifications. Successful technology transfer initiatives frequently involve collaborations between scientists in developing and industrialized nations; several non-profit agencies have evolved to facilitate formation of these partnerships. Capacity building is a core tenet of many such programs, and new paradigms for incorporation of indigenous knowledge at all stages of decision-making are under development. The complex (and sometimes controversial) social and scientific issues associated with the technology notwithstanding, *Agrobacterium*-mediated enhancement of agronomic traits provides novel approaches to address commercial, environmental, and humanitarian goals.

1 INTRODUCTION

Plant biotechnology has had a dramatic impact on agriculture, and on public awareness of the role of the private sector in industrial-scale farming in developed countries. This chapter focuses on the seminal contributions of *Agrobacterium tumefaciens* to this technological revolution, and on the applications of genetic engineering that continue to expand the limits of plant productivity. *Agrobacterium*-mediated transformation has yielded a stunning array of transgenic plants with novel properties ranging from enhanced agronomic performance, nutritional content, and disease resistance to the production of pharmaceuticals and industrially important compounds. Many of these advances have been made possible by creative and elegant methodological innovations that have enabled gene stacking, targeted mutagenesis, and the transformation of previously recalcitrant hosts.

Transgenic plants are not a panacea for global food shortages, distributional failures, or other structural causes of poverty. They can, however, have a positive impact on both human and environmental health. Agricultural biotechnology's image has been tarnished by the perception that it fails to address the needs of the world's hungry, and indeed most of the commercial products to date represent technology that is inappropriate for subsistence farmers (Huang et al., 2002a). As this chapter documents, there is ample potential for genetically modified plants to ameliorate some of the constraints faced by resource-poor farmers. Even modest enhancements of agronomic traits have the potential to help farmers overcome endemic problems such as lack of food security, limited purchasing power, and inadequate access to balanced nutritional resources (Leisinger, 1999). Many of these innovations will come from public sector research, and the vast majority of the applications described herein have in fact emanated from basic investigations and collaborative product-oriented research originating in the non-profit realm. As plant biotechnology research moves forward and outward to include more stakeholders in developing countries, it will continue to complement, rather than to replace, plant breeding (Morandini and Salamini, 2003). Whether these applications will enjoy increased public acceptance depends in large part on whether they progress in a context of sustainable development that incorporates integrated natural resource management and understanding of the socioeconomic realities of small-scale farming (Serageldin, 1999).

2 THE DEVELOPMENT OF *AGROBACTERIUM*-MEDIATED TRANSFORMATION

The first demonstration that *A. tumefaciens* could be used to generate transgenic plants (Barton et al., 1983 and see Chapter 2) heralded the beginning of a new era in agriculture as well as in plant molecular biology research. Plant transformation entails not only delivery and integration of engineered DNA into plant cells, but also the regeneration of transgenic plants from those genetically altered cells. Thus it was no accident that the earliest successes in plant genetic engineering occurred in species (e.g., tobacco, petunia, carrot and sunflower) that were both good hosts for *A. tumefaciens* and for which much was known about the conditions required to regenerate whole plants. Indeed, it has frequently been the plant tissue culture technology, rather than the transformation process itself, that has been the limiting step in achieving efficient genetic modification (Herrera-Estrella et al., 2005). Through extensive experimentation, protocols have been established for *Agrobacterium*-mediated transformation and regeneration of many other host plants including cotton (Umbeck et al., 1987), soybean (Hinchee et al., 1988), sugarbeet (D'Halluin et al., 1992), rice (Hiei et al., 1994), maize (Ishida et al., 1996), sorghum (Nguyen et al., 1996; Zhao et al., 2000), wheat (Cheng et al., 1997), barley (Tingay et al., 1997), papaya (Fitch et al., 1993), banana (May et al., 1995), and cassava (Li et al., 1996). Generation of transgenic monocots using *Agrobacterium*, initially believed to be impossible, is now considered routine for particular cultivars of some monocot species. However, transformation of several agronomically important cereal genotypes still poses significant challenges and represents an area where considerably more research is needed (S.B. Gelvin, personal communication).

2.1 Requirements for generation of transgenic plants

Generally speaking, *Agrobacterium*-mediated transformation involves incubating cells or tissues with bacteria carrying the foreign gene construct of interest, flanked by border sequences. Plant cells in which the foreign DNA has integrated into the genome are selected and propagated via a callus stage before hormone-induced regeneration of a transgenic plant, in which each cell is derived from the genetically altered progenitor cell (Walden and Wingender, 1995). Over the past two decades, a number of techniques have been developed to improve the efficiency of *Agrobacterium*-mediated gene delivery: wounding the plant tissue by sonication of embryonic suspension cultures, by glass beads, or by particle bombardment; bombardment with microprojectiles coated with agrobacteria; and imbibing germinating seeds have all proven successful in at least one host species. Other approaches are summarized in Newell (2000). The totipotency of plant cells has allowed the transformation of many different cell types, although tissues from different plant species respond differently to culture conditions, so optimal culture and regeneration methods must be established for every host tissue and species (Walden and Wingender, 1995). Explants are often used as the target for transformation because they are less prone to changes in DNA methylation status, chromosomal rearrangements and other genetic and epigenetic alterations that occur in plant tissue culture and that result in somaclonal variation (Christou, 1996). Hormone-induced regeneration of transgenic plants from transformed explants can occur via organogenesis (the direct formation of shoots) or somatic embryogenesis (the generation of embryos that can directly germinate into seedlings from somatic tissues). Most economically important plants, especially monocots, are regenerated using the latter approach, since callus is easily initiated from the scutellum of immature embryos (Hansen and Wright, 1999; Zuo et al., 2002). Delivery of the foreign DNA directly into meristematic tissue or immature embryos has also been found to limit somaclonal variation because it minimizes the amount of time in tissue culture (Walden and Wingender, 1995; Christou, 1996). A vacuum infiltration method, in which agrobacteria are applied to entire flowering *Arabidopsis*, was developed to avoid altogether the requirement for plant tissue culture or regeneration (Bechtold et al., 1993). More recently, this approach has been further simplified; in the "floral dip" process only the developing floral tissue is submerged into a solution of agrobacterial cells, and the labor-intensive vacuum infiltration step is eliminated (Clough and Bent, 1998).

In addition to susceptibility to *Agrobacterium* infection and the ability to regenerate whole plants from transformed cells, a third requirement for successful genetic modification is an efficient selection method for plant cells containing integrated trans-DNA (Chung et al., 2006). As described in Chapter 2, the first demonstration that the *Agrobacterium* lifestyle could be exploited to generate transgenic plants relied on a bacterial strain in which the T-DNA was still partially intact. Identification of transformed cells was achieved by screening for the production of nopaline (Barton et al., 1983). Published almost simultaneously, a number of other papers provided several key improvements on this initial transformation system. Foremost among these was the use of T-DNA-derived promoters and 3' regulatory regions (from the nopaline synthase gene) to drive *in planta* transcription of a bacterial antibiotic resistance gene such as chloramphenicol acetyltransferase or neomycin phosphotransferase (*nptII*). Expression of these chimeric genes in the plant allowed the selection of antibiotic-resistant transformed plant cells and hence the elimination of the opine synthesis genes from the transferred DNA (Bevan et al., 1983; Fraley et al., 1983; Herrera-Estrella et al., 1983a; Herrera-Estrella et al., 1983b). Phenotypically normal and fertile plants were regenerated from the resistant calli, and the resistance trait was passed to the progeny in a Mendelian fashion (De Block et al., 1984; Horsch et al., 1984). Two innovations in vector design circumvented the difficulties associated with cloning into the very large Ti plasmid. Zambryski et al. (1983) replaced the entire oncogenic region of the Ti plasmid with the standard cloning vector pBR322; DNA sequences of interest cloned into a pBR vector could thus easily be introduced into the T-region by a single recombination event.

The labs of Schilperoort and Bevan designed binary vector strategies in which one broad-host range replicon carried the DNA to be transferred, while a second, compatible pTi-derived plasmid provided the *vir* functions required for DNA transfer (Hoekema et al., 1983; Bevan, 1984). Both of these systems provided enormous versatility because the DNA to be transferred could be easily manipulated in *E. coli*, and the demonstration that integration of these altered T-DNAs did not interfere with normal plant cell differentiation (Zambryski et al., 1983) opened the floodgates for the wave of plant genetic modifications that followed.

2.2 Binary vectors

In the two decades since their initial development, *Agrobacterium*-mediated transformation systems have undergone a number of refinements. Ease of DNA manipulation in *E. coli* has been achieved by modification of the replication functions on the binary vectors to enhance copy number, reduction in the size of the vectors (Hellens et al., 2000), and incorporation of convenient multiple cloning sites (Komari et al., 2006). The Overdrive sequence adjacent to the right border (RB) enhances T-DNA transfer (Peralta et al., 1986), and some binary vectors include this sequence (Hellens et al., 2000). In addition to the *nptII* gene originally used as the selectable marker, a variety of other selection schemes, including chimeric genes conferring resistance to methotrexate (Eichholtz et al., 1987) and hygromycin (Van de Elzen et al., 1985) have been developed, and several families of binary vectors now provide a choice of marker (Hellens et al., 2000). Many of the early binary vectors carried the selectable marker near the RB, where it would be transferred before the transgene of interest. In contrast, placement of the marker closest to the left border greatly diminishes the chance of selecting transgenic plants resulting from interrupted bacterium-to-plant DNA transfer that carry only the marker (Hellens et al., 2000). This strategy is especially important when introducing very large fragments of foreign DNA into plants. Binary bacterial artificial chromosomes (BIBAC) and transformation-competent bacterial artificial chromosomes (TAC) have been developed that allow the delivery of fragments of at least 80-150 kb (Hamilton et al., 1996; Shibata and Liu, 2000). Such large-capacity vectors are likely to prove particularly useful in identifying and confirming quantitative trait loci (QTLs) controlling agronomically significant characteristics such as crop yield, disease resistance, and stress tolerance (Shibata and Liu, 2000; Salvi and Tuberosa, 2005).

Binary vectors are typically used with so-called "disarmed" *Agrobacterium* strains, in which the virulence functions required for DNA processing and transfer are provided by a modified Ti plasmid lacking oncogenic DNA. Certain strains carrying the "supervirulent" Ti plasmid pTiBo542 exhibit greatly enhanced transformation efficiency (Jin et al., 1987), and the popular transformation strain EHA101 carries a disarmed version of pTiBo542 (Hood et al., 1986). Capitalizing on the discovery of a supervirulent pTi, super-binary vectors carry the *virB*, *virE,* and *virG* genes of pTiBo542 or the Ti plasmid from another supervirulent strain, Chry5 (Torisky et al., 1997). Super-binary vectors have provided critical improvements in transformation efficiency, and were a key factor in extending the host range of *Agrobacterium*-mediated transformation to the cereals in the 1990s (Komari et al., 2006). Practical technical information about binary and super-binary vectors and disarmed strains, along with email addresses and websites of contacts for those who wish to obtain these resources, has been compiled in two recent reviews (Hellens et al., 2000; Komari et al., 2006).

Many binary vectors use the strong constitutive cauliflower mosaic virus (CaMV) 35S promoter to drive expression of the target gene (Chung et al., 2005), although the maize ubiquitin I promoter and the rice actin promoter/intron sequences are more frequently used for expression in monocots (Walden and Wingender, 1995). Alternative promoters exhibiting similarly high or even higher levels of constitutive transcription include a chimera derived from the octopine and mannopine synthase genes (Ni et al., 1995). Inducible and/or tissue-specific promoters provide the possibility of activating a transgene at the most favorable time of development or upon perception of certain environmental cues; use of such promoters can also prevent deleterious effects associated with constitutive production of a toxic product (reviewed in Gelvin, 2003b). A bidirectional promoter, permitting expression of a gene at either end, offers the potential to stack traits (Xie et al., 2001). A completely different strategy for the coordinated production of two proteins makes use of a virally derived polyprotein proteolytic processing peptide. A gene constructed from multiple coding regions separated by this 18-amino acid peptide gives rise to a polyprotein that is co-translationally self-processed to yield stoichiometric amounts of the individual proteins (de Felipe et al., 2006). This approach has even been used successfully to co-produce two proteins targeted to different subcellular compartments (François et al., 2004).

2.3 Transgene stacking

As researchers have moved beyond the introduction of simple traits conferred by a single gene, strategies have been developed that allow the coordinated manipulation of multiple genes in the same plant. The most basic approaches entail sequential sexual crossings or retransformation, and both of these have been used successfully, although they are time-consuming and prone to complications arising from independent segregation in subsequent generations if the introduced genes integrate at different loci (Halpin and Boerjan, 2003). Thus, methods that allow the introduction of multiple genes in one step and their co-integration are more desirable. Somewhat unexpectedly, co-transformation with two T-DNAs on the same or different plasmids within the same bacterium, or even in two different bacterial cultures that are mixed before co-cultivation, can yield remarkably high rates of co-transformation and even, on occasion, co-integration (Halpin and Boerjan, 2003). Such double-T-DNA systems have proven effective in manipulating two or more transgenes at a time in *Arabidopsis*, tobacco, rapeseed, rice, soybean and maize (Slater et al., 1999; Miller et al., 2002; Li et al., 2003). However, engineering of more complex metabolic pathways will require that even more transgenes be stacked. Recently, construction of transformation-ready cassettes was greatly simplified by the advent of binary vectors compatible with the GATEWAY technology, which is based on site-specific recombination between two DNA molecules carrying complementary recombination sites (Invitrogen; http://www.invitrogen. com). The first generation of GATEWAY-compatible destination binary vectors allowed overexpression of a gene, with or without a visible marker, construction of N- or C-terminal Green Fluorescent Protein fusions, or post-transcriptional gene silencing of a target gene (Karimi et al., 2002). This elegant system was subsequently extended to accommodate simultaneous assembly of up to three DNA fragments onto one binary vector (Karimi et al., 2005). Alternatively, as many as six genes can be inserted into a single binary vector containing sites for rare-cutting restriction enzymes (Goderis et al., 2002). Transfer of up to 10 genes into the rice genome has also been achieved using a TAC-based vector, with assembly of the various inserts mediated by the Cre/*lox*P recombination system and homing endonucleases (Lin et al., 2003). Like the GATEWAY-based system, both of these approaches rely on auxillary donor vectors. Perhaps the most advanced system currently available is the pSAT series of vectors, which offers unprecedented versatility in the choice of restriction sites, plant selectable markers, and the possibility of constructing fusions with any of six different autofluorescent tags (Tzfira et al., 2005). An added benefit of the newest pSAT vectors is the opportunity to choose from among a variety of promoter and terminator sequences for combined expression of the target, selection and reporter genes. This is a critical advantage, since diversity among promoters and terminators can reduce the risk of transgene silencing in plants (see section 2.6 and Chung et al., 2005).

2.4 Marker genes and marker-free transformation

In response to concerns about the potential for transfer of antibiotic-resistance genes to gut microbes, a number of antibiotic-free marker systems have been developed. Herbicide resistance genes are also in widespread use as selectable markers [see Hare and Chua (2002) for examples], although they too pose perceived dangers to health and the environment (Hood, 2003). Rather than killing non-transformed cells (negative selection), one can also use a positive selectable marker that confers on transformed cells a growth or metabolic advantage (Hohn et al., 2001). For example, introduction of the phosphomannose isomerase gene rescues plants from the growth inhibition associated with mannose (Negretto et al., 2000). The desire to transform recalcitrant plant species has driven the development of other positive selection markers, including native plant genes conferring resistance to bacterial pathogens (Hood, 2003). Erikson et al. (2004) devised a clever scheme in which introduction of a single gene, encoding a D-amino acid oxidase, allows either positive or negative selection. Selection can be exerted by spraying certain D-amino acids onto soil-grown seedlings; transformed plants exhibit resistance to toxic D-amino acids (e.g. D-alanine or D-serine), whereas only wild-type plants survive exposure to innocuous D-amino acids (D-isoleucine or D-valine) that are converted by the enzyme to toxic keto acids. This dual selection scheme has the distinct advantage of permitting positive selection for transformation, followed by negative selection to identify desired plants that have lost the selectable marker gene (Scheid, 2004). Other positive selection markers, such as the agrobacterial or plant cytokinin synthesis isopentyl transferase (*ipt*) genes, promote regeneration of shoots from transformed calli or explants in the absence of critical growth regulators (Zuo et al., 2002). Inducible expression circumvents the developmental defects associated with constitutive overexpression of *ipt* (Kunkel et al., 1999). However, over-produced cytokinins can cause spurious regeneration of non-transformed neighboring cells. Thus, introduction of cytokinin signal transduction pathway genes may be a preferable selection scheme to avoid non-transgenic escapes (Zuo et al., 2002). Finally, marker-free transformation was achieved in potato using a virulent *A. tumefaciens* and PCR screening for successfully altered shoots (de Vetten et al., 2003).

Non-antibiotic/herbicide resistance markers address concerns about potential health and ecological risks, but they still suffer from other shortcomings: the selection scheme can have negative consequences for plant cell proliferation and differentiation, and multiple transgenes cannot be stacked through sequential retransformations using the same marker gene (Ebinuma et al., 1997). These constraints have spurred the development of various methods to remove the marker gene after transformation. In one such strategy, the selectable marker is inserted into a transposable element, allowing transposition-mediated loss of the marker after selection of the transformed plants (Ebinuma et al., 1997). Alternatively, one can place the transgene of interest on the transposon; in this case, transposition to new sites not only separates the transgene from the marker gene, but also provides an opportunity to obtain a series of plants with varying transgene loci, and potentially differing expression levels, from a single transformant (Hohn et al., 2001). Excision of the marker gene can be achieved by flanking the marker gene with recombination sites and incorporating the cognate site-specific recombinase on the transgenic unit or crossing with a second plant carrying a recombinase-encoding transgene; in either case counter-selectable marker genes can be included within the "elimination cassette" to ensure excision (Hohn et al., 2001). Among the popular recombinase options are the bacteriophage P1 Cre/*lox* system (Dale and Ow, 1991) and the yeast Flp/FRT system (Hare and Chua, 2002). In the simplest case, the marker gene is excised in the F1 generation and the recombinase gene is removed through segregation in the subsequent generations (Gilbertson et al., 2003). Inducible (Zuo et al., 2001) or transient (Hare and Chua, 2002) expression of the recombinase, or transient exposure of the plants to agrobacteria that deliver the recombinase (Vergunst et al., 2000) avoids the need to eliminate the recombinase gene through genetic segregation. Marker excision through recombination has also been achieved using bacteriophage λ *attP* sequences as the flanking DNA (Zubko et al., 2000). Surprisingly, introduction of a recombinase was not required, making this strategy especially attractive for crops that are propagated vegetatively and for which it would therefore be difficult to eliminate the recombinase gene through subsequent crosses (Zubko et al., 2000). Finally, marker genes can be eliminated by co-transforming with tandem marker- and trans-genes, each flanked by its own border sequences, on a single binary vector (Matthews et al., 2001). *Agrobacterium*-mediated delivery of such a construct can lead to independent integration events in the same plant cell, and the marker can therefore be segregated away from the transgene (Hohn et al., 2001).

2.5 Elimination of foreign DNA other than the transgene of interest

Agrobacterium-mediated transformation frequently results in the unintentional introduction of vector backbone sequences (Kononov et al., 1997; Wenck et al., 1997; see also Chapter 12 in this volume for a more detailed description of T-DNA integration patterns). Like marker genes, backbone sequences in a transformed plant are undesirable from a commercial perspective. Incorporation of a lethal gene into the non-T-DNA portion of a binary vector causes a dramatic decrease in the percentage of tobacco, tomato, and grape plants carrying a vector-borne reporter gene, without markedly reducing the overall transformation efficiency (Hanson et al., 1999). Thus, this strategy can be an efficacious way to enrich for T-DNA-only transformants in situations where the presence of vector backbone sequences would be problematic. Alternatively, a systematic comparison of multiple agrobacterial strains and T-DNA origins of replication revealed that integration of "backbone" sequences can almost be eliminated if the border-flanked transgene is located on the bacterial chromosome (H. Oltmanns and S.B. Gelvin, personal communication).

As the preceding discussion implies, the presence of foreign DNA (in addition to the desired transgene itself) may or may not increase health or environmental risks associated with a transgenic plant, but it frequently poses public relations problems, and in fact accounts for much of the dissatisfaction that has led to widespread public rejection of genetically modified crops (Rommens, 2004). In addition to the transgene and the selectable marker, other non plant-derived genetic elements needed for stable transgene expression frequently include promoters, transcriptional terminators, and of course the T-DNA borders. On average, genetically engineered plants approved for commercialization contain ten genetic elements from non-plant sources; typically these have come from bacteria or viruses, or are synthetic sequences. In an effort to decrease dependence on non-plant genetic material, researchers have identified a variety of plant genes associated with agronomically relevant traits such as disease resistance, insect resistance, herbicide tolerance, enhanced storage or nutritional characteristics, and stress tolerance. Additionally, hundreds of plant promoters, both constitutive and tissue-specific, and transcription termination sequences for most important crop species are now available (Rommens, 2004). Rommens et al. (2004) used database searches and PCR to isolate plant sequences that resemble T-DNA borders. Strikingly, these "P-DNA" sequences function to mediate DNA transfer to potato. Using transient expression of a selectable marker carried on a conventional "Life-Support" T-DNA to block proliferation and regeneration of cells that had not received exogenous DNA, these authors were able to document integration events comprised of marker-free P-DNA. Such all-native transformations can be obtained by co-infecting with two *Agrobacterium* strains (one carrying the P-DNA binary and the other providing the Life-Support T-DNA vector), either simultaneously or sequentially, but the frequency of marker-free P-DNA insertion is as high or higher (depending on the host species) if both binaries are present in a single bacterial strain. By selecting against backbone integration events as described above, this approach can be used to generate completely marker-free transgenic plants at a frequency that is consistent with commercial scale production (Rommens et al., 2004).

2.6 Influence of position effects and gene silencing on transgene expression levels

The fact that *Agrobacterium*-mediated DNA integration into the host plant's genome occurs by illegitimate recombination (see Chapter 11) has profound implications for the generation of transgenic plants. Expression levels of the transgene can be dramatically affected by the chromosomal context of the integration site, and insertional disruption of an active host gene can have unintended phenotypic consequences on the resulting plant (Kumar and Fladung, 2001). Targeting the insertion event (see section 2.7) to a specific innocuous, yet transcriptionally active, locus could provide a way to circumvent this variability, particularly if insertions at the same genomic position routinely exhibit similar expression levels (Gilbertson et al., 2003). In at least one study, targeted insertions into the same site did result in reproducible transgene expression levels; however, in nearly half the insertion events, partial or complete silencing of the transgene was observed (Day et al., 2000). Such "position effects" are consistent with our growing appreciation for the striking variability and unpredictable nature of transgene expression levels, a ubiquitous phenomenon in almost all eukaryotes. In the face of repressive influences exerted on transgenes by neighboring genes or the surrounding chromosomal structure, the standard, albeit costly, approach has been to generate enough transgenic plants to find some with the desired level of expression (Hansen and Wright, 1999).

There is some hint that naturally occurring matrix attachment regions (MARs), sequences that associate with the nuclear matrix and mediate looping of DNA, may stabilize expression levels (Han et al., 1997; Iglesias et al., 1997), although the benefit of flanking *Agrobacterium*-delivered transgenes with MARs may be only marginal (Gelvin, 2003b).

In the context of plant transformation, transgene silencing also results from insertion of multiple copies or high-level expression from a constitutive promoter, and an introduced transgene can lead to silencing of a homologous host gene (Vaucheret et al., 1998). Multicopy transgenic loci, particularly those including binary vector sequences, appear prone to transcriptional silencing attributable to meiotically heritable epigenetic modifications, most often methylation and/or condensation of chromatin (Matzke and Matzke, 1998; Vaucheret et al., 1998). Silencing can also occur by a post-transcriptional mechanism termed "cosuppression," in which the formation of double-stranded RNA (dsRNA) results in sequence-specific degradation of homologous RNA molecules (Soosaar et al., 2005). The degree of cosuppression tends to correlate with the strength of the promoter driving the transgene, although reciprocal and synergistic silencing between host genes and transgenes can also result from production of aberrant RNA above a threshold level that activates the RNA degradation machinery (Vaucheret et al., 1998). Conversely, expression of heterologous genes can be stimulated by adjacent ribosomal DNA spacer regions, at least in transgenic tobacco. Strikingly, the enhancement is attributable to amplification of the gene copy number as well as increased transcription, and both changes are stably inherited (Borisjuk et al., 2000).

2.7 Targeting transgene insertions

Gene targeting after *Agrobacterium*-mediated transformation was initially demonstrated as recombination between endogenous or engineered tobacco protoplast sequences and a homologous incoming gene fragment; successful targeting restored a functional selectable marker gene (Lee et al., 1990; Offringa et al., 1990). However, the frequency of such homologous recombination events is relatively low. In contrast, efficient targeted transgene insertion can be achieved by first creating a plant line with a *lox* "target" site; in subsequent transformations of this plant line, incoming DNA carrying a *lox* sequence is specifically and precisely integrated at this chromosomal site via Cre-mediated recombination (Gilbertson et al., 2003). Inclusion of a promoter at the site of integration provides a simple selection scheme for successful insertion of a T-DNA carrying the marker gene (Albert et al., 1995). Targeted transgene integration via site-specific recombination can be combined with a second recombination system that eliminates the selectable marker gene (Srivastava and Ow, 2004). The efficiency of the targeted integration reaction is enhanced when the T-DNA carries two *lox* sites, allowing for formation of the required circular integration substrate (Vergunst et al., 1998). A variety of approaches have been used to stabilize the insertion and prevent subsequent Cre-mediated excision (Gilbertson et al., 2003).

A second important application of homology-directed DNA insertion is gene inactivation via targeted disruption. Although large collections of random T-DNA insertions (e.g., Feldmann, 1991) have proven to be an immensely valuable tool for plant molecular biologists, not all genes are represented and not all alleles are null mutants (Britt and May, 2003). Disruption of a specified locus in *Arabidopsis* can be accomplished by flanking a selectable marker with two genomic fragments from the target gene and screening for a double cross-over event that eliminates another T-DNA-borne marker gene or other T-DNA sequences (Miao and Lam, 1995; Kempin et al., 1997; Hannin et al., 2001). Several refinements of this procedure enabled the first targeted disruption in a monocot, rice (Terada et al., 2002). Those improvements include optimizing the efficiency of the *Agrobacterium*-mediated transformation itself and the use of a stringent PCR screen for true recombinants. A third, and probably critical, factor was the inclusion of toxin-encoding genes at either end of the vector DNA to provide strong counter selection against random integration of the T-DNA elsewhere in the genome. Finally, it is plausible that recombination occurs more readily in the highly proliferative callus tissue typically used in rice transformation than in the plant tissues used in other transformations (Shimamoto, 2002). The gene targeted for disruption in this application was Waxy, which encodes granule-bound starch synthase. Lower Waxy mRNA abundance in Japonica rice accounts for its stickier nature as compared to Indica rice, in which the gene is expressed at higher levels (Hohn and Puchta, 2003). The success of this gene targeting process in rice paves the way for other gene knockouts in this important staple crop to study gene function or to alter nutritional or growth traits.

Conventional approaches to gene targeting appear to be limited by the preference in plants for non-homologous end-joining (NHEJ) over homologous recombination for DNA double-stranded break repair. Recent advances in enhancing targeted mutagenesis have focused on harnessing the NHEJ process and on stimulating homologous recombination by engineering plants to express a yeast recombination gene (Tzfira and White,

2005). NHEJ frequently introduces insertion and/or deletion mutations at double-stranded breaks, thus raising the possibility that targeted mutagenesis could be accomplished by inducing double stranded breaks at the desired locus. Successful implementation of this approach was achieved using *Agrobacterium*-mediated transformation to introduce a synthetic zinc-finger nuclease that then created the double-stranded break (Lloyd et al., 2005). These zinc-finger nucleases consist of custom-made C2H2 zinc fingers, with each finger recognizing a specified three-nucleotide sequence, fused to a non-specific restriction enzyme. Expression of this chimeric gene in a plant allows the targeted digestion of a specific and unique sequence of 18 nucleotides, which then becomes a substrate for error-prone NHEJ-mediated repair (Tzfira and White, 2005). Using a heat-shock promoter to drive production of the zinc-finger nuclease in *Arabidopsis*, Lloyd et al. (2005) demonstrated highly efficient mutagenesis and transmission of the induced mutations, and suggested on theoretical grounds that this technology should be applicable to most plant genes in most plant species. In a second approach to increasing the frequency of directed gene disruption or replacement, Shaked et al. (2005) introduced the yeast chromatin remodeling protein RAD54 into *Arabidopsis* and reported a 10-to-100 fold improvement in homology-based integration efficiency.

2.8 Extending the range of susceptible hosts for *Agrobacterium*-mediated transformation

A variety of factors have been shown to influence the range of hosts that can be transformed by *A. tumefaciens*. On the bacterial side of the interaction, certain virulence loci including *virC* and *virF* are considered host range determinants (Yanofsky et al., 1985; Jarchow et al., 1991; Regensburg-Tuink and Hooykaas, 1993), and constitutive transcription of the virulence genes improves the efficiency of plant transformation in both susceptible and recalcitrant species (Hansen et al., 1994). Genes within the T-region can also affect the range of susceptible host species (Hoekema et al., 1984). Overexpression of certain plant genes, particularly *HTA1* (encoding histone 2A) and *VIP1* (which may facilitate nuclear targeting of the T-complex) can also enhance plant susceptibility (Mysore et al., 2000; Tzfira et al., 2002). The manipulation of host genes to improve transformation frequency is the subject of two recent reviews (Gelvin, 2003a; Gelvin, 2003b). Bacterial and plant contributions to host range are discussed in more detail in Chapters 1 and 13, respectively, in this volume. It is worth noting that there are almost certainly more factors yet to be identified that limit the interaction between *A. tumefaciens* and specific plant species. For example, maize root exudates contain a potent inhibitor of VirA/VirG-mediated signal perception, leading to the possibility that bacterial mutants with enhanced resistance to this inhibition may prove useful in extending the transformation efficiency of maize (Zhang et al., 2000). One approach to circumvent host range limitations involves the use of *Agrobacterium rhizogenes* to generate composite plants, comprised of transgenic roots on wild-type shoots. This system provides a useful method to study transgene activity in the root in the context of a wild-type plant, and has been used successfully in species such as soybean, sweet potato and cassava, that are recalcitrant to *A. tumefaciens* transformation (Taylor et al., 2006).

Somewhat ironically, of all the advances in plant transformation described in this chapter, some of the most pronounced long-term impacts on plant biotechnology may result from an innovation that has the potential to obviate the requirement for *Agrobacterium* as a gene delivery vehicle. Motivated by the desire to "invent around" the myriad intellectual property constraints that limit use of *Agrobacterium*-mediated transformation by the public and the private sector, Broothaerts et al. (2005) successfully modified several species outside the *Agrobacterium* genus to stably transform a variety of plants. (The complex issues surrounding intellectual property in agricultural biotechnology are developed more fully in Chapter 20.) *Rhizobium*, *Sinorhizobium*, and *Mesorhizobium* strains of bacteria endowed with a disarmed Ti plasmid acquired the ability to deliver DNA from a standard binary vector; the vector was modified with a unique tag to facilitate tracking of the provenance of the transferred DNA. Rice, tobacco and *Arabidopsis* were genetically modified to express an intron-containing beta-glucuronidase (GUS) gene, indicating that monocots as well as dicots can serve as recipients with non-*Agrobacterium* bacteria, albeit at frequencies that ranged from 1-40% of that observed with *Agrobacterium*-mediated transformation. (The presence of the intron prevents reporter gene expression in the bacteria, and thus ensures that any observed GUS activity results from expression in the plant cell; Vacanneyt et al., 1990). Various tissues, and hence transformation mechanisms (floral dip for *Arabidopsis*, somatic tissue for tobacco and rice), were utilized in these experiments, and stable integration was confirmed by Southern blotting, sequence analysis of the insertion junctions, and Mendelian transmission of the transgene to progeny. This alternative technology may have profound implications for the plant biotechnology community for two reasons. First, this technology has been configured to be freely accessible and "open-source," with no commercial restrictions other than covenants for sharing improvements, relevant safety information, and regulatory data (http://www.bioforge.net). Second, the exceptionally broad host range of the *Rhizobium* strain used, and the potential to extend the technology to additional bacteria species, make it likely that previously recalcitrant plant species may become transformable. As a plant pathogen, *Agrobacterium* elicits a variety of defense responses that can block any step of the transformation process, thereby limiting its host range. While a better understanding of *Agrobacterium*-triggered defense responses may lead to methods to lower or subvert a plant's natural barriers (Zipfel et al., 2006), the use of non-*Agrobacterium* species as T-DNA delivery systems provides a way for plant biotechnologists to invent around the obstacles erected by both plant evolution and patent lawyers.

2.9 Alternatives to *Agrobacterium*-mediated gene delivery

In the 1980's, the apparent recalcitrance of several agronomically important crop plants, including maize, wheat, barley, and rice, to infection by *A. tumefaciens* drove the development of alternative methods of DNA delivery for genetic engineering. Protoplast transformation, although achievable through electroporation, microinjection, or polyethylene glycol fusion, proved to be inefficient because the regeneration of plants from protoplasts is time-consuming and non-trivial (Newell, 2000). Particle bombardment, in which tungsten or gold microprojectiles are coated with DNA and accelerated into the target plant tissue, has proven highly successful in a wide range of species (Klein et al., 1987), and is the most reliable method by which chloroplasts can be transformed. This biolistic approach presents certain advantages over *Agrobacterium*-mediated gene delivery; many types of explants can be bombarded and yield fertile plants, and the gene to be delivered need not be cloned into a specialized transformation vector (Herrera-Estrella et al., 2005). Nonetheless, particle gun delivery of DNA is generally not the method of choice for a plant species that can be transformed by *Agrobacterium*, as the bombardment process typically results in integration of multiple copies of the DNA, as well as rearranged and/or truncated DNA sequences (Newell, 2000). These complex integration patterns can lead to genetic instability, due to homologous recombination among the identical copies, and/or epigenetic silencing of the transgene (see section 2.6). "Agrolistic" transformation was designed to mitigate these shortcomings by combining the high efficiency of biolistic DNA delivery with the simpler integration pattern characteristic of *Agrobacterium*-mediated DNA transfer. Particle bombardment of the *virD1* and *virD2* genes, under the control of the CaMV35S promoter, with a target plasmid carrying the transgene of interest flanked by T-DNA border sequences, allows transient expression of the *vir* genes in the plant. The insertion events resulting from *in planta* VirD1/2-mediated processing and integration resemble those generated by traditional *Agrobacterium*-mediated transformation (Hansen and Chilton, 1996).

Agrobacterium-mediated and biolistic delivery of foreign DNA are typically used to stably transform plants, although transient expression of genes delivered by *A. tumefaciens* on binary vectors can be used to produce recombinant proteins without the delays and technical barriers associated with stable integration (Chung et al., 2006). Heterologous genes can also be introduced into plants on viral vectors; because of the amplification associated with viral infection, transient expression of the transgenes can yield commercial-scale quantities of pharmaceutical proteins. In a novel hybrid technology, *A. tumefaciens* has been used to expedite the production process by circumventing the need for *in vitro* synthesis of the RNA viral vector. Building on the idea of "agroinfection," in which a viral genome is delivered as a cDNA inserted between border sequences (Grimsley et al., 1986; Grimsley et al., 1987), complete viral replicons have been assembled *in planta* through site-specific recombination among DNA modules delivered by *Agrobacterium* (Marillonnet et al., 2004). Additional refinements of the viral vectors further enhanced the efficiency of the system, which was limited by the low infectivity of viral vectors carrying larger genes and apparently by nuclear processing of a viral transcript that normally never experiences the nuclear milieu (Marillonnet et al., 2005). By infiltrating whole mature plants with a suspension of agrobacteria carrying the encoded viral replicons, the bacteria take on the viral infection function, while the viral vector mediates cell-to-cell dissemination, amplification, and high-level expression of the transgene (Gleba et al., 2005). This "magnifection" process is rapid and scalable; the modular nature of the viral components facilitates adaptation to new transgenes, and the yield can reach 80% of total soluble protein (Marillonnet et al., 2004).

3 APPLICATIONS OF *AGROBACTERIUM*-MEDIATED TRANSFORMATION

3.1 Production of foreign proteins in plant cell cultures

Agrobacterium-mediated transformation has been extensively utilized to engineer plants producing a wide variety of useful, and in many cases clinically relevant, metabolites and exogenous proteins. Most applications to date have focused on field-grown plants, although recombinant proteins and metabolites can also be produced in plant cell cultures. Despite limited commercial use so far, cultured plant cells such as the tobacco-derived BY-2 and NT-1 lines offer several advantages over expression systems in intact plants: they can be maintained in simple media, and are not subject to variations in weather and soil conditions; products can be easily harvested, especially when secreted into the culture medium (Hellwig et al., 2004). In the future, functional genomics and combinatorial biochemistry are likely to increase dramatically the range of products that can be generated in genetically modified plant cell cultures (Oksman-Caldentey and Inze, 2004).

3.2 Genetic modification of plants to generate useful products

3.2.1 Biodegradable plastics

Among the more notable foreign products produced in plants are biodegradable plastics. Drawing on the natural ability of many bacterial species, including *Ralstonia eutropha*, to synthesize carbon storage products with plastic-like properties (Hanley et al., 2000), Chris Somerville's lab first demonstrated poly-3-hydroxybutyrate (PHB) synthesis in *Arabidopsis* by introducing biosynthetic genes from *R. eutropha* (formerly *Alcaligenes eutrophus*) (Poirier et al., 1992). Yields of this simple C4 polymer, which is synthesized from acetyl-CoA by the sequential action of the bacterial *phbA*, *phbB*, and *phbC* gene products, could be increased 100-fold by N-terminal addition of the pea small subunit RUBISCO-transit peptide, thereby targeting the three encoded enzymes to the chloroplast (Nawrath et al., 1994). Further increases in yield, from 14% to as much as 40% of the plant dry weight, were achieved by using gas chromatography-mass spectrometry to screen large numbers of transgenic *Arabidopsis* plants for high levels of production; however, the high producing lines exhibited stunted growth, loss of fertility, and significant alterations in the levels of various amino acids, organic acids, sugars, and sugar alcohols (Bohmert et al., 2000).

Properties of PHB, including brittleness and low-temperature decomposition, preclude its use commercial use. In contrast, the co-polymer poly(3-hydroxybutyrate-co-3-hydroxyvalerate) (PHBV) is considerably more flexible and therefore useful (Hanley et al., 2000). Slater et al. (1999) successfully engineered both *A. thaliana* leaves and seeds of *Brassica napus* (oilseed rape) to synthesize PHBV at a significantly lower cost than the previous industrial-scale bacterial fermentation process (Poirier, 1999). Because PHBV synthesis requires not only the abundant metabolite acetyl-CoA, but also the relatively scarce propionyl-CoA, Slater et al. had to redirect the metabolic flow of two independent pathways to generate a pool of propionyl-CoA in the plastid (Slater et al., 1999).

Finally, Neumann et al. (2005) have recently reported the synthesis in transgenic tobacco and potato plants of cyanophycin, which can be hydrolyzed to yield the soluble, non-toxic, biodegradable plastic-like compound poly-aspartate. Although these transgenic plants exhibit morphological alterations in chloroplast structure and in growth rate, additional engineering of the amino-acid biosynthesis pathways may permit economically viable levels of biodegradable plastic production (Conrad, 2005). If successful, the substitution of a renewable process (solar-driven carbon fixation) for conventional petrochemically derived plastic production technologies would have substantial positive environmental consequences, decreasing our reliance on finite petroleum resources, while reducing the accumulation of indestructible plastics (Poirier, 1999; Conrad, 2005).

3.2.2 Primary and secondary metabolites with desirable properties

Considerable effort has been dedicated to metabolic engineering of terpenoids in plants. Terpenoids, also known as isoprenoids, are a family of more than 40,000 natural compounds, including both primary and secondary metabolites, that are critically important for plant growth and survival. Some of the primary metabolites produced by the terpenoid biosynthetic pathway include phytohormones, pigments involved in photosynthesis, and the ubiquinones required for respiration (Aharoni et al., 2005). Secondary metabolites, including monoterpenoids (C10), sesquiterpenoids (C15), diterpenoids (C20), and triterpenoids (C30), also provide physiological and ecological benefits to plants. Some function as antimicrobial agents, thus contributing to plant disease resistance, while other terpenoid compounds serve to repel pests, attract pollinators, or inhibit growth of neighboring competitor plant species. Additionally, many terpenoids have commercial value as medicinals, flavors, and fragrances. Interest in manipulating the inherent properties of plants (e.g., enhanced aromas of ornamentals, fruits, and vegetables), or in using plants as sources of pharmaceuticals and cosmetics, has driven the development of terpenoid metabolic engineering in a variety of species (Aharoni et al., 2005).

The terpenoid biosynthetic pathway and strategies for its manipulation have been reviewed recently (Mahmoud and Croteau, 2002; Aharoni et al., 2005). A comprehensive listing of transgenic plants with altered terpenoid biosynthetic properties is available elsewhere (Aharoni et al., 2005). Examples include expression of heterologous synthases in tomato, leading to enhanced aroma in ripening fruit (Lewinsohn et al., 2001), reduced production in mint of an undesirable monoterpenoid that promotes off-color and off-flavor (Mahmoud and Croteau, 2001), and the introduction of bacterial genes directing the production of keto-carotenoids, thought to have medicinal value, into tomato and tobacco (Ralley et al., 2004). Other endogenous, plant-derived terpenoids with demonstrated pharmaceutical properties include the anti-malarial agent artemisinin, the diuretic glycyrrhizin, and the cancer drugs Taxol and perilla alcohol. Several of these compounds are currently derived from endangered species in threatened ecosystems (Bouwmeester, 2006), while chemical synthesis of terpenes can be prohibitively costly and inefficient (Wu et al., 2006).

Plants contain two terpene biosynthetic pathways; the mevalonate pathway leads to the synthesis of sesquiterpenes and triterpenes at the level of the endoplasmic reticulum, while the methyl-D-erythritol-4-phosphate pathway functions in the chloroplast to produce monoterpenes, diterpenes, and carotenoids (Aharoni et al., 2005). Most attempts to manipulate the pathways involve introducing terpene synthase genes whose products could divert pathway intermediates towards the production of desired, and in some cases novel, terpenes (Chappell, 2004). To date, generating monoterpenes in transgenic plants has proven easier than modifying the metabolism of longer-chain terpenoids (Aharoni et al., 2005). The complexity of the biosynthetic pathway, giving rise to a vast number of natural products, and the subcellular compartmentalization of the processes pose challenges for terpenoid genetic engineering. Manipulating terpenoid biosynthetic pathways in plant species that produce the same class of terpenes is less problematic, because the plant already has the specialized structures necessary to carry out the storage and transport of volatile, hydrophobic compounds. In contrast, introducing novel pathways into species that lack the secretory structures required may prove to be far more difficult

(Mahmoud and Croteau, 2002). A recent comprehensive evaluation of the factors required for high-level terpene production in tobacco identified several effective strategies for enhancing synthesis as much as 1,000-fold (Wu et al., 2006). By over-producing, in the same subcellular compartment, an enzyme producing an isoprenoid substrate and a terpene synthase that rapidly incorporates this substrate, Wu et al. (2006) have advanced the technology necessary to achieve commercial-scale production of industrially or pharmaceutically relevant terpenes (Bouwmeester, 2006).

3.2.3 Commercially relevant traits in ornamentals and trees

In ornamental plants, flower color, architecture, and post-harvest life are all targets for transgenic modification (Mol et al., 1995). Commercially important traits in trees have also been a focus of recent *Agrobacterium*-mediated transformation (Tzfira et al., 1998). Tree improvement goals include increasing timber yield and decreasing generation time; together, these traits could pave the way for economically viable plantation forests, leading to decreased pressure on natural forests as sources of wood (Fenning and Gershenzon, 2002). In this regard, overexpression of a key enzyme in the gibberellin biosynthetic pathway resulted in enhanced biomass and accelerated growth rate in hybrid aspen, but had a negative effect on rooting. Interestingly, the transgenic trees also exhibited longer and more numerous xylem fibers that could be advantageous in producing stronger paper (Eriksson et al., 2000). Altering plant composition could also enhance the production of bioethanol, a renewable energy source for the transportation sector with substantial positive environmental impact (Boudet et al., 2003). Finally, in poplar and aspen, biotechnology has proven to be an effective way to manipulate levels of the undesirable cell wall component lignin by downregulating the last step of the lignin biosynthetic pathway through an antisense strategy (Baucher et al., 1996; Li et al., 2003); the transgenic trees required fewer chemicals for delignification and yielded more high-quality pulp (Pilate et al., 2002). Since removal of lignin in the paper and pulp industry is an energy-consuming process that requires large amounts of hazardous chemicals, the success of the antisense trees holds promise for more environmentally friendly processing in the future.

3.2.4 Biopharmaceuticals/edible vaccines

Using *Agrobacterium*-mediated transformation, transgenic plants have been engineered to express a wide variety of exogenous proteins, from spider dragline silk (a fiber with high tensile strength and elasticity; Scheller et al., 2001) to vaccines, antibodies, and other life-saving biopharmaceuticals such as anti-coagulants, human epidermal growth factor, and interferon (Giddings et al., 2000). To date, most such clinically relevant proteins have been produced in tobacco, although potatoes, alfalfa, soybean, rice and wheat have also been used successfully. While green tissue has a distinct advantage in terms of productivity, seeds or tubers are most useful for delivery of an edible product such as a vaccine; they can be stored for long periods of time (Daniell et al., 2001) and shipped long distances at ambient temperature (Streatfield et al., 2001).

Edible vaccines may hold considerable promise for the developing world, where refrigeration, sterile syringes and needles, and trained health care personnel are frequently in short supply (Arntzen et al., 2005). Since many pathogens utilize mucosal surfaces as their point of entry, priming the entire mucosal immune system via oral stimulation is an especially attractive mode of immunization (Streatfield et al., 2001). Nonetheless, lack of a profit incentive for private industry, coupled with concerns about inadequate biosafety infrastructure in developing countries and the complexity of government-financed health care delivery systems, have resulted in the development of relatively few products (Ma et al., 2005b) in the 14 years since the first report of an antigen expressed in transgenic plants (Mason et al., 1992). Oral immunization has been achieved using transgenic potatoes expressing antigens including the heat-labile enterotoxin from *E. coli* (Haq et al., 1995; Mason et al., 1998), the Norwalk virus capsid protein (Tacket et al., 2000), and the hepatitis B surface antigen (Richter et al., 2000; Kong et al., 2001), as well as transgenic alfalfa expressing proteins from the foot and mouth disease virus (Dus Santos et al., 2005), among others. Despite these successes, it should be noted that there are no transgenic-plant-derived pharmaceuticals in commercial production (Ma et al., 2005a). This may change in the near future, as a large European consortium with collaborators in South Africa is actively engaged in developing plant-based production platforms for pharmaceuticals targeted to HIV, rabies, tuberculosis and diabetes. This group would be the first to carry out clinical trials of plant-derived candidate pharmaceuticals within the European Union regulatory framework (http://www.pharma-planta.org/).

Plant-derived pharmaceuticals have many potential advantages over those produced in animal cell culture or by microbial fermentation. High yields, favorable economics, existing technologies for harvesting and processing large numbers of plants, and the possibility of expressing proteins in specific subcellular compartments where they may be more stable, all contribute to the choice of transgenic plants over bacterial expression systems for recombinant proteins (Daniell et al., 2001). Like animal cells, plants have the ability to carry out post-translational modifications, and can fold and assemble recombinant proteins using eukaryotic chaperones, but plant expression systems have the added benefit of minimizing the potential for contamination with human pathogens (Woodard et al., 2003; Arntzen et al., 2005). Finally, multimeric protein complexes may be reconstructed in transgenic plants by stacking transgenes through successive crosses among plants resulting from single transformation events (Hiatt et al., 1989; Ma et al., 1995; Ma et al., 2005a). This is a particularly important consideration when producing multimeric secretory antibodies to protect against microbial infection at mucosal sites (Giddings et al., 2000).

One concern about plant-based pharmaceuticals is the potential for non-mammalian glycosylation patterns that might result in immune sensitization or loss of function (Bardor et al., 1999; Giddings et al., 2000). However, at least one plant-derived monoclonal antibody was found to be functional despite differences in N-linked glycosylation (Ko et al., 2003), and stable expression of a human galactosyltransferase in plants has been shown to yield "plantibodies" with mammalian glycosyl modifications (Bakker et al., 2001). Other potential limitations of plant expression systems include low and/or variable yield (Chargelegue et al., 2001), unexpected localization of the expressed protein (Hood, 2004), and, for edible vaccines, induction of oral tolerance and/or gastrointestinal degradation of the antigen (Ma, 2000; Daniell et al., 2001). Finally, contamination of food and feed crops with pharmaceutical crops, either in the field or post-harvest, poses potentially serious health and public relations risks (Ma et al., 2005b).

3.3 Bioremediation

Two classes of transgenic plants have been developed to address the serious risks to human health posed by industrial and naturally occurring environmental pollutants: some serve as biomonitors, detecting the presence of toxic compounds in the environment, while others detoxify contaminated soils. By integrating an engineered marker gene, beta-glucuronidase, Barbara Hohn and coworkers have pioneered a strategy in which transgenic *Arabidopsis* has successfully been used to report enhanced rates of homologous recombination or point mutation due to heavy metal ions (Kovalchuk et al., 2001a; Kovalchuk et al., 2001b), and to ionizing radiation resulting from the Chernobyl accident (Kovalchuk et al., 1998).

Increasing levels of pollution resulting from global industrialization have focused attention on the possibility of phytoremediation: using plants to remove or inactivate pollutants from soil or surface waters. Factors that influence the utility of a plant in phytoremediation include (i) the availability of the trace element in a form that can be taken up by the plant's roots; (ii) the rate of uptake; (iii) the ability of the plant to transform the pollutant into a less toxic, and potentially volatile, compound; and (iv) the movement of the compound from the roots into the shoots (Kramer and Chardonnens, 2001). Theoretically, genetic manipulation of heavy metal accumulation in plants could be used to imbue a plant with any of these traits or to enhance an existing capability (Clemens et al., 2002). Introduction of bacterial genes has enabled the creation of transgenic *Arabidopsis* plants capable of converting the highly toxic contaminant methylmercury to the volatile and much less toxic elemental mercury (Bizily et al., 1999; Bizily et al., 2000). Similar modifications have resulted in *Arabidopsis* and poplar able to process and sequester mercury ion (Rugh et al., 1996; Rugh et al., 1998), Indian mustard that processes selenite (a common contaminant in oil-refinery wastewater) (Pilon-Smits et al., 1999), and tobacco engineered to facilitate degradation of the explosive trinitrotoluene (TNT) (Hannink et al., 2001). To deplete arsenic contamination from groundwater, researchers have introduced bacterial genes that confer on *Arabidopsis* the ability to extract and accumulate in the leaf levels of arsenic that would normally poison the plant (Dhankher et al., 2002). Second generation phytoremediating plants will likely capitalize on the finding that overexpression of a yeast vacuolar transporter in *Arabidopsis* leads to enhanced accumulation, and hence tolerance, of heavy metals such as cadmium and lead (Song et al., 2003b).

3.4 Increasing crop plant productivity by altering plant physiology and photosynthetic capacity

The Green Revolution succeeded in increasing net food productivity per capita in Asia, India, and Latin America by combining, through traditional breeding, high yield and dwarfing traits in several of the world's most important grain crops (Evenson and Gollin, 2003). The advent of basic plant molecular biology, made possible in large part by the availability of *Agrobacterium*-mediated techniques for introducing and knocking out plant genes, has dramatically augmented our understanding of how plant architecture and generation time are regulated, and these discoveries may enable further improvements in yield. For example, manipulating plant brassinosteroid levels resulted in a more erect leaf structure in rice, increasing yield under dense planting conditions (Sakamoto et al., 2006). Tissue-specific modulation of the growth hormone gibberellin catabolism in transgenic rice led to a semi-dwarf phenotype without a loss in grain productivity (Sakamoto et al., 2003). In other cases, yield may be enhanced by decreasing the time required for the plant to produce the edible portion. Exogenous expression of the *Arabidopsis* flower initiation genes LEAFY or APETALA accelerated the generation time of citrus trees (Pena and Seguin, 2001). Dormancy in potatoes was controlled by expressing a bacterial gene that altered sprouting behavior (Farre et al., 2001), while tomatoes with prolonged shelf- and vine-life characteristics were created by manipulating the biosynthesis of the ripening-promoting hormone ethylene (Oeller et al., 1991), or by increasing levels of the anti-ripening polyamines (Mehta et al., 2002), respectively.

Other attempts to increase yield potential have centered on the photosynthetic process, and in particular the inefficiency of the carbon assimilation pathway in C_3 plants, a group that includes many agronomically important crop plants. The alternative C_4 pathway makes use of both altered biochemical pathways and spatial segregation within the plant to concentrate CO_2 for the crucial Calvin-cycle enzyme ribulose 1,5-bisphosphate carboxylase (Rubisco) (Edwards, 1999). Using *Agrobacterium*-mediated transformation, Matsuoka and co-workers have expressed three key C_4 enzymes in rice (a C_3 plant) (Ku et al., 1999; Ku et al., 2001), but it seems likely that successful enhancement of photosynthetic capabilities will require the specialized leaf anatomy of C_4 plants. Another strategy, involving expression in tobacco of a cyanobacterial enzyme, successfully improved photosynthetic capacity and concomitantly increased the plants' biomass (Miyagawa et al., 2001). However, grain production is tightly linked to nitrogen availability, and hence larger plants will not necessarily yield more grain unless soil nitrogen levels are sufficient (Sinclair et al., 2004).

3.5 Enhancing crop productivity by mitigating external constraints

A plant's physiology and its photosynthetic capacity are inherent characteristics, but crop yields can also be limited by many external factors, including inadequate soil fertility, disease, climatic stresses, and/or the presence of soil constituents (e.g., heavy metals) that compromise plants' growth and development. Among the approaches to mitigating these constraints are some that involve genetically modifying the crop plant. It is important to stress that there are also many highly successful non-biotechnological practices that have been in use for centuries, including integrated pest and vector management, crop rotation, dissemination of pathogen-free plant material (Rudolph et al., 2003), and removal of weeds that can serve as reservoirs of infection (Wilson, 1993). Indeed, farming systems that combine careful land management with a diverse array of species and genetic backgrounds within a species can be highly productive even in the absence of modern varieties or biotechnology "improvements" (Brown, 1998). The lessons of such a holistic approach to agriculture are enjoying a resurgence of popularity among small and some medium-scale farmers in the industrialized world; for example, integrated production and organic farming guidelines are in practice on 85% of the farmland in Switzerland (Xie et al., 2002). Nonetheless, the predominant model of agriculture in much of the developed world is one of monocultures grown with high external inputs. At the other end of the spectrum, resource-poor farmers cultivating marginally arable land face myriad environmental constraints which, for a variety of reasons, have proven recalcitrant to the available integrated approaches. The following sections highlight some of the applications of *Agrobacterium*-mediated genetic modification of plants that may address these constraints and/or mitigate negative consequences of the conventional solutions. None of these biotechnological approaches is a panacea. On the other hand, although biotechnology is anathema to most proponents of organic farming practices, it is likely that our ability to meet the growing challenge of adequate food production may benefit from open-mindedness and creative approaches that incorporate the genetic modifications described below into sustainable, ecosystem-centered cultivation systems.

3.5.1 Enhanced nutrient utilization

The negative environmental impacts of inorganic, petroleum-based fertilizers are well-documented, as are the prohibitive costs that preclude their use by subsistence farmers attempting to cultivate depleted soils (Good et al., 2004). Engineering plants with enhanced capabilities to absorb micronutrients from the soil, by over-expressing nitrogen, potassium and phosphorus transporters and/or manipulating their regulation, could decrease the need for fertilizers (Hirsch and Sussman, 1999). For some nutrients, such as iron and phosphorus, the limiting factor is often solubility rather than abundance in the soil. Plants synthesize and secrete a variety of organic acids that can chelate insoluble compounds, allowing uptake of the complex (Guerinot, 2001). Several important grain crops such as rice, maize and sorghum are particularly sensitive to low iron availability in alkaline soils, where iron is less soluble. *Agrobacterium*-mediated introduction of genes conferring enhanced biosynthesis of an iron chelator in rice resulted in improved growth and four-fold higher grain yields under conditions of low iron availability (Takahashi et al., 2001). Finally, it may be possible to engineer plants to secrete nutrients that specifically promote growth of beneficial microbes in the rhizosphere (O'Connell et al., 1996).

3.5.2 Enhanced tolerance to abiotic stress

Solubility of soil constituents is also an important factor influencing a plant's tolerance for metal ions. The abundant metal aluminum normally exists as harmless oxides and aluminosilicates, but in acidic soils it is solubilized into the toxic $Al^{3+,}$ which inhibits root growth. Plants that tolerate otherwise toxic levels of $Al^{3+,}$ do so by secreting organic acids such as citrate or malate at the root apex that chelate the $Al^{3+,}$ in the soil and prevent uptake (Ma et al., 2001). [Other plants accumulate aluminum in the leaves and detoxify it internally by forming organic acid-complexes; the characteristic variation in hydrangea sepals from pink to blue, for example, is determined by the pH-dependent aluminum concentration in the cell sap; Ma et al., (2001).] Attempts to engineer aluminum tolerance by introducing bacterial citrate synthase genes into tobacco and papaya were met with mixed success; enhanced tolerance was reported, but could not be reproduced by another group (de la Fuente et al., 1997; Delhaize et al., 2001). Improved tolerance of zinc in transgenic plants has also been observed (van der Zaal et al., 1999).

Metal contamination in the soil is but one of the abiotic stresses that constrain crop plant productivity. Growing global demand for food continues to force farmers onto marginally arable land where soil salinity, water deficits, and climatic challenges such as low or high temperatures limit cultivation (Bartels, 2001). Strategies to engineer enhanced tolerance to such adverse conditions fall into at least two categories: direct protection from the stressor(s), and enhanced resistance to the physiological damage caused by the stressor. In the latter category, a family of aldose-aldehyde reductases are activated in response to a wide variety of stresses (Bartels, 2001). Ectopic expression of an alfalfa aldose-aldehyde reductase gene via *Agrobacterium*-mediated transformation results in reduced damage upon oxidative stress, apparently by eliminating reactive aldehydes, and increased tolerance to salt, dehydration, or heavy metal stress (Oberschall et al., 2000). Several other transgenic improvements in stress tolerance [e.g., overexpression of glutathione peroxidase (Roxas et al., 1997) and overexpression of superoxide dismutase (McKersie et al., 1996)] likewise function by providing oxidative protection (Zhu, 2001).

Osmolytes also confer stress tolerance by scavenging reactive oxygen species (Zhu, 2001). The non-reducing disaccharide trehalose stabilizes biological structures upon dessication in many bacteria, fungi, and invertebrates, but apparently does not accumulate naturally in plants (Penna, 2003). Transgenic tobacco and rice engineered to produce trehalose exhibit enhanced resistance to drought (Romero et al., 1997; Pilon-Smits et al., 1998), salt, and low-temperature stress (Garg et al., 2002). Production of mannitol results in tobacco with enhanced tolerance to high salinity (Tarczynski et al., 1993). Other low-molecular-weight compatible solutes that accumulate in some plants to protect proteins from stress-induced damage include glycinebetaine, polyols and amino acids. Glycinebetaine accumulation confers on transgenic *Arabidopsis* an increased ability to withstand high temperatures during germination and seedling growth (Alia et al., 1998).

In addition to small osmolytes, a number of proteins have also been shown to have stress protective activity, primarily in response to low temperature. A variety of plants produce antifreeze proteins, as do several fish and insects; these proteins function to inhibit growth of intercellular ice crystals (Griffith and Yaish, 2004). There have been a variety of attempts to introduce a gene encoding one of these anti-freeze proteins into tobacco, tomato, potato, and *Arabidopsis*, with the ultimate goal of lowering the freezing temperature, even by a few degrees, so that the plants could survive a light frost (Griffith and Yaish, 2004). At least one such experiment was successful; although the plant did not exhibit higher rates of survival upon freezing, the freezing temperature was indeed lowered (Huang et al., 2002c). Another transgenic strategy to achieve freezing tolerance involves the introduction of bacterial ice nucleation genes, which permit slow dehydration that minimizes tissue damage (Baertlein et al., 1992).

High soil salinity impedes plant growth, both by creating a water deficit in the soil and within the plant, as sodium ions impinge on many key biochemical processes. Strategies to increase salt tolerance involve limiting exposure of cytoplasmic enzymes to the salt and may include blocking Na^+, influx, increasing Na^+ efflux, and compartmentalizing Na^+ (Zhu, 2001). Successful transgenic approaches are described in detail in Yamaguchi and Blumwald (2005). Many of these entail over-expressing the *Arabidopsis* vacuolar Na^+/H^+ antiporter, which enhances tolerance to soil salinity, with few or no detrimental effects on seed quality or plant growth, in canola (Zhang et al., 2001), *Arabidopsis* (Apse et al., 1999), tomato

(Zhang and Blumwald, 2001), and wheat (Xue et al., 2004). Sequestration of cations in the *Arabidopsis* vacuole, resulting in enhanced salt and drought tolerance was also achieved by overexpressing the vacuolar H^+ pyrophosphatase (Gaxiola et al., 2001). Increased expression of a plasma membrane Na^+/H^+ antiporter augmented salt tolerance by limiting Na^+ accumulation (Shi et al., 2003). Finally, tolerance in rice, resulting from expression of a bacterial Na^+/H^+ antiporter, was accompanied by biosynthetic activation of the osmoregulatory molecule proline (Wu et al., 2004).

Both freezing and high temperature cause damage to plant tissues and proteins, leading to diminished crop yield. A comprehensive listing of attempts to enhance plant thermo-tolerance through genetic modification can be found in Sung et al. (2003). This chapter will highlight some of the most common approaches. One of the earliest reports of altered chilling sensitivity resulted from engineering the degree of fatty acid saturation in tobacco membranes (Murata et al., 1992). Another strategy stems from the identification of the low-temperature transcriptional activator CBF1, which induces expression of multiple cold-regulated (COR) genes associated with cold acclimation (Sarhan and Danyluk, 1998). Using *Agrobacterium*-mediated overexpression of CBF1, Jaglo-Ottosen et al. (1998) successfully mimicked an acclimated state and enhanced the freezing tolerance of *Arabidopsis*. The existence of a more universal transcriptional response that includes *cor* genes, induced by the DREB (dehydration-responsive element binding) transcription factor family (Smirnoff and Bryant, 1999), suggests that there is likely to be extensive cross-talk among the stress-responsive signal transduction pathways (Sung et al., 2003). Indeed, stress-inducible over-expression of DREB1A conferred enhanced tolerance to freezing, water stress, and salinity without affecting plant growth, while increased constitutive expression of DREB1A also caused a significant improvement in stress tolerance but at the expense of severe growth retardation under normal growing conditions (Kasuga et al., 1999). Although the biochemical functions of the encoded stress-induced proteins are unknown, it is worth noting that the effect of DREB1A on freezing tolerance was substantially greater than that of CBF1 (>10°C vs. 1°C) (Zhu, 2001).

In concluding this section on engineering tolerance to environmental constraints, it is important to recognize that reductions in crop viability and yield are compounded by combinations of abiotic stresses. Such combinations can elicit plant responses that are not easily extrapolated from the plant's response to each stress applied individually (Mittler, 2006). Strategies designed to mitigate the effects of combinations of environmental stress conditions might, for example, target stress-responsive signal transduction pathways, which could exhibit synergistic or antagonistic cross-talk. Regardless of the approach taken, lab-based proof-of-concept experiments must be complemented by testing under conditions that mimic the field environment (Mittler, 2006).

3.5.3 Improved disease resistance

Crop productivity is limited by a variety of parasites and pathogens, including fungi, bacteria, viruses, and insects (Baker et al., 1997). In naturally occurring ecosystems, elaborate networks of defenses function at many levels to protect plants from disease (Abramovitch and Martin, 2004). Elucidation of these defense pathways has recently become a particularly active area of research in plant molecular biology, and has led to our growing appreciation for the complex interplay between basal defenses and specific disease resistance (Feys and Parker, 2000). A major contributor to disease susceptibility is the reliance of industrial-scale agriculture on monocultures. Cultivation of plant lines bred for resistance to one or a few pathogens, often conferred by so-called R genes, can lead to the emergence of pathogens that have undergone natural selection to overcome the resistance (Gurr and Rushton, 2005). Despite the potentially short-sighted nature of such agricultural practices, identification of R genes has been the focus of considerable effort over the past decade (Baker et al., 1997; Dangl and Jones, 2001). At least one such gene, the Bs2 gene from pepper, has been used successfully to engineer durable resistance to the agronomically significant bacterial spot disease in tomato (Tai et al., 1999). The Xa21 resistance gene from rice, which provides wide-spectrum resistance to the devasting bacterial blight caused by *Xanthomonas oryzae* pathovar *oryzae*, has been introduced into a variety of rice cultivars using *Agrobacterium*-mediated gene delivery (Wang et al., 2005). Likewise, broad spectrum resistance to potato late blight is conferred by one of four R genes cloned from a wild, highly resistant, potato species (Song et al., 2003a). Pyramiding of multiple R genes can confer resistance to a range of pathovars within a species (e.g., Li et al., 2001), but the introduction of R genes can also result in a substantial fitness cost to the plant (Gurr and Rushton, 2005).

More recently, attention has shifted to the basal or non-host resistance plant defenses, which tend to target entire classes of pathogens. These pathways are generally activated in response to common patterns shared by many pathogens, such as fungal cell walls or bacterial flagellin. Elicitation of defense-related signal transduction pathways can be achieved by introduction or overexpression of receptor-like kinases (Gurr and Rushton, 2005) such as the receptor responsible for perception of the pathogen-associated molecule flagellin (Zipfel et al., 2004). A related strategy involves engineering a plant to express a pathogen-derived elicitor of specific or basal defense responses (Keller et al., 1999). In this case, limiting the expression to sites of infection using pathogen-inducible promoters (Rushton, 2002) is essential, since constitutive activation of defense pathways can lead to reductions in plant health and even cell death (Gurr and Rushton, 2005).

Generic plant defenses include antimicrobial compounds such as defensins and chitinases. Ectopic expression of plant-derived or synthetic antimicrobial peptides in transgenic potatoes provides robust resistance to bacterial and fungal pathogens (Gao et al., 2000; Osusky et al., 2000), although only the former study tested resistance in the more relevant field setting (van der Biezen, 2001). A variety of antibacterial proteins from sources other than plants have been used to confer resistance to bacterial diseases in several transgenic plants (reviewed in Mourgues et al., 1998). *Arabidopsis* plants expressing antifungal peptides fused to a pathogen-specific recombinant antibody derived from chicken exhibited resistance to the fungal pathogen (Peschen et al., 2004). Finally, plant-derived defense molecules including proteinase inhibitors (Urwin et al., 1997) and lectins (Jung et al., 1998) have potential as nematicidal agents.

A third approach to engineering enhanced disease resistance takes advantage of the rapid expansion in our understanding of the pathways downstream of the initial pathogen perception events. Here, targets for genetic manipulation include "master-switch" transcriptional regulators, particularly those that activate local or global resistance networks involving salicylic acid, jasmonate, pathogenesis-related proteins, and the systemic acquired resistance that primes defenses in uninfected areas of the plant (Gurr and Rushton, 2005). For example, overproduction of the transcription factor NPR1 (also known as NIM1) results in enhanced resistance to bacterial and fungal pathogens and enhances the efficacy of fungicides (Cao et al., 1998; Friedrich et al., 2001). Plants engineered to produce elevated levels of salicylic acid also exhibit enhanced disease resistance (Verberne et al., 2000). Finally, appreciation for the involvement of the iron-binding protein ferritin in the oxidative stress response and the central role of oxidative stress in plant defense responses led to the successful demonstration that ectopic expression of ferritin can enhance tolerance to viral and fungal pathogens (Deak et al., 1999). Given the explosion in knowledge of plant defense mechanisms over the past decade, as well as the continued reliance on approaches to industrial-scale cultivation that foment rampant pathogen spread, genetic engineering for disease resistance promises to be a very active area of research in the near future.

The quest to engineer virus resistance in plants stems from the proposal that expression of pathogen-derived genes within a plant can induce resistance to the pathogen in question (Sanford and Johnston, 1985). The first successful validation of this concept was the creation of tobacco mosaic virus-resistant tobacco plants producing the virus coat protein (Powell-Abel et al., 1986). A multitude of virus-resistant plants have since been developed using the same strategy (reviewed in Lomonossoff, 1995 and Wilson, 1993). A markedly effective implementation of coat protein-mediated protection was instrumental in saving the Hawaiian papaya crop from the papaya ringspot virus; the transgenic papaya has been commercialized and efforts are underway to transfer the technology to developing countries, which produce 98% of the world's papaya crop (Gonsalves, 1998). Although this particular application made use of particle bombardment rather than *Agrobacterium* to deliver the transgene, it serves as a convincing illustration of the potential for achieving virus resistance in other highly susceptible crops.

Production of viral proteins generally provides moderate levels of protection to a relatively broad spectrum of related viruses (Lomonossoff, 1995). In several cases, *Agrobacterium*-mediated expression of a viral replicase gene (Baulcombe, 1994) or virus movement proteins (e.g., Beck et al., 1994), rather than the viral coat protein, effectively conferred resistance. Unexpectedly, a number of researchers discovered that in some instances, levels of resistance did not correlate with the amount of foreign protein produced; furthermore, translationally defective genes could also provide protection (reviewed in Lomonossoff, 1995). Taken together, these findings indicated that at least some component of the resistance was attributable to the transgenic RNA, not the protein itself (Lindbo et al., 1993; Pang et al., 1993; Goregaoker et al., 2000; Prins, 2003). These observations coincided roughly with the initial reports of cosuppression (see section 2.6), and contributed to the discovery of post-transcriptional RNA silencing (PTGS) in plants, as well as in fungi and animals (Hannon, 2002). The recognition that the observed RNA-mediated virus resistance was a manifestation of PTGS, in turn, led to the realization that homology-dependent gene silencing is responsible for much of the phenotypic variability observed in transgenic plants (Kooter et al., 1999).

The intracellular series of events by which dsRNA brings about gene silencing in plants has been extensively studied (reviewed in Tenllado et al., 2004), and it is now clear that the process functions as a naturally occurring defense system in plants in response to dsRNA formed during virus replication (Tenllado et al., 2004; Soosaar et al., 2005). RNA-mediated protection tends to provide resistance even to high levels of viral infection, but, as might be expected given the mechanism, is usually very virus-specific (Lomonossoff, 1995). Engineering plants to produce a self-complementary hairpin RNA corresponding to a viral gene target confers virus resistance; notably, the percentage of virus-resistant plants can be increased to almost 100% by including an intron within the hairpin region (Smith et al., 2000). Using *Agrobacterium*-mediated infiltration to deliver hairpin loops of viral RNA, Diaz-Ruiz and colleagues have demonstrated that it is possible to induce virus resistance in plants simply by exogenous exposure to the dsRNA; this work opens the door for future development of field-scale approaches in which bacterial lysates containing dsRNA are sprayed directly on the plants to confer resistance (Tenllado et al., 2004). Endogenous microRNAs, important regulators of gene expression that cause translational repression or cleavage of their target mRNAs, can also be engineered to contain sequences complementary to particular plant viruses. Transgenic plants expressing precursors of these artificial microRNAs exhibit resistance to the targeted viruses, even at temperatures that compromise hairpin dsRNA–mediated silencing (Niu et al., 2006). In contrast with RNA-mediated resistance, artificial microRNAs do not run the risk of complementing or recombining with non-target viruses, and thus pose less of an environmental biosafety threat (Garcia and Simon-Mateo, 2006).

Generally speaking, the mechanism by which expression of viral proteins causes resistance in plants is not as well understood as the process of viral RNA mediated suppression (Uhrig, 2003), and is rather protein-specific (Lomonossoff, 1995). Nonetheless, recent attempts to improve virus resistance in transgenic plants have targeted both protein- and RNA-mediated mechanisms. In several instances, introduction of a defective or truncated protein-coding sequence has proven more effective than expression of an intact, functional version in inducing resistance (Uhrig, 2003). Rudolph et al., (2003) have demonstrated that transgenic expression of a dominant interfering peptide from a viral nucleocapsid protein, identified using the yeast dihybrid assay, is sufficient to bring about virus resistance. *Agrobacterium*-mediated delivery of a ribozyme, a small RNA molecule capable of cleaving RNA, has been successful in conferring at least partial resistance to viruses and viroids in tobacco and potato (de Feyter et al., 1996; Yang et al., 1997).

A significant shortcoming of RNA-mediated virus resistance is the high degree of sequence homology (>90%) required (Prins, 2003), limiting the possibility of engineering resistance to multiple viruses with one transgene. Furthermore, virus resistance achieved in the lab does not always translate into the field, where added environmental stresses compound the plants' susceptibility (Wilson, 1993). With our growing appreciation of PTGS as a natural form of self-protection in plants came the predictable discovery that many viruses produce suppressors of PTGS as a counter-defense strategy (Rovere et al., 2002; Soosaar et al., 2005). This presents a potential problem for the use of silencing-based virus resistance in the field, where a secondary infection with a suppressor-carrying virus could allow the targeted virus to overcome the engineered resistance. Some attempts to stack viral genes have been successful in achieving resistance to multiple, related, viruses (Prins et al., 1995); one logical strategy would entail engineering resistance to possible co-infecting viruses that carry PTGS suppressors as well as the virus of interest (Rovere et al., 2002). Finally, expression of a viral transgene under the control of the 35S CaMV promoter can be substantially attenuated if the plants happen to become infected with CaMV, leading to silencing of the transgene and a loss of immunity (Mitter et al., 2001). Likewise, herbicide resistance, conferred by a 35S CaMV-driven transgene, was rendered ineffective upon CaMV infection (Al-Kaff et al., 2000). These observations suggest that virus-derived suppression of transgene expression, attributable to transcriptional or post-transcriptional gene silencing, may prove to be a significant limitation in maintaining engineered traits in a field setting.

3.6 Reduction in the use of harmful agrochemicals by enhancing plant resistance to herbicides and pests

3.6.1 Herbicide resistance

Much has been written in the popular press about the creation and marketing of herbicide resistant crop plants. The rationale is that these crops allow farmers to eliminate weeds with one broad-spectrum, somewhat less toxic, herbicide without damaging the crop. Two of the most common herbicide/herbicide resistant seed packages involve the herbicides glyphosate (inhibitor of the shikimate pathway for aromatic amino acid biosynthesis; marketed by Monsanto as RoundupTM) and glufosinate ammonium (glutamine synthase inhibitor; Hoechst's trademark BastaTM); others include sulfonylurea (acetolactate synthase inhibitor) and bromoxynil (Nottingham, 1998). In most cases, resistance is conferred by foreign genes encoding enzymes that are not susceptible to the action of the herbicide (Comai et al., 1985), or by overproduction of the target enzyme (Dale et al., 1993). In addition, the bacterial *bar* gene product provides resistance to glufosinate ammonium by detoxifying it (De Block et al., 1987). The most widely planted herbicide-resistant crop plant is Monsanto's Round-Up Ready soybean; other glyphosate-resistant crops include maize, canola, oilseed rape, sugarbeet, tobacco, and cotton (Nottingham, 1998). Although many herbicide-resistant crops were initially developed using *Agrobacterium*-mediated gene delivery, the current method of choice is particle bombardment. For this reason, these plants will not be discussed further here; the reader is referred to Nottingham (1998) for a more complete discussion of the private sector interests responsible for the development of these crops.

3.6.2 Insect resistance

One of the early selling points of transgenic crop plants was the promise of a reduction in the use of hazardous pesticides. By far the most widely used insect resistance traits are conferred by the *cry* genes, encoding toxins derived from the soil bacterium *Bacillus thuringiensis*. Several different Bt toxin gene products have slightly different modes of action and target different orders of insects, but the general strategy is similar: the crystalline toxins bind to the membrane of the larval gut and prevent nutrient uptake (Nottingham, 1998). Bt toxins are considered particularly attractive because of their high specificity, biodegradable nature, and lack of toxicity for humans and other non-target animals. *Agrobacterium* was first used to introduce a Bt gene into tobacco and tomato in 1987 (Vaeck et al., 1987), and the first transgenic plant was commercialized in 1996. The most widely planted Bt crops include maize (resistant to the European corn borer and/or southern corn rootworm), cotton (resistant to the cotton bollworm and the tobacco budworm), and potato (target pest is the Colorado potato beetle) (Shelton et al., 2002). Several other Bt crops, including canola, soybean, tomato, apple, peanuts, and broccoli are under development (Bates et al., 2005). Bt rice may hold considerable promise for Asian agriculture (High et al., 2004). A substantial body of literature exists on the economic impact of Bt crops in industrial and developing countries [see, for example, Morse et al., (2004); for a comprehensive review of ecological, economic, and social consequences, together with risk assessment of Bt crops, the reader is referred to Shelton et al., (2002)]. It should be noted that yield increases due to genetic modifications such as Bt transgenes are likely to be much higher in developing countries than in industrialized nations; this difference is attributable to high pest pressure, and low availability/adoption of chemical alternatives in areas such as south/southeast Asia and Africa, where farmers cannot afford chemical inputs (Qaim and Zilberman, 2003). In a clear validation of the original rationale for insecticide-producing transgenic crops, Huang et al. have documented impressive reductions in pesticide application and in pesticide-related poisoning among Chinese farmers cultivating Bt cotton and rice (Huang et al., 2002b; Huang et al., 2005). Similar decreases in the use of pesticides have also been reported among farmers planting Bt cotton in India (Qaim and Zilberman, 2003).

At the same time, effects such as the long-term regional declines in the pink bollworm population density attributed to the planting of Bt cotton (Carriere et al., 2003), suggest that this technology will have significant and lasting ecological impacts. Concern about the emergence of insect resistance to Bt has led to a variety of insect resistance management strategies including regulation of the toxin dosage, mandated planting of refuge regions, and temporal or tissue-specific toxin expression (Bates et al., 2005). Pyramiding two or more Bt toxin genes in the same transgenic plant has been demonstrated to delay the evolution of resistance (Zhao et al., 2003). However, selection for resistance will continue to occur even in plants with pyramided resistance genes as long as the transgenes are also used singly in other varieties; Pink and Puddephat (1999) have argued instead for plant "multilines" that are heterogeneous with respect to the resistance genes they carry, with the composition of the mixture commensurate with the frequency of the corresponding virulence alleles in the pathogen population. Additional non-Bt proteins that target non-Bt receptors under development include the Vip3A toxin (Moar, 2002) and toxin A from the bacterium *Photorabdus luminescens* (Liu et al., 2003). Other classes of insecticidal proteins are the protease inhibitors, produced by a wide variety of plants to inhibit animal or microbial digestive enzymes, and plant-derived lectins (Nottingham, 1998). *Agrobacterium*-mediated introduction of the cowpea trypsin inhibitor gene has been shown to provide tobacco with increased resistance to the tobacco budworm (Hilder et al., 1987); the same gene in rice also confers greatly enhanced resistance to the rice stem borer (Wang et al., 2005). Pyramiding *cry* genes with genes encoding lectins and/or protease inhibitors is an active area of research in many crops of import to developing world agriculture (see, for example, the report from the Indo-Swiss Collaboration in Biotechnology at http://iscb.epfl.ch).

3.7 Enhanced nutritional content in crop plants

In addition to increasing yields and reducing the use of inputs associated with negative environmental and/or health consequences, genetic modification of food crops offers the possibility of enhancing the nutritional content of the food (Huang et al., 2002a). In some cases, the goal is to improve nutritional value by removing naturally occurring, but harmful, substances. Perhaps the best-known cases are the toxic cyanogens found in the important staple food cassava. Labor-intensive processing is required to remove these cyanide precursors, which pose particular risks to individuals with protein-poor diets, from the tubers. By blocking the synthesis of the cyanogen precursors with antisense constructs, Siritunga and Sayre (2003) achieved a 99% reduction in root cyanogen levels, even though the *Agrobacterium*-mediated transgenic modification targeted the leaf-based biosynthetic pathway.

More frequently, however, nutritional enhancement entails increasing the content of relatively rare constituents and/or creating a more balanced amino acid complement. Rice, for example, is a staple crop for over half the world's population (Wang et al., 2005), yet lacks many essential nutrients (Ye et al., 2000), and loses more nutritional value during processing (Al-Babili and Beyer, 2005). The gene encoding a non-allergenic seed albumin protein with a well-balanced amino acid content was introduced into potato (Chakraborty et al., 2000), while canola and soybean have been modified to augment their notoriously low levels of lysine (Tabe and Higgins, 1998). Transgene-driven biosynthesis of naturally occurring or modified sulfur-rich proteins has been achieved in canola (Altenbach et al., 1992) and could be used to ameliorate low methionine levels in other edible plants; this deficiency is especially pronounced in legume seeds (Tabe and Higgins, 1998). Quantity and quality of starch are other targets of food-crop engineering (Slattery et al., 2000); manipulation of the adenylate pools in potato increased both the starch content and the yield of transgenic potatoes (Regierer et al., 2002). Successful production of health-promoting very long chain polyunsaturated (including omega-3) fatty acids in oilseed crops has recently been reported (Wu et al., 2005). Although accumulation of the desirable fatty acids in linseed is limited by the availability of biosynthetic intermediates, alternative strategies, including engineering fatty acid production in green vegetables, have been proposed (Abbadi et al., 2004). If successful, such genetic modifications hold promise as a sustainable alternative to fish, which are prone to problems including dwindling stocks and contamination with heavy metals and other pollutants (Qi et al., 2004).

As might have been predicted from their roles in cancer prevention, in promoting immunity, and in slowing the progression of several degenerative human diseases (Shintani and DellaPenna, 1998), augmentation of anti-oxidant levels in plants has been another attractive goal of food crop engineering. Biosynthesis of one such group of essential antioxidants, Vitamin E (Sattler et al., 2004), has been significantly enhanced in *Arabidopsis*, corn, and soybean using *Agrobacterium*-mediated redesign of the pertinent pathways (Shintani and DellaPenna, 1998; Cahoon et al., 2003; Van Eenennaam et al., 2003). Production of other potent anti-oxidants including lycopene has been increased through transgenic overexpression of relevant enzymes in tomatoes (Muir et al., 2001; Mehta et al., 2002; Niggeweg et al., 2004). Fruit-specific silencing of the photomorphogenesis gene *DET1* in tomato elevated flux through both the flavonoid and carotenoid biosynthetic pathways, increasing the content of beta-carotene as well as lycopene, without the use of exogenous genes and without negative impacts on fruit yield or quality (Davuluri et al., 2005).

Mineral fortification of crop plants through genetic alteration or selection has been envisioned as a way to address dramatic global dietary deficiencies in iron, zinc, iodine, selenium and several other essential minerals. Identifying genes and conditions that promote mineral accumulation in plants is the focus of the HarvestPlus program within the CGIAR (Consultative Group of International Agricultural Research). In the initial phase of this initiative, six crops (beans, cassava, maize, rice, sweet potato and wheat) are being targeted; an additional 11 subsistence crops will be added in phase 2 (http://www.harvestplus.org/about.html). Transgenic approaches to increase bioavailability have targeted mineral uptake, transport to edible tissues, and augmented levels of organic compounds, including ascorbate and beta-carotene, that promote mineral absorption in humans (White and Broadley, 2005). Expression of the soybean iron storage protein ferritin in rice, for example, resulted in a three-fold rise in seed iron content (Goto et al., 1999). Alternative strategies include engineering plants to express phytase, thereby removing a key impediment in most animals to mineral uptake (Brinch-Pedersen et al., 2002).

3.7.1 "Golden Rice"

Beta-carotene is an essential dietary constituent, required in vertebrates to synthesize Vitamin A, the key visual pigment retinal, and the morphogen retinoic acid (Giuliano et al., 2000). Beta-carotene and other carotenoids are synthesized in plants from phytoene, and introduction of a phytoene synthase or desaturase from bacteria dramatically increased flux through the carotenoid pathway in canola (Shewmaker et al., 1999) and in tomato (Romer et al., 2000), respectively. Vitamin A deficiency is a significant health problem in much of the developing world, leading to an estimated 2 million deaths and 250,000 cases of childhood blindness each year (Ye et al., 2000). A public sector initiative to engineer the beta-carotene biosynthetic pathway into rice endosperm resulted in the development of "golden rice" in 1999. *A. tumefaciens* was used to deliver into rice phytoene synthase and lycopene cyclase genes from daffodil, along with a bacterial phytoene desaturase gene. All three encoded enzymes carried transit peptides targeting them to the plastid, the natural site of synthesis of the phytoene precursor geranylgeranyl diphosphate (Ye et al., 2000). This prototype golden rice contained one tenth of the recommended daily allowance (RDA) of beta-carotene per 300 grams of rice (Giuliano et al., 2000). This relatively low yield, coupled with concerns about the presence of an antibiotic selectable marker, led to the creation of second-generation golden rice in two agronomically important rice cultivars. The details of these modifications, carried out in parallel by both public and private sector researchers, have been summarized in an excellent review by Al-Babili and Beyer (2005). Dramatic improvement in the yield of beta-carotene was achieved by substituting a phytoene synthase gene from maize for that from daffodil (Paine et al., 2005). Using generally accepted conversion factors for bioavailability and processing within the human, it is estimated that this second generation golden rice can provide 50% of the vitamin A RDA for children in a 72 g serving (Al-Babili and Beyer, 2005). Ultimately, however, as with all transformed crop plants, the only relevant value will be the nutritional contribution provided by field-grown, locally adapted varieties. Additional goals for complementary rice improvement include increasing the content of vitamin E to stabilize the beta-carotene, and increasing iron accumulation to address the iron deficiencies often found in the same populations who would benefit from golden rice (Al-Babili and Beyer, 2005).

Golden rice serves as an excellent illustration of the challenges inherent in technology transfer to developing countries. Although the research and development was provided exclusively though the public sector, the project had drawn on a wide variety of patent-protected DNA fragments and technologies, and hence the modified plant was encumbered with no fewer than 70 patent constraints held by 32 different companies and universities (Potrykus, 2001). Through a series of complex negotiations, free licenses were eventually obtained for every intellectual and technical property component (see, for example, Normile, 2000). Current efforts are focused on introducing the engineered traits into as many local adapted varieties and ecotypes as possible. The central player in this phase of the project is the Indo-Swiss Collaboration in Biotechnology, a program funded through the Indian Department of Biotechnology and the Swiss Development Corporation. This collaborative effort, which incorporates studies on biosafety, ecological impact, and socioeconomic considerations, and which is committed to ensuring that the technology reaches the target populations, should serve as a model for future technology transfers (Potrykus, 2001). Public adoption of golden rice will depend on many factors, but as Potrykus (2001) succinctly spells out, this product of *Agrobacterium*-mediated genetic modification fulfills each of the requirements for acceptability put forth by activists opposed to genetically engineered crops.

4 GENE FLOW AND MOLECULAR APPROACHES TO TRANSGENE CONTAINMENT/MONITORING

Despite the panoply of potential benefits associated with plant genetic modification, public enthusiasm for this technology has been far from universal. Concerns range from risks inherent in the technology-such as potential ecological damage resulting from transgene escape to wild plant relatives, or possible adverse health effects of consuming genetically modified (GM) food-to sociopolitical ramifications that transcend the technology. In the latter category, valid questions have been raised about inequitable access to the new crop varieties and the impact that this may have on the distribution of wealth within poor societies. On a global scale, growing disparities in wealth between North and South (industrialized and developing countries) may be exacerbated by the practice referred to as bio-prospecting or bio-piracy (depending on one's perspective), in which genes from landraces and traditional varieties found to confer desirable traits are utilized/appropriated to genetically modify crop plants (Leisinger, 1999). Like the dangers associated with monoculture discussed earlier, these technology-transcending risks are not specific to plant genetic engineering, but they should not be dismissed as irrelevant to the discourse on GM crops.

A detailed discussion of biosafety issues and regulatory considerations associated with agricultural biotechnology is beyond the scope of this chapter. However, in light of the serious nature of the concerns, and the widespread public mistrust of the technology (Kleter et al., 2001), it would be irresponsible not to include a brief overview of the topic. For a more detailed analysis, the reader is referred to chapter 19 in this volume.

Most health-related concerns center on the possibility of transgene transfer to gut microbes. As alluded to in section 2.4, antibiotic-resistance marker genes have come in for special scrutiny in this regard. Many questions remain unanswered concerning the ability of ingested DNA to survive passage through the digestive tract in a biologically active form, the potential for gene flow during silage production using GM crops, and the significance of GM plant-derived antibiotic resistance marker genes in comparison with the rampant dissemination of bacterial resistance attributable to overuse of antibiotics in clinical and livestock settings (Heritage, 2005). The cultivation of plants producing pharmaceutical proteins also presents possible health risks including exposure of non-target organisms and of humans to potential allergens (Peterson and Arntzen, 2004).

Adverse environmental impacts of transgenic plants may arise from toxicity to non-target organisms or from increased selective pressure on target pests, although our ability to predict the evolution of resistance development is limited (Sandermann, 2004). Transgene contamination of plants can occur via cross-pollination or inadvertent dispersal of GM seeds during harvest, transportation, or planting (Smyth et al., 2002). Gene flow from transgenic plants to wild relatives and non-transgenic crop plants has been documented for both Bt and herbicide resistance traits (reviewed in Sandermann, 2004). Contamination of conventional varieties destined for "GM-free" or organic markets represents a serious concern to farmers who have chosen to abstain from cultivating genetically engineered crops (Smyth et al., 2002). Incidents that appear to threaten the livelihood of this cohort of producers or the integrity of the booming organic movement are likely to cause a substantial negative backlash in public perceptions of agricultural biotechnology.

Molecular strategies to limit gene flow include interfering with pollen production and interruption of seed formation; both approaches can rely on *Agrobacterium*-mediated delivery of exogenous genetic material. A third, non-*Agrobacterium* mediated approach- maternal inheritance-involves introducing the transgene into the chloroplast genome to avoid pollen-based gene dissemination (Daniell, 2002). Nuclear-encoded male sterility was first accomplished by Mariani et al. (1990), who expressed an RNase gene under the control of a promoter specific for the tapetum. RNase-induced destruction of the tapetum, one of the specialized tissues in the anther required for pollen development, prevents pollen formation. Restoration of male fertility can be achieved by crossing in the *barstar* gene, encoding an inhibitor of the *barnase* RNase, also under tapetum-specific promoter control (Williams, 1995). Other approaches to conditional male sterility include engineering a plant with a gene or set of genes, under inducible control in male reproductive tissues, that poison the plant cells or that alter the levels of metabolites such as amino acids needed for the production of pollen (Perez-Prat and van Lookeren Campagne, 2002).

Genes required for seed formation have also been targets for gene containment strategies. In the infamous "terminator" technology, inducible expression of Cre recombinase results in the removal of a spacer sequence that otherwise prevents seed-specific production of a cytotoxic ribosome inhibitor protein; application of an exogenous stimulus relieves repression of the *cre* gene and leads to destruction of the seed tissue (Daniell, 2002). Although this technology has significant potential as a built-in safety mechanism to prevent unintended dispersal of GM seed, it gained notoriety as an impediment to growers wishing to save and replant harvested seed containing proprietary alterations. As such, it is perceived as exemplifying the insensitivity of the agricultural biotechnology enterprise to the needs of subsistence farmers, and winning the acceptance of biotechnology skeptics will be a challenge (Smyth et al., 2002).

Several recent reviews focus on monitoring gene flow and on mathematical modeling of risk assessment (Wilkinson et al., 2003; Heinemann and Traavik, 2004; Nielsen and Townsend, 2004; Lee and Natesan, 2006). Transgene presence in living plants can be monitored using fluorescent marker genes (Stewart, 2005). "Bio-barcodes," consisting of uniform recognition sequences flanking a unique variable sequence to facilitate PCR amplification and sequencing of the barcode, could be incorporated into all transgene events; comparison to a universal database of barcode sequences would provide information pertinent to liability claims, intellectual property violations, and dispersal tracing (Gressel and Ehrlich, 2002). Unfortunately, our ability to predict ecological consequences of transgenic crop cultivation still lags far behind the implementation of monitoring technology, and even further behind the development of the crops themselves (Snow, 2002). As the many emerging applications of plant genetic engineering described in section 3 are adapted for novel geographical locations, each will need to be assessed on a case-by-case basis, taking into consideration the particular ecological context in which the plants are to be grown (Dale et al., 2002).

5 GLOBAL STATUS OF AGRICULTURAL BIOTECHNOLOGY AND TECHNOLOGY TRANSFER

A concise summary of the global growth of commercialized GM crops and their economic impact, replete with graphs and figures, is compiled annually by the International Service for the Acquisition of Agribiotech Applications (ISAAA) and can be easily accessed via the internet at http://www.isaaa.org (James, 2005). Additional insights concerning future trends can be gleaned by examining data available on the internet regarding approved field trials, field trial applications, and patent applications in the U.S and internationally (http://www.aphis.usda.gov/brs/brs_charts.html; http://www.nbiap.vt.edu). An analysis of these data from 1987 through 1999 reveals that the early emphasis on single gene traits-primarily herbicide and insect resistance-has now given way to attempts to alter more complex traits, such as nutritional quality and the physiological characteristics that affect crop yield (Dunwell, 2000).

As of 2005, 90 million hectares in 21 countries were planted with approved GM crops; 11 of the 21 nations, producing 38% of the world's biotech crops, are in the developing world (James, 2005). With public funding levels that far exceed those in any other country, China accounts for over half of the plant biotechnology expenditures in lesser-developed nations, with Brazil and India trailing far behind (Huang et al., 2002a). Early claims that plant genetic engineering would help ameliorate food shortages among the world's poorest populations have led to sustained skepticism and even cynicism from biotechnology opponents, in part because the first transgenic crops to be commercialized appear to benefit primarily the agro-chemical industry and corporate-scale farmers in industrialized countries, rather than consumers or subsistence farmers (Vasil, 2003). This picture is changing, however; of the 8.5 million farmers cultivating genetically engineered crops in 2005, 7.7 million of them were poor subsistence farmers. The vast majority of those farmers (6.4 million) live in China (James, 2005), the world's largest producer of rice, and genetic modification of rice is the focus of considerable attention in China's program to develop more sustainable agriculture (Wang et al., 2005). Biosafety procedures in China require multiple levels of testing for environmental release, and rice engineered for resistance to lepidopteran insects or bacterial blight is currently in the final stages of safety trials prior to commercialization (Wang et al., 2005). In addition to rice, the Chinese have placed substantial emphasis on engineering a variety of fruit and vegetable crops in an effort to bolster food security (Huang et al., 2002b). Although herbicide tolerance is still the most prevalent engineered trait worldwide (currently constituting 71% of the global area devoted to GM crops) (James, 2005), over 90% of the field trials in China target insect and disease resistance (Huang et al., 2002b). Pest and pathogen resistant plants are already starting to have a significant impact on productivity and on reducing the environmental impact of pesticide use in China (Huang et al., 2002b; Huang et al., 2005).

Over the past four decades, public sector research institutions in several regions of the developing world have played a pivotal role in the improvement of staple crops through conventional breeding. In addition to the 20 CGIAR centers (http://www.cgiar.org), national agricultural research agencies have contributed to introducing new traits into local varieties and to facilitating distribution and adoption of these varieties by farmers. Collaborations with both academic and corporate plant biotechnology programs in industrialized nations are now beginning to make biotechnology approaches available to these public sector institutions (Toenniessen, 1995). The most successful of these collaborations have as core tenets strong emphases on capacity building, and on sustainable cropping practices that incorporate indigenous knowledge at all levels of decision-making. The following section highlights the goals, participants, and innovative aspects of some of these programs. More information on national and international public-sector research stations, and on international organizations involved in facilitating biotechnology transfer is available elsewhere (Toenniessen, 1995). A detailed investigation of the capacity for biotechnology research in four developing countries-Mexico, Kenya, Indonesia, and Zimbabwe-together with policy recommendations arising from the study, has also been published (Falconi, 2002).

The resource- and knowledge-intensive nature of plant genetic engineering has precluded development of biotechnology research programs by many of the countries that face the most pressing food security issues. Furthermore, with a few notable exceptions (China, Brazil, India and South Africa), national government investment in agricultural research is generally insufficient to maintain programs that could address local constraints and/or transfer modifications developed elsewhere to locally favored varieties (Huang et al., 2002a). Several collaborative initiatives, some including private sector partners, have evolved to meet these challenges; most of the projects undertaken within these collaborations rely on *Agrobacterium*-mediated transformation of target plants. One of the oldest such partnerships is the Indo-Swiss Collaboration in Biotechnology (ISCB), which was established in 1974. During its first two decades, this long-term bilateral program focused on developing a cadre of highly trained Indian scientists and establishing research capacity within the Indian academic sector. In the last few years, the ISCB has promoted research partnerships between Swiss and Indian institutions, with an emphasis on increased productivity of wheat and pulses through enhanced disease resistance; a parallel initiative centers on sustainable management of soil resources (http://iscb.epfl.ch). Other bilateral programs include the Peking-Yale Joint Center for Plant Molecular Genetics and Agrobiotechnology, established in 2000 (Yimin and Mervis, 2002), and a partnership between scientists in Bolivia and those at the University of Leeds, who are developing nematode-resistant potatoes by introducing proteinase-inhibitor genes (Atkinson et al., 2001).

For several years starting in 1992, the Dutch government-funded Special Programme on Biotechnology brought together scientists, farmers, and local leaders in Zimbabwe, India, Kenya, and Colombia to develop local biotechnology agendas that addressed the needs of small scale producers. This project-based program differed from most other collaborations in the primacy it placed on participatory technology development, developing new paradigms for integrating the perspectives of farmers, consumers, and socio-economic policy experts into the process of setting research priorities (Broerse, 1998). Specific research projects included the transformation of cassava, sweet potato, and cowpea to confer virus resistance (Sithole-Niang, 2001). In Zimbabwe, the Dutch program also funded capacity building through a Master's level training program in biotechnology and shorter local training workshops.

Complementing these bilateral models for technology transfer are networks of researchers focused on one crop, as exemplified by the Cassava Biotechnology Network (CBN). Founded in 1988 by two CGIAR centers, the Centro International de Agricultural Tropical (CIAT) and the International Institute of Tropical Agriculture (IITA), in collaboration with several small research institutes in North America and Europe, the goals of the CBN are to develop strategic biotechnology tools and appropriate biotechnology applications for cassava improvement. With support from the Dutch Special Programme for Biotechnology, the network has expanded to over 800 active researchers in 35 countries and includes collaborators focusing on needs assessment, anthropology, plant breeding, and post-harvest issues including market economics. Although cassava is not cultivated in industrialized nations and therefore has not been a target for improvement by the private sector, it is an important source of nutrition and food security in many of the world's least developed areas (Taylor et al., 2004). Several characteristics make cassava a staple for subsistence farmers, a cash crop for local markets, and a reliable source of food and animal feed during periods of famine. It is drought tolerant and grows with low inputs in areas of marginal fertility. The edible roots can be left in the ground for one to two years without decay and the leaves are an important source of protein and vitamins in many parts of Africa (Siritunga and Sayre, 2003). Through direct participation by cassava farmers, the CBN has identified several targets for improvement: resistance to bacterial blight, viral disease, and insect-inflicted damage; reduction of toxic cyanogens; enhanced nutritional value including increased vitamin content, protein content, and quantity and quality of starch in the roots; and stress tolerance (Thro et al., 1999). A recent comprehensive review, describing how each of these goals is being addressed through *Agrobacterium*-mediated transformation of cassava (Taylor et al., 2004), serves to illustrate the potential for improvement in one key crop of central importance to resource-poor farmers.

Several collaborative ventures have involved significant contributions from the private sector. In 1991, the ISAAA was created to build partnerships and to broker transfer of proprietary technology from industrialized countries to developing nations. One model project praised for its inclusion of a substantial training component involved the donation by Monsanto of coat protein genes conferring virus-resistance to Mexican scientists working on potato. The technology was further disseminated to scientists from the Kenyan Agricultural Research Institute (Krattiger, 1999). ISAAA has centers on five continents and is funded by the McKnight Foundation, the Rockefeller Foundation, various bilateral agencies, and the private sector. Like ISAAA, the USAID-funded Agricultural Biotechnology Support Project, based at Michigan State University, was initiated to bring together public sector and commercial research efforts. Between 1991 and 2003 this program funded a number of plant genetic modification projects that were undertaken in collaboration with the Agricultural Genetic Engineering Research Institute in Egypt. Goals included development of resistance to potato tuber moth, drought- and salinity-tolerant tomato and wheat, stem borer resistance in tropical maize, virus resistant tomato and sweet potato, and micropropagation techniques for pineapple and banana (http://www.iia.msu.edu/absp/). As a third example, in 2005 the Bill and Melinda Gates Foundation provided funding through its Grand Challenges in Global Health initiative for the Kenyan-based food organization A Harvest to partner with Pioneer Hi-Bred International and the Council for Scientific and Industrial Research in South Africa to develop a more nutritious and easily digested variety of sorghum (http://www.gcgh.org/).

The future success of these collaborative programs will depend on sustained commitments to their funding, and on a continued recognition of the complementarity between biotechnology and traditional crop breeding programs (Huang et al., 2002a). A key component in many of the examples described above is the emphasis placed on transfer of the technological knowledge and the tools required for scientists in the developing countries to pursue future projects more independently. Such capacity building includes training of research personnel, but also requires the establishment of a regulatory framework that is sensitive to local ecological, legal, and cultural contexts. In contrast with commercial crops that have already been vetted by Western regulatory agencies, novel locally developed crops "pose unique challenges for institutes seeking regulatory approval" (Cohen, 2005). The Swedish Biotechnology Advisory Commission was formed to help developing countries meet the challenges of biosafety capacity building through training, advising, and information exchange (L. Paula, personal communication). Similarly, a core mission of the ISAAA is training, including institutional capacity building in biosafety regulation (Krattiger, 1999). The Biotechnology Service at the International Service for National Agricultural Research (ISNAR) has also provided training in agricultural biotechnology management and performed assessments on intellectual property issues as they related to agricultural biotechnology (http://www.isnar.cgiar.org). In 2004, ISNAR was folded into the International Food Policy Research Institute and is now located in Addis Ababa, Ethiopia (http://www.ifpri.org/divs/isnar.htm).

Finally, a number of public sector research institutes are dedicated to developing and transferring biotechnology knowledge and resources to developing countries. These include the Applied Biotechnology Center at CIMMYT in Mexico City, devoted to genetic engineering of wheat and maize (http://www.cimmyt.org/ABC); the Center for the Application of Molecular Biotechnology to International Agriculture (CAMBIA) in Canberra, focusing on rice transformation using *Agrobacterium* (http://www.cambia.org); and the International Laboratory for Tropical Agricultural Biotechnology, focusing on rice, cassava, and tomato (http://www.danforthcenter.org/iltab).

A recent and highly illuminating survey of the public-sector research pipelines for GM crops in 15 developing countries identified a number of key trends in research agendas and regulatory considerations (Cohen, 2005). In contrast with the worldwide situation, where three crops (soybean, maize, and cotton) account for 95% of the global land area devoted to commercialized GM crops (James, 2005), the 201 genetic transformation events in these 15 countries encompassed no fewer than 45 different crops important to local economies, including chickpeas, cowpeas, lupin, cacao and a wide variety of fruits and vegetables in addition to rice, potato and maize (Cohen, 2005). Somewhat surprisingly, given the potential of the collaborative ventures described in the preceding section, most of the research described in this survey was carried out by single institutions, and the partnerships that did exist most often involved only public-sector institutions within the same country (Cohen, 2005).

The vast majority of the projects surveyed target biotic or abiotic stresses, while others strive to achieve prolonged shelf life or nutritional enhancement; only 5% of the transgenic plants under development are being engineered for herbicide tolerance (Cohen, 2005). This, again, is in stark contrast to the situation in industrialized nations, as described above, and reflects a much more consumer-centric approach to genetic transformation that focuses on local needs in these predominantly poor countries. Indeed, one of the most important observations to be made from this survey is the degree to which the biotechnology research in these countries actually has the potential to realize the oft-touted promise of enhancing human health and reducing poverty. By substantially decreasing the use of pesticides, fungicides, and other harmful agrochemicals, these crops should provide significant environmental and health benefits. Reducing losses attributable to pests can result in less acreage devoted to a single staple or cash crop, thereby contributing to greater biodiversity in a given area (Atkinson et al., 2001)). Enhanced shelf life can diversify a farm family's diet and allow farmers to wait out a glutted supply stream before bringing crops to market, thus increasing the financial return on their investment. Likewise, higher yields due to improved disease, pest, salt and drought tolerance lead to increased food security and more purchasing power, with the potential for "spillover effects" in the local economies (Cohen, 2005). These effects include enhanced educational opportunities for female children, personal hygiene leading to less transmission of communicable disease, and reduced population growth (Rosegrant and Cline, 2003). These, then, are the applications of *Agrobacterium*-mediated transformation that truly reflect a "poverty focus" (Conway, 1997) and that give renewed life to the promise of benefits for resource-poor farmers.

6 ACKNOWLEDGEMENTS

We are grateful to Gape Machao for assistance in assembling this manuscript. This work was supported by grant MCB-0416471 from the National Science Foundation to LMB.

7 REFERENCES

Abbadi A, Domergue F, Bauer J, Napier JA, Welti R, Zahringer U, Cirpus P, Heinz E (2004) Biosynthesis of very-long-chain polyunsaturated fatty acids in transgenic oilseeds: constraints on their accumulation. Plant Cell 16: 2734-2748

Abramovitch RB, Martin GB (2004) Strategies used by bacterial pathogens to suppress plant defenses. Curr Opin Plant Biol 7: 356-364

Aharoni A, Jongsma MA, Bouwmeester HJ (2005) Volatile science? Metabolic engineering of terpenoids in plants. Trends Plant Sci 10: 594-602

Al-Babili S, Beyer P (2005) Golden Rice–five years on the road–five years to go? Trends Plant Sci 10: 565-573

Albert H, Dale EC, Lee E, Ow DW (1995) Site specific integration of DNA into wild-type and mutant *lox* sites placed in the plant genome. Plant J 7: 649-659

Alia, Hayashi H, Sakamoto A, Murata N (1998) Enhancement of the tolerance of Arabidopsis to high temperatures by genetic engineering of the synthesis of glycinebetaine. Plant J 16: 155-161

Al-Kaff NS, Kreike MM, Covey SN, Pitcher R, Page AM, Dale PJ (2000) Plants rendered herbicide-susceptible by cauliflower mosaic virus-elicited suppression of a 35S promoter-regulated transgene. Nat Biotechnol 18: 995-999

Altenbach SB, Kuo CC, Staraci LC, Pearson KW, Wainwright C, Georgescu A, Townsend J (1992) Accumulation of a Brazil nut albumin in seeds of transgenic canola results in enhanced levels of seed protein methionine. Plant Mol Biol 18: 235-245

Apse MP, Aharon GS, Snedden WA, Blumwald E (1999) Salt tolerance conferred by overexpression of a vacuolar Na^+/H^+ antiport in *Arabidopsis*. Science 285: 1256-1258

Arntzen C, Plotkin S, Dodet B (2005) Plant-derived vaccines and antibodies: potential and limitations. Vaccine 23: 1753-1756

Atkinson HJ, Green J, Cowgill S, Levesley A (2001) The case for genetically modified crops with a poverty focus. Trends Biotechnol 19: 91-96

Baertlein DA, Lindow SE, Panopoulos NJ, Lee SP, Mindrinos MN, Chen TH (1992) Expression of a bacterial ice nucleation gene in plants. Plant Physiol 100: 1730-1736

Baker B, Zambryski P, Staskawicz B, Dinesh-Kumar SP (1997) Signaling in plant-microbe interactions. Science 276: 726-733

Bakker H, Bardor M, Molthoff JW, Gomord V, Elbers I, Stevens LH, Jordi W, Lommen A, Faye L, Lerouge P, Bosch D (2001) Galactose-extended glycans of antibodies produced by transgenic plants. Proc Natl Acad Sci USA 98: 2899-2904

Bardor M, Faye L, Lerouge P (1999) Analysis of the N-glycosylation of recombinant glycoproteins produced in transgenic plants. Trends Plant Sci 4: 376-380

Bartels D (2001) Targeting detoxification pathways: an efficient approach to obtain plants with multiple stress tolerance? Trends Plant Sci 6: 284-286

Barton KA, Binns AN, Matzke AJ, Chilton M-D (1983) Regeneration of intact tobacco plants containing full length copies of genetically engineered T-DNA, and transmission of T-DNA to R1 progeny. Cell 32: 1033-1043

Bates SL, Zhao JZ, Roush RT, Shelton AM (2005) Insect resistance management in GM crops: past, present and future. Nat Biotechnol 23: 57-62

Baucher M, Chabbert B, Pilate G, Van Doorsselaere J, Tollier MT, Petit-Conil M, Cornu D, Monties B, Van Montagu M, Inzé D, Jouanin L, Boerjan W (1996) Red xylem and higher lignin extractability by down-regulating a cinnamyl alcohol dehydrogenase in poplar. Plant Physiol 112: 1479-1490

Baulcombe D (1994) Replicase-mediated resistance: a novel type of virus resistance in transgenic plants? Trends Microbiol 2: 60-63

Bechtold N, Ellis J, Pelletier G (1993) *In planta Agrobacterium* mediated gene transfer by infiltration of adult *Arabidopsis thaliana* plants. C R Acad Sci Paris Life Sciences 316: 1194-1199

Beck DL, Van Dolleweerd CJ, Lough TJ, Balmori E, Voot DM, Andersen MT, O'Brien IE, Forster RL (1994) Disruption of virus movement confers broad-spectrum resistance against systemic infection by plant viruses with a triple gene block. Proc Natl Acad Sci USA 91: 10310-10314

Bevan MW (1984) Binary *Agrobacterium* vectors for plant transformation. Nucleic Acids Res 12: 8711-8720

Bevan MW, Flavell RB, Chilton M-D (1983) A chimeric antibiotic resistance gene as a selectable marker for plant transformation. Nature 304: 184-187

Bizily SP, Rugh CL, Meagher RB (2000) Phytodetoxification of hazardous organomercurials by genetically engineered plants. Nat Biotechnol 18: 213-217

Bizily SP, Rugh CL, Summers AO, Meagher RB (1999) Phytoremediation of methylmercury pollution: *merB* expression in *Arabidopsis thaliana* confers resistance to organomercurials. Proc Natl Acad Sci USA 96: 6808-6813

Bohmert K, Balbo I, Kopka J, Mittendorf V, Nawrath C, Poirier Y, Tischendorf G, Trethewey RN, Willmitzer L (2000) Transgenic *Arabidopsis* plants can accumulate polyhydroxybutyrate to up to 4% of their fresh weight. Planta 211: 841-845

Borisjuk N, Borisjuk L, Komarnytsky S, Timeva S, Hemleben V, Gleba Y, Raskin I (2000) Tobacco ribosomal DNA spacer element stimulates amplification and expression of heterologous genes. Nat Biotechnol 18: 1303-1306

Boudet AM, Kajita S, Grima-Pettenati J, Goffner D (2003) Lignins and lignocellulosics: a better control of synthesis for new and improved uses. Trends Plant Sci 8: 576-581

Bouwmeester HJ (2006) Engineering the essence of plants. Nat Biotechnol 24: 1359-1361

Brinch-Pedersen H, Sørensen LD, Holm PB (2002) Engineering crop plants: getting a handle on phosphate. Trends Plant Sci 7: 118-125

Britt AB, May GD (2003) Re-engineering plant gene targeting. Trends Plant Sci 8: 90-95

Broerse JEW (1998) Towards a new development strategy: How to include small-scale farmers in the biotechnological innovation process. Eburon Publishers, The Netherlands

Broothaerts W, Mitchell HJ, Weir B, Kaines S, Smith LM, Yang W, Mayer JE, Roa-Rodriguez C, Jefferson RA (2005) Gene transfer to plants by diverse species of bacteria. Nature 433: 629-633

Brown J (1998) How to feed the world, in two contradictory lessons. Trends Plant Sci 3: 409-410

Cahoon EB, Hall SE, Ripp KG, Ganzke TS, Hitz WD, Coughlan SJ (2003) Metabolic redesign of vitamin E biosynthesis in plants for tocotrienol production and increased antioxidant content. Nat Biotechnol 21: 1082-1087

Cao H, Li X, Dong X (1998) Generation of broad-spectrum disease resistance by overexpression of an essential regulatory gene in systemic acquired resistance. Proc Natl Acad Sci USA 95: 6531-6536

Carrière Y, Ellers-Kirk C, Sisterson M, Antilla L, Whitlow M, Dennehy TJ, Tabashnik BE (2003) Long-term regional suppression of pink bollworm by *Bacillus thuringiensis* cotton. Proc Natl Acad Sci USA 100: 1519-1523

Chakraborty S, Chakraborty N, Datta A (2000) Increased nutritive value of transgenic potato by expressing a nonallergenic seed albumin gene from *Amaranthus hypochondriacus*. Proc Natl Acad Sci USA 97: 3724-3729

Chappell J (2004) Valencene synthase–a biochemical magician and harbinger of transgenic aromas. Trends Plant Sci 9: 266-269

Chargelegue D, Obregon P, Drake PM (2001) Transgenic plants for vaccine production: expectations and limitations. Trends Plant Sci 6: 495-496

Cheng M, Fry JE, Pang S, Zhou H, Hironaka CM, Duncan DR, Conner TW, Wan Y (1997) Genetic transformation of wheat mediated by *Agrobacterium tumefaciens*. Plant Physiol 115: 971-980

Christou P (1996) Transformation technology. Trends Plant Sci 1: 423-431

Chung SM, Frankman EL, Tzfira T (2005) A versatile vector system for multiple gene expression in plants. Trends Plant Sci 10: 357-361

Chung SM, Vaidya M, Tzfira T (2006) *Agrobacterium* is not alone: gene transfer to plants by viruses and other bacteria. Trends Plant Sci 11: 1-4

Clemens S, Palmgren MG, Kramer U (2002) A long way ahead: understanding and engineering plant metal accumulation. Trends Plant Sci 7: 309-315

Clough SJ, Bent AF (1998) Floral dip: a simplified method for *Agrobacterium*-mediated transformation of *Arabidopsis thaliana*. Plant J 16: 735-743

Cohen JI (2005) Poorer nations turn to publicly developed GM crops. Nat Biotechnol 23: 27-33

Comai L, Faccioti D, Hiatt WR, Thompson G, Rose RE, Stalker DM (1985) Expression in plants of a mutant *aroA* gene from *Salmonella typhimurium* confers tolerance to glyphosate. Nature 370: 741-744

Conrad U (2005) Polymers from plants to develop biodegradable plastics. Trends Plant Sci 10: 511-512

Conway G (1997) The doubly green revolution: Food for all in the twenty-first century. Comstock Publishing Associates, Cornell University Press, Ithaca, NY

Dale EC, Ow DW (1991) Gene transfer with subsequent removal of the selection gene from the host genome. Proc Natl Acad Sci USA 88: 10558-10562

Dale PJ, Clarke B, Fontes EM (2002) Potential for the environmental impact of transgenic crops. Nat Biotechnol 20: 567-574

Dale PJ, Irwin JA, Scheffler JA (1993) The experimental and commercial release of transgenic crop plants. Plant Breeding 111: 1-22

Dangl JL, Jones JD (2001) Plant pathogens and integrated defense responses to infection. Nature 411: 826-833

Daniell H (2002) Molecular strategies for gene containment in transgenic crops. Nat Biotechnol 20: 581-586

Daniell H, Streatfield SJ, Wycoff K (2001) Medical molecular farming: production of antibodies, biopharmaceuticals and edible vaccines in plants. Trends Plant Sci 6: 219-226

Davuluri GR, van Tuinen A, Fraser PD, Manfredonia A, Newman R, Burgess D, Brummell DA, King SR, Palys J, Uhlig J, Bramley PM, Pennings HMJ, Bowler C (2005) Fruit-specific RNAi-mediated suppression of DET1 enhances carotenoid and flavonoid content in tomatoes. Nat Biotechnol 23: 890-895

Day CD, Lee E, Kobayashi J, Holappa LD, Albert H, Ow DW (2000) Transgene integration into the same chromosome location can produce alleles that express at a predictable level, or alleles that are differentially silenced. Genes Dev 14: 2869-2880

De Block M, Herrera-Estrella L, Van Montagu M, Schell J, Zambryski P (1984) Expression of foreign genes in regenerated plants and in their progeny. EMBO J 3: 1681-1689

De Block MD, Botterman J, Vandewiele M, Dockx J, Thoen C, Gosselé V, Movva NR, Thompson C, Montagu MV, Leemans J (1987) Engineering herbicide resistance in plants by expression of a detoxifying enzyme. EMBO J 6: 2513-2518

de Felipe P, Luke GA, Hughes LE, Gani D, Halpin C, Ryan MD (2006) *E unum pluribus*: multiple proteins from a self-processing polyprotein. Trends Biotechnol 24: 68-75

de Feyter R, Young M, Schroeder K, Dennis ES, Gerlach W (1996) A ribozyme gene and an antisense gene are equally effective in conferring resistance to tobacco mosaic virus on transgenic tobacco. Mol Gen Genet 250: 329-338

de la Fuente JM, Ramìrez-Rodrìguez V, Cabrera-Ponce JL, Herrera-Estrella L (1997) Aluminum tolerance in transgenic plants by alteration of citrate synthesis. Science 276: 1566-1568

de Vetten N, Wolters AM, Raemakers K, van der Meer I, ter Stege R, Heeres E, Heeres P, Visser R (2003) A transformation method for obtaining marker-free plants of a cross-pollinating and vegetatively propagated crop. Nat Biotechnol 21: 439-442

Deák M, Horváth GV, Davletova S, Török K, Sass L, Vass I, Barna B, Király Z, Dudits D (1999) Plants ectopically expressing the iron-binding protein, ferritin, are tolerant to oxidative damage and pathogens. Nat Biotechnol 17: 192-196

Delhaize E, Hebb DM, Ryan PR (2001) Expression of a *Pseudomonas aeruginosa* citrate synthase gene in tobacco is not associated with either enhanced citrate accumulation or efflux. Plant Physiol 125: 2059-2067

D'Halluin K, Bossut M, Bonne E, Mazur B, Leemans J, Botterman J (1992) Transformation of sugarbeet (*Beta vulgaris* L.) and evaluation of herbicide resistance in transgenic plants. Bio/Technology 10: 309 - 314

Dhankher OP, Li Y, Rosen BP, Shi J, Salt D, Senecoff JF, Sashti NA, Meagher RB (2002) Engineering tolerance and hyperaccumulation of arsenic in plants by combining arsenate reductase and gamma-glutamylcysteine synthetase expression. Nat Biotechnol 20: 1140-1145

Dixon RA (2005) A two-for-one in tomato nutritional enhancement. Nat Biotechnol 23: 825-826

Dunwell JM (2000) Transgenic approaches to crop improvement. J Exp Bot 51: 487-496

Dus Santos MJ, Carrillo C, Ardila F, Ríos RD, Franzone P, Piccone ME, Wigdorovitz A, Borca MV (2005) Development of transgenic alfalfa plants containing the foot and mouth disease virus structural polyprotein gene P1 and its utilization as an experimental immunogen. Vaccine 23: 1838-1843

Ebinuma H, Sugita K, Matsunaga E, Yamakado M (1997) Selection of marker-free transgenic plants using isopentenyl transferase gene. Proc Natal Acad USA 94: 2117-2121

Edwards G (1999) Tuning up crop photosynthesis. Nat Biotechnol 17: 22-23

Eichholtz DA, Rogers SG, Horsch RB, Klee HJ, Hayford M, Hoffmann NL, Braford SB, Fink C, Flick J, O'Connell KM, Fraley RT (1987) Expression of mouse dihydrofolate reductase gene confers methotrexate resistance in transgenic petunia plants. Somat Cell Mol Genet 13: 67-76

Erikson O, Hertzberg M, Näsholm T (2004) A conditional marker gene allowing both positive and negative selection in plants. Nat Biotechnol 22: 455-458

Eriksson ME, Israelsson M, Olsson O, Moritz T (2000) Increased gibberellin biosynthesis in transgenic trees promotes growth, biomass production and xylem fiber length. Nat Biotechnol 18: 784-788

Evenson RE, Gollin D (2003) Assessing the impact of the green revolution, 1960 to 2000. Science 300: 758-762

Falconi CA (2002) Briefing paper 42: Agricultural biotechnology research capacity in for developing countries. International Service for National Agricultural Research (ISNAR).

Farré EM, Bachmann A, Willmitzer L, Trethewey RN (2001) Acceleration of potato tuber sprouting by the expression of a bacterial pyrophosphatase. Nat Biotechnol 19: 268-272

Feldmann KA (1991) T-DNA insertion mutagenesis in *Arabidopsis*: mutational spectrum. Plant J 1: 71-82

Fenning TM, Gershenzon J (2002) Where will the wood come from? Plantation forests and the role of biotechnology. Trends Biotechnol 20: 291-296

Feys BJ, Parker JE (2000) Interplay of signaling pathways in plant disease resistance. Trends Genet 16: 449-455

Fitch MMM, Manshardt RM, Gonsalves D, Slightom JL (1993) Transgenic papaya plants from *Agrobacterium*-mediated transformation of somatic embryos. Plant Cell Rep 12: 245-249

Fraley RT, Rogers SG, Horsch RB, Sanders PR, Flick JS, Adams SP, Bittner ML, Brand LA, Fink CL, Fry JS, Galluppi GR, Goldberg SB, Hoffmann NL, Woo SC (1983) Expression of bacterial genes in plant cells. Proc Natl Acad Sci USA 80: 4803-4807

François IEJA, Van Hemelrijck W, Aerts AM, Wouters PFJ, Proost P, Broekaert WF, Cammue BPA (2004) Processing in *Arabidopsis thaliana* of a heterologous polyprotein resulting in differential targeting of the individual plant defensins. Plant Sci 166: 113-121

Friedrich L, Lawton K, Dietrich R, Willits M, Cade R, Ryals J (2001) *NIM1* overexpression in *Arabidopsis* potentiates plant disease resistance and results in enhanced effectiveness of fungicides. Mol Plant Microbe Interact 14: 1114-1124

Gao AG, Hakimi SM, Mittanck CA, Wu Y, Woerner BM, Stark DM, Shah DM, Liang J, Rommens CM (2000) Fungal pathogen protection in potato by expression of a plant defensin peptide. Nat Biotechnol 18: 1307-1310

Garcia JA, Simon-Mateo C (2006) A micropunch against plant viruses. Nat Biotechnol 24: 1358-1359

Garg AK, Kim JK, Owens TG, Ranwala AP, Choi YD, Kochian LV, Wu RJ (2002) Trehalose accumulation in rice plants confers high tolerance levels to different abiotic stresses. Proc Natl Acad Sci USA 99: 15898-15903

Gaxiola RA, Li J, Undurraga S, Dang LM, Allen GJ, Alper SL, Fink GR (2001) Drought- and salt-tolerant plants result from overexpression of the AVP1 H^+-pump. Proc Natl Acad Sci USA 98: 11444-11449

Gelvin S (2003a) Improving plant genetic engineering by manipulating the host. Trends Biotechnol 21: 95-98

Gelvin SB (2003b) *Agrobacterium*-mediated plant transformation: the biology behind the "gene-jockeying" tool. Microbiol Mol Biol Rev 67: 16-37

Giddings G, Allison G, Brooks D, Carter A (2000) Transgenic plants as factories for biopharmaceuticals. Nat Biotechnol 18: 1151-1155

Gilbertson L (2003) Cre-lox recombination: Cre-ative tools for plant biotechnology. Trends Biotechnol 21: 550-555

Giuliano G, Aquilani R, Dharmapuri S (2000) Metabolic engineering of plant carotenoids. Trends Plant Sci 5: 406-409

Gleba Y, Klimyuk V, Marillonnet S (2005) Magnifection--a new platform for expressing recombinant vaccines in plants. Vaccine 23: 2042-2048

Goderis IJ, De Bolle MF, François IE, Wouters PF, Broekaert WF, Cammue BP (2002) A set of modular plant transformation vectors allowing flexible insertion of up to six expression units. Plant Mol Biol 50: 17-27

Gonsalves D (1998) Control of papaya ringspot virus in papaya: a case study. Annu Rev Phytopathol 36: 415-437

Good AG, Shrawat AK, Muench DG (2004) Can less yield more? Is reducing nutrient input into the environment compatible with maintaining crop production? Trends Plant Sci 9: 597-605

Goregaoker SP, Eckhardt LG, Culver JN (2000) Tobacco mosaic virus replicase-mediated cross-protection: contributions of RNA and protein-derived mechanisms. Virology 273: 267-275

Goto F, Yoshihara T, Shigemoto N, Toki S, Takaiwa F (1999) Iron fortification of rice seed by the soybean ferritin gene. Nat Biotechnol 17: 282-286

Gressel J, Ehrlich G (2002) Universal inheritable barcodes for identifying organisms. Trends Plant Sci 7: 542-544

Griffith M, Yaish MW (2004) Antifreeze proteins in overwintering plants: a tale of two activities. Trends Plant Sci 9: 399-405

Grimsley N, Hohn B, Hohn T, Walden R (1986) "Agroinfection," an alternative route for viral infection of plants by using the Ti plasmid. Proc Natl Acad Sci USA 83: 3282-3286

Grimsley N, Hohn T, Davies JW, Hohn B (1987) *Agrobacterium*-mediated delivery of infectious maize streak virus into maize plants. Nature 325: 177-179

Guerinot ML (2001) Improving rice yields-ironing out the details. Nat Biotechnol 19: 417-418

Gurr SJ, Rushton PJ (2005) Engineering plants with increased disease resistance: what are we going to express? Trends Biotechnol 23: 275-282

Halpin C, Boerjan W (2003) Stacking transgenes in forest trees. Trends Plant Sci 8: 363-365

Hamilton CM, Frary A, Lewis C, Tanksley SD (1996) Stable transfer of intact high molecular weight DNA into plant chromosomes. Proc Natl Acad Sci USA 93: 9975-9979

Han KH, Ma CP, Strauss SH (1997) Matrix attachment regions (MARs) enhance transformation frequency and transgene expression in poplar. Transgenic Res 6: 415-420

Hanley Z, Slabas T, Elborough KM (2000) The use of plant biotechnology for the production of biodegradable plastics. Trends Plant Sci 5: 45-46

Hannin M, Volrath S, Bogucki A, Briker M, Ward E, Paskowski J (2001) Gene targeting in Arabidopsis. Plant J 28: 671-677

Hannink N, Rosser SJ, French CE, Basran A, Murray JA, Nicklin S, Bruce NC (2001) Phytodetoxification of TNT by transgenic plants expressing a bacterial nitroreductase. Nat Biotechnol 19: 1168-1172

Hannon GJ (2002) RNA interference. Nature 418: 244-251

Hansen G, Chilton M-D (1996) "Agrolistic" transformation of plant cells: integration of T-strands generated in planta. Proc Natl Acad Sci USA 93: 14978-14983

Hansen G, Das A, Chilton M-D (1994) Constitutive expression of the virulence genes improves the efficiency of plant transformation by Agrobacterium. Proc Natl Acad Sci USA 91: 7603-7607

Hansen G, Wright MS (1999) Recent advances in the transformation of plants. Trends Plant Sci 4: 226-231

Hanson B, Engler D, Moy Y, Newman B, Ralston E, Gutterson N (1999) A simple method to enrich an *Agrobacterium*-transformed population for plants containing only T-DNA sequences. Plant J 19: 727-734

Haq TA, Mason HS, Clements JD, Arntzen CJ (1995) Oral immunization with a recombinant bacterial antigen produced in transgenic plants. Science 268: 714-716

Hare PD, Chua NH (2002) Excision of selectable marker genes from transgenic plants. Nat Biotechnol 20: 575-580

Heinemann JA, Traavik T (2004) Problems in monitoring horizontal gene transfer in field trials of transgenic plants. Nat Biotechnol 22: 1105-1109

Hellens R, Mullineaux P, Klee H (2000) Technical Focus:a guide to Agrobacterium binary Ti vectors. Trends Plant Sci 5: 446-451

Hellwig S, Drossard J, Twyman RM, Fischer R (2004) Plant cell cultures for the production of recombinant proteins. Nat Biotechnol 22: 1415-1422

Heritage J (2005) Transgenes for tea? Trends Biotechnol 23: 17-21

Herrera-Estrella L, Block MD, Messens E, Hernalsteens JP, Montagu MV, Schell J (1983a) Chimeric genes as dominant selectable markers in plant cells. EMBO J 2: 987-995

Herrera-Estrella L, Depicker A, Van Montagu M, Schell J (1983b) Expression of chimaeric genes transferred into plant cells using a Ti plasmid derived vector. Nature 303: 209-213

Herrera-Estrella L, Simpson J, Martinez-Trujillo M (2005) Transgenic plants: an historical perspective. Methods Mol Biol 286: 3-32

Hiatt A, Cafferkey R, Bowdish K (1989) Production of antibodies in transgenic plants. Nature 342: 76-78

Hiei Y, Ohta S, Komari T, Kumashiro T (1994) Efficient transformation of rice (*Oryza sativa* L.) mediated by *Agrobacterium* and sequence analysis of the boundaries of the T-DNA. Plant J 6: 271-282

High SM, Cohen MB, Shu QY, Altosaar I (2004) Achieving successful deployment of *Bt* rice. Trends Plant Sci 9: 286-292

Hilder VA, Gatehouse AMR, Sheerman SE, Barker RF, Boulter D (1987) A novel mechanism of insect resistance engineered into tobacco. Nature 330: 160-163

Hinchee MAW, Connor-Ward DV, Newell CA, McDonnell RE, Sato SJ, Gasser CS, Fischhoff DA, Re DB, Fraley RT, Horsch RB (1988) Production of transgenic soybean plants using *Agrobacterium*-mediated DNA transfer. Bio/Technology 6: 915-922

Hirsch RE, Sussman MR (1999) Improving nutrient capture from soil by the genetic manipulation of crop plants. Trends Biotechnol 17: 356-361

Hoekema A, de Pater BS, Fellinger AJ, Hooykaas PJ, Schilperoort RA (1984) The limited host range on an *Agrobacterium tumefaciens* strain extended by a cytokinin gene from a wide host range T-region. EMBO J 3: 3043-3047

Hoekema A, Hirsch PR, Hooykaas PJJ, Schilperoort RA (1983) A binary plant vector strategy based on separation of vir and T-region of the *Agrobacterium tumefaciens* Ti plasmid. Nature 303: 179-180

Hohn B, Levy AA, Puchta H (2001) Elimination of selection markers from transgenic plants. Curr Opin Biotechnol 12: 139-143

Hohn B, Puchta H (2003) Some like it sticky: targeting of the rice gene *Waxy*. Trends Plant Sci 8: 51-53

Hood EE (2004) Where, oh where has my protein gone? Trends Biotechnol 22: 53-55

Hood EE (2003) Selecting the fruits of your labors. Trends Plant Sci 8: 357-358

Hood EE, Helmer GL, Fraley RT, Chilton M-D (1986) The hypervirulence of *Agrobacterium tumefaciens* A281 is encoded in a region of pTiBo542 outside of T-DNA. J Bacteriol 168: 1291-1301

Horsch RB, Fraley RT, Rogers SG, Sanders PR, Lloyd A, Hoffmann N (1984) Inheritance of functional foreign genes in plants. Science 223: 496-498

Huang J, Hu R, Rozelle S, Pray C (2005) Insect-resistant GM rice in farmers' fields: assessing productivity and health effects in China. Science 308: 688-690

Huang J, Pray C, Rozelle S (2002a) Enhancing the crops to feed the poor. Nature 418: 678-684

Huang J, Rozelle S, Pray C, Wang Q (2002b) Plant biotechnology in China. Science 295: 674-676

Huang T, Nicodemus J, Zarka DG, Thomashow MF, Wisniewski M, Duman JG (2002c) Expression of an insect (*Dendroides canadensis*) antifreeze protein in *Arabidopsis thaliana* results in a decrease in plant freezing temperature. Plant Mol Biol 50: 333-344

Iglesias V, Moscone E, Papp I, Neuhuber F, Michalowski S, Phelan T, Spiker S, Matzke M, Matzke A (1997) Molecular and cytogenetic analysis of stably and unstably expressed transgene loci in tobacco. Plant Cell 9: 1251-1264

Ishida Y, Saito H, Ohta S, Hiei Y, Komari T, Kumashiro T (1996) High efficiency transformation of maize (*Zea mays* L.) mediated by *Agrobacterium tumefaciens*. Nat Biotechnol 14: 745-750

Jaglo-Ottosen KR, Gilmour SJ, Zarka DG, Schabenberger O, Thomashow MF (1998) *Arabidopsis CBF1* overexpression induces *COR* genes and enhances freezing tolerance. Science 280: 104-106

James C (2005) Executive summary: Brief 34, Global status of commercialized biotech/GM crops: 2005. The International Service for the Acquisition of Agri-biotech Applications (ISAAA), Ithaca NY

Jarchow E, Grimsley NH, Hohn B (1991) *virF*, the host range-determining virulence gene of *Agrobacterium tumefaciens*, affects T-DNA transfer to *Zea mays*. Proc Natl Acad Sci USA 88: 10426-10430

Jin SG, Komari T, Gordon MP, Nester EW (1987) Genes responsible for the supervirulence phenotype of *Agrobacterium tumefaciens* A281. J Bacteriol 169: 4417-4425

Jung C, Cai D, Kleine M (1998) Engineering nematode resistance in crop species. Trends Plant Sci 3: 266-271

Karimi M, De Meyer B, Hilson P (2005) Modular cloning in plant cells. Trends Plant Sci 10: 103-105

Karimi M, Inzé D, Depicker A (2002) GATEWAY™ vectors for *Agrobacterium*-mediated plant transformation. Trends Plant Sci 7: 193-195

Kasuga M, Liu Q, Miura S, Yamaguchi-Shinozaki K, Shinozaki K (1999) Improving plant drought, salt, and freezing tolerance by gene transfer of a single stress-inducible transcription factor. Nat Biotechnol 17: 287-291

Keller H, Pamboukdjian N, Ponchet M, Poupet A, Delon R, Verrier JL, Roby D, Ricci P (1999) Pathogen-induced elicitin production in transgenic tobacco generates a hypersensitive response and nonspecific disease resistance. Plant Cell 11: 223-235

Kempin SA, Liljegren SJ, Block LM, Rounsley SD, Yanofsky MF, Lam E (1997) Targeted disruption in Arabidopsis. Nature 389: 802-803

Klein TM, Wolf ED, Wu R, Sanford JC (1987) High-velocity microprojectiles for delivering nucleic acids into living cells. Nature 327: 70-73

Kleter GA, van der Krieken WM, Kok EJ, Bosch D, Jordi W, Gilissen LJ (2001) Regulation and exploitation of genetically modified crops. Nat Biotechnol 19: 1105-1110

Ko K, Tekoah Y, Rudd PM, Harvey DJ, Dwek RA, Spitsin S, Hanlon CA, Rupprecht C, Dietzschold B, Golovkin M, Koprowski H (2003) Function and glycosylation of plant-derived antiviral monoclonal antibody. Proc Natl Acad Sci USA 100: 8013-8018

Komari T, Takakura Y, Ueki J, Kato N, Ishida Y, Hiei Y (2006) Binary vectors and super-binary vectors. Methods Mol Biol 343: 15-41

Kong Q, Richter L, Yang YF, Arntzen CJ, Mason HS, Thanavala Y (2001) Oral immunization with hepatitis B surface antigen expressed in transgenic plants. Proc Natl Acad Sci USA 98: 11539-11544

Kononov ME, Bassuner B, Gelvin SB (1997) Integration of T-DNA binary vector 'backbone' sequences into the tobacco genome: evidence for multiple complex patterns of integration. Plant J 11: 945-957

Kooter JM, Matzke AM, Meyer P (1999) Listening to the silent genes: transgene silencing, gene regulation and pathogen control. Trends Plant Sci 4: 340-347

Kovalchuk I, Kovalchuk O, Arkhipov A, Hohn B (1998) Transgenic plants are sensitive bioindicators of nuclear pollution caused by the Chernobyl accident. Nat Biotechnol 16: 1054-1059

Kovalchuk I, Kovalchuk O, Hohn B (2001a) Biomonitoring the genotoxicity of environmental factors with transgenic plants. Trends Plant Sci 6: 306-310

Kovalchuk O, Titov V, Hohn B, Kovalchuk I (2001b) A sensitive transgenic plant system to detect toxic inorganic compounds in the environment. Nat Biotechnol 19: 568-572

Krämer U, Chardonnens AN (2001) The use of transgenic plants in the bioremediation of soils contaminated with trace elements. Appl Microbiol Biotechnol 55: 661-672

Krattiger AF (1999) Networking biotechnology solutions with developing countries: the mission and strategy of the International Service for the Acquisition of Agri-biotech Applications. *In* T Hohn, KM Leisinger, eds, Biotechnology of food cropsin developing countries. Springer-Verlag Wien, New York, pp 25-33

Ku MS, Agarie S, Nomura M, Fukayama H, Tsuchida H, Ono K, Hirose S, Toki S, Miyao M, Matsuoka M (1999) High-level expression of maize phosphoenolpyruvate carboxylase in transgenic rice plants. Nat Biotechnol 17: 76-80

Ku MSB, Cho D, Li X, Jiao D, Pinto M, Miyao M, Matsuoka M (2001) Introduction of genes encoding C4 photosynthesis enzymes into rice plants: physiological consequences. *In* Rice biotechnology: improving yield, stress tolerance and grain quality, Wiley, Chichester (Novartis Foundation Symposium 236), pp 100-116

Kumar S, Fladung M (2001) Controlling transgene integration in plants. Trends Plant Sci 6: 155-159

Kunkel T, Niu QW, Chan YS, Chua N-H (1999) Inducible isopentenyl transferase as a high-efficiency marker for plant transformation. Nat Biotechnol 17: 916-919

Lee D, Natesan E (2006) Evaluating genetic containment strategies for transgenic plants. Trends Biotechnol 24: 109-114

Lee KY, Lund P, Lowe K, Dunsmuir P (1990) Homologous recombination in plant cells after *Agrobacterium*-mediated transformation. Plant Cell 2: 415-425

Leisinger KM (1999) The contribution of genetic engineering to the fight against hunger in developing countries. *In* T Hohn, KM Leisinger, eds, Biotechnology of food crops in developing countries. Springer-Verlag Wien, New York, pp 1-19

Lewinsohn E, Schalechet F, Wilkinson J, Matsui K, Tadmor Y, Nam KH, Amar O, Lastochkin E, Larkov O, Ravid U, Hiatt W, Gepstein S, Pichersky E (2001) Enhanced levels of the aroma and flavor compound *S*-linalool by metabolic engineering of the terpenoid pathway in tomato fruits. Plant Physiol 127: 1256-1265

Li HQ, Sautter C, Potrykus I, Puonti-Kaerlas J (1996) Genetic transformation of cassava (*Manihot esculenta* Crantz). Nat Biotechnol 14: 736-740

Li L, Zhou Y, Cheng X, Sun J, Marita JM, Ralph J, Chiang VL (2003) Combinatorial modification of multiple lignin traits in trees through multigene cotransformation. Proc Natl Acad Sci USA 100: 4939-4944

Li ZK, Sanchez A, Angeles E, Singh S, Domingo J, Huang N, Khush GS (2001) Are the dominant and recessive plant disease resistance genes similar? A case study of rice R genes and *Xanthomonas oryzae* pv. *oryzae* races. Genetics 159: 757-765

Lin L, Liu YG, Xu X, Li B (2003) Efficient linking and transfer of multiple genes by a multigene assembly and transformation vector system. Proc Natl Acad Sci USA 100: 5962-5967

Lindbo JA, Silva-Rosales L, Proebsting WM, Dougherty WG (1993) Induction of a Highly Specific Antiviral State in Transgenic Plants: Implications for Regulation of Gene Expression and Virus Resistance. Plant Cell 5: 1749-1759

Liu D, Burton S, Glancy T, Li ZS, Hampton R, Meade T, Merlo DJ (2003) Insect resistance conferred by 283-kDa *Photorhabdus luminescens* protein TcdA in *Arabidopsis thaliana*. Nat Biotechnol 21: 1222-1228

Lloyd A, Plaisier CL, Carroll D, Drews GN (2005) Targeted mutagenesis using zinc-finger nucleases in Arabidopsis. Proc Natl Acad Sci USA 102: 2232-2237

Lomonossoff GP (1995) Pathogen-derived resistance to plant viruses. Annu Rev Phytopathol 33: 323-334

Ma JF, Ryan PR, Delhaize E (2001) Aluminium tolerance in plants and the complexing role of organic acids. Trends Plant Sci 6: 273-278

Ma JK (2000) Genes, greens, and vaccines. Nat Biotechnol 18: 1141-1142

Ma JK, Chikwamba R, Sparrow P, Fischer R, Mahoney R, Twyman RM (2005a) Plant-derived pharmaceuticals--the road forward. Trends Plant Sci 10: 580-585

Ma JK, Drake PM, Chargelegue D, Obregon P, Prada A (2005b) Antibody processing and engineering in plants, and new strategies for vaccine production. Vaccine 23: 1814-1818

Ma JK, Hiatt A, Hein M, Vine ND, Wang F, Stabila P, van Dolleweerd C, Mostov K, Lehner T (1995) Generation and assembly of secretory antibodies in plants. Science 268: 716-719

Mahmoud SS, Croteau RB (2001) Metabolic engineering of essential oil yield and composition in mint by altering expression of deoxyxylulose phosphate reductoisomerase and menthofuran synthase. Proc Natl Acad Sci USA 98: 8915-8920

Mahmoud SS, Croteau RB (2002) Strategies for transgenic manipulation of monoterpene biosynthesis in plants. Trends Plant Sci 7: 366-373

Mariani C, De Beuckeleer M, Truettner J, Leemans J, Goldberg R (1990) Induction of male sterility in plants by a chimaeric ribonuclease gene. Nature 347: 737-741

Marillonnet S, Giritch A, Gils M, Kandzia R, Klimyuk V, Gleba Y (2004) *In planta* engineering of viral RNA replicons: efficient assembly by recombination of DNA modules delivered by *Agrobacterium*. Proc Natl Acad Sci USA 101: 6852-6857

Marillonnet S, Thoeringer C, Kandzia R, Klimyuk V, Gleba Y (2005) Systemic *Agrobacterium tumefaciens*-mediated transfection of viral replicons for efficient transient expression in plants. Nat Biotechnol 23: 718-723

Mason HS, Haq TA, Clements JD, Arntzen CJ (1998) Edible vaccine protects mice against *Escherichia coli* heat-labile enterotoxin (LT): potatoes expressing a synthetic LT-B gene. Vaccine 16: 1336-1343

Mason HS, Lam DM, Arntzen CJ (1992) Expression of hepatitis B surface antigen in transgenic plants. Proc Natl Acad Sci USA 89: 11745-11749

Matthews P, Wang M, Waterhouse P, Thornton S, Fieg S, Gubler F, Jacobsen J (2001) Marker gene elimination from transgenic barley, using co-transformation with adjacent 'twin T-DNAs' on a standard *Agrobacterium transformation* vector. Mol Breed 7: 195-202

Matzke AJ, Matzke MA (1998) Position effects and epigenetic silencing of plant transgenes. Curr Opin Plant Biol 1: 142-148

May G, Afza R, Mason H, Wiecko A, Novak F, Arntzen C (1995) Generation of transgenic banana (*Musa acuminata*) plants via *Agrobacterium*-mediated transformation. Biotechnology 13: 486-492

McKersie BD, Bowley SR, Harjanto E, Leprince O (1996) Water-deficit tolerance and field performance of transgenic alfalfa overexpressing superoxide dismutase. Plant Physiol 111: 1177-1181

Mehta RA, Cassol T, Li N, Ali N, Handa AK, Mattoo AK (2002) Engineered polyamine accumulation in tomato enhances phytonutrient content, juice quality, and vine life. Nat Biotechnol 20: 613-618

Miao ZH, Lam E (1995) Targeted disruption of the TGA3 locus in *Arabidopsis thaliana*. Plant J 7: 359-365

Miller M, Tagliani L, Wang N, Berka B, Bidney D, Zhao ZY (2002) High efficiency transgene segregation in co-transformed maize plants using an *Agrobacterium tumefaciens* 2 T-DNA binary system. Transgenic Res 11: 381-396

Mitter N, Sulistyowati E, Graham MW, Dietzgen RG (2001) Suppression of gene silencing: a threat to virus-resistant transgenic plants? Trends Plant Sci 6: 246-247

Mittler R (2006) Abiotic stress, the field environment and stress combination. Trends Plant Sci 11: 15-19

Miyagawa Y, Tamoi M, Shigeoka S (2001) Overexpression of a cyanobacterial fructose-1,6-/sedoheptulose-1,7-bisphosphatase in tobacco enhances photosynthesis and growth. Nat Biotechnol 19: 965-969

Moar W (2003) Breathing new life into insect-resistant plants. Nat Biotechnol 21: 1152-1154

Mol J, Holton T, Koes R (1995) Floriculture: genetic engineering of commercial traits. Trends Biotechnol 13: 350-355

Morandini P, Salamini F (2003) Plant biotechnology and breeding: allied for years to come. Trends Plant Sci 8: 70-75

Morse S, Bennett R, Ismael Y (2004) Why Bt cotton pays for small-scale producers in South Africa. Nat Biotechnol 22: 379-380

Mourgues F, Brisset MN, Chevreau E (1998) Strategies to improve plant resistance to bacterial diseases through genetic engineering. Trends Biotechnol 16: 203-210

Muir SR, Collins GJ, Robinson S, Hughes S, Bovy A, Ric De Vos CH, van Tunen AJ, Verhoeyen ME (2001) Overexpression of petunia chalcone isomerase in tomato results in fruit containing increased levels of flavonols. Nat Biotechnol 19: 470-474

Murata N, Ishizaki-Nishizawa O, Higashi S, Hayashi H, Tasaka Y, Nishida I (1992) Genetically engineered alteration in the chilling sensitivity of plants. Nature 356: 710-713

Mysore KS, Nam J, Gelvin SB (2000) An *Arabidopsis* histone H2A mutant is deficient in *Agrobacterium* T-DNA integration. Proc Natl Acad Sci USA 97: 948-953

Nawrath C, Poirier Y, Somerville C (1994) Targeting of the polyhydroxybutyrate biosynthetic pathway to the plastids of *Arabidopsis thaliana* results in high levels of polymer accumulation. Proc Natl Acad Sci USA 91: 12760-12764

Negretto DB, Jolley MB, Beer SB, Wenck AR, Hansen G (2000) The use of phosphomannose-isomerase as a selectable marker to recover transgenic maize plants (*Zea mays* L.) via *Agrobacterium transformation.* Plant Cell Rep 19: 798-803

Neumann K, Stephan DP, Ziegler K, Hühns M, Broer I, Lockau W, Pistorius EK (2005) Production of cyanophycin, a suitable source for the biodegradable polymer polyaspartate, in transgenic plants. Plant Biotechnol J 3: 249-258

Newell CA (2000) Plant transformation technology. Developments and applications. Mol Biotechnol 16: 53-65

Nguyen HT, Xu W, Rosenow DT, Mullett JE, McIntyre L (1996) Use of biotechnology in sorghum breeding. *In* Proceedings of the International Conference on Genetic Improvement of Sorghum and Pearl Millet,, Texas, Sept 22-27, 1997 INTSORMIL/ICRISAT, pp 412-424

Ni M, Cui D, Einstein J, Narasimhulu S, Vergara CE, Gelvin SB (1995) Strength and tissue specificity of chimeric promoters derived from the octopine and mannopine synthase genes. Plant J 7: 661-667

Nielsen KM, Townsend JP (2004) Monitoring and modeling horizontal gene transfer. Nat Biotechnol 22: 1110-1114

Niggeweg R, Michael AJ, Martin C (2004) Engineering plants with increased levels of the antioxidant chlorogenic acid. Nat Biotechnol 22: 746-754

Niu QW, Lin SS, Reyes JL, Chen KC, Wu HW, Yeh SD, Chua NH (2006) Expression of artificial microRNAs in transgenic *Arabidopsis thaliana* confers virus resistance. Nat Biotechnol 24: 1420-1428

Normile D (2000) Agricultural biotechnology. Monsanto donates its share of golden rice. Science 289: 843-845

Nottingham S (1998) Eat your genes: how genetically modified food is entering our diet. Zed Books Ltd, New York

Oberschall A, Deák M, Török K, Sass L, Vass I, Kovács I, Fehér A, Dudits D, Horvath GV (2000) A novel aldose/aldehyde reductase protects transgenic plants against lipid peroxidation under chemical and drought stresses. Plant J 24: 437-446

O'Connell K, Goodman R, Handelsmar J (1996) Engineering the rhizosphere: expressing a bias. Trends Biotechnol 14: 83-88

Oeller PW, Min-Wong L, Taylor LP, Pike DA, Theologis A (1991) Reversible inhibition of tomato fruit senescence by antisense RNA. Science 254: 437-439

Offringa R, de Groot MJA, Haagsman HJ, Does MP, van den Elzen PJM, Hooykaas PJJ (1990) Extrachromosomal homologous recombination and gene targeting in plant cells after *Agrobacterium* mediated transformation. EMBO J 9: 3077-3084

Oksman-Caldentey KM, Inzé D (2004) Plant cell factories in the post-genomic era: new ways to produce designer secondary metabolites. Trends Plant Sci 9: 433-440

Osusky M, Zhou G, Osuska L, Hancock RE, Kay WW, Misra S (2000) Transgenic plants expressing cationic peptide chimeras exhibit broad-spectrum resistance to phytopathogens. Nat Biotechnol 18: 1162-1166

Paine JA, Shipton CA, Chaggar S, Howells RM, Kennedy MJ, Vernon G, Wright SY, Hinchliffe E, Adams JL, Silverstone AL, Drake R (2005) Improving the nutritional value of Golden Rice through increased pro-vitamin A content. Nat Biotechnol 23: 482-487

Pang SZ, Slightom JL, Gonsalves D (1993) Different mechanisms protect transgenic tobacco against tomato spotted wilt and impatiens necrotic spot Tospoviruses. Biotechnol 11: 819-824

Peña L, Seguin A (2001) Recent advances in the genetic transformation of trees. Trends Biotechnol 19: 500-506

Penna S (2003) Building stress tolerance through over-producing trehalose in transgenic plants. Trends Plant Sci 8: 355-357

Peralta EG, Hellmiss R, Ream W (1986) *Overdrive*, a T-DNA transmission enhancer on the *A. tumefaciens* tumour-inducing plasmid. EMBO J 5: 1137-1142

Perez-Prat E, van Lookeren Campagne MM (2002) Hybrid seed production and the challenge of propagating male-sterile plants. Trends Plant Sci 7: 199-203

Peschen D, Li HP, Fischer R, Kreuzaler F, Liao YC (2004) Fusion proteins comprising a *Fusarium*-specific antibody linked to antifungal peptides protect plants against a fungal pathogen. Nat Biotechnol 22: 732-738

Peterson RK, Arntzen CJ (2004) On risk and plant-based biopharmaceuticals. Trends Biotechnol 22: 64-66

Pilate G, Guiney E, Holt K, Petit-Conil M, Lapierre C, Leplé JC, Pollet B, Mila I, Webster EA, Marstorp HG, Hopkins DW, Jouanin L, Boerjan W, Schuch W, Cornu D, Halpin C (2002) Field and pulping performances of transgenic trees with altered lignification. Nat Biotechnol 20: 607-612

Pilon-Smits EA, Hwang S, Lytle CM, Zhu Y, Tai JC, Bravo RC, Chen Y, Leustek T, Terry N (1999) Overexpression of ATP sulfurylase in indian mustard leads to increased selenate uptake, reduction, and tolerance. Plant Physiol 119: 123-132

Pilon-Smits EAH, Terry N, Sears T, Kim H, Zayed A, Hwang S, van Dunn K, Voogd E, Verwoerd TC, Krutwagen RWHH, Goddijn OJM (1998) Trehalose-producing transgenic tobacco plants show improved growth performance under drought stress. Plant Physiol J 152: 525-532

Pink D, Puddephat I (1999) Deployment of disease resistance genes by plant transformation – a 'mix and match' approach. Trends Plant Sci 4: 71-75

Poirier Y (1999) Green chemistry yields a better plastic. Nat Biotechnol 17: 960-961

Poirier Y, Dennis DE, Klomparens K, Somerville C (1992) Polyhydroxybutyrate, a biodegradable thermoplasic, produced in transgenic plants. Science 256: 520-523

Potrykus I (2001) Golden rice and beyond. Plant Physiol 125: 1157-1161

Powell-Abel P, Nelson RS, De B, Hoffmann N, Rogers SG, Fraley RT, Beachy RN (1986) Delay of disease development in transgenic plants that express the tobacco mosaic virus coat protein gene. Science 232: 738-743

Prins M (2003) Broad virus resistance in transgenic plants. Trends Biotechnol 21: 373-375

Prins M, de Haan P, Luyten R, van Veller M, van Grinsven MQ, Goldbach R (1995) Broad resistance to tospoviruses in transgenic tobacco plants expressing three tospoviral nucleoprotein gene sequences. Mol Plant Microbe Interact 8: 85-91

Qaim M, Zilberman D (2003) Yield effects of genetically modified crops in developing countries. Science 299: 900-902

Qi B, Fraser T, Mugford S, Dobson G, Sayanova O, Butler J, Napier JA, Stobart AK, Lazarus CM (2004) Production of very long chain polyunsaturated omega-3 and omega-6 fatty acids in plants. Nat Biotechnol 22: 739-745

Ralley L, Enfissi EM, Misawa N, Schuch W, Bramley PM, Fraser PD (2004) Metabolic engineering of ketocarotenoid formation in higher plants. Plant J 39: 477-486

Regensburg-Tuink AJG, Hooykaas PJJ (1993) Transgenic *N. glauca* plants expressing bacterial virulence gene *virF* are converted into hosts for nopaline strains of *A. tumefaciens*. Nature 363: 69-71

Regierer B, Fernie AR, Springer F, Perez-Melis A, Leisse A, Koehl K, Willmitzer L, Geigenberger P, Kossmann J (2002) Starch content and yield increase as a result of altering adenylate pools in transgenic plants. Nat Biotechnol 20: 1256-1260

Richter LJ, Thanavala Y, Arntzen CJ, Mason HS (2000) Production of hepatitis B surface antigen in transgenic plants for oral immunization. Nat Biotechnol 18: 1167-1171

Römer S, Fraser PD, Kiano JW, Shipton CA, Misawa N, Schuch W, Bramley PM (2000) Elevation of the provitamin A content of transgenic tomato plants. Nat Biotechnol 18: 666-669

Romero C, Bellés JM, Vayé JL, Serrano R, Culiáñez-Marciá FA (1997) Expression of the yeast trehalose-6-phosphate synthase gene in transgenic tobacco plants: pleiotropic phenotypes include drought tolerance. Planta 201: 293-297

Rommens C (2004) All-native DNA transformation: a new approach to plant genetic engineering. Trends Plant Sci 9: 457-464

Rommens CM, Humara JM, Ye J, Yan H, Richael C, Zhang L, Perry R, Swords K (2004) Crop improvement through modification of the plant's own genome. Plant Physiol 135: 421-431

Rosegrant MW, Cline SA (2003) Global food security: challenges and policies. Science 302: 1917-1919

Rovere CV, del Vas M, Hopp HE (2002) RNA-mediated virus resistance. Curr Opin Biotechnol 13: 167-172

Roxas VP, Smith RK, Allen ER, Allen RD (1997) Overexpression of glutathione S-transferase/glutathione peroxidase enhances the growth of transgenic tobacco seedlings during stress. Nat Biotechnol 15: 988-991

Rudolph C, Schreier PH, Uhrig JF (2003) Peptide-mediated broad-spectrum plant resistance to tospoviruses. Proc Natl Acad Sci USA 100: 4429-4434

Rugh CL, Senecoff JF, Meagher RB, Merkle SA (1998) Development of transgenic yellow poplar for mercury phytoremediation. Nat Biotechnol 16: 925-928

Rugh CL, Wilde HD, Stack NM, Thompson DM, Summers AO, Meagher RB (1996) Mercuric ion reduction and resistance in transgenic *Arabidopsis thaliana* plants expressing a modified bacterial *merA* gene. Proc Natl Acad Sci USA 93: 3182-3187

Rushton P (2002) Exciting prospects for plants with greater disease resistance. Trends Plant Sci 7: 325

Sakamoto T, Morinaka Y, Ishiyama K, Kobayashi M, Itoh H, Kayano T, Iwahori S, Matsuoka M, Tanaka H (2003) Genetic manipulation of gibberellin metabolism in transgenic rice. Nat Biotechnol 21: 909-913

Sakamoto T, Morinaka Y, Ohnishi T, Sunohara H, Fujioka S, Ueguchi-Tanaka M, Mizutani M, Sakata K, Takatsuto S, Yoshida S, Tanaka H, Kitano H, Matsuoka M (2006) Erect leaves caused by brassinosteroid deficiency increase biomass production and grain yield in rice. Nat Biotechnol 24: 105-109

Salvi S, Tuberosa R (2005) To clone or not to clone plant QTLs: present and future challenges. Trends Plant Sci 10: 297-304

Sandermann H (2004) Molecular ecotoxicology of plants. Trends Plant Sci 9: 406-413

Sanford JC, Johnston SA (1985) The concept of parasite-derived resistance - deriving resistance genes from the parasite's own genome. J Theor Biol 113: 395-405

Sarhan F, Danyluk J (1998) Engineering cold-tolerant crops - throwing the master switch. Trends Plant Sci 3: 289-290

Sattler SE, Cheng Z, DellaPenna D (2004) From *Arabidopsis* to agriculture: engineering improved Vitamin E content in soybean. Trends Plant Sci 9: 365-367

Scheid O (2004) Either/or selection markers for plant transformation. Nat Biotechnol 22: 398-399

Scheller J, Gührs KH, Grosse F, Conrad U (2001) Production of spider silk proteins in tobacco and potato. Nat Biotechnol 19: 573-577

Serageldin I (1999) Biotechnology and food security in the 21st century. Science 285: 387-389

Shaked H, Melamed-Bessudo C, Levy AA (2005) High frequency gene targeting in *Arabidopsis* plants expressing the yeast RAD54 gene. Proc Natl Acad Sci USA 102: 12265-12269

Shelton AM, Zhao JZ, Roush RT (2002) Economic, ecological, food safety, and social consequences of the deployment of Bt transgenic plants. Annu Rev Entomol 47: 845-881

Shewmaker CK, Sheehy JA, Daley M, Colburn S, Ke DY (1999) Seed-specific overexpression of phytoene synthase: increase in carotenoids and other metabolic effects. Plant J 20: 401-412

Shi H, Lee BH, Wu SJ, Zhu JK (2003) Overexpression of a plasma membrane Na+/H+ antiporter gene improves salt tolerance in *Arabidopsis thaliana*. Nat Biotechnol 21: 81-85

Shibata D, Liu YG (2000) *Agrobacterium*-mediated plant transformation with large DNA fragments. Trends Plant Sci 5: 354-357

Shimamoto K (2002) Picking genes in the rice genome. Nat Biotechnol 20: 983-984

Shintani D, DellaPenna D (1998) Elevating the vitamin E content of plants through metabolic engineering. Science 282: 2098-2100

Sinclair TR, Purcell LC, Sneller CH (2004) Crop transformation and the challenge to increase yield potential. Trends Plant Sci 9: 70-75

Siritunga D, Sayre RT (2003) Generation of cyanogen-free transgenic cassava. Planta 217: 367-373

Sithole-Niang I (2001) Future of plant science in Zimbabwe. Trends Plant Sci 6: 493-494

Slater S, Mitsky TA, Houmiel KL, Hao M, Reiser SE, Taylor NB, Tran M, Valentin HE, Rodriguez DJ, Stone DA, Padgette SR, Kishore G, Gruys KJ (1999) Metabolic engineering of *Arabidopsis* and *Brassica* for poly(3-hydroxybutyrate-co-3-hydroxyvalerate) copolymer production. Nat Biotechnol 17: 1011-1016

Slattery CJ, Kavakli IH, Okita TW (2000) Engineering starch for increased quantity and quality. Trends Plant Sci 5: 291-298

Smirnoff N, Bryant JA (1999) DREB takes the stress out of growing up. Nat Biotechnol 17: 229-230

Smith NA, Singh SP, Wang MB, Stoutjesdijk PA, Green AG, Waterhouse PM (2000) Total silencing by intron-spliced hairpin RNAs. Nature 407: 319-320

Smyth S, Khachatourians GG, Phillips PW (2002) Liabilities and economics of transgenic crops. Nat Biotechnol 20: 537-541

Snow A (2002) Transgenic crops – why gene flow matters. Nat Biotechnol 20: 542

Song J, Bradeen JM, Naess SK, Raasch JA, Wielgus SM, Haberlach GT, Liu J, Kuang H, Austin-Phillips S, Buell CR, Helgeson JP, Jiang J (2003a) Gene *RB* cloned from *Solanum bulbocastanum* confers broad spectrum resistance to potato late blight. Proc Natl Acad Sci USA 100: 9128-9133

Song WY, Sohn EJ, Martinoia E, Lee YJ, Yang YY, Jasinski M, Forestier C, Hwang I, Lee Y (2003b) Engineering tolerance and accumulation of lead and cadmium in transgenic plants. Nat Biotechnol 21: 914-919

Soosaar JL, Burch-Smith TM, Dinesh-Kumar SP (2005) Mechanisms of plant resistance to viruses. Nat Rev Microbiol 3: 789-798

Srivastava V, Ow DW (2004) Marker-free site-specific gene integration in plants. Trends Biotechnol 22: 627-629

Stewart CN, Jr. (2005) Monitoring the presence and expression of transgenes in living plants. Trends Plant Sci 10: 390-396

Streatfield SJ, Jilka JM, Hood EE, Turner DD, Bailey MR, Mayor JM, Woodard SL, Beifuss KK, Horn ME, Delaney DE, Tizard IR, Howard JA (2001) Plant-based vaccines: unique advantages. Vaccine 19: 2742-2748

Sung DY, Kaplan F, Lee KJ, Guy CL (2003) Acquired tolerance to temperature extremes. Trends Plant Sci 8: 179-187

Tabe L, Higgins TJV (1998) Engineering plant protein composition for improved nutrition. Trends Plant Sci 3: 282-286

Tacket CO, Mason HS, Losonsky G, Estes MK, Levine MM, Arntzen CJ (2000) Human immune responses to a novel norwalk virus vaccine delivered in transgenic potatoes. J Infect Dis 182: 302-305

Tai TH, Dahlbeck D, Clark ET, Gajiwala P, Pasion R, Whalen MC, Stall RE, Staskawicz BJ (1999) Expression of the Bs2 pepper gene confers resistance to bacterial spot disease in tomato. Proc Natl Acad Sci USA 96: 14153-14158

Takahashi M, Nakanishi H, Kawasaki S, Nishizawa NK, Mori S (2001) Enhanced tolerance of rice to low iron availability in alkaline soils using barley nicotianamine aminotransferase genes. Nat Biotechnol 19: 466-469

Tarczynski MC, Jensen RG, Bohnert HJ (1993) Stress protection of the transgenic tobacco by production of the osmolyte mannitol. Science 259: 508-510

Taylor CG, Fuchs B, Collier R, Lutke WK (2006) Generation of composite plants using *Agrobacterium rhizogenes*. Methods Mol Biol 343: 155-167

Taylor N, Chavarriaga P, Raemakers K, Siritunga D, Zhang P (2004) Development and application of transgenic technologies in cassava. Plant Mol Biol 56: 671-688

Tenllado F, Llave C, Díaz-Ruíz JR (2004) RNA interference as a new biotechnological tool for the control of virus diseases in plants. Virus Res 102: 85-96

Terada R, Urawa H, Inagaki Y, Tsugane K, Iida S (2002) Efficient gene targeting by homologous recombination in rice. Nat Biotechnol 20: 1030-1034

Thro AM, Fregene M, Taylor N, Raemakers KCJJM, Puonit-Kaerlas J, Schöpke C, Visser R, Potrykus I, Fauquet C, Roca W, Hershey C (1999) Genetic biotechnologies and cassava-based development. *In* T Hohn, KM Leisinger, eds, Biotechnology of food crops in developing countries. Springer-Verlag Wien, New York, pp 143-185

Tingay S, McElroy D, Kalla R, Fieg S, Wang M, Thornton S, Brettell R (1997) *Agrobacterium tumefaciens*-mediated barley transformation. Plant J 11: 1369-1376

Toenniessen G (1995) Plant biotechnology and developing countries. Trends Biotechnol 13: 404-409

Torisky RS, Kovacs L, Avdiushko S, Newman JD, Hunt AG, Collins GB (1997) Development of a binary vector system for plant transformation base on the supervirulent *Agrobacterium tumefaciens* strain Chry 5. Plant Cell Rep 17: 102-108

Tzfira T, Tian G-W, Lacroix B, Vyas S, Li J, Leitner-Dagan Y, Krichevsky A, Taylor T, Vainstein A, Citovsky V (2005) pSAT vectors: a modular series of plasmids for autofluorescent protein tagging and expression of multiple genes in plants. Plant Mol Biol 57: 503-516

Tzfira T, Vaidya M, Citovsky V (2002) Increasing plant susceptibility to *Agrobacterium* infection by overexpression of the *Arabidopsis* nuclear protein VIP1. Proc Natl Acad Sci USA 99: 10435-10440

Tzfira T, White C (2005) Towards targeted mutagenesis and gene replacement in plants. Trends Biotechnol 23: 567-569

Tzfira T, Zuker A, Altman A (1998) Forest-tree biotechnology: genetic transformation and its application to future forests. Trends Biotechnol 16: 439-446

Uhrig JF (2003) Response to Prins: broad virus resistance in transgenic plants. Trends Biotechnol 21: 376-377

Umbeck P, Johnson G, Barton K, Swain W (1987) Genetically transformed cotton (*Gossypium Hirsutum* L.) plants. Bio/Technology 5: 263-266

Urwin PE, Lilley CJ, McPherson MJ, Atkinson HJ (1997) Resistance to both cyst and root-knot nematodes conferred by transgenic *Arabidopsis* expressing a modified plant cystatin. Plant J 12: 455-461

Vaeck M, Reynaerts A, Höfte H, Jansens S, De Beuckeleer M, Dean C, Zabeau M, Van Montagu M, Leemans J (1987) Transgenic plants protected from insect attack. Nature 328: 33-37

Van de Elzen PJM, Townsend J, Lee KY, Bedbrook JR (1985) A chimaeric hygromycin resistance gene as a selectable marker in plant cells. Plant Mol Biol 5: 299-302

van der Biezen EA (2001) Quest for antimicrobial genes to engineer disease-resistant crops. Trends Plant Sci 6: 89-91

van der Zaal BJ, Neuteboom LW, Pinas JE, Chardonnens AN, Schat H, Verkleij JA, Hooykaas PJJ (1999) Overexpression of a novel *Arabidopsis* gene related to putative zinc-transporter genes from animals can lead to enhanced zinc resistance and accumulation. Plant Physiol 119: 1047-1055

Van Eenennaam AL, Lincoln K, Durrett TP, Valentin HE, Shewmaker CK, Thorne GM, Jiang J, Baszis SR, Levering CK, Aasen ED, Hao M, Stein JC, Norris SR, Last RL (2003) Engineering vitamin E content: from Arabidopsis mutant to soy oil. Plant Cell 15: 3007-3019

Vancanneyt G, Schmidt R, O'Connor-Sanchez A, Willmitzer L, Rocha-Sosa M (1990) Construction of an intron-containing marker gene: splicing of the intron in transgenic plants and its use in monitoring early events in *Agrobacterium*-mediated plant transformation. Mol Gen Genet 220: 245-250

Vasil IK (2003) The science and politics of plant biotechnology--a personal perspective. Nat Biotechnol 21: 849-851

Vaucheret H, Béclin C, Elmayan T, Feuerbach F, Godon C, Morel JB, Mourrain P, Palauqui JC, Vernhettes S (1998) Transgene-induced gene silencing in plants. Plant J 16: 651-659

Verberne MC, Verpoorte R, Bol JF, Mercado-Blanco J, Linthorst HJ (2000) Overproduction of salicylic acid in plants by bacterial transgenes enhances pathogen resistance. Nat Biotechnol 18: 779-783

Vergunst AC, Jansen LE, Hooykaas PJ (1998) Site-specific integration of *Agrobacterium* T-DNA in *Arabidopsis thaliana* mediated by Cre recombinase. Nucleic Acids Res 26: 2729-2734

Vergunst AC, Schrammeijer B, den Dulk-Ras A, de Vlaam CMT, Regensburg-Tuink TJ, Hooykaas PJJ (2000) VirB/D4-dependent protein translocation from *Agrobacterium* into plant cells. Science 290: 979-982

Walden R, Wingender R (1995) Gene-transfer and plant-regeneration techniques. Trends Biotechnol 13: 324-331

Wang Y, Xue Y, Li J (2005) Towards molecular breeding and improvement of rice in China. Trends Plant Sci 10: 610-614

Wenck A, Czako M, Kanevski I, Marton L (1997) Frequent collinear long transfer of DNA inclusive of the whole binary vector during *Agrobacterium*-mediated transformation. Plant Mol Biol 34: 913-922

White PJ, Broadley MR (2005) Biofortifying crops with essential mineral elements. Trends Plant Sci 10: 586-593

Wilkinson MJ, Sweet J, Poppy GM (2003) Risk assessment of GM plants: avoiding gridlock? Trends Plant Sci 8: 208-212

Williams M (1995) Genetic engineering for pollination control. Trends Biotechnol 13: 344-349

Wilson TM (1993) Strategies to protect crop plants against viruses: pathogen-derived resistance blossoms. Proc Natl Acad Sci USA 90: 3134-3141

Woodard SL, Mayor JM, Bailey MR, Barker DK, Love RT, Lane JR, Delaney DE, McComas-Wagner JM, Mallubhotla HD, Hood EE, Dangott LJ, Tichy SE, Howard JA (2003) Maize (*Zea mays*)-derived bovine trypsin: characterization of the first large-scale, commercial protein product from transgenic plants. Biotechnol Appl Biochem 38: 123-130

Wu L, Fan Z, Guo L, Li Y, Chen Z, Qu L (2004) Over-expression of the bacterial *nha*A gene in rice enhances salt and drought tolerance. Plant Sci 168: 297-302

Wu G, Truksa M, Datla N, Vrinten P, Bauer J, Zank T, Cirpus P, Heinz E, Qiu X (2005) Stepwise engineering to produce high yields of very long-chain polyunsaturated fatty acids in plants. Nat Biotechnol 23: 1013-1017

Wu S, Schalk M, Clark A, Miles RB, Coates R, Chappell J (2006) Redirection of cytosolic or plastidic isoprenoid precursors elevates terpene production in plants. Nat Biotechnol 24: 1441-1447

Xie M, He Y, Gan S (2001) Bidirectionalization of polar promoters in plants. Nat Biotechnol 19: 677-679

Xie ZP, Auberson-Huang L, Malnoë P, Yao H, Kaeppeli O (2002) Comparison of driving forces in sustainable food production and the future of plant biotechnology in Switzerland and China. Trends Plant Sci 7: 416-418

Xue Z, Zhi D, Xue G, Zhang H, Zhao Y, Xia G (2004) Enhanced salt tolerance of transgenic wheat (*Tritivum aestivum* L.) expressing a vacuolar Na^+/H^+ antiporter gene with improved grain yields in saline soils in the field and a reduced level of leaf Na^+. Plant Sci 167: 849-859

Yamaguchi T, Blumwald E (2005) Developing salt-tolerant crop plants: challenges and opportunities. Trends Plant Sci 10: 615-620

Yang X, Yie Y, Zhu F, Liu Y, Kang L, Wang X, Tien P (1997) Ribozyme-mediated high resistance against potato spindle tuber viroid in transgenic potatoes. Proc Natl Acad Sci USA 94: 4861-4865

Yanofsky M, Lowe B, Montoya A, Rubin R, Krul W, Gordon M, Nester E (1985) Molecular and genetic analysis of factors controlling host range in *Agrobacterium tumefaciens*. Mol Gen Genet 201: 237-348

Ye X, Al-Babili S, Klöti A, Zhang J, Lucca P, Beyer P, Potrykus I (2000) Engineering the provitamin A (β-carotene) biosynthetic pathway into (carotenoid-free) rice endosperm. Science 287: 303-305

Yimin D, Mervis J (2002) Transgenic crops. China takes a bumpy road from the lab to the field. Science 298: 2317-2319

Zambryski PC, Joos H, Genetello C, Leemans J, Van Montagu M, Schell J (1983) Ti plasmid vector for the introduction of DNA into plant cells without alteration of their normal regeneration capacity. EMBO J 2: 2143-2150

Zhang HX, Blumwald E (2001) Transgenic salt-tolerant tomato plants accumulate salt in foliage but not in fruit. Nat Biotechnol 19: 765-768

Zhang HX, Hodson JN, Williams JP, Blumwald E (2001) Engineering salt-tolerant *Brassica* plants: characterization of yield and seed oil quality in transgenic plants with increased vacuolar sodium accumulation. Proc Natl Acad Sci USA 98: 12832-12836

Zhang J, Boone L, Kocz R, Zhang C, Binns AN, Lynn DG (2000) At the maize/*Agrobacterium* interface: natural factors limiting host transformation. Chem Biol 7: 611-621

Zhao JZ, Cao J, Li Y, Collins HL, Roush RT, Earle ED, Shelton AM (2003) Transgenic plants expressing two *Bacillus thuringiensis* toxins delay insect resistance evolution. Nat Biotechnol 21: 1493-1497

Zhao ZY, Cai T, Tagliani L, Miller M, Wang N, Pang H, Rudert M, Schroeder S, Hondred D, Seltzer J, Pierce D (2000) *Agrobacterium*-mediated sorghum transformation. Plant Mol Biol 44: 789-798

Zhu JK (2001) Plant salt tolerance. Trends Plant Sci 6: 66-71

Zipfel C, Kunze G, Chinchilla D, Caniard A, Jones JD, Boller T, Felix G (2006) Perception of the bacterial PAMP EF-Tu by the receptor EFR restricts *Agrobacterium*-mediated transformation. Cell 125: 749-760

Zipfel C, Robatzek S, Navarro L, Oakeley EJ, Jones JD, Felix G, Boller T (2004) Bacterial disease resistance in *Arabidopsis* through flagellin perception. Nature 428: 764-767

Zubko E, Scutt C, Meyer P (2000) Intrachromosomal recombination between *attP* regions as a tool to remove selectable marker genes from tobacco transgenes. Nat Biotechnol 18: 442-445

Zuo J, Niu QW, Ikeda Y, Chua NH (2002) Marker-free transformation: increasing transformation frequency by the use of regeneration-promoting genes. Curr Opin Biotechnol 13: 173-180

Zuo J, Niu QW, Møller SG, Chua NH (2001) Chemical-regulated, site-specific DNA excision in transgenic plants. Nat Biotechnol 19: 157-161

Chapter 4

THE *AGROBACTERIUM TUMEFACIENS* C58 GENOME

Steven C. Slater[1], Brad W. Goodner[2], João C. Setubal[3], Barry S. Goldman[4], Derek W. Wood[5,6] and Eugene W. Nester[5]

[1]The Biodesign Institute and Department of Applied Biological Sciences, Arizona State University, Mesa, AZ 85212, USA; [2]Department of Biology, Hiram College, Hiram, OH 44234, USA; [3]Virginia Bioinformatics Institute and Department of Computer Science, Virginia Polytechnic Institute and State University, Blacksburg, VA 24060, USA; [4]Monsanto Company, 800 North Lindbergh Boulevard, St. Louis, MO 63167, USA; [5]Department of Biology, Seattle Pacific University, Seattle, WA 98119, USA; [6]Department of Microbiology, University of Washington, Seattle WA 98195, USA

Abstract. *Agrobacterium* is a bacterial plant pathogen capable of transferring a specific fragment of DNA, called the T-DNA, into plants and other organisms. Once in a eukaryotic cell, the T-DNA moves to the nucleus and integrates into the genome at an essentially random location. T-DNA integration generally leads to tumor formation in the plant host, and *Agrobacterium's* ability to transfer DNA has been adapted as an important tool for mutagenesis and genetic engineering of plants and fungi. *Agrobacterium tumefaciens* C58 was the first species of *Agrobacterium* to have a fully-sequenced genome, and the sequence data are catalyzing expansion of *A. tumefaciens* research beyond its traditional focus on plant pathogenesis and T-DNA transfer. This chapter reviews many of the findings of the original genome publications and discusses many new insights derived from the availability of the genome sequence.

1 INTRODUCTION

In 2001 the journal *Science* published two papers back-to-back on the genome of the *Agrobacterium* biovar I organism *A. tumefaciens* C58 (Goodner et al., 2001; Wood et al., 2001). Two different teams of scientists had raced to complete and publish this genome, only becoming aware of the other's efforts near the end of the projects. After contacting each other, and thanks to the vision of *Science* editors, both teams were able to publish their results simultaneously. An interesting account of this race was published several years later in *Nature Biotechnology* (Harvey and McMeekin, 2004). The principle members of both groups have now combined efforts and, in addition to authoring this chapter, have completed the genome sequences of representative *Agrobacterium* strains from biovars II and III (Wood D, Burr T, Farrand S, Goldman B, Nester E, Setubal J and Slater S, unpublished data).

The two original *Science* papers, although covering a lot of common ground, were surprisingly complementary. Over 250 manuscripts have used the data from the original C58 genome sequences. The types of manuscripts fall into three basic categories: (i) those that use the sequence as part of genome-scale comparative analyses, (ii) those that simply cite the identification of an ortholog of their gene of interest in *A. tumefaciens*, and (iii) those that follow-up on specific genes in *A. tumefaciens* after identifying them in the genome sequence. The last category contains about 20% of these manuscripts. Here we present a description of the C58 genome that combines the findings of both teams, and summarizes many new results on *A. tumefaciens* biology that have been enabled by the *A. tumefaciens* C58 genome sequence. *Table 4-1* lists all genes discussed herein and their designations by the original genome publications (Goodner et al., 2001; Wood et al., 2001). To harmonize nomenclature as we continue our annotation of the *Agrobacterium* genomes, we have chosen to use the gene designations and style of Wood et al. (2001).

2 GENERAL FEATURES OF THE GENOME

The 5.67-Mb genome of *A. tumefaciens* C58 (Hamilton and Fall, 1971) is comprised of four replicons (Allardet-Servent et al., 1993): a circular chromosome, a linear chromosome and the pAtC58 and pTiC58 plasmids. The original sequences generated by the two groups had only 38 potential sequence discrepancies. Re-sequencing of discrepant regions showed 15 verified sequence differences between the two isolates, with the remaining differences being due to sequencing errors by one team or the other (Slater, S, Burr T, Farrand S, Goldman B, Kaul R, Nester E, Setubal J, Wood D and Zhao Y, unpublished data). Thus, the overall sequencing error rate was well below the 1 in 10^5 required by the Bermuda Standard (Wellcome Trust, 1997).

Gene density is very similar between the two chromosomes. However, genes involved in most essential processes are significantly over-represented on the circular chromosome (Goodner et al., 1999; Goodner et al., 2001; Wood et al., 2001). This asymmetry is consistent with direct descent of the circular chromosome from the primordial α-proteobacterial genome, with a minority of essential genes moving to the linear chromosome. Consistent with lateral transfer between chromosomes, the overall dinucleotide signatures (Karlin, 2001) of the two chromosomes are essentially identical, but are significantly different from those of the two plasmids. The dinucleotide signatures of the two plasmids are quite similar to each other and to related plasmids from other members of the Rhizobiaceae family (Goodner et al., 2001). Several other bacterial species have true multipartite genomes (that is, essential genes on more than one chromosome), including *Vibrio cholerae, Rhodobacter sphaeroides*, and organisms in the *Burkholderia pseudomallei* complex (Heidelberg et al., 2000; Holden et al., 2004; Copeland et al., 2005). Other members of the Rhizobiaceae also have large plasmids with many genes that are critical for bacteria-plant interaction, although not necessarily for survival (Capela et al., 2001; Galibert et al., 2001; Kaneko et al., 2002; Gonzalez et al., 2003; Gonzalez et al., 2006).

In contrast to the pSymB plasmid of *S. meliloti* (Galibert et al., 2001), both *A. tumefaciens* plasmids contain all the necessary machinery for conjugation, but do not contain essential genes. A new conjugal transfer system belonging to the Type IV secretion family (AvhB) was identified on pAtC58, and was shown to be required for the conjugal transfer of pAtC58 following the original publications (Chen et al., 2002). Recent work in *Rhizobium etli* has shown that a similar system is under the control of the regulatory pair RctA and RctB, both of which have orthologs in C58 (Perez-Mendoza et al., 2005).

More than 6000 bp of near perfect sequence identity extend across the two rRNA gene clusters on each of the two chromosomes (Goodner et al., 2001; Wood et al., 2001). The chromosomes also share some shorter regions of greater than 90% sequence identity with pAtC58. The overall GC content of the *A. tumefaciens* genome is 58%. The TiC58 plasmid has two regions of distinctive GC content, the T-DNA (46%) and the *vir* region (54%). Low GC content was noted previously in the T-DNA of a related Ti plasmid (Suzuki et al., 2000). Reduced GC content (53%) is also seen within a 24-kb segment of pAtC58. This region includes seventeen conserved hypothetical or hypothetical genes, an ATP-dependent DNA helicase and an IS element. These genes are flanked by a phage integrase and a second IS element. The genes in these three regions have a distinct codon usage as compared to the rest of the genome, providing evidence for their recent evolutionary acquisition.

A. tumefaciens contains four rRNA operons and transcription of all these gene clusters is oriented away from the DNA replication origins, with those on the linear chromosome being in the same orientation. The genome contains 53 tRNAs that represent all 20 amino acids. These tRNAs are distributed unevenly between the circular and linear chromosomes. Transfer RNA species corresponding to the most frequently represented alanine, glutamine and valine codons are found only on the linear replicon.

The genome contains 25 predicted insertion sequence (IS) elements representing 8 different families. The largest is the IS3 family comprising 10 IS elements. The IS elements are not equally distributed among the replicons, but are located preferentially on the linear chromosome and pAtC58. The adjacent virH1 and virH2 genes of the Ti plasmid, encoding p450 mono-oxygenases, are flanked by IS elements, suggesting that they arrived in *A. tumefaciens* as part of a compound transposon. Twelve genes of probable phage origin were identified, most of which are on the circular chromosome. Many of these genes cluster in two discrete regions suggesting that they represent prophage remnants. None of these clustered phage-related genes are shared with *S. meliloti*, implying they were lost from *S. meliloti* or entered the *A. tumefaciens* genome after these organisms evolutionarily diverged.

3 THE LINEAR CHROMOSOME

Historically, genetic research on *A. tumefaciens* focused on its virulence mechanism, and almost all of the early mapped mutations affecting virulence were localized to the Ti plasmid (Binns and Thomashow, 1988). However, a few chromosomal virulence genes were found at around that same time, and several labs constructed genetic maps of the *A. tumefaciens* C58 "chromosome" (Hooykaas et al., 1982; Pischl and Farrand, 1984;

Miller et al., 1986; Robertson et al., 1988; Cooley and Kado, 1991). These efforts involved mobilization of mutations and transposable element insertions via R factor-mediated chromosome transfer between *Agrobacterium* strains. All of the early papers supported a single circular chromosome in C58, in line with the *E. coli* standard for bacterial chromosomes.

The first chink in the single chromosome model for C58 came from a combination of physical mapping by pulsed-field gel electrophoresis and Southern blotting (Allardet-Servent et al., 1993). This work showed that there were clearly two distinct mega-DNA molecules in C58 and one of them was linear. Moreover, rRNA operons were found on both molecules, supporting a multi-chromosome genome architecture. The move from a one chromosome model to a two chromosome model was solidified by two independent efforts. In one, additional biovar I strains of *A. tumefaciens* and *A. rubi* were shown to have two mega-DNA molecules, one being linear, with each bearing one or more rRNA operons (Jumas-Bilak et al., 1998). Interestingly, the grape-limited biovar III strains of *Agrobacterium* also have two chromosomes, but both of them are circular. In the second parallel effort, the C58 genome was subjected to extensive genetic and physical mapping which further confirmed the linear nature of the second chromosome and identified several auxotrophic markers on the linear chromosome (Goodner et al., 1999). This work also showed that the linear chromosome was not "invisible" to the R factor-mediated chromosome transfer technique. The most likely explanation lies in the relative rarity of auxotrophic markers on the linear chromosome as compared to the circular chromosome. The idea that the linear chromosome originated from a breaking/rejoining event of the circular chromosome seemed improbable given the stable nature of the linear chromosome, which implied the presence of stable telomeres. The remaining hypotheses required some level of "foreign" DNA insertion into the linear chromosome, the real question being: How much?

Linear replicons are the norm in eukaryotes, but only a few examples are known in prokaryotes and viruses (Casjens et al., 1997; Goshi et al., 2002; Bao and Cohen, 2003; Ravin et al., 2003; Chaconas, 2005). The chromosome ends, or telomeres, of these known examples fall into two groups; those with proteins attached to free DNA ends, such as in *Streptomyces* species, and those with covalently closed hairpin loops, such as *Borrelia* species and numerous viruses (e.g., phage N15). As the C58 genome projects got underway, experiments were done to analyze the nature of the linear chromosome telomeres. Comparisons of pulsed-field gel electrophoresis done with and without proteolysis showed that there are no large proteins attached to the telomeres. Rather, experiments that tested for the snap-back characteristics of hairpin loops demonstrated the presence of hairpin loops at the telomeres (Goodner et al., 2001). Neither of the original genome projects clearly defined the hairpin loop sequences, but not for a lack of trying. Since then, the sequences of the hairpin loops have been determined; one telomere had been sequenced around the hairpin loop in the original genome projects but not enough of the stem structure was present to catch the eyes of researchers (Huang et al., 2006). The hairpin loop sequence is highly conserved between the two telomeres, reflecting the fact that they both are substrates of the same protelomerase, TelA, encoded by the Atu2523/AGR_C_4584 gene. The TelA protein is distantly related to the protelomerases found in *Borrelia* species and in numerous viruses (Huang et al., 2006). In contrast to the conservation of the hairpin loops, the sequences proximal to the telomeres are not similar in sequence even though they both are rich in IS elements and potential secondary structures.

The complete sequence of the C58 linear chromosome answered the questions raised by the earlier genetic and physical mapping as to its origin. Sitting almost perfectly in the middle of the linear replicon is an intact *repABC* operon, the key element involved in the replication and segregation of almost all plasmids known in the family Rhizobiaceae. The position of the *repABC* operon coincides with a GC-skew inversion indicative of bi-directional replication typical of circular DNA molecules. In addition to the *repABC* operon, other indications of a plasmid origin for the linear chromosome include the presence of genes for the conjugation proteins TraA, TraG, and MobC. Based on these and other data, both Goodner et al. (Goodner et al., 2001) and Wood et al. (Wood et al., 2001) proposed that the linear chromosome is evolutionarily derived from a plasmid, although the replication mechanism remains to be experimentally verified for the C58 linear chromosome. The plasmid origin of an "extra" chromosome in proteobacteria had been predicted for multi-chromosome genomes of the α-proteobacteria (Moreno, 1998).

How then did a plasmid become a chromosome? Syntenic analysis of the two chromosomes of C58 in comparison with the single chromosome of *Sinorhizobium meliloti* provided a big clue (Goodner et al., 2001; Wood et al., 2001; Wood, 2002). The C58 circular chromosome shared large stretches of synteny with the *S. meliloti* chromosome broken by some gaps. In several cases, sequences absent from the C58 circular chromosome are present on the linear chromosome. Thus, the second chromosome in biovar I strains of *Agrobacterium* originated from a *repABC*-type plasmid through the intragenomic transfer of several chunks of sequence from the ancestral chromosome to this new chromosome. It is valid to call the linear chromosome a chromosome because several of those intragenome transfer events involved genes essential for prototrophic growth.

4 PHYLOGENY AND WHOLE-GENOME COMPARISON

The two original genome papers show the close similarities that exist between the *A. tumefaciens* genome and those of two rhizobial species, *Sinorhizobium meliloti* and *Mesorhizobium loti*. The circular chromosomes of all three organisms show extensive nucleotide colinearity and gene order conservation, which can be readily seen in pairwise whole-replicon alignments (Goodner et al., 2001; Wood et al., 2001). Since 2001 several additional α-proteobacterial genomes have been sequenced, and chromosome colinearity can also be readily detected against *Brucella* and *Bartonella* (our unpublished data). Thus it seems that chromosomal gene order in this subgroup of α-proteobacteria has been under selective pressure to be maintained, even though the group includes organisms with diverse lifestyles, such as plant and animal pathogens, and plant symbionts. An interesting perspective on the evolution of α-proteobacteria is given by Boussau et al. (2004). According to this study *A. tumefaciens* has undergone genome reduction whereas its relatives *S. meliloti* and *M. loti* have undergone genome expansions.

Contrasting with the large scale conservation of the chromosomal backbones, we do not see significant colinearities between other replicons of C58 and those of *S. meliloti* and *M. loti*. The vast majority of protein-coding genes in the three smaller C58 replicons do have orthologs in the other two rhizobial species, but it appears that widespread gene shuffling has taken place since divergence. It should be noted that IS elements are relatively rare in C58 and therefore cannot by themselves explain the highly distributed nature of orthologous genes in the smaller replicons. One notable exception to the shuffling exists in the C58 linear chromosome. It has two regions that exhibit significant conservation of gene order with a segment of the *S. meliloti* chromosome. The first is comprised of 46 genes (44 kb) and the second contains 65 genes (89 kb). If portions of the linear chromosome arose via an excision event from the ancestral chromosome, the excision may have originated in these regions, with subsequent insertions moving particular sections apart.

5 DNA REPLICATION AND THE CELL CYCLE

The circular chromosome contains a putative origin of replication (C*ori*) similar to the known C*ori* of *Caulobacter crescentus* (Brassinga et al., 2002). The linear chromosome, as mentioned above, has a plasmid-type replication system of the same type found on the two plasmids. This system, encoded by the *repABC* genes, expresses a pair of segregation proteins (RepA, RepB) and an origin-binding replication initiation protein (RepC) (Li and Farrand, 2000). A number of additional papers published in the last several years have added to our knowledge of *repABC* origins, and support the annotation of these regions provided by the original C58 genome papers (Bartosik et al., 2002; Cevallos et al., 2002; Pappas and Winans, 2003; Soberon et al., 2004; Venkova-Canova et al., 2004; Cho and Winans, 2005; MacLellan et al., 2005; MacLellan et al., 2006).

Coordinating replication and segregation of multiple chromosomes is critical for cell survival. The topic was initially addressed in *A. tumefaciens* by Kahng and Shapiro (2001) who showed cell-cycle regulation by CcrM DNA adenine methyltransferase and replication of the *A. tumefaciens att* locus in coordination with the cell cycle. CcrM is also critical for cell cycle regulation in *Caulobacter crescentus* and appears to be a general mechanism for cell cycle control in the α–proteobacteria (Marczynski and Shapiro, 2002). At the time of the Kahng and Shapiro study (Kahng and Shapiro, 2001), the *att* locus was thought to be located on the *A. tumefaciens* chromosome. Their interpretation was slightly modified and expanded by Goodner et al. (2001) after the genome sequence revealed *att* to be located on pTiC58. It appears that both the chromosomes and plasmids are replicated synchronously, although the means for coordinating the two types of replication origins remain unclear. Kahng and Shapiro later demonstrated (2003) that the DNA replication origins in both *A. tumefaciens* and *S. meliloti* are localized at the cell poles, although their precise locations don't necessarily overlap. The multipartite genome of *V. cholerae* is also replicated in a coordinated manner (Egan et al., 2004), and some of the issues surrounding replication of multipartite genomes have been recently reviewed (Egan et al., 2005).

Processive DNA replication is performed by DNA Polymerase III (Pol III); the *A. tumefaciens* genome carries four paralogs of the *dnaE* gene encoding the Pol III α (polymerase) subunit. Goodner et al. (2001) demonstrated that these *dnaE* genes fall into two distinct sequence families, designated as categories A and B. The category A gene of the circular chromosome is conserved in all sequenced α-proteobacteria and probably encodes the primary replication enzyme. Each of the *A. tumefaciens repABC* replicons (linear chromosome, pTiC58 and pAtC58) encodes a Category B *dnaE* gene within an operon containing two other conserved genes. The operon was found to be present in all fully sequenced α-proteobacteria, except the *Rickettsia* species, and was hypothesized by Goodner et al. (2001) to encode a novel DNA polymerase complex. In the past several years, a broader description of these genes has been published (Abella et al., 2004) and the operon in *Caulobacter crescentus* has been characterized (Galhardo et al., 2005). As predicted, these genes form an auxiliary DNA polymerase complex involved in repair of damaged DNA. It is still not clear why *A. tumefaciens* carries three copies of these operons, or whether all are functional.

6 GENUS-SPECIFIC GENES

The original analyses assigned the *A. tumefaciens* predicted proteins to about 500 paralogous families containing between two and 206 members. The two largest families are composed of genes belonging to the ATPase and membrane-spanning components of the ATP Binding Cassette (ABC) transport family. Comparison of the genomes of *A. tumefaciens, S. meliloti* and *M. loti* identified genes in each organism that likely contribute to genus-specific biology. Of the 5,419 originally-predicted *A. tumefaciens* proteins, 853 (16%) are not found in these other organisms. Of these, 97 have a proposed function, whereas 756 are hypothetical or conserved hypothetical. The predicted products of these genes are diverse, and include proteins involved in cellulose production, plasmid maintenance, cell growth, transcriptional regulation and cell wall synthesis. Several additional proteins are predicted to catabolize plant cell wall materials, sugars and exudates. These include polygalacturonases, a glycosidase, an endoglucanase, a myo-inositol catabolism protein and a cell wall lysis associated protein. Additional specific genes, predictably found on the Ti plasmid, include those encoding virulence, T-DNA and conjugal transfer associated proteins. Conversely, nitrogen-fixing genes and others associated with symbiosis are specific to *S. meliloti* and *M. loti* when compared to *A. tumefaciens.* With 756 *Agrobacterium*-specific ORFs yet to be characterized, much remains to be elucidated regarding the genetic distinction between *A. tumefaciens* and its rhizobial relatives.

7 PLANT TRANSFORMATION AND TUMORIGENESIS

Genes involved in plant transformation and tumorigenesis are located on all four genetic elements. The circular chromosome harbors the well-studied *chvAB* genes required for synthesis and transport of the extracellular β-1,2-glucan involved in binding to plant cells, the *chvG/I, chvE*, and *ros* genes involved in regulation of Ti plasmid vir genes, as well as the *chvD, chvH*, and *acvB* genes. The linear chromosome harbors the *exoC* (*pgm*) gene required for synthesis of the extracellular β-1,2-glucan and succinoglucan polysaccharides, and the cellulose synthesis (*cel*) genes. The *cel* region sequence produced by the genome projects differed from the published sequence (Matthysse et al., 1995), and the locus was reannotated by both groups, resulting in several changes to predicted proteins.

An early interaction of *Agrobacterium* with its plant hosts is mediated by several attachment-related genes (Matthysse and Kijne, 1998). The *att* (attachment) genes were reported to be involved in initial specific attachment of the bacterium to plant cells (Binns and Thomashow, 1988). The genome sequence showed the *att* genes to be present on the pAtC58 replicon, rather than the bacterial chromosome, as was originally reported. pAtC58 also has a second, partial *att* locus. Since the pAtC58 replicon is dispensable for virulence (Rosenberg and Huguet, 1984), a reevaluation of the *att* genes was required. Nair et al. (2003) took on the problem and determined that neither pAtC58 nor two specific *att* genes (*attR* and *attD*) are required for T-DNA transfer to plants. They also showed that pAtC58 has a positive effect on *vir* gene induction and T-DNA transfer, but did not define the precise manner in which this positive effect is produced.

The importance of small heat-shock proteins (sHsps) in bacteria-plant interactions has recently been identified, both in *Sinorhizobium* and *Agrobacterium* (Natera et al., 2000; Baron et al., 2001; Balsiger et al., 2004). Balsiger et al. (2004) identified four sHsp proteins in *A. tumefaciens* C58 and investigated their expression. One gene (*hspC*) was located on the circular chromosome, one (*hspL*) on the linear chromosome, and two (*hspAT1* and *hspAT2*) on the pAT plasmid. They found that while *hspC* was poorly or never expressed under their growth conditions, *hspL* is part of the RpoH regulon and induced by heat stress. The two *hsp* genes on pAtC58 are regulated independently of RpoH, via ROSE sequences in their 5' regions (Nocker et al., 2001a; Nocker et al., 2001b).

Our original genomic analysis identified genes whose products are similar to plant pathogen virulence proteins required for host cell wall degradation (Goodner et al., 2001; Wood et al., 2001). These include pectinase (*kdgF*), ligninase (*ligE*) and xylanase as well as regulators of pectinase and cellulase production (*pecS/M*). *A. tumefaciens* may use such enzymes to breach the cell wall of its host before T-DNA transfer. In addition, we identified numerous orthologs of animal virulence genes. Examples include those involved in host survival, such as the bacA locus of *Brucella* and *S. meliloti*, and two members of the widely conserved HtrA family of serine proteases implicated in response to oxidative stress in *Salmonella* and *Yersinia*. Invasion-related homologs include the *ialA* and *ialB* genes of *Bartonella henselae* as well as five hemolysin-like proteins with associated type I secretion systems. The highly conserved *mviN* gene, implicated in *Salmonella* virulence, is also present. Two autotransporting virulence factor family members are encoded by pAtC58. Such proteins cross the plasma membrane via the signal peptide-dependent pathway and self-insert into the outer membrane. Typically, a large extracellular domain is exposed, where it modifies cell adhesion or host cell functions. Other genes similar to known virulence factors include putative adhesins, *icmF* (macrophage killing in *Rickettsia*), and as many as six different iron uptake systems.

Another potential virulence locus includes the genes Atu4334, Atu4337, Atu4340, Atu4341, and Atu4343. This locus encodes proteins that are similar to members of the IcMF-associated homologous protein (IAHP) group (Das and Choudhuri, 2003). Orthologous genes in *Vibrio cholerae* and *Pseudomonas aeruginosa* are protein exporters that play a role in pathogenesis of mammalian hosts (Mougous et al., 2006; Pukatzki et al., 2006). However, this locus has not yet been analyzed in *A. tumefaciens*.

8 TRANSPORT

Transporters comprise 15% of the *A. tumefaciens* genome (Goodner et al., 2001; Wood et al., 2001). *A. tumefaciens* possesses broad capabilities for the transport of common nutrients found in the rhizosphere including sugars, amino acids and peptides. Overall, our analyses indicate that *A. tumefaciens* and the other sequenced members of the Rhizobiaceae have similar transport capabilities.

ABC transporters in *A. tumefaciens* constitute 60% of its total transporter complement. There are 153 complete systems plus additional "orphan" subunits. Predicted substrates of these ABC transport systems include sugars (53 systems), amino acids (29 systems) and peptides (25 systems). As we speculated in the original papers, the large number of ABC transporters in *A. tumefaciens* may reflect a need for high-affinity uptake systems to facilitate the acquisition of nutrients in the highly competitive soil and rhizosphere environments.

Subsequent to publication of the original genome analyses the twin-arginine targeting system that mediates the secretion of folded proteins in a sec-independent manner has been shown to be involved in virulence of *A. tumefaciens*, symbiotic interactions of *S. meliloti* and pathogenesis of a variety of bacterial pathogens (Ding and Christie, 2003; Meloni et al., 2003; Lee and Bostock, 2006). Further work on this and other secretion systems will help shed light on how this organism mediates interactions within the complex rhizosphere environment.

9 REGULATION

Our original analyses of *A tumefaciens* indicated that at least 9% of its genome was dedicated to regulation. This is consistent with observations by Stover et al. (2000) who suggest that bacteria that inhabit diverse environments tend to have large complements of regulatory genes. This regulatory capacity likely facilitates survival of *A. tumefaciens* within the dynamic soil and rhizosphere environments. The availability of the genome has facilitated numerous large scale studies to address regulatory issues.

An oligonucleotide-based microarray is available from Agilent biotechnologies and has been used to identify genes under the control of the master VirG virulence regulon (Wood DW, Monks D, Houmiel K, Monks S, Tompa M, Bumgamer R, Slater SC and Nester EW, unpublished data). Preliminary data from this array indicated that the *repABC* system required for replication of the Ti plasmid was under the control of VirG. Subsequent work confirmed this observation and described additional novel genes controlled by acetosyringone induction on both the A6 and C58 Ti plasmids (Cho and Winans, 2005). Interestingly, a recent report by Liu and Nester suggests that indole acetic acid produced by the tumor effectively represses *vir* gene expression (Liu and Nester, 2006). These results suggest an opposing regulatory mechanism that functions to shut off the *vir* system following tumor formation. A number of proteomics analyses have subsequently identified C58 gene products induced by heat shock and plant signal molecules (Rosen et al., 2002; Rosen et al., 2003; Rosen et al., 2004).

Karlin and colleagues have also used the availability of the C58 genome sequence to identify highly transcribed genes from C58 and other α-proteobacteria, among which were genes of the TCA cycle, aerobic respiration and ribosomal function (Karlin et al., 2003). These studies were later extended to other plant associated bacteria including *Ralstonia solanacearum* and *Pseudomonas syringae* (Fu et al., 2005).

We noted the presence of numerous nucleotide cyclases in the plant symbionts *S. meliloti* and *M. loti* and in the evolutionarily distinct human pathogen, *Mycobacterium tuberculosis* in our original analysis of the genome. These nucleotide cyclases have been noted previously in *S. meliloti* and were postulated to function in signal transduction (Galibert et al., 2001). Contrary to our expectation, only three proteins of this class were identified in *A. tumefaciens*. It is unclear at this point why the nitrogen-fixing plant symbionts share similarly large numbers of nucleotide cyclases with a human pathogen, whereas few such genes are found in the evolutionarily related *A. tumefaciens*. Since that observation at least one additional nucleotide cyclase, conserved in C58, has been examined in *Rhizobium etli* (Tellez-Sosa et al., 2002). While it is able to functionally complement a *cya* mutant of *E. coli*, it has no discernable role in *R. etli*. The authors speculate that given the large number of nucleotide cyclases, there may be extensive functional redundancy among these systems. More detailed analyses are required in order to tease out the specific functionality of these systems in the Rhizobiaceae.

The genome of *A. tumefaciens* encodes two new bacterial phytochromes. Typically, bacterial phytochromes contain the sensory portion of the protein, including the tetrapyrrole chromophore-binding site, attached to a histidine kinase domain. One of the *A. tumefaciens* phytochromes has this structure and its gene is in a putative operon with a partner response regulator. The other phytochrome is itself a response regulator. The *Agrobacterium* bacteriophytochromes have since become the subject of intensive study (Lamparter et al., 2002; Karniol and Vierstra, 2003; Lamparter et al., 2003; Karniol and Vierstra, 2004; Lamparter, 2004; Borucki et al., 2005; Inomata et al., 2005; Lamparter and Michael, 2005; Lamparter, 2006; Oberpichler et al., 2006). These studies have characterized in detail the crystal structure, chromophore binding, spectral properties, and histidine kinase properties of these proteins. However, phenotypes associated with deletion mutants have not yet been identified.

10 RESPONSE TO PLANT DEFENSES

A recent study (Chevrot et al., 2006) showed that γ-aminobutryic acid (GABA) produced by the plant in wounded tissue can affect virulence of *A. tumefaciens*. GABA is a well-known signaling molecule in bacteria, plants and animals that is involved in cell-cell interactions in development and response to stress. The Chevrot study showed for the first time that GABA can signal between a bacterial pathogen and its host. The bacterial quorum-sensing signal homoserine lactone (OC8-HSL) is inactivated via the *attM* pathway in a GABA-responsive manner. Induction of the *attKLM* operon was demonstrated and a moderate decrease in tumorigenicity was documented in transgenic tobacco plants that overproduce GABA. The authors propose several models by which inactivation of OC8-HSL might affect *A. tumefaciens* interaction with its environment, including reduction of Ti plasmid conjugation, suppression of quorum sensing by potential competitor bacteria, and reduction of HSL-induced plant defense responses.

A major plant defense against bacterial infection is production of oxidative bursts (primarily H_2O_2) designed to cripple the invading organisms. Several recent studies have focused on *Agrobacterium*'s defense against oxidation agents. Ceci et al. (2003) solved the crystal structure of the *A. tumefaciens* Dps protein, which protects DNA from damage by oxidation. This protein was discovered in *E. coli* and is required for viability in stationary phase. While the protective qualities of the *E. coli* protein were attributed to DNA-binding activity and oxidation of bound iron (Almiron et al., 1992; Zhao et al., 2002), Ceci et al. (2003) showed that the *Agrobacterium* enzyme protects against oxidation without binding DNA.

Several additional studies have focused on regulation of the oxidative response in *A. tumefaciens*, and the role of catalase in resistance to oxidative stress (Ceci et al., 2003; Eiamphungporn et al., 2003; Nakjarung et al., 2003; Prapagdee et al., 2004a; Prapagdee et al., 2004b; Chuchue et al., 2006). These studies disrupted the regulatory genes *oxyR* and *oxyS*, plus the *katA* and *catE* genes that encode a bifunctional catalase-peroxidase and monofunctional catalase, respectively. They demonstrate that both the *katA* and *catE* genes are induced by superoxide via the OxyR protein, that the KatA protein is primarily responsible for resistance to H_2O_2, and that the CatE protein serves a supplementary role to KatA. A mutation in the *katA* gene results in an avirulent strain (Xu et al., 2001).

Chuchue et al. (2006) analyzed *A. tumefaciens ohr*, a gene originally identified in *Xanthomonas campestris* as an organic hydroperoxidase resistance protein (Mongkolsuk et al., 1998). They determined that *ohr* performs the same function in *A. tumefaciens*, and that it is regulated by the adjacent gene, *ohrR*. They also disrupted five additional genes predicted to have a similar activity, demonstrating that *ohr* is *A. tumefaciens*' primary (but not sole) means of resistance to organic hydroperoxides. Since their work did not assign a specific activity to any of the five additional genes, all remain candidates as additional enzymes capable of degrading hydroperoxides. It is possible that these genes may have overlapping functions.

11 GENERAL METABOLISM

Annotation of the *A. tumefaciens* genome provides broad insight into prototrophic growth mechanisms (Goodner et al., 1999; Goodner et al., 2001; Wood et al., 2001). Entry ways for inorganic and organic sources of nitrogen, sulfur, and phosphate are present. C58 was known to lack catechol and hydroxamate-type siderophores (Penyalver et al., 2001). The genome sequence suggests that the bacterium can scavenge iron from other organisms by transport of iron-chelate complexes such as ferric citrate (Page and Dale, 1986). More recently, a huge gene cluster has been analyzed that encodes a novel hybrid nonribosomal peptide-polyketide siderophore produced by C58 under iron limitation (Rondon et al., 2004). An unrelated siderophore identified in *A. tumefaciens* MAFF0301001 by Sonoda et al. (2002) is not present in *A. tumefaciens* C58.

Complete biosynthetic pathways for amino acids, nucleotides, lipids, vitamins, and cofactors are encoded by chromosomal genes. One interesting sidelight is that C58 uses only the vitamin B12-dependent branch of methionine synthesis involving the MetH protein, however the organism can synthesize vitamin B12 itself. In terms of central metabolism, *A. tumefaciens* has the enzymatic machinery for the Embden-Meyeroff (glycolysis), pentose phosphate, and Entner-Doudoroff pathways, but prefers the Entner-Doudoroff pathway for glucose catabolism (Fuhrer et al., 2005). The Embden-Meyeroff and pentose phosphate pathways may be more important for intermediary metabolism leading to biosynthetic pathways and for scavenging biological forms of sulfur (Roy et al., 2003). *A. tumefaciens* C58 is known to grow on glycerol or ethanol, so gluconeogenesis is also a possible role for the Embden-Meyeroff pathway. However, the genome lacks a homolog for all seven known fructose-1,6-bisphosphatase types suggesting a different enzyme is involved in this gluconeogenic step (Csonka et al., 2005). Based on codon usage, the genes encoding components of the TCA cycle and aerobic respiration are predicted to be highly expressed, well in line with the strong aerobic growth of C58 (Karlin et al., 2003). There is considerable gene redundancy for TCA enzymes in C58 with 4 citrate synthase homologs, 4 malate dehydrogenase homologs, and 2 aconitases, however evidence is now available that there is not functional redundancy for citrate synthase and aconitase (Suksomtip et al., 2005; Whiteside D, Johnson T, Collins J, Kuhns JM, Ohlin V, Law T, Yasin R, Dottle G, Livingston L, Wheeler C and Goodner B, unpublished data). C58 does contain the *fixNOQP* operon which should allow it to grow under microaerobic conditions (Lopez et al., 2001). Under anaerobic conditions, the only well established growth route is anaerobic respiration using nitrate as the terminal electron acceptor (see below). C58 does contain enzymes, such as lactate dehydrogenases, that might allow for some fermentation. While there is some evidence for fermentation under certain conditions in related organisms, there is no experimental evidence for fermentation in *Agrobacterium* (Sardesai and Babu, 2000).

Carbon metabolism is very broad in *A. tumefaciens* and the C58 genome sequence predicts many routes including those for the plant-derived sugars fructose, sucrose, ribose, xylose, xylulose and lactose as well as compounds such as myo-inositol, hydantoin, urea and glycerol. The capacity to metabolize glucuronate, galactonate, galactarate, gluconate, ribitol, glycogen, quinate, L-idonate, creatinine, stachydrine, ribosylnicotinamide and 4-hydroxymandelate is also implied by the genome content. The ability to break down and possibly metabolize plant-derived polymers such as hemicellulose, pectin, lignin, and tannin has some bioinformatics support as well. Chemotaxis systems corresponding to many of these compounds have been experimentally verified in *A. tumefaciens* (Ashby et al., 1988). Sucrose is a preferred growth substrate and chemoattractant, and there are at least 4 putative enzymatic routes for its degradation – 2 α-glycosidases, a sucrose hydrolase, and a novel route involving the oxidation of sucrose to 3-ketosucrose (Schuerman et al., 1997; Willis and Walker, 1999). Evidence exists that there is some functional redundancy for sucrose degradation, but in addition sucrose metabolism is linked to other cellular processes such as osmoregulation (Smith et al., 1990; Goodner B, Hardesty J, Edwards J, Reed A, Mateo V, Shelton B and Wheeler C, unpublished data). Experimental evidence has now linked C58 genes directly or indirectly to palatinose, rhamnose, Amadori compounds, and alginate catabolism (De Costa et al., 2003; Richardson et al., 2004; Baek et al., 2005; Baek and Shapleigh, 2005; Ochiai et al., 2006a; Ochiai et al., 2006b). Other carbon metabolic routes that have been experimentally tested since the C58 genome was published include the conversion of fructose to psicose, glycogen synthesis, hydantoin racemization, methyl and the erythritol phosphate pathway (Ugalde et al., 2003; Martinez-Rodriguez et al., 2004; Kim et al., 2006; Lherbet et al., 2006).

Nitrogen metabolism in *Agrobacterium* has long been overshadowed by its N-fixing cousins *Rhizobium*, *Sinorhizobium*, and *Mesorhizobium*. However, it is clear that nitrogen metabolism in *Agrobacterium* is not a minor story and may have some real goldmines of basic knowledge and commercial application (Cheneby et al., 2004). For example, the C58 genome contains the genes for two nitrate reductases, one an assimilatory type (NAS) and the other a potential bifunctional type (NAP) (Richardson et al., 2001). The extent to which these two nitrate reductases divide up or share the nitrogen assimilation and anaerobic respiration duties is currently under investigation (Baek and Shapleigh, 2005; Abraham N, Bennett I, Wheeler C and Goodner B, unpublished data). Recently, nitrate and nitrite have been shown to be chemoattractants for *A. tumefaciens* (Lee et al., 2002). As another example, the C58 genome encodes six different glutamine synthetases, two different glutamate synthases plus an orphan glutamate synthase large subunit, and a glutamate dehydrogenase. As a final example, biovar I strains of *A. tumefaciens* can grow on media lacking an exogenous nitrogen source, especially under low oxygen tensions (Abraham N, Bennett I, Wheeler C and Goodner B, unpublished data). This growth can occur repeatedly through serial transfer and it is enough growth to entice those interested in nitrogen fixation (Kanvinde and Sastry, 1990). Given that the C58 genome lacks genes for nitrogenase, it seems likely that this is an example of high efficiency scavenging of nitrogenous compounds. There are orthologs of a high affinity urea ABC-type transporter (*urtABCDE*, Valladares et al., 2002), a high affinity ammonium transporter (*amtB*, Meletzus et al., 1998), and a formate dehydrogenase and urease that might also serve as a methylenediurease system for the degradation of methyleneureas (Jahns et al., 1998). There are also weak orthologs of several enzymes that comprise the cyanide oxygenase complex in *Pseudomonas flourescens* (Fernandez and Kunz, 2005). That said, it remains possible that a novel nitrogen fixation system exists in *Agrobacterium*, but highly efficient nitrogen scavenging is the hypothesis of choice for the time being.

Once a virulent *A. tumefaciens* strain, such as C58, initiates tumor formation, the bacterium can also benefit from the proprietary carbon and nitrogen sources called opines that are produced by the transformed plant cells (Bevan and Chilton, 1982). In C58, the Ti plasmid pTiC58 carries most of the genes necessary for utilizing nopaline an agrocinopine.

However, it is known that C58 derivatives lacking the pTiC58 replicon can take up octopine and nopaline without subsequent hydrolysis. Spontaneous mutants have been found that now express the ability to utilize octopine or mannopine/mannopinic acid (LaPointe et al., 1992). The C58 genome contains many ABC-type transport systems for amino acids and their derivatives that may allow for the uptake of opines. Both the linear chromosome and the cryptic plasmid pAtC58 harbor strong homologs of the *agaE* gene that encodes an enzyme involved in mannopine metabolism (Lyi et al., 1999).

12 CONCLUSIONS

The most striking finding from our original analyses was the extensive similarity of the circular chromosomes of *A. tumefaciens* and the plant symbiont *S. meliloti*, supporting the view that these bacteria originated from a recent common ancestor. The mosaic structure of the *A. tumefaciens* linear chromosome and plasmids, predominantly composed of orthologs found on each of the *S. meliloti* replicons, suggests that these organisms diverged following acquisition of the pSymA and pSymB ancestral molecules by this progenitor. Subsequent sequencing of the Brucella suis genome by Paulsen et al. (2002) extended these analyses and showed similar synteny between these organisms and chromosome I of *B. suis* and *B. melitensis*. These findings, along with information on the gene content of these three organisms, suggest that they all arose from a common plant-associated ancestor. The intriguing divergence of this ancestor along distinctly different lineages dedicated to plant symbiosis, plant pathogenesis and animal pathogenesis provides a fascinating model system from which to investigate bacterial evolution. Ongoing work by us and others on the genome analysis of *Agrobacterium* biovars I, II and III representatives, plus the recently published genomes of *Rhizobium etli* (Gonzalez et al., 2006) and *Rhizobium leguminosarum* (Young et al., 2006) should present a wealth of new data for this investigation.

A significant body of literature has been built upon the foundation of the *Agrobacterium* genome sequence. This is representative of the explosion of literature that accompanies the completion of most model genomes and strongly supports the need for continued research into new and more efficient sequencing technologies. Methods to elucidate the functional and evolutionary relationships of these genome systems must expand as well. Such functional analyses, on both the global and individual gene scales, are critical to expand our understanding of key biological systems and represent the next challenge of the genomic revolution.

Table 4-1. Agrobacterium tumefaciens C58 genes discussed in this chapter

Gene or Protein Name	Atu Number[a]	AGR_Number[b]
new Type IV secretion system on pAtC58 (*avhB* operon)	5162-5172	pAT_218-pAT_bx65
rctA	5160	pAT_217
rctB	5116	pAT_169
virH1	6150	pTi_272
virH2	6151	pTi_273
telA	2523	C_4584
chromosome II *repABC* operon	3924-3922	L_1843-L_1847
chromosome II *traA*	4855	L_66
chromosome II *traG*	4858	L_58
chromosome II *mobC*	4857	L_60
chromosome I *dnaE*	1292	C_2379
chromosome II *dnaE*	3228	L_3173
pAtC58 *dnaE*	5100	pAT_bx5
pTiC58 *dnaE*	6093	pTi_175
chvA	2728	C_4944
chvB	2730	C_4949
chvGI operon	0033-0034	C_52-C_54
chvE	2348	C_4267
ros	916	C_1669
chvD	2125	C_3855
chvH	2553	C_4625
acvB	2522	C_4582
exoC	4074	L_1564
cel gene cluster	3302-3309	L_3032-L_3021
hspC	375	C_657
hspL	3887	L_1921
hspAT1	5052	pAT_69
hspAT2	5449	pAT_660
rpoH	2445	C_4439
kdgF	3145	L_3329
ligE	1121	C_2076
endo-1,4-beta-xylanase	2371	C_4304
pecSM operon	0272-0273	C_466-C_468
bacA	2304	C_4191
htrA family member	1915	C_3507
htrA family member	2043	C_3700
ialA	2772	C_5030
ialB	3275	L_3087
hemolysin	359	C_627
hemolysin	736	C_1334

(*Continued*)

Gene or Protein Name	Atu Number[a]	AGR_Number[b]
hemolysin	3732	L_2203
mviN	347	C_608
autotransporting virulence factor family member	5354	pAT_511
autotransporting virulence factor family member	5364	pAT_528
icmF	4332	L_1062
virG	6178	pTi_15
adenylate cyclase	1149	C_2127
adenylate cyclase	2277	C_4137
adenylate cyclase	2580	C_4673
adenylate cyclase	4013	L_1679
bacterial phytochrome	1990	C_3620
bacterial phytochrome	2165	C_3927
attKLM operon	5137-5139	pAT_197-pAT_200
dps	2477	C_4495
oxyR	4641	L_484
katA	4642	L_481
catE	5491	pAT_722
ohr	847	C_1547
ohrR	846	C_1544
siderophore biosynthetic gene cluster	3668-3691	L_2335-L_2292
metH	2155	C_3907
verified citrate synthase	1392	C_2572
citrate synthase-related gene	4851	L_71
citrate synthase-related gene	5306	pAT_439
citrate synthase-related gene	5307	pAT_441
putative malate dehydrogenase	164	C_268
putative malate dehydrogenase	2639	C_4782
putative malate dehydrogenase	3208	L_3209
putative malate dehydrogenase	4676	L_410
aconitase A	2685	C_4866
aconitase B	4734	L_294
fixNOQP operon	1537-1534	C_2835-C_2829
NAS-type nitrate reductase operon	3900-3899	L_1895-L_1897
NAP-type nitrate reductase operon	4405-4410	L_921-L_913
α-glucosidase (with associated transport operon)	594	C_1051
α-glucosidase (orphan)	2295	C_4169
sucrose hydrolase	944	C_1721
putative glutamine synthetase	193	C_326
putative glutamine synthetase	602	C_1068
putative glutamine synthetase I	1770	C_3253

Gene or Protein Name	Atu Number[a]	AGR_Number[b]
putative glutamine synthetase	2142	C_3883
putative glutamine synthetase II	2416	C_4385
putative glutamine synthetase III	4230	L_1262
orphan glutamate synthase large subunit	145	C_235
bacterial-type glutamate synthase operon	3783-3784	L_2104-L2101
archaeal-type glutamate synthase operon	4227-4229	L_1268-L_1265
glutamate dehydrogenase	2766	C_5015
urtABCDE operon	2414-2410	C_4380-C_4373
amtB	2758	C_5001
urease operon	2401-2408	C_4357-C_4369
putative *agaE*	3050	L_3510
putative *agaE*	3414	L_2815
putative *agaE*	5003	pAT_5

[a] Gene numbering system used by Wood et al. (2001) and adopted for our subsequent genome consortium reannotation of the *A. tumefaciens* C58 genome.
[b] Gene numbering system used by Goodner et al. (2001).

13 ACKNOWLEDGEMENTS

This work has been supported by grant number 0333297 from the National Science Foundation to the authors, Grant number 52005125 from the Howard Hughes Medical Institute to B.G. at Hiram College, and by Monsanto Company.

14 REFERENCES

Abella M, Erill I, Jara M, Mazon G, Campoy S, Barbe J (2004) Widespread distribution of a *lexA*-regulated DNA damage-inducible multiple gene cassette in the *Proteobacteria phylum*. Mol Microbiol 54: 212-222

Allardet-Servent A, Michaux-Charachon S, Jumas-Bilak E, Karayan L, Ramuz M (1993) Presence of one linear and one circular chromosome in the *Agrobacterium tumefaciens* C58 genome. J Bacteriol 175: 7869-7874

Almiron M, Link AJ, Furlong D, Kolter R (1992) A novel DNA-binding protein with regulatory and protective roles in starved *Escherichia coli*. Genes Dev 6: 2646-2654

Ashby AM, Watson MD, Loake GJ, Shaw CH (1988) Ti plasmid-specified chemotaxis of *Agrobacterium tumefaciens* C58C1 toward vir-inducing phenolic compounds and soluble factors from monocotyledonous and dicotyledonous plants. J Bacteriol 170: 4181-4187

Baek CH, Farrand SK, Park DK, Lee KE, Hwang W, Kim KS (2005) Genes for utilization of deoxyfructosyl glutamine (DFG), an amadori compound, are widely dispersed in the family Rhizobiaceae. FEMS Microbiol Ecol 53: 221-233

Baek SH, Shapleigh JP (2005) Expression of nitrite and nitric oxide reductases in free-living and plant-associated *Agrobacterium tumefaciens* C58 cells. Appl Environ Microbiol 71: 4427-4436

Balsiger S, Ragaz C, Baron C, Narberhaus F (2004) Replicon-specific regulation of small heat shock genes in *Agrobacterium tumefaciens*. J Bacteriol 186: 6824-6829

Bao K, Cohen SN (2003) Recruitment of terminal protein to the ends of *Streptomyces* linear plasmids and chromosomes by a novel telomere-binding protein essential for linear DNA replication. Genes Dev 17: 774-785

Baron C, Domke N, Beinhofer M, Hapfelmeier S (2001) Elevated temperature differentially affects virulence, VirB protein accumulation, and T-pilus formation in different *Agrobacterium tumefaciens* and *Agrobacterium vitis* strains. J Bacteriol 183: 6852-6861

Bartosik D, Baj J, Piechucka E, Waker E, Wlodarczyk M (2002) Comparative characterization of *repABC*-type replicons of *Paracoccus pantotrophus* composite plasmids. Plasmid 48: 130-141

Bevan MW, Chilton M-D (1982) T-DNA of the *Agrobacterium* Ti and Ri plasmids. Annu Rev Genet 16: 357-384

Binns AN, Thomashow MF (1988) Cell biology of *Agrobacterium* infection and transformation of plants. Annu Rev Microbiol 42: 575-606

Borucki B, von Stetten D, Seibeck S, Lamparter T, Michael N, Mroginski MA, Otto H, Murgida DH, Heyn MP, Hildebrandt P (2005) Light-induced proton release of phytochrome is coupled to the transient deprotonation of the tetrapyrrole chromophore. J Biol Chem 280: 34358-34364

Boussau B, Karlberg EO, Frank AC, Legault B-A, Andersson SG (2004) Computational inference of scenarios for alpha-proteobacterial genome evolution. Proc Natl Acad Sci USA 101: 9722-9727

Brassinga AKC, Siam R, McSween W, Winkler H, Wood D, Marczynski GT (2002) Conserved response regulator CtrA and IHF binding sites in the alpha-proteobacteria Caulobacter crescentus and *Rickettsia prowazekii* chromosomal replication origins. J Bacteriol 184: 5789-5799

Capela D, Barloy-Hubler F, Gouzy J, Bothe G, Ampe F, Batut J, Boistard P, Becker A, Boutry M, Cadieu E, Dreano S, Gloux S, Godrie T, Goffeau A, Kahn D, Kiss E, Lelaure V, Masuy D, Pohl T, Portetelle D, Puhler A, Purnelle B, Ramsperger U, Renard C, Thebault P, Vandenbol M, Weidner S, Galibert F (2001) Analysis of the chromosome sequence of the legume symbiont *Sinorhizobium meliloti* strain 1021. Proc Natl Acad Sci USA 98: 9877-9882

Casjens S, Murphy M, DeLange M, Sampson L, van Vugt R, Huang WM (1997) Telomeres of the linear chromosomes of Lyme disease spirochaetes: nucleotide sequence and possible exchange with linear plasmid telomeres. Mol Microbiol 26: 581-596

Ceci P, Ilari A, Falvo E, Chiancone E (2003) The Dps protein of *Agrobacterium tumefaciens* does not bind to DNA but protects it toward oxidative cleavage: x-ray crystal structure, iron binding, and hydroxyl-radical scavenging properties. J Biol Chem 278: 20319-20326

Cevallos MA, Porta H, Izquierdo J, Tun-Garrido C, Garcia-de-los-Santos A, Davila G, Brom S (2002) Rhizobium etli CFN42 contains at least three plasmids of the repABC family: a structural and evolutionary analysis. Plasmid 48: 104-116

Chaconas G (2005) Hairpin telomeres and genome plasticity in *Borrelia*: all mixed up in the end. Mol Microbiol 58: 625-635

Chen LS, Chen YC, Wood DW, Nester EW (2002) A new type IV secretion system promotes conjugal transfer in *Agrobacterium tumefaciens*. J Bacteriol 184: 4838-4845

Cheneby D, Perrez S, Devroe C, Hallet S, Couton Y, Bizouard F, Iuretig G, Germon JC, Philippot L (2004) Denitrifying bacteria in bulk and maize-rhizospheric soil: diversity and N2O-reducing abilities. Can J Microbiol 50: 469-474

Chevrot R, Rosen R, Haudecoeur E, Cirou A, Shelp BJ, Ron E, Faure D (2006) GABA controls the level of quorum-sensing signal in *Agrobacterium tumefaciens*. Proc Natl Acad Sci USA 103: 7460-7464

Cho H, Winans SC (2005) VirA and VirG activate the Ti plasmid *repABC* operon, elevating plasmid copy number in response to wound-released chemical signals. Proc Natl Acad Sci USA 102: 14843-14848

Chuchue T, Tanboon W, Prapagdee B, Dubbs JM, Vattanaviboon P, Mongkolsuk S (2006) ohrR and ohr are the primary sensor/regulator and protective genes against organic hydroperoxide stress in *Agrobacterium tumefaciens*. J Bacteriol 188: 842-851

Cooley MB, Kado CI (1991) Mapping of the *ros* virulence regulatory gene of *A. tumefaciens*. Mol Gen Genet 230: 24-27

Copeland A, Lucas S, Lapidus A, Barry K, Detter JC, Glavina T, Hammon N, Israni S, Pitluck S, Richardson P, Mackenzie C, Choudhary M, Larimer F, Hauser LJ, Land M, Donohue TJ, Kaplan S (2005) Complete Sequence of Chromosome 1 of Rhodobacter sphaeroides 2.4.1. Unpublished, but available via GenBank at http://www.ncbi.nlm.nih.gov/entrez/query.fcgi?db=genome&cmd=Retrieve&dopt=Overview&list_uids=18843

Csonka LN, O'Connor K, Larimer F, Richardson P, Lapidus A, Ewing AD, Goodner BW, and Oren A (2005) What we can deduce about metabolism in the moderate halophile *Chromohalobacter salexigens* from its genomic sequence? In NA Oren, A Plemenita, eds, Adaptation To Life At High Salt Concentrations In Archaea, Bacteria, and Eukarya. Springer, Dordrecht

Das S, Choudhuri K (2003) Identification of a unique IAHP (IcmF-associated homologous proteins) cluster in *Vibrio cholerae* and other proteobacteria through *in silico* analysis. In Silico Biol 3: 287-300

De Costa DM, Suzuki K, Yoshida K (2003) Structural and functional analysis of a putative gene cluster for palatinose transport on the linear chromosome of *Agrobacterium tumefaciens* MAFF301001. J Bacteriol 185: 2369-2373

Ding Z, Christie PJ (2003) *Agrobacterium tumefaciens* twin-arginine-dependent translocation is important for virulence, flagellation, and chemotaxis but not type IV secretion. J Bacteriol 185: 760-771

Egan ES, Fogel MA, Waldor MK (2005) Divided genomes: negotiating the cell cycle in prokaryotes with multiple chromosomes. Mol Microbiol 56: 1129-1138

Egan ES, Lobner-Olesen A, Waldor MK (2004) Synchronous replication initiation of the two *Vibrio cholerae* chromosomes. Curr Biol 14: R501-502

Eiamphungporn W, Nakjarung K, Prapagdee B, Vattanaviboon P, Mongkolsuk S (2003) Oxidant-inducible resistance to hydrogen peroxide killing in *Agrobacterium tumefaciens* requires the global peroxide sensor-regulator OxyR and KatA. FEMS Microbiol Lett 225: 167-172

Fernandez RF, Kunz DA (2005) Bacterial cyanide oxygenase is a suite of enzymes catalyzing the scavenging and adventitious utilization of cyanide as a nitrogenous growth substrate. J Bacteriol 187: 6396-6402

Fu QS, Li F, Chen LL (2005) Gene expression analysis of six GC-rich Gram-negative phytopathogens. Biochem Biophys Res Commun 332: 380-387

Fuhrer T, Fischer E, Sauer U (2005) Experimental identification and quantification of glucose metabolism in seven bacterial species. J Bacteriol 187: 1581-1590

Galhardo RS, Rocha RP, Marques MV, Menck CF (2005) An SOS-regulated operon involved in damage-inducible mutagenesis in *Caulobacter crescentus*. Nucleic Acids Res 33: 2603-2614

Galibert F, Finan TM, Long SR, Puhler A, Abola P, Ampe F, Barloy-Hubler F, Barnett MJ, Becker A, Boistard P, Bothe G, Boutry M, Bowser L, Buhrmester J, Cadieu E, Capela D, Chain P, Cowie A, Davis RW, Dreano S, Federspiel NA, Fisher RF, Gloux S, Godrie T, Goffeau A, Golding B, Gouzy J, Gurjal M, Hernandez-Lucas I, Hong A, Huizar L, Hyman RW, Jones T, Kahn D, Kahn ML, Kalman S, Keating DH, Kiss E, Komp C, Lelaure V, Masuy D, Palm C, Peck MC, Pohl TM, Portetelle D, Purnelle B, Ramsperger U, Surzycki R, Thebault P, Vandenbol M, Vorholter F-J, Weidner S, Wells DH, Wong K, Yeh K-C, Batut J (2001) The composite genome of the legume symbiont *Sinorhizobium meliloti*. Science 293: 668-672

Gonzalez V, Bustos P, Ramirez-Romero MA, Medrano-Soto A, Salgado H, Hernandez-Gonzalez I, Hernandez-Celis JC, Quintero V, Moreno-Hagelsieb G, Girard L, Rodriguez O, Flores M, Cevallos MA, Collado-Vides J, Romero D, Davila G (2003) The mosaic structure of the symbiotic plasmid of *Rhizobium etli* CFN42 and its relation to other symbiotic genome compartments. Genome Biol 4: R36

Gonzalez V, Santamaria RI, Bustos P, Hernandez-Gonzalez I, Medrano-Soto A, Moreno-Hagelsieb G, Janga SC, Ramirez MA, Jimenez-Jacinto V, Collado-Vides J, Davila G (2006) The partitioned *Rhizobium etli* genome: genetic and metabolic redundancy in seven interacting replicons. Proc Natl Acad Sci USA 103: 3834-3839

Goodner B, Hinkle G, Gattung S, Miller N, Blanchard M, Qurollo B, Goldman BS, Cao YW, Askenazi M, Halling C, Mullin L, Houmiel K, Gordon J, Vaudin M, Iartchouk O, Epp A, Liu F, Wollam C, Allinger M, Doughty D, Scott C, Lappas C, Markelz B, Flanagan C, Crowell C, Gurson J, Lomo C, Sear C, Strub G, Cielo C, Slater S (2001) Genome sequence of the plant pathogen and biotechnology agent *Agrobacterium tumefaciens* C58. Science 294: 2323-2328

Goodner BW, Markelz BP, Flanagan MC, Crowell CB, Jr., Racette JL, Schilling BA, Halfon LM, Mellors JS, Grabowski G (1999) Combined genetic and physical map of the complex genome of *Agrobacterium tumefaciens*. J Bacteriol 181: 5160-5166

Goshi K, Uchida T, Lezhava A, Yamasaki M, Hiratsu K, Shinkawa H, Kinashi H (2002) Cloning and analysis of the telomere and terminal inverted repeat of the linear chromosome of *Streptomyces griseus*. J Bacteriol 184: 3411-3415

Hamilton RC, Fall MZ (1971) The loss of tumor initiating ability in *Agrobacterium tumefaciens* by incubation at high temperature. Experientia 27: 229-230

Harvey M, McMeekin A (2004) Public-private collaborations and the race to sequence *Agrobacterium tumefaciens*. Nat Biotechnol 22: 807-810

Heidelberg JF, Eisen JA, Nelson WC, Clayton RA, Gwinn ML, Dodson RJ, Haft DH, Hickey EK, Peterson JD, Umayam L, Gill SR, Nelson KE, Read TD, Tettelin H, Richardson D, Ermolaeva MD, Vamathevan J, Bass S, Qin H, Dragoi I, Sellers P, McDonald L, Utterback T, Fleishmann RD, Nierman WC, White O, Salzberg SL, Smith HO, Colwell RR, Mekalanos JJ, Venter JC, Fraser CM (2000) DNA sequence of both chromosomes of the cholera pathogen *Vibrio cholerae*. Nature 406: 477-483

Holden MTG, Titball RW, Peacock SJ, Cerdeno-Tarraga AM, Atkins T, Crossman LC, Pitt T, Churcher C, Mungall K, Bentley SD, Sebaihia M, Thomson NR, Bason N, Beacham IR, Brooks K, Brown KA, Brown NF, Challis GL, Cherevach I, Chillingworth T, Cronin A, Crossett B, Davis P, DeShazer D, Feltwell T, Fraser A, Hance Z, Hauser H, Holroyd S, Jagels K, Keith KE, Maddison M, Moule S, Price C, Quail MA, Rabbinowitsch E, Rutherford K, Sanders M, Simmonds M, Songsivilai S, Stevens K, Tumapa S, Vesaratchavest M, Whitehead S, Yeats C, Barrell BG, Oyston PCF, Parkhill J (2004) Genomic plasticity of the causative agent of melioidosis, *Burkholderia pseudomallei*. Proc Natl Acad Sci USA 101: 14240-14245

Hooykaas PJJ, Peerbolte R, Regensburg-Tuink AJ, de Vries P, Schilperoort RA (1982) A chromosomal linkage map of *Agrobacterium tumefaciens* and a comparison with the maps of *Rhizobium* spp. Mol Gen Genet 188: 12-17

Huang WM, Davis J, Ruan Q, Aron J, Goodner B, Pride N, Henry E, Sabo A, Telepak E, Joss L and Casjens S (2006) Linear chromosome end generating system of *Agrobacterium tumefaciens* C58. Submitted

Inomata K, Hammam MAS, Kinoshita H, Murata Y, Khawn H, Noack S, Michael N, Lamparter T (2005) Sterically locked synthetic bilin derivatives and phytochrome Agp1 from *Agrobacterium tumefaciens* form photoinsensitive Pr- and Pfr-like adducts. J Biol Chem 280: 24491-24497

Jahns T, Schepp R, Siersdorfer C, Kaltwasser H (1998) Microbial urea-formaldehyde degradation involves a new enzyme, methylenediurease. Acta Biol Hung 49: 449-454

Jumas-Bilak E, Michaux-Charachon S, Bourg G, Ramuz M, Allardet-Servent A (1998) Unconventional genomic organization in the alpha subgroup of the Proteobacteria. J Bacteriol 180: 2749-2755

Kahng LS, Shapiro L (2001) The CcrM DNA methyltransferase of *Agrobacterium tumefaciens* is essential, and its activity is cell cycle regulated. J Bacteriol 183: 3065-3075

Kahng LS, Shapiro L (2003) Polar localization of replicon origins in the multipartite genomes of *Agrobacterium tumefaciens* and *Sinorhizobium meliloti*. J Bacteriol 185: 3384-3391

Kaneko T, Nakamura Y, Sato S, Minamisawa K, Uchiumi T, Sasamoto S, Watanabe A, Idesawa K, Iriguchi M, Kawashima K, Kohara M, Matsumoto M, Shimpo S, Tsuruoka H, Wada T, Yamada M, Tabata S (2002) Complete genomic sequence of nitrogen-fixing symbiotic bacterium *Bradyrhizobium japonicum* USDA110. DNA Res 9: 189-197

Kanvinde L, Sastry GR (1990) *Agrobacterium tumefaciens* is a diazotrophic bacterium. Appl Environ Microbiol 56: 2087-2092

Karlin S (2001) Detecting anomalous gene clusters and pathogenicity islands in diverse bacterial genomes. Trends Microbiol 9: 335-343

Karlin S, Barnett MJ, Campbell AM, Fisher RF, Mrazek J (2003) Predicting gene expression levels from codon biases in alpha-proteobacterial genomes. Proc Natl Acad Sci USA 100: 7313-7318

Karniol B, Vierstra RD (2003) The pair of bacteriophytochromes from *Agrobacterium tumefaciens* are histidine kinases with opposing photobiological properties. Proc Natl Acad Sci USA 100: 2807-2812

Karniol B, Vierstra RD (2004) The HWE histidine kinases, a new family of bacterial two-component sensor kinases with potentially diverse roles in environmental signaling. J Bacteriol 186: 445-453

Kim JG, Park BK, Kim SU, Choi D, Nahm BH, Moon JS, Reader JS, Farrand SK, Hwang I (2006) Bases of biocontrol: sequence predicts synthesis and mode of action of agrocin 84, the Trojan horse antibiotic that controls crown gall. Proc Natl Acad Sci USA 103: 8846-8851

Lamparter T (2004) Evolution of cyanobacterial and plant phytochromes. FEBS Lett 573: 1-5

Lamparter T (2006) A computational approach to discovering the functions of bacterial phytochromes by analysis of homolog distributions. BMC Bioinformatics 7: 141

Lamparter T, Michael N (2005) *Agrobacterium* phytochrome as an enzyme for the production of ZZE bilins. Biochemistry 44: 8461-8469

Lamparter T, Michael N, Caspani O, Miyata T, Shirai K, Inomata K (2003) Biliverdin binds covalently to *Agrobacterium* phytochrome Agp1 *via* its ring A vinyl side chain. J Biol Chem 278: 33786-33792

Lamparter T, Michael N, Mittmann F, Esteban B (2002) Phytochrome from *Agrobacterium tumefaciens* has unusual spectral properties and reveals an N-terminal chromophore attachment site. Proc Natl Acad Sci USA 99: 11628-11633

LaPointe G, Nautiyal CS, Chilton WS, Farrand SK, Dion P (1992) Spontaneous mutation conferring the ability to catabolize mannopine in *Agrobacterium tumefaciens*. J Bacteriol 174: 2631-2639

Lee DY, Ramos A, Macomber L, Shapleigh JP (2002) Taxis response of various denitrifying bacteria to nitrate and nitrite. Appl Environ Microbiol 68: 2140-2147

Lee MH, Bostock RM (2006) *Agrobacterium* T-DNA-mediated integration and gene replacement in the brown rot pathogen *Monilinia fructicola*. Curr Genet 49: 309-322

Lherbet C, Pojer F, Richard SB, Noel JP, Poulter CD (2006) Absence of substrate channeling between active sites in the *Agrobacterium tumefaciens* IspDF and IspE enzymes of the methyl erythritol phosphate pathway. Biochemistry 45: 3548-3553

Li PL, Farrand SK (2000) The replicator of the nopaline-type Ti plasmid pTiC58 is a member of the *repABC* family and is influenced by the TraR-dependent quorum-sensing regulatory system. J Bacteriol 182: 179-188

Liu P, Nester EW (2006) Indoleacetic acid, a product of transferred DNA, inhibits *vir* gene expression and growth of *Agrobacterium tumefaciens* C58. Proc Natl Acad Sci USA 103: 4658-4662

Lopez O, Morera C, Miranda-Rios J, Girard L, Romero D, Soberon M (2001) Regulation of gene expression in response to oxygen in *Rhizobium etli*: role of FnrN in fixNOQP expression and in symbiotic nitrogen fixation. J Bacteriol 183: 6999-7006

Lyi SM, Jafri S, Winans SC (1999) Mannopinic acid and agropinic acid catabolism region of the octopine-type Ti plasmid pTi15955. Mol Microbiol 31: 339-347

MacLellan SR, Smallbone LA, Sibley CD, Finan TM (2005) The expression of a novel antisense gene mediates incompatibility within the large *repABC* family of alpha-proteobacterial plasmids. Mol Microbiol 55: 611-623

MacLellan SR, Zaheer R, Sartor AL, MacLean AM, Finan TM (2006) Identification of a megaplasmid centromere reveals genetic structural diversity within the *repABC* family of basic replicons. Mol Microbiol 59: 1559-1575

Marczynski GT, Shapiro L (2002) Control of chromosome replication in *Caulobacter crescentus*. Annu Rev Microbiol 56: 625-656

Martinez-Rodriguez S, Las Heras-Vazquez FJ, Clemente-Jimenez JM, Rodriguez-Vico F (2004) Biochemical characterization of a novel hydantoin racemase from *Agrobacterium tumefaciens* C58. Biochimie 86: 77-81

Matthysse AG, Kijne JW (1998) Attachment of Rhizobiaceae to plant cells. In HP Spaink, A Kondorosi, PJJ Hooykaas, eds, The Rhizobiaceae: Molecular Biology of Model Plant-Associated Bacteria. Kluwer Academic Publishers, Dordrecht/Boston/London, pp 235-249

Matthysse AG, White S, Lightfoot R (1995) Genes required for cellulose synthesis in *Agrobacterium tumefaciens*. J Bacteriol 177: 1069-1075

Meletzus D, Rudnick P, Doetsch N, Green A, Kennedy C (1998) Characterization of the *glnK-amtB* operon of *Azotobacter vinelandii*. J Bacteriol 180: 3260-3264

Meloni S, Rey L, Sidler S, Imperial J, Ruiz-Argueso T, Palacios JM (2003) The twin-arginine translocation (Tat) system is essential for Rhizobium-legume symbiosis. Mol Microbiol 48: 1195-1207

Miller IS, Fox D, Saeed N, Borland PA, Miles CA, Sastry GR (1986) Enlarged map of *Agrobacterium tumefaciens* C58 and the location of the chromosomal regions which affect tumorigenicity. Mol Gen Genet 205: 153-159

Mongkolsuk S, Praituan W, Loprasert S, Fuangthong M, Chamnongpol S (1998) Identification and characterization of a new organic hydroperoxide resistance (ohr) gene with a novel pattern of oxidative stress regulation from *Xanthomonas campestris* pv. *phaseoli*. J Bacteriol 180: 2636-2643

Mougous JD, Cuff ME, Raunser S, Shen A, Zhou M, Gifford CA, Goodman AL, Joachimiak G, Ordonez CL, Lory S, Walz T, Joachimiak A, Mekalanos JJ (2006) A virulence locus of *Pseudomonas aeruginosa* encodes a protein secretion apparatus. Science 312: 1526-1530

Moreno E (1998) Genome evolution within the alpha *Proteobacteria*: why do some bacteria not possess plasmids and others exhibit more than one different chromosome? FEMS Microbiol Rev 22: 255-275

Nair GR, Liu ZU, Binns AN (2003) Re-examining the role of the cryptic plasmid pAtC58 in the virulence of *Agrobacterium tumefaciens* strain C58. Plant Physiol 133: 989-999

Nakjarung K, Mongkolsuk S, Vattanaviboon P (2003) The *oxyR* from *Agrobacterium tumefaciens*: evaluation of its role in the regulation of catalase and peroxide responses. Biochem Biophys Res Commun 304: 41-47

Natera SH, Guerreiro N, Djordjevic MA (2000) Proteome analysis of differentially displayed proteins as a tool for the investigation of symbiosis. Mol Plant Microbe Interact 13: 995-1009

Nocker A, Hausherr T, Balsiger S, Krstulovic NP, Hennecke H, Narberhaus F (2001a) A mRNA-based thermosensor controls expression of rhizobial heat shock genes. Nucleic Acids Res 29: 4800-4807

Nocker A, Krstulovic NP, Perret X, Narberhaus F (2001b) ROSE elements occur in disparate rhizobia and are functionally interchangeable between species. Arch Microbiol 176: 44-51

Oberpichler I, Molina I, Neubauer O, Lamparter T (2006) Phytochromes from *Agrobacterium tumefaciens*: difference spectroscopy with extracts of wild type and knockout mutants. FEBS Lett 580: 437-442

Ochiai A, Hashimoto W, Murata K (2006a) A biosystem for alginate metabolism in *Agrobacterium tumefaciens* strain C58: Molecular identification of Atu3025 as an exotype family PL-15 alginate lyase. Res Microbiol 157: 642-649

Ochiai A, Yamasaki M, Mikami B, Hashimoto W, Murata K (2006b) Crystallization and preliminary X-ray analysis of an exotype alginate lyase Atu3025 from *Agrobacterium tumefaciens* strain C58, a member of polysaccharide lyase family 15. Acta Crystallograph Sect F Struct Biol Cryst Commun 62: 486-488

Page WJ, Dale PL (1986) Stimulation of *Agrobacterium tumefaciens* growth by *Azotobacter vinelandii* Ferrisiderophores. Appl Environ Microbiol 51: 451-454

Pappas KM, Winans SC (2003) The RepA and RepB autorepressors and TraR play opposing roles in the regulation of a Ti plasmid *repABC* operon. Mol Microbiol 49: 441-455

Paulsen IT, Seshadri R, Nelson KE, Eisen JA, Heidelberg JF, Read TD, Dodson RJ, Umayam L, Brinkac LM, Beanan MJ, Daugherty SC, Deboy RT, Durkin AS, Kolonay JF, Madupu R, Nelson WC, Ayodeji B, Kraul M, Shetty J, Malek J, Van Aken SE, Riedmuller S, Tettelin H, Gill SR, White O, Salzberg SL, Hoover DL, Lindler LE, Halling SM, Boyle SM, Fraser CM (2002) The *Brucella suis* genome reveals fundamental similarities between animal and plant pathogens and symbionts. Proc Natl Acad Sci USA 99: 13148-13153

Penyalver R, Oger P, Lopez MM, Farrand SK (2001) Iron-binding compounds from *Agrobacterium* spp.: biological control strain *Agrobacterium rhizogenes* K84 produces a hydroxamate siderophore. Appl Environ Microbiol 67: 654-664

Perez-Mendoza D, Sepulveda E, Pando V, Munoz S, Nogales J, Olivares J, Soto MJ, Herrera-Cervera JA, Romero D, Brom S, Sanjuan J (2005) Identification of the *rctA* gene, which is required for repression of conjugative transfer of rhizobial symbiotic megaplasmids. J Bacteriol 187: 7341-7350

Pischl DL, Farrand SK (1984) Characterization of transposon Tn5-facilitated donor strains and development of a chromosomal linkage map for *Agrobacterium tumefaciens*. J Bacteriol 159: 1-8

Prapagdee B, Eiamphungporn W, Saenkham P, Mongkolsuk S, Vattanaviboon P (2004a) Analysis of growth phase regulated KatA and CatE and their physiological roles in determining hydrogen peroxide resistance in *Agrobacterium tumefaciens*. FEMS Microbiol Lett 237: 219-226

Prapagdee B, Vattanaviboon P, Mongkolsuk S (2004b) The role of a bifunctional catalase-peroxidase KatA in protection of *Agrobacterium tumefaciens* from menadione toxicity. FEMS Microbiol Lett 232: 217-223

Pukatzki S, Ma AT, Sturtevant D, Krastins B, Sarracino D, Nelson WC, Heidelberg JF, Mekalanos JJ (2006) Identification of a conserved bacterial protein secretion system in *Vibrio cholerae* using the *Dictyostelium* host model system. Proc Natl Acad Sci USA 103: 1528-1533

Ravin NV, Kuprianov VV, Gilcrease EB, Casjens SR (2003) Bidirectional replication from an internal ori site of the linear N15 plasmid prophage. Nucleic Acids Res 31: 6552-6560

Richardson DJ, Berks BC, Russell DA, Spiro S, Taylor CJ (2001) Functional, biochemical and genetic diversity of prokaryotic nitrate reductases. Cell Mol Life Sci 58: 165-178

Richardson JS, Hynes MF, Oresnik IJ (2004) A genetic locus necessary for rhamnose uptake and catabolism in *Rhizobium leguminosarum* bv. *trifolii*. J Bacteriol 186: 8433-8442

Robertson JL, Holliday T, Matthysse AG (1988) Mapping of *Agrobacterium tumefaciens* chromosomal genes affecting cellulose synthesis and bacterial attachment to host cells. J Bacteriol 170: 1408-1411

Rondon MR, Ballering KS, Thomas MG (2004) Identification and analysis of a siderophore biosynthetic gene cluster from *Agrobacterium tumefaciens* C58. Microbiology 150: 3857-3866

Rosen R, Buttner K, Becher D, Nakahigashi K, Yura T, Hecker M, Ron EZ (2002) Heat shock proteome of *Agrobacterium tumefaciens*: evidence for new control systems. J Bacteriol 184: 1772-1778

Rosen R, Matthysse AG, Becher D, Biran D, Yura T, Hecker M, Ron EZ (2003) Proteome analysis of plant-induced proteins of *Agrobacterium tumefaciens*. FEMS Microbiol Ecol 44: 355-360

Rosen R, Sacher A, Shechter N, Becher D, Buttner K, Biran D, Hecker M, Ron EZ (2004) Two-dimensional reference map of *Agrobacterium tumefaciens* proteins. Proteomics 4: 1061-1073

Rosenberg C, Huguet T (1984) The pAtC58 plasmid of *Agrobacterium tumefaciens* is not essential for tumour induction. Mol Gen Genet 196: 533-536

Roy AB, Hewlins MJ, Ellis AJ, Harwood JL, White GF (2003) Glycolytic breakdown of sulfoquinovose in bacteria: a missing link in the sulfur cycle. Appl Environ Microbiol 69: 6434-6441

Sardesai N, Babu CR (2000) Cold stress induces switchover of respiratory pathway to lactate glycolysis in psychrotrophic *Rhizobium* strains. Folia Microbiol (Praha) 45: 177-182

Schuerman PL, Liu JS, Mou H, Dandekar AM (1997) 3-Ketoglycoside-mediated metabolism of sucrose in E. coli as conferred by genes from *Agrobacterium tumefaciens*. Appl Microbiol Biotechnol 47: 560-565

Smith LT, Smith GM, Madkour MA (1990) Osmoregulation in *Agrobacterium tumefaciens*: accumulation of a novel disaccharide is controlled by osmotic strength and glycine betaine. J Bacteriol 172: 6849-6855

Soberon N, Venkova-Canova T, Ramirez-Romero MA, Tellez-Sosa J, Cevallos MA (2004) Incompatibility and the partitioning site of the *repABC* basic replicon of the symbiotic plasmid from *Rhizobium etli.* Plasmid 51: 203-216

Sonoda H, Suzuki K, Yoshida K (2002) Gene cluster for ferric iron uptake in *Agrobacterium tumefaciens* MAFF301001. Genes Genet Syst 77: 137-146

Stover CK, Pham XQ, Erwin AL, Mizoguchi SD, Warrener P, Hickey MJ, Brinkman FSL, Hufnagle WO, Kowalik DJ, Lagrou M, Garber RL, Goltry L, Tolentino E, Westbrock-Wadman S, Yuan Y, Brody LL, Coulter SN, Folger KR, Kas A, Larbig K, Lim R, Smith K, Spencer D, Wong GKS, Wu Z, Paulsen IT, Reizer J, Saier MH, Hancock REW, Lory S, Olson MV (2000) Complete genome sequence of *Pseudomonas aeruginosa* PA01, an opportunistic pathogen. Nature 406: 959-964

Suksomtip M, Liu P, Anderson T, Tungpradabkul S, Wood DW, Nester EW (2005) Citrate synthase mutants of *Agrobacterium* are attenuated in virulence and display reduced *vir* gene induction. J Bacteriol 187: 4844-4852

Suzuki K, Hattori Y, Uraji M, Ohta N, Iwata K, Murata K, Kato A, Yoshida K (2000) Complete nucleotide sequence of a plant tumor-inducing Ti plasmid. Gene 242: 331-336

Tellez-Sosa J, Soberon N, Vega-Segura A, Torres-Marquez ME, Cevallos MA (2002) The *Rhizobium etli* cyaC product: characterization of a novel adenylate cyclase class. J Bacteriol 184: 3560-3568

Trust W (1997) Summary of the Report of the Second International Strategy Meeting on Human Genome Sequencing.

Ugalde JE, Parodi AJ, Ugalde RA (2003) De novo synthesis of bacterial glycogen: *Agrobacterium tumefaciens* glycogen synthase is involved in glucan initiation and elongation. Proc Natl Acad Sci USA 100: 10659-10663

Valladares A, Montesinos ML, Herrero A, Flores E (2002) An ABC-type, high-affinity urea permease identified in cyanobacteria. Mol Microbiol 43: 703-715

Venkova-Canova T, Soberon NE, Ramirez-Romero MA, Cevallos MA (2004) Two discrete elements are required for the replication of a *repABC* plasmid: an antisense RNA and a stem-loop structure. Mol Microbiol 54: 1431-1444

Willis LB, Walker GC (1999) A novel Sinorhizobium meliloti operon encodes an alpha-glucosidase and a periplasmic-binding-protein-dependent transport system for alpha-glucosides. J Bacteriol 181: 4176-4184

Wood DW, Setubal JC, Kaul R, Monks DE, Kitajima JP, Okura VK, Zhou Y, Chen L, Wood GE, Almeida Jr. NF, Woo L, Chen Y, Paulsen IT, Eisen JA, Karp PD, Bovee Sr. D, Chapman P, Clendenning J, Deatherage G, Gillet W,

Grant C, Kutyavin T, Levy R, Li MJ, McClelland E, Palmieri P, Raymond C, Rouse R, Saenphimmachak C, Wu Z, Romero P, Gordon D, Zhang S, Yoo H, Tao Y, Biddle P, Jung M, Krespan W, Perry M, Gordon-Kamm B, Liao L, Kim S, Hendrick C, Zhao ZY, Dolan M, Chumley F, Tingey SV, Tomb JF, Gordon MP, Olson MV, Nester EW (2001) The genome of the natural genetic engineer *Agrobacterium tumefaciens* C58. Science 294: 2317-2323

Wood TK (2002) Active expression of soluble methane monooxygenase from *Methylosinus trichosporium* OB3b in heterologous hosts. Microbiology 148: 3328-3329

Xu XQ, Li LP, Pan SQ (2001) Feedback regulation of an *Agrobacterium* catalase gene *katA* involved in *Agrobacterium*-plant interaction. Mol Microbiol 42: 645-657

Young JP, Crossman LC, Johnston AW, Thomson NR, Ghazoui ZF, Hull KH, Wexler M, Curson AR, Todd JD, Poole PS, Mauchline TH, East AK, Quail MA, Churcher C, Arrowsmith C, Cherevach I, Chillingworth T, Clarke K, Cronin A, Davis P, Fraser A, Hance Z, Hauser H, Jagels K, Moule S, Mungall K, Norbertczak H, Rabbinowitsch E, Sanders M, Simmonds M, Whitehead S, Parkhill J (2006) The genome of *Rhizobium leguminosarum* has recognizable core and accessory components. Genome Biol 7: R34

Zhao G, Ceci P, Ilari A, Giangiacomo L, Laue TM, Chiancone E, Chasteen ND (2002) Iron and hydrogen peroxide detoxification properties of DNA-binding protein from starved cells. A ferritin-like DNA-binding protein of *Escherichia coli*. J Biol Chem 277: 27689-27696

Chapter 5

AGROBACTERIUM—TAXONOMY OF PLANT-PATHOGENIC *RHIZOBIUM* SPECIES

John M. Young

Landcare Research, Private Bag 92170, Auckland, New Zealand

Abstract. Traditionally, *Agrobacterium* spp. have been regarded as unique, predominantly soil-inhabiting, oncogenic plant pathogenic bacteria, thus justifying their inclusion in a single genus that encompassed species allocated according to the nature of symptoms produced. Tumorigenic strains have been included in *A. tumefaciens* and rhizogenic strains in *A. rhizogenes*; each species having a wide host plant range. Other species (*A. larrymoorei, A. rubi* and *A. vitis* and) have relatively restricted host ranges. Non-pathogenic, exclusively soil-inhabiting strains have been allocated to *A. radiobacter*. From its inception, the authenticity of *Agrobacterium* was questioned because of its possible synonymy with *Rhizobium*, a genus until recently considered to be represented only by bacteria forming nodulating, nitrogen-fixing, symbiotic relationships with legume plants. Accumulated phenotypic and molecular evidence now shows that these two genera can be circumscribed as single taxon. Furthermore, *Agrobacterium* pathogenicity and *Rhizobium* nodulation characters are plasmid-borne and interchangeable between individual species and between members of the two genera. This evidence militates against stable nomenclature based on pathogenic characters for the genus, *Agrobacterium*, or for its species. According to modern approaches to classification of the two genera, all *Agrobacterium* spp. should be allocated to the genus *Rhizobium*, natural species being distinguished on the basis of phenotypic and genomic data. Differences in pathogenicity can be accommodated by nomenclature referring to the presence or absence of different oncogenic plasmids. In this chapter, the classification and nomenclature of *Agrobacterium* is chronicled in relation to the evolution of bacterial taxonomy as a discipline intended to inform natural relationships. The 'agrobacteria' are considered in the context of the diversity of related soil-inhabiting 'rhizobia' of which they form a sub-population.

1 INTRODUCTION

The classification of bacteria at generic and specific levels has been subject to repeated amendment, with frequent revisions made to keep nomenclature in line with contemporary taxonomic approaches. The genus *Agrobacterium* Conn 1942 is an exception. Although they had their origins in diverse genera, the plant pathogenic bacteria associated with oncogenic symptoms, commonly called 'crown gall' and 'hairy root', and other more rece ntly identified oncogenic pathogens, have been recognized as distinct species in the genus *Agrobacterium* since the genus was established (Kersters and De Ley, 1984).

Classification of the genus *Agrobacterium* and of its species has been based on its once-puzzling oncogene pathogenicity, which was the defining character of the genus (Kersters and De Ley, 1984). This was paralleled in the genus *Rhizobium* Frank 1889, originally reserved for bacteria with the capacity to form symbiotic nitrogen-fixing symbioses with legume species. For both genera, their distinctive generic characteristics are now known to be the result of the presence or absence of interchangeable conjugative plasmids that confer specific oncogenic or nodulating capabilities. However, a character that is the result of arbitrary acquisition or loss of a plasmid is obviously unstable and cannot form the basis of formal nomenclature. Although comparative phenotypic and genetic studies of *Agrobacterium* spp. and *Rhizobium* spp. have failed to confirm differentiation into separate genera based on oncogenicity and nitrogen-fixation respectively (Young et al., 2001), an element of the bacteriological community has continued to support a special-purpose nomenclature based on pathogenicity alone.

Pathogenicity was also used as the single defining character of individual *Agrobacterium* species (Kersters and De Ley, 1984) although, following comprehensive genetic and phenotypic studies, the genus has been revised with the recognition of natural species (Holmes and Roberts, 1981) to accord with current interpretations of bacterial taxonomy. Nomenclature reflecting species epithets based on pathogenicity alone has also continued to be strongly supported.

Although the nomenclature of the genus has been in question since its inception (Pribram, 1933; Conn, 1942; Graham, 1964; Allen and Holding, 1974; Kersters and De Ley, 1984), formal nomenclatural revisions (Keane et al., 1970; Holmes and Roberts, 1981) have not generally been adopted. The account presented here of the taxonomy of oncogenic agrobacteria details the history of their classification in relation to the evolution of bacterial systematic, and the implications that this has on their nomenclature. [It is customary in papers on taxonomy to provide full author citations where the name of a validly published taxon is mentioned for the first time. Full citation includes the names of authors together with the date of publication of the proposal of the taxon, although the reference is not usually recorded in literature cited. This is largely uninformative except to indicate valid publication, and it breaks the flow of text. Hereafter, this treatment reports only authorities for names in *Agrobacterium* and *Rhizobium* that are validly published (Lapage et al., 1992). The validity of other names can be assumed unless otherwise indicated. Historic names that are not valid are identified with a superscript (^{NV}) to indicate that they have not been validly published in the Approved Lists of Bacterial Names (Skerman et al., 1980) or more recent proposals of new taxa in the *International Journal of Systematic and Evolutionary Microbiology* (*IJSEM*) or in Validation Lists in *IJSEM* of names published elsewhere. Euzéby (1997–2006) offers a comprehensive reference list of published names].

2 HISTORICAL PERSPECTIVE—ORIGINS

2.1 Taxonomy, classification and nomenclature

Taxonomy, as it relates to the systematic study of bacteria, has three main components – classification, nomenclature and identification (Young et al., 1992) – of which the following are considered here:

1. *Classification*, the grouping of bacteria in taxa in a hierarchy based on some principle and methodology (Young et al., 1992; Goodfellow and O'Donnell, 1993; Young, 2001; Brenner et al., 2005).
2. *Nomenclature*, the application of names to these taxa. Refinements of classification have usually improved understanding of bacterial rela-

tionships and, when expressed in formal nomenclature, names help to conceptualize those relationships. Adopting any particular nomenclature should imply acceptance of a particular classification (Goodfellow and O'Donnell, 1993; Sneath, 2005). Nomenclature therefore has the capacity to illuminate classification but, if not strictly applied, can mislead.

The evolution of bacterial taxonomy can be divided into three periods (Young et al., 1992):

1. In the 19th and early 20th centuries (up to 1940), bacterial nomenclature was notable for the proliferation of species names, in a period when the principles and practices of bacterial taxonomy were relatively undeveloped and when bacteria were regularly given names that reflected particular characters regarded as important in areas of human endeavour (e.g. agriculture, medicine). The eight editions of *Bergey's Manual of Determinative Bacteriology* (1923–1974) were prepared on the basis that they provided a determinative system for bacterial identification. Although some tests (e.g. morphology, Gram's reaction, flagellar insertion, fermentation) were considered to represent taxonomically significant bacterial characters, it gradually became clear that much information (colony growth, broth turbidity) was inadequate for reliable classification of bacteria.

2. In the period 1940–1975, with progressive expansion of phenotypic databases (Stanier et al., 1966) and the introduction of numerical computer-based analysis, demonstrations of taxa based on overall phenotypic differences using numerical analysis (Sokal and Sneath, 1963) supported a concept of natural classification. Colwell (1970a; 1970b) introduced the term 'polyphasic' for natural classification based on all available phenotypic data. Such natural classifications based on phenotypic comparisons were considered to allow predictions about the characteristics of populations; at any taxonomic level, bacteria in the same taxon were expected to have more attributes in common than with bacteria in other taxa at the same level. Such general purpose or natural classifications can be contrasted with special purpose or artificial classifications (Sneath, 2005) often framed around individual characters of significant interest in areas of human endeavour (Lelliott, 1972).

3. Since 1975, molecular methods have been used increasingly to establish classification and to generate nomenclature (Young, 2001; Young

et al., 1992). Early studies entailed comparisons of complex compounds such as peptides and fatty acids, using SDS-PAGE and FAME, respectively. However these have become secondary to genetic methods, especially those based on comparisons of PCR-amplified DNA sequences. As attention turned to phylogenetic approaches in taxonomy (Sneath, 1988), the claim that comparative analyses of 16S rDNA would give 'true' phylogenies (Stackebrandt and Woese, 1984; Woese, 1987) and the ease with which sequences could be obtained and analysed using tree-building software, has seen proliferation of names based on phylogenetic inference. In a refinement, Vandamme et al. (1996) proposed a revised polyphasic approach derived from Colwell (1970a; 1970b) that entailed investigating strain diversity by a variety of methods in order to establish natural taxa based on overall similarities (Gillis et al., 2005). An essential additional component in the new approach was a requirement to include inference of phylogenetic relationships at taxonomic levels above species, based on comparative analysis of 16S rDNA sequences. This is how the terms 'polyphasic taxonomy' and 'polyphasic classification' are used today.

2.2 Early days of bacterial taxonomy

Before 1940, classification of bacteria relied on structural descriptions based on cell morphology and colony growth on different media. For plant pathogens, specific epithets usually referred to host species or to distinct symptoms, it being assumed that pathogenicity represented the expression of a major component of the underlying phenotype. Generic names were proposed and revised, sometimes without explanation, on the basis of limited investigations of what would now be seen as ephemeral or inadequate criteria.

Subsequently, taxa established according to these criteria were often amalgamated. However, when these bacteria were investigated in more detail, the extent of their biochemical diversity became apparent, and the heterogeneity of named genera came to be recognized as concealing recognizable taxa based both on morphological and physiological characters.

2.3 The genus *Agrobacterium*

In the period around 1940 generic classifications were reassessed on the basis of more detailed investigations of morphology, cell wall structure (Gram's reaction), flagellar insertion, and a relatively small number of physiological reactions considered to represent fundamental metabolic processes; these are now regarded as the classic methods by which genera were differentiated. That approach allowed the redistribution of most pathogenic species according to broad similarity groups, to *Corynebacterium*, *Erwinia*, *Pseudomonas*, and *Xanthomonas*. A new genus, *Agrobacterium*, was proposed by Conn (1942) to include two pathogenic species previously allocated to *Phytomonas*; *A. tumefaciens* and *A. rhizogenes* and a non-pathogenic species, *A. radiobacter*. The first description of the genus Conn (1942) was: '*Agrobacterium*: small, short, non-spore-forming rods, which are typically motile with 1–4 peritrichous flagella (if only one flagellum, lateral attachment is as common as polar). Occur primarily in soil or as pathogens attacking roots or producing hypertrophies on the stems of plants. Are ordinarily gram-negative. Do not produce acid or gas in glucose-peptone media, although a certain amount of acid is evident in synthetic media; this latter observation is ordinarily due merely to the presence of CO_2 which may be produced in considerable abundance. Liquefy gelatine slowly or not at all.' A slightly expanded description of the genus was provided by Allen and Holding (1974). Although this classification is now understood to reflect only a part of generic diversity, it allowed more systematic comparative examination of relatively similar organisms.

2.4 History of species allocated to *Agrobacterium*

2.4.1 Species transferred when Agrobacterium was first proposed

1. *Agrobacterium radiobacter* (Beijerinck and van Delden 1902) Conn 1942
   A. radiobacter was originally proposed as *Bacillus radiobacter*^{NV} by Beijerinck and van Delden (1902) in a study of soil bacteria associated with nitrogen utilization. The species was not recognized as a plant pathogen. Subsequently, the species was reclassified as *Bacterium radiobacter*^{NV}, *Rhizobium radiobacter*, *Achromobacter radiobacter*^{NV} and *Alcaligenes radiobacter*^{NV}. Eventually, Conn (1942) proposed reclassification as *Agrobacterium radiobacter*.

2. *Agrobacterium tumefaciens* (Smith and Townsend 1907) Conn 1942
   Crown gall, the unregulated growth of plant tissue of many plant species, usually occurring in the roots and crown, has probably been known from antiquity. Proof that crown gall was a disease caused by a

bacterial pathogen was made by Smith and Townsend (1907), who named the organism *Bacterium tumefaciens* with *Chrysanthemum frutescens* as host plant but suggesting a wider host range. This was one of the early demonstrations of bacterial pathogenicity to plants when that concept was still contentious. The crown gall pathogen was renamed *Pseudomonas tumefaciens*^{NV} by Duggar (1909), who confirmed a wide host range for this pathogen and then *Phytomonas tumefaciens*^{NV} before Conn (1942) created the combination *Agrobacterium tumefaciens*. Subsequently, bacteria were classified as *A. tumefaciens* solely on the basis of their tumorigenic pathogenicity.

3. *Agrobacterium rhizogenes* (Riker et al. 1930) Conn 1942
   Phytomonas rhizogenes^{NV} was the name proposed by Riker et al. (1930) for bacteria causing the 'hairy root' or rhizogenic symptom in a range of plant species. (So confused was bacterial systematics at that time that Riker et al. (1930) also proposed *Bacterium rhizogenes*^{NV} and *Pseudomonas rhizogenes*^{NV} as synonyms to ensure recognition of the species). Conn (1942) transferred the species to *Agrobacterium*. Subsequently, bacteria were classified as *A. rhizogenes* solely on the basis of their rhizogenic pathogenicity.

2.4.2 Additional species allocated to Agrobacterium after Conn proposed the genus

1. *Agrobacterium rubi* (Hildebrand 1940) Starr and Weiss 1943
   Hildebrand (1940) proposed *Phytomonas rubi*^{NV} for a tumorigenic pathogen, similar in character to *Phytomonas tumefaciens* but which he considered to be specific to *Rubus*. Subsequently, Starr and Weiss (1943) transferred the species to *Agrobacterium*.

2. *Agrobacterium vitis* Ophel and Kerr 1990
   Ophel and Kerr (1990) re-examined a sub-population of tumorigenic strains isolated from grape previously described as *A. radiobacter* biotype 3 of Keane et al. (1970), biovar 3 of Kersters and De Ley (1984), Kerr and Panagopoulos (1977), Süle (1978) and Panagopoulos et al. (1978) and allocated them to a new species, *A. vitis*.

3. *Agrobacterium larrymoorei* Bouzar and Jones 2001
   A tumorigenic pathogen isolated from aerial tumours in *Ficus benjamina* (Bouzar et al., 1995) was named *A. larrymoorei* by Bouzar and Jones (2001).

4. Incidental reference has been made to *A. albertimagni*[NV] (Salmassi et al., 2001; Han et al., 2005; Liu et al., 2005) but the epithet has not been proposed legitimately and is not validly published.

2.5 Phenotypic species classification

2.5.1 Pathogenic species

After 1940 *Agrobacterium* classification developed within the framework of understanding of plant pathogenic bacteria, that '.... where there is a true difference in pathogenic ability, some other type of difference .... cultural, biochemical, metabolic, serological, or some other category .... should be demonstrable' (Burkholder and Starr, 1948). For pathogens in *Agrobacterium*, as for pathogenic species in other genera, it was taken for granted that pathogenicity was the expression of significant and substantial phenotypic, and therefore genetic, differences, and that improved methods would eventually sustain their characterization.

The probable synonymy of *A. radiobacter* and *A. tumefaciens* has been repeatedly noted (Heberlein et al., 1967; Moffett and Colwell, 1968; Graham, 1976) but species nomenclature was not revised (Allen and Holding, 1974) (*Table 5-1*), remaining essentially the same as described by Bergey et al. (1939) in spite of mounting circumstantial evidence that it expressed an over-simple, not to say mistaken, interpretation of relationships. This uncertainty was acknowledged to be the result of the small number of strains available and comparatively few established discriminating examined (Allen and Holding, 1974).

2.5.2 Comparative studies of Agrobacterium species

Numerical analysis of phenotypic characteristics (White, 1972; Kersters et al., 1973); biochemical and physiological tests (Keane et al., 1970; Kersters et al., 1973; Kerr and Panagopoulos, 1977; Sule, 1978) DNA-DNA reassociation (De Ley, 1972, 1974; De Ley et al., 1973), and comparison of soluble proteins (Kersters and De Ley, 1975) indicated four genetically and phenotypically distinct groups or clusters that were unrelated to pathogenic capability. Keane et al. (1970) were first to propose that speciation based on pathogenicity was untenable for the agrobacteria. They proposed that all pathogens be included in a single species, *A. radiobacter*, with biotype nomenclature to differentiate physiological differences, and variety nomenclature to distinguish between tumorigenic, rhizogenic and non-pathogenic states (Keane et al., 1970) (*Table 5-1*). Allen and Holding (1974) maintained species nomenclature based on pathogenicity and although they did note the work of Keane et al. (1970) incidentally, for whatever reason, they did not appreciate the full implications of that report.

2.6 The approved lists and *Agrobacterium* nomenclature

Recognition of the uncertainties in *Agrobacterium* classification and nomenclature occurred contemporaneously with the development (Lapage et al., 1975) and publication of the Approved Lists of Bacterial Names ['the Approved Lists' (Skerman et al., 1980)]. Realization that a high proportion (>90%) of species names of bacteria were synonyms or were illegitimate indicated the need for nomenclatural revision. The Approved Lists included only species names recognized as valid where there was a modern description and at least one extant strain which could be accepted as the type, or name-bearing, strain (Lapage et al., 1975). Descriptions based on pathogenic characterization alone would not justify inclusion of species in the Approved Lists (Lapage et al., 1975; Young et al., 1978) or in or in subsequent lists of validly published taxa (Lapage et al., 1992). Species epithets based solely on this criterion were excluded. Little public consideration had been given to revision of *Agrobacterium* nomenclature and, in spite of a proposal by Keane et al. (1970) for a natural classification, and a proposal based on application of pathovar nomenclature for *A. radiobacter* (Kerr et al., 1978), neither of these nomenclatural alternatives was adopted. Notwithstanding the doubtful status of *A. radiobacter* and *A. tumefaciens* as independent species, as well as the uncertainty of the standing of species based on plasmid-borne pathogenicity (Kerr et al., 1978), *A. radiobacter*, *A. rhizogenes*, *A. rubi* and *A. tumefaciens* were included in the Approved Lists.

2.6.1 Pathogenicity is plasmid-borne

Genetic studies of *Agrobacterium* spp. showed that pathogenicity, as expressed by tumorigenic capability of *A. tumefaciens*, and by rhizogenic capability of *A. rhizogenes*, could be transferred between strains of *Agrobacterium*, or be lost (Kerr, 1969b). Subsequently, this behavior was shown to be derived from transfer of plasmids as conjugative elements (Genetello et al., 1977). Tumorigenic pathogenicity of *A. tumefaciens* depends on acquisition of a Ti plasmid (Van Larebeke et al., 1975; Watson et al., 1975) and rhizogenic pathogenicity of *A. rhizogenes* depends on the acquisition of an Ri plasmid (Willmitzer et al., 1980; Tepfer, 1984). It is generally assumed that the genes for pathogenicity are plasmid-borne in all *Agrobacterium* spp.

Recognition that tumorigenic or rhizogenic capability was mediated by genes on transmissible plasmids carried by *A. tumefaciens* and *A. rhizogenes* had important implications for species classification although these were not immediately understood. Tumorigenic and rhizogenic populations could not be circumscribed as species in formal nomenclature because acquisition, exchange, or loss, of one of these plasmids by a bacterial strain would lead to a change in its species identity (Kersters and De Ley, 1984; Kerr, 1992; Young et al., 2001). Furthermore, the small genetic and phenotypic contribution of plasmids to the phenotype and genotype of bacteria was believed to be insignificant in terms of differentiation of species.

Pathogenic *Agrobacterium* spp. are represented by strains that may be either tumorigenic, rhizogenic or non-pathogenic according to their plasmid complement. Pathogenicity characters are mobile. For this reason, the epithets *tumefaciens, rhizogenes* and *radiobacter,* if restricted to use for populations defined by their pathogenicity or lack thereof, could not refer to stable taxa (Kersters and De Ley, 1984; Young et al., 2001).

Oncogenic activity is also associated with all *Agrobacterium* spp. and there may be several additional bacterial populations that may merit classification as species (Sawada and Ieki, 1992b).

2.7 Natural *Agrobacterium* species

Based on their own numerical analysis of data, and on previous studies described above, Holmes and Roberts (1981) proposed a natural classification for *A. tumefaciens* and *A. rhizogenes* (*Table 5-1*):

1. *A. tumefaciens* corresponded to biotype 1 (Keane et al., 1970), group I of White (1972), *Agrobacterium* cluster 1 of Kersters et al. (1973), biovar 1 of Kersters and De Ley (1984), of Willems and Collins (1993), and of Sawada et al. (1993). The species included the type strains of both *A. tumefaciens* and *A. radiobacter* and, because *A. tumefaciens* is type species of the genus, Holmes and Roberts (1981) considered that the epithet '*tumefaciens*' took priority as name of the species.

2. *A. rhizogenes* corresponds to biotype 2 of Keane et al. (1970), group III of White (1972), cluster 2 of Kersters et al. (1973), biovar 2 of Kersters and De Ley (1984), of Willems and Collins (1993) and of Sawada et al. (1993).

Table 5-1. Comparison of nomenclature used for plant pathogenic *Rhizobium* spp., previously allocated to the genus *Agrobacterium*

Species names based on natural classification			Species names based on pathogenicity	
After Young et al. (2001)	After Holmes and Roberts (1981); Bradbury (1986); Holmes (1988); Moore et al. (2001)	After Keane et al. (1970); Kerr and Panagopoulos (1977); Panagopoulos et al. (1978)	After Allen and Holding (1974); Skerman et al. (1980)	After Kersters and De Ley (1984)
R. larrymoorei[a]	*A. larrymoorei*[b]	NR[c]	NR	NR
R. radiobacter (Ti)	*A. tumefaciens*[d] (tumorigenic)	*A. radiobacter* biovar *tumefaciens* (biotype 1)	*A. tumefaciens*	*A. tumefaciens* (biovar 1)
R. radiobacter (Ti)	*A. tumefaciens*[d] (tumorigenic)	*A. radiobacter* biovar *tumefaciens* (biotype 1)	*A. tumefaciens*	*A. tumefaciens* (biovar 1)
R. radiobacter (Ri)	*A. tumefaciens*[d] (rhizogenic)	*A.* radiobacter biovar *rhizogenes* (biotype 1)	*A. rhizogenes*	*A. rhizogenes* (biovar 1)
R. radiobacter	*A. tumefaciens* (non-pathogenic)	*A. radiobacter* biovar *radiobacter* (biotype 1)	*A. radiobacter*	*A. radiobacter* (biovar 1)
R. rhizogenes (Ti)	*A. rhizogenes* (tumorigenic)	*A. radiobacter* biovar *tumefaciens* (biotype 2)	*A. tumefaciens*	*A. tumefaciens* (biovar 2)
R. rhizogenes (Ri)	*A. rhizogenes* (rhizogenic)	*A. radiobacter* biovar *rhizogenes* (biotype 2)	*A. rhizogenes*	*A. rhizogenes* (biovar 2)
R. rhizogenes	*A. rhizogenes* (non-pathogenic)	*A. radiobacter* biovar *radiobacter* (biotype 2)	*A. radiobacter*	*A. radiobacter* (biovar 2)
R. rubi[b]	*A. rubi*[b]	*A. radiobacter* biovar *tumefaciens* (biotype 2)	*A. rubi*	*A. rubi*
R. vitis (Ti)	*A. vitis* (tumorigenic)[e]	*A. radiobacter* biovar *tumefaciens* (biotype 3)	NR	*A. tumefaciens* (biovar 3)
R. vitis	*A. vitis* (non-pathogenic)	NR	NR	NR

[a] Oncogene designations are indicated when necessary for clarity; [b] Only tumorigenic (Ti) capability has been reported for this species; [c] NR = not recorded; [d] The correct name for this species is *A. radiobacter* (see Young et al., 2006; Sawada et al., 1993); [e] Moore et al. (2001)

Popoff et al. (1984) supported this natural classification, including *A. rubi*. They discriminated nine genomic species within *A. tumefaciens* (Holmes and Roberts, 1981) based on DNA-DNA reassociation. Six of these corresponded to phenotypic groups in *A. tumefaciens* reported in the numerical analysis of Kersters et al. (1973).

Kersters and De Ley (1984) acknowledged the compelling evidence based on the earlier studies for recognition of natural species of *Agrobacterium*. However, they stated without explanation, and apparently without considering the possibility of emendation of species descriptions, that acceptance of a classification based on natural species would require a change of names of *A. tumefaciens* and *A. rhizogenes* (*Table 5-1*). They perceived a requirement to use the names *A. tumefaciens* and *A. rhizogenes* but, notwithstanding Holmes and Roberts' (1981) proposal, considered that these were unacceptable as epithets in a natural classification. Kersters and De Ley (1984) therefore elected to follow the earlier nomenclature based on pathogenicity (Allen and Holding, 1974).

Classification and nomenclature of natural *Agrobacterium* species is (*Table 5-1*):

1. *Agrobacterium radiobacter* as described by Holmes and Roberts (1981) includes the type strains of oncogenic *A. tumefaciens* and of non-pathogenic *A. radiobacter*. [The species epithet '*radiobacter*' takes precedence when *A. radiobacter* and *A. tumefaciens* are amalgamated because it was the first published name. The correct name of the species is *A. radiobacter*; *A. tumefaciens* retains its nomenclatural status as type species of the genus as a junior heterotypic (subjective) synonym of *A. radiobacter* (Young et al., 2006a). An anticipatory correction is made here.]

2. *Agrobacterium rhizogenes* as described by Holmes and Roberts (1981).

3. *Agrobacterium rubi* is characterized in genotypic and phenotypic terms and is usually isolated from above-ground cane galls of *Rubus* spp., although strains have been isolated from other hosts (Bradbury, 1986) and are capable of infecting other plant hosts (Anderson and Moore, 1979; Sawada and Ieki, 1992b).

4. *Agrobacterium vitis* Ophel and Kerr (1990) was proposed as the name for biotype 3 of Keane et al. (1970), biovar 3 of Kersters and De Ley (1984), Kerr and Panagopoulos (1977), Süle (1978) and Panagopoulos

et al. (1978). The predominant *Agrobacterium* species in grape is *A. vitis* but strains of this species have also been isolated from *Actinidia* (Sawada and Ieki, 1992a).

Following the report of Holmes and Roberts (1981) of a natural classification for *Agrobacterium* spp. the application of species epithets has become ambiguous. The names may be used either in a natural classification (Bradbury, 1986; Moore et al., 2001), or in special-purpose nomenclature based on differences in pathogenicity (Kersters and De Ley, 1984). Preliminary clarification is therefore always necessary when they are used now.

2.7.1 Pathogenic designations

Several approaches have been made to describing pathogenic strains as part of natural species classification (Kerr et al., 1978) and subsequently (Kersters and De Ley, 1984) proposed the application of pathovar nomenclature in terms of the International Standards for Naming Pathovars (Dye et al., 1980). However, the fact that pathogenicity genes are carried on plasmids means that the pathogenic character of any strain is unstable. This lack of stability would make uncertain the application of pathovar names to particular strains, most notably to pathotype strains. For pathogenic agrobacterial strains, therefore, formal pathovar nomenclature seems inappropriate. Holmes and Roberts (1981) recognized pathogenic strains within species according to their 'tumorigenic', 'rhizogenic' or 'non-pathogenic' states. Species comprising pathogenic or non-pathogenic strains can also be reported as tumorigenic ('Ti'), as rhizogenic ('Ri'), or as non-pathogenic strains of the species, where relevant and necessary. There is no taxonomic basis for according pathogenicity of strains greater nomenclatural formality.

3 *AGROBACTERIUM—RHIZOBIUM* RELATIONSHIPS

Agrobacterium has long been recognized as closely related to *Rhizobium*. Without explanation, Pribram (1933) proposed the combination *Rhizobium radiobacter*, anticipating the debate concerning the common generic relationship of *Agrobacterium* spp. to *Rhizobium*. In proposing the new combination *Achromobacter radiobacter*, Bergey et al. (1934) noted that the species was indistinguishable from *Rhizobium* spp. except for a few characters that would not now be recognized as adequate generic determinants. Conn (1942) also noted the close relationship of *Agrobacterium* spp. to known *Rhizobium* spp. but, on advice, proposed *Agrobacterium* as new genus that included tumorigenic and soil-inhabiting bacteria.

3.1 Phenotypic comparisons of *Agrobacterium* and *Rhizobium*

In almost every discussion of *Agrobacterium* and *Rhizobium* the close similarity of their descriptions has been pointed out, and amalgamation of the genera has often been suggested (Bonnier, 1953; Graham, 1964; Heberlein et al., 1967; De Ley, 1968; White, 1972; Graham, 1976; Kerr et al., 1978; Kerr, 1992). The only systematic difference recorded between the genera has been their oncogenic (*Agrobacterium*) or symbiotic (*Rhizobium*) interactions with plants. Allen and Allen (1950) listed a number of tests that they claimed discriminated the two genera. However, their study compared only a few of the *Rhizobium* species recognized today with *A. radiobacter* alone, and none of the differentiating tests described are acknowledged as significant in the literature today. Recent comprehensive studies of phenotypic data (De Lajudie et al., 1994) and a study of fatty acid profiles by Tighe et al. (2000) confirm the integrity of individual agrobacterial and rhizobial species, but neither study supports differentiation of *Agrobacterium* and *Rhizobium* as separate taxa.

4 GENOTYPIC RELATIONSHIPS

4.1 Comparative molecular analysis of *Agrobacterium*

4.1.1 16S rDNA

Early studies based on comparative molecular analyses were made at a time when 16S rDNA was considered to provide a reliable means of establishing accurate phylogenetic inferences, following the work of Stackebrandt and Woese (1984) and Woese (1987). Comparative analyses of 16S rDNA sequences of *Agrobacterium*, *Rhizobium* and related genera showed that the two genera were not distinguished as separate monophyletic clades (reviewed in Young et al., 2001). These studies were based on comparisons of 16S rDNA sequences from type strains only, in which minor

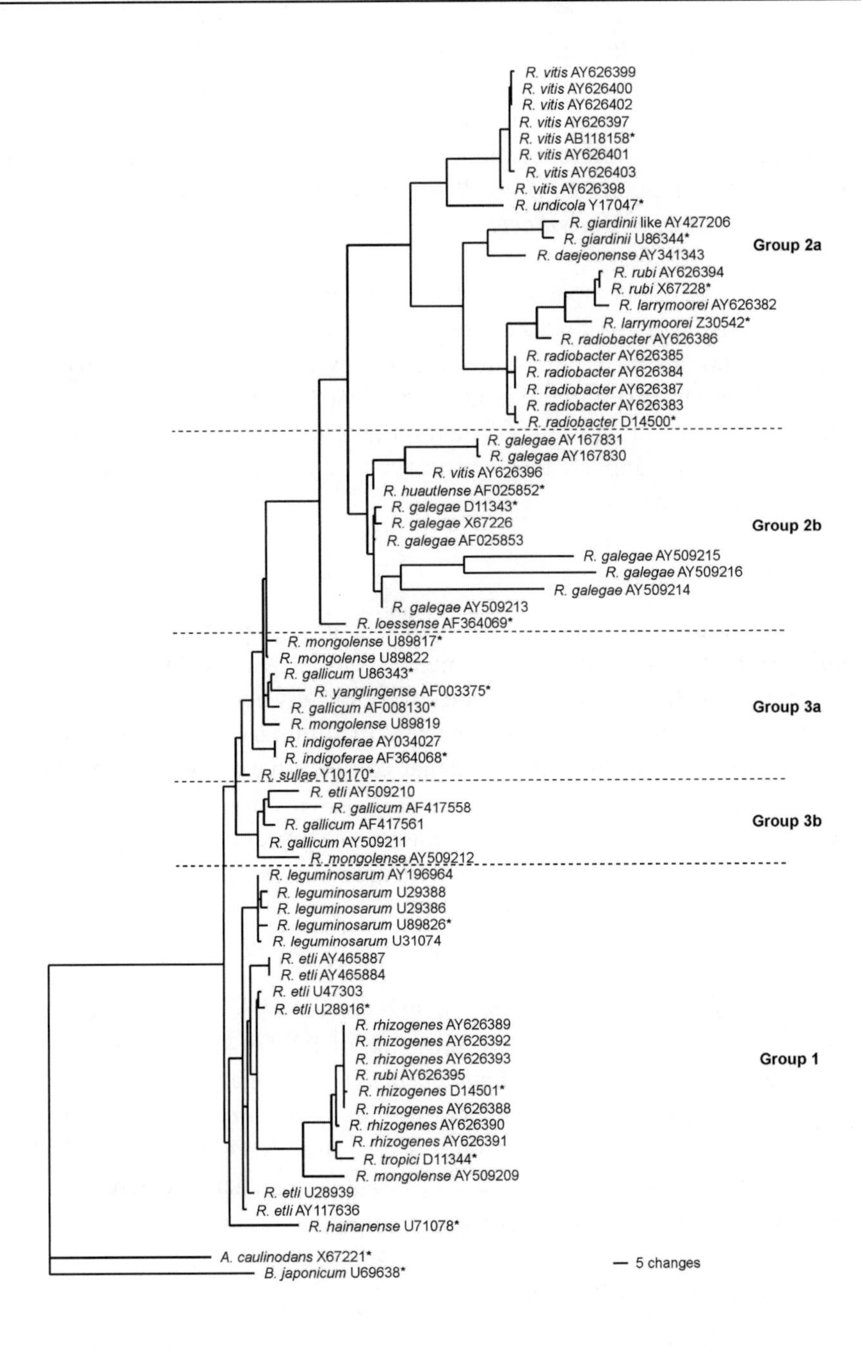

Figure 5-1. Relationships of *Rhizobium* species, including oncogenic (*Agrobacterium*) species, inferred from a comparative analyses of 16S rDNA using Maximum Likelihood (from Young et al., 2004). Sequences from type strains are marked *.

variations in indicated relationships were the result of different methods of analysis. Young et al. (2004) obtained the same result when 16S rDNA sequences of several strains from each species were compared using a selection of phylogenetic inference models. Considered as rooted trees, analyses expressed *R. undicola, A. vitis, R. giardinii, R. daejoenense, A. rubi, A. larrymoorei* and *A. radiobacter* as intermingled species in *Rhizobium* (*Figure 5-1*).

Dependence on comparisons of this sequence is questionable (Young 2001; Young et al., 1992; Zeigler, 2003) because ribosomal DNA, although ubiquitous in living organisms, is only privileged in the sense that all bases are conserved; there is no third position redundancy as occurs in codons of open reading frames. The moiety is subject to the same selective pressures that are applied to all conserved sequences (Ueda et al., 1999; Eardly et al., 2005).

4.1.2 Other sequences

In a comparative analysis of housekeeping sequences (16S rRNA, *atpD*, *recA*), Gaunt et al. (2001) obtained similar results for each moiety, indicating that *Agrobacterium* species were closely related to, or intermingled with, nodulating *Rhizobium* species. In summary, there was a period after 1980 when it seemed possible that comparative sequence analysis of 16S rDNA, then seen as the touchstone of generic relationships (Ludwig and Schleifer, 1999), or other comparative sequence analyses might demonstrate generic differences between oncogenic *Agrobacterium* and nodulating *Rhizobium* species. However, such comparative sequence analyses have given no support for differentiation and hope of such an outcome in future is speculative.

In their study of rhizobium-specific intergenic mosaic elements (RIMEs), Østerås et al. (1995) reported the presence of RIMEs in *Rhizobium* (now *Sinorhizobium*) *meliloti, R. leguminosarum* and unassigned *Rhizobium* spp., as well as in *A. rhizogenes*, but not in *A tumefaciens*. However, because they presented data from only a few strains, Østerås et al. (1995) make a limited contribution to the discussion of generic differences between *Agrobacterium* and *Rhizobium.*

4.1.3 Genomic comparisons

A genomic comparison of *Agrobacterium, Rhizobium* and *Sinorhizobium* by Jumas-Bilak et al. (1998) indicated a high level of diversity without demonstrating a systematic difference between these genera. A recent study (Young et al., 2006b), comparing *A. radiobacter* (as *A. tumefaciens*), *R. leguminosarum*, *Sinorhizobium meliloti*, *Mesorhizobium meliloti*, *Brucella melitensis* and *Bradyrhizobium japonicum* indicated the same relationships as those inferred by 16S rDNA sequence analyses. This study also indicated extensive structural variation between chromosomes, and that a significant number of nitrogen fixing symbiotic species share numbers of genes not found in *A. radiobacter.*

5 PLASMID TRANSFER AND GENUS RECLASSIFICATION

5.1 Transfer of oncogenic Ti and nodulating Sym plasmids

Intergeneric transmissibility of Ti and nodulating plasmids has been demonstrated from nodulating *Rhizobium* to tumorigenic *Agrobacterium* species (Martínez-Romero et al., 1987; Brom et al., 1988; Abe et al., 1998), and from *Agrobacterium* species to *Rhizobium* species (Hooykaas et al., 1977) and occurs in nature (see section 6.1).

A further confusion arises because of transfer of Sym plasmids to genera outside the *Rhizobiales*. Sawada et al. (2003) record nodulation of legumes by *Blastobacter, Burkholderia, Devosia, Methylobacterium, Photorhizobium* and *Ralstonia*, indicating the probable horizontal transfer of oncogenic plasmids between these genera and members of the *Rhizobiales*. The need for rigour in the correct application of nomenclature for strains named on the basis of plasmid-borne characters is becoming increasingly important.

5.2 Revision of oncogenic *Rhizobium* species

5.2.1 Plant pathogenic Rhizobium species

As the extent of species diversity became increasingly clear so the case for retaining the separate genera has become more obscure (Young et al., 2001, 2003; Kuykendall et al., 2005; Young et al., 2005) [The manuscript for Young et al. (2005) was submitted in 2000]. Indeed it is hard to find any justification in the literature for separation of *Agrobacterium* from *Rhizobium* as a genus representing the pathogenic populations. Comparison of the reported phenotypic characters of the two genera indicates that only nodulating and oncogenic behaviour differentiate the two genera and, as noted, these characters are plasmid borne.

Species nomenclature of these strains based on symbiosis and pathogenicity fails to indicate their underlying relationships. Because there was no support in any systematic studies for considering them as separate genera, Young et al. (2001) proposed that *Agrobacterium* and *Rhizobium* species be amalgamated in a single genus, *Rhizobium*, comprising pathogenic, symbiotic nitrogen-fixing, and unspecialized soil populations. An emended description of the genus *Rhizobium* that includes *Agrobacterium* is described in *Box 5-1* and *Rhizobium* spp. known to include oncogenic strains are listed in *Box 5-2*. Future studies might justify the division of the genus comprising *Rhizobium* and *Agrobacterium* species, but there is no basis for thinking that such a division would lead to the reinstatement of *Agrobacterium* based on pathogenic species.

6 DIVERSITY WITHIN *RHIZOBIUM*

6.1 Symbiotic *Agrobacterium* and oncogenic *Rhizobium* (and other genera)

Recent advances in rapid molecular methods have resulted in more extensive surveys of reliably identified rhizobial and agrobacteria populations than was possible in the past. There are now several reports of nodulating rhizobia belonging to the 16S rDNA clade associated with gall-forming agrobacteria, as members of '*Agrobacterium tumefaciens*' (Han et al., 2005; Liu et al., 2005) or closely related to this or other *Agrobacterium* spp. (Chen et al., 2000; Gao et al., 2001; Hungria et al., 2001; Kwon et al., 2005; Wolde-Meskel et al., 2005). A strain (TAL 1145) nodulating *Leucaena leucocephala*, *Phaseolus vulgaris* and a wide range of tropical tree legumes (George et al., 1994) is in the '*Agrobacterium rhizogenes*' clade (B.S. Weir, Landcare Research, Auckland, New Zealand, pers. comm.). Rhizogenic strains of *Rhizobium* spp., *Ochrobactrum* spp., and *Sinorhizobium* sp. have been isolated from hydroponicly grown cucumber exhibiting hairy root (Weller et al., 2004). Velázquez et al. (2005) recorded strains identified as *Rhizobium* (*Agrobacterium*) *rhizogenes* with both oncogenic and nodulating capabilities. As further investigations are made of nodulating rhizobia of the more than 14 000 known legume species (Jordan,

Genus *Rhizobium* Frank 1889

= *Agrobacterium* Conn 1942; *Allorhizobium* de Lajudie, Laurent-Fulele, Willems, Torck, Coopman, Collins, Kersters, Dreyfus & Gillis 1998)

Rhi.zo'bi.um. Gr. n. *rhiza* a root; Gr. n. *bios* life; M.L. neut. n. *Rhizobium* that which lives in a root.

Rods 0.5-1.0 x 1.2-3.0 µm. Non-spore-forming. Gram-negative. Motile by 1-6 flagella. Insertion usually peritrichous, or peritrichous/sub-polar. Fimbriae have been described on some strains. Aerobic, possessing a respiratory type of metabolism with oxygen as the terminal electron acceptor. Optimum temperature, 25-30°C some species can grow at temperatures >40°C. Optimum pH, 6-7; range pH 4-10. Generation times of *Rhizobium* strains are 1.5-3.0 h. Colonies are usually white or beige, circular, convex, semi-translucent or opaque, raised and mucilaginous, usually 2-4 mm in diameter within 3-5 days on yeast-mannitol-mineral salts agar (YMA). Growth on carbohydrate media is usually accompanied by copious amounts of extracellular polysaccharide slime. Chemo-organotrophic, utilizing a wide range of carbohydrates and salts of organic acids as sole carbon sources, without gas formation. Cellulose and starch are not utilized. Produce an acidic reaction in mineral-salts medium containing mannitol or other carbohydrates. Ammonium salts, nitrate, nitrite and most amino acids can serve as nitrogen sources. Strains of some species will grow in a simple mineral salts medium with vitamin-free casein hydrolysate as the sole source of both carbon and nitrogen, but strains of many species require one or more growth factors such as biotin, pantothenate or nicotinic acid. Casein, starch, chitin and agar are not hydrolyzed. Members of *Rhizobium* are distinguished from those in the related genera, *Mesorhizobium* and *Phyllobacterium*, by differences in growth rate, fatty acid profiles and 16S rDNA sequence. Closely related in terms of 16S rDNA sequence similarity, all known *Rhizobium* species include strains which induce hypertrophisms in plants. Hypertrophisms in most species are either root nodules with or without symbiotic nitrogen fixation while in other species they occur as unregulated oncogenic (tumorigenic or rhizogenic) growths. Some cells of symbiotic bacterial species enter root hair cells of leguminous plants (Family Leguminosae) and elicit the production of root nodules wherein the bacteria may engage as intracellular symbionts to fix nitrogen. Many well-defined nodulation (*nod*) and nitrogen fixation (*nif*) genes are clustered on large or megaplasmids (pSyms). Nod factors produced. Strains of plant pathogenic *Rhizobium* (previously *Agrobacterium*) species invade the crown, roots and stems of many dicotyledonous and some gymnospermous plants, via wounds. Self-proliferating tumors are induced by the genetic transfer of a small DNA region carried on large tumor-inducing Ti, or hairy root-inducing Ri, plasmids into the host plant genome. Plasmid transfer between species results in the expression of the particular plant-interactive properties of the plasmid-donor species.

The mol% G + C of the DNA is 57-66 (T_m).

Type species: *Rhizobium leguminosarum* (Frank 1879) Frank 1889

Box 5-1. The emended description of the *Rhizobium* as proposed in Young et al. (2001).

1. *Rhizobium larrymoorei* (Bouzar and Jones 2001) Young 2004
= *Agrobacterium larrymoorei* Bouzar and Jones 2001
Type strain: ATCC 51759; CFBP 5473; ICMP 14256; LMG 21410; NCPPB 4096

2. *Rhizobium radiobacter* (Beijerinck and van Delden 1902) Young et al. 2001
= *Agrobacterium radiobacter* (Beijerinck and van Delden 1902) Conn 1942
= *Agrobacterium tumefaciens* (Smith and Townsend 1907) Conn 1942
Type strain: ATCC 19358; DSMZ 30147; ICMP 5785; LMG 140; NCPPB 3001

3. *Rhizobium rhizogenes* (Riker et al. 1930) Young et al. 2001
= *Agrobacterium rhizogenes* (Riker et al. 1930) Conn 1942
Type strain: ATCC 11325; DSMZ 30148; ICMP 5794; LMG 150

4. *Rhizobium rubi* (Hildebrand 1940) Young et al. 2001
= *Agrobacterium rubi* (Hildebrand 1940) Starr and Weiss 1943
Type strain: ATCC 13335; CFBP 1317; ICMP 6428; LMG 156; NCPPB 1854

5. *Rhizobium vitis* (Ophel and Kerr 1990) Young et al. 2001
= *Agrobacterium vitis* Ophel and Kerr 1990
Type strain: ATCC 49767; ICMP 10752; LMG 8750; NCPPB 3554

Culture Collection Abbreviations:

ATCC	American Type Culture Collection
CFBP	Collection Française de Bactèries Phytopathogènes
DSMZ	Deutsche Sammlung von Mikroorganismen und Zellkulturen
ICMP	International Collection of Micro-organisms from Plants
LMG	Collection of the Laboratorium voor Microbiologie en Microbiele Genetica
NCPPB	National Collection of Plant Pathogenic Bacteria

Box 5-2. Rhizobium spp. that include oncogenic strains

1984; Lindström et al., 1998) it seems likely that more examples will arise of *Agrobacterium* spp. with nodulating capabilities and more strains with oncogenic capabilities will be isolated from rhizobial species hitherto associated with nodulating capabilities. These reports indicate functional diversity that will be difficult, if not impossible, to express in formal nomenclature, and will be misleading, if binomial nomenclature based on pathogenicity and symbiosis is used.

6.2 Clinical '*Agrobacterium*' species

Based on DNA-DNA reassociation, Popoff et al. (1984) demonstrated that *A. radiobacter* is represented by nine genomic species, three of which comprised non-oncogenic strains isolated from clinical human sources.

These genomic characterizations are supported by a study using AFLP (Mougel et al., 2002). More detailed studies are needed to investigate the internal structure of this species, and to authenticate these species as formal taxa. These species have an equal claim for recognition as *Rhizobium* spp. and indicate a level of diversity in '*Agrobacterium*' not easily reconciled within present concepts of the genus or its species as being specifically associated with plant oncogenicity.

6.3 Soil agrobacteria

Interest in the distribution of agrobacteria from sources other than plants has largely been confined to oncogenic plasmid-bearing strains in soils (Sadowsky and Graham, 1998). However, avirulent agrobacterial populations have long been known to be widely distributed in soils (Conn, 1942; Kerr, 1969a; Panagopoulos and Psallidas, 1973; Sule, 1978; Burr et al., 1987; Sadowsky and Graham, 1998) but only recently have been the subject of more detailed study. It is now clear that these act as recipients of Ti- or Ri-plasmids resulting in oncogenic sub-populations that strains fluctuate according to the presence or absence of susceptible host plants. Apparently, some sub-populations, as *Agrobacterium* spp., are more likely to act as recipients of oncogenic plasmids but they can also act as recipients of Sym plasmids (Chen et al., 2000; Gao et al., 2001; Hungria et al., 2001; Kwon et al., 2005; Velazquez et al., 2005; Wolde-Meskel et al., 2005). Recognition that agrobacteria are sub-populations of a wider rhizobial population has been slow and to some extent is inhibited by a divisive nomenclature. All available evidence points to the identity of the rhizobia as a diverse population of soil-inhabiting bacteria with the capacity to exchange plasmids that confer oncogenic and nodulating capabilities (Segovia et al., 1991). It can also be expected that novel species of *Rhizobium* will be identified that are uncharacteristic of the genus as it is identified as present. Salmassi et al. (2001) record the isolation and characterization of *Agrobacterium albertimagni*, as an arsenite oxidizing bacterium. The record of a sulfur-oxidizing chemolithoautotrophic *Mesorhizobium* spp. from legume rhizosphere soil (Ghosh and Roy, 2006) is an early indication of greater phenotypic diversity in the rhizobia.

7 REVISION OF *AGROBACTERIUM* NOMENCLATURE

7.1 Why is the revision of *Agrobacterium* nomenclature controversial?

Following the proposal of Young et al. (2001), Farrand et al. (2003) responded in a note co-signed by 83 individuals, objecting to inclusion of *Agrobacterium* spp. in *Rhizobium*. Farrand et al. (2003) argued that Young et al. (2001) misinterpreted earlier data, and that data not considered by them were adequate support for continued recognition of *Agrobacterium* as a distinct genus. Concerns indicated by Farrand et al. (2003) have already been addressed (Young et al., 2003), though they may not have been satisfied. *Agrobacterium* nomenclature remains in vogue, especially in the large literature of biotechnical reports of plant transformation, a field in which oncogenic *Agrobacterium* plays a major part. It continues to be used in reports specifically discussing the characteristics of the oncogenic pathogens. This is not unexpected because in most studies of this kind, binomial nomenclature is redundant; names being treated as special-purpose nomenclature, with no requirement to indicate relationships to other taxa. On the other hand, *Rhizobium* nomenclature encompassing *Agrobacterium* is increasingly accepted for reporting ecological and taxonomic studies, and for cataloguing culture collections (e.g. ATCC, DSMZ, ICMP, LMG).

Since the Approved Lists were published, all other genera containing plant pathogenic species have been the subject of substantial revision. For instance, plant pathogenic *Pseudomonas* spp. have been transferred to *Acidovorax*, *Burkholderia* and *Ralstonia*. *Erwinia* spp. have been allocated to *Brenneria*, *Dickeya*, *Pantoea*, *Pectobacterium* and *Samsonia*. *Corynebacterium* spp. have been transferred to *Clavibacter*, *Curtobacterium* and *Leifsonia*. The taxonomic basis of some of these proposals is questionable, yet none has raised such a controversy as the proposal of *Agrobacterium–Rhizobium* amalgamation. As indicated, there is little or no taxonomic data to support *Agrobacterium* as a genus separate from *Rhizobium*. Why then, has this proposal proved so controversial? Perhaps it arises from a misunderstanding of the nature of names in taxonomy, from over-emphasis given to the particular implied characters of the two genera, and from a habit of thought.

7.1.1 Names are not descriptive

The tradition of proposing bacterial names that describe a significant character of the taxon has generated a perception that names express a particular meaning and descriptive intention. Taken literally, the etymology of the name '*Rhizobium*' implies an association of the bacterium with plant roots; however subsequent use has resulted in the name being applied to nitrogen-fixing symbionts of legumes. The etymology of the name '*Agrobacterium*' implies an association of the bacterium with soil, but subsequent use has resulted in the named being applied to oncogenic pathogens. To apply the epithets '*tumefaciens*' to strains inducing crown-gall and the epithet '*rhizogenes*' to strains that induce hairy root, and to refer all oncogenic stains to '*Agrobacterium*' would make common-sense if it were not for the fact that binomial nomenclature has the intention of indicating a hierarchy of natural relationships (Sneath, 1988; Goodfellow and O'Donnell, 1993).

Descriptive terms necessarily refer to one or a few characters, often regulated by a few genes that may not be present in all members of a taxon, especially after revision in classification (Young, 2000b; Young et al., 2003). As taxa are redefined to include or exclude populations that do not conform to all the characters of the original description, so names lose their descriptive relevance (Young, 2001; Sneath, 2005). The etymology of generic names *Agrobacterium* and *Rhizobium* is not indicative of their current use, nor can the epithets *tumefaciens* and *rhizogenes* be applied descriptively to species.

7.1.2 Binomial names should indicate natural relationships

The task of modern systematics has been to determine natural relationships that are indicated by application of binomial names. In cases where genera have been divided, or species distributed into genera unfamiliar to a scholar, the application of novel binomials has created little tension, as when comprehensive revisions of *Pseudomonas* has resulted in transfer of species to genera distributed across the *Proteobacteria* (Kersters et al., 1996), or when plant pathogenic *Corynebacterium* spp. were transferred to *Clavibacter* and *Curtobacterium*. However, in this particular case, two genera with popular and long-standing names, *Agrobacterium* and *Rhizobium*, have been amalgamated. This poses a burden on translation from the old to the new nomenclature for those who have a developed familiarity with *Agrobacterium* nomenclature, but it is not insuperable, and is unlikely to pose difficulties for those who approach bacterial nomenclature for the first time. The reward is a nomenclature that allocates due weight to taxonomic and pathogenic differences.

7.2 The status of *Agrobacterium* nomenclature

7.2.1 Species

Before 1981, application of *Agrobacterium* species names was universally applied to strains based on pathogenicity. Following publication of nomenclature based on phenotypic species (Holmes and Roberts, 1981) and support for this concept (Bradbury, 1986; Moore et al., 2001), species circumscriptions could no longer be assumed. Application of names, whether as references to pathogenic (Kersters and De Ley, 1984) or natural (Holmes and Roberts, 1981; Bradbury, 1986; Young et al., 2005) species, must always be made explicit.

The taxonomic literature contains nomenclature based on distinct pathogenic characters (Jarvis et al., 1966; Sawada et al., 1993; Weibgen et al., 1993; Bouzar, 1994; Broughton, 2003; Farrand et al., 2003; Weller et al., 2004, and many recent reports of plant conjugation using the tumorigenic capabilities of agrobacterial strains carrying a Ti plasmid), as well as nomenclature reflecting natural classification (Bradbury, 1986; Nour et al., 1995; Rome et al., 1996; Tan et al., 1997; De Lajudie et al., 1998a; 1998b; Moore et al., 2001; Young et al., 2001; Eardly et al., 2005; Kwon et al., 2005; Young et al., 2005). Both forms of *Agrobacterium* nomenclature are now used.

7.2.2 Genus

The genus *Agrobacterium* and its species are validly published. Amalgamation with the genus *Rhizobium* does not affect the validity of the earlier nomenclature and the relevant nomenclature can still be used, although the special purpose nature of the nomenclature should not be lost sight of. A possible complication would only arise if in future a novel oncogenic pathogen was proposed as an *Agrobacterium* sp. In such a case, as well as a circumscription of the novel species, it might be required of proposers of novel *Agrobacterium* species that they produce a circumscription to justify *Agrobacterium* as a genus distinct from *Rhizobium*.

7.2.3 Vernacular alternative

Alternative to using genus and species names as binomials for the pathogens, which are both misleading and ambiguous, could be to use vernacular names. Rather than applying *Agrobacterium* spp. epithets to pathogenic strains in the sense proposed by Kersters and De Ley (1984), '*Agrobacterium* crown gall bacterium' and '*Agrobacterium* hairy root bacterium' could be terms used that accurately describe particular pathogens.

The term 'agrobacteria' can be used with little ambiguity to refer specifically to bacteria with oncogenic capability based on the expression of Ti or Ri plasmids irrespective of taxon, in a way analogous to the application of 'rhizobia'. The Subcommittee on the Taxonomy of *Agrobacterium* and *Rhizobium* of the International Committee on Systematics of Prokaryotes suggest that '.... the word rhizobium (plural rhizobia) can be used as a common name for legume-nodulating, nitrogen-fixing bacteria irrespective of genus' (Lindström and Martínez-Romero, 2005). The term agrobacteria will need to be carefully applied in future if more widespread examples are discovered of *Agrobacterium* species with legume-nodulating, nitrogen-fixing capabilities, or of *Rhizobium* species with oncogenic capabilities, or of '*Agrobacterium*' species from other environments (Popoff et al., 1984).

8 RELATIONSHIP OF *RHIZOBIUM* TO OTHER MEMBERS OF THE RHIZOBIACEAE

In the past, attempts to establish bacterial taxa above the level of genera foundered on the lack of common characters for comparison. Conn (1938) proposed the bacterial family, *Rhizobiaceae*, which originally included *Rhizobium*, *Chromobacterium* and *Alcaligenes*. Subsequently it included *Agrobacterium* and *Rhizobium* (Jordan and Allen, 1974), and *Agrobacterium*, *Rhizobium* and *Bradyrhizobium*.

Most recently, genera included in the family *Rhizobiaceae* are *Rhizobium, Agrobacterium, Allorhizobium*, *Carbophilus*, *Chelatobacter,* and *Ensifer*. The family as now defined is a phenotypically heterogeneous assemblage of aerobic, Gram-negative rod-shaped bacteria and is based solely on a 16S rRNA sequence analysis (Kuykendall, 2005). Young et al. (2001) noted that *Allorhizobium* was proposed solely on the basis of a 16S rDNA sequence comparison, but the analysis did not differentiate it from a clade comprising *R. galegae, R. huautlense*, *A. radiobacter, A. rubi,* and *A. vitis*. Willems et al. (2003) reported the synonymy of *Ensifer* and

Sinorhizobium Chen et al. (1988) and proposed that the name *Sinorhizobium* take priority although it was the later published name. Because the stability of names depends on priority of publication, a central principle of the International Code of Nomenclature of Prokaryotes (formerly the International Code of Nomenclature of Bacteria – (Lapage et al., 1975; 1992), this proposal was not accepted by the Judicial Commission of the International Committee on the Systematics of Prokaryotes (B. Tindall, pers. comm.). *Carbophilus* Meyer et al. (1994), isolated from soil, is strictly aerobic with facultative chemolithotrophic capacity. It does not fix nitrogen (Meyer, 2005). Members of *Chelatobacter* are obligately aerobic, nitrilotriacetate (NTA)-utilizing bacteria. The relationships of these genera are likely to be modified by more detailed studies based on a wider selection of sequence data that include more representatives of related taxa.

As noted, caution is in order when classification is based solely on a single sequence comparison because analyses give differing results depending on the chosen algorithm and, most particularly, on the selection of included sequences as shown by comparison of inferred phylogenies. Strains representing *Bartonella*, *Brucella*, *Blastobacter*, *Phyllobacterium* and *Mesorhizobium* have been reported as interspersed between the members of the *Rhizobiaceae* (Young and Huakka, 1996; De Lajudie et al., 1998; Young et al., 2001). Expanded studies can be expected to resolve these anomalies.

9 OTHER '*AGROBACTERIUM*' SPECIES

Based on phenotypic characterizations of bacteria isolated from marine and brackish environments, Rüger and Höfle (1992) proposed new species, *Agrobacterium atlanticum* and *A. meteori*, and reinstatement of *A. ferrugineum*, *A. gelatinovorum* and *A. stellulatum*. *A. atlanticum* has since been reclassified as *Ruegeria atlantica*, and *A. stellulatum* as *Stappia stellulata* by Uchino et al. (1998), who also proposed that *A. meteori* is a synonym of *A. atlanticum*. *A. ferrugineum* has been reclassified as *Pseudorhodobacter ferrugineus* by Uchino et al. (2002), and subsequently as *Hoeflea marina* by Peix et al. (2005). *A. gelatinovorum* has been reclassified as *Ruegeria gelatinovorans* by Uchino et al. (1998), and subsequently as *Thalassobius gelatinovorus* by Arahal et al. (2005).

10 SUMMARY

For most of the history of the genus, the unusual symptoms, aetiology, and genetics of pathogenicity shaped classification and nomenclature of *Agrobacterium* spp. However, as now understood, the distribution, diversity and systematics of these pathogenic bacteria is similar to those of other bacteria. They are small but significant populations of the soil microflora that comprise closely related bacteria with the capacity to exchange characteristic plasmids that usually confer oncogenic capabilities affecting plants, but can also form symbiotic nitrogen-fixing associations with legume plants. Clinical isolates have also been reported. The oncogenic species are members of the genus *Rhizobium*, whose species, until now, have largely been characterized on the basis of their symbiotic nitrogen-fixing associations with legume plants. However, the present record of characterized species is strongly biased in favour of organisms of anthropocentric interest and there is little basis for believing even that nitrogen-fixing or oncogenic strains are the predominant representatives of species with which they are associated; these nitrogen-fixing strains almost certainly represent only a small part of the greater diversity of soil bacteria potentially identifiable with this genus. The past literature that has separated these similar bacteria into distinct taxa has been an obvious hindrance to conceptualizing their ecology. If formal nomenclature is to serve the purpose of indicating natural bacterial relationships then oncogenic strains must be identified in *Rhizobium*. It can be expected that novel species of *Rhizobium* will be identified that are uncharacteristic of the genus as now understood, and in these circumstances, names established and maintained as keys to characters such as tumorigenic capabilities or nitrogen fixation can be expected only to become more confusing.

11 ACKNOWLEDGEMENTS

The New Zealand Foundation for Research Science and Technology for financial support under contract CO9X0001. S.R. Pennycook, Landcare Research, and D.R.W. Watson, Plant Diseases Division, DSIR, Auckland (retired) are thanked for their critical reading of the manuscript.

12 REFERENCES

Abe M, Kawamura R, Higashi S, Mori S, Shibata M, Uchiumi T (1998) Transfer of the symbiotic plasmid from *Rhizobium leguminosarum* biovar trifolii to *Agrobacterium tumefaciens*. J Gen Appl Microbiol 44: 65-74

Allen EK, Allen ON (1950) Biochemical and symbiotic properties of the rhizobia. Bacteriol Rev 14: 273-330

Allen ON, Holding AJ (1974) Genus II. *Agrobacterium* Conn. 1942. 359. *In* RE Buchanan, NE Gibbons, eds, Bergey's Manual of Determinative Bacteriology, 8th ed. The Williams and Wilkins Co., Baltimore, pp 264-267

Anderson AR, Moore LW (1979) Host specificity of the genus *Agrobacterium*. Phytopathol 69: 320-323

Arahal DR, Macián MC, Garay E, Pujalte MJ (2005) *Thalassobius mediterraneus* gen. nov., sp. nov., and reclassification of *Ruegeria gelatinovorans* as *Thalassobius gelatinovorus* comb. nov. Int J Syst Evol Microbiol 55: 2371-2376

Beijerinck MW, van Delden A (1902) Ueber die Assimilation des freien Stickstoffs durch Bakterien. Zentralbl Bakteriol Parasitenk Infektionskr Hyg Abt II 9: 3-43

Bergey DH, Breed RS, Hammer BW, Huntoon FM, Murray EGD, Harrison FC (1934) Bergey's Manual of Determinative Bacteriology, 4th ed. The Williams & Wilkins Co., Baltimore

Bergey DH, Breed RS, Murray EGD, Hitchens AP (1939) Bergey's Manual of Determinative Bacteriology, 5th ed. The Williams & Wilkins Co., Baltimore

Bonnier C (1953) Classification et spécificité de l'hôte, dans le genre *Rhizobium*. Atti VI Cong Int Microbiol 6: 325-327

Bouzar H (1994) Request for a judicial opinion concerning the type species of *Agrobacterium*. Int J Syst Bacteriol 44: 373-374

Bouzar H, Chilton WS, Nesme X, Dessaux Y, Vaudequin V, Petit A, Jones JB, Hodge NC (1995) A new *Agrobacterium* strain isolated from aerial tumors on *Ficus benjamina* L. Appl Environ Microbiol 61: 65-73

Bouzar H, Jones JB (2001) *Agrobacterium larrymoorei* sp. nov., a pathogen isolated from aerial tumours of *Ficus benjamina*. Int J Syst Evol Microbiol 51: 1023-1026

Bradbury JF (1986) Guide to Plant Pathogenic Bacteria. CAB International Mycological Institute, London

Brom S, Martínez E, Palacios R (1988) Narrow- and broad-host-range symbiotic plasmids of *Rhizobium* spp. strains that nodulate *Phaseolus vulgaris*. Appl Environ Microbiol 54: 1280-1283

Broughton WJ (2003) Roses by other names: taxonomy of the *Rhizobiaceae*. J Bacteriol 185: 2975-2979

Brenner DJ, Staley JT, Krieg NR (2005) Classification of prokaryotic organisms and the concept of bacterial speciation. *In* DJ Brenner, NR Krieg, JT Staley, eds, Bergey's Manual of Systematic Bacteriology. The Proteobacteria, Vol. 2A, 2nd ed. Springer-Verlag, New York, USA, pp 27-32

Burkholder WH, Starr MP (1948) The generic and specific characters of phytopathogenic species of *Pseudomonas* and *Xanthomonas*. Phytopathol 38: 494-502

Burr TJ, Katz BH, Bishop AL (1987) Populations of *Agrobacterium* in vineyard and non vineyard soils and grape roots in vineyards and nurseries. Plant Disease 71: 617-620

Chen LS, Figueredo A, Pedrosa FO, Hungria M (2000) Genetic characterization of soybean rhizobia in Paraguay. Appl Environ Microbiol 66: 5099-5103

Colwell RR (1970a) Polyphasic taxonomy of bacteria. *In* H Iizuka, T Hasegawa, eds, Culture Collections of Microorganisms. University Park Press, Baltimore, pp 421-426

Colwell RR (1970b) Polyphasic taxonomy of the genus *vibrio*: numerical taxonomy of *Vibrio cholerae*, *Vibrio parahaemolyticus*, and related *Vibrio* species. J Bacteriol 104: 410-433

Conn HJ (1938) Taxonomic relationships of certain non-sporeforming rods in soil. J Bacteriol 36: 320-321

Conn HJ (1942) Validity of the genus *Alcaligenes*. J Bacteriol 44: 353-360

De Lajudie P, Laurent-Fulele E, Willems A, Torck U, Coopman R, Collins MD, Kersters K, Dreyfus B, Gillis M (1998a) Description of *Allorhizobium undicola* gen. nov., sp. nov. for nitrogen-fixing bacteria efficiently nodulating *Neptunia natans* in Senegal. Int J Syst Bacteriol 48: 1277-1290

De Lajudie P, Willems A, Nick G, Moreira F, Molouba F, Hoste B, Torck U, Neyra M, Collins MD, Lindström K, Dreyfus B, Gillis M (1998b) Characterization of tropical tree rhizobia and description of *Mesorhizobium plurifarium* sp. nov. Int J Syst Bacteriol 48 Pt 2: 369-382

De Lajudie P, Willems A, Pot B, Dewettinck D, Maestrojuan G, Neyra M, Collins MD, Dreyfus B, Kersters K, Gillis M (1994) Polyphasic taxonomy of rhizobia: emendation of the genus Sinorhizobium and description of *Sinorhizobium meliloti* comb. nov., *Sinorhizobium saheli* sp. nov. and *Sinorhizobium teranga* sp. nov. Int J Syst Bacteriol 44: 715-733

De Ley J (1968) DNA base composition and hybridization in the taxonomy of the phytopathogenic bacteria. Annu Rev Phytopathol 6: 63-90

De Ley J (1972) *Agrobacterium*: intrageneric relationships and evolution. *In* HP Maas Geesteranus, ed, Plant Pathogenic Bacteria 1971, Proc 3rd Int Conf Pl Path Bact. Centre for Agricultural Publishing and Documentation, Wageningen, pp 251-259

De Ley J (1974) Phylogeny of the prokaryotes. Taxon 23: 291-300

De Ley J, Tijtgat R, De Smedt J, Michiels M (1973) Thermal stability of DNA: DNA hybrids within the genus *Agrobacterium*. J Gen Microbiol 78: 241-252

Duggar BM (1909) Fungous diseases of plants. Ginn & Co., Boston

Dye DW, Bradbury JF, Goto M, Hayward AC, Lelliott RA, Schroth MN (1980) International standards for naming pathovars of phytopathogenic bacteria and list of pathovar names and pathotype strains. Rev Pl Pathol 59: 153-168

Eardly BD, Nour SM, van Berkum P, Selander RK (2005) Rhizobial 16S rRNA and *dnaK* genes: mosaicism and the uncertain phylogenetic placement of *Rhizobium galegae*. Appl Environ Microbiol 71: 1328-1335

Euzéby JP (1997–2006) List of bacterial names with standing in nomenclature: a folder available on the internet (updated: March, 2006; revised URL: http://www.bacterio.cict.fr. Int J Syst Evol Microbiol 47: 590-592

Farrand SK, Van Berkum PB, Oger P (2003) *Agrobacterium* is a definable genus of the family *Rhizobiaceae*. Int J Syst Evol Microbiol 53: 1681-1687

Gao J, Terefework Z, Chen W-X, Lindstrom K (2001) Genetic diversity of rhizobia isolated from *Astragalus adsurgens* growing in different geographical regions of China. J Biotechnol 91: 155-168

Gaunt MW, Turner SL, Rigottier-Gois L, Lloyd-Macgilp SA, Young JPW (2001) Phylogenies of *atpD* and *recA* support the small subunit rRNA-based classification of rhizobia. Int J Syst Evol Microbiol 51: 2037-2048

Genetello C, Van Larebeke N, Holsters M, De Picker A, Van Montagu M, Schell J (1977) Ti plasmids of *Agrobacterium* as conjugative plasmids. Nature 265: 561-563

George ML, Young JPW, Borthakur D (1994) Genetic characterization of *Rhizobium* sp. strain TAL1145 that nodulates tree legumes. FEMS Microbiol Lett 40: 208-215

Ghosh W, Roy P (2006) *Mesorhizobium thiogangeticum* sp. nov., a novel sulfur-oxidizing chemolithoautotroph from rhizosphere soil of an Indian tropical leguminous plant. Int J Syst Evol Microbiol 56: 91-97

Gillis M, Vandamme P, De Vos P, Swings J, Kersters K (2005) Polyphasic taxonomy. *In* DJ Brenner, NR Krieg, JT Staley, eds, Bergey's Manual of Systematic Bacteriology. The Proteobacteria, Vol. 2A, 2nd ed. Springer-Verlag, New York, USA, pp 43-48

Goodfellow M, O'Donnell AG (1993) Roots of bacterial systematics. *In* M Goodfellow, AG O'Donnell, eds, Handbook of New Bacterial Systematics. Academic Press, London, San Diego, pp 3-54

Graham PH (1964) The application of computer techniques to the taxonomy of the root nodule bacteria of legumes. J Gen Microbiol 35: 511-517

Graham PH (1976) Identification and classification of root nodule bacteria. *In* Symbiotic Nitrogen Fixation in Plants. International Biological Programme 7. University Press, Cambridge, pp 99-112

Han S-Z, Wang E-T, Chen W-X (2005) Diverse bacteria isolated from root nodules of Phaseolus vulgaris and species within the genera Campylotropis and Cassia grown in China. Syst Appl Microbiol 28: 265-276

Heberlein GT, De Ley J, Tijtgat R (1967) Deoxyribonucleic acid homology and taxonomy of *Agrobacterium*, *Rhizobium*, and *Chromobacterium*. J Bacteriol 94: 116-124

Hildebrand EM (1940) Cane gall of brambles caused by *Phytomonas rubi* n. sp. J Agr Res 61: 685-696

Holmes B (1988) The taxonomy of *Agrobacterium*, Acta Hort 225: 47–52

Holmes B, Roberts P (1981) The classification, identification and nomenclature of agrobacteria. J Appl Bacteriol 50: 443-467

Hooykaas PJ, Klapwijk PM, Nuti MP, Schilperoort RA, Rörsch A (1977) Transfer of the *Agrobacterium tumefaciens* Ti plasmid to avirulent agrobacteria and to *Rhizobium ex planta*. J Gen Microbiol 98: 477-484

Hungria M, Campio RJ, Chueire LMO, Grange L, Megias M (2001) Symbiotic effectiveness of fast-growing rhizobial strains isolated from soybean nodules in Brazil. Biol Fert Soils 33: 387-394

Jarvis BDW, Sivakumaran SW, Tighe SW, Gillis M (1966) Identification of *Agrobacterium* and *Rhizobium* species based on cellular fatty acid composition. Pl Soil 184: 143-158

Jordan DC (1984) Genus1. *Rhizobium* Frank 1889, 339^{AL}. *In* NR Krieg, JG Holt, eds, Bergey's Manual of Systematic Bacteriology, 7th ed. The Williams & Wilkins Co., Baltimore, pp 235-242

Jordan DC, Allen ON (1974) Family III. *Rhizobiaceae* Conn 1938, 321, *In* RE Buchanan, NE Gibbons, eds, Bergey's Manual of Determinative Bacteriology, 8th ed. The Williams & Wilkins Co., Baltimore, pp 261-262

Jumas-Bilak E, Michaux-Charachon S, Bourg G, Ramuz M, Allardet-Servent A (1998) Unconventional genomic organization in the alpha subgroup of the *Proteobacteria*. J Bacteriol 180: 2749-2755

Keane PJ, Kerr A, New PB (1970) Crown gall of stone fruit. II. Identification and nomenclature of *Agrobacterium* isolates. Aus J Biol Sci 23: 585-595

Kerr A (1969a) Crown gall of stonefruit. I. Isolation of *Agrobacterium tumefaciens* and related fruits. Aus J Biol Sci 22: 111-116

Kerr A (1969b) Transfer of virulence between isolates of *Agrobacterium*. Nature 223: 1175-1176

Kerr A (1992) The genus *Agrobacterium. In* A Balows, HG Trüper, M Dworkin, W Harder, K-H Schleifer, eds, The Prokaryotes – a Handbook on the Biology of Bacteria, Vol. 3, 2ed ed. Springer-Verlag, Heidelberg, pp 2214-2235

Kerr A, Panagopoulos CG (1977) Biotypes of *Agrobacterium radiobacter* var. *tumefaciens* and their biological control. Phytopath Z 90: 172-179

Kerr A, Young JM, Panagopoulos CG (1978) Genus II. *Agrobacterium* Conn 1942. *In* Young JM, Dye DW, Bradbury JF, Panagopoulos CG, Robbs CF, eds, A proposed nomenclature and classification for plant pathogenic bacteria. NZ J Agric Res 21: 153-177

Kersters K, De Ley J (1975) Identification and grouping of bacteria by numerical analysis of their electrophoretic protein patterns. J Gen Microbiol 87: 333-342

Kersters K, De Ley J (1984) Genus III. *Agrobacterium* Conn. *In* NR Krieg, JG Holt, eds, Bergey's Manual of Systematic Bacteriology, Vol. 1, 8th ed. The Williams and Wilkins Co., Baltimore-London, pp 244-254

Kersters K, De Ley J, Sneath PHA, Sackin M (1973) Numerical taxonomic analysis of *Agrobacterium*. J Gen Microbiol 78: 227-239

Kersters K, Ludwig W, Vancanneyt M, De Vos P, Gillis P, Schleifer K-H (1996) Recent changes in the classification of pseudomonads. Syst App Microbiol 19: 465-477

Kuykendall LD (2005) Family *Rhizobiaceae* Conn 1938, 321^{AL}. *In* DJ Brenner, NR Krieg, JT Staley, GM Garrity, eds, Bergey's Manual of Systematic Bacteriology. The Proteobacteria, Vol. 2C, 2ed ed. Springer-Verlag, New York, pp 324-325

Kuykendall LD, Young JM, Martínez-Romero E, Kerr A, Sawada H (2005) Genus *Rhizobium* Frank 1889, 338^{AL} emend. *In* DJ Brenner, NR Krieg, JT Staley, GM Garrity, eds, Bergey's Manual of Systematic Bacteriology. The Proteobacteria, Vol. 2C, 2ed ed. Springer-Verlag, New York, pp 325-340

Kwon SW, Park JY, Kim JS, Kang JW, Cho YH, Lim CK, Parker MA, Lee GB (2005) Phylogenetic analysis of the genera *Bradyrhizobium*, *Mesorhizobium*, *Rhizobium* and *Sinorhizobium* on the basis of 16S rRNA gene and internally transcribed spacer region sequences. Int J Syst Evol Microbiol 55: 263-270

Lapage SP, Sneath PHA, Lessel EF, Skerman VBD, Seeliger HPR, Clark WA (1975) International Code of Nomenclature of Bacteria – Bacteriological Code, 1976 Revision. American Society of Microbiology, Washington: 1-180

Lapage SP, Sneath PHA, Lessel EF, Skerman VBD, Seeliger HPR, Clark WA (1992) International Code of Nomenclature of Bacteria – Bacteriological Code, 1990 Revision. American Society of Microbiology, Washington: 1-189

Lelliott RA (1972) The genus *Xanthomonas*. *In* HP Maas Geesteranus, ed, Plant Pathogenic Bacteria 1971, Proc 3rd Int Conf Pl Path Bact. Centre for Agricultural Publishing and Documentation, Wageningen, pp 269-271

Lindström K, Laguerre G, Normand P, Rassmussen U, Heulin T, De Lajudie P, Martínez-Romero E, Chen W-X (1998) Taxonomy and phylogeny of diazotrophs. *In* C Elmerich, A Kondorosi, WE William, eds, Biological Nitrogen Fixation for the 21st Century: Proceedings of the 11th International Congress on Nitrogen Fixation Institut Pasteur, Paris, France, 1997, July 20-25. Kluwer Academic, Dordrecht, pp 559-570.

Lindström K, Martínez-Romero ME (2005) International Committee on Systematics of Prokaryotes; Subcommittee on the taxonomy of *Agrobacterium* and *Rhizobium*. Minutes of the meeting, 26 July 2004, Toulouse, France. Int J Syst Evol Microbiol 55: 1383

Liu J, Wang ET, Chen WX (2005) Diverse rhizobia associated with woody legumes *Wisteria sinensis, Cercis racemosa* and *Amorpha fruticosa* grown in the temperate zone of China. Syst Appl Microbiol 28: 465-477

Ludwig W, Schleifer K-H (1999) Phylogeny of bacteria beyond the 16S rDNA standard. ASM News 65: 752-757

Martínez-Romero E, Palacios R, Sánchez F (1987) Nitrogen-fixing nodules induced by *Agrobacterium tumefaciens* harboring *Rhizobium phaseoli* plasmids. J Bacteriol 169: 2828-2833

Meyer OO (2005) Genus IV. Carbophilus Meyer, Stackebrandt and Auling,G, 1994, 182^{VP}. *In* DJ Brenner, NR Krieg, JT Staley, GM Garrity, eds, Bergey's Manual of Systematic Bacteriology. The Proteobacteria, Vol. 2C, 2nd ed. Springer-Verlag, New York, pp 346-347

Moffett ML, Colwell RR (1968) Adansonian analysis of the *Rhizobiaceae*. J Gen Microbiol 51: 245-266

Moore LW, Bouzar H, Burr TJ (2001) *Agrobacterium*. *In* NW Schaad, JB Jones, W Chun, eds, Laboratory Guide for Identification of Plant Pathogenic Bacteria. American Phytopathological Society Press, St. Paul, Minnesota, pp 17-33

Mougel C, Thioulouse J, Perrière G, Nesme X (2002) A mathematical method for determining genome divergence and species delineation using AFLP. Int J Syst Evol Microbiol 52: 573-586

Nour SM, Cleyet-Marel JC, Normand P, Fernandez MP (1995) Genomic heterogeneity of strains nodulating chickpeas (*Cicer arietinum* L.) and description of *Rhizobium mediterraneum* sp. nov. Int J Syst Bacteriol 45: 640-648

Ophel K, Kerr A (1990) *Agrobacterium vitis* sp. nov. for strains of *Agrobacterium* biovar 3 from grapevines. Int J Syst Bacteriol 40: 236-241

Østerås M, Stanley J, Finan TM (1995) Identification of *Rhizobium*-specific intergenic mosaic elements within an essential two-component regulatory system of *Rhizobium* species. J Bacteriol 177: 5485-5494

Panagopoulos CG, Psallidas PG (1973) Characteristics of Greek Isolates of *Agrobacterium tumefaciens* (E. F. Smith & Townsend) Conn. J Appl Bacteriol 36: 233-240

Panagopoulos CG, Psallidas PG, Alivizatos AS (1978) Studies on biotype 3 of *Agrobacterium radiobacter* var. *tumefaciens*. *In* Station de Pathologie Végétale et Phytobactériologie, ed., Plant Pathogenic Bacteria Proc 4th Internat Conf Plant Path Bact, Angers, France, pp 221-228

Peix A, Rivas R, Trujillo ME, Vancanneyt M, Velázquez E, Willems A (2005) Reclassification of *Agrobacterium ferrugineum* LMG 128 as *Hoeflea marina* gen. nov., sp. nov. Int J Syst Evol Microbiol 55: 1163-1166

Popoff MY, Kersters K, Kiredjian M, Miras I, Coynault C (1984) Position taxonomique de souches de *Agrobacterium* d'origine hospitalière. Ann Microbiol 135A: 427-442

Pribram E (1933) Klassification der Shizomyceten, F. Deuticke, Leipzig

Riker AJ, Banfield WM, Wright WH, Keitt GW, Sagen HE (1930) Studies on infectious hairy root of nursery trees of apples. J Agr Res 41: 507-540

Rome S, Fernandez MP, Brunel B, Normand P, Cleyet-Marel JC (1996) *Sinorhizobium medicae* sp. nov., isolated from annual *Medicago* spp. Int J Syst Bacteriol 46: 972-980

Rüger HJ, Höfle MG (1992) Marine star-shaped-aggregate-forming bacteria: *Agrobacterium atlanticum* sp. nov.; *Agrobacterium meteori* sp. nov.; *Agrobacterium ferrugineum* sp. nov., nom. rev.; *Agrobacterium gelatinovorum* sp. nov., nom. rev.; and *Agrobacterium stellulatum* sp. nov., nom. rev. Int J Syst Bacteriol 42: 133-143

Sadowsky MJ, Graham PH (1998) Soil biology of the *Rhizobiaceae*. *In* HP Spaink, A Kondorosi, PJJ Hooykaas, eds, The *Rhizobiaceae*: Molecular Biology of Model Plant-Associated Bacteria. Kluwer Academic Publications, Dordrecht, pp 155-172

Salmassi TM, Venkateswaren K, Satomi M, Newman DK, Nealson KH, Hering JG (2001) Isolation and characterization of *Agrobacterium albertimagni*, strain AOL15, sp. nov., a new arsenite oxidizing bacterium isolated from Hot Creek. Cited in EMBL Nucleotide Sequence Database: Accession AF316615

Sawada H, Ieki H (1992a) Crown gall of kiwifruit caused by *Agrobacterium tumefaciens* in Japan. Plant Dis 76: 212

Sawada H, Ieki H (1992b) Phenotypic characteristics of the genus *Agrobacterium*. Ann Phytopathol Soc Jpn 58: 37-45

Sawada H, Ieki H, Oyaizu H, Matsumoto S (1993) Proposal for rejection of *Agrobacterium tumefaciens* and revised descriptions for the genus *Agrobacterium* and for *Agrobacterium radiobacter* and *Agrobacterium rhizogenes*. Int J Syst Bacteriol 43: 694-702

Sawada H, Kuykendall LD, Young JM (2003) Changing concepts in the systematics of bacterial nitrogen-fixing legume symbionts. J Gen Appl Microbiol 49: 155-179

Segovia L, Piñero D, Palacios R, Martinez-Romero E (1991) Genetic structure of a soil population of nonsymbiotic *Rhizobium leguminosarum*. Appl Environ Microbiol 57: 426-433

Skerman VBD, McGowan V, Sneath PHA (1980) Approved lists of bacterial names. Int J Syst Bacteriol 30: 225-420

Smith EF, Townsend CO (1907) A plant tumor of bacterial origin. Science 25: 671-673

Sneath PHA (1984) Bacterial nomenclature. *In* NR Krieg, JG Holt, eds, Bergey's Manual of Systematic Bacteriology. The Williams & Wilkins, Baltimore, pp 19-23

Sneath PHA (1988) The phenetic and cladistic approaches. *In* DL Hawksworth, ed, Prospects in Systematics. Clarendon Press, Oxford, pp 252-273

Sneath PHA (2005) Bacterial nomenclature. *In* DJ Brenner, NR Krieg, JT Staley, eds, Bergey's Manual of Systematic Bacteriology. The Proteobacteria, Vol. 2A, 2nd. Springer-Verlag, New York, USA, pp 83-88

Sokal RR, Sneath PHA (1963) Principles of Numerical Taxonomy. W.H. Freeman, San Francisco

Stackebrandt E, Woese CR (1984) The phylogeny of prokaryotes. Microbiol Sci 1: 117-122

Stanier RY, Palleroni NJ, Doudoroff M (1966) The aerobic pseudomonads: a taxonomic study. J Gen Microbiol 43: 159-271

Starr MP, Weiss JE (1943) Growth of phytopathogenic bacteria in a synthetic asparagine medium. Phytopathol 33: 313-318

Süle S (1978) Biotypes of *Agrobacterium tumefaciens* in Hungary. J Appl Bacteriol 44: 207-213

Tan ZY, Xu XD, Wang ET, Gao JL, Martinez-Romero E, Chen WX (1997) Phylogenetic and genetic relationships of *Mesorhizobium tianshanense* and related rhizobia. Int J Syst Bacteriol 47: 874-879

Tepfer D (1984) Transformation of several species of higher plants by *Agrobacterium rhizogenes*: sexual transmission of the transformed genotype and phenotype. Cell 37: 959-967

Tighe SW, de Lajudie P, Dipietro K, Lindström K, Nick G, Jarvis BDW (2000) Analysis of cellular fatty acids and phenotypic relationships of *Agrobacterium, Bradyrhizobium, Mesorhizobium, Rhizobium* and *Sinorhizobium* species using the Sherlock Microbial Identification System. Int J Syst Evol Microbiol 50: 787-801

Uchino Y, Hamada T, Yokota A (2002) Proposal of *Pseudorhodobacter ferrugineus* gen. nov., comb. nov., for a non-photosynthetic marine bacterium, *Agrobacterium ferrugineum*, related to the genus *Rhodobacter*. J Gen Appl Microbiol 48: 309-319

Uchino Y, Hirata A, Yokota A, Sugiyama J (1998) Reclassification of marine *Agrobacterium* species: proposals of *Stappia stellulata* gen. nov., comb. nov., *Stappia aggregata* sp. nov., nom. rev., *Ruegeria atlantica* gen. nov., comb. nov., *Ruegeria gelatinovora* comb. nov., *Ruegeria algicola* comb. nov., and *Ahrensia kieliense* gen. nov., sp. nov., nom. rev. J Gen Appl Microbiol 44: 201-210

Ueda K, Seki T, Kudo T, Yoshida T, Kataoka M (1999) Two distinct mechanisms cause heterogeneity of 16S rRNA. J Bacteriol 181: 78-82

Van Larebeke N, Genetello C, Schell J, Schilperoort RA, Hermans AK, Van Montagu M, Hernalsteens JP (1975) Acquisition of tumour-inducing ability by non-oncogenic agrobacteria as a result of plasmid transfer. Nature 255: 742-743

Vandamme P, Pot B, Gillis M, de Vos P, Kersters K, Swings J (1996) Polyphasic taxonomy, a consensus approach to bacterial systematics. Microbiol Rev 60: 407-438

Velázquez E, Peix A, Zurdo-Piñeiro JL, Palomo JL, Mateos PF, Rivas R, Muñoz-Adelantado E, Toro N, García-Benavides P, Martínez-Molina E (2005) The coexistence of symbiosis and pathogenicity-determining genes in *Rhizobium rhizogenes* strains enable them to induce nodules and tumors or hairy roots in plants. Mol Plant Microbe Interact 18: 1325-1332

Watson B, Currier TC, Gordon MP, Chilton M-D, Nester EW (1975) Plasmid required for virulence of *Agrobacterium tumefaciens*. J Bacteriol 123: 255-264

Weibgen U, Russa R, Yokota A, Mayer H (1993) Taxonomic significance of the lipopolysaccharide composition of the three biovars of *Agrobacterium tumefaciens*. Syst Appl Microbiol 16: 177-182

Weller SA, Stead DE, Young JPW (2004) Acquisition of an *Agrobacterium* Ri plasmid and pathogenicity by other α-proteobacteria in cucumber and tomato crops affected by root mat. Appl Environ Microbiol 70: 2778-2785

White LO (1972) The taxonomy of the crown-gall organism *Agrobacterium tumefaciens* and its relationship to rhizobia and other agrobacteria. J Gen Microbiol 72: 565-574

Willems A, Collins MD (1993) Phylogenetic analysis of rhizobia and agrobacteria based on 16S rRNA gene sequences. Int J Syst Bacteriol 43: 305-313

Willems A, Fernandez-Lopez M, Munoz-Adelantado E, Goris J, De Vos P, Martinez-Romero E, Toro N, Gillis M (2003) Description of new *Ensifer* strains from nodules and proposal to transfer *Ensifer adhaerens* Casida 1982 to *Sinorhizobium* as *Sinorhizobium adhaerens* comb. nov. Request for an opinion. Int J Syst Evol Microbiol 53: 1207-1217

Willmitzer L, De Beuckeleer M, Lemmers M, van Montagu M, Schell J (1980) DNA from Ti plasmid present in nucleus and absent from plastids of crown gall plants. Nature 287: 359-361

Woese CR (1987) Bacterial evolution. Microbiol Rev 51: 221-271

Wolde-Meskel E, Terefework Z, Frostegård A, Lindström K (2005) Genetic diversity and phylogeny of rhizobia isolated from agroforestry legume species in southern Ethiopia. Int J Syst Evol Microbiol 55: 1439-1452

Young JM (2000a) Recent systematics developments and implications for plant pathogenic bacteria. *In* FG Priest, M Goodfellow, eds, Applied Microbial Systematics. Kluwer, Dordrecht, pp 133-160

Young JM (2000b) Suggestions for avoiding on-going confusion from the Bacteriological Code. Int J Syst Evol Microbiol 50: 1687-1698

Young JM (2001) Implications of alternative classifications and horizontal gene transfer for bacterial taxonomy. Int J Syst Evol Microbiol 51: 945-953

Young JM, Dye DW, Bradbury JF, Panagopoulos CG, Robbs CF (1978) A proposed nomenclature and classification for plant pathogenic bacteria. NZ J Agric Res 21: 153-177

Young JM, Kerr A, Sawada H (2005) Genus *Agrobacterium* Conn 1942, 359[AL]. *In* DJ Brenner, NR Krieg, JT Staley, GM Garrity, eds, Bergey's Manual of Systematic Bacteriology. The Proteobacteria, Vol. 2C, 2ed ed. Springer-Verlag, New York, pp 340-345

Young JM, Kuykendall LD, Martinez-Romero E, Kerr A, Sawada H (2001) A revision of *Rhizobium* Frank 1889, with an emended description of the genus, and the inclusion of all species of *Agrobacterium* Conn 1942 and *Allorhizobium undicola* de Lajudie et al. 1998 as new combinations: *Rhizobium radiobacter*, *R. rhizogenes*, *R. rubi*, *R. undicola* and *R. vitis*. Int J Syst Evol Microbiol 51: 89-103

Young JM, Kuykendall LD, Martinez-Romero E, Kerr A, Sawada H (2003) Classification and nomenclature of *Agrobacterium* and *Rhizobium* – a reply to Farrand, van Berkum and Oger. Int J Syst Evol Microbiol 53: 1689-1695

Young JM, Park DC, Weir BS (2004) Diversity of 16S rDNA sequences of *Rhizobium* spp. implications for species determinations. FEMS Microbiol Lett 238: 125-131

Young JM, Pennycook SR, Watson DRW (2006a) Proposal that *Agrobacterium radiobacter* has priority over *Agrobacterium tumefaciens*. Request for an opinion. Int J Syst Evol Microbiol 56: 491-493

Young JM, Takikawa Y, Gardan L, Stead DE (1992) Changing concepts in the taxonomy of plant pathogenic bacteria. Annu Rev Phytopathol 30: 67-105

Young JPW, Huakka KE (1996) Diversity and phylogeny of rhizobia. New Phytolol 133: 87-94

Young JPW, Crossman LC, Johnston AWB, Thomson NR, Ghazoui ZF, Hull KH, Wexler M, Curson ARJ, Todd JD, Poole PS, Mauchline TH, East AK, Quail MA, Churcher C, Arrowsmith C, Cherevach I, Chillingworth T, Clarke K, Cronin A, Davis P, Fraser A, Hance Z, Hauser H, Jagels K, Moule S, Mungall K, Norbertczak H, Rabbinowitsch E, Sanders M, Simmonds M, Whitehead S, Parkhill J (2006b) The genome of *Rhizobium leguminosarum* has recognizable core and accessory components. Genome Biology 7: R34, pp 1-26 [http://genomebiology.com/2006/7/4/R34]

Zeigler DR (2003) Gene sequences useful for predicting relatedness of whole genomes in bacteria. Int J Syst Evol Microbiol 53: 1893-1900

Chapter 6

THE INITIAL STEPS IN *AGROBACTERIUM TUMEFACIENS* PATHOGENESIS: CHEMICAL BIOLOGY OF HOST RECOGNITION

Yi-Han Lin[1], Andrew N. Binns[2] and David G. Lynn[1]

[1]Center for Fundamental and Applied Molecular Evolution, Departments of Chemistry and Biology, Emory University, Atlanta, GA 30322, USA; [2]Department of Biology, Plant Sciences Institute, University of Pennsylvania, Philadelphia, PA 19104, USA

Abstract. The biology of host recognition in *Agrobacterium tumefaciens* has set the tone for host interactions and xenognosis for several decades, and the twists and turns of the discoveries provide many valuable lessons and insights. From transposon mutagenesis enabling discovery of the initial chemical exchanges to two-component signal transduction and receptor identification, this organism continues to enrich our understanding of chemical ecology and pathogenic strategies. The complexity of the host commitment and the intricate nature of the evolved machinery remains awe inspiring. This system is now poised with the necessary chemical and biological resources, for both host and parasite, to reveal the detailed chemical biology that occurs within the host tissues. Here we review our current understanding of the signal exchanges, and highlight the many questions that remain to be addressed. We use this perspective to set the stage for the rich chemical biology this organism continues to offer.

1 INTRODUCTION

Transformation of plants by wild type strains of *Agrobacterium tumefaciens* results from the transfer of the Ti plasmid's T-DNA into host cells where it is ultimately integrated into chromosomal DNA and expressed (see other chapters in this volume). The virulence (*vir*) genes of the Ti plasmid required for virulence (Klee et al., 1983; Stachel and Nester, 1986) encode, for example, proteins involved in the processing, transport and ultimate integration of the T-DNA in the host (see other chapters). The resultant 'crown gall' tumors potentially yield great benefits to the infecting bacteria in the form of opines produced via enzymes encoded on the T-DNA (De Greve et al., 1982), yet the process requires significant energy expenditures by the bacterium and, accordingly, should be tightly regulated. In agreement with this hypothesis is the finding that the virulence genes are essentially silent unless the bacteria are exposed to a plant or plant derived molecules (Stachel et al., 1985b; Stachel et al., 1986). Activation of the genes in response to the host or host derived signals was first shown via experiments exploiting *vir::lacZ* fusions (Stachel et al., 1985a), and further experiments, importantly, showed that two virulence proteins encoded on the Ti plasmid, VirA and VirG, were required for the host-induced expression of the *vir* genes (Stachel and Zambryski, 1986; Engstrom et al., 1987; Winans et al., 1988).

Early studies of VirA and VirG demonstrated that they were related to the just discovered class of bacterial regulatory 'two component' systems (TCS) (Winans, 1991; Charles et al., 1992). TCS are comprised, minimally, of a histidine autokinase (often called sensor kinase) that responds, either directly or indirectly to environmental input, and a response regulator that is phosphorylated by its cognate histidine kinase (Robinson et al., 2000; Stock et al., 2000; West and Stock, 2001). Often, but not exclusively, the response regulator controls transcription of sets of genes via binding to specific regions of promoters and recruiting the RNA polymerase (Makino et al., 1993; Kenney et al., 1995). The phosphorylation status of the response regulator, which can also be affected by phosphatase activities of the sensor kinase as well as other proteins (Perego and Hoch, 1996), determines its activity. VirA is a membrane bound sensor kinase that has a large periplasmic domain (Melchers et al., 1989b; Chang and Winans, 1992), and VirG is the response regulator that binds to *vir* operon promoters resulting in their activation (Jin et al., 1990; Charles et al., 1992). Together, these proteins form the central control unit governing *vir* gene expression, though their activity is modulated by a series of other proteins encoded on the chromosome, most notably ChvE, a periplasmic protein with significant homology to periplasmic sugar binding proteins from a variety of bacteria (Cangelosi et al., 1990).

As noted above, the *vir* genes are essentially silent unless the bacterium is exposed to plants or plant derived molecules. Stachel et al (1985b) demonstrated that a particular phenol, 3,5-dimethoxy acetophenone (acetosyringone (AS)), isolated from medium in which plant roots or leaves had been cultured, is capable of inducing expression of the *vir* genes as reported by *vir::lacZ* expression. A wide variety of other related phenols were shown to be *vir* gene inducers (Stachel et al., 1985b; Melchers et al., 1989a; Duban et al., 1993; Lee, 1997; Peng et al., 1998), and the mechanistic and biological significance of this diversity will be considered in detail below. Soon after the discovery of the role played the phenols, several other conditions were found to be critical for optimal induction – low pH, low phosphate, temperature <30°C and sugars (Stachel et al., 1986; Winans et al., 1988; Cangelosi et al., 1990; Chang and Winans, 1992) – though phenols appear to be the only signal that is absolutely required. A variety of studies indicate that the response to each of these conditions is mediated by the VirA/VirG system, though in some cases they do so in concert with other gene products (e.g. ChvE).

The objectives of this review are to examine the diversity of signals and control mechanisms involved in *vir* gene expression from two different perspectives. First, what, exactly, are the signals, how are they recognized and what is the functional significance of the diversity? Second, how are the diverse signals integrated by the recognition system(s) to control response regulator activity? In relation to this question, we will explore the possible role played by other control systems, as well as understand how the induction of *vir* gene expression may have a more global (though possibly indirect) affect on bacterial gene expression in general.

2 SIGNAL DIVERSITY

2.1 Discovery of signals

The first demonstration of plant-induced *vir* gene expression came as a result of the development of the Tn*3*HoHo1 transposon (Stachel et al., 1985a). When inserted into an operon, in the correct orientation, the transposon creates a transcriptional or translational fusion, and, thus, could be used to monitor expression of the operon. Cosmids carrying portions of the *vir* region of the Ti plasmid were mutagenized with Tn*3*HoHo1 and then moved into a strain carrying a wild type Ti plasmid. The fundamental observation was that the vast majority of these insertions were silent, but activity was observed in response to co-cultivation of bacteria with plant cells (Stachel et al., 1985a; Stachel et al., 1986). Quickly it became apparent that conditioned medium from the cultured plant cells could induce expression of the *vir* genes, and the role of phenols in this process was discovered (Stachel et al., 1985b; Stachel et al., 1986). In these reports, the diversity of phenols was noted, including the critical role of the hydroxyl group and the ortho methoxy substituents (see below for more detail). Additionally, Stachel et al. (1986) reported a distinct pH optimum, and above pH 6.0, very little activity was observed. Intriguingly, low pH has a role in both AS-independent expression of *virG* (Winans et al., 1988), as well as in the VirA/VirG mediated control of phenol dependent vir gene expression (Melchers et al., 1989a). Examination of factors released by wheat seedlings that could induce *vir* gene expression, as well as the characterization of *Agrobacterium* chromosomal mutants that were deficient in virulence, lead to the discovery that certain monosaccharides could enhance the *vir* inducing activity of the phenols through the activity of a periplasmic protein, ChvE, which has homology to many known bacterial periplasmic sugar binding proteins (Cangelosi et al., 1990; Shimoda et al., 1993). Here we examine the chemical features of the signals and what that suggests about activity requirements, their location(s) in the host plant, and the possible relevance of the chemical and spatial diversity of the signals.

2.2 Structural class and diversity

Signaling in *Agrobacterium* pathogenesis is unique among TCS for many reasons, but central among them is the inherent interplay between the specificity and generality that forms our current understanding of signaling function. The ability of *Agrobacterium* to recognize and respond to seemingly all dicotyledonous plants must underpin the success of this multi-host pathogen. Consistent with this hypothesis, the VirA/VirG system responds to four distinct classes of molecular signals – phenols, sugars, phosphate and H^+ – and the magnitude of the response depends on integration across many of the inputs. For example, while the phenol is necessary and sufficient for VirA/VirG activation, both the sensitivity and the maximal response to the phenol are significantly enhanced in the presence of sugar and low pH (Chang and Winans, 1992).

Even more mechanistically interesting are the broad structural requirements within many of these molecular classes. Even though the sugar response requires reducing hexoses, several of the hexose diastereomers mediate similar responses. Even more remarkable, almost 70 different phenols have been reported to be active inducers (Palmer et al., 2004). While ortho-methoxy substituents do enhance activity, many other substitutions on the ring are compatible with the inducing activity. Therefore, not only is the response to many different classes of signal molecules a hallmark of *Agrobacterium* pathogenicity, but within each structural class, a broad range of structures can be accommodated to mediate the response.

2.3 Signal landscape

As described above, the host-sensing system of *Agrobacterium* recognizes and integrates at least 4 different signals, raising the critical question of why such a complex recognition landscape may have evolved. A logical hypothesis is that the presence of all the signals at some specific location, and/or developmental state, indicates an organ, tissue or cell type that is maximally susceptible to the pathogen. Interestingly, little is known about the specifics of this 'signal landscape' in the plants and how it might change during, for example, development, environmental stress or infection by *Agrobacterium* or by other pathogens. We do know that, for example, root exudates, conditioned culture medium and extracts of seedlings and plants are sources of *vir* inducing signals. Yet thorough qualitative and quantitative information is surprisingly scarce. An example of what might be occurring in relation to the *vir* inducing phenols is reflected in the composition of lignin during development of some plants. In many cases, the relative ratio of monomethoxy- (guaicyl) vs dimethoxy- (syringyl) phenols found in lignin varies significantly within different developmental stages of the plant (Dixon et al., 2001). This diversity is likely to reflect the availability of the phenolic monomers that are used in the biosynthesis of this polymer rather than post-polymerization modification of the constituents. The observation that the monomethoxyphenols are less efficient inducers of *vir* induction than the dimethoxy derivatives (Melchers et al., 1989a; Duban et al., 1993) suggests that different regions or cell types in the plant are likely to present different inducing environments as they relate to phenols. It is also likely that similar variation in the amounts and/or types of available sugars varies significantly throughout the plant.

Wounds on host plants are common sites of transformation by *A. tumefaciens*. While the wound may simply be a 'portal of entry', other specific processes may occur at the wound site and facilitate transformation (Braun, 1952). Identification of the *vir* inducing signals described above has provided one likely explanation for the importance of the wound site: high activity of the phenylpropanoid pathway, low pH, and sugars associated with cell wall synthesis are routinely associated with wound repair (Matsuda et al., 2003). Even at this most widely hypothesized source, the actual distribution of the signal molecules is not known. For example, are all wounds equal in their capacity to produce quantities (and types) of phenols and sugars; are they sufficiently acidified to efficiently induce *vir* gene expression? While the wound site is of obvious importance in *vir* gene induction and tumorigenesis, two studies have established that transformation can occur in unwounded tobacco seedlings and in one case, *vir* gene expression could be observed in the bacteria in the absence of wounding (Escudero and Hohn, 1997; Brencic et al., 2005). These observations raise important questions concerning the infection process and the role of wounding. What is the relative efficiency of *vir* induction, for example, at wounded vs. non-wounded tissues of the same tissue type? How does the signal landscape differ in these two cases? And, importantly, is the efficiency of transformation the same or different at such sites and is this efficiency related only to *vir* induction? Answering these questions will provide not only information about the specifics of *vir* induction, but also let us determine whether the signal landscape is monitored by the bacterium in order to provide information about the competence of the plant cells for transformation.

3 SIGNAL RECOGNITION, INTEGRATION AND TRANSMISSION

3.1 Signal recognition

As noted above, numerous signals regulate the VirA/VirG system. While the identification of host signals involved in controlling pathogenesis has been accomplished, only in one case – the sugars – is there good evidence for the means by which the signal is recognized. The physical basis of signal perception, integration and transmission will be impossible to understand if the sites of signal perception are not defined. Here we will review the progress towards identifying specific regions of the VirA/VirG control system that are involved in signal recognition and the means by which physical proof of this might be accomplished.

3.1.1 Phenols

Biochemical evidence now exists for the interaction of the inducing phenols with a receptor. The chirality of the *para*-substituent is critical for activity (Campbell et al., 2000); consistent with the three dimensional structure being recognized by a "receptor" and necessary for function. The *ortho*-methoxy groups and *para*-substituents all potentially contribute to binding affinity. On the basis of the broad structural and physical attributes of the phenols necessary for induction and the general correlation with phenol pKa, a 'proton transfer model' of signal recognition and receptor activation has been put forward (Palmer et al., 2004). This model holds that activation involves donation of a proton to a basic site on the receptor, initiating allosteric changes and activation via the ultimate phosphotransfer to VirG.

While the chemistry of the inducing phenols has led to models suggesting particular features of a receptor as described above, there remains no evidence for a physical interaction between these compounds and any of the components of the VirA/VirG system. Affinity labeling and affinity chromatography protocols identified several proteins in *Agrobacterium* that can interact with phenols (Lee et al., 1992; Dye and Delmotte, 1997), but VirA was not amongst them, nor is there evidence that the identified phenol binding proteins are required for *vir* induction. Despite this paucity of physical evidence, genetic evidence strongly suggests that VirA is, indeed, the phenol perceiving element. The clearest genetic evidence comes from studies in which a phenol responsive VirA/VirG mediated signal transduction system could be reconstructed in *E. coli* (Lohrke et al., 2001). To be successful, the RpoA alpha-subunit of RNA polymerase from *A. tumefaciens* had to be present in *E. coli* as well as VirA and VirG – VirG-P apparently requires RpoA to initiate transcription at the *vir* promoter (Lohrke et al., 1999).

Strong evidence now exists that phenol perception can be moderated by a variety of factors that are not part of the phenol binding site. One example is mutations in the receiver domain of VirA that have been reported to broaden the range of phenols capable of activating *vir* gene transcription, despite the fact that this domain is not necessary for the phenol response (Chang et al., 1996). A similar case is seen in the capacity to recognize sugars via ChvE (see below) and alter phenol specificity (Peng et al., 1998). In each of these cases, one could envision a binding site, or 'activation energy', necessary for phenol mediated signaling being altered as a result of changes in the VirA conformation. Yet another case is the report of mutations in *A. tumefaciens* that alter sensitivity to phenols but that do not map to the Ti plasmid or *virA* (Campbell et al., 2000). The genes responsible for this phenotype have, however, not been isolated. Thus, the picture of how phenols are perceived by *A. tumefaciens* is not clear and remains a fundamental goal of current research.

3.1.2 Sugars

The sugar environment is critical in terms of *vir* gene activation through the VirA/VirG system, and two major types of responses to these sugars have been noted. First, the sensitivity towards the inducing phenolics is greatly increased in the presence of sugar. For example, the dose of AS required for half-maximal *vir* inducing activity can be 20-50 fold lower in the presence of 'inducing' sugars (Cangelosi et al., 1990; Shimoda et al., 1990). Second, the maximal level of *vir* gene induction in response to phenols is 10-20 fold higher in the presence of such sugars. Both of these phenotypes require the presence of ChvE, a chromosomally encoded protein, as well as VirA and VirG. ChvE is an abundant periplasmic protein and most homologous to a series of periplasmic sugar-binding proteins that are involved in sugar transport (via ABC transporters) and chemotaxis to sugars. Of these, ChvE is most similar to the ribose-binding protein (RBP) of *E. coli* (Gao et al., 2006). X-ray crystal derived structures of several such proteins, in the presence or absence of sugars, has revealed the sugar binding site (Ricagno et al., 2006; Tremblay et al., 2006). When the ChvE sequence is modeled via threading onto this structure, the sugar-binding site is apparent (Gao and Lynn, 2005). Genetic and physical analysis of ChvE and its interaction with VirA remains critical to an understanding of how this interaction is transmitted through the inner membrane to regulate phenol perception. Early models of this transfer of information are just now emerging through investigations of different alleles of VirA (Gao and Lynn, 2007).

3.1.3 pH

A mildly acidic pH (optimum pH 5.5) has a marked affect on *vir* gene expression and does so through several mechanisms. *virG* expression is induced by low pH through activities on the "P2" portion of the *virG* promoter, and this regulation is independent of VirA (Mantis and Winans, 1992; Chang and Winans, 1996). The means by which regulation is controlled is not known, though one possibility is via the ChvG/ChvI two component system which appears to be involved in the regulation of numerous genes via acidic pH (Li et al., 2002). The sensor kinase ChvG is required for this regulation, but the means by which it is responding to pH is unknown. Beyond its affects on *virG* expression, low pH affects *vir* gene expression through VirA. Expression of *virG* from promoters (P_{lac} or P_{N25}) that are not sensitive to pH results in *Agrobacterium* strains that remain pH sensitive (Chang and Winans, 1996; Gao and Lynn, 2005): 10-20 fold increase in the expression of VirA dependent *vir*-reporter fusions occur at pH 5.5 in comparison to pH 7.0. The genetic basis of this regulation is complex and involves both the periplasmic domain of VirA and ChvE (Gao and Lynn, 2005). When ChvE is absent, the VirA/VirG system is responsive to phenols but, neither sugar or pH (pH5-7) affect the that response (Gao and Lynn, 2005). Additionally, a large deletion of the periplasmic domain (Melchers et al., 1989b) or small insertions at numerous locations across the periplasmic domain (Nair GR and Binns AN, unpublished), results in the capacity of those forms of VirA to support *vir* gene expression at pH 7. An important interpretation of these results is that pH5.5, along with ChvE, is required to relieve a repressive influence of the periplasmic domain of VirA. Intriguingly, two mutant forms of VirA– $VirA^{\Delta 242\text{-}257}$ and $VirA^{E255L}$ result in strains that respond poorly if at all to sugar, but continue to exhibit pH regulation, thereby uncoupling these activities (Gao and Lynn, 2005).

3.2 Signal integration and transmission

3.2.1 HK/RR structures and transmission

As shown in *Figure 6-1*, the structures from several histidine kinases (HK) and response regulators (RR) have been reported recently (Robinson et al., 2000; Stock et al., 2000; West and Stock, 2001). The data suggest two conserved domains in the HKs: the dimerization domain, a four-helix bundle containing the conserved histidine reside, and the ATP-binding phosphotransfer domain. The RRs contain a five strand α/β fold with the conserved aspartate residue located in an acidic pocket (*Figure 6-1b*) (Robinson et al., 2003). Recently, the crystal structure of the HK from *Thermotoga maritima* was solved (*Figure 6-1a*) (Marina et al., 2005), positioning each domain and limiting the possibilities for inter-domain phosphotransfer as well as the inter-molecular association of HK and RR proteins. Both HK and RR are generally highly conserved in primary sequence, and yet the real power of the TCSs resides in their broad scope and utility. From nutrients such as amino acids, monosaccharides and oxygen to osmoregulation, pathogenicity and cell density, bacteria rely on signal perception and functional integration through these elements. Most of the signal sensing domains reside at the N terminus of HK (*in cis*) and regulate HK activity intramolecularly. An example is the heme-binding oxygen sensing PAS domain in FixL of *Bradyrhizobium japonicum* (*Figure 6-1c*) (Gong et al., 1998). The HK activity turns 'ON' without oxygen binding to the sensing domain, and turns 'OFF' when oxygen binds. On the other hand, some of the HKs have the signal sensing domain separated, in *trans*, and regulate kinase activity intermolecularly. The most well-known example is the signal sensing domains of *E. coli* chemotaxis histidine protein kinase CheA (Falke and Hazelbauer, 2001; Wadhams and Armitage, 2004).

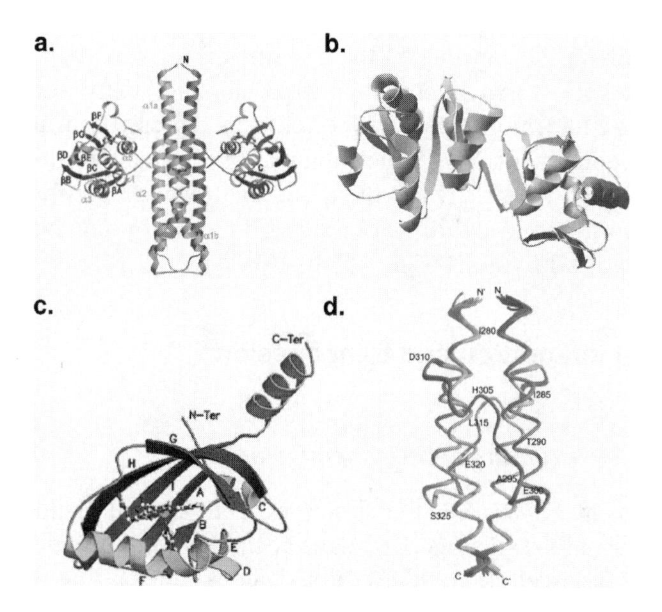

Figure 6-1. Structures of some functional and regulatory domains of TCS. (a) Crystal structure of the *Thermotoga maritima* HK0853 showing a dimer of HK0853 with each subunit containing dimerization and ATP-binding domains. (b) Crystal structure of the response regulator DrrB of *T. maritima* with the receiver domain colored in blue and yellow, the conserved Asp residue in red, and the DNA binding domain in gray. (c) The crystal structure of the PAS domain of *Bradyrhizobium japonicum* oxygen sensing protein FixL. The bound heme cofactor is shown as a ball-and-stick representation. (d) NMR structure of a HAMP domain of Af1503 from *Archaeoglobus fulgidus*. The dimeric domain maintains a coiled-coil structure.

CheA responds to various chemoattractants by coordinating with different transmembrane signal binding proteins, for example Tar for aspartate sensing, Trg for ribose and galactose sensing, and Tap for dipeptide sensing. Since the signal sensing domain controls the conserved kinase activity, understanding the signal sensing domain structure becomes critical to elucidating the regulation mechanisms.

Two of the most commonly seen domains in the signal sensing region of two-component systems are the PAS *(Per, ARNT, Sim)* and the HAMP *(histidine kinases, adenylyl cyclases, methyl-binding proteins and phosphatases')* domains (Ponting and Aravind, 1997; Aravind and Ponting, 1999; Williams and Stewart, 1999). In addition to the oxygen binding PAS domain of FixL (*Figure 6-1c*), *B. subtilis* KinA, *B. bronchiseptica* BvgS, *E. coli* ArcB and NtrB all contain one or more PAS domains, but few of their signals have been identified (Ponting and Aravind, 1997). The HAMP domain, which is ~50aa in length, is also observed in histidine kinases and methyl-accepting proteins (Aravind and Ponting, 1999; Williams and Stewart, 1999). This domain usually presents as a linkage between the signal perception region and the kinase domain, and has a coiled-coil like structure (*Figure 6-1d*) with both rotation and piston motions possible for signal transmission (Ottemann et al., 1999; Hulko et al., 2006).

3.2.2 Model for signal integration in VirA/VirG

While several complex phosphorelay systems exist, including sporulation in *Bacillus subtilis* and osmosensing in *Saccaromyces cerevisiae*, and rely on more than one pair of HK/RR for signal transmission (Varughese, 2002; Stephenson and Lewis, 2005), the basic chemical phosphotransfer steps are expected to provide critical points for the regulation signal input. At this point, studies of the signal-mediated kinase activity have been limited by knowledge of the signal. In contrast, the signals for the VirA/VirG system in *Agrobacterium tumefaciens* are well characterized, and the accumulated evidence points toward specific domains responsible for signal sensing (Chang and Winans, 1992). The limitation in the VirA/VirG system is how multiple input domains, the periplasmic domain responsible for pH and monosaccharide sensing, the cytoplasmic linker domain for phenol signaling, and the receiver domain, which is highly homologous with the dimerization domain of most response regulators, all function cooperatively to optimize output.

These cooperative functions certainly appear intricate and tightly integrated. As mentioned above, the presence of sugar and low pH not only increase the sensitivity of the system for the phenol but also increase the level of the response. The previously assigned repressive role of the receiver domain (Chang et al., 1996) now appears to be a function of the cytosolic VirG concentration; at low VirG levels the receiver domain functions as an activator (Fang F, Lynn DG, Binns AN, Wise AA, submitted). As discussed above, there has been significant debate as to whether auxiliary proteins are responsible for phenol binding, similar to those seen for ChvE and sugar binding (Lee et al., 1992; Lee et al., 1995, 1996; Campbell et al., 2000; Lohrke et al., 2001; Joubert et al., 2002), and a precise role for these proteins has yet to be assigned.

Recently predicted secondary structure analyses of the linker domain finds homology with GAF domains (*Figure 6-2*) (Gao and Lynn, 2007). Like the PAS domains, the (cGMP-specific and -stimulated phosphodiesterase, *Anabaena* adenylate cyclase, and *E. coli* FhlA) GAF domain is a small molecule binding element which usually localizes at the N terminus of functional proteins, including histidine kinases (Aravind and Ponting, 1997; Ho et al., 2000). These elements appear to regulate functional activity

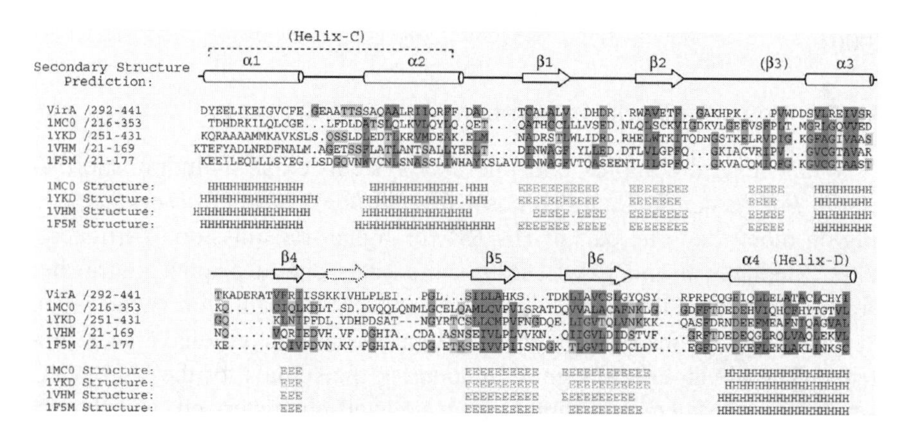

Figure 6-2. Alignment of VirA linker with known GAF structures. Sequence alignment is from PFAM database and refined based on structures (protein data bank ID: 1MC0, 1YKD, 1VHM and 1F5M). H and E represent α-helices and β-strands observed in the structure, respectively. The secondary structure of VirA linker was predicted by SAM-T02 method. Predicted α-helices and β-strands are marked as cylinders and arrows. The dotted arrow indicates a β-strand but not conserved among GAF structures. Residues with remote homology were colored as the following: blue, hydrophobic and aromatic residues (L, I, V, M, C, A, F, W); red, charged residues (D, E, K, R); orange, G; yellow, P; green, (S, T, N, Q). Helix C refers to the region of α1 and α2, while helix D is α4.

by small molecule binding, however, the mechanism of regulation remains unclear. The VirA linker domain regulates the kinase activity upon phenol binding, and the signal transmission is proposed to be through two amphipathic helixes, helix C – located at the N terminus of the linker domain connected to TM2, and helix D – located at the C terminus of the linker domain and connected to K domain (*Figure 6-2*). These helices organize in an anti-parallel dimer in the GAF structural predictions.

Fusing a GCN4 helix at the N terminus of either linker helix successfully 'ratchets ON' kinase activity in the absence of phenols (*Figure 6-3*) (Wang et al., 2002). GCN4 provides a strong leucine-zipper homo-dimerization interface, so that fusing GCN4 in front of the amphipathic helix C

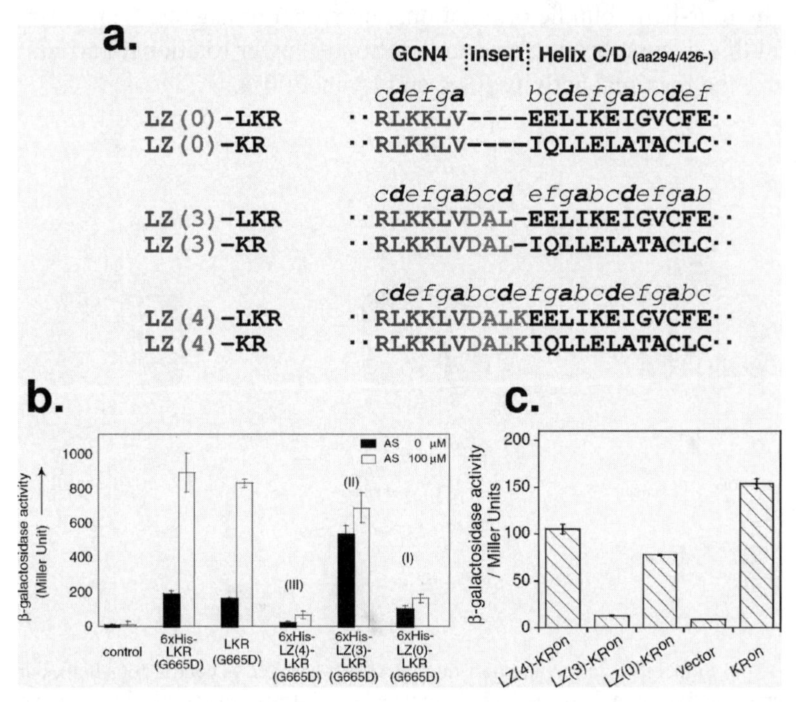

Figure 6-3. N'-fused GCN4 leucine zipper of helix C/D of the linker domain. (a) The heptad repeats registry of LZ(0/3/4) chimeras. Fused at aa294 is the helix C fusion (LZ-LKR), while fusion made at aa426 is the helix D fusion (LZ-KR). LZ residues are shown in blue, the inserting amino acids between LZ and VirA are in red, and VirA sequence in black. (b) β-galactosidase activity of different LZ-LKR (helix C) fusions. G665D is a constitutive on mutant of VirA, using this high activity mutant simplified the activity measurement. (c) β-galactosidase activity of different LZ-KR (helix D) fusions with ON mutant also at G665D (see Gao and Lynn, 2007).

or D, with different amino acid insertions at the fusion point, could change the registry of the heptad repeats of each helix (*Figure 6-3a*), mimicking a rotational motion of the coiled-coil. Successfully engineering the kinase locked in 'ON' or 'OFF' positions by GCN4 is consistent with a model in which phenol binding induces a rotational motion, a motion proposed in other signaling events (Kwon et al., 2003; Hulko et al., 2006), and this information is propagated through a four-helix bundle model for signal transmission to the kinase (*Figure 6-4*) (Gao and Lynn, 2007). In the absence of phenol, VirA is at the 'OFF' state; upon phenol binding, the rotation of the helices switch the protein to the 'ON' state. In this model, the sugar and pH sensed by the periplasmic domain would effectively fine-tune the four-helix bundle orientation, and recent mutagenesis studies have argued this occurs through a piston motion to lower rotational barriers and enhance the maximal activity (Gao and Lynn, 2007).

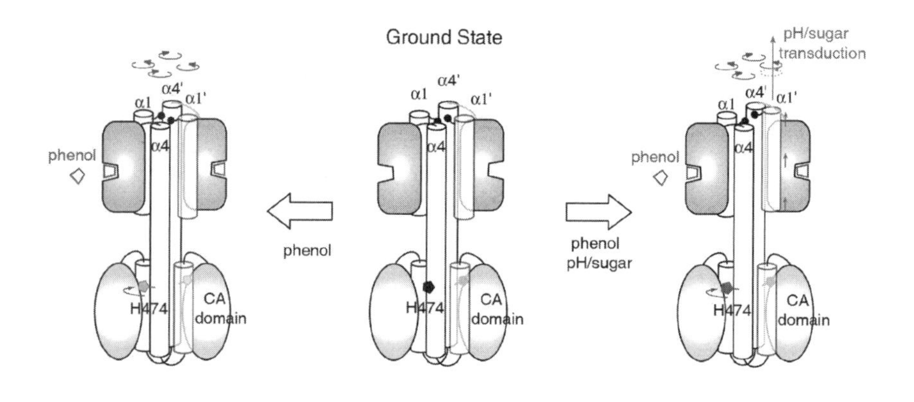

Figure 6-4. Signal integration and transduction of VirA linker. A central four-helix bundle formed by Helix-C and D (α1 and α4 in predicted GAF structure) is critical for both periplasmic and cytoplasmic signaling. Helix-D is directly connected to the histidine containing helix of kinase and the rotational motion modulates the phophosrylation of the histidine residue (pentagon). Phenol perception by the linker domain is postulated to initiate the rotation within the four-helix bundle. The periplasmic sensing of pH/sugar is proposed to induce a sliding of the signaling helices, thus enhancing the phenol response.

The secondary structure predictions and the crystal structure of the histidine kinase of *Thermotoga maratima* (Marina et al., 2005) suggest that Helix D is directly connected to the major helix of the K domain and rotation may well position the critical histidine for phosphorylation. Like other kinases, the VirA dimer appears to mediate phosphotransfer intermolecularly between the two subunits of this dimer (Pan et al., 1993; Brencic et al., 2004; Wise et al., 2005), arguing that both kinase and linker regions exist as a four-helix bundle. Although mutageneses studies have identified several residues which abolish phenol sensing and support this model (Toyoda-Yamamoto et al., 2000), acquiring physical evidence for the rotational motion of the four-helix bundles, in both the linker and the kinase, stands as the critical challenge for future experiments.

Unlike other histidine kinases, VirA maintains an extra receiver domain at its C terminus. When tested in otherwise wild type strains, this domain is required for phenol mediated activation in the absence of sugar, and is critical for maximal activation by phenols in the presence of sugar (Fang F, Lynn DG, Binns AN, Wise, AA, submitted). Earlier studies indicating that the receiver domain was repressive were all done in the presence of constitutively expressed VirG (Chang et al., 1996), further enriching our understanding of the role played be the receiver. Moreover, the homology of the VirA receiver domain with the N-terminal of VirG has suggested that phosphorylation at D766 may be critical to this regulatory activity, but *in vivo* phosphorylation analysis suggested that the receiver is not phosphorylated at an aspartic acid residue. It may well be that the R domain functions as a guide for VirG phosphorylation at low VirG concentrations and this function is disrupted as the VirG concentration is elevated. Accordingly, the physiological significance of the regulatory role of the receiver domain is just now emerging.

In contrast to the R domain, VirG is readily phosphorylated in *in vivo* experiments (Mukhopadhyay et al., 2004). These experiments suggest that the accumulation of phosphate on VirA occurs in the absence of phenol, and that phosphoryltransfer to VirG takes place when phenol is added. Moreover, the conserved aspartate at position 52 directly accepts the phosphate from VirA (Jin et al., 1990). The molecular mechanism of switching VirG 'ON' was deconvoluted with two AS-independent alleles, I77V/D52E and N54D (Scheeren-Groot et al., 1994; Gubba et al., 2005; Gao et al., 2006). An 'aromatic switch' mechanism for response regulator dimerization was proposed for VirG (Gao et al., 2006), but how this VirG phosphorylation is regulated by the extra R domain in VirA, and the physiological significance of the VirG concentration difference has not been resolved.

4 SUMMARY

Taken together, the VirA/VirG system presents a model for our understanding of pathogenesis signaling and highlights the highly modular protein nature of TCS. That said, major questions remain as to how the signals are actually perceived and how the information is integrated and transmitted to output. Maybe even more important are how these precise signaling networks exploit the biological matrix in which pathogenesis has evolved. While we know the signals and the protein components involved in signal transmission, the signaling landscape within the host and the regulation that enables *Agrobacterium tumefaciens* to function as a successful multi-host pathogen is only now emerging. We hope that this review sets these questions in clear contrast for those designing experiments to resolve the critical chemical events occurring at the host/pathogen interface.

5 ACKNOWLEDGEMENTS

We gratefully acknowledge the NIH RO1 GM047369 for support. We are very grateful to the present and past members of the Binns and Lynn labs for many valuable contributions, entertaining discussions, and the opportunity to work in a warmly collaborative and cohesive environment.

6 REFERENCES

Aravind L, Ponting CP (1997) The GAF domain: an evolutionary link between diverse phototransducing proteins. Trends Biochem Sci 22: 458-459

Aravind L, Ponting CP (1999) The cytoplasmic helical linker domain of receptor histidine kinase and methyl-accepting proteins is common to many prokaryotic signalling proteins. FEMS Microbiol Lett 176: 111-116

Braun AC (1952) Conditioning of the host cell as a factor in the transformation process in crown gall. Growth 16: 65-74

Brencic A, Angert ER, Winans SC (2005) Unwounded plants elicit *Agrobacterium vir* gene induction and T-DNA transfer: transformed plant cells produce opines yet are tumor free. Mol Microbiol 57: 1522-1531

Brencic A, Xia Q, Winans SC (2004) VirA of *Agrobacterium tumefaciens* is an intradimer transphosphorylase and can actively block *vir* gene expression in the absence of phenolic signals. Mol Microbiol 52: 1349-1362

Campbell AM, Tok JB, Zhang J, Wang Y, Stein M, Lynn DG, Binns AN (2000) Xenognosin sensing in virulence: is there a phenol receptor in *Agrobacterium tumefaciens*? Chemistry & Biology 7: 65-76

Cangelosi GA, Ankenbauer RG, Nester EW (1990) Sugars induce the *Agrobacterium* virulence genes through a periplasmic binding protein and a transmembrane signal protein. Proc Natl Acad Sci USA 87: 6708-6712

Chang CH, Winans SC (1992) Functional roles assigned to the periplasmic, linker, and receiver domains of the *Agrobacterium tumefaciens* VirA protein. J Bacteriol 174: 7033-7039

Chang CH, Winans SC (1996) Resection and mutagenesis of the acid pH-inducible P2 promoter of the *Agrobacterium tumefaciens virG* gene. J Bacteriol 178: 4717-4720

Chang CH, Zhu J, Winans SC (1996) Pleiotropic phenotypes caused by genetic ablation of the receiver module of the *Agrobacterium tumefaciens* VirA protein. J Bacteriol 178: 4710-4716

Charles TC, Jin S, Nester EW (1992) Two-component sensory transduction systems in *Phytobacteria*. Ann Rev Phyto 30: 463-484

De Greve H, Dhaese P, Seurinck J, Lemmers M, Van Montagu M, Schell J (1982) Nucleotide sequence and transcript map of the *Agrobacterium tumefaciens* Ti plasmid-encoded octopine synthase gene. J Mol Appl Genet 1: 499-511

Dixon RA, Chen F, Guo D, Parvath K (2001) The biosynthesis of monolignols: a metabolic grid or independent pathways to guaiacyl and syringal units. Phytochemistry 57: 1069

Duban ME, Lee KH, Lynn DG (1993) Strategies in pathogenesis: mechanistic specificity in the detection of generic signals. Mol Microbiol 7: 637-645

Dye F, Delmotte FM (1997) Purification of a protein from *Agrobacterium tumefaciens* strain A348 that binds phenolic compounds. Biochem J 321: 319-324

Engstrom P, Zambryski P, Van Montagu M, Stachel SE (1987) Characterization of *Agrobacterium tumefaciens* virulence proteins induced by the plant factor acetosyringone. J Mol Biol 197: 635-645

Escudero J, Hohn B (1997) Transfer and integration of T-DNA without cell injury in the host plant. Plant Cell 9: 2135-2142

Falke JJ, Hazelbauer GL (2001) Transmembrane signaling in bacterial chemoreceptors. Trends Biochem Sci 26: 257-265

Gao R, Lynn DG (2005) Environmental pH sensing: resolving the VirA/VirG two-component system inputs for *Agrobacterium* pathogenesis. J Bacteriol 187: 2182-2189

Gao R, Lynn DG (2007) Integration of rotation and piston motions in coiled-coil signal transduction. J Bacteriol 189: 6048-6056

Gao R, Mukhopadhyay A, Fang F, Lynn DG (2006) Constitutive activation of two-component response regulators: characterization of VirG activation in *Agrobacterium tumefaciens*. J Bacteriol 188: 5204-5211

Gong W, Hao B, Mansy SS, Gonzalez G, Gilles-Gonzalez MA, Chan MK (1998) Structure of a biological oxygen sensor: a new mechanism for heme-driven signal transduction. Proc Natl Acad Sci USA 95: 15177-15182

Gubba S, Xie Y, Das A (2005) Regulation of *Agrobacterium tumefaciens* virulence gene expression: isolation of a mutation that restores *virGD52E* function. Mol Plant Microbe Interact 8: 788-791

Ho Y-S, Burden LM, Hurley JH (2000) Structure of the GAF domain, a ubiquitous signaling motif and a new class of cyclic GMP receptor. EMBO J 19: 5288-5299

Hulko M, Berndt F, Gruber M, Linder JU, Truffault V, Schultz A, Martin J, Schultz JE, Lupas AN, Coles M (2006) The HAMP domain structure implies helix rotation in transmembrane signaling. Cell 126: 929-940

Jin SG, Roitsch T, Christie PJ, Nester EW (1990) The regulatory VirG protein specifically binds to a *cis*-acting regulatory sequence involved in transcriptional activation of *Agrobacterium tumefaciens* virulence genes. J Bacteriol 172: 531-537

Joubert P, Beaupere D, Lelievre P, Wadouachi A, Sangwan RS, Sangwan-Norreel BS (2002) Effects of phenolic compounds on *Agrobacterium vir* genes and gene transfer induction – a plausible molecular mechanism of phenol binding protein activation. Plant Sci 162: 733-743

Kenney LJ, Bauer MD, Silhavy TJ (1995) Phosphorylation-dependent conformational changes in OmpR, an osmoregulatory DNA-binding protein of *Escherichia coli*. Proc Natl Acad Sci USA 92: 8866-8870

Klee HJ, White FF, Iyer VN, Gordon MP, Nester EW (1983) Mutational analysis of the virulence region of an *Agrobacterium tumefaciens* Ti plasmid. J Bacteriol 153: 878-883

Kwon O, Georgellis D, Lin ECC (2003) Rotational on-off switching of a hybrid membrane sensor kinase Tar-ArcB in *Escherichia coli.* J Biol Chem 278: 13192-13195

Lee K (1997) A structure-based activation model of phenol-receptor protein interactions. Bull Korean Chem Soc 18: 18-23

Lee K, Dudley MW, Hess KM, Lynn DG, Joerger RD, Binns AN (1992) Mechanism of activation of *Agrobacterium* virulence genes: Identification of phenol-binding proteins. Proc Natl Acad Sci USA 89: 8666-8670

Lee Y-W, Jin S, Sim W-S, Nester EW (1995) Genetic evidence for direct sensing of phenolic compounds by the VirA protein of *Agrobacterium tumefaciens*. Proc Natl Acad Sci USA 92: 12245-12249

Lee Y-W, Jin S, Sim WS, Nester EW (1996) The sensing of plant signal molecules by *Agrobacterium*: genetic evidence for direct recognition of phenolic inducers by the VirA protein. Gene 179: 83-88

Li L, Jia Y, Hou Q, Charles TC, Nester EW, Pan SQ (2002) A global pH sensor: *Agrobacterium* sensor protein ChvG regulates acid-inducible genes on its two chromosomes and Ti plasmid. Proc Natl Acad Sci USA 99: 12369-12374

Lohrke SM, Nechaev S, Yang H, Severinov K, Jin SJ (1999) Transcriptional activation of *Agrobacterium tumefaciens* virulence gene promoters in *Escherichia coli* requires the *A. tumefaciens* RpoA gene, encoding the alpha subunit of RNA polymerase. J Bacteriol 181: 4533-4539

Lohrke SM, Yang H, Jin S (2001) Reconstitution of acetosyringone-mediated *Agrobacterium tumefaciens* virulence gene expression in the heterologous host *Escherichia coli*. J Bacteriol 183: 3704-3711

Makino K, Amemura M, Kim SK, Nakata A, Shinagawa H (1993) Role of the sigma 70 subunit of RNA polymerase in transcriptional activation by activator protein PhoB in *Escherichia coli*. Genes Dev 7: 149-160

Mantis NJ, Winans SC (1992) The *Agrobacterium tumefaciens vir* gene transcriptional activator *virG* is transcriptionally induced by acid pH and other stress stimuli. J Bacteriol 174: 1189-1196

Marina A, Waldburger CD, Hendrickson WA (2005) Structure of the entire cytoplasmic portion of a sensor histidine-kinase protein. EMBO J 24: 4247-4259

Matsuda F, Morino K, Miyashita M, Miyagawa H (2003) Metabolic flux analysis of the phenylpropanoid pathway in wound-healing potato tuber tissue using stable isotope-labeled tracer and LC-MS spectroscopy. Plant Cell Physiol 44: 510-517

Melchers LS, Regensburg-Tuink AJG, Schilperoort RA, Hooykaas PJJ (1989a) Specify of signal molecules in the activation of *Agrobacterium* virulence gene-expression. Molecular Microbiology 3: 969-977

Melchers LS, Regensburg-Tuink TJ, Bourret RB, Sedee NJ, Schilperoort RA, Hooykaas PJ (1989b) Membrane topology and functional analysis of the sensory protein VirA of *Agrobacterium tumefaciens*. EMBO J 8: 1919-1925

Mukhopadhyay A, Gao R, Lynn DG (2004) Integrating input from multiple signals: the VirA/VirG two-component system of *Agrobacterium tumefaciens*. Chembiochem 5: 1535-1542

Ottemann KM, Xiao W, Shin YK, Koshland DE, Jr. (1999) A piston model for transmembrane signaling of the aspartate receptor. Science 285: 1751-1754

Palmer AG, Gao R, Maresh J, Erbil WK, Lynn DG (2004) Chemical biology of multi-host/pathogen interactions: chemical perception and metabolic complementation. Annu Rev Phytopathol 42: 439-464

Pan SQ, Charles T, Jin S, Wu ZL, Nester EW (1993) Preformed dimeric state of the sensor protein VirA is involved in plant-*Agrobacterium* signal transduction. Proc Natl Acad Sci USA 90: 9939-9943

Peng WT, Lee YW, Nester EW (1998) The phenolic recognition profiles of the *Agrobacterium tumefaciens* VirA protein are broadened by a high level of the sugar binding protein ChvE. J Bacteriol 180: 5632-5638

Perego M, Hoch JA (1996) Protein aspartate phosphatases control the output of two-component signal transduction systems. Trends Genet 12: 97-101

Ponting CP, Aravind L (1997) PAS: a multifunctional domain family comes to light. Curr Biol 7: R674-677

Ricagno S, Campanacci V, Blangy S, Spinelli S, Tremblay D, Moineau S, Tegoni M, Cambillau C (2006) Crystal structure of the receptor-binding protein head domain from Lactococcus lactis phage bIL170. J Virol 80: 9331-9335

Robinson VL, Buckler DR, Stock AM (2000) A tale of two components: a novel kinase and a regulatory switch. Nat Struct Biol 7: 626-633

Robinson VL, Wu T, Stock AM (2003) Structural analysis of the domain interface in DrrB, a response regulator of the OmpR/PhoB subfamily. J Bacteriol 185: 4186-4194

Scheeren-Groot EP, Rodenburg KW, den Dulk-Ras A, Turk SC, Hooykaas PJ (1994) Mutational analysis of the transcriptional activator VirG of *Agrobacterium tumefaciens*. J Bacteriol 176: 6418-6426

Shimoda N, Toyoda-Yamamoto A, Aoki S, Machida Y (1993) Genetic evidence for an interaction between the VirA sensor protein and the ChvE sugar-binding protein of *Agrobacterium*. J Biol Chem 268: 26552-26558

Shimoda N, Toyoda-Yamamoto A, Nagamine J, Usami S, Katayama M, Sakagami Y, Machida Y (1990) Control of expression of *Agrobacterium vir* genes by synergistic actions of phenolic signal molecules and monosaccharides. Proc Natl Acad Sci USA 87: 6684-6688

Stachel SE, An G, Flores C, Nester EW (1985a) A Tn3 *lacZ* transposon for the random generation of beta-galactosidase gene fusions: application to the analysis of gene expression in *Agrobacterium*. EMBO J 4: 891-898

Stachel SE, Messens E, Van Montagu M, Zambryski PC (1985b) Identification of the signal molecules produced by wounded plant cell that activate T-DNA transfer in *Agrobacterium tumefaciens*. Nature 318: 624-629

Stachel SE, Nester EW (1986) The genetic and transcriptional organization of the *vir* region of the A6 Ti plasmid of *Agrobacterium tumefaciens*. EMBO J 5: 1445-1454

Stachel SE, Nester EW, Zambryski PC (1986) A plant cell factor induces *Agrobacterium tumefaciens vir* gene expression. Proc Natl Acad Sci USA 83: 379-383

Stachel SE, Zambryski PC (1986) *virA* and *virG* control the plant-induced activation of the T-DNA transfer process of *A. tumefaciens*. Cell 46: 325-333

Stephenson K, Lewis RJ (2005) Molecular insights into the initiation of sporulation in Gram-positive bacteria: new technologies for an old phenomenon. FEMS Microbiol Rev 29: 281-301

Stock AM, Robinson VL, Goudreau PN (2000) Two-component signal transduction. Annu Rev Biochem 69: 183-215

Toyoda-Yamamoto A, Shimoda N, Machida Y (2000) Genetic analysis of the signal-sensing region of the histidine protein kinase VirA of *Agrobacterium tumefaciens*. Mol Gen Genet 263: 939-947

Tremblay DM, Tegoni M, Spinelli S, Campanacci V, Blangy S, Huyghe C, Desmyter A, Labrie S, Moineau S, Cambillau C (2006) Receptor-binding protein of *Lactococcus lactis* phages: identification and characterization of the saccharide receptor-binding site. J Bacteriol 188: 2400-2410

Varughese KI (2002) Molecular recognition of bacterial phosphorelay proteins. Curr Opin Microbiol 5: 142-148

Wadhams GH, Armitage JP (2004) Making sense of it all: bacterial chemotaxis. Nat Rev Cell Biol 5: 1024-1037

Wang Y, Gao R, Lynn DG (2002) Ratcheting up *vir* gene expression in *Agrobacterium tumefaciens*: coiled coils in histidine kinase signal transduction. Chembiochem 3: 311-317

West AH, Stock AM (2001) Histidine kinases and response regulator proteins in two-component signaling systems. Trends Biochem Sci 26: 369-376

Williams SB, Stewart V (1999) Functional similarities among two-component sensors and methyl-accepting chemotaxis proteins suggest a role for linker region amphipathic helices in transmembrane signal transduction. Molecular Microbiology 33: 1093-1102

Winans SC (1991) An *Agrobacterium* two-component regulatory system for the detection of chemicals released from plant wounds. Mol Microbiol 5: 2345-2350

Winans SC, Kerstetter RA, Nester EW (1988) Transcriptional regulation of the *virA* and *virG* genes of *Agrobacterium tumefaciens*. J Bacteriol 170: 4047-4054

Wise AA, Voinov L, Binns AN (2005) Intersubunit complementation of sugar signal transduction in VirA heterodimers and posttranslational regulation of VirA activity in *Agrobacterium tumefaciens*. J Bacteriol 187: 213-223

Chapter 7

AGROBACTERIUM-HOST ATTACHMENT AND BIOFILM FORMATION

Clay Fuqua

Department of Biology, Indiana University, Bloomington, IN 47405, USA

Abstract. Physical association with host plant tissue is a prerequisite to *Agrobacterium tumefaciens* infection and subsequent disease. Mechanisms of tissue adherence have been extensively studied in mammalian pathogens, but less so in plant-associated bacteria. Cells of *A. tumefaciens* often attach to plant tissue by a single pole. In the appropriate environment, these attached bacteria eventually develop into multicellular assemblies called biofilms, enmeshed within exopolymeric material produced by the bacteria and possibly the plant host. It remains unclear whether all modes of plant attachment can lead to interkingdom gene transfer, or whether the conformation of the infecting agrobacterial population influences this process. A two-step model was proposed in which the bacterium initially attaches reversibly by way of interactions between a bacterial adhesin structure(s) and a plant receptor(s), followed by a more tenacious attachment coincident with production of cellulose fibrils. This adherence model, while potentially still valid, remains largely untested. Possible *A. tumefaciens* adherence functions, including lipopolysaccharides and cyclic β-1,2-glucans have been identified, but none has been definitively shown to mediate productive attachment to plants. Similarly, despite some promising leads, no confirmed plant receptor candidates have been identified. *A. tumefaciens* forms biofilms on a variety of surfaces including but not restricted to plant tissues. Studies of biofilm formation by *A. tumefaciens* on model surfaces have revealed a degree of structural and functional overlap with plant association, including several common cell surface structures and key regulatory pathways.

1 INTRODUCTION

Agrobacterium tumefaciens attaches to plant tissues during initial stages of crown gall pathogenesis and this physical interaction is required for subsequent DNA transfer (Lippincott and Lippincott, 1969). Adherent *A. tumefaciens* can accumulate on these plant tissue surfaces to form aggregates and biofilms, and similarly adhere to abiotic surfaces in the terrestrial environment (*Figure 7-1*). This chapter will address the process of attachment and subsequent biofilm formation by *A. tumefaciens*, examine the molecular requirements for these processes, and consider their impact on plant disease.

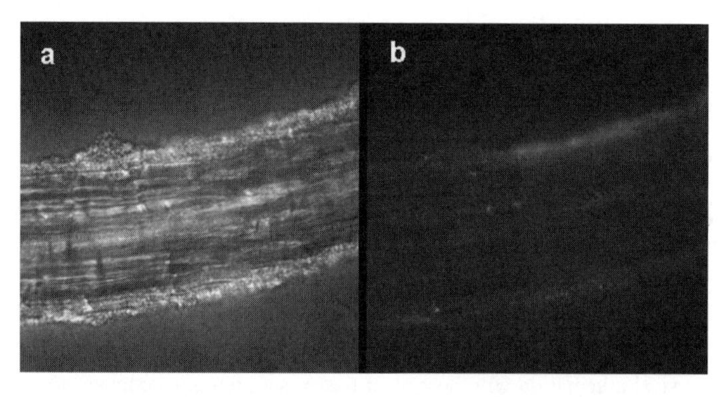

Figure 7-1. Plant tissue attachment and biofilm formation. Adherent *A. tumefaciens* C58 harboring GFP on *Arabidopsis thaliana* WS seedling root. (a) Bright field microscopy, (b) Fluorescence microscopy, bacteria are pseudocolored red. Images captured on a Deltavision deconvolution microscope.

Bacterial attachment to host tissues is an obligatory first step to disease progression for many plant and animal pathogens. Host binding and recognition has been intensively explored for several mammalian pathogens. In these systems, attachment can be highly specific, often mediated through receptors that decorate the exterior of host cells and the extracellular matrix (Boyle and Finlay, 2003; Pizarro-Cerda and Cossart, 2006). For several well-studied systems, such as enteropathogenic *Escherichia coli* (EPEC), host association is comprised of multiple steps, initiating with surface engagement and followed by a more intimate interaction in which receptors on the target cells are recognized and tightly complexed (Nougayrede et al., 2003). Binding of these receptors by infecting pathogens often causes profound alterations in cytoskeletal elements, disruption of internal signaling pathways or uptake of the bacteria into the target cells. Adhesins are cell surface structures produced by the infecting bacteria that engage host cells, often via specific receptors, and promote intimate association of the pathogen with its target cells. In mammalian pathogens, adhesins are strictly defined as those cell surface structures including pili, flagella, or other surface proteins, that directly engage host receptors (Nougayrede et al., 2003). Other surface structures may act to promote physical interaction between microbes and their hosts, but are considered to be accessory adherence functions.

Colonization of host tissue may lead to establishment of localized, adherent populations that share many attributes with environmental biofilms, sessile populations of bacteria, associated with the surface and with each other through an extracellular matrix material usually produced by the bacteria themselves (Parsek and Fuqua, 2004). Several mammalian pathogens clearly proceed through a biofilm state during disease progression (Parsek and Singh, 2003). For example, uropathogenic *E. coli* (UPEC) can enter into a persistent infective state in which they form dense biofilm-type populations within cells that line the bladder (Justice et al., 2004). For many other mammalian pathogens, the link between biofilm formation and infection is less established, but the persistence of these microbes within environmental reservoirs involves residence within biofilms.

In contrast to animal pathogens, far less is known regarding attachment of plant pathogens to their hosts and the role of biofilm formation. Only a few potential bacterial adhesins have been identified in plant-associated bacteria, and even fewer have been functionally evaluated (Rojas et al., 2002; Guilhabert and Kirkpatrick, 2005; Laus et al., 2006). In the association of rhizobial species with legumes host specificity at the level of bacterial attachment is mediated in part through plant-produced sugar-binding lectins which presumably recognize the appropriate rhizobial symbiotic partner (Hirsch et al., 2001). Plant lectins are likely common targets for bacterial attachment, although no other plant attachment systems are known to this level of detail. Furthermore, biofilm formation among plant-associated bacteria, while a common observation in microscopic studies, is only now being examined for its role in plant tissue interactions during disease and symbiosis, and as a mechanism for persistence within environmental reservoirs (Ramey et al., 2004a).

1.1 A simple model for agrobacterial attachment to plants?

The attachment of *Agrobacterium tumefaciens* to plant tissues during crown gall pathogenesis has been the subject of study for decades. Despite years of work and an exquisite level of detail on plant-microbe signaling and cross-kingdom DNA transfer, there is a very limited understanding of the recruitment and attachment processes that bring forth *A. tumefaciens* from the soil into contact with the plant to initiate pathogenesis. Furthermore, the structure and complexity of the agrobacterial population that forms on plant tissues during benign and pathogenic interactions, and in response to infected tissue, has never been systematically examined.

Studies of microbial surface interactions in many different environments have led to the concept that bacteria attach to abiotic surfaces in two discrete stages, first via relatively weak interactions that comprise a reversible stage, followed by a stronger, relatively irreversible stage (Marshall et al., 1971; Fletcher, 1996). Reversible attachment of motile bacteria in aqueous environments is often mediated by flagellar locomotion overcoming repulsive forces at the surface. The irreversible stage of attachment involves inhibition of motility and synthesis of extracellular polymeric substances, including polysaccharide, protein and DNA that act to hold the bacteria in place.

Matthysse (1983) proposed a dual-stage model for *A. tumefaciens* attachment that shares some, but not all of the features of the general two-step model. In this model, it was proposed that *A. tumefaciens* attached via an interaction with plant cell receptors and bacterial adhesins. This stage was considered reversible because cells could be removed from plant tissues with washing or vortexing. Certain avirulent *A. tumefaciens* mutants were reported to be deficient specifically at the reversible stage of attachment, lending support to the importance of this step. The second, irreversible binding stage was proposed to be concomitant with synthesis of cellulose fibrils by the bacteria, that appeared to be induced in response to plant-released signals (Matthysse et al., 1981). Electron micrographs of plant cell bound bacteria revealed the presence of cellulose fibrils, presumably anchoring cells to the infected tissue (Matthysse, 1983). Cellulose-deficient *A. tumefaciens* mutants were somewhat attenuated for virulence and are more readily washed from wound sites than the wild type (Matthysse, 1983; Minnemeyer et al., 1991). The observation that these mutants remained virulent was interpreted to indicate that the irreversible stage of attachment was dispensable for pathogenesis, although whether the virulence assays employed would reflect conditions *in situ* can be debated. It may be that irreversible attachment is required for virulence under natural infection conditions or perhaps a subset of these conditions.

There is no reason to question the general framework of this two-step attachment model, particularly given its facile similarity to the more generally supported two step models of bacterial sorption to surfaces. However, in the more than 20 years since the attachment model was proposed for *A. tumefaciens*, the details that would validate and provide mechanistic insights into this process have remained elusive. Despite some tantalizing leads, the bacterial adhesins and attachment factors involved in the presumptive early interactions with plant receptors have not been definitively identified, and the interpretation of mutant phenotypes that seemed deficient at this stage are confounded by pleiotropic effects. The production of cellulose fibrils following initial attachment, although an appealing observation, has not been substantiated by identification of relevant regulatory pathways or by additional mechanistic insights. Meanwhile, the importance of cellulose in more general bacterial attachment and biofilm formation by diverse microbes has gained tremendous experimental support (Romling, 2002). Substantial progress has been made in understanding plant functions involved in interkingdom genetic exchange, most notably through the use of *Arabidopsis thaliana*, but even here, the *A. tumefaciens* attachment receptor or receptors have not been identified (Zhu et al., 2003). In short, our understanding of attachment processes leading to T-DNA transfer and otherwise, remains at a relatively rudimentary level, and this area warrants significant attention

2 PRESUMPTIVE ADHERENCE FACTORS

Many different approaches have been employed to identify *A. tumefaciens* functions required for plant attachment. Lippincott and Lippincott (Lippincott and Lippincott, 1969) reasoned that lipopolysaccharide (LPS) from *A. tumefaciens* would contact the plant surface, and evaluated the effect of adding purified LPS preparations during *A. tumefaciens* infection on tumorigenesis (Lippincott and Lippincott, 1969; Whatley et al., 1976). Other studies, including the analysis of the presumptive adhesin called rhicadhesin, adopted a similar approach (Smit et al., 1989). Douglas et al. (1982) identified an avirulent mutant in a gene they designated *chvB* (chromosomal virulence gene B), encoding a β-1,2-glucan biosynthetic function, and subsequently concluded that it manifested an attachment deficiency. Matthysse used binding of *A. tumefaciens* to carrot tissue culture cells as a direct attachment assay, and in so doing identified the Att genes (Matthysse, 1987). More recently, *A. tumefaciens* mutants with deficiencies in attachment to model surfaces (PVC plastic and glass) have been identified and subsequently screened for plant attachment deficiencies (Ramey et al., 2004b). Although it is possible and perhaps likely that these and similar approaches have identified some of the important adherence functions, it remains unclear whether any of these are primary adhesins responsible for attachment processes that lead to T-DNA transfer (*Table 7-1*). Support of major roles in attachment for several of these identified functions has diminished, complicated by complex phenotypes, or has been refuted by more recent work.

2.1 Flagellar motility and chemotaxis

Passive deposition of bacteria onto root surfaces may foster limited colonization, but this process is greatly enhanced by swimming, swarming and gliding motility. Flagellar-based locomotion, including swimming and swarming, is a well established factor in the colonization of plant tissues by bacteria and aflagellate mutants often manifest deficiencies in attachment processes (Burdman et al., 2000; Lugtenberg et al., 2002). In addition, flagella may also function as adhesins, directly contacting surfaces and promoting cellular association. *A. tumefaciens* elaborates several flagella, arranged circumthecally towards one pole of the cell, and exhibits swimming, but not swarming motility (Kado, 1992). There are four different presumptive flagellin genes, *flaA*, *flaB*, *flaC* and *flaD* in *A. tumefaciens* C58 (Deakin et al., 1999). Nonmotile *A. tumefaciens* transposon mutants were deficient in root colonization (Shaw et al., 1991). Analysis of a "bald" mutant, with defined deletions for three of the four flagellins (Δ*flaABC*) revealed a modest deficiency in tumor size when manually inoculated into wounds on several different plant hosts (Chesnokova et al., 1997). Aflagellate pseudomonads only reveal significant plant colonization deficiencies when examined in competition with motile bacteria (Lugtenberg et al., 2002). *A. tumefaciens* motility mutants have not been examined using more quantitative assays or in competition, and it is unclear whether the manual inoculation virulence assays would reveal more subtle attachment deficiencies.

Directed motility through chemotaxis and aerotaxis is also very likely to play a role in plant colonization in the environment. *A. tumefaciens* is reported to chemotax towards plant-released compounds including *vir*-inducing phenolic compounds and opines (Ashby et al., 1988; Kim and

Farrand, 1998). There are as many as 20 different methyl-accepting chemotaxis protein (MCP) homologues annotated in the *A. tumefaciens* C58 genome sequence suggesting diverse chemotactic behavior (Goodner

Table 7-1. Identified *A. tumefaciens* attachment factors

Factor	Gene(s)	Relevant Characteristics	Standing	References
Rhicadhesin	Unidentified	Small exocellular Ca^{++}-binding protein, crude preparations inhibit attachment to pea root hairs, no gene(s) or mutant(s) isolated.	Unclear	(Smit et al., 1989)
Lipopolysaccharide (LPS)	*rfx* genes	Outer membrane components.	Unclear	(Whatley et al., 1976)
Cellulose	*celABCDE*	Attenuated virulence and reduced attachment. Easily dislodged.	Validated	(Matthysse, 1983)
β-1,2 glucans	*chvAB*	Mutants avirulent, but pleiotropic defects. Non-attaching on plants, mammalian cells and abiotic surfaces.	Validated	(Douglas et al., 1982)
"Att" functions	Att cluster	C58 mutants reported to be avirulent and nonattaching. Gene cluster is located on pAtC58, which is dispensable for virulence, attachment and biofilm formation.	Contradictory reports	(Matthysse, 1987)
Motility	*fla, fli, flg, mot*	Nonmotile and reduced virulence.	Validated	(Chesnokova et al., 1997; Shaw et al., 1991)
T-pilus	*virB2*	Polarly localized, coincident with fibrils attached to *Streptomyces* cells during T-DNA transfer.	Speculative	(Kelly and Kado, 2002; Lai et al., 2000)

et al., 2001; Wood et al., 2001). Chemotaxis mutants have not been thoroughly tested for plant interactions. Similar to other soil microbes, it seems virtually certain that chemotaxis plays a role in recruiting agrobacteria from the soil environment into the rhizosphere, and that these functions may also have a more direct impact in surface colonization and perhaps attachment.

2.2 Lipopolysaccharide (LPS)

Studies in the late 1960s and 1970s by Barbara and James Lippincott and colleagues examined the effect of LPS preparations on crown gall tumorigenesis, using the rationale that externalized *A. tumefaciens* LPS might be the molecule in most intimate contact with plant tissue during attachment (Whatley et al., 1976). The recognized importance of LPS in animal defense responses also provided significant precedence for this work. Interestingly, crude envelope preparations and purified LPS from virulent strains were effective at inhibiting tumor formation on pinto bean leaves (Whatley et al., 1976). Similar preparations from at least some avirulent *A. tumefaciens* derivatives were noninhibitory. The interpretation was that free LPS was binding to receptors on the plant surface, and thus blocking binding of agrobacterial cells. Treatment with LPS was much less effective when administered after *A. tumefaciens* was provided a short prebinding period. These studies did not directly evaluate binding to the leaf tissue, but rather measured binding indirectly as formation of tumors on the infected tissue after seven days. Subsequent work has not further implicated LPS as an important *A. tumefaciens* attachment factor, although there is significant debate whether rhizobial LPS might function during legume symbiosis (Noel and Duelli, 2000). It remains possible that LPS plays a role in *A. tumefaciens* attachment. The preparations used in these early studies were however very likely to have had impurities. Such impurities, including abundant cellular components such as EF-Tu and even LPS itself, are now known to elicit plant basal defense response in plants and can effectively reduce tumorigenesis (Zipfel et al., 2006).

2.3 Rhicadhesin

A promising candidate for a bacterial adhesin involved in plant attachment has been called rhicadhesin. Examination of the calcium (Ca^{++}) dependence of rhizobial attachment to pea root hairs led to the identification of rhicadhesin, a small Ca^{++}-binding protein, that could block root attachment when added in semi-purified form to attachment assays (Smit et al., 1989). Similar proteins have been reported for other members of the *Rhizobiaceae*, including *A. tumefaciens*, but have not been identified outside of this group (Dardanelli et al., 2003). Rhicadhesin preparations from *A. tumefaciens* and other rhizobia share the Ca^{++}-dependent ability to inhibit bacterial attachment to pea roots. An *A. tumefaciens chvB* mutant, deficient for attachment and synthesis of β-1,2-glucans (see below), was unable to synthesize detectable rhicadhesin, and addition of rhicadhesin corrected its attachment deficiency, suggesting a connection between these functions (Swart et al., 1993). Cell-surface Ca^{++} along with rhicadhesin, is released at pH < 6.5, and therefore rhicadhesin has been proposed to function in plant attachment specifically under non-acidic conditions (Swart et al., 1993; Laus et al., 2006). These observations all are supportive of rhicadhesin functioning to foster early stage plant interactions, but the experiments rely on observations in which semi-purified protein is added to plant binding assays.

Despite the availability of several rhizobial and agrobacterial genome sequences, the gene(s) encoding rhicadhesin and its elaboration has not been identified nor have rhicadhesin-deficient mutants been isolated. Therefore the simple experiment of asking whether rhicadhesin is required for productive plant attachment, has never been performed, and its true role, if any, in attachment has never been confirmed. In promising recent work, additional studies in the rhizobia have identified several secreted Ca^{++} binding proteins presumptively called Rap adhesins, one of which may be rhicadhesin (Ausmees et al., 2001a; Russo et al., 2006). Several Rap protein amino acid sequences were determined, but none of these matched sequences in the *A. tumefaciens* C58 genome (Fuqua C, unpublished data). It is unclear what role the presumptive rhicadhesin might play in crown gall disease. Conditions known to induce the Vir regulon at wounds sites include acidic pH, and these would apparently promote the loss of rhicadhesin from the cell surface (Winans, 1992). It is possible that rhicadhesin functions at attachment sites other than wounds. It was recently shown that T-DNA transfer and subsequent opine production does not require wounding (Brencic et al., 2005).

2.4 ChvA/B and cyclic β-1,2-glucans

In the early 80s Carl Douglas, Eugene Nester and colleagues developed a plant attachment assay that utilized adherence to *Zinnia* leaf mesophyll cells to screen a series of avirulent transposon mutants (Garfinkel and

Nester, 1980; Douglas et al., 1982). A non-attaching mutant was isolated with clearly diminished binding to *Zinnia* tissue relative to the parent strain. Subsequent genetic mapping and sequence analysis revealed a transposon insertion within a two gene operon designated *chvAB* (Douglas et al., 1985; Zorreguieta et al., 1988). Many studies since this time have repeatedly validated the avirulent, non-attaching phenotype of the *A. tumefaciens chvAB* mutant, both for T-DNA transfer to plants and engineered, *Agrobacterium*-dependent DNA transfer to human tissue culture cells (Swart et al., 1993; Kunik et al., 2001). The *chvAB* genes encode synthesis and export of cyclic β-1,2-glucans in *A. tumefaciens* (Puvanesarajah et al., 1985; Cangelosi et al., 1989). This polysaccharide in *A. tumefaciens* and in numerous rhizobia can be periplasmic or secreted and is typically cyclized with 17-40 sugar residues (Breedveld and Miller, 1994). Although the historical pedigree of the ChvAB proteins and β-1,2-glucans linking virulence and attachment in *A. tumefaciens* seems promising, a precise role for these gene products in these processes has never been defined. Rather, the primary function ascribed to β-1,2-glucans is as periplasmic osmoregulators, controlling the movement of water and protecting against osmotic shock (Breedveld and Miller, 1998). The changes in periplasmic osmolarity lead to a variety of pleiotropic effects in *chvAB* mutants including reduced numbers of flagella, increased antibiotic sensitivity, differences in cell surface proteins, and increased exopolysaccharide synthesis (Breedveld and Miller, 1998). It is therefore difficult to distinguish between a direct role for *chvAB* in plant association, or such significant alteration of cell surface properties that mutants in these genes are dysfunctional in localization or elaboration of other attachment factors, or simply elevated sensitivity to the rhizosphere environment. The attachment deficiencies and the avirulent phenotype, as well as several pleiotropic cell surface properties are however, reported to be corrected at lower temperatures (Bash and Matthysse, 2002). The *chvB* mutant does not produce the presumptive attachment protein rhicadhesin, perhaps due to osmotic stress, and is corrected for attachment deficiencies by addition of exogenous rhicadhesin (Swart et al., 1993). These observations and better evidence for the adherence function of rhicadhesin, could clarify the underlying cause of the *chvAB* mutant phenotypes, although a conservative assessment at this juncture is that reduced attachment and avirulence are the indirect consequence of misregulated osmolarity and resulting changes in cell surface properties.

2.5 The attachment (*Att*) genes—not required for attachment?

Recognizing the limitations of assays that inferred attachment efficiency through effects on virulence, Matthysse employed a painstaking microscopic screening method to isolate mutants of *A. tumefaciens* with decreased attachment to carrot tissue culture cells (Matthysse, 1987). Several *A. tumefaciens* C58 transposon mutants that did not attach to carrot cells were isolated and found to have insertions within a common EcoRI restriction enzyme cleavage fragment of 12 kb. These attachment or Att mutants were reported to be avirulent when manually inoculated onto *Bryophyllum* (*Kalanchoe*) *diagremontiana* leaves. The attachment and virulence deficiencies were complemented with a pair of overlapping cosmids from an *A. tumefaciens* C58 genomic library (Matthysse et al., 1996; Matthysse et al., 2000). Transposon insertions throughout these cosmids abrogated complementation and allelic replacement mutants generated with these insertions and by other means resulted in non-attaching, avirulent *A. tumefaciens* mutants.

The DNA sequence of the 29 kilobases region spanned by these overlapping cosmids, and subsequently designated the Att region, revealed genes with a wide range of predicted functions including an ABC-type transporter system, polysaccharide synthesis and modification enzymes, peptidases, Mg^{++} transporters and transcription regulators (Matthysse et al., 1996; Matthysse et al., 2000). Mutations in most of the Att genes resulted in loss of attachment and virulence, while consistent with earlier reports, mutations in one subregion (*atrA-attG*), could be rescued by addition of conditioned medium derived from *A. tumefaciens*-plant co-culture (Matthysse, 1994). The *attR* gene was the most extensively studied *att* gene, with the most consistent attachment defect (Matthysse and McMahan, 2001). The *attR* gene product is a predicted transacetylase and was demonstrated to be required for the synthesis of an acidic polysaccharide, consistent with a surface structure that might promote attachment (Reuhs et al., 1997).

The concept of a large genetic cluster devoted to attachment was quite intriguing, but it was difficult to envision how genes of such diverse predicted functions might all impinge upon the attachment process. The attachment mutants described above were isolated from and characterized in *A. tumefaciens* C58. The complete genome sequence of C58 revealed a multipartite composition with a circular (2.84 Mb) and a linear chromosomes (2.07 Mb), the Ti plasmid pTiC58 (0.21 Mb), and another large plasmid (0.54 Mb) pAtC58 (Goodner et al., 2001; Wood et al., 2001). The genome sequence revealed that the Att gene cluster resides on pAtC58, and that there are several apparent *att* gene copies elsewhere in the genome (e.g. *attH*). This was a puzzling result, as several earlier studies had suggested that pAtC58 was dispensable for virulence (Hooykaas et al., 1977; Rosenberg and Huguet, 1984; Hynes et al., 1985). A recent study using isogenic derivatives of C58 carefully examined the effect of this plasmid on virulence and found that although it was not required, it did have a modest positive impact on tumor size and induction of *vir* genes (Nair et al., 2003). In this same study a targeted disruption of the *attR* gene on pAtC58 did not influence virulence, and did not abolish the positive impact of the pAtC58 on virulence and *vir* gene induction. It remains unclear what gene(s) on pAtC58 is responsible for the enhanced *vir* regulon induction. These results strikingly contradict the earlier work on the *att* genes. If pAtC58 is dispensable for virulence, an observation verified by several labs in multiple publications, why do transposon insertions in the *att* genes lead to avirulent and non-attaching mutants? Additionally, several published studies suggested that the *attR* mutant was avirulent and manifested a strong attachment deficiency, while the more recent work found no virulence role for *attR* (Matthysse and McMahan, 1998, 2001; Nair et al., 2003).

Several other genes in the Att region, *attJ* and *attKLM*, are now known to direct the degradation of acylhomoserine lactone (AHL) quorum sensing signals and thereby modulate cell-cell signaling (so-called quorum-quenching) (Zhang et al., 2002). AttM is a lactonase enzyme that cleaves the lactone ring on the AHL signal molecule, AttJ and AttK may assist in further degradation, and AttJ is a transcriptional regulator of the *attKLM* genes. Mutations in these *A. tumefaciens* genes had previously been reported to result in avirulent, non-attaching derivatives (Matthysse et al., 2000). There is no evidence linking the quorum sensing process in *A. tumefaciens* to plant attachment or virulence, and it is therefore difficult to reconcile the now established biochemical activity of these proteins with their previously proposed role in attachment.

The role of genes within the Att region in the process of attachment for which they were named, now seems tenuous at best. It seems plausible that some of the original transposon mutants might have generated dominant-negative alleles that interfered with attachment and virulence. The number of different Att mutant derivatives, the uniformity of the reported phenotypes, and the effective complementation results reported with cosmids and smaller plasmids however, make this possibility much less likely (Matthysse et al., 1996; Matthysse et al., 2000). Perhaps more plausible is the possibility that the C58 derivative in which these were first isolated possessed a second site mutation, that indirectly affected the attachment process, and that this was aggravated by the mutations in the Att region. Either way, the observations that pAtC58 is dispensable for virulence and manifests no attachment defect argues strongly that the Att genes are not directly required for these processes. In the end, a comprehensive re-analysis of the so-called *att* genes on pAtC58, in a *bona fide* wild type C58 genetic background, is required in order to better elucidate a function for these genes.

2.6 Synthesis of cellulose fibrils and irreversible attachment

The production of cellulose fibrils is often cited as the visual indication that *A. tumefaciens* and other rhizobia have transitioned to the irreversible stage of attachment (Matthysse and Kijne, 1998). Elaboration of these fibrils is observed to be induced during interaction with plant tissue surfaces and cells. These fibrils are not observed in electron micrographs of cellulose synthesis (Cel$^-$) mutants (Matthysse et al., 1981). The Cel- mutants are more easily washed from inoculation sites and exhibit attenuated virulence, suggesting a role in adhesion. In *A. tumefaciens,* cellulose production requires genes encoded within the *celABCG* and *celDE* operons on the C58 linear chromosome (*Figure 7-2*) (Matthysse et al., 1995b). Mutations in *celA, celB*, *celC*, *celD* and *celE* abolish cellulose biosynthesis, while disruption of *celG* results in its overproduction (Matthysse et al., 2005). Based largely on analysis of homologous systems, the Cel proteins are thought to form a membrane-associated complex that directs cellulose synthesis and export (*Figure 7-2*). CelA is a membrane-associated cellulose synthase (CS) enzyme, utilizing the precursor UDP-glucose. Homologues of CelB bind cyclic diguanosine monophosphate (c-di-GMP), an allosteric regulator of CS activity (see below), and physically interact with CS in the membrane (Romling, 2002). CelC is a secreted protein similar to endoglucanases, and CelDE are cytoplasmic proteins that may be required for lipid carrier activity (Matthysse et al., 1995a). Regulation of *cel* gene expression is not well understood, but recently a cellulose-overproducing mutant has been identified with a lesion in a gene designated *celI* (cellulose synthesis Inhibitor) encoding a MarR/ArsR type repressor protein (Matthysse et al., 2005).

The CS activity of *A. tumefaciens* is allosterically regulated by the intracellular signal molecule c-di-GMP, which strongly stimulates cellulose synthesis in cell extracts (Amikam and Benziman, 1989). Originally identified as an allosteric regulator of cellulose synthase in *Gluconacetobacter xylinus*, the role of c-di-GMP as a cellular signal is an emerging theme in bacterial physiology (D'Argenio and Miller, 2004). Synthesis of c-di-GMP is catalyzed by proteins with a so-called GGDEF domain (also called DUF1), and conversely c-di-GMP turnover is mediated through proteins that share the EAL signature motif (also called DUF2). The same proteins may often contain both motifs (Paul et al., 2004). Bacterial genome sequencing has revealed a large number of GGDEF and EAL proteins in bacteria, commonly multiple different derivatives encoded within the same genome. The GGDEF and EAL domains appear to be highly modular and are often associated with other recognized motifs involved in signal perception, such as PAS and HAMP domains (Jenal, 2004). Although cellulose synthesis is a confirmed target for c-di-GMP, there are clearly other processes under its control. It is not known how multiple signaling systems directing synthesis of the same compound, c-di-GMP, would impart specific responses. The current view is that GGDEF/EAL proteins provide environmentally-responsive control over cell surface properties through modulating cellular pools of c-di-GMP.

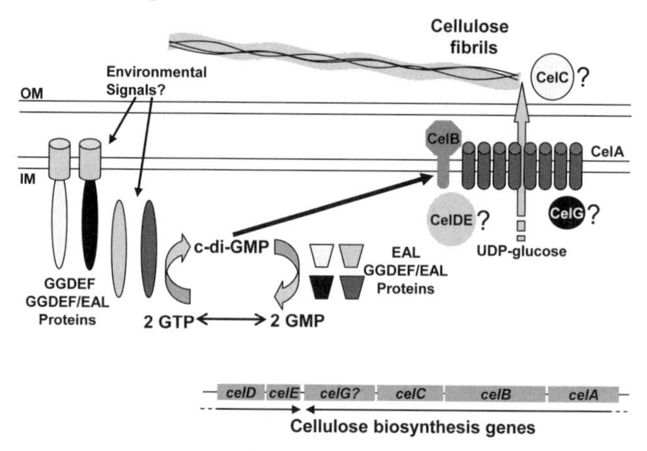

Figure 7-2. Cellulose biosynthesis in *A. tumefaciens*. Gene map of cellulose biosynthesis operons, model of membrane-associated cellulose synthase complex, and a depiction of c-di-GMP synthesis and turnover by GGDEF/EAL proteins.

Analysis of the *A. tumefaciens* C58 genome reveals 31 gene products with GGDEF domains, 16 of which also have EAL domains (Goodner et al., 2001; Wood et al., 2001). It is clear that c-di-GMP influences *A. tumefaciens* CS activity in vitro and that ectopic expression of a heterologous GGDEF protein enhances cellulose synthesis (Amikam and Benziman, 1989; Ausmees et al., 2001b). Therefore, we hypothesize that one or more of the *A. tumefaciens* GGDEF proteins regulates cellulose biosynthesis, thereby affecting plant attachment and biofilm formation (see *Figure 7-2* and below). It is plausible that the induction of cellulose fibrils upon plant interactions may require c-di-GMP signaling. Other *A. tumefaciens* GGDEF proteins are likely to control different cell surface features.

2.7 Plant attachment *via* the T-pilus?

The VirB gene products plus the VirD4 protein comprise a Type IV Secretion (T4S) system that transfers the T-DNA as a nucleoprotein complex into targeted plant cells and independently also introduces several other proteins, including VirE2 and VirF (Christie et al., 2005). A subset of the eleven VirB gene products are involved in elaboration of an extracellular pilus structure, called the T-pilus, visible in electron micrographs of cells grown under *vir*-inducing conditions (Fullner et al., 1996). The VirB2 protein is the propilin protein, and is cyclized in the process of polymerization into the T-pilus (Lai et al., 2002). Recent work suggests that VirB proteins are localized to a single pole of the *A. tumefaciens* cell, consistent with the observation that the T-pilus is also elaborated from a pole (Lai et al., 2000; Judd et al., 2005). Although it has not been experimentally proven, it seems reasonable to speculate that the pole to which the T-pilus and the VirB proteins localize is the same, and that this is also the end of the cell that contacts the plant surface during infection. During adhesion of *E. coli* to the mammalian intestinal tract during EPEC infection, an extracellular filament comprised of the EspA protein and perhaps EspB and EspD, elaborated through the type III secretion system, acts as an adhesin (Knutton et al., 1998). Similarly, it seems plausible that a component of the T-pilus functions as an adhesin prior to and perhaps during T-DNA transfer. Electron micrographs of *A. tumefaciens* associated with *Streptomyces lividans* cells reveals a filamentous structure that bridges between the pole of the *A. tumefaciens* cell and the *S. lividans* hyphae (Kelly and Kado, 2002). T-DNA is successfully transferred from *A. tumefaciens* to *S. lividans*. Although it is not certain whether this filament is synthesized by *A. tumefaciens* or *S. lividans*, it was absent in *vir* gene mutants nor under non-inducing conditions. Given these observations, it seems possible that components of the T-DNA transfer machinery act as attachment or adhesion factors DNA transfer to *S. lividans* and by extension, to plants. However, avirulent *Agrobacterium* species with no functional Vir system and *A. tumefaciens vir* mutants that do not elaborate a T-pilus attach efficiently to plants, so it appears that any role in attachment for the Vir T4S system is either ancillary or restricted to sites at which T-DNA transfer occurs.

3 PLANT RECEPTORS RECOGNIZED DURING *A. TUMEFACIENS* INFECTION

Does *A. tumefaciens* recognize specific structures on the plant tissue surface? *A. tumefaciens* has a strikingly wide host range and can infect a variety of tissues including roots, stems and leaves (De Cleene and De Ley, 1976). For rhizobial systems, legume surface-localized lectins impart a signicant portion of host specificity, presumably through productive binding of the rhizobial cell to root hairs (Hirsch et al., 2001). The wide host range of *A. tumefaciens* however argues for the recognition of multiple structures or a more general feature of plants. Appropriately induced *A. tumefaciens* also can productively attach and transfer DNA to other bacteria, fungi and even mammalian cells (Kunik et al., 2001; Kelly and Kado, 2002; Lacroix et al., 2006). This impressive host range suggests that a specific structure may not be required for T-DNA transfer or that the requisite structure is conserved among the major domains of life. *A. tumefaciens* may still however interact with specific plant surface components during attachment, even if these are not absolutely required for transfer to all hosts.

Several mutants of *Arabidopsis thaliana* resistant to *A. tumefaciens* (*rat* mutants) appear to be colonized poorly by *A. tumefaciens* (Zhu et al., 2003). One of the *rat* mutants (*rat1*), blocked at an early stage of the host-microbe interaction, carries a T-DNA insertion in a gene required for synthesis of arabinogalactans, polymers that localize to the plant cell wall surface. Another mutant (*rat4*), also blocked at an early step, is disrupted in a cellulose-synthase like protein, again related to plant surface functions. Screens for plant proteins that interact directly with *A. tumefaciens* VirB2, the T-pilin, have identified additional candidate receptors (Hwang and Gelvin, 2004). Three VirB2-interacting proteins (BTIs) with no known function were identified, as well as a membrane-associated GTPase. Tagged versions of these proteins physically associate with VirB2 and localize proximally to the plant cell wall in transgenic *Arabidopsis*. Furthermore, inhibition of their expression leads to plants that are poorly infected by *A. tumefaciens*, and elevated expression of at least one BTI protein (BTI1) enhances transformation by *A. tumefaciens*. All of these properties are consistent with BTI proteins functioning in recognition or productive interactions with the *A. tumefaciens* T-pilus, and they are promising candidates for plant features recognized by the T-pilus during *Arabidopsis* infection. Whether the BTI proteins function at early stages of plant interaction, including attachment, or in later stages following attachment is yet to be determined. Additionally, it is not clear how uniformly these proteins are conserved among other plants.

4 BIOFILM FORMATION BY *A. TUMEFACIENS*

A natural consequence of bacterial attachment to surfaces is the formation of multicellular adherent populations collectively called biofilms. Biofilm formation has received increasing attention as the ubiquity of these structures and their importance to medicine, industry and agriculture have become apparent (Hall-Stoodley et al., 2004; Parsek and Fuqua, 2004). In general, biofilms are surface-associated microbial populations in which the individual cells are affixed to surfaces and cohered to each other through an extracellular polymeric matrix, often produced by the bacteria themselves. Biofilms can range from relatively flat, featureless films to highly structured, discontinuous and porous complexes. The point at which adherent cells may be considered biofilms varies widely between different investigators with some considering any adherent cells to be a biofilm, while others only classifying adherent populations as biofilms when they have reached some minimum level of structure. Conceptually, the point at which the presence of multiple cells adhered to the surface changes the attributes of the population as a whole, can arguably be considered a biofilm. Operationally this point can be difficult to define, and varies among different microbes and different environments.

4.1 Adherent bacterial populations on plants and in the rhizosphere

Biofilms have been most extensively studied on abiotic surfaces in aquatic environments. More recently, biofilms that form on living tissues during interactions with metazoan host organisms have gained attention, and the role of biofilm formation in pathogenesis has become an active area of research (Parsek and Singh, 2003). Bacterial adherence to plant surfaces, in both commensal and pathogenic relationships, shares many features with adherence on abiotic surfaces. In both cases, the bacterial populations form complex, structured assemblages (Morris and Monier,

2003; Ramey et al., 2004a). Large and small clusters of microbes along roots, stems, leaves, and within the plant vasculature have been variably described as aggregates, microcolonies, symplasmata and biofilms (Morris and Monier, 2003). Many plant-associated bacteria also reside as saprophytes in the terrestrial environment, adhered to soil particles and decaying plant matter. Metabolically active plant tissue presents a unique surface that can vary among different plants and for different tissues of the same plant. Gradients of nutrients, balanced by sequestration and exudation make the plant surface a dynamic environment (Walker et al., 2003). In general, plant-associated bacteria must recognize, adapt to and interact with both living and inert surfaces, and these interactions are critical features of their life cycles.

4.2 Biofilm formation and structure

A. tumefaciens forms architecturally complex biofilms on host tissues, as well as model abiotic surfaces (*Figure 7-3*). Confocal laser scanning microscopy has revealed that *A. tumefaciens* biofilms on model surfaces are characterized by densely packed, but relatively shallow layers of cells along the surface, punctuated by larger, globular aggregates of 20-30 cells in depth (Danhorn et al., 2004; Ramey et al., 2004b). Early stages of biofilm formation exhibit a large proportion of cells attached to surfaces via their poles (*Figure 7-3*). In contrast to other well-studied biofilm forming bacteria, *A. tumefaciens* cells remain attached by single poles, consistent with the manner in which they bind to root tissues (Douglas et al., 1982; Pueppke and Hawes, 1985; Hinsa et al., 2003). As the biofilm matures, more complex cellular arrangements emerge. On plant root surfaces, the adherent biomass is somewhat more heterogeneous, but quite substantial, sharing many of the structural features observed on abiotic surfaces (Matthysse et al., 1995b; Ramey et al., 2004b).

It is clear that *A. tumefaciens* cells can and do form biofilms on a variety of surfaces. As with other bacteria, the ability to form a biofilm is likely to enhance nutrient acquisition, provide protection from dessication and predation, and improve tolerance towards chemical and physical stress (Hall-Stoodley et al., 2004). Does biofilm formation influence the process of pathogenesis? T-DNA transfer is recognized as a relatively inefficient event. Although single cells are capable of T-DNA transfer, it is much more common for aggregates and other multicellular assemblies to form at the site of infection (Escudero and Hohn, 1997; Brencic et al., 2005). In practical transformation applications, huge numbers of *A. tumefaciens* cells are inoculated onto plants and other hosts, compensating for overall inefficiency. It seems intuitive, that the larger number of bacteria in physical proximity to the infected tissue afforded through biofilm formation, the greater the likelihood of successful transformation. Additionally, the biofilm may promote productive *in situ* activation of the *vir* genes by concentrating or slowing the diffusion of phenolics and other *vir* inducers. Biofilms might also provide protection or overall resistance to the plant basal defense response.

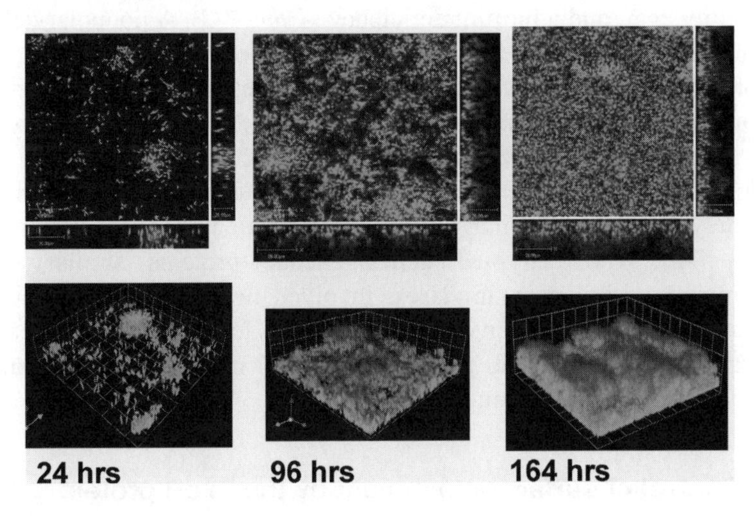

Figure 7-3. A. tumefaciens biofilm formation. Adherent *A. tumefaciens* C58 harboring GFP on a glass surface over time. Images were acquired using a confocal laser scanning microscopy and Volocity software to render the image. In collaboration with Dingding An and Matthew Parsek.

4.3 Mutations that diminish biofilm formation and plant attachment

Genetic screens of *A. tumefaciens* C58 transposon mutant libraries have lead to identification of several functions important for biofilm formation (Ramey, 2004 and *Table 7-2*). These screens were performed using a modification of the O'Toole and Kolter approach, isolating mutants with reduced surface adherence to polyvinylchloride (PVC) 96-well microtitre plates (O'Toole et al., 1999). Several of the functions identified thus far were also implicated in previous work on *A. tumefaciens* plant association. An existing mutant for the *chvB* gene and therefore unable to make cyclic β-1,2 glucans, did not adhere efficiently to PVC (Danhorn T and Fuqua C, unpublished data). A *chvA* transposon mutant was isolated from the biofilm screen and manifested the same attachment deficient phenotype as the *chvB* mutant (*Table 7-2*). This suggests that mutants unable to synthesize cyclic β-1,2-glucans are generally deficient for adherence, irrespective of the surface. Nonmotile transposon mutants (disruptions in *flgD* and *fliR*) were also biofilm deficient, consistent with reduced virulence for a nonmotile mutant (Δ*flaABC*) in a previous study (Chesnokova et al., 1997). A transposon insertion in the *tlpA* gene, the first gene in the *A. tumefaciens* C58 chemotaxis operon, results in a motile, but nonchemotactic phenotype on motility agar, and a biofilm deficiency (*Table 7-2*). A nonpolar deletion within the *cheA* gene of the same operon, encoding the two-component sensor kinase known to interact with methyl-accepting chemotaxis proteins and in turn control flagellar rotation, manifests the identical phenotype as the original *tlpA* mutant (Merritt PM and Fuqua C, unpublished data).

Transposon insertions in a gene homologous to the *amiA* gene and separately to a gene similar to *sodB* also result in biofilm deficiencies (*Table 7-2*). The disrupted genes encode proteins similar to *N*-acetylmuramoyl-l-alanine amidases, involved in cell wall synthesis and superoxide dismutase, converting superoxide to peroxide, respectively. The underlying reasons that the *amiA* and *sodB* mutants are compromised for biofilm formation remain unclear.

4.4 Control of surface attachment by the ExoR protein

Two independent biofilm deficient derivatives were isolated with transposon insertions in an *A. tumefaciens* gene highly similar to *exoR* from *Sinorhizobium meliloti* (Reed et al., 1991). These two mutants manifest a dramatic inability to colonize surfaces, with no significant biofilm formation (Tomlinson et al., in preparation). Additionally, the *A. tumefaciens exoR* mutants are quite mucoid and brightly fluorescent when plated on medium containing Calcoflour, a polysaccharide β-linkage-specific dye. These features are consistent with the *S. meliloti exoR* mutant, in which the *exo* genes encoding synthesis and export of succinoglycan (SCG), a differentially modified exopolysaccharide required for plant nodulation, are derepressed (Reed et al., 1991). *A. tumefaciens* also synthesizes SCG and has homologues of the *exo* genes (Cangelosi et al., 1987). Deletion of the *exoA* gene, directing the first unique step of SCG biosynthesis, reduced the mucoidy and Calcofluor staining in the wild type and *exoR* mutant backgrounds (Becker and Puhler, 1998). The *A. tumefaciens exoA* SCG$^-$ mutant forms biofilms qualitatively similar to wild type, suggesting that this exopolysaccharide is not required for biofilm formation. It seemed likely that the biofilm formation deficiency in the *exoR* mutant was due to elevated SCG production. The *exoAexoR* double mutant, although unable to synthesize SCG, exhibits the identical biofilm deficiency as the *exoR* mutant, indicating that SCG overproduction is not responsible for this phenotype (Tomlinson AD, Ramey BE, Day TW, Rodriguez JL, Lawler ML Fuqua C, unpublished data). It was recently reported that the *S. meliloti exoR* mutant is aflagellate and nonmotile (Yao et al., 2004). Although the *A. tumefaciens exoR* mutant and the *exoAexoR* mutant remain motile, they are clearly less active on motility agar than the wild type (Tomlinson AD, Ramey BE, Day TW, Rodriguez JL, Lawler ML Fuqua C, unpublished data). Flagellar staining suggests that *A. tumefaciens exoR* mutants produce or retain fewer flagella than wild type. Given the established relationship between motility and adherence in *A. tumefaciens* this may explain the *exoR* phenotype on abiotic surfaces, although the severity of the biofilm deficiency suggests additional problems.

Table 7-2. A. tumefaciens C58 biofilm mutants

Gene name	Atu number[a]	Presumptive function	Reference
chvA	Atu2728	β-1,2-glucan synthesis; chromosomal virulence factor	(Douglas et al., 1982)
flgD	Atu0579	Flagellar synthesis; hook protein	(Ohnishi et al., 1994)
fliR	Atu0582	Flagellar synthesis; hook/basal body	(Armitage et al., 1997)
tlpA	Atu0514	Cytoplasmic methyl-accepting chemotaxis protein	(Kawagishi et al., 1992)
sodB	Atu4726	Iron superoxide dismutase	(Cortez et al., 1998)
amiA	Atu1340	*N*-acetylmuramoyl-$_{L}$-alanine amidases	(Langaee et al., 2000)
exoR	Atu1715	Negative regulator of exopolysaccharide production	(Reed et al., 1991)
sinR	Atu2394	FNR homologue, DNR subfamily	(Ramey et al., 2004)

[a]Atu gene designation as available through http://depts.washington.edu/agro/genomes/ c58/ c58homeF.htm

The *exoR* mutant is virulent when manually inoculated into wound sites on the stems of cowpeas and on potato disks (Tomlinson AD, Ramey BE, Day TW, Rodriguez JL, Lawler ML Fuqua C, unpublished data). In contrast, the mutant manifests a striking binding deficiency on *Arabidopsis* roots, with very few attached cells, similar to its phenotype on abiotic surfaces. Surprisingly, the inability to bind *Arabidopsis* roots is corrected by the *exoA* deletion, suggesting that the binding deficiency on plant roots is due to SCG overproduction, opposite to the findings on abiotic surfaces. It is interesting that a disruption of *exoR* leads to a virtually identical nonadherent phenotype on biotic and abiotic surfaces, but that the cause of these phenotypes is different depending upon the surface. It is therefore apparent that ExoR regulates multiple surface properties relevant to surface interactions.

ExoR is not a standard regulatory protein but rather contains an N-terminal secretion signal, a possible trans-membrane segment, and a single tetratricopeptide repeat sequence, a motif known to promote protein-protein interactions (Blatch and Lassle, 1999). ExoR is therefore likely to reside within the periplasm, perhaps associated with the cytoplasmic membrane, interacting with an as yet unidentified signal transduction system(s) that directly regulates target functions such as *exo* gene expression. Perhaps in this location ExoR can influence adaptation to surfaces and the transition from planktonic to sessile life styles.

4.5 Control of biofilm maturation by an FNR homologue

One of the *A. tumefaciens* biofilm deficient transposon mutants identified was proficient in initial attachment to surfaces, but never attained the density or surface coverage of wild-type (Ramey et al., 2004b). This mutant was disrupted for an FNR-type transcription factor designated SinR (surface interaction Regulator), and expression of the *sinR* gene from a plasmid not only corrected the deficiency, but accelerated and exaggerated the density of the biofilm. FNR is an oxygen-responsive regulatory protein best studied in *E. coli* where it controls expression of a wide range of genes involved in the switch from aerobic to anaerobic growth (Lazazzera et al., 1996). FNR is the founding member of a large family of regulators, often, but not always involved in oxygen responsive gene regulation (Korner et al., 2003). SinR lacks the conserved cysteine residues found in the amino terminus of the FNR, and thus is unlikely to coordinate an [4Fe-4S] cluster and provide direct oxygen-responsiveness. The presence of a canonical FNR binding site centered at -42 relative to the *sinR* transcription start hinted that this gene is itself regulated by an FNR-type protein. The *A. tumefaciens* FNR orthologue FnrN, is required to activate *sinR* expression under oxygen limitation and within biofilms (Ramey et al., 2004b). FnrN does have the four conserved cysteines and regulates a number of *A. tumefaciens* genes involved in the oxygen limitation response (Li P, Ramey BE and Fuqua C, unpublished data). SinR also regulates its own expression. It is as yet unclear whether SinR is ligand-responsive or simply constitutive in activity. We hypothesize that SinR functions to promote the spread of the biofilm along surfaces in response to the oxygen limitation that occurs as a consequence of oxygen utilization within the biofilm. The *sinR* mutant also exhibits a maturation defect on *Arabidopsis* roots, and when the regulator is overexpressed, forms strikingly dense biofilms (Ramey et al., 2004b). Tumorigenesis assays on tobacco leaf cuttings revealed a modest, but significant deficiency for the *sinR* mutant, whereas overexpression increases the efficiency of tumor formation (Ramey, 2004).

4.6 Phosphorus limitation stimulates biofilm formation

The nutrient composition of the environment is known to have a profound effect on surface interactions. Phosphorous limitation is known to augment the virulence of *A. tumefaciens* in part through increased expression of the *virG* transcription factor (Winans, 1990). Limiting the source of inorganic phosphorous (P_i) enhanced biofilm formation in *A. tumefaciens* (Danhorn et al., 2004). Despite significant reductions in planktonic culture density, biofilms observed under Pi limitation are as much as 4-fold more dense than those formed in Pi replete conditions, with much greater overall surface coverage. The enhanced biofilm formation was designated the Sin^{PL} phenotype (surface interactions under P_i limitation). This observation was in contrast to those of Monds and colleagues with *Pseudomonas aureofaciens* in which Pi-limitation (simulated by a *pstC* mutation) reduced biofilm formation (Monds et al., 2001).

The increase in *A. tumefaciens* biofilm formation was coincident with induction of alkaline phosphatase activity, the standard indicator of the Pho regulon. This suggested that the PhoR-PhoB two-component system might be responsible for the enhanced biofilm formation (Wanner, 1995). Surprisingly it was discovered that the *phoR* and *phoB* genes are essential even under Pi-replete conditions in *A. tumefaciens*, as the chromosomal copies of either of these regulators could only be disrupted if an intact copy was present (Danhorn et al., 2004). A recent publication has suggested that the essentiality for *phoR* and *phoB* might be due to a nonfunctional low affinity, high capacity PO_4^- transporter (Pit) and a PhoB requirement to express the high affinity P_i transport systems PstSCAB and PhoCDET (Yuan et al., 2006). To circumvent the problem of essentiality, a tightly controlled *phoB* expression plasmid was introduced into *A. tumefaciens*. In the presence of the *phoB* expression plasmid, a chromosomal *phoB* disruption was generated (Danhorn et al., 2004). Induction of the plasmid-borne *phoB* gene under P_i-limiting and P_i-replete conditions elevated alkaline phosphatase activity and reproduced the Sin^{PL} phenotype.

Examination of Pho regulon induction during early stages of surface attachment on abiotic surfaces revealed significantly greater numbers of adherent cells under Pho-inducing conditions compared to wild type (Danhorn T and Fuqua C, unpublished data). Strikingly, two-to-three fold more cells were attached by their poles to the surface, and polarly aggregated, in the Pho-inducing conditions. These patterns were observed with wild type *A. tumefaciens* C58 under Pi limitation, and when *phoB* expression was induced under P_i-replete conditions. Increased polar adherence is likely to lead to the increased biomass we observe in mature Sin^{PL} biofilms, but there may be other PhoB-regulated phenomena that also come into play. There are uniformly P_i-limiting conditions in the soil environment and even greater P_i depletion through plant sequestration from the rhizosphere (Holford, 1997). It is therefore highly likely that the Sin^{PL} phenotype is engaged during plant interactions, and thereby contributes to plant adherence.

5 A MODEL FOR ADHERENCE AND BIOFILM FORMATION

Although the requisite components and sub-processes of the adherence process are only partially defined, we can expand upon the original two-step model for *A. tumefaciens* attachment originally proposed by Matthysse (Matthysse, 1983). Contact with the surface is enhanced by active flagellar motility and chemotaxis, perhaps simply through increasing the chances of collision (*Figure 7-4*). Binding the surface often occurs on a single pole, perhaps through the function of a cell surface protein such as rhicadhesin. Phosphorous limitation, as sensed via the PhoR-PhoB system increases the efficiency of this polar adherence. Cyclic β-1,2-glucans are required for this process, although these may be indirectly involved by their function as periplasmic osmoregulators. ExoR is required to control several processes relevant to attachment including but not restricted to motility and synthesis of exopolysaccharide(s). On or in close proximity to plant tissues, Vir-inducing conditions can stimulate initiation of T-DNA processing and transfer, and these conditions do not necessarily require wounding of the plant tissue. The T-pilus may contribute to intimate association with plant tissues, functioning as an additional adhesin. The transition from reversible adherence to irreversible binding, has been defined by visible synthesis of cellulose fibrils. Other polysaccharides and biofilm matrix components are also likely to be produced during this time.

Subsequent clonal growth of attached cells and additional colonization from the planktonic phase can result in formation of a biofilm. SinR and perhaps FnrN are involved in late maturation of the biofilm. Although this model is deliberately generalized to include abiotic and biotic surfaces, there are clearly aspects of each specific surface that are unique, and processes that are specifically adapted to that surface (*Figure 7-4*). The best examples of this are the induction of *vir* genes by plant released signals and the presence of specific receptors on the plant surface, but there are certain to be many others awaiting discovery.

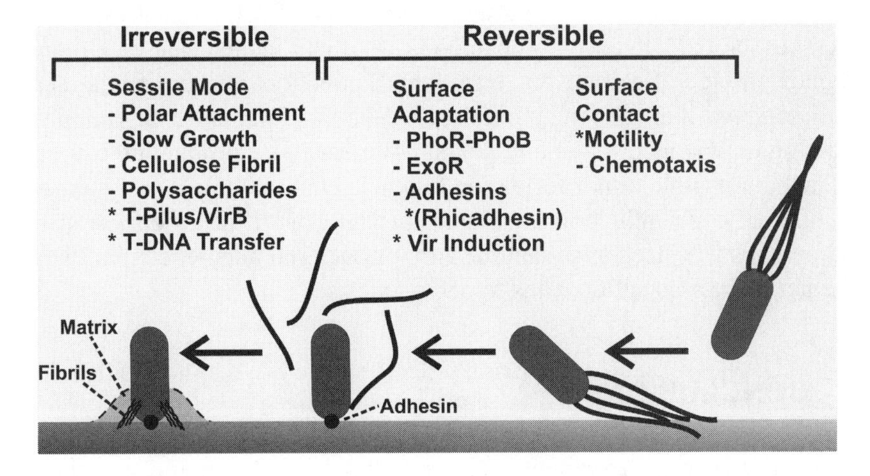

Figure 7-4. Current model for surface interactions by *A. tumefaciens*. Reversible and irreversible stages are depicted. Features marked with an asterisk are specific to plant surfaces.

6 A WIDE RANGE OF SURFACE INTERACTIONS

Many agrobacteria benignly reside on the surfaces of plants and soils as saprophytes, and are not directly engaged in pathogenesis (Bouzar and Moore, 1987; Burr et al., 1987). In fact, soils that have never demonstrated crown gall infections can carry high numbers of Ti+ and Ti- (avirulent) *Agrobacterium* species. As with other soil bacteria, agrobacteria associate with inert material of biotic and abiotic origin in the soil environment (Mills and Powelson, 1996). The soil is the ultimate reservoir for both pathogenic and non-pathogenic agrobacteria and the distribution of *A. tumefaciens* in the terrestrial environment largely determines whether introduced plants will acquire crown gall (Burr et al., 1987). *A. tumefaciens* can be divided into two functional populations: those engaged in pathogenesis and actively inciting crown gall, and those associated with the soil or plant tissues but existing as saprophytes or commensals. As with other pathogens, these bacteria are likely to be in perpetual flux as sub-populations that can mobilize from benign environmental reservoirs to cause disease in response to environmental conditions and host susceptibility. Interactions of *A. tumefaciens* with plant tissue surfaces that result in pathogenesis may therefore be viewed as a subset of the larger group of interactions of the pathogen with a variety of biotic and abiotic surfaces in the terrestrial environment. *A. tumefaciens* is well adapted to colonization of the plant surface, and both virulent and a virulent isolates are effective plant-associated microbes. Not all attached agrobacteria can or will incite disease. Although wounds provide effective sites from which infections can initiate, a recent study suggests that *vir* gene induction, T-DNA transfer and integration can occur on unwounded tissue and that subsequent opine production can happen without visible tumor formation (Brencic et al., 2005). It is certain that in addition to *vir* induction there are other aspects of the colonization site and perhaps the mechanism and density of bacterial adherence that dictate whether plant-association leads to pathogenesis.

7 CONCLUSIONS

A large body of excellent work in many laboratories has provided a relatively sophisticated understanding of many aspects of *A. tumefaciens* pathogenesis including plant-microbe signaling, interkingdom DNA transfer, T-DNA integration, tumorigenesis, opine production and bacterial cell-cell communication. Despite years of work however, the processes leading to productive physical association with host plant tissues remain largely undefined. With all of the presumptive adherence and colonization functions identified to this point, there are ambiguities as to their identity, their role in the process, or their general function. Perhaps this complex picture simply reflects that bacterial adherence and attachment are not a single process, but multiple processes that can dramatically differ between hosts and surfaces, and are highly sensitive to prevailing environmental conditions? The recognition that biofilm formation and structured adherent populations may also strongly influence the outcome of plant-microbe interactions, adds additional potential complexity to understanding initial steps in pathogenic as well as benign associations. Future work will need to determine whether there is a primary underlying process common to all surface interactions with baroque modifications adapted to specific conditions, or whether there are truly discrete and separable mechanisms that are largely independent of each other. There remains a great deal to be done in defining this area *A. tumefaciens* biology, with the promise, as in so many other areas of study on this fascinating and adaptable microbe, that this work will also illuminate other areas of prokaryotic biology and host-microbe associations.

8 ACKNOWLEDGEMENTS

Research on agrobacterial adherence and biofilm formation in the author's laboratory is supported by the United States Department of Agriculture (CRI 2002-35319-12636) and the Indiana University META-Cyt program.

9 REFERENCES

Amikam D, Benziman M (1989) Cyclic diguanylic acid and cellulose synthesis in *Agrobacterium tumefaciens*. J Bacteriol 171: 6649-6655

Armitage JP, Schmitt R (1997) Bacterial chemotaxis: *Rhodobacter sphaeroides* and *Sinorhizobium meliloti* – variations on a theme? Microbiology 143: 3671-3682

Ashby AM, Watson MD, Loake GJ, Shaw CH (1988) Ti plasmid-specified chemotaxis of *Agrobacterium tumefaciens* C58C1 toward *vir*-inducing phenolic compounds and soluble factors from monocotyledonous and dicotyledonous plants. J Bacteriol 170: 4181-4187

Ausmees N, Jacobsson K, Lindberg M (2001a) A unipolarly located, cell-surface-associated agglutinin, RapA, belongs to a family of *Rhizobium*-adhering proteins (Rap) in *Rhizobium leguminosarum* bv. *trifolii*. Microbiology 147: 549-559

Ausmees N, Mayer R, Weinhouse H, Volman G, Amikam D, Benziman M, Lindberg M (2001b) Genetic data indicate that proteins containing the GGDEF domain possess diguanylate cyclase activity. FEMS Microbiol Lett 204: 163-167

Bash R, Matthysse AG (2002) Attachment to roots and virulence of a *chvB* mutant of *Agrobacterium tumefaciens* are temperature sensitive. Mol Plant-Microbe Interact 15: 160-163

Becker A, Puhler A (1998) Production of expolysaccharides. In HP Spaink, A Kondorosi, PJJ Hooykaas, eds, The Rhizobeaceae: Molecular Biology of Model Plant-Associated Bacteria. Kluwer Academic Publishers, Boston, pp 97-118

Blatch GL, Lassle M (1999) The tetratricopeptide repeat: a structural motif mediating protein-protein interactions. Bio Essays 21: 932-939

Bouzar H, Moore LW (1987) Isolation of different *Agrobacterium* biovars from a natural oak savanna and tallgrass prairie. Appl Environm Microbiol 53: 717-721

Boyle EC, Finlay BB (2003) Bacterial pathogenesis: exploiting cellular adherence. Curr Opin Cell Biol 15: 633-639

Breedveld MW, Miller KJ (1994) Cyclic beta-glucans of members of the family Rhizobiaceae. Microbiol Rev 58: 145-161

Breedveld MW, Miller KJ (1998) Cell-surface beta-glucans. In HP Spaink, A Kondorosi, PJJ Hooykaas, eds, The Rhizobiaceae: Molecular Biology of Model Plant-Associated Bacteria. Kluwer Academic Publishers, Boston, pp 81-96

Brencic A, Angert ER, Winans SC (2005) Unwounded plants elicit *Agrobacterium vir* gene induction and T-DNA transfer: transformed plant cells produce opines yet are tumor free. Mol Microbiol 57: 1522-1531

Burdman S, Okon Y, Jurkevitch E (2000) Surface characteristics of *Azospirillum brasilense* in relation to cell aggregation and attachment to plant roots. Crit Rev Microbiol 26: 91-110

Burr TJ, Katz BH, Bishop AL (1987) Populations of *Agrobacterium* in vineyard and nonvineyard soils and grape roots in vineyards and nurseries. Plant Dis 71: 617-620

Cangelosi GA, Hung L, Puvanesarajah V, Stacey G, Ozga DA, Leigh JA, Nester EW (1987) Common loci for *Agrobacterium tumefaciens* and *Rhizobium meliloti* exopolysaccharide synthesis and their roles in plant interactions. J Bacteriol 169: 2086-2091

Cangelosi GA, Martinetti G, Leigh JA, Lee CC, Theines C, Nester EW (1989) Role for *Agrobacterium tumefaciens* ChvA protein in export of beta-1,2-glucan. J Bacteriol 171: 1609-1615

Chesnokova O, Coutinho JB, Khan IH, Mikhail MS, Kado CI (1997) Characterization of flagella genes of *Agrobacterium tumefaciens*, and the effect of a bald strain on virulence. Mol Microbiol 23: 579-590

Christie PJ, Atmakuri K, Krishnamoorthy V, Jakubowski S, Cascales E (2005) Biogenesis, architecture, and function of bacterial type IV secretion systems. Annu Rev Microbiol 59: 451-485

Cortez N, Carrillo N, Pasternak C, Balzer A, Klug G (1998) Molecular cloning and expression analysis of the *Rhodobacter capsulatus sodB* gene, encoding an iron superoxide dismutase. J Bacteriol 180: 5413-5420.

Danhorn T, Hentzer M, Givskov M, Parsek MR, Fuqua C (2004) Phosphorus limitation enhances biofilm formation of the plant pathogen *Agrobacterium tumefaciens* through the PhoR-PhoB regulatory system. J Bacteriol 186: 4492-4501

Dardanelli M, Angelini J, Fabra A (2003) A calcium-dependent bacterial surface protein is involved in the attachment of rhizobia to peanut roots. Can J Microbiol 49: 399-405

D'Argenio DA, Miller SI (2004) Cyclic di-GMP as a bacterial second messenger. Microbiology 150: 2497-2502

De Cleene M, De Ley J (1976) The host range of crown gall. Bot Rev 42: 389-466

Deakin WJ, Parker VE, Wright EL, Ashcroft KJ, Loake GJ, Shaw CH (1999) *Agrobacterium tumefaciens* possesses a fourth flagelin gene located in a large gene cluster concerned with flagellar structure, assembly and motility. Microbiology 145: 1397-1407

Douglas CJ, Halperin W, Nester EW (1982) *Agrobacterium tumefaciens* mutants affected in attachment to plant cells. J Bacteriol 152: 1265-1275

Douglas CJ, Staneloni RJ, Rubin RA, Nester EW (1985) Identification and genetic analysis of an *Agrobacterium tumefaciens* chromsomal virulence region. J Bacteriol 161: 850-860

Escudero J, Hohn B (1997) Transfer and integration of T-DNA without cell injury in the host plant. Plant Cell 9: 2135-2142

Fletcher M (1996) Bacterial attachment in aquatic environments: a diversity of surfaces and adhesion strategies. In M Fletcher, ed, Bacterial Adhesion: Molecular and Ecological Diversity. Wiley-Liss, New York, pp 1-24

Fullner KJ, Lara JC, Nester EW (1996) Pilus assembly by *Agrobacterium* T-DNA transfer genes. Science 273: 1107-1109

Garfinkel DJ, Nester EW (1980) *Agrobacterium tumefaciens* mutants affected in crown gall tumorigenesis and octopine catabolism. J Bacteriol 144: 732-743

Goodner B, Hinkle G, Gattung S, Miller N, Blanchard M, Qurollo B, Goldman BS, Cao Y, Askenazi M, Halling H, Mullin L, Houmiel K, Gordon J, Vaudin M, Iartchouk O, Epp A, Liu F, Wollam C, Allinger M, Doughty D, Scott C, Lappas C, Markelz B, Flanagan C, Crowell C, Gurson J, Lomo C, Sear C, Strub G, Cielo C, Slater S (2001) Genome sequence of the plant pathogen and biotechnology agent *Agrobacterium tumefaciens* C58. Science 294: 2323-2328

Guilhabert MR, Kirkpatrick BC (2005) Identification of *Xylella fastidiosa* antivirulence genes: hemagglutinin adhesins contribute a biofilm maturation to *X.*

fastidios and colonization and attenuate virulence. Mol Plant Microbe Interact 18: 856-868

Hall-Stoodley L, Costerton JW, Stoodley P (2004) Bacterial biofilms: from the natural environment to infectious diseases. Nat Rev Microbiol 2: 95-108

Hinsa SM, Espinosa-Urgel M, Ramos JL, O'Toole GA (2003) Transition from reversible to irreversible attachment during biofilm formation by *Pseudomonas fluorescens* WCS365 requires an ABC transporter and a large secreted protein. Mol Microbiol 49: 905-918

Hirsch AM, Lum MR, Downie JA (2001) What makes the rhizobia-legume symbiosis so special? Plant Physiol 127: 1484-1492

Holford ICR (1997) Soil phosphorous: its measurement and its uptake by plants. Aus J Biol Res 35: 227-239

Hooykaas PJ, Klapwijk PM, Nuti MP, Schilperoort RA, Rorsch A (1977) Transfer of the *Agrobacterium tumefaciens* Ti plasmid to avirulent agrobacteria and to *Rhizobium ex planta*. J Gen Microbiol 98: 477-484

Hwang HH, Gelvin SB (2004) Plant proteins that interact with VirB2, the *Agrobacterium tumefaciens* pilin protein, mediate plant transformation. Plant Cell 16: 3148-3167

Hynes MF, Simon R, Puhler A (1985) The development of plasmid-free strains of *Agrobacterium tumefaciens* by using incompatibility with a *Rhizobium meliloti* plasmid to eliminate pAtC58. Plasmid 13: 99-105

Jenal U (2004) Cyclic di-guanosine-monophosphate comes of age: a novel secondary messenger involved in modulating cell surface structures in bacteria? Curr Opin Microbiol 7: 185-191

Judd PK, Kumar RB, Das A (2005) Spatial location and requirements for the assembly of the *Agrobacterium tumefaciens* type IV secretion apparatus. Proc Natal Acad Sci USA 102: 11498-11503

Justice SS, Hung C, Theriot JA, Fletcher DA, Anderson GG, Footer MJ, Hultgren SJ (2004) Differentiation and developmental pathways of uropathogenic *Escherichia coli* in urinary tract pathogenesis. Proc Natl Acad Sci USA 101: 1333-1338

Kado CI (1992) Plant pathogenic bacteria. In A. I. Balows, H. G. Truper, M. Dworkin, W. Harder, K-H Schleifer, eds, The Prokaryotes, 2 ed, vol 1. Springer-Verlag, New York

Kawagishi I, Muller V, Williams AW, Irikura VM, Macnab RM (1992) Subdivision of flagellar region III of the *Escherichia coli* and *Salmonella typhimurium* chromosomes and identification of two additional flagellar genes. J Gen Microbiol 138: 1051-1065

Kelly BA, Kado CI (2002) *Agrobacterium*-mediated T-DNA transfer and integration into the chromosome of *Streptomyces lividans*. Mol Plant Pathol 3: 125-134

Kim H, Farrand SK (1998) Opine catabolic loci from *Agrobacterium* plasmids confer chemotaxis to their cognate substrates. Mol Plant Microbe Interact 11: 131-143

Knutton S, Rosenshine I, Pallen MJ, Nisan I, Neves BC, Bain C, Wolff C, Dougan G, Frankel G (1998) A novel EspA-associated surface organelle of enteropathogenic *Escherichia coli* involved in protein translocation into epithelial cells. Embo J 17: 2166-2176

Korner H, Sofia HJ, Zumft WG (2003) Phylogeny of the bacterial superfamily of Crp-Fnr transcription regulators: exploiting the metabolic spectrum by controlling alternative gene programs. FEMS Microbiol Rev 27: 559-592

Kunik T, Tzfira T, Kapulnik Y, Gafni Y, Dingwall C, Citovsky V (2001) Genetic transformation of HeLa cells by *Agrobacterium*. Proc Natl Acad Sci USA 98: 1871-1876

Lacroix B, Tzfira T, Vainstein A, Citovsky V (2006) A case of promiscuity: *Agrobacterium*'s endless hunt for new partners. Trends Genet 22: 29-37

Lai EM, Chesnokova O, Banta LM, Kado CI (2000) Genetic and environmental factors affecting T-pilin export and T-pilus biogenesis in relation to flagellation of *Agrobacterium tumefaciens*. J Bacteriol 182: 3705-3716

Lai EM, Eisenbrandt R, Kalkum M, Lanka E, Kado CI (2002) Biogenesis of T pili in *Agrobacterium tumefaciens* requires precise VirB2 propilin cleavage and cyclization. J Bacteriol 184: 327-330

Langaee TY, Gagnon L, Huletsky A (2000) Inactivation of the *ampD* gene in *Pseudomonas aeruginosa* leads to moderate-basal-level and hyperinducible AmpC beta-lactamase expression. Antimicrob Agents Chemother 44: 583-589

Laus MC, Logman TJ, Lamers GE, Van Brussel AA, Carlson RW, Kijne JW (2006) A novel polar surface polysaccharide from *Rhizobium leguminosarum* binds host plant lectin. Mol Microbiol 59: 1704-1713

Lazazzera BA, Beinert H, Khoroshilova N, Kennedy MC, Kiley PJ (1996) DNA binding and dimerization of the Fe-S-containing FNR protein from Escherichia coli are regulated by oxygen. J Biol Chem 271: 2762-2768

Lippincott BB, Lippincott JA (1969) Bacterial attachment to a specific wound site as an essential stage in tumor initiation by *Agrobacterium tumefaciens*. J Bacteriol 97: 620-628

Lugtenberg BJJ, Chin-A-Woeng TF, Bloemberg GV (2002) Microbe-plant interactions: principles and mechanisms. Antonie Van Leeuwenhoek 81: 373-383

Marshall KC, Stout R, Mitchell R (1971) Mechanisms of the initial events in the sorption of marine bacteria to surfaces. J Gen Microbiol 68: 337-348

Matthysse AG (1983) Role of bacterial cellulose fibrils in *Agrobacterium tumefaciens* infection. J Bacteriol 154: 906-915

Matthysse AG (1987) Characterization of nonattaching mutants of *Agrobacterium tumefaciens*. J Bacteriol 169: 313-323

Matthysse AG (1994) Conditioned medium promotes the attachment of *Agrobacterium tumefaciens* strain NT1 to carrot cells. Protoplasma 183: 131-136

Matthysse AG, Holmes KV, Gurlitz RHG (1981) Elaboration of cellulose fibrils by *Agrobacterium tumefaciens* during attachment to carrot cells. J Bacteriol 145: 583-595

Matthysse AG, Kijne JW (1998) Attachment of Rhizobiaceae to plant cells. In HP Spaink, A Kondorosi, PJJ Hooykaas, eds, The Rhizobiaceae: Molecular Biology of Model Plant-Associated Bacteria. Kluwer Academic Publishers, Dordrecht/Boston/London, pp 235-249

Matthysse AG, Marry M, Krall L, Kaye M, Ramey BE, Fuqua C, White AR (2005) The effect of cellulose overproduction on binding and biofilm formation on roots by *Agrobacterium tumefaciens*. Mol Plant Microbe Interact 18: 1002-1010

Matthysse AG, McMahan S (1998) Root colonization by *Agrobacterium tumefaciens* is reduced in *cel, attB, attD,* and *attR* mutants. Appl Environ Microbiol 64: 2341-2345

Matthysse AG, McMahan S (2001) The effect of the *Agrobacterium tumefaciens attR* mutation on attachment and root colonization differs between legumes and other dicots. Appl Environ Microbiol 67: 1070-1075

Matthysse AG, Thomas DL, White AR (1995a) Mechanism of cellulose synthesis in *Agrobacterium tumefaciens*. J Bacteriol 177: 1076-1081

Matthysse AG, White S, Lightfoot R (1995b) Genes required for cellulose synthesis in *Agrobacterium tumefaciens*. J Bacteriol 177: 1069-1075

Matthysse AG, Yarnall H, Boles SB, McMahan S (2000) A region of the *Agrobacterium tumefaciens* chromosome containing genes required for virulence and attachment to host cells. Biochim Biophys Acta 1490: 208-212

Matthysse AG, Yarnall HA, Young N (1996) Requirement for genes with homology to ABC transport systems for attachment and virulence of *Agrobacterium tumefaciens*. J Bacteriol 178: 5302-5308

Mills AL, Powelson DK (1996) Bacterial interactions with surfaces in soils. In M Fletcher, ed, Bacterial Adhesion: Molecular and Ecological Diversity. Wiley-Liss, New York, pp 25-57

Minnemeyer SL, Lightfoot R, Matthysse AG (1991) A semiquantitative bioassay for relative virulence of *Agrobacterium tumefaciens* strains on *Bryophyllum daigremontiana*. J Bacteriol 173: 7723-7724

Monds RD, Silby MW, Mahanty HK (2001) Expression of the Pho regulon negatively regulates biofilm formation by *Pseudomonas aureofaciens* PA147-2. Mol Microbiol 42: 415-426

Morris CE, Monier JM (2003) The ecological significance of biofilm formation by plant-associated bacteria. Annu Rev Phytopathol 41: 429-453

Nair GR, Liu Z, Binns AN (2003) Re-examining the role of the accessory plasmid pAtC58 in the virulence of *Agrobacterium tumefaciens* strain C58. Plant Physiol 133: 989-999

Noel KD, Duelli DM (2000) Rhizobium lipopolysacchride and its role in symbiosis. In EW Triplett, ed, Prokaryotic Nitrogen Fixation: A Model System for the Analysis of a Biological Process. Horizon Scientific Press,, Norfolk, UK, pp 415-431

Nougayrede JP, Fernandes PJ, Donnenberg MS (2003) Adhesion of enteropathogenic *Escherichia coli* to host cells. Cell Microbiol 5: 359-372

Ohnishi K, Ohto Y, Aizawa S-I, Macnab RM, Iino T (1994) FlgD is a scaffolding protein needed for flagellar hook assembly in *Salmonella typhimurium*. J Bacteriol 176: 2272-2281

O'Toole GA, Pratt LA, Watnick PI, Newman DK, Weaver VB, Kolter R (1999) Genetic approaches to study of biofilms. Methods Enzymol 310: 91-109

Parsek MR, Fuqua C (2004) Biofilms 2003: emerging themes and challenges in studies of surface-associated microbial life. J Bacteriol 186: 4427-4440

Parsek MR, Singh PK (2003) Bacterial biofilms: an emerging link to disease pathogenesis. Annu Rev Microbiol 57: 677-701

Paul R, Weiser S, Amiot NC, Chan C, Schirmer T, Giese B, Jenal U (2004) Cell cycle-dependent dynamic localization of a bacterial response regulator with a novel di-guanylate cyclase output domain. Genes Dev 18: 715-727

Pizarro-Cerda J, Cossart P (2006) Bacterial adhesion and entry into host cells. Cell 124: 715-727

Pueppke SG, Hawes MC (1985) Understanding the binding of bacteria to plant surfaces. Trends Biotechnol 3: 310-313

Puvanesarajah V, Schell FM, Stacey G, Douglas CJ, Nester EW (1985) Role for 2-linked-beta-D-glucan in the virulence of *Agrobacterium tumefaciens*. J Bacteriol 164: 102-106

Ramey BE (2004) Biofilm formation by *Agrobacterium tumefaciens* and its role in plant interactions. Doctoral dissertation, Indiana University, Bloomington

Ramey BE, Koutsoudis M, von Bodman SB, Fuqua C (2004a) Biofilm formation in plant-microbe associations. Curr Opin Microbiol 7: 602-609

Ramey BE, Matthysse AG, Fuqua C (2004b) The FNR-type transcriptional regulator SinR controls maturation of *Agrobacterium tumefaciens* biofilms. Mol Microbiol 52: 1495-1511

Reed JW, Glazebrook J, Walker GC (1991) The *exoR* gene of *Rhizobium meliloti* affects RNA levels of other *exo* genes but lacks homology to known transcriptional regulators. J Bacteriol 173: 3789-3794

Reuhs BL, Kim JS, Matthysse AG (1997) Attachment of *Agrobacterium tumefaciens* to carrot cells and *Arabidopsis* wound sites is correlated with the presence of a cell-associated, acidic polysaccharide. J Bacteriol 179: 5372-5379

Rojas CM, Ham JH, Deng WL, Doyle JJ, Collmer A (2002) HecA, a member of a class of adhesins produced by diverse pathogenic bacteria, contributes to the attachment, aggregation, epidermal cell killing, and virulence phenotypes of *Erwinia chrysanthemi* EC16 on *Nicotiana clevelandii* seedlings. Proc Natl Acad Sci USA 99: 13142-13147

Romling U (2002) Molecular biology of cellulose production in bacteria. Res Microbiol 153: 205-212

Rosenberg C, Huguet T (1984) The pAtC58 plasmid of *Agrobacterium tumefaciens* is not essential for tumour induction. Mol Gen Genet 196: 533-536

Russo DM, Williams A, Edwards A, Posadas DM, Finnie C, Dankert M, Downie JA, Zorreguieta A (2006) Proteins exported *via* the PrsD-PrsE type I secretion system and the acidic exopolysaccharide are involved in biofilm formation by *Rhizobium leguminosarum*. J Bacteriol 188: 4474-4486

Shaw CH, Loake GJ, Brown AP, Garrett CS, Deakin W, Alton G, Hall M, Jones SA, Oleary M, Primavesi L (1991) Isolation and characterization of behavioral mutants and genes of *Agrobacterium tumefaciens*. J Gen Microbiol 137: 1939-1953

Smit G, Logman TJJ, Boerrigter METI, Kijne JW, Lugtenberg BJJ (1989) Purification and partial characterization of the *Rhizobium leguminosarum* biovar viciae Ca2+-dependent adhesin, which mediates the first step in attachment of cells of the family Rhizobiaceae to plant root hair tips. J Bacteriol 171: 4054-4062

Swart S, Smit G, Lugtenberg BJJ, Kijne JW (1993) Restoration of attachment, virulence and nodulation of *Agrobacterium tumefaciens chvB* mutants by rhicadhesin. Mol Microbiol 10: 597-605

Walker TS, Bais HP, Grotewold E, Vivanco JM (2003) Root exudation and rhizosphere biology. Plant Physiol 132: 44-51

Wanner BL (1995) Signal transduction and cross regulation in the *Escherichia coli* phosphate regulaon by PhoR, CreC, and acetyl phosphate. In JA Hoch, TJ Silhavy, eds, Two-Component Signal Transduction. ASM Press, Washington D.C., pp 203-221

Whatley MH, Bodwin JS, Lippincott BB, Lippincott JA (1976) Role of *Agrobacterium* cell envelope lipopolysaccharide in infection site attachment. Infect Immun 13: 1080-1083

Winans SC (1990) Transcriptional induction of an *Agrobacterium* regulatory gene at tandem promoters by plant-released phenolic compounds, phosphate starvation, and acidic growth media. J Bacteriol 172: 2433-2438

Winans SC (1992) Two-way chemical signalling in *Agrobacterium*-plant interactions. Microbiol Rev 56: 12-31

Wood DW, Setubal JC, Kaul R, Monks DE, Kitajima JP, Okura VK, Zhou Y, Chen L, Wood GE, Almeida Jr. NF, Woo L, Chen Y, Paulsen IT, Eisen JA, Karp PD, Bovee Sr. D, Chapman P, Clendenning J, Deatherage G, Gillet W, Grant C, Kutyavin T, Levy R, Li MJ, McClelland E, Palmieri P, Raymond C, Rouse R, Saenphimmachak C, Wu Z, Romero P, Gordon D, Zhang S, Yoo H, Tao Y, Biddle P, Jung M, Krespan W, Perry M, Gordon-Kamm B, Liao L, Kim S, Hendrick C, Zhao ZY, Dolan M, Chumley F, Tingey SV, Tomb JF, Gordon MP, Olson MV, Nester EW (2001) The genome of the natural genetic engineer *Agrobacterium tumefaciens* C58. Science 294: 2317-2323

Yao SY, Luo L, Har KJ, Becker A, Ruberg S, Yu GQ, Zhu JB, Cheng HP (2004) *Sinorhizobium meliloti* ExoR and ExoS proteins regulate both succinoglycan and flagellum production. J Bacteriol 186: 6042-6049

Yuan ZC, Zaheer R, Finan TM (2006) Regulation and properties of PstSCAB, a high-affinity, high-velocity phosphate transport system of *Sinorhizobium meliloti*. J Bacteriol 188: 1089-1102

Zhang HB, Wang LH, Zhang LH (2002) Genetic control of quorum-sensing signal turnover in *Agrobacterium tumefaciens*. Proc Natl Acad Sci USA 99: 4638-4643

Zhu Y, Nam J, Humara JM, Mysore KS, Lee LY, Cao H, Valentine L, Li J, Kaiser AD, Kopecky AL, Hwang HH, Bhattacharjee S, Rao PK, Tzfira T, Rajagopal J, Yi H, Veena, Yadav BS, Crane YM, Lin K, Larcher Y, Gelvin MJ, Knue M, Ramos C, Zhao X, Davis SJ, Kim SI, Ranjith-Kumar CT, Choi YJ, Hallan VK, Chattopadhyay S, Sui X, Ziemienowicz A, Matthysse AG, Citovsky V, Hohn B, Gelvin SB (2003) Identification of *Arabidopsis rat* mutants. Plant Physiol 132: 494-505

Zipfel C, Kunze G, Chinchilla D, Caniard A, Jones JD, Boller T, Felix G (2006) Perception of the bacterial PAMP EF-Tu by the receptor EFR restricts *Agrobacterium*-mediated transformation. Cell 125: 749-760

Zorreguieta A, Geremia RA, Cavaignac S, Cangelosi GA, Nester EW, Ugalde RA (1988) Identification of the product of an *Agrobacterium tumefaciens* chromosomal virulence gene. Mol Plant Microbe Interact 1: 121-127

Color Plates

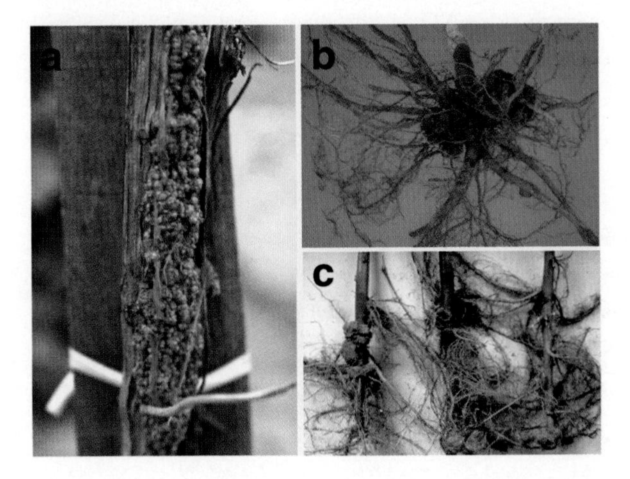

Figure 1-1. Natural crown galls on different hosts. a, grapevine (*Vitis vinifera*, cv. 'Ezerfürtü'); b, raspberry; c, apple. (photo a was kindly provided by Jozsef Mikulas, Kecskemét, Hungary, photo b by Thomas Burr, Cornell University, Geneva and photo c by Christine Blaser, University of Guelph, Laboratory Services, Pest Diagnostic Clinic.)

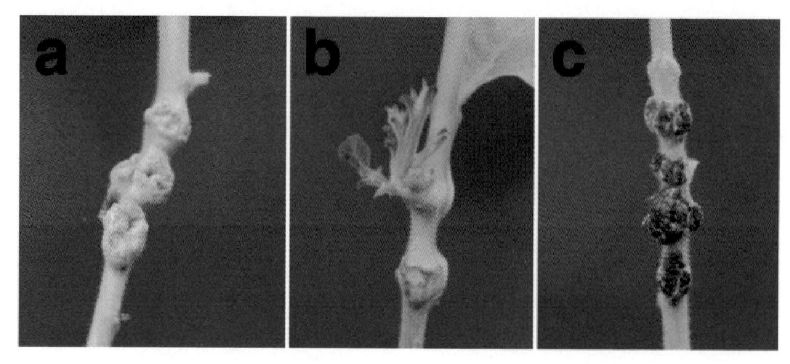

Figure 1-2. Experimental infections of *Nicotiana tabacum* cv. *Samsun* with different *Agrobacterium vitis* strains, showing differences in crown gall morphology. a, strain AB4, undifferentiated tumors; b, strain AT1, shooty tumors (teratomata); c, Tm4, necrotic tumors. (Photos by Ernö Szegedi).

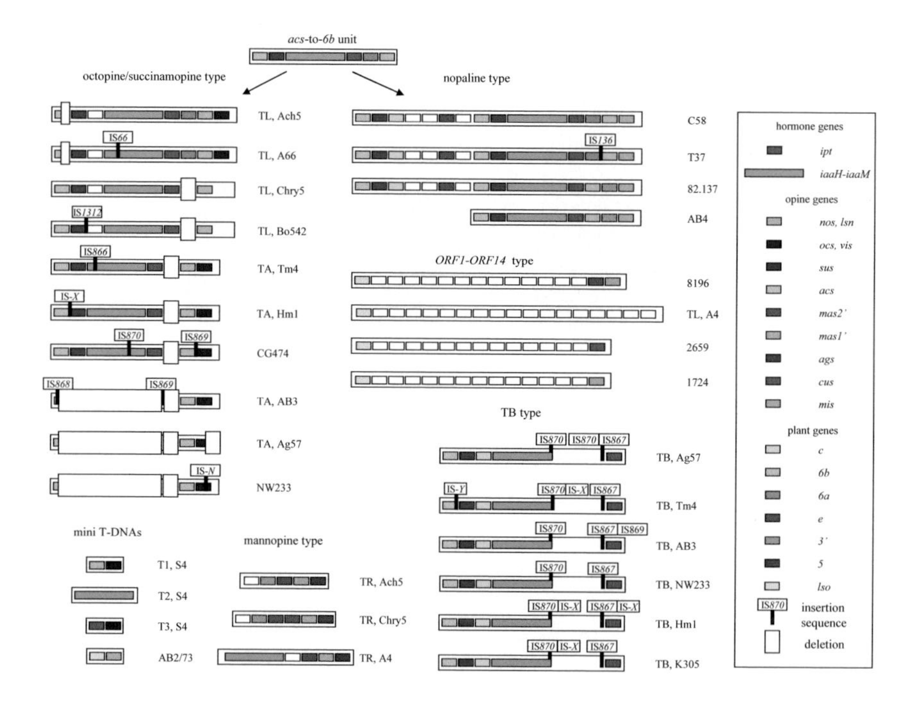

Figure 1-3. Schematic maps of different T-DNA structures. Genes discussed in the text are marked in color. During T-DNA evolution, genes and gene groups were combined in different ways. Maps are not to scale.

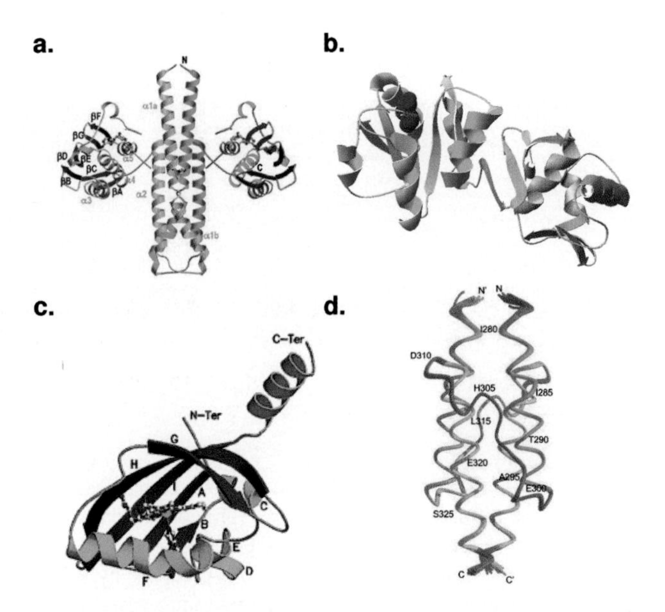

Figure 6-1. Structures of some functional and regulatory domains of TCS. (a) Crystal structure of the *Thermotoga maritima* HK0853 showing a dimer of HK0853 with each subunit containing dimerization and ATP-binding domains. (b) Crystal structure of the response regulator DrrB of *T. maritima* with the receiver domain colored in blue and yellow, the conserved Asp residue in red, and the DNA binding domain in gray. (c) The crystal structure of the PAS domain of *Bradyrhizobium japonicum* oxygen sensing protein FixL. The bound heme cofactor is shown as a ball-and-stick representation. (d) NMR structure of a HAMP domain of Af1503 from *Archaeoglobus fulgidus*. The dimeric domain maintains a coiled-coil structure.

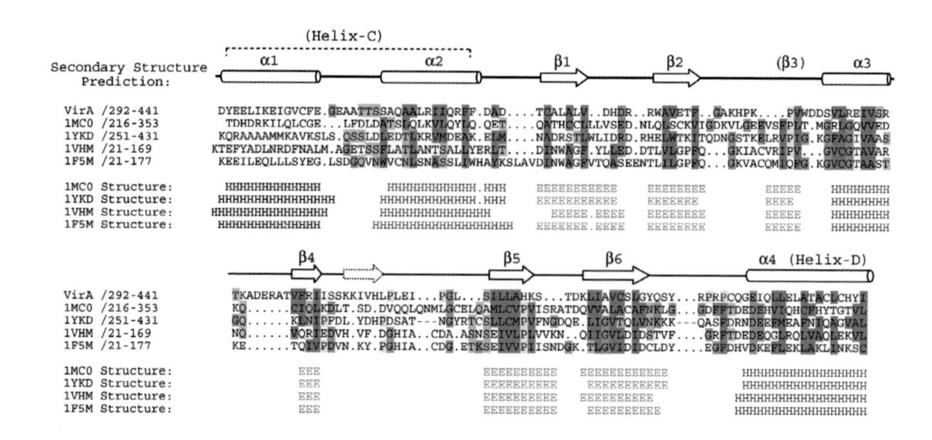

Figure 6-2. Alignment of VirA linker with known GAF structures. Sequence alignment is from PFAM database and refined based on structures (protein data bank ID: 1MC0, 1YKD, 1VHM and 1F5M). H and E represent α-helices and β-strands observed in the structure, respectively. The secondary structure of VirA linker was predicted by SAM-T02 method. Predicted α-helices and β-strands are marked as cylinders and arrows. The dotted arrow indicates a β-strand but not conserved among GAF structures. Residues with remote homology were colored as the following: blue, hydrophobic and aromatic residues (L, I, V, M, C, A, F, W); red, charged residues (D, E, K, R); orange, G; yellow, P; green, (S, T, N, Q). Helix C refers to the region of α1 and α2, while helix D is α4.

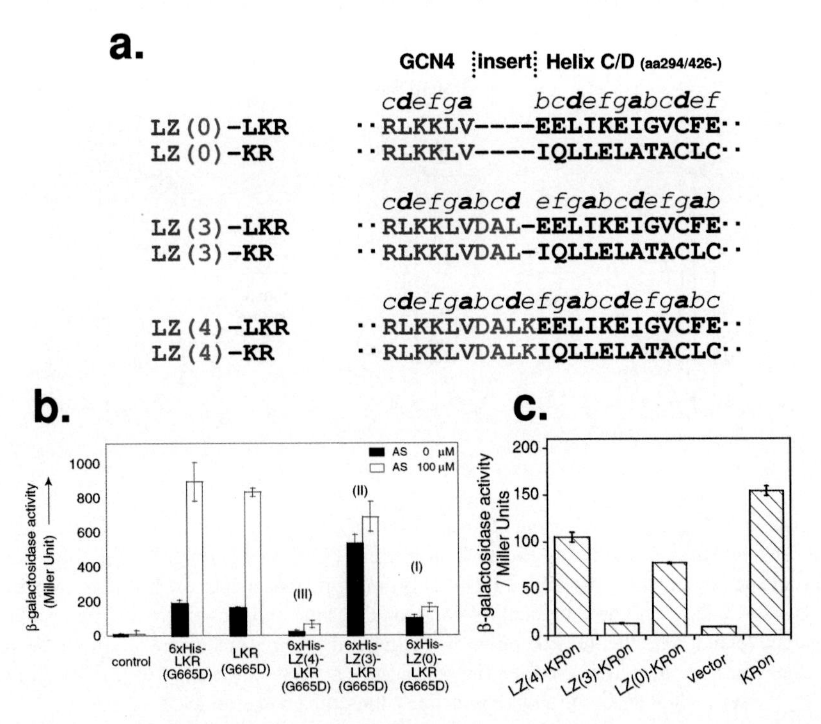

Figure 6-3. N'-fused GCN4 leucine zipper of helix C/D of the linker domain. (a) The heptad repeats registry of LZ(0/3/4) chimeras. Fused at aa294 is the helix C fusion (LZ-LKR), while fusion made at aa426 is the helix D fusion (LZ-KR). LZ residues are shown in blue, the inserting amino acids between LZ and VirA are in red, and VirA sequence in black. (b) β-galactosidase activity of different LZ-LKR (helix C) fusions. G665D is a constitutive on mutant of VirA, using this high activity mutant simplified the activity measurement. (c) β-galactosidase activity of different LZ-KR (helix D) fusions with ON mutant also at G665D (see Gao and Lynn, 2007).

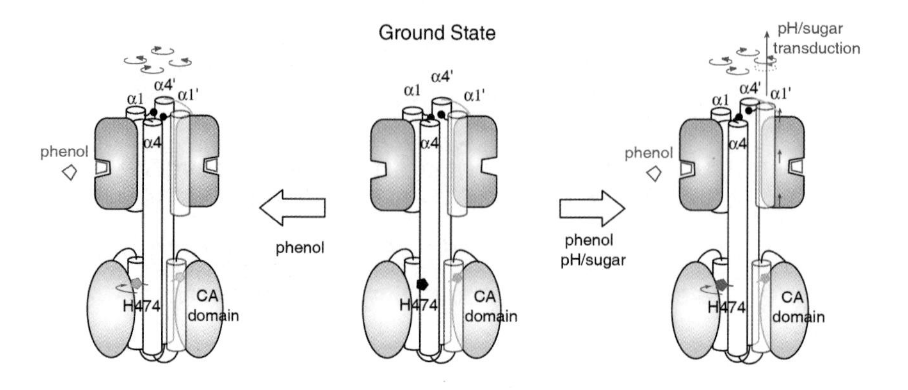

Figure 6-4. Signal integration and transduction of VirA linker. A central four-helix bundle formed by Helix-C and D (α1 and α4 in predicted GAF structure) is critical for both periplasmic and cytoplasmic signaling. Helix-D is directly connected to the histidine containing helix of kinase and the rotational motion modulates the phophosrylation of the histidine residue (pentagon). Phenol perception by the linker domain is postulated to initiate the rotation within the four-helix bundle. The periplasmic sensing of pH/sugar is proposed to induce a sliding of the signaling helices, thus enhancing the phenol response.

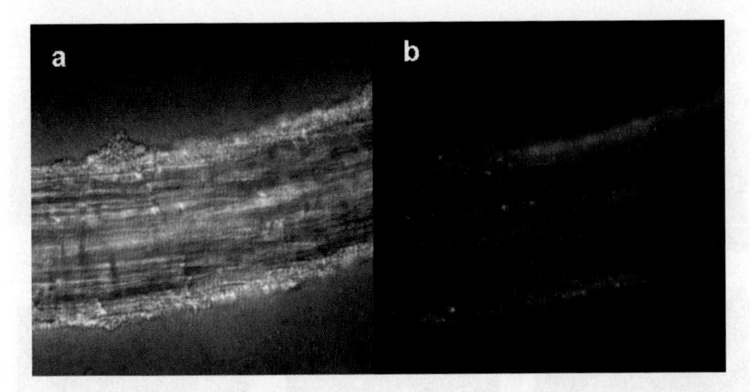

Figure 7-1. Plant tissue attachment and biofilm formation. Adherent *A. tumefaciens* C58 harboring GFP on *Arabidopsis thaliana* WS seedling root. (a) Bright field microscopy, (b) Fluorescence microscopy, bacteria are pseudocolored red. Images captured on a Deltavision deconvolution microscope.

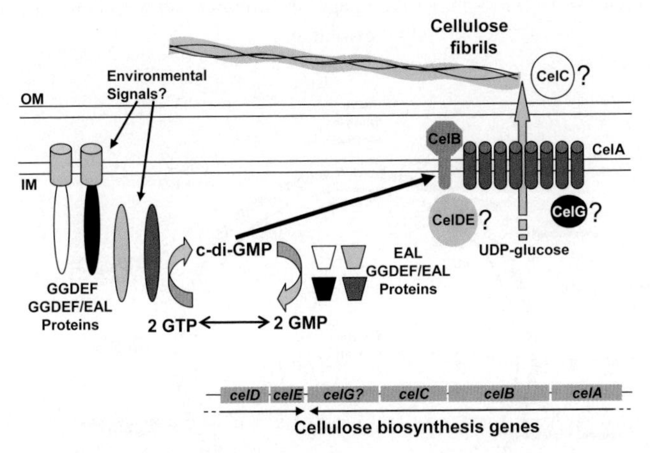

Figure 7-2. Cellulose biosynthesis in *A. tumefaciens*. Gene map of cellulose biosynthesis operons, model of membrane-associated cellulose synthase complex, and a depiction of c-di-GMP synthesis and turnover by GGDEF/EAL proteins.

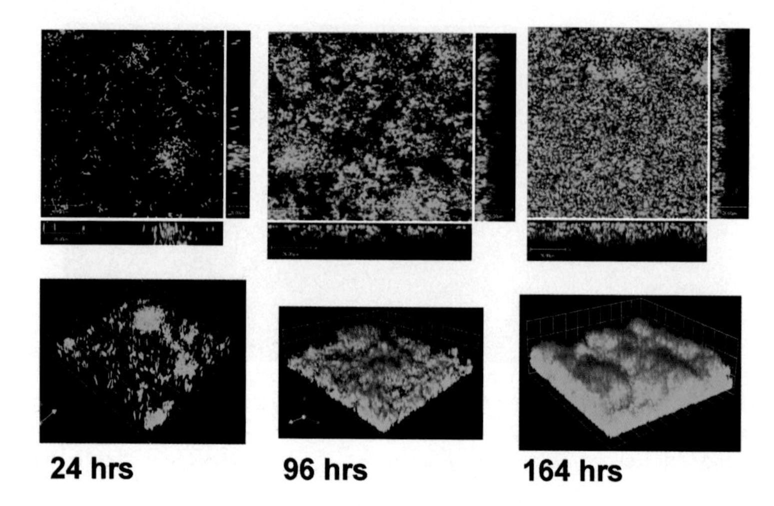

Figure 7-3. A. tumefaciens biofilm formation. Adherent *A. tumefaciens* C58 harboring GFP on a glass surface over time. Images were acquired using a confocal laser scanning microscopy and Volocity software to render the image. In collaboration with Dingding An and Matthew Parsek.

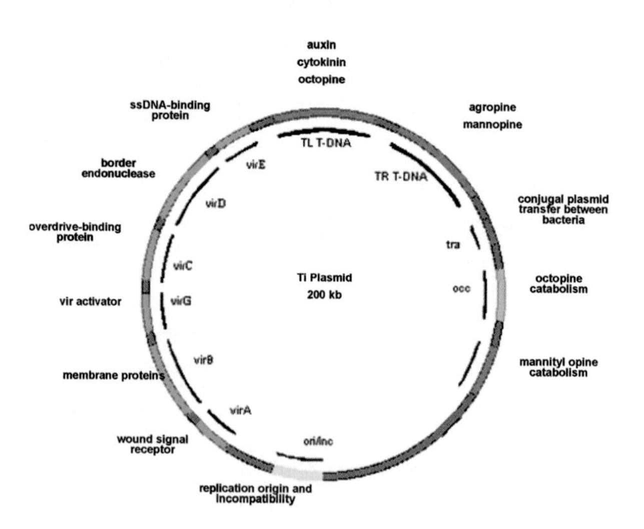

Figure 8-1. Map of an octopine-type tumor-inducing plasmid. The octopine-type Ti plasmid contains two separate T-DNAs, shown in red. TL, the oncogenic T-DNA, encodes proteins for synthesis of two phytohormones, auxin and cytokinin, as well as octopine, which is derived from arginine and pyruvate. TR does not promote tumor growth, but it encodes proteins for synthesis of the mannityl opines (agropine and mannopine), which are derived from mannose and glutamine.

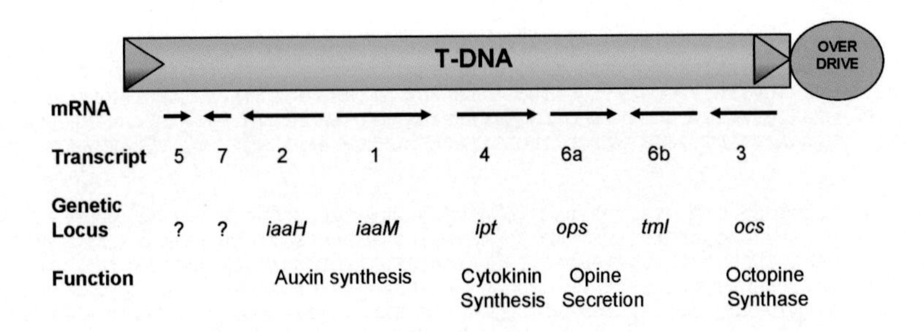

Figure 8-2. Genetic map of the TL transferred DNA (T-DNA) of an octopine type Ti plasmid. The arrowheads in the box represent the left and right borders of the T-DNA, and they indicate the direction of T-DNA transfer, from right to left. *iaa*H (indoleacetamide hydrolase), *iaa*M (tryptophan monooxygenase), *ipt* (isopentenyl transferase), *ops* (octopine secretion), *tml* (tumor morphology large), *ocs* (octopine synthase). Arrows indicate the direction and length of T-DNA encoded mRNAs produced by transformed plant cells.

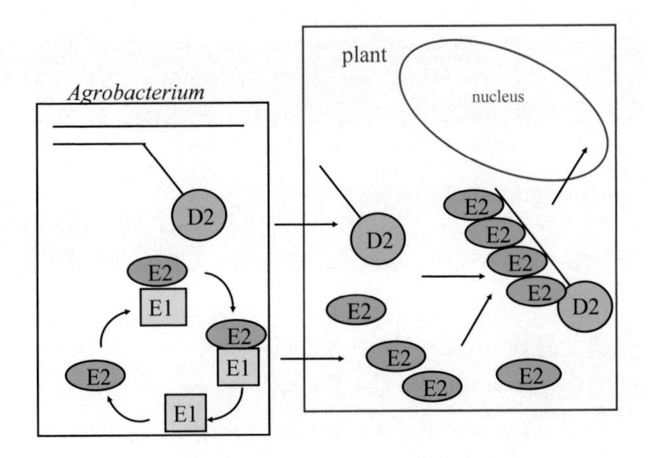

Figure 8-3. The separate export model.

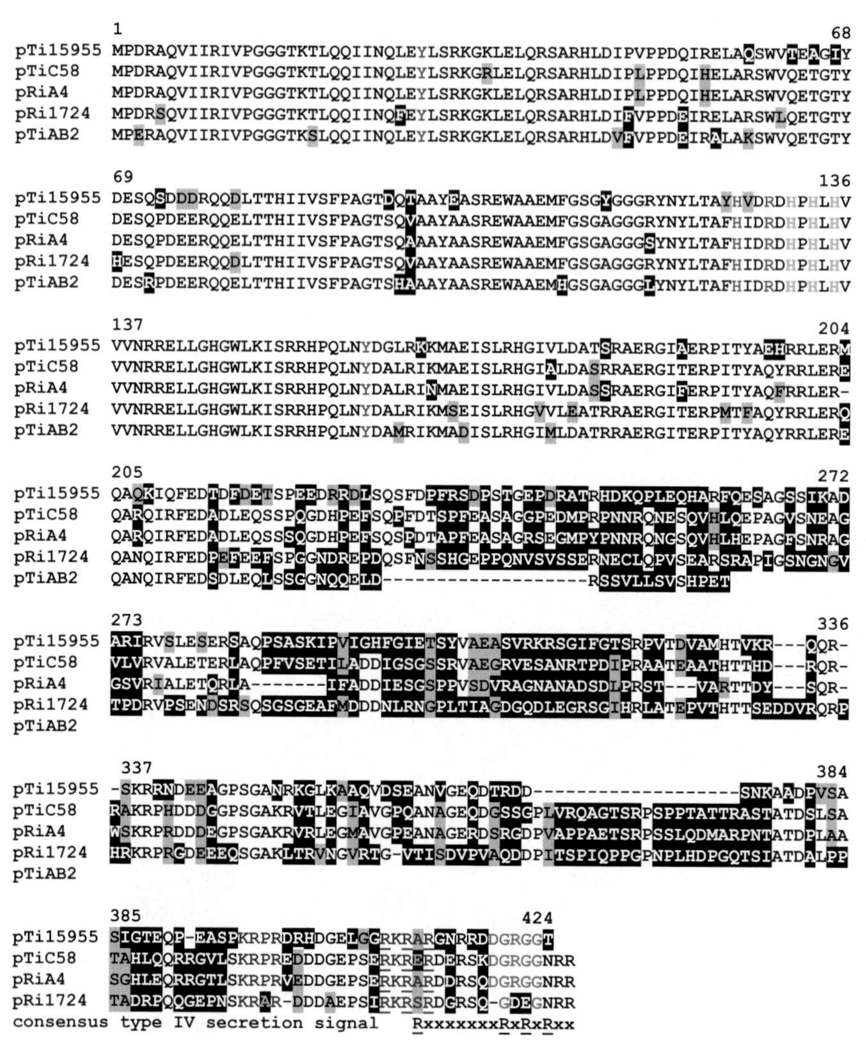

Figure 8-4. Clustal W alignment of VirD2 amino acid sequences. Black letters on a white background indicate amino acids that are identical in at least three VirD2 proteins. Shaded boxes indicate similar amino acids. Groups of amino acids considered similar in this analysis were: I, L, M, and V; A, G, and S; H, K, and R; D and E; N and Q; F, W, and Y; and S and T. Solid boxes (white letters on a black background) indicate nonconserved amino acids. Dashes indicate gaps placed in the sequences to maximize alignment. Numbers indicate the positions of amino acids in VirD2 from pTi15955. The red Y at position 29 indicates the conserved tyrosine that forms a covalent bond with the 5' end of the T-strand (Vogel and Das, 1992b). Conserved histidine (residue 126), arginine (residue 129), and tyrosine (residue 160) residues shown in pink may correspond to a "HRY motif" found in seven phage-encoded site specific integrase proteins (Argos et al., 1986) and in the TraI protein encoded by conjugative plasmid RP4 (Lessl and Lanka, 1994). Three conserved histidine residues (positions 131, 133, and 135) shown in green comprise a "histidine triad" motif (HxHxH), which may mediate the

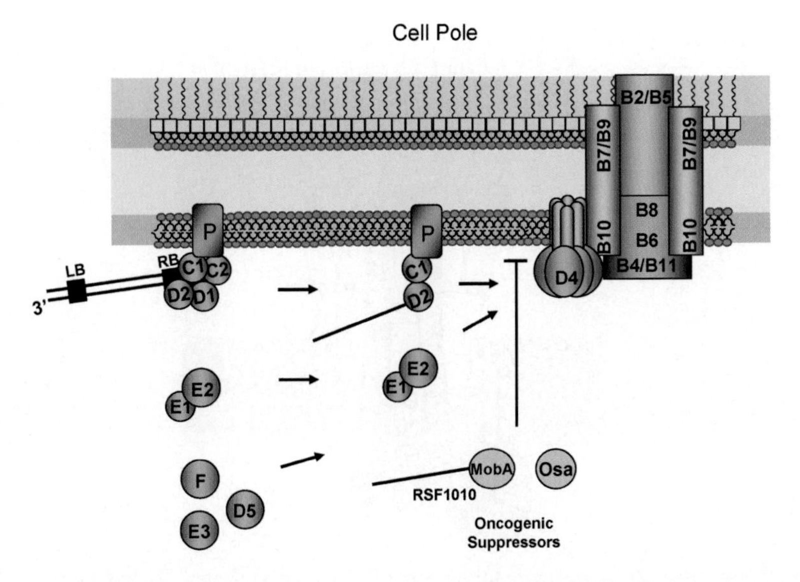

Figure 9-1. Processing of substrates for transfer through the *A. tumefaciens* VirB/D4 type IV secretion system. The Dtr processing factors VirC1, VirC2, VirD1, and VirD2 assemble at the border repeats flanking the T-DNA and VirD2 relaxase generates a nick on the strand destined for transfer (T-strand). ParA-like VirC1 functions both in VirD2 recruitment to the T-DNA borders and to the *A. tumefaciens* cell poles to coordinate substrate docking with the VirD4 receptor. VirE2 is bound by the VirE1 chaperone and is recruited via a C-terminal domain to the polar-localized VirD4 receptor. Two oncogenic suppressors, the RSF1010 transfer intermediate and pSA-encoded Osa, block substrate docking with VirD4. See text for further details and references.

interaction of VirD2 with a nuclear protein kinase, cyclin-dependent kinase-activating kinase (CAK2Ms) (Bako et al., 2003). Amino acids comprising the nuclear localization sequence (KRPRDRHDGELGGRKRAR) are shown in blue. Arginine residues that correspond to the type IV secretion signal consensus sequence (RxxxxxxxRxRxRxx) are underlined. The C-terminal omega sequence (DGRGG) is shown in orange (Shurvinton et al., 1992). The endonuclease domain spans amino acids 1-262. The cyclophilin CypA binds amino acids 274-337, whereas Roc1 binds amino acids 174-337. A deletion that removes amino acids 338-356 does not affect VirD2 function (Shurvinton et al., 1992).

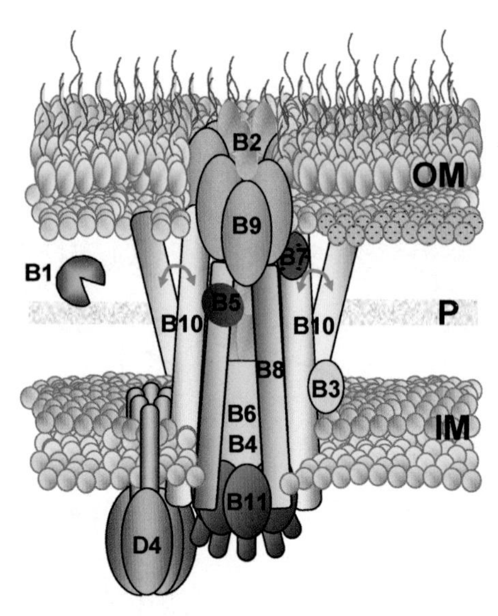

Figure 9-5. A proposed architecture for the VirB/D4 secretion channel. VirB10 senses ATP energy use by VirD4 and VirB11 and, in turn, stably interacts with the outer membrane-associated VirB7-VirB9 complex. The model depicts the VirB/D4 T4S as structurally dynamic, wherein signals including substrate binding, ATP energy, and target cell contacts trigger structural transitions between a quiescent transenvelope complex and an active transport channel.

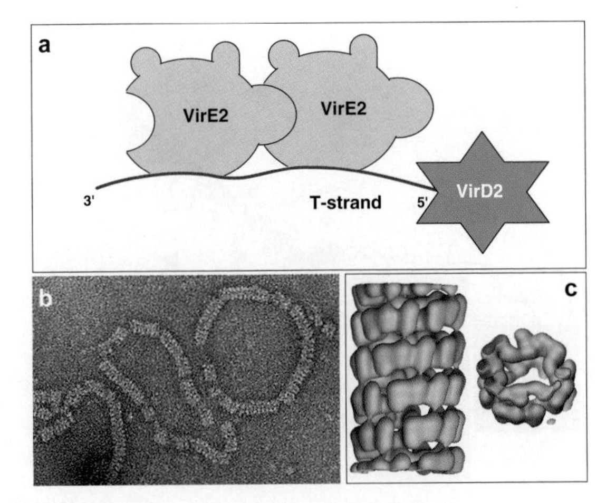

Figure 10-1. The T-complex structure. (a) The T-complex is composed of the T-strand, a single VirD2 which is bound to its 5' end, and numerous VirE2 molecules coating its entire length. (b) A complex of VirE2 proteins with ssDNA molecules, as seen by transmission electron microscopy. (c) A three-dimensional reconstruction of VirE2-ssDNA complexes showing the outer (left) and inner (right) structure of the solenoidal complex. Panels b and c reproduced with permission from (Abu-Arish et al., 2004).

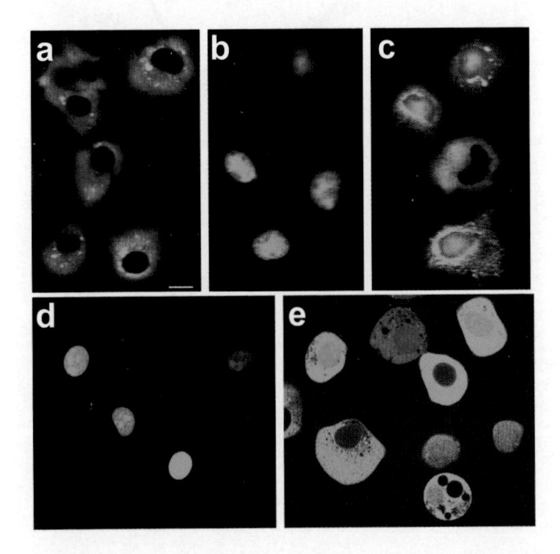

Figure 10-3. Host and bacterial proteins facilitate nuclear import of VirE2 in mammalian cells. (a) GFP-VirE2. (b) GFP-VIP1. (c) GFP-VirE2 + VIP1. (d) GFP-VirE3. (e) GFP-VirE2 + VirE3. Panels a, b, and c reproduced with permission from (Tzfira et al., 2001); panels d and e reproduced with permission from (Lacroix et al., 2005).

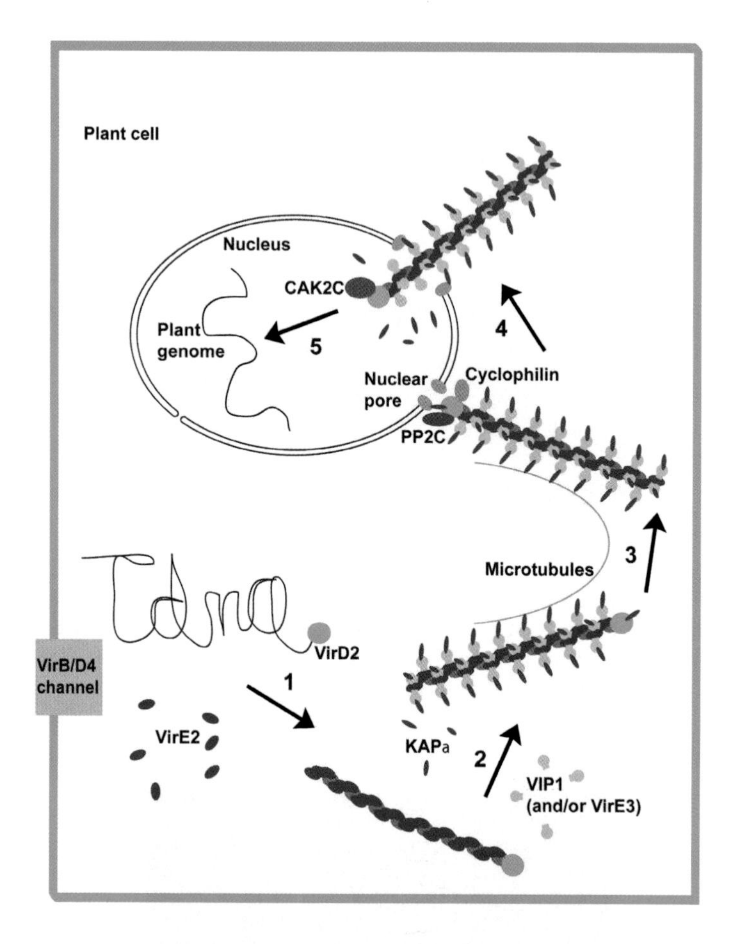

Figure 10-4. A model for T-complex cellular transport (see text for additional details).

(a)

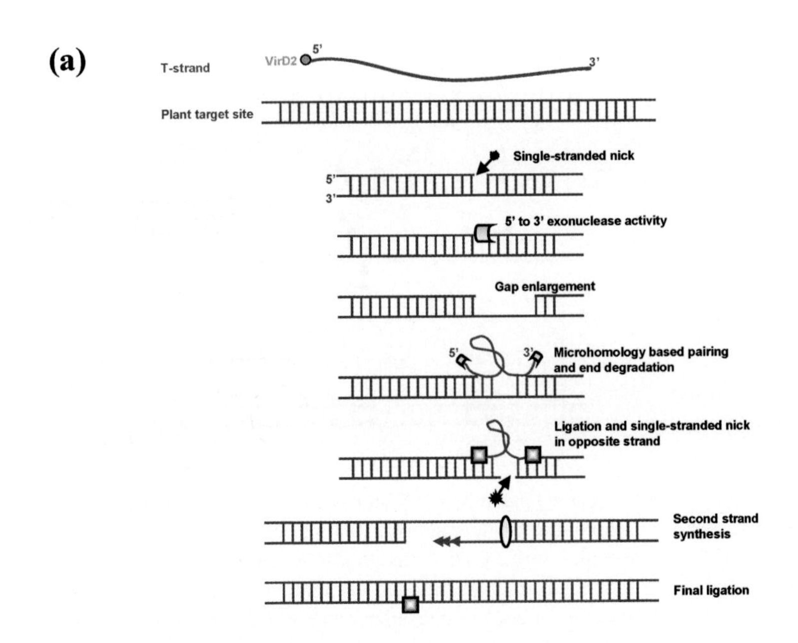
T-strand
VirD2
5'
3'
Plant target site
Single-stranded nick
5'
3'
5' to 3' exonuclease activity
Gap enlargement
5'
3'
Microhomology based pairing
and end degradation
Ligation and single-stranded nick
in opposite strand
Second strand
synthesis
Final ligation

(b)

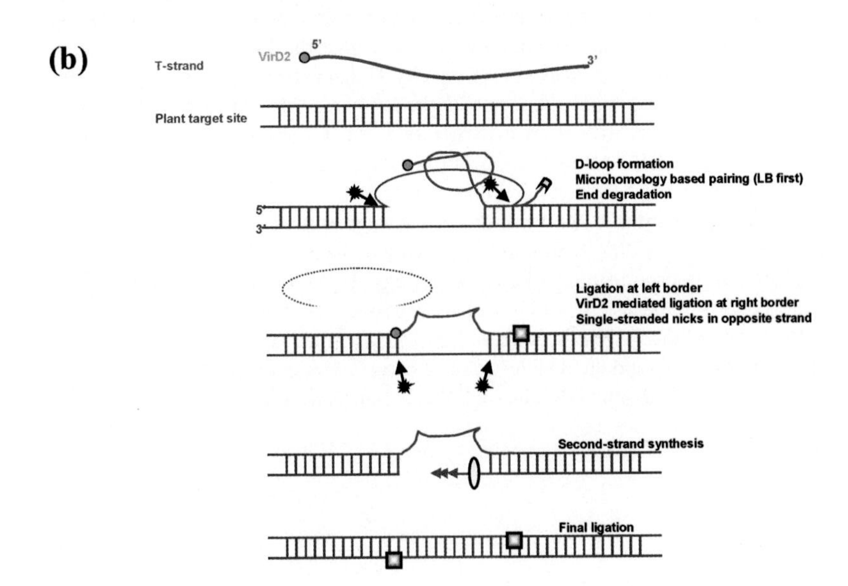
T-strand
VirD2
5'
3'
Plant target site
D-loop formation
Microhomology based pairing (LB first)
End degradation
5'
3'
Ligation at left border
VirD2 mediated ligation at right border
Single-stranded nicks in opposite strand
Second-strand synthesis
Final ligation

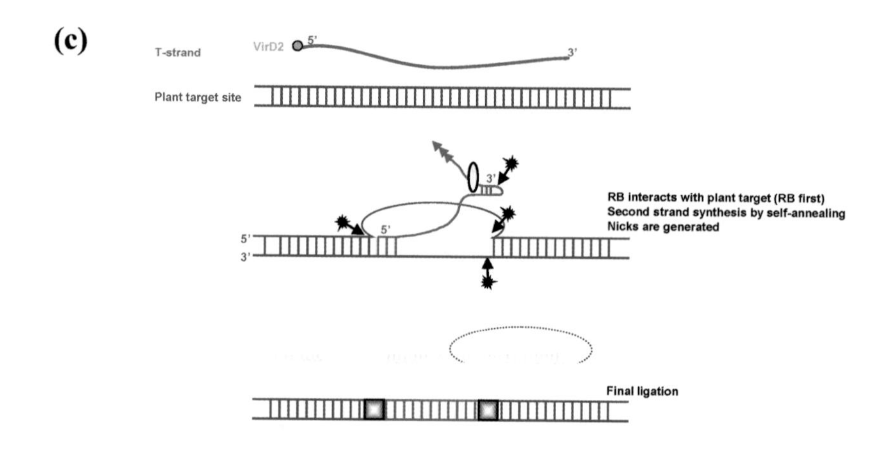

Figure 12-1. Single-stranded gap repair (SSGR) as a model for T-DNA integration. (a) Schematic representation of the single-stranded gap repair model for T-DNA integration as proposed by Gheysen et al. (1991). T-DNA integration is initiated at a single-stranded nick (arrowed asterisk). Upon gap enlargement, due to exonucleolytic degradation, a T-DNA with single-stranded ends and a body that is either single- or double-stranded interacts with the plant DNA target. This initial interaction is stabilized by means of microhomology pairing. Subsequently, the T-DNA ends become ligated to the plant target by the coordinated action of DNA ligases (purple square). A single-stranded nick is induced in the opposite strand of the plant target and polymerase activity copies the T-DNA into the plant genome. (b) T-DNA integration by SSGR as proposed by Tinland (1996). This model is based on the fact that T-DNA integration commences at the left border. The orange circle represents the VirD2 protein, whereas the yellow oval represents a DNA polymerase. (c) Schematic representation of the SSGR "RB first" model for T-DNA integration as proposed by Meza et al. (2002). This model is based on microhomologies between T-DNA and genomic DNA near the T-DNA right border. After the 3' end of the single-stranded T-DNA loops back and anneals to it self, the T-DNA that is still single stranded near the 5' end, invades the genomic DNA and finds a short stretch of microhomology. Nicks will occur in the genomic DNA and the double-stranded 3' end loop will be degraded resulting in a left end deletion. VirD2 and the 5' end of the T-DNA upstream of the microhomology are removed, single-stranded gaps are repaired, and the double-stranded break between the T-DNA left end and the genomic DNA as well, thereby generating filler DNA.

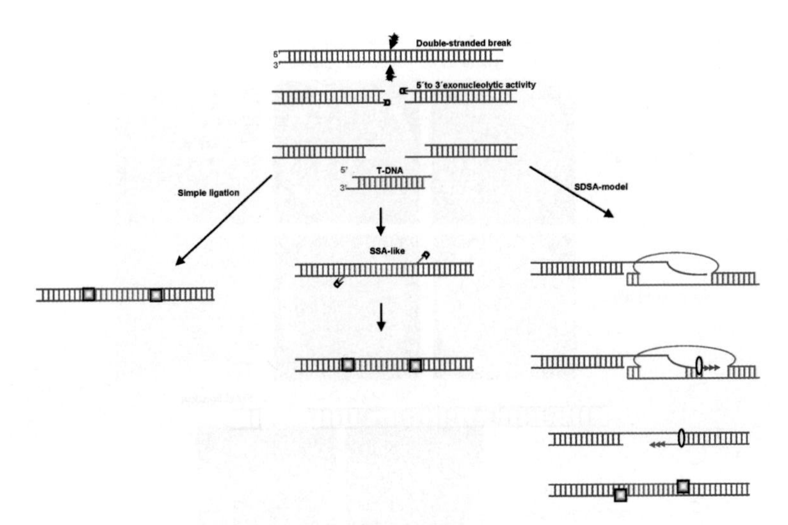

Figure 12-2. Double-stranded break repair (DSBR) as a pathway for T-DNA integration. A double-stranded break (DSB), generated at the plant DNA target site is required for T-DNA integration. After 5' to 3' exonucleolytic degradation of the 5' termini of the DSB, 3' sticky ends are available. A double-stranded DNA is the preferred substrate for integration. Subsequently, three different pathways are represented for resolving a DSB by an incoming T-DNA: simple ligation, a single-strand annealing (SSA)-like pathway, and the SDSA are shown. Simple ligation involves the ligation of the T-DNA ends to the plant DNA ends without deletion of the termini involved and without insertion of filler sequences. An SSA-like mechanism combines microhomology pairing between plant DNA and T-DNA ends, followed by resection of non-complementary overhangs and end ligation. Finally, the most complex mechanism involves SDSA. In these junctions, filler insertions are observed at T-DNA/plant DNA junctions. They originate when the initial interaction between the invading T-DNA end and the plant target is not stabilized yet, but landing and taking off of the T-DNA is repeated regularly. Upon landing, the free 3' end is taken as primer for simultaneous template-based DNA synthesis. Different regions, indicated by green numbers, such as surroundings of the T-DNA target site and the T-DNA plasmid sequence can be invaded, resulting in patchwork-like filler sequence. A reactive plant DNA end invades the double-stranded T-DNA and the T-DNA is copied into the plant genome. Finally, T-DNA and plant DNA are ligated by DNA ligase activity. DNA ligases, purple squares; DNA polymerase, yellow oval (see text, for more details).

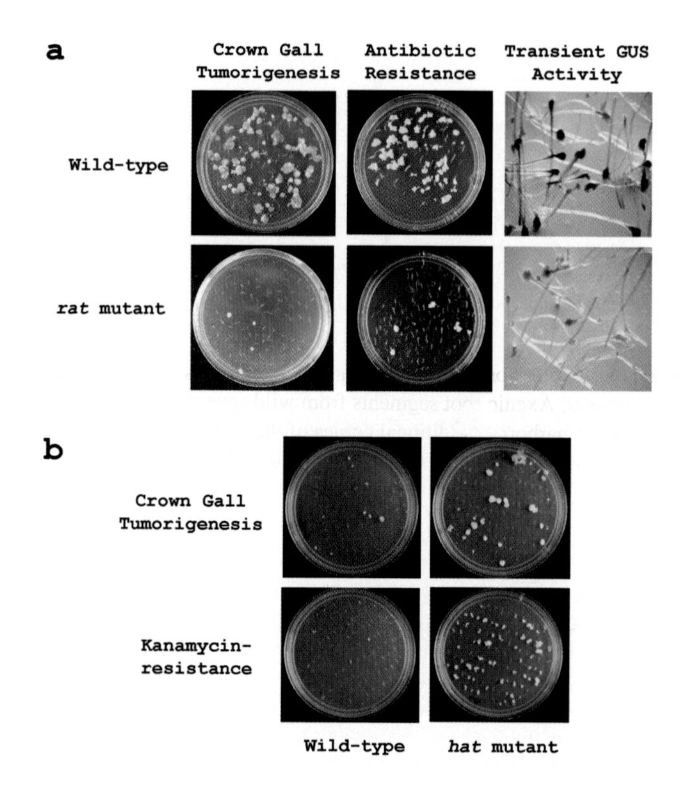

Figure 13-1. Identification of *Arabidopsis rat* (a) and *hat* (b) mutants. (a) For identification of *rat* mutants, axenic root segments of wild-type or mutant *Arabidopsis* plants were inoculated with various *Agrobacterium* strains (10^8 cells/ml) and examined either one month (crown gall tumorigenesis and antibiotic resistance) or six days later (for transient GUS activity). (b) For identification of *hat* mutants, axenic root segments of *Arabidopsis* wild-type or mutant plants containing a T-DNA "activation tag" were inoculated with a low concentration (10^5-10^6 cells/ml) of *A. tumefaciens* cells and examined one month (crown gall tumorigenesis and antibiotic resistance) later. For Crown Gall tumorigenesis assays, the root segments were inoculated with *A. tumefaciens* A208, a tumorigenic strain, and incubated on MS medium lacking phytohormones. For antibiotic resistance, the segments were inoculated with *A. tumefaciens* At849 (containing plant-active *nptII* and *gusA*-intron genes) and incubated on callus inducing medium (CIM) containing 50 mg/l kanamycin. For transient GUS activity, the root segments were incubated on CIM for six days, and then stained with X-gluc.

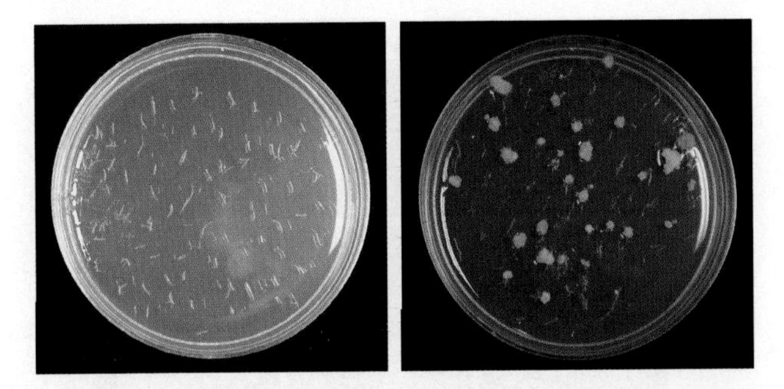

Figure 13-2. Over-expression of the AtAGP17 gene increases the re-transformation frequency of *Arabidopsis*. Axenic root segments from wild-type plants (ecotype Bl-1) (left) or Bl-1 transgenic plants harboring additional copies of the AtAGP17 gene (right) were inoculated with the tumorigenic strain *A. tumefaciens* A208. The segments were incubated on MS medium lacking phytohormones and photographed after one month.

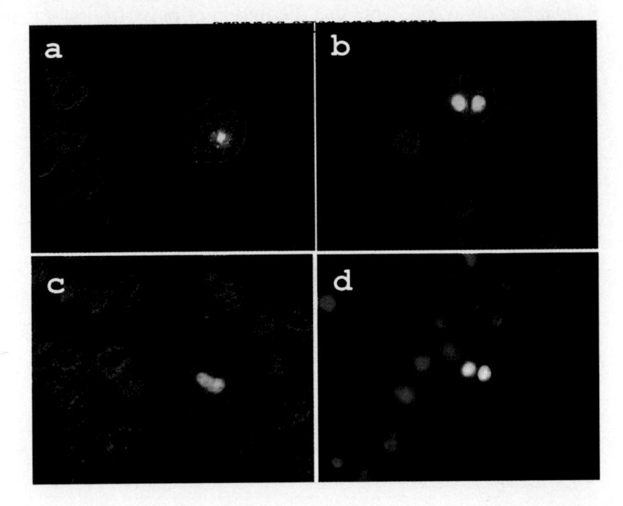

Figure 13-3. Bimolecular fluorescence complementation reveals that VirD2 interacts with several different importin alpha proteins in plant cells. Tobacco BY-2 protoplasts were co-electroporated with constructions containing nYFP-VirD2 and one of four importin alpha-cYFP proteins: AtKAPα (a), AtImpa-4 (b), AtImpa-7 (c) or AtImpa-9 (d). After 24 hours, the cells were visualized using an epifluorescence microscope. For interaction with At-KAPα, AtImpa-4, and AtImpa-7, overlay images are shown with the YFP signal (yellow) in the nucleus and the bright-field image of the cells pseudo-colored in blue. For the interaction with AtImpa-9, an overlay image is shown with the YFP signal (yellow) in the nucleus and with nuclei stained blue with Hoechst 33242.

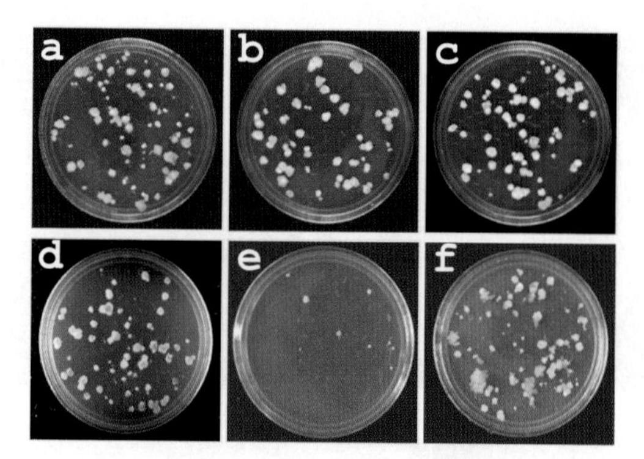

Figure 13-4. Mutation of the *AtImpa-4* gene, but not other importin *a* genes, results in a *rat* phenotype. Axenic root segments of wild-type (a) plants and T-DNA insertions into *At-Kapα* (b), *AtImpa-2* (c), *AtImpa-3* (d) and *AtImpa-4* (e) were inoculated with the tumorigenic strain *A. tumefaciens* A208. Note that disruption of only *AtImpa-4* resulted in a rat phenotype. The right panel (f) shows that complementation of the *AtImpa-4* mutant with an *AtImpa-4* cDNA, under the control of a CaMV 35S promoter, restores transformation susceptibility, indicating that the rat phenotype of the *AtImpa-4* mutant results from disruption of the *AtImpa-4* gene.

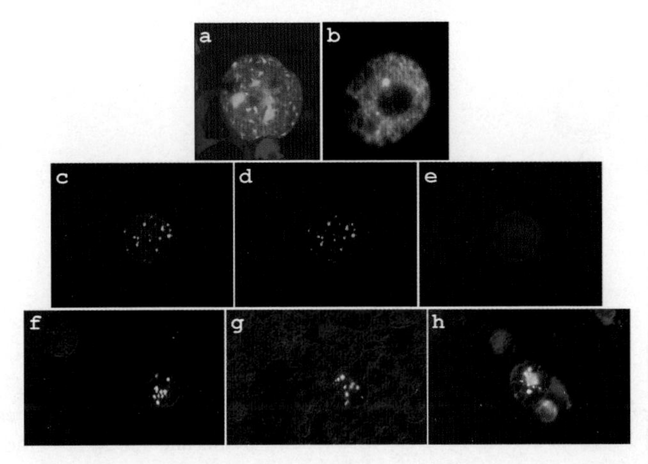

Figure 13-5. Localization and interactions of VirE2. (a, b) VirE2 localizes to the cytoplasm, not the nucleus, of tobacco cells. (c-h) VirE2 interacts with several different importin α proteins in plant cells using the bimolecular fluorescence complementation assay. Tobacco BY-2 protoplasts were electroporated with a construct expressing VirE2-YFP (a), VirE2-cYFP and VirE2-nYFP (b) VirE2-nYFP and AtImpα-4-cYFP (c-e), or VirE2-nYFP with AtKAPα-cYFP (f), AtImpα-7-cYFP (g) or AtImpα-9-cYFP (h). After 24 hours, the cells were visualized using an epifluorescence microscope. Panel a shows an overlay image of YFP fluorescence (yellow) with nuclei stained blue with Hoechst 33242 and panel b shows an overlay image of YFP fluorescence (yellow) with a bright-field image pseudo-colored in blue. The center set of panels show the YFP image (d) the RFP image of the entire cell (e), and the overlay images (c). For interaction with AtKAPα (f), AtImpα-7 (g), and AtImpα-9(h), overlay images are shown with the YFP signal (yellow) in the cytoplasm and the bright-field image of the cells pseudo-colored in blue.

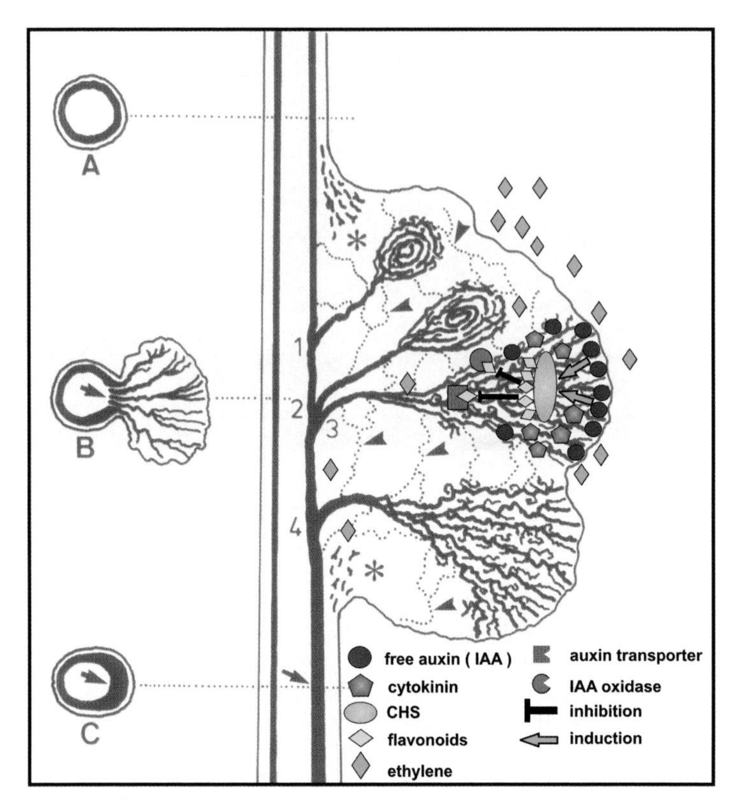

Figure 15-1. Longitudinal diagram with vascularization pattern (thick lines) and phloem anastomoses (dotted lines and arrowheads) in crown gall tumor and host stem. The three diagrams of cross sections illustrate the symmetric structure of a healthy vascular stem system above the tumor (A), the pathological host xylem (arrow) with multiseriate rays in a median position (B), and the asymmetrically enhanced vascular differentiation below the tumor (C). 1 and 2 mark the connecting sites of globular vascular bundles to the host vascular system, while 3 and 4 mark the base of tree-like branched bundles. Asterisks mark regenerative phloem fibres restricted to the upper and lower basal regions of the tumor. Color symbols indicate how T-DNA-elevated IAA concentration leads to enhanced chalcone synthase (*CHS*) expression, which intensifies the synthesis of flavonoid aglycones. These inhibit IAA oxidase. Flavonoids and IAA plus cytokinin (CK) –triggered ethylene impair the basipetal IAA transport (within the same cell as flavonoid synthesis) to regulate sufficiently high free-auxin concentration for tumor vascularization and continuous proliferation. (Follows Aloni et al., 1995, Figure. 1 and Schwalm et al., 2003)

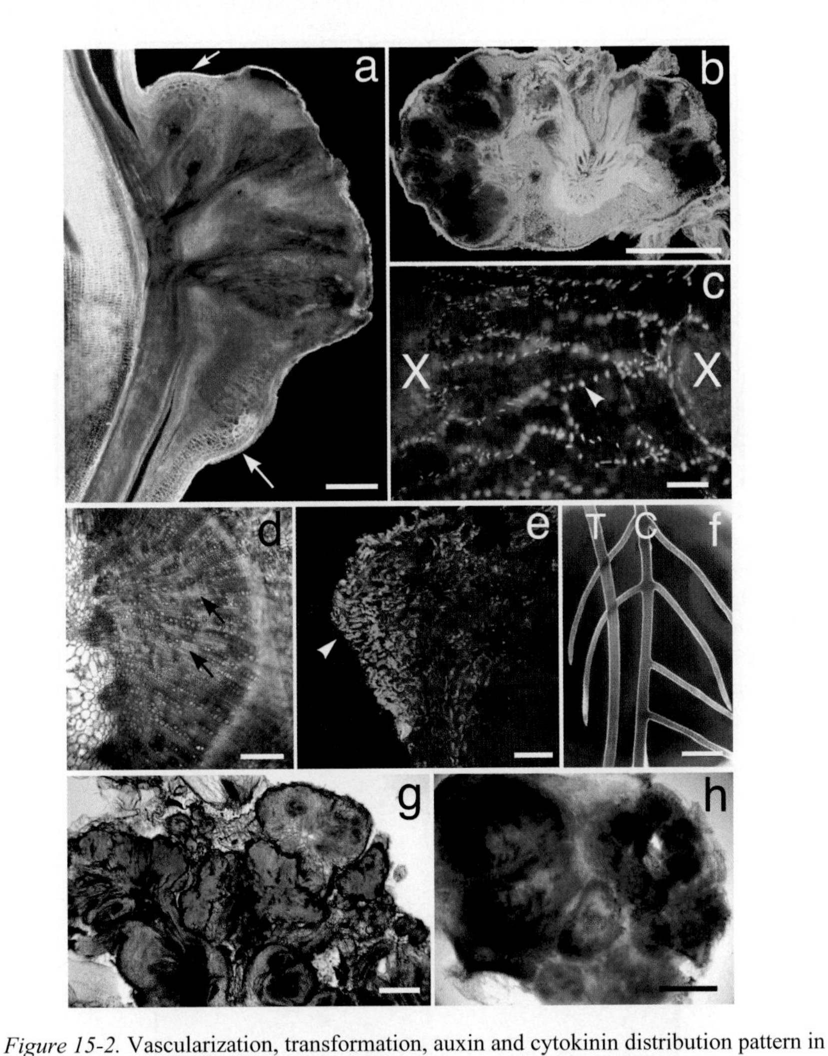

Figure 15-2. Vascularization, transformation, auxin and cytokinin distribution pattern in crown gall tumors. (**a**) Vascularized tumor regions rapidly proliferate (dark blue), distinct aerenchyma development in non-transformed tissue in *Ricinus* (arrows); lacmoid staining. (**b**) Highly transformed 7-week-old *Arabidopsis* rosette tumor, revealed by GUS expression (blue) of the wild-type T-DNA (A281 p35S gus int). (**c**) Tumor bundles (X) are interconnected by a dense net of phloem anastomoses (arrowhead) in *Ricinus*; aniline blue staining. (**d**) 5-week-old *Ricinus* tumor with unlignified multiseriate rays (arrows) and increased number of small vessels (blue); toluidine blue staining. (**e**) Immunolocalization of total auxin in an 8-week-old *Trifolium* tumor, revealing an auxin gradient with the intensity decreasing from the periphery (arrowhead) inwards; red autofluorescence identifies vessels. (**f**) Lack of lateral roots in tumorized (T) in comparison with non-infected control *Ricinus* (C) grown under identical conditions. (**g**) Auxin-induced *DR5::GUS* and (**h**) Cytokinin-induced *ARR5::GUS* expression (blue) in proliferating 5-week-old *Arabidopsis* tumors. Bars = 75 μm (e), 250 μm (c), 500 μm (d,g,h), 2 mm (a,b,f). (From (Aloni et al., 1995; Wächter et al., 1999; Mistrik et al., 2000; Ullrich and Aloni, 2000; Schwalm et al., 2003).

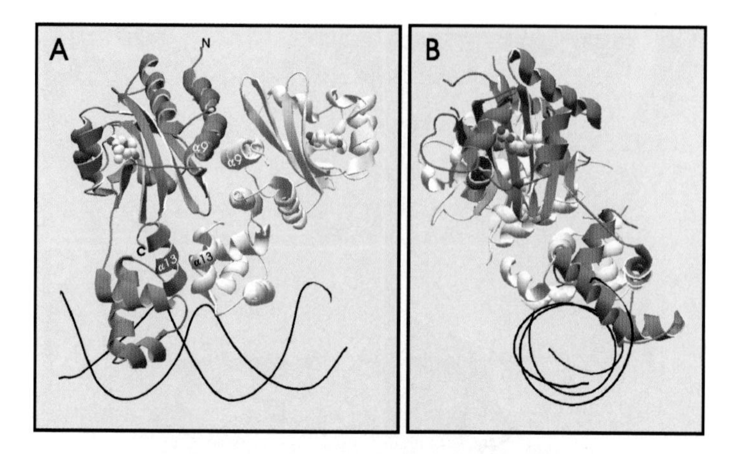

Figure 16-4. Ribbon model of a TraR-OOHL dimer bound to DNA. One monomer is light gray, the other is dark gray, and the backbone of the DNA is shown in black. The OOHL, bound in the N-terminal domain of each monomer, is in space-fill and CPK colors. **a**. The N-terminus (N) and C-terminus (C) of the left monomer are shown for reference, and α helix 9 (α9) and 13 (α13) of each dimer are also marked. These helices are involved in dimerization of the N-terminal domains and the C-terminal domains respectively. **b**. View of the same model, but along the long axis of the DNA to highlight the overall asymmetry of the structure (Vannini et al., 2002; Zhang et al., 2002b).

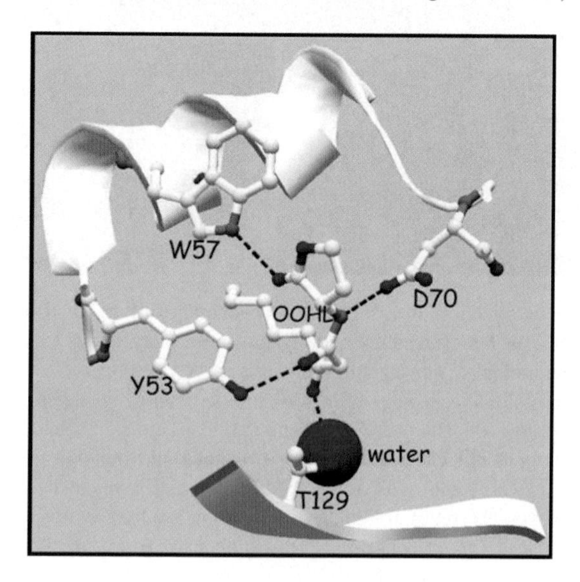

Figure16-5. Model of OOHL in the binding pocket of the TraR N-terminal domain. A portion of an α helix and β sheet are shown for reference. The OOHL and the four residues of TraR that have hydrogen bonds (dashed lines) with OOHL are shown in CPK colors. A water molecule mediates the hydrogen bond between T129 and the 3-oxo group of OOHL (Vannini et al., 2002; Zhang et al., 2002b).

Figure 17-3. Genetic tumors on the stem of aged plant of *N. langsdorffii* × *N. glauca* hybrid (Courtesy of Dr. K. Syono).

Figure17-5. General views of untransformed and pRi-transgenic plants of *Ajuga reptans* cultured for 1 month in a green house (Tanaka and Matsumoto, 1993).

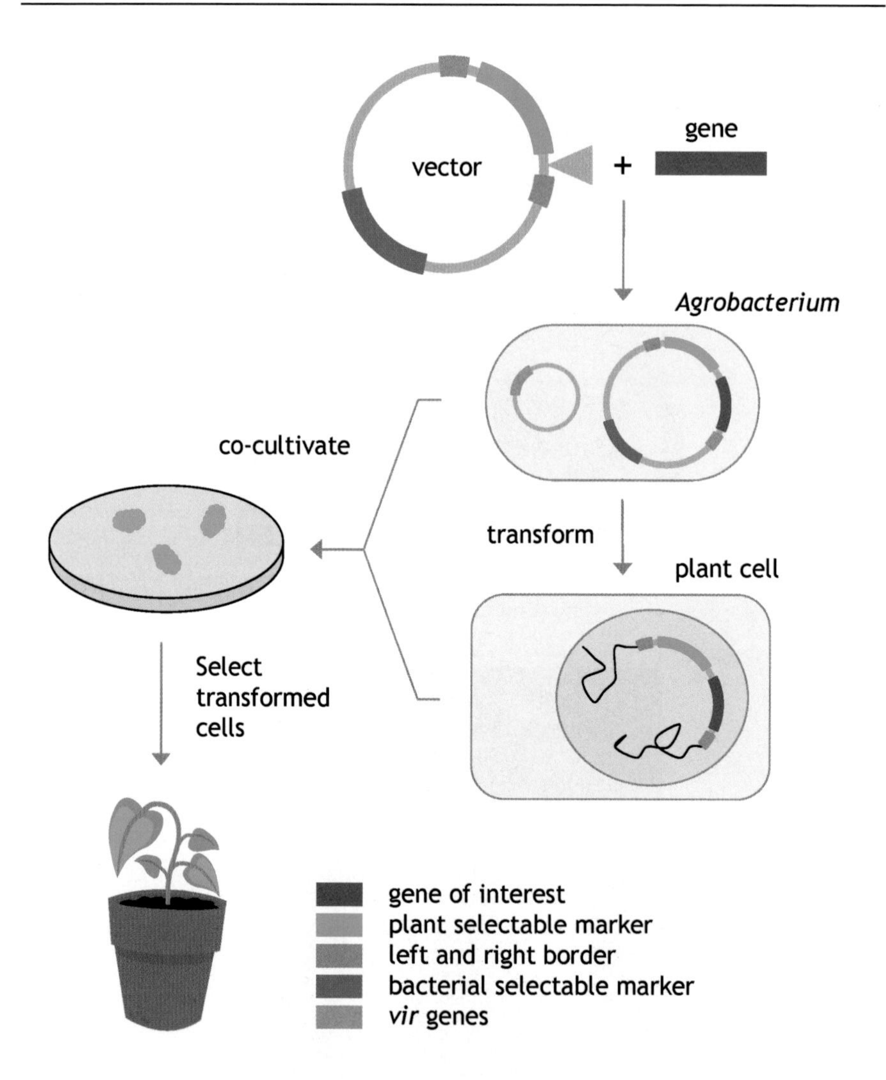

Figure 20-3. A lawyer's view of the transformation process

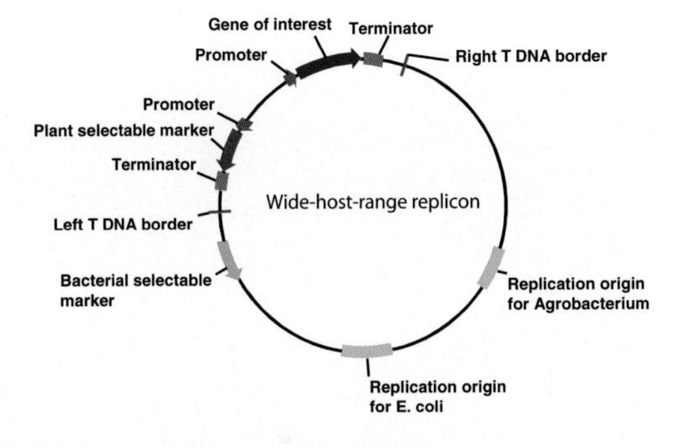

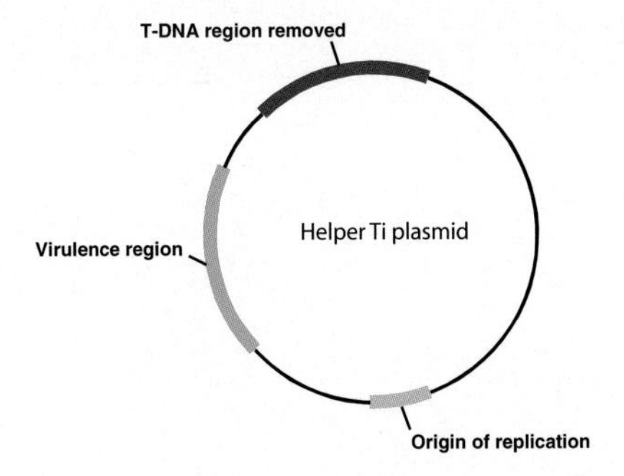

Figure 20-4. Binary vector system

Chapter 8

PRODUCTION OF A MOBILE T-DNA BY *AGROBACTERIUM TUMEFACIENS*

Walt Ream

Department of Microbiology, Oregon State University, Corvallis, OR 97331, USA

Abstract. *Agrobacterium tumefaciens* transfers tumor-inducing (Ti) plasmid-encoded genes and virulence (Vir) proteins into plant cells, where this DNA stably integrates into the plant nuclear genome. The transferred DNA (T-DNA) region of the Ti plasmid is stably inherited and expressed in plant cells, causing crown gall tumors. DNA transfer from *A. tumefaciens* into plant cells resembles plasmid conjugation; single-stranded DNA (ssDNA) is exported from the bacteria via a type IV secretion system (T4SS) comprised of VirB1-VirB11 and VirD4. The bacteria also secrete certain Vir proteins into plant cells through this system. VirD2 (together with VirD1) nicks border sequences at the T-DNA ends and attaches covalently to the 5' end of the nicked strand. The VirB/VirD4 secretion system exports the VirD2-T-DNA complex (T-complex) as well as VirE2 single-stranded DNA-binding protein and ancillary virulence proteins VirF and VirE3. VirE2 and VirF are required only in plant cells. Nuclear localization signals (NLS) in VirD2 and VirE2 target the T-complex into the nucleus where T-DNA integrates into the genome. T-DNA transfer and integration does not require tumorigenesis or T-DNA encoded proteins. This fact has allowed genetic engineers to use *A. tumefaciens* to transfer beneficial genes into plants in place of the T-DNA oncogenes.

1 INTRODUCTION

Crown gall tumors form on most dicotyledonous plants (De Cleene and De Ley, 1976) when virulent strains of *Agrobacterium tumefaciens*, containing a 200-kilobase-pair (kb) Ti plasmid (*Figure 8-1*), infect wounded plant tissue. A specific segment of the Ti plasmid, the T-DNA (*Figure 8-2*), enters plant cells and stably integrates into plant nuclear DNA (Chilton et al., 1977; Chilton et al., 1980). The T-DNA encodes enzymes for biosynthesis of plant growth hormones indole acetic acid (IAA, an auxin) from tryptophan and isopentenyl adenosine monophosphate (ipA, a cytokinin) from adenosine monophosphate, thereby causing transformed cells to grow as crown gall tumors (Binns, 2002). However, T-DNA transfer and integration does not require tumorigenesis or T-DNA encoded proteins (Hoekema et al., 1983; Ream et al., 1983). This fact has allowed genetic engineers to use *A. tumefaciens* to transfer beneficial genes into plants in place of the T-DNA oncogenes (Gelvin, 2003).

2 *A. TUMEFACIENS*—NATURE'S GENETIC ENGINEER

Agrobacterium tumefaciens is nature's genetic engineer. The first genetically modified plant cells were not produced by humans. Instead, plants were first engineered by *A. tumefaciens* (Furner et al., 1986). These bacteria genetically transform host cells with genes that cause rapid growth and production of large quantities of opines, which are used as nutrients by the tumor-inducing bacteria (Guyon et al., 1980; Petit et al., 1983). Many opines are derived from sugars and amino acids, which provide both carbon and nitrogen to the bacteria (Winans, 1992). Transformed plant cells synthesize and secrete significant quantities of specific opines, and the tumor-inducing bacteria carry genes (outside the T-DNA and usually on the Ti plasmid) required to catabolize the same opines synthesized by the tumor. More than 20 different opines exist, and each strain induces and catabolizes a specific set of opines. Generally, each *A. tumefaciens* strain catabolizes only the opines synthesized by tumors it induces. In addition, some opines induce conjugal transfer of self-transmissible Ti plasmids between strains of *Agrobacterium* (Petit et al., 1978; Ellis et al., 1982), thereby conferring on other strains the ability to catabolize extant opines. Apparently, *A. tumefaciens* strains create a niche (a crown gall tumor synthesizing particular opines) that offers an environment favorable for growth of the inducing strain.

3 INTERKINGDOM GENE TRANSFER

3.1 Overview

Agrobacterium tumefaciens transfers the T-DNA portion of its Ti plasmid and virulence proteins VirD2, VirD5, VirE2, VirE3, and VirF into host cells during crown gall tumorigenesis (Sheng and Citovsky, 1996; Zhu et al., 2000; Christie, 2004; Vergunst et al., 2005). VirD2 nicks border sequences at the T-DNA ends and attaches covalently to the 5' end of the nicked strand (Yanofsky et al., 1986; Wang et al., 1987; Herrera-Estrella et al., 1988; Ward and Barnes, 1988; Young and Nester, 1988; Durrenberger et al., 1989; Howard et al., 1989). A type IV secretion system encoded by the *virB* operon and *virD4* (Christie, 2004) mediates export to plant cells of the VirD2-T-DNA complex (T-complex) as well as VirE2 single-stranded DNA (ssDNA) binding protein (SSB) (Gietl et al., 1987; Christie et al., 1988; Citovsky et al., 1988; Das, 1988; Citovsky et al., 1989; Sen et al., 1989) and ancillary virulence proteins VirD5, VirF, and VirE3 (Vergunst et al., 2005). VirE2 and VirF function in plant cells; transgenic plants that express either of these proteins produce tumors when inoculated with *A. tumefaciens* mutants that lack intact copies of the corresponding *vir* gene (Citovsky et al., 1992; Regensburg-Tuink and Hooykaas, 1993). Nuclear localization signals in VirD2 and VirE2 target the T-complex into the nucleus where T-DNA integrates into the genome (Citovsky et al., 1994; Sheng and Citovsky, 1996; Zupan et al., 1996; Citovsky et al., 1997).

3.2 Key early experiments

A series of key observations led to the discovery of interkingdom gene transfer. Crown gall tumor cells continue to proliferate and produce opines even after the tumor-inducing bacteria are killed with antibiotics (Braun, 1958). These observations suggested that *A. tumefaciens* transmits genes for tumor maintenance and opine synthesis to plant cells and that, once established, these genes encode all the functions necessary to confer the transformed phenotype. Because virulence depends on the presence of a Ti plasmid (Van Larebeke et al., 1974; Watson et al., 1975), this extrachromosomal element seemed likely to carry the oncogenes. Hybridization between specific Ti plasmid sequences and DNA isolated from axenic (bacteria-free) tumor cells proved this hypothesis; DNA from nontransformed plant cells did not hybridize (Chilton et al., 1977). Subsequent work established that tumor cells often contain a specific portion of the Ti plasmid, called the T-DNA, integrated into the nuclear DNA of the host (Chilton et al., 1980; Thomashow et al., 1980; Willmitzer et al., 1980), and tumor

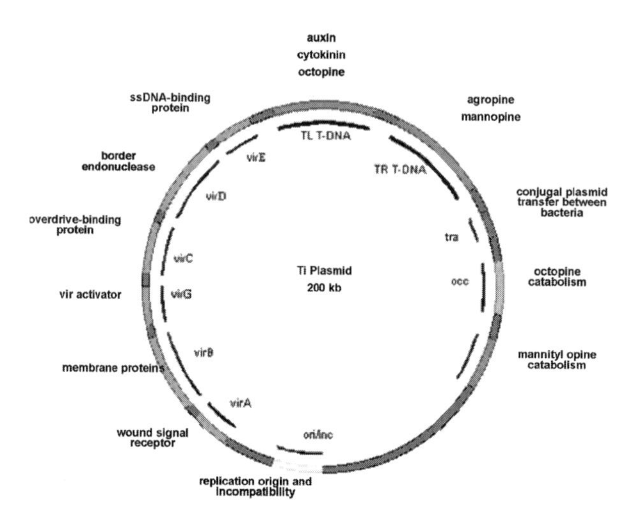

Figure 8-1. Map of an octopine-type tumor-inducing plasmid. The octopine-type Ti plasmid contains two separate T-DNAs, shown in red. TL, the oncogenic T-DNA, encodes proteins for synthesis of two phytohormones, auxin and cytokinin, as well as octopine, which is derived from arginine and pyruvate. TR does not promote tumor growth, but it encodes proteins for synthesis of the mannityl opines (agropine and mannopine), which are derived from mannose and glutamine.

T-DNA

OVER DRIVE

mRNA	→	←	←	→	→	→	←	←
Transcript	5	7	2	1	4	6a	6b	3
Genetic Locus	?	?	*iaaH*	*iaaM*	*ipt*	*ops*	*tml*	*ocs*
Function			Auxin synthesis		Cytokinin Synthesis	Opine Secretion		Octopine Synthase

Figure 8-2. Genetic map of the TL transferred DNA (T-DNA) of an octopine type Ti plasmid. The arrowheads in the box represent the left and right borders of the T-DNA, and they indicate the direction of T-DNA transfer, from right to left. *iaa*H (indoleacetamide hydrolase), *iaa*M (tryptophan monooxygenase), *ipt* (isopentenyl transferase), *ops* (octopine secretion), *tml* (tumor morphology large), *ocs* (octopine synthase). Arrows indicate the direction and length of T-DNA encoded mRNAs produced by transformed plant cells.

cells express genes responsible for the transformed phenotype (Garfinkel and Nester, 1980; Garfinkel et al., 1981; Willmitzer et al., 1982). We now have a reasonably detailed understanding of this gene transfer system, and recently the complete DNA sequences of Ti and Ri plasmids (Zhu et al., 2000; Moriguchi et al., 2001) and an *A. tumefaciens* genome were completed (Goodner et al., 2001; Wood et al., 2001). The remainder of this chapter will cover T-DNA transfer: genetic analysis of *cis*-acting T-DNA border (origin of transfer) sequences and biochemical characterization of VirD2 protein, which interacts with border sequences, becomes covalently attached to T-DNA, pilots T-DNA into plant cells, and helps target T-DNA to the nucleus.

3.3 Protein secretion apparatus

Export of the T-DNA-VirD2 complex and other virulence proteins (VirE2, VirE3, and VirF) requires at least twelve membrane-associated proteins: eleven encoded by the *virB* operon and another encoded by *virD4* (Cascales and Christie, 2003; Ding et al., 2003; Christie, 2004; Li et al., 2005). The VirB proteins and VirD4 belong to a family of type IV secretion systems, which includes the *Bordetella* pertussis toxin liberation (Ptl) proteins (Covacci and Rappuoli, 1993; Weiss et al., 1993), *Legionella pneumophila vir* homologs (Lvh) and some Icm/Dot proteins (Segal et al., 1999), *Helicobacter pylori* Cag proteins (Tummuru et al., 1995; Censini et al., 1996), *Rickettsia prowazekii* VirB proteins (Andersson et al., 1998), and conjugation proteins (Trb, Tra, and Trw) from IncPα plasmid RP4 (Lessl et al., 1992), IncN plasmid pKM101 (Pohlman et al., 1994), and IncW plasmid R388 (Kado, 1994). Thus, type IV secretion systems facilitate two important processes: 1) secretion of virulence factors from pathogen to host, and 2) promiscuous (broad-host-range) conjugation of plasmid DNA. The *A. tumefaciens* VirB/VirD4 transporter is the most versatile of these systems and mediates both promiscuous gene transfer and export of virulence proteins.

The genes that encode type IV secretion systems of *A. tumefaciens*, *L. pneumophila*, *H. pylori*, *B. pertussis*, *R. prowazekii*, and plasmids RP4, pKM101, and R388 share sequence similarities and similar arrangements within operons (Segal et al., 1999). Functional similarities also exist. For example, both the *A. tumefaciens* VirB/VirD4 and *L. pneumophila* Icm/Dot systems mediate conjugation of IncQ plasmid RSF1010 between bacteria, and the presence of RSF1010 abolishes the virulence of both pathogens (Binns et al., 1995; Segal and Shuman, 1998; Stahl et al., 1998;

Vogel et al., 1998). In addition, VirD4 from *A. tumefaciens* can substitute for TraG from pTiC58 in plasmid conjugation (Hamilton et al., 2000). The involvement of closely related type IV secretion systems in both conjugation and protein export suggests that conjugation may be a specialized form of protein export in which the exported protein – the DNA-nicking protein VirD2 in this case – is covalently attached to DNA.

3.4 The conjugation model of T-DNA transfer

3.4.1 Promiscuous conjugation

In many ways T-DNA transfer from *A. tumefaciens* to plant cells resembles broad-host-range plasmid conjugation between bacteria. In each case, a multi-subunit endonuclease binds an origin of transfer (*oriT*) sequence forming a relaxosome, and one endonuclease subunit (relaxase) nicks the DNA and covalently attaches to the 5' end (Herrera-Estrella et al., 1988; Ward and Barnes, 1988; Young and Nester, 1988; Howard et al., 1989; Pansegrau et al., 1993; Jasper et al., 1994; Lessl and Lanka, 1994). The donor transfers a single DNA strand, together with the bound protein, to the recipient via a type IV secretion system.

The conjugation model of T-DNA transfer gained strong support from an unexpected quarter. The *A. tumefaciens* VirB/VirD4 system can transfer a broad-host-range mobilizable bacterial plasmid (RSF1010), which does not contain a T-DNA border sequence, into plant cells where the plasmid DNA integrates into the nuclear genome (Buchanan-Wollaston et al., 1987). In addition to the VirB/VirD4 secretion system, interkingdom transfer of RSF1010 requires the *oriT* sequence and mobilization (*mob*) proteins, which create a site-specific nick within *oriT*. The ability of *A. tumefaciens* to mobilize a broad-host-range plasmid into plant cells supports the conjugation model and opens the possibility that plants potentially receive a great variety of information from many species of gram-negative bacteria. The *A. tumefaciens* VirB/VirD4 system also promotes conjugation of plasmid DNA into other bacteria (Gelvin and Habeck, 1990; Beijersbergen et al., 1992; Fullner et al., 1996; Fullner and Nester, 1996a; Fullner and Nester, 1996b), plants (Buchanan-Wollaston et al., 1987), fungi (Bundock et al., 1995; de Groot et al., 1998)}, or human cells (Kunik et al., 2001) indicating that type IV secretion systems can export a variety of proteins and protein-DNA complexes in to a broad range of recipient cells.

At least one *A. tumefaciens vir* protein can function as part of a different conjugation system. An essential conjugation protein, known as the coupling protein, appears to link the relaxosome to the transmembrane DNA/protein secretion apparatus, which is also called the mating pair formation system. Several coupling proteins, for example, TraG of plasmid RP4 and TraD of F, show limited sequence similarity to *A. tumefaciens* VirD4 (Lessl and Lanka, 1994). In fact, VirD4 can substitute for its pTiC58 homolog, TraG, during conjugation of RSF1010 via the pTiC58 *trb*-encoded mating bridge, thereby proving that VirD4 is a conjugation protein (Hamilton et al., 2000).

3.4.2 Border sequences

T-DNAs from several different *Agrobacterium* strains have very similar 23-base-pair (bp) border sequences at each T-DNA end (*Table 8-1*). T-DNA transfer requires the right-hand border in its wild-type orientation (Shaw et al., 1984; Wang et al., 1984; Peralta and Ream, 1985). Inversion of the right border reduces virulence drastically (Wang et al., 1984; Peralta and Ream, 1985), and the rare tumors that develop contain most or all of the 200-kb Ti plasmid (Miranda et al., 1992). Deletion of the right border abolishes tumorigenesis (Hepburn and White, 1985), whereas removal of the left border does not affect virulence (Joos et al., 1983), indicating that T-DNA transfer begins at the right border, moves leftward through the T-DNA, and (sometimes) terminates at the left border. T-DNA borders share both sequence and functional similarities with the *oriT* of broad-host-range conjugative plasmid RP4 (Lessl and Lanka, 1994). Thus, T-DNA transfer from *A. tumefaciens* into plants strongly resembles plasmid conjugation between bacteria.

Another *cis*-acting sequence, called *overdrive*, flanks right-hand (but not left-hand) border sequences and stimulates T-DNA transfer several hundredfold (*Table 8-1*) (Peralta et al., 1986). Unlike the border sequence, *overdrive* functions in either orientation and at considerable distances on either side of the right-hand border sequence (Ji et al., 1988). Efficient T-DNA transfer requires only two *cis*-acting sequences: the right-hand border sequence and *overdrive*. The following paragraphs will explore the interaction of these DNA sequences with virulence proteins.

Table 8-1. T-DNA Border and Overdrive Sequences

Plasmid/ T-DNA	Border Repeat	Spacer	Overdrive
consensus	TGGCAGGATATAT-CTGTTGTAAAT		T---TC-CTGTGTATGTTTGTTTG
	A T		
pTiA6 TL right	TGGCAGGATATATACCGTTGTAATT	14 bp	TAAGTCGCTGTGTATGTTTGTTTG
pTiA6 TR right	TGGCAGGATATATGCGGTTGTAATT	13 bp	TAAATTTCTGTATTTGTTTGTTTG
pTiAB3 TA rt	TGACAGGATATATACCGTTGTAATT	14 bp	TAAATCGCTGTGTATGTTTGTTTG
pRiA4 TR right	TGACAGGATATATCTTGTGGTCAGG	8 bp	TTTGTGAGGAGGTATGTTTGTTTA
pRiA4 TL right	TGACAGGATATATGTTCCTGTCATG	-3 bp	-------------ATGTTTGTTCA
		77 bp	TTTTAAAAATAGTATGTTTGACTG
pTiT37 right	TGACAGGATATATTGGCGGGTAAAC	62 bp	TTCGTCCATTTGTATGTGCATGCC
pTiA6 TL left	CGGCAGGATATATTCAATTGTAAAT		none
pTiA6 TR left	TGGCAGGATATATCGAGGTGTAAAA		none
pRiA4 TL left	TGGCAGGATATATTGTGATGTAAAC		none
pRiA4 TR left	TGGCAGGATATATGCCAACGTAAAA		none
pTiT37 left	TGGCAGGATATATTGTGGTGTAAAC		none

Strongly conserved bases are underlined; moderately conserved bases are shaded.

3.4.3 The relaxosome

Several *vir* operons encode proteins that participate in DNA-protein interactions necessary for T-DNA transfer. The first two genes of the *virD* operon (*virD1* and *virD2*) encode a site-specific nicking enzyme that nicks the bottom strand of T-DNA border sequences between the third and fourth base (Yanofsky et al., 1986; Wang et al., 1987; Wang et al., 1990). [For *vir* operons containing more than one gene, the number that follows the gene name indicates the position of the gene in the operon rather than an allele number for a specific mutation.] Direct interaction between VirD1 and VirD2 was shown using a novel protein interaction assay in mammalian cells; this study also showed that VirD2 interacts with itself (Relic et al., 1998). VirD2 protein attaches covalently to the 5' end of the nicked DNA (Herrera-Estrella et al., 1988; Ward and Barnes, 1988; Young and Nester, 1988; Durrenberger et al., 1989; Howard et al., 1989) via a phosphodiester bond with a specific tyrosine near the amino terminus (codon 29 in VirD2 encoded by pTiA6) (Vogel and Das, 1992b). The nick within the right border sequence initiates production of T-strands, which are full-length single-stranded copies of the bottom strand of the T-DNA (Stachel et al., 1986; Albright et al., 1987; Jayaswal et al., 1987; Stachel et al., 1987) that the bacteria export into plant cells (Stachel and Zambryski, 1986; Tinland et al., 1994; Yusibov et al., 1994). Thus, early events in T-DNA transfer resemble those in bacterial conjugation: a multiple-subunit nicking enzyme binds an *oriT* sequence forming a DNA-protein complex called the relaxosome, which creates a site-specific nick in one strand of the *oriT* sequence. During this process, one endonuclease subunit covalently binds the nicked DNA.

Other relaxosomes contain more than two mobilization (Mob) proteins. For example, RSF1010 encodes three relaxosome components, MobA, MobB, and MobC (Scholz et al., 1989). Although the *virD* operon contains five genes, only VirD1 and VirD2 are required for border nicking and T-strand production. VirD3 is poorly conserved and not required for T-DNA transmission (Vogel and Das, 1992a). VirD4 is a coupling protein that connects the relaxosome to the mating bridge, and VirD5 is an ancillary protein (Stachel and Nester, 1986) that contains a type IV secretion signal (Vergunst et al., 2005). The VirC1 protein binds a sequence (*overdrive*) adjacent to right borders, and VirC2 may also interact with this complex. The VirC/*overdrive* complex may interact with the relaxosome, but these proteins are not essential for border nicking and T-strand production.

The *overdrive* sequence lies near the right-hand T-DNA border (*Table 8-1*) and, together with the VirC1 and VirC2 proteins, stimulates tumorigenesis several hundredfold (Peralta et al., 1986). VirC1 binds *overdrive*, and the VirD2 nicking protein also interacts with this sequence (or with bound VirC1) (Toro et al., 1989). Although the precise role of *overdrive* and the *virC*-encoded proteins in the relaxosome remain unknown, they appear to distinguish the right- and left-hand border sequences (the origin and terminus of T-DNA transfer). Because T-DNA transfer is unidirectional (Miranda et al., 1992), plant cells will receive the oncogenes and opine synthesis genes only if transfer begins at the right border. In order to avoid unproductive transfer events that begin at the left border, the relaxosome must distinguish between right- and left-hand border sequences, which are functionally equivalent in their interaction with the VirD1/ VirD2 nicking enzyme (Yanofsky et al., 1986; Albright et al., 1987). Apparently, *overdrive* allows the transfer apparatus to recognize the right-hand border as the origin of transfer, perhaps by helping to tether the relaxosome to the VirB/VirD4 secretion/mating bridge apparatus. Indirect support for this idea comes from the observation that mutations in the *virC* operon stimulate (>threefold) VirB/VirD4-dependent conjugation of an RSF1010 plasmid from T-DNA-containing *A. tumefaciens* (Fullner and Nester, 1996b). RSF1010 interferes with secretion of T-DNA from *A. tumefaciens* into plants (Binns et al., 1995; Stahl et al., 1998; Cascales et al., 2005), apparently through competition with VirE2 and VirD2-T-strand complexes for the VirB/VirD4 secretion system. The interactions between the *overdrive* sequence and VirC1 and VirD2 may help localize the T-DNA origin of transfer (right border + *overdrive*) and associated proteins (VirC1, VirC2, VirD1, VirD2) to the VirB/VirD4 pore, competing with RSF1010 for access. Loss of the VirC proteins may greatly reduce the ability of the T-DNA origin of transfer and associated Vir proteins to bind the VirB/VirD4 pore, thereby allowing greater access for the RSF1010 *oriT*-Mob complex.

3.4.4 T-strands

Induction of *vir* expression and border nicking leads to formation of intermediates in T-DNA transfer (T-strands), which are full-length, linear, single-stranded DNA molecules comprised of the bottom strand of the T-DNA (*Figure 8-3*) (Stachel et al., 1986, 1987; Veluthambi et al., 1988). Following a proteinase digestion, DNA isolated from *vir*-induced *A. tumefaciens* cells and subjected to agarose gel electrophoresis can be transferred to nitrocellulose filters by blotting without denaturation. These conditions permit hybridization of only DNAs with single-stranded regions to complementary labeled DNA or RNA probes. Only probes corresponding to the top strand of the T-DNA anneal to T-strands, proving that T-strands are derived from the bottom strand. Treatment with an endonuclease (S1) or exonuclease (*E. coli* exonuclease VII or T4 DNA polymerase) specific for single-stranded DNA destroys T-strands and demonstrates the presence of a free 3' end (exo VII or T4 DNA polymerase) and possibly a free 5' end (exo VII).

Mutations in *virA*, *virG*, *virD1*, and *virD2* abolish T-strand production (Stachel et al., 1987; Veluthambi et al., 1988); *virA* and *virG* are required because these genes control expression of the *virD* operon. T-strand displacement, which occurs 5' to 3', likely requires helicase activity and may be accompanied by synthesis of a new copy of the bottom strand, although this has not been shown. The genes that encode other proteins that are probably involved in T-strand production, for example helicase, DNA polymerase, and topoisomerase, have not been identified and may lie on one of the two *A. tumefaciens* chromosomes.

3.4.5 Secreted single-stranded DNA-binding protein: VirE2

The *virE* operon encodes two proteins: VirE1 (65 amino acids) and VirE2 (533 amino acids; (Winans et al., 1987). The single-stranded DNA-binding (SSB) activity of VirE2 does not depend on VirE1 (Christie et al., 1988; Citovsky et al., 1988; Das, 1988; Citovsky et al., 1989; Sen et al., 1989). In *E. coli* that contain the Lon protease, VirE1 protein stabilizes VirE2 (McBride and Knauf, 1988), suggesting that these proteins interact physically. Indeed, protein interaction cloning (yeast two hybrid) studies showed that VirE2 contains two separable domains that bind VirE1 (Sundberg et al., 1996; Sundberg and Ream, 1999). In *A. tumefaciens*, VirE2 is equally stable with or without VirE1 (Sundberg et al., 1996). VirE1, a secretory chaperone, promotes transport of VirE2 protein into plant cells via the VirB/VirD4 secretion system (Deng et al., 1999; Sundberg and Ream, 1999; Zhou and Christie, 1999). Thus, both proteins are essential for tumorigenesis (Sundberg et al., 1996).

VirE2 likely has multiple roles in T-DNA transfer. VirE2 binding protects single-stranded DNA from nuclease attack *in vitro* (Gietl et al., 1987; Citovsky et al., 1989; Sen et al., 1989) and inside plant cells (Yusibov et al., 1994; Rossi et al., 1996); however, absence of VirE2 does not diminish T-strand accumulation in *A. tumefaciens* (Stachel et al., 1987; Veluthambi et al., 1988). VirE2 may also form a transmembrane channel that helps move T-strand DNA across membranes (Dumas et al., 2001). The presence of nuclear localization signals (NLS) in VirE2 (Citovsky et al., 1992) suggests that it enters plant nuclei during infection. The two NLSs of VirE2 continue to function when VirE2 is bound to ssDNA (Zupan et al., 1996), even though the NLS domains overlap regions of VirE2 involved in cooperativity and ssDNA binding (Citovsky et al., 1992). Fluorescein-labeled ssDNA bound by VirE2 enters the nucleus of plant cells injected with the complex, whereas in the absence of VirE2, the DNA remains cytoplasmic (Zupan et al., 1996). However, in an in vitro assay for nuclear import, VirE2 lost the ability to enter the nucleus upon binding to single-stranded DNA (Ziemienowicz et al., 2001). Differences between the assays used in these studies may explain the contradictory results; neither assay mimics transfer of T-strands and VirE2 from *Agrobacterium* to plant cells during infection. Although this apparent discrepancy remains unresolved, inside plant cells, VirE2 likely shields T-strands from nuclease attack and may target them to the nucleus.

VirE2 is exported from bacterial cells and is required only inside plant cells. VirE2 does not coat T-strands in bacterial cells due to the presence of VirE1, a secretory chaperone that occupies the DNA-binding domain of VirE2 (Dombek and Ream, 1997; Sundberg and Ream, 1999). This was shown directly using the highly sensitive T-DNA immunoprecipitation (TrIP) assay (Cascales and Christie, 2004). T-strand DNA is chemically crosslinked to associated Vir proteins, the complexes are precipitated with specific antibodies, and polymerase chain reaction (PCR) is used to detect T-strand DNA (Cascales and Christie, 2004). Antibodies specific for VirE2 did not precipitate T-strand DNA in this assay, indicating that T-strands are not bound by VirE2 protein in *A. tumefaciens* (Cascales and Christie, 2004).

Interaction cloning experiments identified protein contacts between VirE2 and VirE1 (Sundberg et al., 1996; Deng et al., 1999; Sundberg and Ream, 1999; Zhou and Christie, 1999). VirE1 binds VirE2 domains involved in binding ssDNA and self association, and VirE1 facilitates VirE2 export by preventing VirE2 aggregation and premature binding of VirE2 to ssDNA (Deng et al., 1999; Sundberg and Ream, 1999). Instead, VirE2 is exported separately to plant cells, where it binds T-strands (the Separate Export Model) (Sundberg and Ream, 1999) (*Figure 8-3*). VirE2 is necessary only in plant cells; transgenic plant cells that express VirE2 produce tumors when inoculated with *virE2*-mutant *A. tumefaciens* (Citovsky et al., 1992). Coinoculation of a *virE2* mutant and a *virE+* strain lacking T-DNA results in tumor formation, even though each strain alone is avirulent and unable to exchange genes by conjugation (Otten et al., 1984). This "complementation" by mixed infection requires the VirB/VirD4 transporter (Otten et al., 1984; Christie et al., 1988), and both strains must bind to plant cells (Christie et al., 1988). T-strands accumulate to normal levels in bacterial cells without VirE2 (Stachel et al., 1987; Veluthambi et al., 1988), and *A. tumefaciens* can transfer these uncoated T-strands into plant cells (Citovsky et al., 1992; Sundberg and Ream, 1999). T-strands were detected inside wild-type plant cells infected by a *virE2* mutant (Yusibov et al., 1994). Export of VirE2, but not of T-strands, from *A. tumefaciens* requires VirE1 (Sundberg et al., 1996). Thus, *A. tumefaciens* exports VirE2 protein and uncoated VirD2-T-strand DNA complexes independently into plant cells; one does not depend on the other for transfer (Sundberg et al., 1996; Sundberg and Ream, 1999). T-strand transfer resembles plasmid conjugation in many ways (Kado, 1994; Lessl and Lanka, 1994; Baron and Zambryski, 1996; Firth et al., 1996; Sheng and Citovsky, 1996; Winans et al., 1996; Christie, 1997; Zupan and Zambryski, 1997; Christie, 2004);

conjugal DNA metabolism appears to occur at the transmembrane export channel, with ssDNA transferred directly to the recipient (Firth et al., 1996). Indeed, proper contact between recipient and donor cells triggers conjugal DNA processing (Kingsman and Willetts, 1978; Ou and Reim, 1978; Firth et al., 1996). T-strand production likely occurs in a similar manner, at the VirB/VirD4 macromolecule export apparatus, which is localized to the poles of the bacterial cell (Kumar and Das, 2002; Atmakuri et al., 2003). During a normal infection, T-strands may leave bacterial cells as they are displaced from the Ti plasmid, without significant exposure to the bacterial cytoplasm. The Separate Export model explains several observations: (i) VirE2-producing plant cells restore pathogenicity to *virE*-mutant *A. tumefaciens* (Citovsky et al., 1992), (ii) VirE2 made in one *A. tumefaciens* strain can interact productively with T-strands generated in another (during mixed infections) (Otten et al., 1984; Christie et al., 1988), (iii) export of VirE2 requires VirE1 (Sundberg et al., 1996) and VirB1 (K.J. Fullner, personal communication), whereas T-strand transfer does not, (iv) the presence of plasmid RSF1010 in *A. tumefaciens* blocks VirE2 export but merely reduces T-strand transfer (Binns et al., 1995), and (v) expression of the Osa (oncogenesis suppressing activity) protein in *A. tumefaciens* prevents export of VirE2 but not of T-strands (Lee et al., 1999). From these studies we know *A. tumefaciens* can export VirE2 and T-strand DNA separately under special circumstances. The inability of VirE2-specific antibodies to precipitate T-strand DNA from bacterial cells shows directly that *A. tumefaciens* transfers VirE2 and T-strand DNA into plant cells separately (Cascales and Christie, 2004).

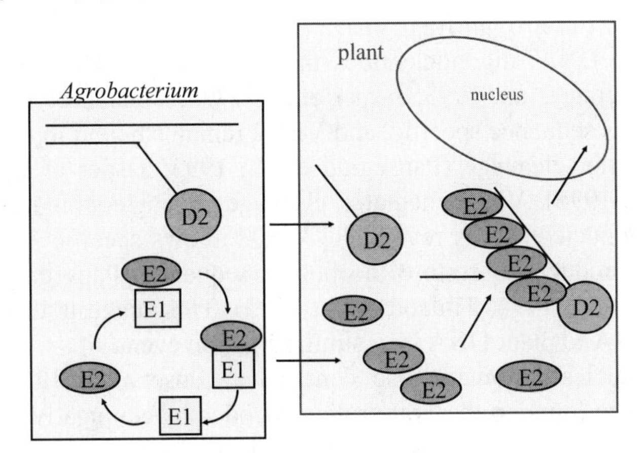

Figure 8-3. The separate export model.

3.4.6 A pilot protein: VirD2

VirD2 endonuclease attaches covalently to T-strand DNA through a phosphodiester bond between the 5' end of the T strand and a conserved tyrosine residue (tyrosine 29; *Figure 8-4*) in VirD2 (Vogel and Das, 1992b) [(reviewed in Sheng and Citovsky, 1996)]. T strands apparently enter plant cells via the VirB/VirD4 secretion system by virtue of this attachment. T-DNA border-specific endonuclease activity lies entirely within the N-terminal half of VirD2 (Ward and Barnes, 1988). Conserved residues at the C-terminus of VirD2 are important for tumorigenesis (Shurvinton et al., 1992; Rossi et al., 1993), although they are not needed for T strand production. Instead, this domain contains a bipartite NLS that helps target T strands to plant nuclei (Howard et al., 1992; Shurvinton et al., 1992; Tinland et al., 1992; Rossi et al., 1993). The carboxy terminus of VirD2 also contains a type IV secretion signal necessary for export from bacterial cells (Vergunst et al., 2005). This secretion signal mediates export of VirD2, even in the absence of T-strand DNA (Vergunst et al., 2005). Thus, VirD2 pilots T-strands into plant cells, and VirD2 may also lead T-strands through the nuclear pore.

3.4.7 Functional domains of VirD2

VirD2 contains several known functional domains. The endonuclease domain is highly conserved (amino acids 1-262; *Figure 8-4*) and includes the active site tyrosine (position 29). This domain is sufficient (together with VirD1) for nicking T-DNA border sequences and for T-strand production in vivo. In vitro, purified VirD2 (in the absence of VirD1) cleaves single-stranded DNA oligonucleotides that contain a T-DNA border sequence (Pansegrau et al., 1993; Jasper et al., 1994; Tinland et al., 1995). This cleavage is sequence specific, and VirD2 remains bound to the 5' end of the DNA after cleavage (Pansegrau et al., 1993; Jasper et al., 1994; Tinland et al., 1995). VirD2-mediated cleavage of single-stranded border sequence oligonucleotides is reversible; VirD2 allows specific ligation of the cleavage products to restore the border sequence (Pansegrau et al., 1993; Jasper et al., 1994; Tinland et al., 1995). This suggests that VirD2 may join T-DNA to plant DNA by a similar ligation event.

The endonuclease domain also contains a conserved "HRY" motif (*Figure 8-4*) also found in the TraI protein encoded by conjugative plasmid RP4 (Lessl and Lanka, 1994) and in phage-encoded site specific integrase proteins (Argos et al., 1986). The conserved arginine (at position 129) is crucial for T-strand production in vivo; conversion of this residue to glycine (an R129G mutation) reduces T-strand levels to <1% of wild-type (Tinland et al., 1995). However, this mutation does not affect the ability of VirD2 to cleave, bind, and religate single-stranded border sequence oligonucleotides in vitro (Tinland et al., 1995). The R129G mutation also affects the precision of T-DNA integration. Normally, junctions between plant DNA and the right-hand T-DNA border occur precisely at the site of the VirD2-mediated nick that initiates T-strand production. In contrast, T-DNAs produced by the mutant VirD2R129G protein all suffer deletions of T-DNA sequences at their right-hand ends (Tinland et al., 1995). Thus, the HRY motif of VirD2 plays a role in precise integration of T-DNA into the plant genome (Tinland et al., 1995).

Except for three conserved C-terminal sequences, the C-terminal half of VirD2 (amino acids 235-395) is not conserved (*Figure 8-4*), and deletion of part of this region (amino acids 338-356) has no effect on T-DNA transmission (transfer and integration) (Shurvinton et al., 1992). However, several functional domains near the C-terminus are necessary for T-DNA transmission to plant cells, including a nuclear localization sequence, a type IV secretion signal, and the omega sequence (*Figure 8-4*). A mutation in VirD2 that removed the omega sequence reduced (fivefold) transient expression (in plant cells) of a T-DNA-borne reporter gene, whereas stable incorporation of T-DNA diminished 25 fold (Mysore et al., 1998). Amino acids within the omega sequence likely contribute to the activity of the adjacent type IV secretion signal, which may explain the fivefold reduction in transient T-DNA transfer. However, stable T-DNA integration is reduced to a greater extent than expected based on the effect of this mutation on VirD2-mediated transport into plant cells. These observations suggest that the omega sequence plays a role in both T-DNA transfer and subsequent integration.

3.4.8 Gateway to the pore: VirD4 coupling protein

VirD4 resembles TraG of RP4 and TraD of F (Lessl and Lanka, 1994; Firth et al., 1996); these proteins appear to connect relaxosomes with membrane-associated secretion/mating bridge systems (Lessl and Lanka, 1994; Firth et al., 1996; Cabezon et al., 1997). VirD4 is similar enough to pTi-encoded TraG that it can substitute for TraG, allowing conjugal transfer of RSF1010 through the pTi Trb pilus into recipient bacteria (Hamilton et al., 2000). Unlike most VirB proteins, a VirD4 analog has not been reported in the *B. pertussis* Ptl toxin export system. However, other type IV secretion systems, including those in *H. pylori*, *L. pneumophila*, and *R. prowazekii*, contain a protein similar to VirD4 (Segal et al., 1999). Thus, VirD4 and most VirB proteins have homologs not only among conjugation proteins but also among proteins devoted solely to toxin secretion.

The function of VirD4 appears more elaborate than an interface between relaxosome and transmembrane pore. Export of VirE2 (in the absence of T-strand DNA) requires VirD4, establishing its involvement in protein transport (Vergunst et al., 2003). Direct physical interaction between VirD4 and the C-terminus of VirE2 occurs in *Agrobacterium* cells (Atmakuri et al., 2003), in which VirD4 localizes to the poles of the rod-shaped bacterial cells (Kumar and Das, 2002). In addition, formation of the VirB pilus requires VirD4 (Fullner et al., 1996), indicating that it participates in translocation of pilus proteins as well. In this respect, VirD4 differs significantly from TraD, which is not needed for F pilus production (Firth et al., 1996).

4 VirD2 INTERACTS WITH HOST PROTEINS

4.1 Nuclear targeting: importin-α proteins

Upon entry into plant cells, VirD2 interacts with host-encoded importin proteins involved in importin-mediated NLS-dependent docking at the nuclear pore. Yeast two-hybrid protein interaction screens identified an *Arabidopsis* importin, AtKAPα (importin α1), that interacts with VirD2 in an NLS-specific manner (Ballas and Citovsky, 1997). AtKAPα shares sequence similarities with yeast NLS-binding proteins, and AtKAPα mediates import of VirD2 into the nuclei of yeast cells (Ballas and Citovsky, 1997). Additional studies used yeast two-hybrid screens, coimmunoprecipitation, and bimolecular fluorescence complementation to show that VirD2 also interacts with importin α-2, 3, 4, 7, and 9 (L.-Y. Lee and S. B. Gelvin, personal communication). These investigators also found that a mutation affecting importin α4 conferred a resistant to *Agrobacterium* transformation (*rat*) phenotype to *Arabidopsis thaliana* root explants (Zhu et al., 2003). Because importin α4 interacts with both VirD2 and VirE2, the Rat phenotype may result from disruption of nuclear targeting of both proteins.

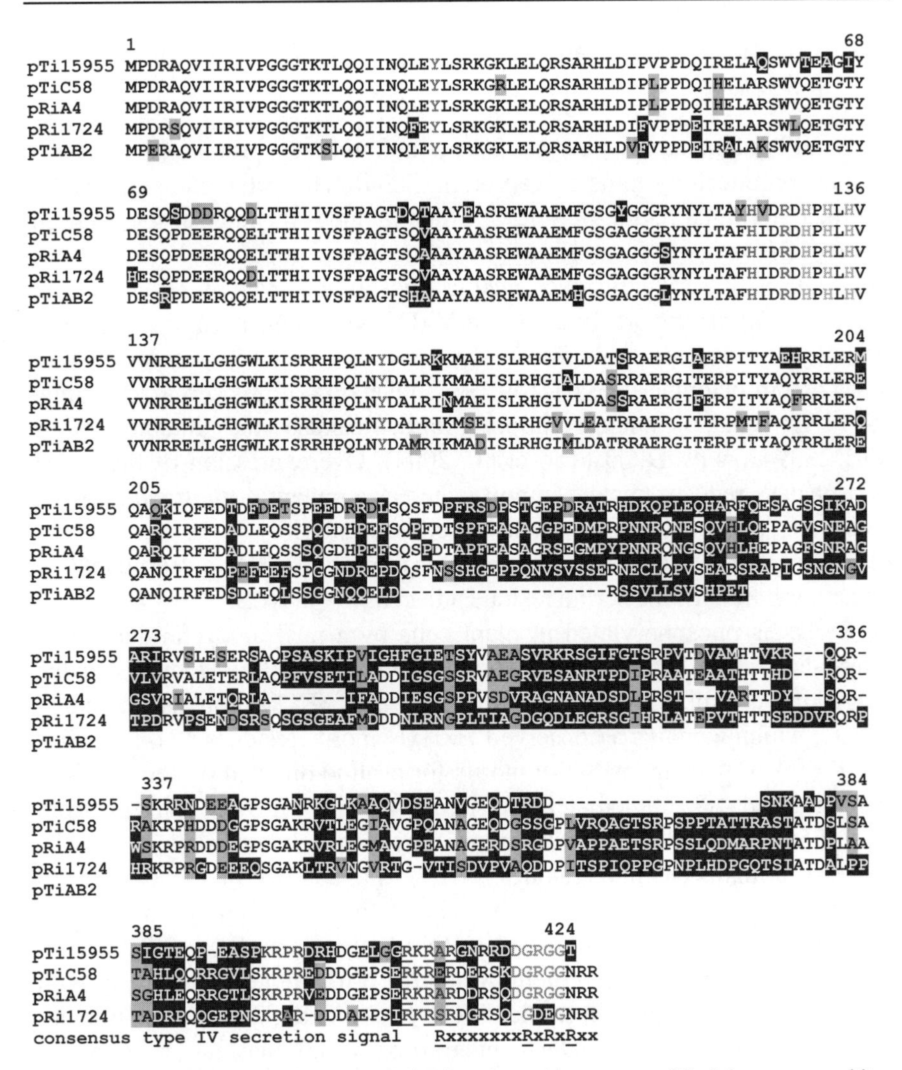

Figure 8-4. Clustal W alignment of VirD2 amino acid sequences. Black letters on a white background indicate amino acids that are identical in at least three VirD2 proteins. Shaded boxes indicate similar amino acids. Groups of amino acids considered similar in this analysis were: I, L, M, and V; A, G, and S; H, K, and R; D and E; N and Q; F, W, and Y; and S and T. Solid boxes (white letters on a black background) indicate nonconserved amino acids. Dashes indicate gaps placed in the sequences to maximize alignment. Numbers indicate the positions of amino acids in VirD2 from pTi15955. The red Y at position 29 indicates the conserved tyrosine that forms a covalent bond with the 5' end of the T-strand (Vogel and Das, 1992b). Conserved histidine (residue 126), arginine (residue 129), and tyrosine (residue 160) residues shown in pink may correspond to a "HRY motif" found in seven phage-encoded site specific integrase proteins (Argos et al., 1986) and in the TraI protein encoded by conjugative plasmid RP4 (Lessl and Lanka, 1994). Three conserved histidine residues (positions 131, 133, and 135) shown in green comprise a "histidine triad" motif (HxHxH), which may mediate the

4.2 Protein phosphatase, kinase, and TATA box-binding proteins

VirD2 interacts with a type 2C serine/threonine protein phosphatase (DIG3) produced by tomato (Tao et al., 2004). This interaction is specific for the C-terminal region of VirD2, which includes the nuclear localization sequence (*Figure 8-4*), and it may affect nuclear import of VirD2. Phosphorylation of serine residues near a nuclear localization sequence can simulate nuclear import. Each of the VirD2 NLS sequences in *Figure 8-4* has a serine residue immediately upstream of the NLS (residue 394 in VirD2 from octopine-type pTi15955). Conversion of this serine to alanine (S394A) reduces nuclear import and weakens the interaction of VirD2S394A with DIG3 (Tao et al., 2004). Overexpression of the DIG3 phosphatase reduces nuclear import, whereas a mutant *A. thaliana* line that lacks this phosphatase shows increased susceptibility to *A. tumefaciens* (Tao et al., 2004), suggesting that nuclear import of VirD2 is stimulated by phosphorylation of the serine residue adjacent to the NLS.

VirD2 is phosphorylated in plant cells by a nuclear cyclin-dependent kinase-activating kinase (CAK2Ms) (Bako et al., 2003). Members of the CAK-kinase family interact with histidine triad (HIT) proteins, including VirD2, which contains a conserved HxHxH motif (residues 131-135; *Figure 8-4*). VirD2 contains S/T-P motifs for proline-directed protein kinases, including an SP immediately upstream of the NLS in VirD2 (from pTi15955; *Figure 8-4*). VirD2 is phosphorylated in two regions, one in the relaxase domain (within residues 2-109) and another in the C-terminal domain (residues 248-447), which includes the NLS (Bako et al., 2003). Normally, CAK2Ms phosphorylates RNA polymerase II, which then recruits TATA box-binding protein. Similarly, VirD2 interacts strongly with TATA box-binding protein in plant cells (Bako et al., 2003). This interaction may play a role in T-DNA integration: VirD2 may target bound T-strands to free 3' ends associated with damaged DNA (Mayerhofer et al., 1991). The observation that TATA box-binding protein can bind to lesions

interaction of VirD2 with a nuclear protein kinase, cyclin-dependent kinase-activating kinase (CAK2Ms) (Bako et al., 2003). Amino acids comprising the nuclear localization sequence (KRPRDRHDGELGGRKRAR) are shown in blue. Arginine residues that correspond to the type IV secretion signal consensus sequence (RxxxxxxxRxRxRxx) are underlined. The C-terminal omega sequence (DGRGG) is shown in orange (Shurvinton et al., 1992). The endonuclease domain spans amino acids 1-262. The cyclophilin CypA binds amino acids 274-337, whereas Roc1 binds amino acids 174-337. A deletion that removes amino acids 338-356 does not affect VirD2 function (Shurvinton et al., 1992).

in DNA and initiate transcription-coupled repair (Vichi et al., 1997; Coin et al., 1998) raises the possibility that VirD2 may direct bound T-strands to free 3' ends in host DNA through its interaction with TATA box-binding protein (Bako et al., 2003).

4.3 Cyclophilins

The central domain of VirD2 interacts with plant cyclophilin proteins CypA and Roc1 (Deng et al., 1998). The cellular function of these cyclophilins is not known, but they may act as chaperones that assist protein folding. The importance of the interaction between VirD2 and cyclophilins is also unknown. CypA interacts strongly with residues 274-337 of VirD2, whereas Roc1 interacts weakly with residues 174-337 (Deng et al., 1998). This region excludes the NLS, omega sequence, type IV secretion signal, and most of the relaxase domain, including the active site, HRY motif, and HIT triad (*Figure 8-4*). In addition, most of this region is very poorly conserved among different VirD2 proteins (*Figure 8-4*), and it adjoins a region not required for VirD2 function (residues 338-356) (Shurvinton et al., 1992). Cyclosporin A, which disrupts the interaction of VirD2 with cyclophilins, also inhibits *Agrobacterium*-mediated transformation of plant cells (Deng et al., 1998), but it is not clear whether this effect is due to disruption of the VirD2-cyclophilin interaction.

5 T-DNA INTEGRATION

5.1 Integration products

Structures of integrated T-DNAs contributed to our understanding of the transformation process. However, the results require cautious interpretation because T-DNAs in established tumor lines may have rearranged subsequent to the initial integration events. Thus, the structures examined may not resemble the initial integration products, although numerous T-DNAs remained stable after lengthy propagation of transformed tissue (Van Lijsebettens et al., 1986). Indeed, non-oncogenic T-DNAs in transgenic plants remain stable through meiosis and mitosis (Barton et al., 1983). Many integrated T-DNAs do not undergo obvious rearrangements and remain colinear with the corresponding portion of the Ti plasmid.

T-DNAs reside at a variety of locations in the plant genome, often as single copies or short tandem arrays (direct or inverted repeats) (Lemmers et al., 1980; Thomashow et al., 1980; Zambryski et al., 1980; Peerbolte et al., 1987), and separate T-DNAs can integrate independently at different locations in the genome of a single plant cell (Chyi et al., 1986).

Alterations of the host target site that accompany T-DNA insertion suggest that this process follows a complex pathway. In one case, T-DNA integration produced a 158-base-pair direct repeat of host sequences as well as a base change, a small deletion, and "filler" DNA of unknown origin that resembles nearby host sequences (Gheysen et al., 1987; Gheysen et al., 1991). Other transformed plant cells contained a particular truncated T-DNA integrated at several different locations in the genome. Apparently the plant cell copied and aberrant T-DNA before integrating the copies at multiple sites. Thus, T-DNA integration occurs via a complex mechanism with several steps, including replication events and ligation of T-DNA to plant DNA.

5.2 The role of VirD2 in T-DNA integration

VirD2 participates in T-DNA integration, as indicated by several observations. First, the left and right ends of T-DNAs are joined to plant DNA via different mechanisms. Sequences of plant-to-T-DNA junctions indicate that right-hand ends of integrated T-DNAs very often correspond exactly to the base at which T strands attach to VirD2; in contrast, the left end of the T-DNA varies by hundreds of bases (Matsumoto et al., 1990; Gheysen et al., 1991; Mayerhofer et al., 1991). This preservation of right-hand T-DNA ends suggests that VirD2 protects them from nuclease attack (Jasper et al., 1994) and may ligate the 5' ends of T strands to plant DNA (Pansegrau et al., 1993; Jasper et al., 1994). Second, specific mutations in *virD2* either reduce integration (but not nuclear entry) of T-DNA (Narasimhulu et al., 1996) or result in T-DNAs with aberrant right-hand ends (Tinland et al., 1995), indicating that proper joining of the VirD2-bound end of a T strand to plant DNA requires wild-type VirD2. Third, VirD2 has ligase activity in some assays. Purified VirD2 cleaves ssDNAs containing the bottom strand of the T-DNA border sequence (Pansegrau et al., 1993; Jasper et al., 1994), and VirD2 can ligate these cut ssDNA molecules to reform the original substrate (Jasper et al., 1994) or join the VirD2-bound portion to another oligonucleotide in a sequence-specific manner (Csonka and Clark, 1979). This same ligase activity may join the

5' end of T strands to plant DNA, although another in vitro study suggests that plant proteins are involved in this step (Ziemienowicz et al., 2000).

6 PLANT GENETIC ENGINEERING

6.1 *Agrobacterium* virulence proteins help preserve T-DNA structure

A. tumefaciens-mediated gene transfer is the preferred method to create transgenic plants. The proteins associated with T-DNA, VirD2 and VirE2, help maintain the integrity of the integrated genes and reduce the frequency of duplications and rearrangements, which often affect transgene expression. For example, transcription through inverted repeat copies of a transgene will produce double-stranded RNA. This aberrant RNA may trigger post-transcriptional silencing of the transgene through systemic sequence-specific degradation of the RNA (Fire et al., 1998; Hamilton and Baulcombe, 1999). Electroporation, transformation, or microprojectile bombardment can introduce "naked" DNA into plant cells, but integration is inefficient and almost always results in multiple tandem copies that suffer frequent rearrangements.

6.2 "Agrolistic" transformation

A novel approach combines the "biolistic" or microprojectile bombardment method with the integration-promoting activity of VirD2 protein. "Agrolistic" transformation uses bombardment with tungsten particles coated with double-stranded T-DNA to introduce DNA into plant cells (Hansen and Chilton, 1996; Hansen et al., 1997). In contrast to standard biolistic transformation, the T-DNA includes border sequences as well as *virD1* and *virD2* fused to plant promoters. Plant cells that receive the T-DNA via particle bombardment produce VirD1 and VirD2 prior to integration of the incoming DNA. The VirD1/VirD2 enzyme nicks the border sequences, which yields an integrated T-DNA with a precise right end (Hansen and Chilton, 1996; Hansen et al., 1997). This approach is important for species recalcitrant to regeneration from tissue culture, because cells within embryos can be transformed.

6.3 Use of the VirD2 omega mutant to create marker-free transgenic plants

Deletion of the omega sequence at the C-terminus of VirD2 (Shurvinton et al., 1992) reduces stable T-DNA integration more profoundly than it reduces T-4DNA transfer and transient expression of genes located on the T-DNA (Mysore et al., 1998). This provides an opportunity to obtain marker-free transgenic plants by transient selection of kanamycin-resistant plant cells (Rommens et al., 2004). Host cells are cocultivated with two disarmed *A. tumefaciens* strains: a gene encoding kanamycin resistance is delivered from a strain with the VirD2-omega mutation, and the desired transgene (not linked to a selectable marker) is delivered from a separate strain with wild-type VirD2 (Rommens et al., 2004). Plant cells that express the kanamycin resistance gene transiently (but fail to inherit this gene stably) can be selected by temporary growth on medium containing kanamycin, and a significant fraction of these cells are stably transformed with the desired transgene (Rommens et al., 2004).

6.4 Efficient transgene targeting by homologous recombination is still elusive in plants

All plant transformation methods, including *A. tumefaciens*-mediated transfer, suffer from one serious limitation: the inability to target transgenes to specific chromosomal locations at a useful frequency. In bacteria, fungi, and mammalian cells, transgenes flanked on both sides with host chromosomal DNA can be introduced into specific locations via homologous recombination. Although such homologous recombination events can occur during plant transformation, they constitute a very small fraction of the total number of integration events. The chromosomal location of a transgene can affect expression of both the transgene and chromosomal genes at the site of insertion. Usually, plant scientists must examine hundreds (or thousands) of transformed plants to find one that exhibits appropriate transgene expression without affecting other important agronomic traits. An efficient integration system based on homologous recombination would allow engineers to place transgenes at specific chromosomal locations that allow good transgene expression without affecting host genes. For these reasons, an efficient transformation method that allows control over transgene insertion site is an important tool that plant genetic engineers currently lack.

7 ACKNOWLEDGEMENTS

Agrobacterium research is my laboratory has been supported by the National Science Foundation, the US Department of Agriculture, the National Institutes of Health, and the American Cancer Society. Larry Hodges and Claire Shurvinton conducted the studies in my lab on VirD2.

8 REFERENCES

Albright LM, Yanofsky MF, Leroux B, Ma DQ, Nester EW (1987) Processing of the T-DNA of *Agrobacterium tumefaciens* generates border nicks and linear, single-stranded T-DNA. J Bacteriol 169: 1046-1055

Andersson SGE, Zomorodipour A, Andersson JO, Sicheritz-Ponten T, Alsmark UCM, Podowski RM, Naslund AK, Eriksson AS, Winkler HH, Kurland CG (1998) The genome sequence of *Rickettsia prowazekii* and the origin of mitochondria. Nature 396: 133-140

Argos P, Landy A, Abremski K, Egan JB, Haggard-Ljungquist E, Hoess RH, Kahn ML, Kalionis B, Narayana SVL, Pierson LS, Sternberg N, Leong JM (1986) The integrase family of site-specific recombinases: regional similarities and global diversity. EMBO J 5: 433-440

Atmakuri K, Ding Z, Christie PJ (2003) VirE2, a type IV secretion substrate, interacts with the VirD4 transfer protein at cell poles of *Agrobacterium tumefaciens*. Mol Microbiol 49: 1699-1713

Bako L, Umeda M, Tiburcio AF, Schell J, Koncz C (2003) The VirD2 pilot protein of *Agrobacterium*-transferred DNA interacts with the TATA box-binding protein and a nuclear protein kinase in plants. Proc Natl Acad Sci USA 100: 10108-10113

Ballas N, Citovsky V (1997) Nuclear localization signal binding protein from *Arabidopsis* mediates nuclear import of *Agrobacterium* VirD2 protein. Proc Natl Acad Sci USA 94: 10723-10728

Baron C, Zambryski PC (1996) Plant transformation: a pilus in *Agrobacterium* T-DNA transfer. Curr Biol 6: 1567-1569

Barton KA, Binns AN, Matzke AJ, Chilton MD (1983) Regeneration of intact tobacco plants containing full length copies of genetically engineered T-DNA, and transmission of T-DNA to R1 progeny. Cell 32: 1033-1043

Beijersbergen A, Dulk-Ras AD, Schilperoort RA, Hooykaas PJJ (1992) Conjugative transfer by the virulence system of *Agrobacterium tumefaciens*. Science 256: 1324-1327

Binns AN (2002) T-DNA of *Agrobacterium tumefaciens*: 25 years and counting. Trends Plant Sci 7: 231-233

Binns AN, Beaupre CE, Dale EM (1995) Inhibition of VirB-mediated transfer of diverse substrates from *Agrobacterium tumefaciens* by the IncQ plasmid RSF1010. J Bacteriol 177: 4890-4899

Braun AC (1958) A physiological basis for the autonomous growth of the crown gall tumor cell. Proc Natl Acad Sci USA 44: 344-349

Buchanan-Wollaston V, Passiatore JE, Cannon F (1987) The mob and oriT mobilization functions of a bacterial plasmid promote its transfer to plants. Nature 328: 172-175

Bundock P, den Dulk-Ras A, Beijersbergen A, Hooykaas PJJ (1995) Trans-kingdom T-DNA transfer from *Agrobacterium tumefaciens* to *Saccharomyces cerevisiae*. EMBO J 14: 3206-3214

Cabezon E, Sastre JI, de la Cruz F (1997) Genetic evidence of a coupling role for the TraG protein family in bacterial conjugation. Mol Gen Genet 254: 400-406

Cascales E, Atmakuri K, Liu Z, Binns AN, Christie PJ (2005) *Agrobacterium tumefaciens* oncogenic suppressors inhibit T-DNA and VirE2 protein substrate binding to the VirD4 coupling protein. Mol Microbiol 58: 565-579

Cascales E, Christie PJ (2003) The versatile bacterial type IV secretion systems. Nat Rev Microbiol 1: 137-149

Cascales E, Christie PJ (2004) Definition of a bacterial type IV secretion pathway for a DNA substrate. Science 304: 1170-1173

Censini S, Lange C, Xiang Z, Crabtree JE, Ghiara P, Borodovsky M, Rappuoli R, Covacci A (1996) *cag*, a pathogenicity island of *Helicobacter pylori*, encodes type I-specific and disease-associated virulence factors. Proc Natl Acad Sci USA 93: 14648-14653

Chilton M-D, Drummond MH, Merio DJ, Sciaky D, Montoya AL, Gordon MP, Nester EW (1977) Stable incorporation of plasmid DNA into higher plant cells: the molecular basis of crown gall tumorigenesis. Cell 11: 263-271

Chilton M-D, Saiki RK, Yadav N, Gordon MP, Quetier F (1980) T-DNA from *Agrobacterium* Ti plasmid is in the nuclear DNA fraction of crown gall tumor cells. Proc Natl Acad Sci USA 77: 4060-4064

Christie PJ (1997) *Agrobacterium tumefaciens* T-complex transport apparatus: a paradigm for a new family of multifunctional transporters in eubacteria. J Bacteriol 179: 3085-3094

Christie PJ (2004) Type IV secretion: the Agrobacterium VirB/D4 and related conjugation systems. Biochim Biophys Acta 1694: 219-234

Christie PJ, Ward JE, Winans SC, Nester EW (1988) The *Agrobacterium tumefaciens virE2* gene product is a single-stranded-DNA-binding protein that associates with T-DNA. J Bacteriol 170: 2659-2667

Chyi YS, Jorgensen RA, Goldstein D, Tanksley SD, Loaiza-Figueroa F (1986) Locations and stability of *Agrobacterium*-mediated T-DNA insertions in the *Lycopersicon* genome. Mol Gen Genet 204: 64-69

Citovsky V, De Vos G, Zambryski P (1988) Single-stranded DNA binding protein encoded by the *virE* locus of *Agrobacterium tumefaciens*. Science 240: 501-504

Citovsky V, Guralnick B, Simon MN, Wall JS (1997) The molecular structure of *Agrobacterium* VirE2-single stranded DNA complexes involved in nuclear import. J Mol Biol 271: 718-727

Citovsky V, Warnick D, Zambryski P (1994) Nuclear import of *Agrobacterium* VirD2 and VirE2 proteins in maize and tobacco. Proc Natl Acad Sci USA 91: 3210-3214

Citovsky V, Wong ML, Zambryski P (1989) Cooperative interaction of *Agrobacterium* VirE2 protein with single stranded DNA: implications for the T-DNA transfer process. Proc Natl Acad Sci USA 86: 1193-1197

Citovsky V, Zupan J, Warnick D, Zambryski P (1992) Nuclear localization of *Agrobacterium* VirE2 protein in plant cells. Science 256: 1802-1805

Coin F, Frit P, Viollet B, Salles B, Egly JM (1998) TATA binding protein discriminates between different lesions on DNA, resulting in a transcription decrease. Mol Cell Biol 18: 3907-3914

Covacci A, Rappuoli R (1993) Pertussis toxin export requires accessory genes located downstream from the pertussis toxin operon. Mol Microbiol 8: 429-434

Csonka LN, Clark AJ (1979) Deletions generated by the transposon Tn*10* in the *srl recA* region of the *Escherichia coli* K-12 chromosome. Genetics 93: 321-343

Das A (1988) *Agrobacterium tumefaciens virE* operon encodes a single-stranded DNA-binding protein. Proc Natl Acad Sci USA 85: 2909-2913

De Cleene M, De Ley J (1976) The host range of crown gall. Bot Rev 42: 389-466

de Groot MJ, Bundock P, Hooykaas PJJ, Beijersbergen AG (1998) *Agrobacterium tumefaciens*-mediated transformation of filamentous fungi. Nat Biotechnol 16: 839-842

Deng W, Chen L, Peng WT, Liang X, Sekiguchi S, Gordon MP, Comai L, Nester EW (1999) VirE1 is a specific molecular chaperone for the exported single-stranded-DNA-binding protein VirE2 in *Agrobacterium*. Mol Microbiol 31: 1795-1807

Deng W, Chen L, Wood DW, Metcalfe T, Liang X, Gordon MP, Comai L, Nester EW (1998) *Agrobacterium* VirD2 protein interacts with plant host cyclophilins. Proc Natl Acad Sci USA 95: 7040-7045

Ding Z, Atmakuri K, Christie PJ (2003) The outs and ins of bacterial type IV secretion substrates. Trends Microbiol 11: 527-535

Dombek P, Ream W (1997) Functional domains of *Agrobacterium tumefaciens* single-stranded DNA-binding protein VirE2. J Bacteriol 179: 1165-1173

Dumas F, Duckely M, Pelczar P, Van Gelder P, Hohn B (2001) An *Agrobacterium* VirE2 channel for transferred-DNA transport into plant cells. Proc Natl Acad Sci USA 98: 485-490

Durrenberger F, Crameri A, Hohn B, Koukolikova-Nicola Z (1989) Covalently bound VirD2 protein of *Agrobacterium tumefaciens* protects the T-DNA from exonucleolytic degradation. Proc Natl Acad Sci USA 86: 9154-9158

Ellis JG, Kerr A, Petit A, Tempe J (1982) Conjugal transfer of nopaline and agropine Ti-plasmids – the role of agrocinopines. Mol Gen Genet 186: 269-274

Fire A, Xu S, Montgomery MK, Kostas SA, Driver SE, Mello CC (1998) Potent and specific genetic interference by double-stranded RNA in *Caenorhabditis elegans*. Nature 391: 806-811

Firth N, Ippen-Ihler K, Skurray RA (1996) Structure and function of the F factor and mechanism of conjugation. *In* FC Neidhardt, Curtiss III R, Ingraham JL, Lin ECC, Low KB, Magasanik B, Rsznikoff WS, Riley M, Schaechter M, Umbarger HC, eds, *Escherichia coli* and *Salmonella*: Cellular and Molecular Biology. American Society for Microbiology, Washington, D.C., pp 2377-2401

Fullner KJ, Lara JC, Nester EW (1996) Pilus assembly by *Agrobacterium* T-DNA transfer genes. Science 273: 1107-1109

Fullner KJ Nester EW (1996a) Temperature affects the T-DNA transfer machinery of *Agrobacterium tumefaciens*. J Bacteriol 178: 1498-1504

Fullner KJ, Nester EW (1996b) Environmental and genetic factors affecting RSF1010 mobilization between strains of *Agrobacterium tumefaciens*. *In* W Ream, SB Gelvin, eds, Crown Gall: Advances in Understanding Interkingdom Gene Transfer. American Phytopathological Society, St. Paul., pp 15-29

Furner IJ, Huffman GA, Amasino RM, Garfinkel DJ, Gordon MP, Nester EW (1986) An *Agrobacterium* transformation in the evolution of the genus *Nicotiana*. Nature 319: 422-427

Garfinkel DJ, Nester EW (1980) *Agrobacterium tumefaciens* mutants affected in crown gall tumorigenesis and octopine catabolism. J Bacteriol 144: 732-743

Garfinkel DJ, Simpson RB, Ream LW, White FF, Gordon MP, Nester EW (1981) Genetic analysis of crown gall: fine structure map of the T-DNA by site-directed mutagenesis. Cell 27: 143-153

Gelvin SB (2003) *Agrobacterium*-mediated plant transformation: the biology behind the "gene-jockeying" tool. Microbiol Mol Biol Rev 67: 16-37

Gelvin SB, Habeck LL (1990) *vir* genes influence conjugal transfer of the Ti plasmid of *Agrobacterium tumefaciens*. J Bacteriol 172: 1600-1608

Gheysen G, Van Montagu M, Zambryski P (1987) Integration of *Agrobacterium tumefaciens* transfer DNA (T-DNA) involves rearrangements of target plant DNA. Proc Natl Acad Sci USA 84: 6169-6173

Gheysen G, Villarroel R, Van Montagu M (1991) Illegitimate recombination in plants: a model for T-DNA integration. Genes Dev 5: 287-297

Gietl C, Koukolikova-Nicola Z, Hohn B (1987) Mobilization of T-DNA from *Agrobacterium* to plant cells involves a protein that binds single-stranded DNA. Proc Natl Acad Sci USA 84: 9006-9010

Goodner B, Hinkle G, Gattung S, Miller N, Blanchard M, Qurollo B, Goldman BS, Cao Y, Askenazi M, Halling H, Mullin L, Houmiel K, Gordon J, Vaudin M, Iartchouk O, Epp A, Liu F, Wollam C, Allinger M, Doughty D, Scott C, Lappas C, Markelz B, Flanagan C, Crowell C, Gurson J, Lomo C, Sear C, Strub G, Cielo C, Slater S (2001) Genome sequence of the plant pathogen and biotechnology agent *Agrobacterium tumefaciens* C58. Science 294: 2323-2328

Guyon P, Chilton M-D, Petit A, Tempe J (1980) Agropine in "null-type" crown gall tumors: evidence for the generality of the opine concept. Proc Natl Acad Sci USA 77: 2693-2697

Hamilton AJ, Baulcombe DC (1999) A species of small antisense RNA in post-transcriptional gene silencing in plants. Science 286: 950-952

Hamilton CM, Lee H, Li PL, Cook DM, Piper KR, von Bodman SB, Lanka E, Ream W, Farrand SK (2000) TraG from RP4 and VirD4 from Ti plasmids confer relaxosome specificity to the conjugal transfer system of pTiC58. J Bacteriol 182: 1541-1548

Hansen G, Chilton M-D (1996) "Agrolistic" transformation of plant cells: integration of T-strands generated in planta. Proc Natl Acad Sci USA 93: 14978-14983

Hansen G, Shillito RD, Chilton M-D (1997) T-strand integration in maize protoplasts after codelivery of a T-DNA substrate and virulence genes. Proc Natl Acad Sci USA 94: 11726-11730

Hepburn A, White J (1985) The effect of right terminal repeat deletion on the oncogenicity of the T-region of pTiT37. Plant Mol Biol 5: 3-11

Herrera-Estrella A, Chen Z, Van Montagu M, Wang K (1988) VirD proteins of *Agrobacterium tumefaciens* are required for the formation of a covalent DNA protein complex at the 5' terminus of T-strand molecules. EMBO J 7: 4055-4062

Hoekema A, Hirsch PR, Hooykaas PJJ, Schilperoort RA (1983) A binary plant vector strategy based on separation of vir and T-region of the *Agrobacterium tumefaciens* Ti plasmid. Nature 303: 179-180

Howard EA, Winsor BA, De Vos G, Zambryski PC (1989) Activation of the T-DNA transfer process in *Agrobacterium* results in the generation of a T-strand protein complex: tight association of VirD2 with the 5' ends of T-strands. Proc Natl Acad Sci USA 86: 4017-4021

Howard EA, Zupan JR, Citovsky V, Zambryski PC (1992) The VirD2 protein of *A. tumefaciens* contains a C-terminal bipartite nuclear localization signal: implications for nuclear uptake of DNA in plant cells. Cell 68: 109-118

Jasper F, Koncz C, Schell J, Steinbiss H-H (1994) *Agrobacterium* T-strand production *in vitro*: sequence-specific cleavage and 5' protection of single-stranded DNA templates by purified VirD2 protein. Proc Natl Acad Sci USA 91: 694-698

Jayaswal RK, Veluthambi K, Gelvin SB, Slightom JL (1987) Double-stranded cleavage of T-DNA and generation of single-stranded T-DNA molecules in *Escherichia coli* by a *virD*-encoded border-specific endonuclease from *Agrobacterium tumefaciens*. J Bacteriol 169: 5035-5045

Ji JM, Martinez A, Dabrowski M, Veluthambi K, Gelvin SB, Ream W (1988) The overdrive enhancer sequence stimulates production of T-strands from the *Agrobacterium tumefaciens* tumor-inducing plasmid. *In* B Staskawicz, P Ahlquist, O Yoder, eds, Molecular Biology of Plant-Pathogen Interactions; UCLA Symposia on Molecular and Cellular Biology Alan R Liss, New York, p 229

Joos H, Timmerman B, Van Montagu M, Schell J (1983) Genetic analysis of transfer and stabilisation of *Agrobacterium* DNA in plant cells. EMBO J 2: 2151-2160.

Kado CI (1994) Promiscuous DNA transfer system of *Agrobacterium tumefaciens*: role of the *virB* operon in sex pilus assembly and synthesis. Mol Microbiol 12: 17-22

Kingsman A, Willetts N (1978) The requirements for conjugal DNA synthesis in the donor strain during F*lac* transfer. J Mol Biol 122: 287-300

Kumar RB, Das A (2002) Polar location and functional domains of the *Agrobacterium tumefaciens* DNA transfer protein VirD4. Mol Microbiol 43: 1523-1532

Kunik T, Tzfira T, Kapulnik Y, Gafni Y, Dingwall C, Citovsky V (2001) Genetic transformation of HeLa cells by *Agrobacterium*. Proc Natl Acad Sci USA 98: 1871-1876

Lee LY, Gelvin SB, Kado CI (1999) pSa causes oncogenic suppression of *Agrobacterium* by inhibiting VirE2 protein export. J Bacteriol 181: 186-196

Lemmers M, De Beuckeleer M, Holsters M, Zambryski P, Depicker A, Hernalsteens JP, Van Montagu M, Schell J (1980) Internal organization, boundaries and integration of Ti-plasmid DNA in nopaline grown gall tumours. J Mol Biol 144: 355-378

Lessl M, Balzer D, Pansegrau W, Lanka E (1992) Sequence similarities between the RP4 Tra2 and the Ti VirB region strongly support the conjugation model for T-DNA transfer. J Biol Chem 267: 20471-20480

Lessl M, Lanka E (1994) Common mechanisms in bacterial conjugation and Ti-mediated transfer to plant cells. Cell 77: 321-324

Li J, Wolf SG, Elbaum M, Tzfira T (2005) Exploring cargo transport mechanics in the type IV secretion systems. Trends Microbiol 13: 295-298

Matsumoto M, Ito Y, Hosoi T, Takahashi Y, Machida Y (1990) Integration of *Agrobacterium* T-DNA into a tobacco chromosome: possible involvement of DNA homology between T-DNA and plant DNA. Mol Gen Genet 224: 309-316

Mayerhofer R, Koncz-Kalman Z, Nawrath C, Bakkeren G, Crameri A, Angelis K, Redei GP, Schell J, Hohn B, Koncz C (1991) T-DNA integration: a mode of illegitimate recombination in plants. EMBO J 10: 697-704

McBride KE, Knauf VC (1988) Genetic analysis of the *virE* operon of the *Agrobacterium* Ti plasmid pTiA6. J Bacteriol 170: 1430-1437

Miranda A, Janssen G, Hodges L, Peralta EG, Ream W (1992) *Agrobacterium tumefaciens* transfers extremely long T-DNAs by a unidirectional mechanism. J Bacteriol 174: 2288-2297

Moriguchi K, Maeda Y, Satou M, Hardayani NS, Kataoka M, Tanaka N, Yoshida K (2001) The complete nucleotide sequence of a plant root-inducing (Ri) plasmid indicates its chimeric structure and evolutionary relationship between tumor-inducing (Ti) and symbiotic (Sym) plasmids in Rhizobiaceae. J Mol Biol 307: 771-784

Mysore KS, Bassuner B, Deng X-B, Darbinian NS, Motchoulski A, Ream W, Gelvin SB (1998) Role of the *Agrobacterium tumefaciens* VirD2 protein in T-DNA transfer and integration. Mol Plant-Microbe Interact 11: 668-683

Narasimhulu SB, Deng X-B, Sarria R, Gelvin SB (1996) Early transcription of *Agrobacterium* T-DNA genes in tobacco and maize. Plant Cell 8: 873-886

Otten L, DeGreve H, Leemans J, Hain R, Hooykass P, Schell J (1984) Restoration of virulence of *vir* region mutants of *Agrobacterium tumefaciens* strain B6S3 by coinfection with normal and mutant *Agrobacterium* strains. Mol Gen Genet 195: 159-163

Ou JT, Reim RL (1978) F- mating materials able to generate a mating signal in mating with HfrH *dnaB*(Ts) cells. J Bacteriol 133: 442-445

Pansegrau W, Schoumacher F, Hohn B, Lanka E (1993) Site-specific cleavage and joining of single-stranded DNA by VirD2 protein of *Agrobacterium tumefaciens* Ti plasmids: analogy to bacterial conjugation. Proc Natl Acad Sci USA 90: 11538-11542

Peerbolte R, te Lintel-Hekkert W, Barfield DG, Hoge JHC, Wullems GL, Schilperoort RA (1987) Structure, organization and expression of transferred DNA in *Nicotiana plumbaginifolia* crown gall tissues. Planta 171: 393-405

Peralta EG, Hellmiss R, Ream W (1986) *Overdrive*, a T-DNA transmission enhancer on the *A. tumefaciens* tumour-inducing plasmid. EMBO J 5: 1137-1142

Peralta EG, Ream LW (1985) T-DNA border sequences required for crown gall tumorigenesis. Proc Natl Acad Sci USA 82: 5112-5116

Petit A, David C, Dahl GA, Ellis JG, Guyon P, Casse-Delbert F, Tempe J (1983) Further extension of the opine concept: plasmids in *Agrobacterium rhizogenes* cooperate for opine degradation. Mol Gen Genet 190: 204-414

Petit A, Tempe J, Kerr A, Holsters M, Van Montagu M, Schell J (1978) Substrate induction of conjugative activity of *Agrobacterium tumefaciens* Ti plasmids. Nature 271: 570-571

Pohlman RF, Genetti HD, Winans SC (1994) Common ancestry between IncN conjugal transfer genes and macromolecular export systems of plant and animal pathogens. Mol Microbiol 14: 655-668

Ream LW, Gordon MP, Nester EW (1983) Multiple mutations in the T-region of the *Agrobacterium tumefaciens* tumor-inducing plasmid. Proc Natl Acad Sci USA 80: 1660-1664

Regensburg-Tuink AJG, Hooykaas PJJ (1993) Transgenic *N. glauca* plants expressing bacterial virulence gene *virF* are converted into hosts for nopaline strains of *A. tumefaciens*. Nature 363: 69-71

Relic B, Andjelkovic M, Rossi L, Nagamine Y, Hohn B (1998) Interaction of the DNA modifying proteins VirD1 and VirD2 of *Agrobacterium tumefaciens*: analysis by subcellular localization in mammalian cells. Proc Natl Acad Sci USA 95: 9105-9110

Rommens CM, Humara JM, Ye J, Yan H, Richael C, Zhang L, Perry R, Swords K (2004) Crop improvement through modification of the plant's own genome. Plant Physiol 135: 421-431

Rossi L, Hohn B, Tinland B (1993) The VirD2 protein of *Agrobacterium tumefaciens* carries nuclear localization signals important for transfer of T-DNA to plant. Mol Gen Genet 239: 345-353

Rossi L, Hohn B, Tinland B (1996) Integration of complete transferred DNA units is dependent on the activity of virulence E2 protein of *Agrobacterium tumefaciens*. Proc Natl Acad Sci USA 93: 126-130

Scholz P, Haring V, Wittmann-Liebold B, Ashman K, Bagdasarian M, Scherzinger E (1989) Complete nucleotide sequence and gene organization of the broad-host-range plasmid RSF1010. Gene 75: 271-288

Segal G, Russo JJ, Shuman HA (1999) Relationships between a new type IV secretion system and the *icm/dot* virulence system of *Legionella pneumophila*. Mol Microbiol 34: 799-809

Segal G, Shuman HA (1998) Intracellular multiplication and human macrophage killing by *Legionella pneumophila* are inhibited by conjugal components of IncQ plasmid RSF1010. Mol Microbiol 30: 197-208

Sen P, Pazour GJ, Anderson D, Das A (1989) Cooperative binding of *Agrobacterium tumefaciens* VirE2 protein to single-stranded DNA. J Bacteriol 171: 2573-2580

Shaw CH, Watson MD, Carter GH (1984) The right hand copy of the nopaline Ti-plasmid 25 bp repeat is required for tumour formation. Nucleic Acids Res 12: 6031-6041

Sheng J, Citovsky V (1996) *Agrobacterium*-plant cell interaction: have virulence proteins - will travel. Plant Cell 8: 1699-1710

Shurvinton CE, Hodges L, Ream W (1992) A nuclear localization signal and the C-terminal omega sequence in the *Agrobacterium tumefaciens* VirD2 endonuclease are important for tumor formation. Proc Natl Acad Sci USA 89: 11837-11841

Stachel SE, Nester EW (1986) The genetic and transcriptional organization of the *vir* region of the A6 Ti plasmid of *Agrobacterium tumefaciens*. EMBO J 5: 1445-1454

Stachel SE, Timmerman B, Zambryski PC (1986) Generation of single-stranded T-DNA molecules during the initial stages of T-DNA transfer for *Agrobacterium tumefaciens* to plant cells. Nature 322: 706-712

Stachel SE, Timmerman B, Zambryski PC (1987) Activation of *Agrobacterium tumefaciens vir* gene expression generates multiple single-stranded T-strand molecules from the pTiA6 T-region: requirement for 5' *virD* gene products. EMBO J 6: 857-863

Stachel SE, Zambryski PC (1986) *Agrobacterium tumefaciens* and the susceptible plant cell: a novel adaptation of extracellular recognition and DNA conjugation. Cell 47: 155-157

Stahl LE, Jacobs A, Binns AN (1998) The conjugal intermediate of plasmid RSF1010 inhibits *Agrobacterium tumefaciens* virulence and VirB-dependent export of VirE2. J Bacteriol 180: 3933-3939

Sundberg C, Meek L, Carrol K, Das A, Ream W (1996) VirE1 protein mediates export of single-stranded DNA binding protein VirE2 from *Agrobacterium tumefaciens* into plant cells. J Bacteriol 178: 1207-1212

Sundberg CD, Ream W (1999) The *Agrobacterium tumefaciens* chaperone-like protein, VirE1, interacts with VirE2 at domains required for single-stranded DNA binding and cooperative interaction. J Bacteriol 181: 6850-6855

Tao Y, Rao PK, Bhattacharjee S, Gelvin SB (2004) Expression of plant protein phosphatase 2C interferes with nuclear import of the *Agrobacterium* T-complex protein VirD2. Proc Natl Acad Sci USA 101: 5164-5169

Thomashow MF, Nutter R, Montoya AL, Gordon MP, Nester EW (1980) Integration and organization of Ti plasmid sequences in crown gall tumors. Cell 19: 729-739

Tinland B, Hohn B, Puchta H (1994) *Agrobacterium tumefaciens* transfers single-stranded transferred DNA (T-DNA) into the plant cell nucleus. Proc Natl Acad Sci USA 91: 8000-8004

Tinland B, Koukolikova-Nicola Z, Hall MN, Hohn B (1992) The T-DNA-linked VirD2 protein contains two distinct nuclear localization signals. Proc Natl Acad Sci USA 89: 7442-7446

Tinland B, Schoumacher F, Gloeckler V, Bravo-Angel AM, Hohn B (1995) The *Agrobacterium tumefaciens* virulence D2 protein is responsible for precise integration of T-DNA into the plant genome. EMBO J 14: 3585-3595

Toro N, Datta A, Carmi OA, Young C, Prusti RK, Nester EW (1989) The *Agrobacterium tumefaciens virC1* gene product binds to Overdrive, a T-DNA transfer enhancer. J Bacteriol 171: 6845-6849

Tummuru MK, Sharma SA, Blaser MJ (1995) *Helicobacter pylori picB*, a homologue of the *Bordetella pertussis* toxin secretion protein, is required for induction of IL-8 in gastric epithelial cells. Mol Microbiol 18: 867-876

Van Larebeke N, Engler G, Holsters M, Van den Elsacker S, Zaenen I, Schilperoort RA, Schell J (1974) Large plasmid in *Agrobacterium tumefaciens* essential for crown gall-inducing ability. Nature 252: 169-170

Van Lijsebettens M, Inze D, Schell J, Van Montagu M (1986) Transformed cell clones as a tool to study T DNA integration mediated by *Agrobacterium tumefaciens*. J Mol Biol 188: 129-145

Veluthambi K, Ream W, Gelvin SB (1988) Virulence genes, borders, and overdrive generate single-stranded T-DNA molecules from the A6 Ti plasmid of *Agrobacterium tumefaciens*. J Bacteriol 170: 1523-1532

Vergunst AC, van Lier MCM, den Dulk-Ras A, Stüve TAG, Ouwehand A, Hooykaas PJJ (2005) Positive charge is an important feature of the C-terminal transport signal of the VirB/D4-translocated proteins of *Agrobacterium*. Proc Natl Acad Sci USA 102: 832-837

Vergunst AC, van Lier MCM, den Dulk-Ras A, Hooykaas PJJ (2003) Recognition of the *Agrobacterium* VirE2 translocation signal by the VirB/D4 transport system does not require VirE1. Plant Physiol 133: 978-988

Vichi P, Coin F, Renaud J-P, Vermeulen W, Hoeijmakers JH, Moras D, Egly J-M (1997) Cisplatin- and UV-damaged DNA lure the basal transcription factor TFIID/TBP. EMBO J 16: 7444-7456

Vogel AM, Das A (1992a) The *Agrobacterium tumefaciens virD3* gene is not essential for tumorigenicity on plants. 174: 5161-5164

Vogel AM, Das A (1992b) Mutational analysis of *Agrobacterium tumefaciens virD2*: tyrosine 29 is essential for endonuclease activity. J Bacteriol 174: 303-308

Vogel JP, Andrews HL, Wong SK, Isberg RR (1998) Conjugative transfer by the virulence system of *Legionella pneumophila*. Science 279: 873-876

Wang K, Herrera-Estrella A, Van Montagu M (1990) Overexpression of *virD1* and *virD2* genes in *Agrobacterium tumefaciens* enhances T-complex formation and plant transformation. J Bacteriol 172: 4432-4440

Wang K, Herrera-Estrella L, Van Montagu M, Zambryski PC (1984) Right 25 bp terminus sequence of the nopaline T-DNA is essential for and determines direction of DNA transfer from *Agrobacterium* to the plant genome. Cell 38: 455-462

Wang K, Stachel SE, Timmerman B, Van Montagu M, Zambryski PC (1987) Site-specific nick in the T-DNA border sequence as a result of *Agrobacterium vir* gene expression. Science 235: 587-591

Ward ER, Barnes WM (1988) VirD2 protein of *Agrobacterium tumefaciens* very tightly linked to the 5' end of T-strand DNA. Science 242: 927-930

Watson B, Currier TC, Gordon MP, Chilton M-D, Nester EW (1975) Plasmid required for virulence of *Agrobacterium tumefaciens*. J Bacteriol 123: 255-264

Weiss AA, Johnson FD, Burns DL (1993) Molecular characterization of an operon required for pertussis toxin secretion. Proc Natl Acad Sci USA 90: 2970-2974

Willmitzer L, De Beuckeleer M, Lemmers M, Van Montagu M, Schell J (1980) DNA from Ti plasmid present in nucleus and absent from plastids of crown gall plants. Nature 287: 359-361

Willmitzer L, Simons G, Schell J (1982) The TL-DNA in octopine crown-gall tumours codes for seven well-defined polyadenylated transcripts. EMBO J 1: 139-146

Winans SC (1992) Two-way chemical signalling in *Agrobacterium*-plant interactions. Microbiol Rev 56: 12-31

Winans SC, Allenza P, Stachel SE, McBride KE, Nester EW (1987) Characterization of the *virE* operon of the *Agrobacterium* Ti plasmid pTiA6. Nucleic Acids Res 15: 825-837

Winans SC, Burns DL, Christie PJ (1996) Adaptation of a conjugal transfer system for the export of pathogenic macromolecules. Trends Microbiol 4: 64-68

Wood DW, Setubal JC, Kaul R, Monks DE, Kitajima JP, Okura VK, Zhou Y, Chen L, Wood GE, Almeida Jr. NF, Woo L, Chen Y, Paulsen IT, Eisen JA, Karp PD, Bovee Sr. D, Chapman P, Clendenning J, Deatherage G, Gillet W, Grant C, Kutyavin T, Levy R, Li MJ, McClelland E, Palmieri P, Raymond C, Rouse R, Saenphimmachak C, Wu Z, Romero P, Gordon D, Zhang S, Yoo H, Tao Y, Biddle P, Jung M, Krespan W, Perry M, Gordon-Kamm B, Liao L, Kim S, Hendrick C, Zhao ZY, Dolan M, Chumley F, Tingey SV, Tomb JF, Gordon MP, Olson MV, Nester EW (2001) The genome of the natural genetic engineer *Agrobacterium tumefaciens* C58. Science 294: 2317-2323

Yanofsky MF, Porter SG, Young C, Albright LM, Gordon MP, Nester EW (1986) The *virD* operon of *Agrobacterium tumefaciens* encodes a site-specific endonuclease. Cell 47: 471-477

Young C, Nester EW (1988) Association of the VirD2 protein with the 5' end of T-strands in *Agrobacterium tumefaciens*. J Bacteriol 170: 3367-3374

Yusibov VM, Steck TR, Gupta V, Gelvin SB (1994) Association of single-stranded transferred DNA from *Agrobacterium tumefaciens* with tobacco cells. Proc Natl Acad Sci USA 91: 2994-2998

Zambryski P, Holsters M, Kruger K, Depicker A, Schell J, Van Montagu M, Goodman HM (1980) Tumor DNA structure in plant cells transformed by *A. tumefaciens*. Science 209: 1385-1391

Zhou XR, Christie PJ (1999) Mutagenesis of the *Agrobacterium* VirE2 single-stranded DNA-binding protein identifies regions required for self-association and interaction with VirE1 and a permissive site for hybrid protein construction. J Bacteriol 181: 4342-4352

Zhu J, Oger PM, Schrammeijer B, Hooykaas PJJ, Farrand SK, Winans SC (2000) The bases of crown gall tumorigenesis. J Bacteriol 182: 3885-3895

Zhu Y, Nam J, Humara JM, Mysore KS, Lee LY, Cao H, Valentine L, Li J, Kaiser AD, Kopecky AL, Hwang HH, Bhattacharjee S, Rao PK, Tzfira T, Rajagopal J, Yi H, Veena, Yadav BS, Crane YM, Lin K, Larcher Y, Gelvin MJ, Knue M, Ramos C, Zhao X, Davis SJ, Kim SI, Ranjith-Kumar CT, Choi YJ, Hallan VK, Chattopadhyay S, Sui X, Ziemienowicz A, Matthysse AG, Citovsky V, Hohn B, Gelvin SB (2003) Identification of Arabidopsis *rat* mutants. Plant Physiol 132: 494-505

Ziemienowicz A, Merkle T, Schoumacher F, Hohn B, Rossi L (2001) Import of *Agrobacterium* T-DNA into plant nuclei: Two distinct functions of VirD2 and VirE2 proteins. Plant Cell 13: 369-384

Ziemienowicz A, Tinland B, Bryant J, Gloeckler V, Hohn B (2000) Plant enzymes but not *Agrobacterium* VirD2 mediate T-DNA ligation in vitro. Mol Cell Biol 20: 6317-6322

Zupan J, Citovsky V, Zambryski PC (1996) *Agrobacterium* VirE2 protein mediates nuclear uptake of single-stranded DNA in plant cells. Proc Natl Acad Sci USA 93: 2392-2397

Zupan J, Zambryski PC (1997) The *Agrobacterium* DNA transfer complex. Crit Rev Plant Sci 16: 279-295

Chapter 9

TRANSLOCATION OF ONCOGENIC T-DNA AND EFFECTOR PROTEINS TO PLANT CELLS

Krishnamohan Atmakuri and Peter J. Christie

Department of Microbiology and Molecular Genetics, University of Texas Medical School at Houston, 6431 Fannin, Houston, TX 77030, USA

Abstract. *Agrobacterium tumefaciens* has evolved as a phytopathogen by adapting a DNA conjugation system for the novel purpose of delivering oncogenic T-DNA and protein substrates to susceptible plant cells. This transfer system is a member of a large family of translocation systems termed the type IV secretion (T4S) systems. The T4S systems are structurally complex machines assembled from a dozen or more membrane proteins often in response to environmental signals. In *A. tumefaciens* and other Gram-negative bacteria, the T4S machines assemble as a cell-envelope spanning secretion channel and an extracellular pilus. Recent studies of the *A. tumefaciens* VirB/D4 T4S system and closely related systems have advanced our understanding of T4S secretion in several fundamental areas, including: (i) T-DNA processing reactions and requirements for T-DNA and protein substrate recruitment, (ii) stages leading to assembly and polar positioning of the transfer apparatus, (iii) VirB subunit membrane topologies and structures and transfer channel architecture, (iv) energetic contributions to machine assembly and function, and (v) the T-DNA translocation route through the VirB/D4 transfer channel. These studies are generating a picture of the VirB/D4 T4S system as multifunctional and structurally dynamic. The wealth of information generated by many laboratories in recent years has established the *A. tumefaciens* VirB/D4 T4S system as an important paradigm for unraveling the mechanistic details of DNA and protein trafficking between diverse cell types.

1 INTRODUCTION

The ability of *Agrobacterium tumefaciens* to transfer oncogenic T-DNA as well as effector proteins across kingdom boundaries is a unique aspect and hallmark feature of *A. tumefaciens* phytopathogenesis. The cell surface machine mediating translocation of DNA and protein macromolecules is composed of proteins encoded by the *virB* and *virD* operons, specifically, VirB1 through VirB11 and VirD4. Besides elaborating a transenvelope channel allowing substrate transfer to plants and other cell types, the VirB proteins mediate production of an extracellular pilus for establishing contact with target cells. The VirB/D4 machine is now recognized as a member of a growing family of translocation systems termed the type IV secretion systems (T4S systems or T4SS). Both in terms of their mechanism of action and their broad biological functions, the T4S systems are a fascinatingly diverse group of translocation systems required for infection by many agriculturally and medically important bacterial pathogens. Today, the VirB/D4 T4S system is recognized as an important model for the T4S superfamily, due to the dedicated efforts of many laboratories intent on defining the molecular details underlying transkingdom T-DNA transfer. In this chapter, we will summarize recent progress in our understanding of the VirB/D4 T4S system with emphasis on the mechanistic and structural features of the translocation channel. Information will be derived mainly from studies of VirB/D4 machines encoded by the octopine-type pTiA6NC and nopaline-type pTiC58 plasmids.

2 A HISTORICAL OVERVIEW

We will first highlight pioneering studies that led to the discovery of the T-DNA transfer system as an adapted conjugation machine, and discoveries that resulted in renaming conjugation and related protein trafficking systems as type IV secretion (T4S) systems.

2.1 Discovery of the VirB/D4 transfer system

Stachel and Nester delimited the boundaries of the *virB* gene cluster and established its importance for virulence by Tn3HoHo1 transposon mutagenesis (Stachel and Nester, 1986). Nearly all transposon insertions within an ~11-kilobase region located between *virA* and *virG* abolished virulence. The transposon insertions generated transcriptional and translational fusions to *lacZ* and supplied evidence for a single operon which, like the flanking *vir* operons, was strongly induced upon co-cultivation of *A. tumefaciens* with plant cells, or exposure to plant cell exudates or specific classes of purified plant phenolic compounds, e.g., acetosyringone (Bolton et al., 1986; Engstrom et al., 1987; Melchers et al., 1989).

Four laboratories originally sequenced *virB* operons from octopine (pTiA6NC, pTi15955) and nopaline (pTiC58) Ti plasmids (Thompson et al., 1988; Ward et al., 1988; Kuldau et al., 1990; Shirasu et al., 1990; Ward et al., 1990). The *virB* operon from these plasmids encodes 11 VirB proteins, VirB1 through VirB11. The VirB proteins from the octopine pTiA6/pTi15955 and the nopaline pTiC58 plasmids are highly related, with sequence identities ranging from 71% for VirB1 and VirD4 to over 90% for VirB2. Despite this high sequence conservation and the fact that the VirB/D4 systems from the pTiA6NC and pTiC58 plasmids are studied in the same strain C58 genetic background, there have been several experimental findings distinguishing the two systems that unlikely are due to differences in experimental approaches; these are mentioned in the appropriate sections below.

In the early days, computer analyses yielded only a few clues as to possible functions of the VirB proteins for virulence (see above refs. and Beijersbergen et al., 1994). Most notably, characteristic signal sequences were identified at the N termini of VirB1, VirB2, VirB5, VirB7 and VirB9 suggestive of translocation across the inner membrane. Potential transmembrane (TM) domains near the N termini of VirB3, VirB8, and VirB10 were suggestive of bitopic inner membrane configurations oriented with C-terminal domains in the periplasm. Multiple potential TM segments in VirB6 were suggestive of a polytopic configuration, whereas the absence of characteristic TM segments suggested VirB4 and VirB11 reside predominantly or exclusively in the cytoplasm. Consistent with their predicted cellular locations, VirB4 and VirB11 were shown to carry conserved Walker A motifs suggestive of NTP binding activities important for energizing reactions associated with machine biogenesis and/or translocation. Finally, VirB7 was shown to possess a signal sequence characteristic of lipoproteins.

Stachel and Zambryski proposed that the T-DNA transfer process might be an adapted form of bacterial conjugation (Stachel and Zambryski, 1986). Strong support for this hypothesis emerged with the discoveries that *A. tumefaciens* can deliver the mobilizable IncQ plasmid RSF1010 to plant cells (Buchanan-Wollaston et al., 1987). VirD2 and T-DNA border sequences respectively were shown to resemble proteins termed relaxases and cognate origin-of-transfer (*oriT*) sequences of several self-transmissible plasmids as well as the mobilizable IncQ plasmid, RSF1010 (Lessl et al., 1992a; Waters and Guiney, 1993). Sequence analyses of the transfer (*tra*) regions of several conjugative plasmids identified many VirB and VirD4 homologs (Lessl et al., 1992b; Shirasu and Kado, 1993). Notably, the *tra* regions of both pKM101 and R388 were found to encode subunits homologous to all 11 VirB subunits and the VirD4 subunit, whereas the RP4 *tra* region encodes at least 7 discernible VirB/D4-like subunits plus several other ancestrally unrelated components. Following the demonstration of interkingdom transfer of RSF1010, Hooykaas and colleagues showed that the VirB and VirD4 proteins mediate transfer of RSF1010 from agrobacterial donor to recipient cells (Beijersbergen et al., 1992). Ward and colleagues then reported the interesting finding that wild-type *A. tumefaciens* cells carrying RSF1010 display attenuated virulence, whereas the overexpression of three *virB* genes, *virB9*, *virB10*, and *virB11*, restored virulence to wild-type levels (Ward et al., 1991). These investigators proposed that RSF1010 and T-DNA compete for available VirB/D4 transfer machines, with VirB9, VirB10, and VirB11 being rate limiting for machine assembly. Thus, by the mid-1990's, it was widely accepted that *A. tumefaciens* had appropriated an ancestral conjugation system for the purpose of delivering oncogenic T-DNA to plant cells.

2.2 Renaming the mating pore as a type IV translocation channel

The next significant conceptual advance in this field derived from a sequence analysis of the *Bordetella pertussis ptl* gene cluster (Shirasu and Kado, 1993; Weiss et al., 1993). The *ptl* genes mediate export of the 6-subunit pertussis toxin across the outer membrane of *B. pertussis*, and the sequence studies identified the *ptl* gene products as VirB homologs. The *ptl* genes are collinear with the *virB* genes in the respective operons and encode homologs for all but the VirB1 and VirB5 proteins. These findings established for the first time an evolutionary link between a conjugation system and a dedicated protein secretion machine. Soon afterward, a new nomenclature was proposed for macromolecular trafficking systems displaying ancestral relatedness to the *A. tumefaciens* VirB/D4 and related conjugation systems. These systems were designated as type IV secretion systems (T4SS, or T4S systems) in order to distinguish them from the type I (T1S; ATP-binding cassette or ABC transporters), type II (T2S; terminal branch of the general secretory pathway), III (T3S; ancestry-related to bacterial flagella) systems, type V (T5S; autotransporter) and, most recently, type VI (T6S) systems (Economou et al., 2006). The VirB/D4-like systems are also designated type IVA, as a means of distinguishing this subfamily from the type IVB systems. The types IVA and IVB components are unrelated with the exception of subunits homologous to VirB10, VirB11, and VirD4 (Christie and Vogel, 2000).

Throughout the last 10 years, many T4S systems have been identified through genome sequencing projects or screens for virulence factors (Cascales and Christie, 2003; Christie et al., 2005). In the latter screens, mutations introduced into T4S machine subunits invariably abolished virulence, establishing the importance of these machines for one or more stages of infection. Where investigated, these systems have been shown to translocate effector proteins to the cytosols of eukaryotic cells, leading to their classification as 'effector translocation' systems. Most T4S machines likely contribute to pathogenesis through delivery of proteins to target eukaryotic cells, but recent studies also suggest the T4S machines might promote infection in various other ways. For example, in *Helicobacter pylori*, the CagA T4SS is implicated in translocation of peptidoglycan fragments to mammalian cells, resulting in induction of a host defense response involving the protein Nod1 (Viala et al., 2004). Some T4S systems, including the VirB/D4 transfer system of *A. tumefaciens*, the *tra* system of the conjugative plasmid RP4, the *dot/icm* system of *Legionella pneumophila*, and a T4S system of *Bartonella henselae*, deliver DNA to various eukaryotic cells at least in specific laboratory conditions (Beijersbergen et al., 1992; Bates et al., 1998; Vogel et al., 1998; Kunik et al., 2001; Waters, 2001 and C. Dehio, personal communication). At this time, the *A. tumefaciens* VirB/D4 system remains the only T4S system for which interkindgom DNA transfer is an essential feature of phytopathogenicity, but it is intriguing to speculate that other T4S systems also deliver DNA effector molecules during the course of infection. Finally, the T4S systems might enhance virulence by mechanisms other than interkingdom effector translocation. For example, these systems elaborate surface adhesins or extracellular pili which can aid in colonization and establishment of biofilms on host tissues (Ghigo, 2001; Reisner et al., 2006). Also, the Cag T4S of

H. pylori induces IL-8 secretion in mammalian cells even in the absence of detectable substrate transfer, suggesting that the T4S machine itself triggers this response through binding a host cell surface receptor(s) (Selbach et al., 2002).

3 *A. TUMEFACIENS* VIRB/D4 SECRETION SUBSTRATES

Before discussing the *A. tumefaciens* VirB/D4 system in structural and mechanistic detail, it is necessary to present available information pertaining to the macromolecular substrates of this transfer system. It is now widely appreciated that *A. tumefaciens* incites Crown Gall disease by translocating oncogenic T-DNA to susceptible plants cells. In the past 15 years or so, studies also have identified a number of protein substrates whose translocation to plant cells is necessary for tumor formation. In general, these effector proteins are thought to promote transmission of the T-DNA transfer intermediate through the plant cytoplasm to the nucleus. However, their specific functions in the plant are beyond the scope of this Chapter; the reader is referred to chapter 13 for current information. Here, we will summarize recent progress in our understanding of how the DNA and protein substrates of the VirB/D4 system are recognized and recruited to the transfer apparatus.

3.1 T-DNA processing and recruitment to the VirB/D4 channel

The T-DNA is processed from its position on the Ti plasmid into a translocation-competent substrate through the action of the DNA processing factors acting at T-DNA border repeat sequences (*Figure 9-1*). The catalytic subunit, VirD2, is a member of a large family of transesterases that generate single-stranded nicks at origin-of-transfer (*oriT*) sequences of conjugative plasmids (Pansegrau et al., 1993; Scheiffele et al., 1995). Upon recruitment to *oriT*-like sequences within the T-DNA borders, VirD2 nicks both borders, and remains covalently bound to the 5' end of the transferred strand (T-strand). Purified VirD2 catalyzes nicking of T-DNA substrates *in vitro*, but border cleavage *in vivo* requires accessory proteins including VirD1 and the VirC1 and VirC2 proteins (Veluthambi et al., 1988; De Vos and Zambryski, 1989 Pansegrau, 1993 #687; Scheiffele et al., 1995). VirC1 binds the *overdrive* sequence flanking the right border repeat sequences of octopine-type Ti plasmids (Toro et al.,

1988; Toro et al., 1989), and this binding reaction stimulates T-DNA processing (Atmakuri et al., 2007). Interestingly, VirC1 is related to the ParA family of ATPases, which mediate partitioning of chromosomes and plasmids during cell division (Zhu et al., 2000).

Recent work advanced our understanding of how VirC1 and VirC2 contribute to the efficiency of plant transformation (Atmakuri et al., 2007). First, a quantitative analysis confirmed and extended early studies by showing that wild-type *A. tumefaciens* cells accumulate approximately 12-14 molecules of processed T-strand per Ti plasmid within 24 h of induction of the virulence (*vir*) genes with the plant phenolic acetosyringone (AS). Both VirC1 and VirC2 were shown to be required to stimulate this high level of T-strand production, and a mutation in an invariant Lys residue in the Walker A nucleotide triphosphate binding motif of VirC1 (VirC1K15Q) abolished the stimulatory effect indicating the importance of ATP binding or hydrolysis. Very intriguingly, VirC1, VirC2, and VirD1 localize predominantly at one pole of *A. tumefaciens* cells, as shown by immunofluoresence microscopy (Atmakuri et al., 2007). Each of these processing factors localizes at the cell poles independently of each other as well as the VirB/D4 T4S machine components (which also localizes at *A. tumefaciens* cell poles - see below). VirC1 polar localization serves to recruit the VirD2 relaxase to the cell membrane as well as to the cell pole, and again studies of the VirC1K15Q mutant established that ATP binding or hydrolysis is important for polar recruitment of the relaxase. By adapting the fluorescence *in situ* hybridization (FISH) assay for detection of single-stranded DNA in a cell, Atmakuri and colleagues further showed that VirC1 positions the T-strand at cell poles by an ATP-dependent mechanism. The results strongly indicate that VirC1 functions to recruit the processed VirD2-T-strand transfer intermediate to the cell pole.

Adding to the above, Atmakuri and colleagues presented evidence that VirC1 interacts with VirD4, the substrate receptor for the VirB/D4 T4S translocation system (see below). The data therefore suggest that *A. tumefaciens* evolved Par-like proteins for two novel purposes associated with conjugation: (i) VirC1 functions together with VirC2 to stimulate DNA processing through recruitment of VirD2 to the T-DNA borders and (ii) VirC1 recruits the T-complex from a cytosolic pool to coordinate substrate docking with the polar-localized VirD4 substrate receptor (Atmakuri et al., 2007). It is important to note that *virC* mutants can still transform certain plant species. The novel functions ascribed to VirC1 and VirC2 thus serve principally to enhance the efficiency of the *A. tumefaciens* infection process and broaden its host range.

3.2 Processing and recruitment of protein substrates

The VirB/D4 T4S system also translocates multiple protein substrates (*Figure 9-1*). Early studies suggested the VirE2 (SSB) associates noncovalently with T-strand to form a VirD2-T-strand-VirE2 particle (Christie et al., 1988). While formation of a VirD2-T-strand-VirE2 particle, termed the T-complex, is still considered essential for T-DNA transmission to the plant nucleus, several lines of evidence now argue strongly that the VirD2-T-strand particle and VirE2 are exported separately across the *A. tumefaciens* envelope and assemble within the plant cell (Stahl et al., 1998; Vergunst et al., 2000; Simone et al., 2001).

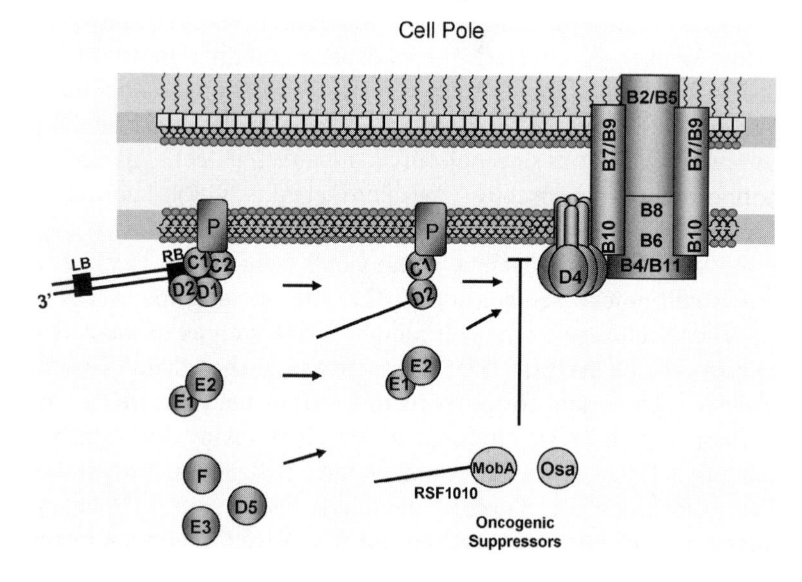

Figure 9-1. Processing of substrates for transfer through the *A. tumefaciens* VirB/D4 type IV secretion system. The Dtr processing factors VirC1, VirC2, VirD1, and VirD2 assemble at the border repeats flanking the T-DNA and VirD2 relaxase generates a nick on the strand destined for transfer (T-strand). ParA-like VirC1 functions both in VirD2 recruitment to the T-DNA borders and to the *A. tumefaciens* cell poles to coordinate substrate docking with the VirD4 receptor. VirE2 is bound by the VirE1 chaperone and is recruited via a C-terminal domain to the polar-localized VirD4 receptor. Two oncogenic suppressors, the RSF1010 transfer intermediate and pSA-encoded Osa, block substrate docking with VirD4. See text for further details and references.

Most convincingly, Vergunst and colleagues demonstrated that VirE2 fused to the Cre recombinase mediates transfer of Cre to plant cells independently of the T-DNA, as assessed by recombinase activity at *lox* target sites (Vergunst et al., 2000). This assay, termed CrAFT, has now been widely used to screen for effectors of the *A. tumefaciens* VirB/D4 system as well as several other T4S systems. Coupled with this finding, an *A. tumefaciens virE2* mutant will induce tumor formation on transgenic plants engineered to produce VirE2, establishing the importance of VirE2 for T-DNA stability or translocation in the plant (Citovsky and Zambryski, 1993). *virE2* is spatially juxtaposed to the upstream *virE1*, and both genes must be coexpressed from the native *virE* promoter to complement a *virE2* null mutation suggestive of translational coupling. As expected for products of translationally coupled genes, VirE1 and VirE2 form a stable complex *in vivo* as shown by two-hybrid assays and biochemical screens (Deng et al., 1999; Sundberg and Ream, 1999; Zhao et al., 2001). VirE1 is a small protein with physical properties resembling secretion chaperones required for substrate transfer via type III secretion systems (Deng et al., 1999; Sundberg and Ream, 1999). Consistent with a proposed chaperone function, VirE1 prevents VirE2 from premature self-association (Zhao et al., 2001; Frenkiel-Krispin et al., 2006), and a very recent study showed that chaperone binding prevents VirE2 from binding the T-DNA transfer intermediate in the bacterium (Frenkiel-Krispin et al., 2006). VirE1 is not required for VirE2 docking with the VirB/D4 T4S system (Atmakuri et al., 2003; Vergunst et al., 2003), consistent with findings that VirE2 carries C-terminal signals conferring substrate recognition (see below).

As noted above, VirE1 is the only identified secretion chaperone in *A. tumefaciens*, and it seems specifically adapted for VirE2 export. Other protein substrates include VirE3, VirF, and VirD5, and these likely are exported independently of cognate secretion chaperones (Vergunst et al., 2000; Vergunst et al., 2005). These proteins are not implicated in binding the T-strand and various physical properties, e.g., folding kinetics, monomeric status, and accessibility of secretion signals, might obviate the need for a secretion chaperone. Recent work also has shown that the VirD2 relaxase possesses secretion signals recognizable by the transfer machinery and, when fused to Cre, mediates transfer of the recombinase to target cells independently of an association with the T-strand (Vergunst et al., 2005).

Ream et al. discovered a protein termed GALLS that is required for T-DNA transfer from *Agrobacterium rhizogenes* to plants. Intriguingly, expression of the *GALLS* gene can complement a *virE2* mutation in *A. tumefaciens*, suggesting that GALLS is translocated to plant cells where it substitutes for VirE2 function (Hodges et al., 2004). GALLS shows no sequence similarities with VirE2, but contains several domains for NTP binding, nuclear localization in the plant, and T4S secretion (see below). Recently, Ream et al. confirmed that GALLS indeed is translocated to plant cells by use of the CrAFT assay (Hodges et al., 2006). See chapter 10 for updated information on this interesting protein.

In addition to protein effectors substrates that contribute to the infection process – VirE2, VirE3, VirF, and VirD5 – the *A. tumefaciens* VirB/D4 T4S system also translocates GALLS, the VirD2 and RSF1010 MobA relaxases independently of associated DNA, and Msi059 and Msi061, two substrates of a related T4S system carried by *Mesorhizobium meliloti* (Hubber et al., 2004).

3.3 Secretion signals

What is the nature of the secretion signal conferring substrate recognition by the VirB/D4 T4S system? A common feature among the above substrates is the presence of C-terminal clusters of positively-charged amino acids, e.g., Arg-x-Arg motifs (Vergunst et al., 2005). These residues are important for transfer, as shown by analyses of various substrate mutants by CrAFT (Vergunst et al., 2005). Further, the C-terminal 11 residues of GALLS and 10 residues of VirF have been shown to confer weak transfer of Cre, whereas the last 27 residues of both proteins constituted strong transport signals (Vergunst et al., 2005; Hodges et al., 2006). These are the smallest sequences identified to date that confer detectable substrate transfer. Consistent with these findings, an *A. tumefaciens* strain producing a VirE2 variant deleted of C-terminal residues is avirulent, yet a *virE2* mutant strain can nevertheless incite tumor formation on transgenic plants engineered to produce such VirE2 truncation derivatives (Simone et al., 2001). These findings suggest that the C-terminal residues are required for translocation but not for VirE2 function in the plant. It was also shown that as little as 100 C-terminal residues of VirE2 is sufficient for recruitment to the polar-localized VirD4 receptor (see below), as monitored with a GFP tag (Atmakuri et al., 2003). With the exception of positive-charge residues, the putative C-terminal secretion signals do not possess other discernible sequence signatures. Thus, the VirB/D4 T4S system very likely recognizes secretion substrates through C-terminal charge-based interactions.

For several T4S systems, there is accumulating evidence that additional motifs located elsewhere on native protein substrates also are important for transfer. Such motifs might participate in substrate recognition or serve as discrimination signals for controlling the relative amounts or the temporal order of substrate transfer. As described below, VirB9 is postulated to comprise a distal portion of the secretion channel. By mutational analysis, it was shown that VirB9 has the capacity to discriminate between different

DNA substrates (Jakubowski et al., 2005). One class of VirB9 mutations was identified that selectively block translocation of the VirD2-T-strand and not RSF1010 transfer intermediate, whereas another class exerted the opposite effect. Both types of substrate discrimination mutations mapped predominantly to the N-terminal third of VirB9. Both VirD2 and MobA carry similar C-terminal Arg clusters, suggesting that this region of VirB9 selectively regulates passage of these two substrates through recognition of other motifs carried by these relaxases.

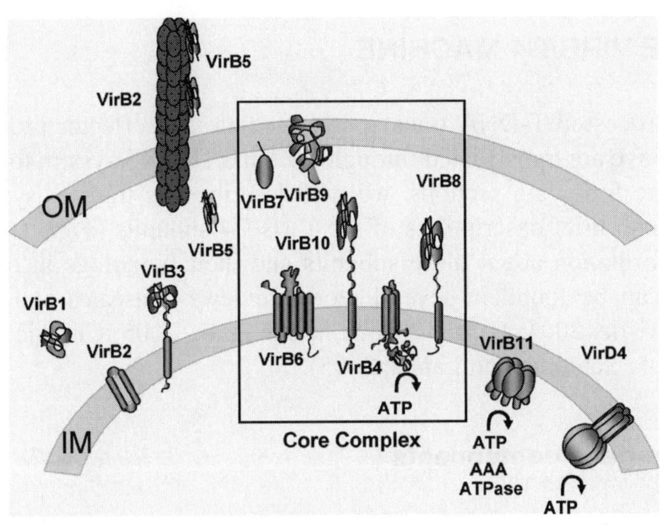

Figure 9-2. Topologies and cellular locations of the VirB/D4 T4S subunits at the *A. tumefaciens* cell envelope. Proteins that are highly conserved and form a stable assembly intermediate, are boxed and denoted as a 'core complex'. All proteins are required for channel activity, and only the VirB proteins are required for polymerization of the VirB2 pilin as the extracellular T-pilus. X-ray structures are known for homologs of VirD4, the VirB11 AAA ATPase, VirB5, and soluble domains of VirB8 and VirB10. An NMR structure has been developed for a co-complex of a VirB7 ortholog and the C-terminal domain of a VirB9 homolog. IM, inner membrane; OM, outer membrane. See text for details.

3.4 Inhibitors of VirB/D4-mediated substrate translocation

Two factors have been shown to suppress virulence of *A. tumefaciens*, the Osa fertility inhibition factor of IncW plasmid pSa and the MobA-R-strand transfer intermediate of IncQ plasmid RSF1010 (Ward et al., 1991; Chen and Kado, 1994; Binns et al., 1995; Chen and Kado, 1996; Lee et al., 1999). Recently, a combination of biochemical and cytological evidence was presented indicating that both inhibitors act by blocking VirD4 receptor access to both T-DNA and protein substrates (*Figure 9-1*; see below and Cascales et al., 2005). RSF1010 is a substrate of this system and, indeed, processing to form the MobA-R-strand is necessary for the 'oncogenic suppression' (Binns et al., 1995). RS1010 therefore probably competes with the native substrates of this system. Osa is not a secretion substrate, but still might exert its inhibitory effect through binding VirD4 at the substrate interface, though other mechanisms are possible.

4 THE VIRB/D4 MACHINE

The processed T-DNA transfer intermediate and effector proteins discussed above are translocated through the VirB/D4 T4S system to the plant cell. In the following sections, we will describe this transfer system, beginning with brief descriptions of the VirB/D4 subunits (*Figure 9-2*). Detailed information about these subunits and their homologs in other T4S systems can be found in several recent reviews (Cascales and Christie, 2003; Christie, 2004; Baron, 2005; Christie et al., 2005; Christie and Cascales, 2005; Schroder and Lanka, 2005).

4.1 Energetic components

Three subunits, VirD4, VirB11, and VirB4, carry conserved Walker A motifs required for function. All three subunits bind ATP, and VirB11 and VirD4 or homologs of these subunits also hydrolyze ATP (Christie et al., 1989; Krause et al., 2000b; Tato et al., 2005). These subunits are postulated to supply the energy required for channel or pilus assembly or delivery of secretion substrates to the cell surface.

4.1.1 VirD4

VirD4 is a member of a family of ATPases related to the SpoIIIE and FtsK DNA translocases (Gomis-Ruth et al., 2004). Farrand et al. constructed chimeric T4S systems composed of homologs of VirD4 from one T4S system and VirB-like components from a second T4S system (Hamilton et al., 2000). Such chimeric T4S systems were shown to be functional, and, furthermore, these systems exported substrates characteristically translocated by the system from which the VirD4-like protein was derived. These findings strongly suggest that VirD4 and its homologs link the Dtr processing proteins bound at *oriT* - the relaxosome - to the T4S system. Consequently, these proteins have been termed 'coupling proteins' or 'substrate receptors'.

VirD4 homologs have been purified and shown to bind DNA *in vitro* (Moncalian et al., 1999; Hormaeche et al., 2002; Schroder et al., 2002; Schroder and Lanka, 2003). Single-stranded (ss)-DNA is preferred over double-stranded (ds)-DNA, yet binding is also sequence non-specific, suggesting that these proteins recognize DNA substrates by virtue of interactions with the relaxase or other relaxosomal subunits bound at *oriT* or the T-DNA border sequences. VirD4 probably interacts with VirD2, although as discussed above ParA-like VirC1 mediates this interaction. The role of VirD4 as a substrate receptor has been confirmed for both T-DNA and protein substrates (see below).

Structures of soluble domains of two VirD4-like proteins have now been solved by X-ray crystallography, one of TrwB encoded by plasmid R388 and one of *E. coli* FtsK. TrwB presents as six equivalent protomers assembled as a spherical particle of overall dimensions 110 Å in diameter and 90 Å in height. This ring-like structure possesses a central channel of 20 Å in diameter, constricted to 8 Å at the entrance facing the cytoplasm (Gomis-Ruth et al., 2001). This channel traverses the structure, possibly connecting cytoplasm with periplasm. An appendix corresponding to the N-terminal TM domain was discernible by image averaging of electron micrographs. This overall structure bears a striking resemblance to the F1-ATPase $\alpha3\beta3$ heterohexamer, whereas the structure of the soluble domain closely resembles DNA ring helicases and other proteins, such as FtsK, that translocate along DNA. The FtsK structure is slightly larger with an outer diameter of 120 Å and a central annulus of 30 Å (Massey et al., 2006). The predicted structure is a dodecamer composed of two hexamers stacked in a head-to-head arrangement. As shown by electron microscopy imaging, double-stranded DNA runs through the FtsK annulus, providing a structural view of a previously described ATP-dependent translocase activity (Saleh et al., 2004).

Thus, VirD4 functions as a receptor for the T-DNA and protein substrates of the VirB/D4 T4S system, and it might also function as an inner membrane translocase. Another interesting feature of VirD4 is that it localizes at the cell poles of *A. tumefaciens*; this will be discussed in more detail below.

4.1.2 VirB11

VirB11 is a member of a large family of ATPases associated with systems dedicated to secretion of macromolecules (Krause et al., 2000a;

Planet et al., 2001). Purified homologs TrbB, TrwC, *H. pylori* HP0525, and *B. suis* VirB11 assemble as homohexameric rings discernible by electron microscopy, and the last two also by X-ray crystallography (Krause et al., 2000a; Yeo et al., 2000; Savvides et al., 2003; Hare et al., 2006). These structures present as double-stacked rings formed by the N- and C-terminal halves and a central cavity of ~50 Å in diameter. *A. tumefaciens* VirB11 was modeled on HP0525, but recent evidence suggests it is more structurally similar to *B. suis* VirB11 (Hare et al., 2006). The most recent update to this structural information is that *B. suis* VirB11 is configured such that the nucleotide binding site is composed of the N-terminal domain of one monomer and the C-terminal domain of the next monomer in the hexamer. This domain swap likely ensures a coordination of ATP utilization among the subunits of the hexamer to drive machine assembly or activity (Hare et al., 2006). VirB11 associates peripherally but tightly with the inner membrane of *A. tumefaciens*. Mutants of VirB11 bearing Walker A nucleotide-binding motif substitutions bind the inner membrane more tightly than the wild-type protein, suggestive of an ATP-regulated membrane interaction (Rashkova et al., 1997; Rashkova et al., 2000). Despite the accumulation of structure – function data over the past few years, the role of VirB11 in T4S is still fundamentally unknown.

4.1.3 VirB4

VirB4 subunits are large inner membrane proteins with consensus Walker A and B nucleoside triphosphate-binding domains (Rabel et al., 2003). VirB4 homologs are extensively distributed among T4S systems of Gram-negative and Gram-positive bacteria. A VirB4 topology model was generated by extensive reporter fusion and protease susceptibility studies; this model depicts VirB4 as predominantly cytoplasmic with possible periplasmic loops, one near the N terminus and a second just N-terminal to the Walker A motif (Dang and Christie, 1997). These experimental findings are consistent with computer-based predictions and results of fractionation studies of the TrbE homolog from RP4 (Rabel et al., 2003). These findings are also consistent with an *in silico* analysis of the VirB4 structure, based on observed sequence similarities between the C-terminal residues 426 to 787 of VirB4 and TrwB of plasmid R388 (see below), that placed VirB4 at the entrance to the VirB/D4 channel (Middleton et al., 2005). However, recently, a completely different model based primarily on yeast two-hybrid interaction data placed VirB4 almost entirely in the periplasm (Draper et al., 2006). This seems improbable both on grounds of the reporter fusion protein data and the fact that no other protein with conserved nucleotide binding motifs have yet been shown to be located in the periplasm.

4.2 Inner-membrane translocase components

Several VirB proteins are postulated to assemble with VirD4 to form the inner membrane translocase.

4.2.1 VirB6

VirB6 is a highly hydrophobic inner membrane proteins with multiple predicted TMS. A combination of reporter fusion and Cys accessibility studies of functional substitution mutants support a topology model consisting of a periplasmic N terminus, five TMS, and a cytoplasmic C terminus (Jakubowski et al., 2004). A particularly notable feature of the topology model is loop P2, a large central periplasmic loop whose secondary structure appears important for DNA substrate translocation (see below). Homologs of VirB6 display relatively low overall sequence similarities with the exception of a conserved region corresponding to residues ~170 to 205 that includes an invariant Trp residue required for protein function (Judd et al., 2005c). VirB6 has been shown to stabilize other VirB proteins, notably, VirB3, VirB5, and a VirB7 homodimer species, and it is also participates in some way in the formation of an outer membrane-associated VirB7-VirB9 heterodimer (Hapfelmeier et al., 2000; Jakubowski et al., 2003). Two mutational analyses have begun to define domains and residues required for VirB6 function (Jakubowski et al., 2004; Judd et al., 2005c). Results of these analyses suggest VirB6 is part of the inner membrane translocation channel.

4.2.2 VirB8

VirB8 is a bitopic inner membrane protein with an N-proximal TMS. X-ray structures of periplasmic domains of the *A. tumefaciens* and *B. suis* VirB8 subunits have been solved and both present as a large extended β-sheet of four antiparallel strands juxtaposed to five α-helices, giving rise to an overall globular fold (Terradot et al., 2005; Bailey et al., 2006; Paschos et al., 2006). Conserved residues important for protein function are buried in the hydrophobic core, where they are predicted to contribute to VirB8 structural integrity. Other conserved residues are surface exposed and might mediate contacts with VirB8 partner subunits. VirB8 assembles as a homodimer, and also interacts with several other VirB subunits, including

VirB4, VirB5, VirB9, and VirB10 (Kumar and Das, 2001; Ward et al., 2002).

4.2.3 VirB10

VirB10 also is a bitopic inner membrane protein situated with the bulk of the protein in the periplasm. After the TMS, most homologs possess a Pro-rich region, which is predicted to form an extended structure in the periplasm. A crystal structure is available for a periplasmic fragment corresponding to residues 146 to 376 of the *H. pylori* ComB10 subunit (Terradot et al., 2005). The structure presents as an extensively modified β-barrel with an α-helix projecting off one side and a second, flexible helix-loop-helix of 70 Å in length projecting off the top. This structure is compatible with recent evidence that VirB10 senses ATP energy use by the inner membrane proteins VirD4 and VirB11 for a dynamic association with the outer membrane protein VirB9 (Cascales and Christie, 2004a). Like VirB8, VirB10 establishes multiple contacts with several T4S channel subunits, including VirD4, VirB4, VirB8, and VirB9 (Beaupre et al., 1997; Das and Xie, 2000; Ding et al., 2002; Ward et al., 2002; Llosa et al., 2003; Atmakuri et al., 2004; Jakubowski et al., 2005).

4.2.4 VirB3

VirB3 is a short polypeptide with one or two predicted transmembrane domains (TM's) near the N terminus. An early study reported that the VirB4 ATPase contributes to the localization of VirB3 at the outer membrane (Jones et al., 1994). However, VirB3 lacks an N-terminal signal sequence and, furthermore, a BLAST search identified a phylogenetic clade in a subset of T4S systems in which VirB3 and VirB4 are fused as a single polypeptide. (Christie et al., 2005). It is intriguing to speculate that VirB3 interacts with and transduces ATP energy from the cytosolic VirB4 ATPase into the periplasm for some aspect of pilus biogenesis or substrate transfer.

4.3 Periplasmic/outer-membrane channel components

Several subunits are exported to the periplasm or outer membrane where they likely form the distal portion of the secretion channel.

4.3.1 VirB1

VirB1 is a member of a large family of subunits commonly associated with macromolecular surface structures, including the T2S, T3S, and T4S systems, type IV pili and flagella, DNA-uptake systems, and bacteriophage entry systems (Koraimann, 2003). The signature for this protein family is a lysozyme-like structural fold. In *A. tumefaciens*, VirB1 is not essential for T-DNA transfer, though it does augment transfer efficiencies and is also important for production of the T-pilus (Berger and Christie, 1994; Lai et al., 2000). The dispensibility of VirB1 for substrate transfer suggests the VirB/D4 channel can assemble through holes in the peptidoglycan generated by alternative murein hydrolases or during remodeling of the peptidoglycan, presumably during a specific phase of cell growth. Supporting the former possibility, VirB1 orthologs from *Brucella suis* and pKM101 complement a *virB1* null mutation, restoring T-pilus production and T-DNA transfer to WT levels (Hoppner et al., 2004). Interestingly, VirB1 from the nopaline Ti plasmid appears to be proteolytically processed, and the C-terminal 73 residues, termed VirB1*, is released to the exterior of the cell (Llosa et al., 2000). The N- and C-terminal portions of VirB1 were reported to independently enhance tumorigenesis of strain C58. By contrast, VirB1 from the octopine TiA6 plasmid is not proteolytically cleaved and no processed form of VirB1 is detected in the extracellular fraction (P. J. Christie, unpublished data). The function of nopaline VirB1* is not known, nor is the basis for the strain-specificity of the putative VirB1 processing reaction.

4.3.2 VirB5

VirB5 subunits are exported to the periplasm, and they also localize extracellularly as components of the pilus (Schmidt-Eisenlohr et al., 1999). An X-ray crystallography structure of one family member, TraC of pKM101, presents as a single-domain protein with a mostly α-helical, elongated structure (Yeo et al., 2003). Evidence from studies of VirB5 and homologs in other systems suggest the pilus-associated forms of VirB5-like subunits might contribute to target cell attachment (Schmidt-Eisenlohr et al., 1999; Schmid et al., 2004). In addition to its extracellular function, the periplasmic form of VirB5 is required for T-DNA translocation to the cell surface (see below).

4.3.3 VirB2

VirB2 is the major pilin subunit of the *A. tumefaciens* VirB/D4 T-pilus and an essential component of the secretion channel (Jones et al., 1996; Lai and Kado, 1998). VirB2 is a small, hydrophobic subunit with an unusually long signal sequence and two hydrophobic stretches. Recent studies have described two fundamentally important properties of VirB2. First, both VirB2 and a homolog, TrbC encoded by plasmid RP4, are processed to form cyclic polypeptides (Eisenbrandt et al., 1999). TrbC is cyclized through the action of the serine protease TraR, and VirB2 is cyclized by an unknown chromosomal enzyme (Eisenbrandt et al., 2000). Second, specific mutations in the VirB11 ATPase (see below) and VirB9 have been shown to block polymerization of VirB2 pilin monomers to form the extracellular T-pilus, yet these mutations do not abolish substrate transfer through the VirB/D4 T4S channel (Zhou and Christie, 1997; Sagulenko et al., 2001a; Jakubowski et al., 2005). In follow-up studies, it was shown strains producing the so-called "uncoupling" mutations still require VirB2 for substrate transfer, e.g., these strains do not bypass the requirement for VirB2 pilin for substrate transfer (Jakubowski et al., 2005). Isolation of these mutations thus constitute an important line of evidence that VirB2 alternatively polymerizes as the T-pilus and as a component of the secretion channel (see below).

4.3.4 VirB7

VirB7 is a small lipoprotein required for assembly of the VirB/D4 T4S system. VirB7 localizes predominantly at the outer membrane, although both inner-membrane-associated and extracellular forms also have been detected (Fernandez et al., 1996a). Extracellular VirB7 copurifies with the T-pilus but is also recovered from the supernatant of pilus-minus cells (Sagulenko, 2001b). VirB7 stabilizes other VirB proteins, in part through formation of a disulfide bridge with the outer membrane-associated VirB9 subunit (Anderson et al., 1996; Fernandez et al., 1996b; Spudich et al., 1996; Baron et al., 1997). An NMR structure recently was presented for a co-complex consisting of homologs of VirB7 and VirB9 from the plasmid pKM101 (Bayliss et al., 2007). This structure and other experimental findings strongly suggest VirB7 and the C-terminal region of the VirB9 are situated at the inner leaflet of the outer membrane.

4.3.5 VirB9

VirB9 is generally hydrophilic and contains a number of predicted β-strands. Phylogenetic and mutational analyses supplied evidence that VirB9 is composed of three functional domains, each approximately 80 to 100 residues (Jakubowski et al., 2005). The above-mentioned NMR structure shows that VirB9 forms a β-sandwich around which VirB7 winds (Bayliss et al., 2007). Also, a 3-stranded β-appendage appears to protrude extracellularly, as judged by results of Cys accessibility and immunofluorescence microscopy assays. Whether other N-proximal regions of VirB9 also protrude across the outer membrane remains to be tested. Computer searches have identified sequence similarities between VirB9 and outer membrane pore-forming proteins termed secretins in the types II and III secretion systems (Cao and Saier, 2001; Thanassi, 2002; Lawley et al., 2003). These observations suggest the possiblity that VirB9 oligomerizes to form ring-shaped pores through which protein substrates or the T-pilus pass to the cell surface.

5 VIRB/D4 MACHINE ASSEMBLY AND SPATIAL POSITIONING

Newly-synthesized VirB and VirD4 proteins nucleate to form the secretion channel and T-pilus, and this process is very likely ordered in space and time. Studies exploring the VirB/D4 assembly pathway have developed along several lines. First, early analyses of mutants deleted of a single *virB* gene led to the discovery that certain VirB subunits have destabilizing effects on other VirB proteins (Berger and Christie, 1994; Fernandez et al., 1996b; Beaupre et al., 1997; Hapfelmeier et al., 2000; Jakubowski et al., 2003). Further studies exploring these stabilizing interactions led to formulation of a stabilization pathway that might represent discrete stages of machine assembly at the cell envelope (*Figure 9-3*). Second, capitalizing on findings that the VirB/D4 transfer system assembles at *A. tumefaciens* cell poles (Lai and Kado, 2000), an assembly pathway was developed based on VirB subunit requirements for spatial positioning of individual VirB subunits . Finally, studies exploring the contributions of ATP energy to biogenesis have identified contributions of all three ATPases of this system – VirB4, VirB11, and VirD4 – to assembly of the T-pilus and the secretion channel (Cascales and Christie, 2004a; Yuan et al., 2005).

5.1 A VirB/D4 stabilization pathway

Deletions of several VirB proteins result in destabilization of other VirB proteins (Berger and Christie, 1994). Most strikingly, however, a *virB7* deletion correlated with the absence or striking reduction in levels of most VirB proteins (Fernandez et al., 1996b). VirB7 synthesis was most strongly correlated with stabilization of VirB9, and subsequent studies established that VirB7 interacts with VirB9 through formation of a stabilizing disulfide bridge (Anderson et al., 1996; Spudich et al., 1996; Baron et al., 1997; Beaupre et al., 1997). Assembly of the VirB7-VirB9 dimer or higher-order species at the outer membrane in turn was shown to be important for stabilization of other VirB channel components, including several at the inner membrane, e.g., VirB4, VirB10, and to a lesser extent VirB11 (Fernandez et al., 1996b). Subsequently, studies of native and mutant forms of VirB6 supplied evidence that VirB6 contributes to stabilization of other VirB proteins including VirB3 and VirB5, and also participates in assembly of a VirB7 homodimer species and the VirB7-VirB9 heterodimer (Hapfelmeier et al., 2000; Jakubowski et al., 2003). These and other findings resulted in a stabilization pathway depicted in *Figure 9-3*, In this model, VirB6 promotes assembly of the VirB7-VirB9 dimer and this in turn stabilizes VirB4, VirB8, and VirB10 in the inner membrane. To this 'core' complex, VirB2, VirB3, and VirB5 are added to complete assembly of a transenvelope structure. The VirB11 ATPase, whose production does not affect stabilities of the other VirB proteins, is postulated to interact with the transenvelope structure to initiate polymerization of the T-pilus across the cell envelope and beyond. The VirD4 receptor is not required for stability of VirB proteins, nor assembly of the T-pilus. Hence, the proposed pathway bifurcates so that the core transenvelope structure is used to build the T-pilus, or by addition of VirD4 the secretion channel.

5.2 Polar localization of the T-DNA transfer system

A. tumefaciens cells attach at their poles to plant cells (Matthysse, 1987). The T-pili also can be detected at cell poles, suggesting that the VirB/D4 transfer apparatus assembles at cell poles (Lai and Kado, 2000). Through a combination of electron and immunofluorescence (IFM) microscopy and analyses with the green fluorescent protein (GFP), studies have confirmed that VirB proteins localize at cell poles, though several proteins also form foci distributed around the cell surface (Kumar et al., 2000; Kumar and Das, 2002; Jakubowski et al., 2004; Judd et al., 2005b). A recent study identified a dependency of 6 VirB proteins - VirB1, VirB5-VirB7, VirB9, and VirB10 - on production of VirB8 for polar localization, whereas VirB3, VirB4, and VirB11 localize at cell poles independently of VirB8 (Judd et al., 2005a). The VirB4 and VirB11 ATPases are not required for polar targeting of other VirB proteins and VirB4 and VirB11 Walker A mutants display WT localization patterns, suggesting that nucleation of the VirB proteins at the cell pole does not require ATP energy. Taken together, data acquired through studies exploring the requirements for polar targeting of the VirB protein support an assembly pathway for the T4S apparatus depicted in *Figure 9-3*. The two assembly pathways, postulated on the basis of stabilization data and positional information, share the feature that a core apparatus composed of VirB6 or VirB7 through VirB10 assembles without a requirement for ATP binding or hydrolysis by VirB4 or VirB11 (*Figure 9-3*). The pathways differ in the specifics of subsequent reactions, but it should also be kept in mind that once the core complex is formed, this transenvelope structure might simply represent a nucleation center for other VirB subunits without a specific temporal order.

As noted above, assembly of the secretion channel requires formation of productive contacts between VirD4 and the VirB complex. VirD4 localizes at cell poles, independently of the VirB proteins (Kumar and Das, 2002; Atmakuri et al., 2003). Moreover, as noted above, the VirC proteins are polarly-localized, also independently of VirD4 or the VirB proteins. Thus, at least three protein complexes must be spatially positioned for T-DNA transfer through the VirB/D4 transfer apparatus: (i) the relaxosome bound at T-DNA border sequences and comprised of VirD1, VirD2, VirC1, and VirC2, (ii) the VirD4 substrate receptor, and (iii) the VirB channel complex (*Figure 9-3*). Adding to this picture, Kahng and Shapiro reported that the Ti plasmid itself localizes at or near the cell poles of *A. tumefaciens* vegetative cells (Kahng and Shapiro, 2003) and it was recently confirmed that the Ti plasmid also is localized at the poles of AS-induced cells (Atmakuri et al., 2007). It will be interesting to decipher the underlying targeting systems for the Ti plasmid, the T-DNA processing proteins, the VirD4 receptor and the VirB channel components, and to define how these various cellular constituents coordinate their activities in space and time to achieve T-DNA transfer.

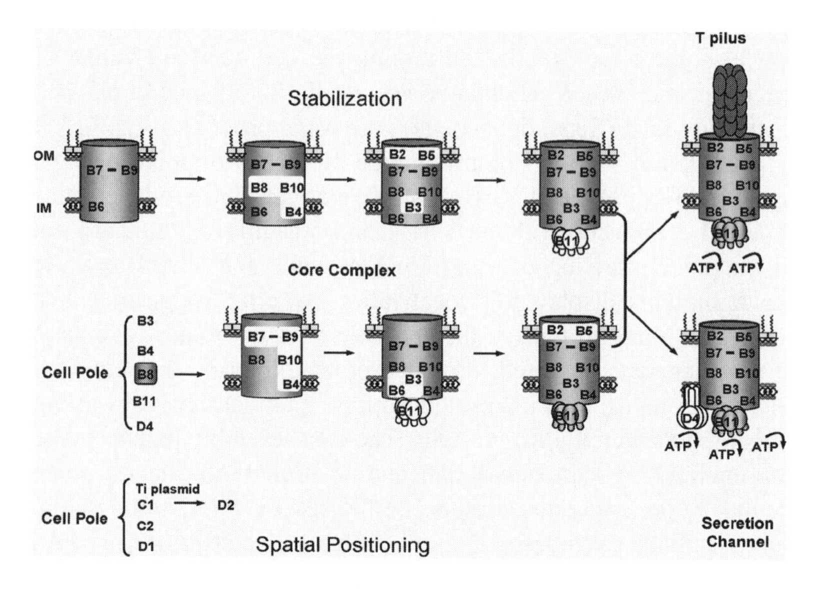

Figure 9-3. Assembly pathways for the VirB/D4 T4S machine based on stabilization data or requirements for localization of VirB subunits at cell poles. Five channel subunits listed at left localize intrinsically at the cell pole. VirB8 (shaded) is thought to nucleate assembly of the secretion system through recruitment of subunits highlighted in gray. These subunits in turn recruit others, again shaded in gray, in staged reactions. Both assembly pathways depict a late-stage bifurcation such that the VirB transenvelope complex elaborates either the extracellular T-pilus through ATP binding or hydrolysis of VirB4 and VirB11 or the secretion channel through engagement with VirD4 and ATP energy use by VirB4, VirB11, and VirD4. In addition to the polar localization of VirB and VirD4 T4S subunits, the Ti plasmid and the VirC and VirD processing factors localize at cell poles; ParA-like VirC1 recruits VirD2 to the cell poles.

5.3 Latter-stage reactions required for machine assembly and substrate transfer

Once the VirB subunits assemble as a stable transenvelope structure at the cell pole, ATP energy is needed to drive biogenesis of the T-pilus and the secretion channel. Two ATPases, VirB4 and VirB11, mediate assembly of the T-pilus, whereas VirB4, VirB11 and VirD4 mediate assembly of the secretion channel. These different energetic requirements suggest that at a late stage in the assembly pathway the membrane-spanning VirB complex serves as a platform for pilus production or alternatively is configured as the secretion channel (*Figure 9-3*).

5.3.1 *VirB4 and VirB11 mediate T-pilus assembly*

Both VirB4 and VirB11 are required for biogenesis of the T-pilus. Mutations in the VirB11 Walker A motif invariably abolish pilus biogenesis. A couple of mutations elsewhere in the protein also abolish pilus biogenesis but do not block translocation of T-DNA or protein substrates (Rashkova et al., 1997; Sagulenko et al., 2001a). Interestingly, a substitution of Arg for the invariant Lys residue in the Walker A motif of VirB4 from the nopaline Ti plasmid did not abolish T-pilus formation (Yuan et al., 2005). The conservative nature of the K439R substitution suggests the mutant protein might still bind but not hydrolyze ATP. Other Walker A mutations, including a three-residue deletion, almost certainly abolish both ATP binding and hydrolysis, and such mutations completely block pilus production (Berger and Christie, 1993). These findings raise the possibility that VirB4 ATP binding but not hydrolysis might be necessary for pilus production. However, further studies are needed before such a conclusion can be drawn, because studies in our laboratory determined that the native form of VirB4 from the nopaline pTiC58 plasmid – but not the K439R mutant – fully complements a *virB4* null mutation in the octopine pTiA6 plasmid of strain A348 with respect both to secretion activity and T-pilus biogenesis (L. Coutte and P. J. Christie, unpublished data). For reasons to be explored, the K439R mutant protein supports T-pilus production only in a nopaline Ti plasmid genetic background.

5.3.2 *VirD4 and VirB11 induce assembly of a stable VirB10-VirB9-VirB7 channel complex*

As discussed above, *A. tumefaciens* VirB10 is a bitopic inner membrane protein whose large C-terminal domain resides in the periplasm and interacts with the OM-associated VirB9-VirB7 dimer complex. VirB10 was shown to undergo a structural transition in response to ATP utilization by VirD4 and the VirB11 ATPase (Cascales and Christie, 2004a). VirB10 interacts with VirD4 independently of ATP energy, but must undergo the structural transition to stably interact with the VirB7-VirB9 multimer. Mutations that abolish the VirB10-VirB9-VirB7 interaction fail to translocate DNA (Cascales and Christie, 2004a and V. Krishnamoorthy and P.J. Christie, unpublished data). VirB10 energization by VirD4 and VirB11 thus appears necessary for a late stage of machine biogenesis required for T-DNA substrate passage to the cell exterior. VirB10 resembles the TonB/TolA proteins in overall structure and membrane topology, and in sensing energy through contacts with inner membrane partner proteins.

TonB transduces energy from the electrochemical gradient to gate outer membrane transporters; similarly, VirB10 might sense and transduce ATP energy to trigger assembly or gating of an outer membrane pore composed of VirB7 and VirB9.

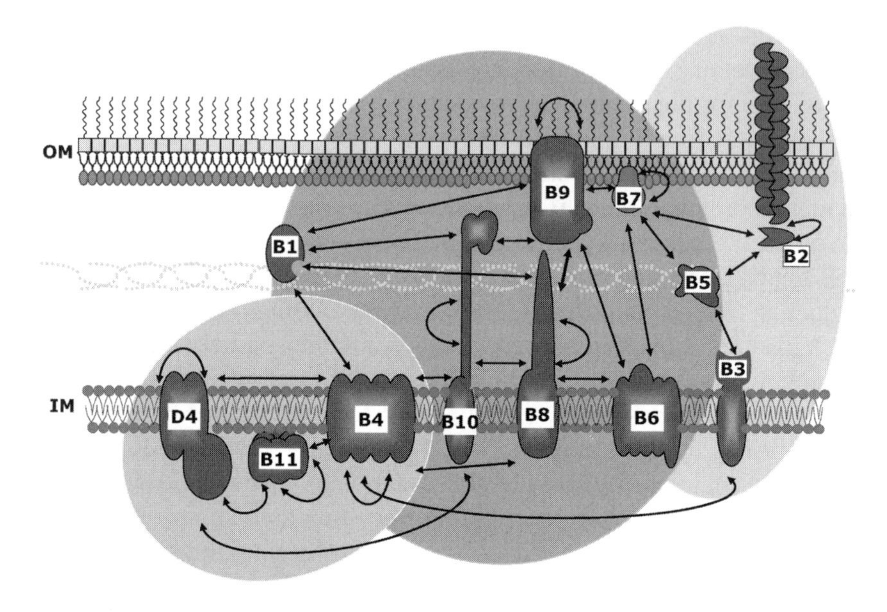

Figure 9-4. Protein interaction network among the VirB/D4 T4S subunits. The localization and topologies of the VirB/D4 subunits are represented. Shaded areas correspond to sub-complexes – energy, core, pilus – described in the text. Arrows indicate interactions detected among the subunits by biochemical/structural approaches and yeast or bacterial two-hybrid screens. See text for references.

5.4 Interactions among the VirB/D4 T4S subunits

Many subunit – subunit interactions have been reported with yeast two hybrid screens (Das et al., 1997; Ward et al., 2002; Draper et al., 2006), and though some pairwise contacts have been substantiated with independent genetic or biochemical approaches other putative interactions need to be experimentally confirmed. When complementary approaches have been used to characterize the *A. tumefaciens* VirB/D4 system and the related *B. suis* VirB system (Anderson et al., 1996; Baron et al., 1997; Beaupre et al., 1997; Dang et al., 1999; Schmidt-Eisenlohr et al., 1999; Yeo et al., 2003; Yuan et al., 2005; Paschos et al., 2006), results suggest the channel subunits form an interaction network as depicted in *Figure 9-4*. Of course, molecular details of these interactions will continue to be described at increasing resolution, and further work also should reveal the structurally dynamic nature of the machine as a function of substrate binding or contact with the target cell.

6 VIRB/D4 CHANNEL/PILUS ARCHITECTURE

With the available information on the VirB channel membrane topologies and structures, stabilizing and spatial positioning interrelationships, energy-mediated structural changes, and subunit – subunit interactions, the architecture of the VirB/D4 T4S system can be depicted as shown in *Figure 9-5*. According to this model, the secretion channel consists of three ATPases, at least two of which are structurally configured as homohexameric complexes, positioned at the cytoplasmic face of the channel entrance. These subunits interact with polytopic VirB6 and the bitopic inner membrane proteins VirB8, VirB10 and, possibly, VirB3. These subunits might also interact with the inner membrane form of VirB2 pilin. The periplasmic domains of the bitopic proteins and a central loop domain of VirB6, together with VirB2 and VirB5, comprise the portion of the channel spanning the periplasm. The bitopic subunits establish contact with a periplasmic domain(s) of VirB9, which, together with the VirB7 lipoprotein form an outer membrane-spanning complex. The composition of channel spanning the outer membrane is not known and here is postulated to consist of VirB7 and VirB9.

An interesting feature of the model, based in part on the discovery that VirB10 is an ATP energy sensor, is that both the secretion channel and the T-pilus are structurally dynamic surface organelles (*Figure 9-5*). For the secretion channel, at two signals – substrate and host cell binding - might regulate structural transitions. For example, in the absence of available substrate or host cell binding the channel would exist in a quiescent state. In this configuration, the channel consists of stable inner and outer membrane machine subassemblies that are only loosely associated with each other. Structural flexibility of a transenvelope macromolecular complex might be important for cell envelope remodeling during the cell growth/division cycle. T-DNA substrate docking with VirD4 and possibly also host cell binding would trigger a cascade of events that include: (i) VirD4 and VirB11 ATP binding/hydrolysis, (ii) VirB10 energy sensing, and (iii) stabilization of the transenvelope VirB complex with accompanying structural changes required for channel assembly or gating. Following translocation, another signal related to substrate passage or dissociation of the mating junction, would trigger conversion of the channel to the quiescent complex. This overall model of the T4S channel existing in a dynamic equilibrium between quiescent and activated conformations is reminiscent of a model postulated for ABC transporters on the basis of findings that substrate binding stimulates late-stage stabilization and contacts between the inner and outer membrane ABC components as a prerequisite to substrate transfer (Letoffe et al., 1996). Though there presently is no evidence for substrate-induced assembly of the VirB/D4 system, ssDNA has been shown to stimulate ATP binding and hydrolysis of R388 TrwB, a homolog of the VirD4 substrate receptor (Tato et al., 2005). T-DNA substrate docking with VirD4 might stimulate ATP hydrolysis and in turn initiate the above morphogenetic reactions.

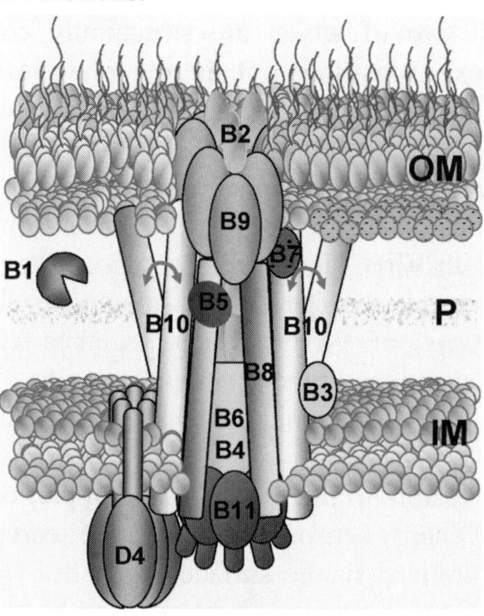

Figure 9-5. A proposed architecture for the VirB/D4 secretion channel. VirB10 senses ATP energy use by VirD4 and VirB11 and, in turn, stably interacts with the outer membrane-associated VirB7-VirB9 complex. The model depicts the VirB/D4 T4S as structurally dynamic, wherein signals including substrate binding, ATP energy, and target cell contacts trigger structural transitions between a quiescent transenvelope complex and an active transport channel.

For the T-pilus, we suggest a VirB inner membrane complex serves as a platform for repetitive cycles of T-pilus polymerization and sloughing. One signal, possibly associated with ATP binding and/or hydrolysis by the VirB4 and VirB11 ATPases, would trigger polymerization of the T-pilus from a pool of pilin monomers accumulated either in the inner membrane or the periplasm. Another signal would induce T-pili sloughing to the extracellular milieu, where the pilus fragments would promote *A. tumefaciens* aggregation and attachment to plant cells through hydrophobic interactions. In view of the proposed dynamic structure of the T4S secretion channel and T-pilus, it is interesting that stress-induced factors are involved in T4S machine assembly and function. For example, an *A. tumefaciens lon* mutant displays highly-attenuated virulence suggestive of a T-DNA transfer defect and does elaborate T-pili (Su et al., 2006 and S. Su and S. Farrand, personal communication). Additionally, it was recently shown that assembly of the T4S system of plasmid R16 correlates with activation of cytoplasmic and extracytoplasmic stress responses (Zahrl et al., 2006). Whether stress-activated proteases or other factors might regulate structural transitions associated with T4S machine function is an intriguing question for further study.

7 T-DNA TRANSLOCATION ACROSS THE CELL ENVELOPE

Progress toward understanding how the VirB/D4 channel functions has been made through development of the transfer DNA immunoprecipitation (TrIP) assay (Cascales and Christie, 2004b). This assay was adapted from the chromatin immunoprecipitation (ChIP) technique and as mentioned above involves formaldehyde treatment of intact cells to crosslink channel subunits to the DNA substrate as it exits the cell. The crosslinked substrate – channel subunit complexes are then recovered from detergent-solubilized cell extracts by immunoprecipitation, and PCR amplification is used to detect the precipitated DNA substrate. With this assay, it was shown that a DNA substrate forms close contacts with 6 of the 12 subunits of the *A. tumefaciens* VirB/D4 T4S machine, including the VirD4 receptor, VirB11 ATPase, VirB6, VirB8, VirB2 pilin and VirB9 (*Figure 9-6*). These subunits are spatially positioned from the cytoplasmic face of the inner membrane to the cell surface, prompting a model that the T-DNA translocates through a transenvelope channel composed of these 6 subunits. The remaining VirB proteins do not form close contacts with the T-DNA but nevertheless are required for substrate transfer; these subunits could function as structural scaffolds for the channel. Further TrIP studies with various mutant strains led to formulation of a translocation pathway, defined as an ordered series of substrate contacts with each of the presumptive channel subunits. Below, we summarize the requirements identified to date for formation of T-DNA substrate contacts with VirD4, VirB11, VirB6 and VirB8, and VirB2 and VirB9.

7.1 Substrate recruitment to the T4S system

VirD4 retains WT T-strand binding activity in a *virB* operon mutant strain, suggesting that receptor activity is not modulated through interactions with the VirB channel subunits (Cascales and Christie, 2004b). VirD4K152Q bearing a substitution of the invariant Lys residue within the Walker A motif also retains WT receptor activity (Atmakuri et al., 2003; Atmakuri et al., 2004), indicating that initial substrate binding is not dependent on ATP utilization. This is especially interesting in view of the finding that VirD4K152Q is distributed at sites around the membrane as opposed to the cell pole (K. Atmakuri and P. J. Christie, unpublished data). The result suggests that polar positioning of the substrate receptor is not a prerequisite for initial substrate binding. As described in Section 2.4, two oncogenic suppressors have been shown to block substrate export through the VirB/D4 channel. With the TrIP assay, it was possible to demonstrate that both the RSF1010 MobA-R-strand transfer intermediate and the pSA-encoded Osa fertility inhibitor protein block T-DNA binding to the VirD4 receptor (Cascales et al., 2005). In addition, both oncogenic suppressors block access of VirE2 to VirD4, as shown by a combination of biochemical and cytology-based assays (Atmakuri et al., 2003). Both oncogenic suppressors thus exert their effects by inhibiting substrate access to the VirD4 receptor (*Figure 9-1*).

7.2 Transfer to the VirB11 hexameric ATPase

VirB11 is the second machine component to interact with the T-DNA substrate. Very interestingly, the Walker A mutant forms of VirD4 and VirB11 retain WT T-DNA binding activity, suggesting that ATP binding or hydrolysis is dispensable both for VirD4 receptor activity and for substrate delivery to VirB11. Reconstitution studies further showed that specific subsets of VirB/D4 subunits are required for T-DNA transfer from VirD4 to VirB11. These include a combination of VirB7, VirB8 and VirB9 or, alternatively, VirB7 and VirB10 (Atmakuri et al., 2004). VirD4 therefore delivers the DNA substrate to VirB11 only in the context of a subset of core subunits, but further studies are needed to understand how different combinations of VirB subunits promote a productive VirD4 – VirB11 interaction.

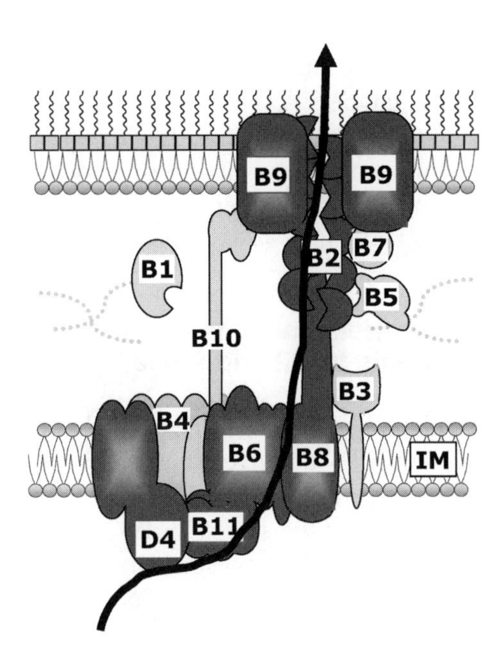

Figure 9-6. Translocation pathway for the VirD2-T-strand transfer intermediate as defined by TrIP. The arrow denotes the transfer route on the basis of close contacts identified between the T-strand and the 6 subunits shaded in dark gray. Other subunits shaded in light gray do not form close substrate contacts but nevertheless are important for substrate transfer; these subunits are proposed to form contacts required for assembly or stability of the secretion channel.

7.3 Transfer to the integral inner membrane proteins VirB6 and VirB8

The TrIP studies place VirB6 and VirB8 at an intermediate point in the postulated DNA translocation pathway, dispensable for DNA transfer to VirD4 and VirB11 but necessary for transfer to VirB2 and VirB9 (Cascales and Christie, 2004b). VirB6 and VirB8 functionally interact, as evidenced by the finding that null mutations in either virB6 or virB8 block substrate transfer to VirB8 or VirB6, respectively. However, certain mutations in VirB6 were shown to permit T-DNA contacts with VirB6 but block contacts with VirB8 (Jakubowski et al., 2004). Hence, the translocation pathway depicts a T-DNA substrate contact first with VirB6 and then with VirB8. Each of the three energetic components – VirD4, VirB4, and VirB11 – each with intact Walker A motifs – are required for substrate transfer to VirB6 and VirB8, suggestive of a coordination of ATP binding or hydrolysis activities for this transfer step (Atmakuri et al., 2004). In addition to VirB6, VirB8, and the 3 ATPases, other subunits including VirB7 and either VirB9 or VirB10 are required for this transfer step. Again, these 'core' subunits probably promote formation of critical contacts between the IM channel subunits, but further studies are needed to decipher the molecular details of these interactions.

7.4 Transfer to the periplasmic and outer-membrane-associated proteins VirB2 and VirB9

The TrIP studies place VirB2 and VirB9 at the distal end of the proposed DNA translocation pathway, given that *virB2* and *virB9* null mutations do not affect DNA transfer to VirD4, VirB11, VirB6, and VirB8. The *virB2* mutation arrests substrate transfer to VirB9 and, conversely, a *virB9* mutation blocks transfer to VirB2, suggesting that VirB2 and VirB9 function together to mediate the latter step(s) of substrate transfer (Cascales and Christie, 2004b). The association of the DNA substrate with a pilin subunit is of special interest. Although this finding could be an indication of substrate transfer through the pilus, the isolation of "uncoupling" mutations (see above) suggests instead that the pilin monomer is both a component of the pilus and the portion of the secretion channel that extends across the periplasm (Sagulenko et al., 2001a; Jakubowski et al., 2005). Recently, the contribution of VirB9 to substrate transfer was assessed by examining the phenotypic consequences of small 2-residue insertion mutations distributed along the length of the polypeptide. Very interestingly, mutations mapping to conserved N- and C-terminal domains blocked substrate transfer to both VirB2 and VirB9, suggesting that these domains might form critical contacts with VirB2 and other channel subunits necessary for latter stages of transfer (Sagulenko et al., 2001a).

7.5 The transfer route

Despite the wealth of information about the VirB/D4 proteins and recent evidence for T-DNA substrate contacts with putative channel subunits, it is still fundamentally unknown how the T-complex and protein substrate traverse the cell envelope and how ATP energy drives translocation. Three models have been presented to describe the substrate transfer route through the VirB/D4 channel (Christie et al., 2005). All models envisage that VirD4 functions as a receptor for the T-DNA and protein substrates. Other proteins might function to couple substrates with the receptor, e.g., ParA-like VirC1 for spatial positioning of the T-complex and VirE1 chaperone for maintaining VirE2 in a translocation-competent conformation. However, the available data indicate that all substrates enter the VirB/D4 transfer system via initial engagement with VirD4.

The first model, mechanistically the simplest, envisages that VirD4 binds the T-complex and then delivers the substrate to the VirB11 ATPase for passage through an inner membrane VirB translocase composed of the VirB6 and VirB8 channel components and VirB4, VirB10, and VirB3 structural scaffolds. The first two reactions, receptor docking and substrate transfer to VirB11, do not require ATP energy. Rather, VirD4, VirB11, and VirB4 together would coordinate their ATP binding/hydrolysis activities to unfold the relaxase and drive the translocation-competent substrate through the VirB translocation channel. This model accommodates most of the available data, though the TrwB F1-ATPase like structure and the recent evidence for DNA protrusion through the central channel of the FtsK hexamer (see above) favors a more active translocase function for the VirD4-like proteins.

The second model, termed the 'shoot-and-pump' model, was set forth by Llosa and colleagues (Llosa et al., 2003). As adapted for the VirB/D4 T4SS in accordance with the TrIP data, this model postulates that VirD4 binds the T-complex, but transfers only the relaxase component of the transfer intermediate to VirB11. VirB11 then unfolds the relaxase through a chaperone activity for delivery through the inner membrane channel composed of the VirB subunits listed above. Simultaneously, VirD4 uses ATP energy to drive the T-strand across the membrane. The complicating aspect of this model is that a single substrate – relaxase covalently bound to T-strand – is delivered across the inner membrane through two distinct translocases. However, it should be noted that the VirD4-like proteins appear to exist as both monomers and higher-order complexes in cells and there is also evidence for DNA-dependent oligomerization. Thus, VirD4 might bind and transfer the T-complex as a monomer and then form a hexameric structure around the T-strand to energize its transfer across the inner membrane simultaneous with relaxase transfer through the VirB channel.

In the third model, termed the 'ping-pong' model, VirD4 binds the T-complex and transfers the relaxase component to the VirB11 chaperone while remaining bound to the T-strand (ping). Next, VirB11 unfolds the relaxase and delivers it back to VirD4 (pong). VirD4 then uses ATP energy to translocate the DNA-protein substrate across the inner membrane. The role of VirB4 ATP energy might be to coordinate substrate transfer from VirD4 to channel subunits in the periplasm. According to this model, the inner membrane VirB channel components would form a structural platform for a VirB secretion channel in the periplasm and also would coordinate substrate transfer from VirD4 into the channel (Atmakuri et al., 2004).

Each of these models also could explain the translocation routes for protein substrates such as VirE2, VirE3, and VirF. Indeed, the latter models also can explain earlier findings that trace amounts of protein substrates can be recovered from periplasmic fractions (Pantoja et al., 2002); following VirD4-mediated translocation across the inner membrane the substrates might pass transiently through the periplasm before gaining access to the VirB channel entrance. Upon transfer across the inner membrane, all models envision that the T-complex and protein substrates enter a secretion channel composed of periplasmic domains of VirB8, VirB9, and VirB2 for transit through the periplasm. The precise route of transfer across the outer membrane is not yet known, though two possibilities include passage through a secretin-like VirB9 channel or the lumen of the T-pilus.

7.6 More jobs than two?

The above discussion presents the VirB proteins as building blocks for a secretion channel or the T-pilus. It is interesting to note, however, that most VirB proteins localize not only at the cell poles but also at discrete sites around the cell surface. The channel subunits might simply accumulate at these sites as dead-end products or assembly intermediates. Another possibility is the surface distributed proteins carry out functions other than those envisioned so far. For example, mating *E. coli* cells examined by transmission electron microscopy have the appearance of being tightly joined not at a specific focus as might be expected if a single T4S channel bound a receptor on the recipient cell surface. Rather, the mating junction extends sometimes up to half the cell lengths. An electron-dense layer can be seen linking the stiffly parallel outer membranes in the junction zone, but there are no cytoplasmic bridges nor apparent breaks in the cell walls or membranes (Samuels et al., 2000). These images suggest that extensive remodeling of the cell envelope accompanies formation of the mating junction. It is interesting to speculate that the nonpolar VirB foci could play a role in remodeling of the *A. tumefaciens* membrane to form extended junctions with bacterial or eukaryotic target cells.

In this context, it is interesting that agrobacterial donor cells carrying the intact VirB/D4 T4S system transfer the IncQ plasmid RSF1010 to agrobacterial recipients at significantly higher frequencies when recipient cells produce a subset of VirB proteins (Bohne et al., 1998). Production of the VirB7 – VirB10 core complex slightly enhances RSF1010 acquisition, but the additional co-production of VirB1 – VirB4 enhance transfer frequencies by 3 to 4 orders of magnitude (Liu and Binns, 2003). Interestingly, this stimulatory effect is not restricted to the *A. tumefaciens* VirB proteins; production of the plasmid RP4 Tra proteins (A. N. Binns, personal communication) or the *B. suis* VirB proteins (Carle et al., 2006) in *A. tumefaciens* recipients also stimulates RSF1010 acquisition by several log orders. These T4S components might provide a conduit for translocation of the RSF1010 transfer intermediate across the recipient cell envelope, but given the apparent non-specificity of the phenomenon with respect to the T4S components, it seems more likely the VirB subunits promote a general remodeling of the recipient cell envelope to favor DNA acquisition. Electron microscopy studies should provide an indication of whether the mating junctions of agrobacterial donor cells with recipients carrying or lacking VirB proteins show any morphological differences.

It has also been shown that the VirB7 lipoprotein can be isolated as a component of a high molecular weight complex exceeding 450 kilodaltons from the extracellular milieu (Sagulenko et al., 2001b). Some VirB7 is associated with the T-pilus in the milieu, but exocellular VirB7 is recovered even from pilus-minus strains. VirB7 is the only VirB protein other than the T-pilus constituents released into the milieu and it is interesting to speculate that this form of VirB7 has an important biological activity during the infection process.

8 THE *AGROBACTERIUM* – PLANT CELL INTERFACE

Currently, there is little information describing the *A. tumefaciens* – plant cell interface. As summarized below, some environmental factors modulate the interkingdom contact through effects on VirB/D4 machine assembly or function. A few candidate plant receptors for the T-pilus have been identified, but clearly further work is needed to understand the mechanism(s) by which T-DNA and protein substrates exit the bacterial cell surface and translocate across the plant plasmamembrane.

8.1 Environmental factors

It has long been known that *A. tumefaciens* VirA/VirG-mediated sensory recognition of plant phenolics and sugars in a low phosphate and acidic (pH=5.5) environment induces the *vir* regulon and synthesis of the proteins mediating T-DNA processing and translocation (McCullen and Binns, 2006). More recently, these environmental conditions were shown to repress flagellum production (Lai et al., 2000). As proposed for *Bordetella bronchiseptica* and other mammalian pathogens, *A. tumefaciens* likely uses flagellar-based motility to locate favorable environmental niches in the rhizosphere, then shuts down motility while inducing the *vir* regulon. Repression of the flagellar genes is achieved through the VirA/VirG two-component system, but details of the regulatory circuitry are unknown. Additionally, environmental conditions leading to *vir* gene induction also upregulate the Ti plasmid *repABC* gene cluster, leading to a 4- to 5-fold increase in plasmid copy number (Cho and Winans, 2005). This effectively increases *vir* gene dosage, which correlates with enhanced cellular accumulation of the processed T-strand. Elevated Ti plasmid copy number might also increase the number of available transport machineries per cell, resulting in enhanced virulence potential.

Temperature is another environmental factor affecting T-DNA transfer. The optimum temperature range for assembly of the VirB/D4 T4S system, as monitored by VirB protein stabilities, T-pilus production, and T-DNA transfer to plants and IncQ plasmid RSF1010 transfer to agrobacterial recipients is between 18 and 23°C. Between 23 and 28°C, *A. tumefaciens* cells show enhanced degradation of the VirB proteins and diminished T-pilus production and substrate transfer. At 28°C, the nopaline strain C58 shows considerable reductions in VirB protein content and T4S function though this is not the case for the octopine strain A348 (Jakubowski et al., 2003) and (Jakubowski et al., 2003 and S. Jakubowski and P. J. Christie, unpublished data), but at or above 30°C, all tested *A. tumefaciens* strains are nonpiliated and avirulent or nearly so. How temperature exerts its effects on the VirB/D4 T4S machine is presently unknown. One possibility is that high temperature elicits an extracytoplasmic stress response, which, in turn leads to degradation of nonessential surface organelles such the T4S machine. As mentioned above, it was recently reported that the *A. tumefaciens* Lon protease contributes to *A. tumefaciens* virulence and T-pilus production. Lon is a member of the heat-shock regulon in *A. tumefaciens,* and thus might be responsible for degradation of off-pathway T4S products, e.g., nonstoichiometrically-produced T4S subunits or dead-end complexes formed at elevated temperatures. Another recent study reported that phosphatidylcholine (PC), which is present in the membranes of *A. tumefaciens* and other alphaproteobacteria, is essential for assembly of the VirB/D4 T4S system (Wessel et al., 2006). PC mutants show defects in *vir* gene expression and assembly of the VirB/D4 machine, and they fail to incite tumor formation on *K. daigremontiana.* In *Yersinia pseudotuberculosis* and *Bacillus cereus*, there are documented effects of temperature on phospholipid composition (Bakholdina et al., 2004; Haque and Russell, 2004). In a similar vein, temperature effects on the physicochemical properties of *A. tumefaciens* membrane lipids might impact biogenesis of the VirB/D4 channel or T-pilus.

It has been assumed that the signal molecules, e.g., plant phenolics and sugars, potentiating expression of the *vir* regulon and T-DNA transfer are released from wounded plant cells. However, it was recently shown that *A. tumefaciens* can induce the *vir* genes and transfer T-DNA to plant cells even in the apparent absence of wounding (Brencic et al., 2005). These studies did not exclude the possibility that *A. tumefaciens* itself causes microscopic wounds through release of plant cell wall degrading enzymes; nevertheless, the data suggest *A. tumefaciens* can sense very low levels of inducing signals and respond by delivering T-DNA at levels below that needed to incite detectable tumors. These observations raise intriguing questions regarding the extent in nature that *A. tumefaciens* transforms plant cells without discernible phenotypic consequences, and they also suggest the possibility that *A. tumefaciens* can establish productive contacts with a much wider range of plant species and cell types than previously envisaged.

8.2 Roles of the T-pilus and plant receptors

The role of the T-pilus in VirB/D4-mediated translocation is not well understood. As in the case of bacterial conjugation systems, the T-pilus likely initiates contact with target cells and thus with specific plant cell receptors. General screens for mutations affecting *A. tumefaciens* transformation efficiencies have identified several candidate genes whose products might represent surface receptors for the T-pilus. For example, *A. tumefaciens* was found to bind poorly to an *Arabidopsis* strain with a mutation in the arabinogalactan protein, a likely constituent of the plant cell wall (Zhu et al., 2003). More recently, a two-hybrid screen for plant proteins that interact with the VirB2 pilin protein identified four candidates, BT1, BT2, and BT3 of unknown function and a membrane-associated GTPase

(Hwang and Gelvin, 2004). Such proteins could serve as T-pilus receptors that are important for the initial bacterial – host contact.

Again by analogy with bacterial conjugation systems, following a T pilus-mediated loose association with the target cell, a tight junction forms between the *A. tumefaciens* outer membrane and the plant plasmamembrane. Several chromosomally–encoded proteins are implicated in mediating tight binding, including *chvA*, *chvB*, and *pscA*(*exoC*) genes whose products are important for synthesis of periplasmic β1,2-glucan and surface factors (Cangelosi et al., 1987; Zorreguieta et al., 1988). However, beyond evidence that *A. tumefaciens* cells attach by their cell poles to plant target cells (Lai et al., 2000; Matthysse, 1987), little else is known about this tight contact. At this junction, surface-displayed VirB channel components might interact with specific plant receptors and/or the bacterial and plant membranes might fuse together. A pore probably forms in the plant membrane, though it is also possible that the VirB/D4 channel elaborates a needle-like structure similar to the injectisomes of T3S machines that penetrate and inject substrates across mammalian cell membranes (Galan, 2001). An interesting candidate pore-forming protein is the VirE2 SSB, which has been shown to form pores in lipid bilayers *in vitro* (Duckely and Hohn, 2003). VirE2 is not required for transfer of the VirD2-T-strand complex to plants, but nevertheless might form pores for its own passage and, possibly, other protein substrates. Clearly, the *A. tumefaciens* – plant cell interface and the mechanism of substrate delivery across target cell membranes are rich areas for further investigation.

9 SUMMARY AND PERSPECTIVES

Since the early sequence analyses of the *A. tumefaciens virB* operons from the pTiA6NC, pTi1955 and pTiC58 plasmids, studies of the now-named VirB/D4 T4S system have led the T4S field in many areas of characterization. As has been the case since the discovery of T-DNA transfer to plant cells, studies of the *A. tumefaciens* VirB/D4 T4S system have generated a number of noteworthy "firsts" by many prominent investigators in the field. With respect to substrate recognition, the elegant studies by Hooykaas, Vergunst and colleagues delineated a C-terminal recognition motif that appears to be common to the VirB/D4 and possibly many other T4S systems. Binns, Kado, and colleagues discovered the first inhibitors of the VirB/D4 T4S system; these since have been shown to inhibit other T4S systems. Das and colleagues pioneered spatial studies of the VirB/D4 system and their findings stimulated further work defining spatial positioning of other components of the VirB/D4 system as well as other T4S systems. Christie, Atmakuri and colleagues determined that ParA-like VirC1 and VirC2 processing factors localize at the cell pole and that VirC1 recruits the VirD2 relaxase to this site, providing the first evidence that the conjugative processing reaction occurs at specific sites in a cell. With respect to the VirB/D4 machine architecture, the VirB proteins themselves have been problematic to work with *in vitro*, but studies by Baron, O' Callaghan and colleagues with the *B. suis* homologs have identified important subunit – subunit interactions, and their creative use of *A. tumefaciens* as a surrogate host supplied the first assay for monitoring assembly of the *B. suis* VirB complex. The assay was adapted from the discovery by Binns and colleagues that production of a subset of VirB proteins in agrobacterial recipients stimulates plasmid acquisition in matings with agrobacterial donors. Kado, Lanka and colleagues supplied the first evidence for novel head-to-tail cyclization of the VirB2 and RP4 TrbB pilin subunits. Though only one X-ray structure of a soluble domain of an *A. tumefaciens* VirB domain was reported so far, the pioneering studies by Gomis-Ruth, Coll, de la Cruz and colleagues, and by Waksman, Bayliss and colleagues have provided the entire T4S field with important X-ray structures of several VirB protein homologs as well as an NMR structure of a VirB co-complex. These structural prototypes are invaluable for understanding the architectures of T4S machines of Gram-negative bacteria and they provide important clues about specific subunit contributions to substrate transfer or T-pilus production. Development of the TrIP assay by Cascales and Christie for the first time defined the likely VirB/D4 channel subunits and specified contributions of the remaining subunits to stages of the translocation pathway. The so-called TrIP translocation pathway represents an important blueprint for further mechanistic studies of the VirB/D4 organelle. Finally, recent progress has been made by several groups toward defining VirB, VirC, and VirD subunit – subunit interactions and contributions of the three ATPases of this system to machine biogenesis and function. This is only a partial list of important discoveries, but it suffices to portray the extreme diversity of approaches being taken to study this fascinating transfer system. It also highlights the importance and power of creative new technologies developed to answer mechanistic questions of the VirB/D4 system that heretofore could not be experimentally addressed; such technologies have already proven invaluable in studies of other T4S machines.

10 ACKNOWLEDGEMENTS

We thank laboratory members for helpful discussions. We acknowledge financial support by the NIH Grant no. GM48746 for studies of the A. tumefaciens VirB/D4 T4S system.

11 REFERENCES

Anderson LB, Hertzel AV, Das A (1996) *Agrobacterium tumefaciens* VirB7 and VirB9 form a disulfide-linked protein complex. Proc Natl Acad Sci USA 93: 8889-8894

Atmakuri K, Cascales E, Burton OT, Banta LM and Christie PJ (2007) *Agrobacterium* ParA-like VirC1 spatially coordinates early conjugative DNA transfer reactions. EMBO J 26 :2540-2551

Atmakuri K, Cascales E, Christie PJ (2004) Energetic components VirD4, VirB11 and VirB4 mediate early DNA transfer reactions required for bacterial type IV secretion. Mol Microbiol 54: 1199-1211

Atmakuri K, Ding Z, Christie PJ (2003) VirE2, a type IV secretion substrate, interacts with the VirD4 transfer protein at cell poles of *Agrobacterium tumefaciens*. Mol Microbiol 49: 1699-1713

Bailey S, Ward D, Middleton R, Grossmann JG, Zambryski PC (2006) *Agrobacterium tumefaciens* VirB8 structure reveals potential protein-protein interaction sites. Proc Natl Acad Sci USA 103: 2582-2587

Bakholdina SI, Sanina NM, Krasikova IN, Popova OB, Solov'eva TF (2004) The impact of abiotic factors (temperature and glucose) on physicochemical properties of lipids from *Yersinia pseudotuberculosis*. Biochimie 86: 875-881

Baron C (2005) From bioremediation to biowarfare: on the impact and mechanism of type IV secretion systems. FEMS Microbiol Lett 253: 163-170

Baron C, Llosa M, Zhou S, Zambryski PC (1997) VirB1, a component of the T-complex transfer machinery of *Agrobacterium tumefaciens*, is processed to a C-terminal secreted product, VirB1*. J Bacteriol 179: 1203-1210

Bates S, Cashmore AM, Wilkins BM (1998) IncP plasmids are unusually effective in mediating conjugation of *Escherichia coli* and Saccharomyces *cerevisiae*: involvement of the tra2 mating system. J Bacteriol 180: 6538-6543

Bayliss R, Harris R, Coutte L, Monier A, Fronzes R, Christie PJ, Driscoll P, Waksman G (2007) NMR structure of a complex between the VirB9/VirB7 interaction domains of the pKM101 type IV secretion system. Proc Natl Acad Sci USA 104: 1673-1678

Beaupre CE, Bohne J, Dale EM, Binns AN (1997) Interactions between VirB9 and VirB10 membrane proteins involved in movement of DNA from *Agrobacterium tumefaciens* into plant cells. J Bacteriol 179: 78-89

Beijersbergen A, Dulk Ras AD, RAS, Hooykaas PJ (1992) Comjugative transfer by the virulence system of *Agrobacterium tumefaciens*. Science 256: 1324-1327

Beijersbergen A, Smith SJ, Hooykaas PJJ (1994) Localization and topology of VirB proteins of *Agrobacterium tumefaciens*. Plasmid 32: 212-218

Berger BR, Christie PJ (1993) The *Agrobacterium tumefaciens virB4* gene product is an essential virulence protein requiring an intact nucleoside triphosphate-binding domain. J Bacteriol 175: 1723-1734

Berger BR, Christie PJ (1994) Genetic complementation analysis of the *Agrobacterium tumefaciens vir*B operon: *vir*B2 through *vir*B11 are essential virulence genes. J Bacteriol 176: 3646-3660

Binns AN, Beaupre CE, Dale EM (1995) Inhibition of VirB-mediated transfer of diverse substrates from *Agrobacterium tumefaciens* by the IncQ plasmid RSF1010. J Bacteriol 177: 4890-4899

Bohne J, Yim A, Binns AN (1998) The Ti plasmid increases the efficiency of *Agrobacterium tumefaciens* as a recipient in *virB*-mediated conjugal transfer of an IncQ plasmid. Proc Natl Acad Sci USA 95: 7057-7062

Bolton GW, Nester EW, Gordon MP (1986) Plant phenolic compounds induce expression of the *Agrobacterium tumefaciens* loci needed for virulence. Science 232: 983-985

Brencic A, Angert ER, Winans SC (2005) Unwounded plants elicit *Agrobacterium vir* gene induction and T-DNA transfer: transformed plant cells produce opines yet are tumour free. Mol Microbiol 57: 1522-1531

Buchanan-Wollaston V, Passiatore JE, Cannon F (1987) The mob and oriT mobilization functions of a bacterial plasmid promote its transfer to plants. Nature 328: 172-175

Cangelosi GA, Hung L, Puvanesarajah V, Stacey G, Ozga DA, Leigh JA, Nester EW (1987) Common loci for *Agrobacterium tumefaciens* and *Rhizobium meliloti* exopolysaccharide synthesis and their roles in plant interactions. J Bacteriol 169: 2086-2091

Cao TB, Saier MH, Jr. (2001) Conjugal type IV macromolecular transfer systems of Gram-negative bacteria: organismal distribution, structural constraints and evolutionary conclusions. Microbiology 147: 3201-3214

Carle A, Hoppner C, Ahmed Aly K, Yuan Q, den Dulk-Ras A, Vergunst A, O'Callaghan D, Baron C (2006) The *Brucella suis* type IV secretion system assembles in the cell envelope of the heterologous host *Agrobacterium tumefaciens* and increases IncQ plasmid pLS1 recipient competence. Infect Immun 74: 108-117

Cascales E, Atmakuri K, Liu Z, Binns AN, Christie PJ (2005) *Agrobacterium tumefaciens* oncogenic suppressors inhibit T-DNA and VirE2 protein substrate binding to the VirD4 coupling protein. Mol Microbiol 58: 565-579

Cascales E, Christie PJ (2003) The versatile bacterial type IV secretion systems. Nat Rev Microbiol 1: 137-149

Cascales E, Christie PJ (2004a) *Agrobacterium* VirB10, an ATP energy sensor required for type IV secretion. Proc Natl Acad Sci USA 101: 17228-17233

Cascales E, Christie PJ (2004b) Definition of a bacterial type IV secretion pathway for a DNA substrate. Science 304: 1170-1173

Chen CY, Kado CI (1994) Inhibition of *Agrobacterium tumefaciens* oncogenicity by the *osa* gene of pSa. J Bacteriol 176: 5697-5703

Chen CY, Kado CI (1996) Osa protein encoded by plasmid pSa is located at the inner membrane but does not inhibit membrane association of VirB and VirD virulence proteins in *Agrobacterium tumefaciens*. FEMS Microbiol Lett 135: 85-92

Cho H, Winans SC (2005) VirA and VirG activate the Ti plasmid *repABC* operon, elevating plasmid copy number in response to wound-released chemical signals. Proc Natl Acad Sci USA 102: 14843-14848

Christie PJ (2004) Bacterial type IV secretion: the *Agrobacterium* VirB/D4 and related conjugation systems. Biochim Biophys Acta 1694: 219-234

Christie PJ, Atmakuri K, Krishnamoorthy V, Jakubowski S, Cascales E (2005) Biogenesis, architecture, and function of bacterial type IV secretion systems. Annu Rev Microbiol 59: 451-485

Christie PJ, Cascales E (2005) Structural and dynamic properties of bacterial type IV secretion systems (review). Mol Membr Biol 22: 51-61

Christie PJ, Vogel JP (2000) Bacterial type IV secretion: conjugation systems adapted to deliver effector molecules to host cells. Trends Microbiol 8: 354-360

Christie PJ, Ward JE Jr, Gordon MP, Nester EW (1989) A gene required for transfer of T-DNA to plants encodes an ATPase with autophosphorylating activity. Proc Natl Acad Sci USA 86: 9677-9681

Christie PJ, Ward JE, Winans SC, Nester EW (1988) The *Agrobacterium tumefaciens virE2* gene product is a single-stranded-DNA-binding protein that associates with T-DNA. J Bacteriol 170: 2659-2667

Citovsky V, Zambryski PC (1993) Transport of nucleic acids through membrane channels: snaking through small holes. Annu Rev Microbiol 47: 167-197

Dang TA, Christie PJ (1997) The VirB4 ATPase of *Agrobacterium tumefaciens* is a cytoplasmic membrane protein exposed at the periplasmic surface. J Bacteriol 179: 453-462

Dang TA, Zhou XR, Graf B, Christie PJ (1999) Dimerization of the *Agrobacterium tumefaciens* VirB4 ATPase and the effect of ATP-binding cassette mutations on the assembly and function of the T-DNA transporter. Mol Microbiol 32: 1239-1253

Das A, Anderson LB, Xie YH (1997) Delineation of the interaction domains of *Agrobacterium tumefaciens* VirB7 and VirB9 by use of the yeast two-hybrid assay. J Bacteriol 179: 3404-3409

Das A, Xie YH (2000) The *Agrobacterium* T-DNA transport pore proteins VirB8, VirB9, and VirB10 interact with one another. J Bacteriol 182: 758-763

De Vos G, Zambryski PC (1989) Expression of *Agrobacterium* nopaline specific VirD1, VirD2, and VirC1 proteins and their requirement for T-strand production in *E. coli*. Mol Plant-Microbe Interact 2: 43-52

Deng W, Chen L, Peng WT, Liang X, Sekiguchi S, Gordon MP, Comai L, Nester EW (1999) VirE1 is a specific molecular chaperone for the exported single-stranded-DNA-binding protein VirE2 in *Agrobacterium*. Mol Microbiol 31: 1795-1807

Ding Z, Zhao Z, Jakubowski SJ, Krishnamohan A, Margolin W, Christie PJ (2002) A novel cytology-based, two-hybrid screen for bacteria applied to protein-protein interaction studies of a type IV secretion system. J Bacteriol 184: 5572-5582

Draper O, Middleton R, Doucleff M, Zambryski PC (2006) Topology of the VirB4 C-terminus in the *Agrobacterium tumefaciens* VirB/D4 type IV secretion system. J Biol Chem 281: 37628-37635

Duckely M, Hohn B (2003) The VirE2 protein of *Agrobacterium tumefaciens*: the Yin and Yang of T-DNA transfer. FEMS Microbiol Lett 223: 1-6

Economou A, Christie PJ, Fernandez RC, Palmer T, Plano GV, Pugsley AP (2006) Secretion by numbers: Protein traffic in prokaryotes. Mol Microbiol 62: 308-319

Eisenbrandt R, Kalkum M, Lai EM, Lurz R, Kado CI, Lanka E (1999) Conjugative pili of IncP plasmids, and the Ti plasmid T pilus are composed of cyclic subunits. J Biol Chem 274: 22548-22555

Eisenbrandt R, Kalkum M, Lurz R, Lanka E (2000) Maturation of IncP pilin precursors resembles the catalytic Dyad-like mechanism of leader peptidases. J Bacteriol 182: 6751-6761

Engstrom P, Zambryski P, Van Montagu M, Stachel S (1987) Characterization of *Agrobacterium tumefaciens* virulence proteins induced by the plant factor acetorsyingone. J Mol Microbiol 4: 635-645

Fernandez D, Dang TAT, Spudich GM, Zhou X-R, Berger BR, Christie PJ (1996a) The *Agrobacterium tumefaciens* VirB7 gene product, a proposed component of the T-complex transport apparatus, is a membrane-associated lipoprotein exposed at the periplasmic surface. J Bacteriol 178: 3156-3167

Fernandez D, Spudich GM, Zhou X-R, Christie PJ (1996b) The *Agrobacterium tumefaciens* VirB7 lipoprotein is required for stabilization of VirB proteins during assembly of the T-complex transport apparatus. J Bacteriol 178: 3168-3176

Frenkiel-Krispin D, Grayer Wolf S, Albeck S, Unger T, Peleg Y, Jacobovitch J, Michael Y, Daube S, Sharon M, Robinson CV, Svergun DI, Fass D, Tzfira T, Elbaum M (2006) Plant transformation by *Agrobacterium tumefaciens*: Modulation of ssDNA-VirE2 complex assembly by VirE1. J Biol Chem (in press)

Galan JE (2001) *Salmonella* interactions with host cells: type III secretion at work. Annu Rev Cell Dev Biol 17: 53-86

Ghigo JM (2001) Natural conjugative plasmids induce bacterial biofilm development. Nature 412: 442-445

Gomis-Ruth FX, Moncalian G, Perez-Luque R, Gonzalez A, Cabezon E, de la Cruz F, Coll M (2001) The bacterial conjugation protein TrwB resembles ring helicases and F1-ATPase. Nature 409: 637-641

Gomis-Ruth FX, Sola M, de la Cruz F, Coll M (2004) Coupling factors in macromolecular type-IV secretion machineries. Curr Pharm Des 10: 1551-1565

Hamilton CM, Lee H, Li PL, Cook DM, Piper KR, von Bodman SB, Lanka E, Ream W, Farrand SK (2000) TraG from RP4 and TraG and VirD4 from Ti plasmids confer relaxosome specificity to the conjugal transfer system of pTiC58. J Bacteriol 182: 1541-1548

Hapfelmeier S, Domke N, Zambryski PC, Baron C (2000) VirB6 is required for stabilization of VirB5 and VirB3 and formation of VirB7 homodimers in *Agrobacterium tumefaciens*. J Bacteriol 182: 4505-4511

Haque MA, Russell NJ (2004) Strains of *Bacillus cereus* vary in the phenotypic adaptation of their membrane lipid composition in response to low water activity, reduced temperature and growth in rice starch. Microbiology 150: 1397-1404

Hare S, Bayliss R, Baron C, Waksman G (2006) A large domain swap in the VirB11 ATPase of *Brucella suis* leaves the hexameric assembly intact. J Mol Biol 360: 56-66

Hodges LD, Cuperus J, Ream W (2004) *Agrobacterium rhizogenes* GALLS protein substitutes for Agrobacterium tumefaciens single-stranded DNA-binding protein VirE2. J Bacteriol 186: 3065-3077

Hodges LD, Vergunst AC, Neal-McKinney J, den Dulk-Ras A, Moyer DM, Hooykaas PJ, Ream W (2006) *Agrobacterium rhizogenes* GALLS protein contains domains for ATP binding, nuclear localization, and type IV secretion. J Bacteriol 188: 8222-8230

Hoppner C, Liu Z, Domke N, Binns AN, Baron C (2004) VirB1 orthologs from *Brucella suis* and pKM101 complement defects of the lytic transglycosylase required for efficient type IV secretion from *Agrobacterium tumefaciens*. J Bacteriol 186: 1415-1422

Hormaeche I, Alkorta I, Moro F, Valpuesta JM, Goni FM, De La Cruz F (2002) Purification and properties of TrwB, a hexameric, ATP-binding integral membrane protein essential for R388 plasmid conjugation. J Biol Chem 277: 46456-46462

Hubber A, Vergunst AC, Sullivan JT, Hooykaas PJ, Ronson CW (2004) Symbiotic phenotypes and translocated effector proteins of the *Mesorhizobium loti* strain R7A VirB/D4 type IV secretion system. Mol Microbiol 54: 561-574

Hwang HH, Gelvin SB (2004) Plant proteins that interact with VirB2, the *Agrobacterium tumefaciens* pilin protein, mediate plant transformation. Plant Cell 16: 3148-3167

Jakubowski SJ, Cascales E, Krishnamoorthy V, Christie PJ (2005) *Agrobacterium tumefaciens* VirB9, an outer-membrane-associated component of a type IV secretion system, regulates substrate selection and T-pilus biogenesis. J Bacteriol 187: 3486-3495

Jakubowski SJ, Krishnamoorthy V, Cascales E, Christie PJ (2004) *Agrobacterium tumefaciens* VirB6 domains direct the ordered export of a DNA substrate through a type IV secretion System. J Mol Biol 341: 961-977

Jakubowski SJ, Krishnamoorthy V, Christie PJ (2003) *Agrobacterium tumefaciens* VirB6 protein participates in formation of VirB7 and VirB9 complexes required for type IV secretion. J Bacteriol 185: 2867-2878

Jones AL, Lai EM, Shirasu K, Kado CI (1996) VirB2 is a processed pilin-like protein encoded by the *Agrobacterium tumefaciens* Ti plasmid. J Bacteriol 178: 5706-5711

Jones AL, Shirasu K, Kado CI (1994) The product of the *virB4* gene of *Agrobacterium tumefaciens* promotes accumulation of VirB3 protein. J Bacteriol 176: 5255-5261

Judd PK, Kumar RB, Das A (2005a) Spatial location and requirements for the assembly of the *Agrobacterium tumefaciens* type IV secretion apparatus. Proc Natl Acad Sci USA 102: 11498-11503

Judd PK, Kumar RB, Das A (2005b) The type IV secretion apparatus protein VirB6 of *Agrobacterium tumefaciens* localizes to a cell pole. Mol Microbiol 55: 115-124

Judd PK, Mahli D, Das A (2005c) Molecular characterization of the *Agrobacterium tumefaciens* DNA transfer protein VirB6. Microbiology 151: 3483-3492

Kahng LS, Shapiro L (2003) Polar localization of replicon origins in the multipartite genomes of *Agrobacterium tumefaciens* and *Sinorhizobium meliloti*. J Bacteriol 185: 3384-3391

Koraimann G (2003) Lytic transglycosylases in macromolecular transport systems of Gram-negative bacteria. Cell Mol Life Sci 60: 2371-2388

Krause S, Barcena M, Pansegrau W, Lurz R, Carazo JM, Lanka E (2000a) Sequence-related protein export NTPases encoded by the conjugative transfer region of RP4 and by the *cag* pathogenicity island of *Helicobacter pylori* share similar hexameric ring structures. Proc Natl Acad Sci USA 97: 3067-3072

Krause S, Pansegrau W, Lurz R, de la Cruz F, Lanka E (2000b) Enzymology of type IV macromolecule secretion systems: the conjugative transfer regions of plasmids RP4 and R388 and the *cag* pathogenicity island of *Helicobacter pylori* encode structurally and functionally related nucleoside triphosphate hydrolases. J Bacteriol 182: 2761-2770

Kuldau GA, De Vos G, Owen J, McCaffrey G, Zambryski P (1990) The *virB* operon of *Agrobacterium tumefaciens* pTiC58 encodes 11 open reading frames. Mol Gen Genet 221: 256-266

Kumar RB, Das A (2001) Functional analysis of the *Agrobacterium tumefaciens* T-DNA transport pore protein VirB8. J Bacteriol 183: 3636-3641

Kumar RB, Das A (2002) Polar location and functional domains of the *Agrobacterium tumefaciens* DNA transfer protein VirD4. Mol Microbiol 43: 1523-1532

Kumar RB, Xie YH, Das A (2000) Subcellular localization of the *Agrobacterium tumefaciens* T-DNA transport pore proteins: VirB8 is essential for the assembly of the transport pore. Mol Microbiol 36: 608-617

Kunik T, Tzfira T, Kapulnik Y, Gafni Y, Dingwall C, Citovsky V (2001) Genetic transformation of HeLa cells by *Agrobacterium*. Proc Natl Acad Sci USA 98: 1871-1876

Lai EM, Chesnokova O, Banta LM, Kado CI (2000) Genetic and environmental factors affecting T-pilin export and T-pilus biogenesis in relation to flagellation of *Agrobacterium tumefaciens*. J Bacteriol 182: 3705-3716

Lai EM, Kado CI (1998) Processed VirB2 is the major subunit of the promiscuous pilus of *Agrobacterium tumefaciens*. J Bacteriol 180: 2711-2717

Lawley TD, Klimke WA, Gubbins MJ, Frost LS (2003) F factor conjugation is a true type IV secretion system. FEMS Microbiol Lett 224: 1-15

Lee L-Y, Gelvin SB, Kado CI (1999) pSa causes oncogenic suppression of *Agrobacterium* by inhibiting VirE2 protein export. J Bacteriol 181: 186-196

Lessl M, Balzer D, Pansegrau W, Lanka E (1992a) Sequence similarities between the RP4 Tra2 and the Ti VirB region strongly support the conjugation model for T-DNA transfer. J Biol Chem 267: 20471-20480

Lessl M, Pansegrau W, Lanka E (1992b) Relationship of DNA-transfer-systems: essential transfer factors of plasmids RP4, Ti and F share common sequences. Nucleic Acids Res 20: 6099-6100

Letoffe S, Delepelaire P, Wandersman C (1996) Protein secretion in gram-negative bacteria: assembly of the three components of ABC protein-mediated exporters is ordered and promoted by substrate binding. EMBO J 15: 5804-5811

Liu Z, Binns AN (2003) Functional subsets of the *virB* type IV transport complex proteins involved in the capacity of *Agrobacterium tumefaciens* to serve as a recipient in *virB*-mediated conjugal transfer of plasmid RSF1010. J Bacteriol 185: 3259-3269

Llosa M, Zunzunegui S, de la Cruz F (2003) Conjugative coupling proteins interact with cognate and heterologous VirB10-like proteins while exhibiting specificity for cognate relaxosomes. Proc Natl Acad Sci USA 100: 10465-10470

Llosa M, Zupan J, Baron C, Zambryski PC (2000) The N- and C-terminal portions of the *Agrobacterium* VirB1 protein independently enhance tumorigenesis. J Bacteriol 182: 3437-3445

Massey TH, Mercogliano CP, Yates J, Sherratt DJ, Lowe J (2006) Double-stranded DNA translocation: structure and mechanism of hexameric FtsK. Mol Cell 23: 457-469

Matthysse AG (1987) Characterization of nonattaching mutants of *Agrobacterium tumefaciens*. J Bacteriol 169: 313-323

McCullen CA, Binns AN (2006) *Agrobacterium tumefaciens* plant cell interactions and activities required for interkingdom macromolecular transfer. Annu Rev Cell Dev Biol 22: 101-127

Melchers LS, Regensburg-Tuink AJ, Schilperoort RA, Hooykaas PJ (1989) Specificity of signal molecules in the activation of *Agrobacterium* virulence gene expression. Mol Microbiol 3: 969-977

Middleton R, Sjolander K, Krishnamurthy N, Foley J, Zambryski P (2005) Predicted hexameric structure of the *Agrobacterium* VirB4 C terminus suggests VirB4 acts as a docking site during type IV secretion. Proc Natl Acad Sci USA 102: 1685-1690

Moncalian G, Cabezon E, Alkorta I, Valle M, Moro F, Valpuesta JM, Goni FM, de La Cruz F (1999) Characterization of ATP and DNA binding activities of TrwB, the coupling protein essential in plasmid R388 conjugation. J Biol Chem 274: 36117-36124

Pansegrau W, Schoumacher F, Hohn B, Lanka E (1993) Site-specific cleavage and joining of single-stranded DNA by VirD2 protein of *Agrobacterium tumefaciens*

Ti plasmids: analogy to bacterial conjugation. Proc Natl Acad Sci USA 90: 11538-11542

Pantoja M, Chen L, Chen Y, Nester EW (2002) *Agrobacterium* type IV secretion is a two-step process in which export substrates associate with the virulence protein VirJ in the periplasm. Mol Microbiol 45: 1325-1335

Paschos A, Patey G, Sivanesan D, Gao C, Bayliss R, Waksman G, O'Callaghan D, Baron C (2006) Dimerization and interactions of *Brucella suis* VirB8 with VirB4 and VirB10 are required for its biological activity. Proc Natl Acad Sci USA 103: 7252-7257

Planet PJ, Kachlany SC, DeSalle R, Figurski DH (2001) Phylogeny of genes for secretion NTPases: identification of the widespread *tadA* subfamily and development of a diagnostic key for gene classification. Proc Natl Acad Sci USA 98: 2503-2508

Rabel C, Grahn AM, Lurz R, Lanka E (2003) The VirB4 family of proposed traffic nucleoside triphosphatases: common motifs in plasmid RP4 TrbE are essential for conjugation and phage adsorption. J Bacteriol 185: 1045-1058

Rashkova S, Spudich GM, Christie PJ (1997) Characterization of membrane and protein interaction determinants of the *Agrobacterium tumefaciens* VirB11 ATPase. J Bacteriol 179: 583-591

Rashkova S, Zhou XR, Chen J, Christie PJ (2000) Self-assembly of the *Agrobacterium tumefaciens* VirB11 traffic ATPase. J Bacteriol 182: 4137-4145

Reisner A, Krogfelt KA, Klein BM, Zechner EL, Molin S (2006) *In vitro* biofilm formation of commensal and pathogenic *Escherichia coli* strains: impact of environmental and genetic factors. J Bacteriol 188: 3572-3581

Sagulenko E, Sagulenko V, Chen J, Christie PJ (2001a) Role of *Agrobacterium* VirB11 ATPase in T-pilus assembly and substrate selection. J Bacteriol 183: 5813-5825

Sagulenko V, Sagulenko E, Jakubowski S, Spudich E, Christie PJ (2001b) VirB7 lipoprotein is exocellular and associates with the *Agrobacterium tumefaciens* T pilus. J Bacteriol 183: 3642-3651

Saleh OA, Perals C, Barre FX, Allemand JF (2004) Fast, DNA-sequence independent translocation by FtsK in a single-molecule experiment. EMBO J 23: 2430-2439

Samuels AL, Lanka E, Davies JE (2000) Conjugative junctions in RP4-mediated mating of *Escherichia coli*. J Bacteriol 182: 2709-2715

Savvides SN, Yeo HJ, Beck MR, Blaesing F, Lurz R, Lanka E, Buhrdorf R, Fischer W, Haas R, Waksman G (2003) VirB11 ATPases are dynamic hexameric assemblies: new insights into bacterial type IV secretion. EMBO J 22: 1969-1980

Scheiffele P, Pansegrau W, Lanka E (1995) Initiation of *Agrobacterium tumefaciens* T-DNA processing. Purified proteins VirD1 and VirD2 catalyze site- and strand-specific cleavage of superhelical T-border DNA *in vitro*. J Biol Chem 270: 1269-1276

Schmid MC, Schulein R, Dehio M, Denecker G, Carena I, Dehio C (2004) The VirB type IV secretion system of *Bartonella henselae* mediates invasion, proinflammatory activation and antiapoptotic protection of endothelial cells. Mol Microbiol 52: 81-92

Schmidt-Eisenlohr H, Domke N, Angerer C, Wanner G, Zambryski PC, Baron C (1999) Vir proteins stabilize VirB5 and mediate its association with the T pilus of *Agrobacterium tumefaciens*. J Bacteriol 181: 7485-7492

Schroder G, Krause S, Zechner EL, Traxler B, Yeo HJ, Lurz R, Waksman G, Lanka E (2002) TraG-like proteins of DNA transfer systems and of the *Helicobacter pylori* type IV secretion system: inner membrane gate for exported substrates? J Bacteriol 184: 2767-2779

Schroder G, Lanka E (2003) TraG-like proteins of type IV secretion systems: functional dissection of the multiple activities of TraG (RP4) and TrwB (R388). J Bacteriol 185: 4371-4381

Schroder G, Lanka E (2005) The mating pair formation system of conjugative plasmids-A versatile secretion machinery for transfer of proteins and DNA. Plasmid 54: 1-25

Selbach M, Moese S, Meyer TF, Backert S (2002) Functional analysis of the *Helicobacter pylori cag* pathogenicity island reveals both VirD4-CagA-dependent and VirD4-CagA-independent mechanisms. Infect Immun 70: 665-671

Shirasu K, Kado CI (1993) The *virB* operon of the *Agrobacterium tumefaciens* virulence regulon has sequence similarities to B, C and D open reading frames downstream of the pertussis toxin-operon and to the DNA transfer-operons of broad-host-range conjugative plasmids. Nucl Acids Res 21: 353-354

Shirasu K, Morel P, Kado CI (1990) Characterization of the *virB* operon of an *Agrobacterium tumefaciens* Ti plasmid: nucleotide sequence and protein analysis. Mol Microbiol 4: 1153-1163

Simone M, McCullen CA, Stahl LE, Binns AN (2001) The carboxy-terminus of VirE2 from *Agrobacterium tumefaciens* is required for its transport to host cells by the *vir*B-encoded type IV transport system. Mol Microbiol 41: 1283-1293

Spudich GM, Fernandez D, Zhou XR, Christie PJ (1996) Intermolecular disulfide bonds stabilize VirB7 homodimers and VirB7/VirB9 heterodimers during biogenesis of the *Agrobacterium tumefaciens* T-complex transport apparatus. Proc Natl Acad Sci USA 93: 7512-7517

Stachel SE, Nester EW (1986) The genetic and transcriptional organization of the *vir* region of the A6 Ti plasmid of *Agrobacterium tumefaciens*. EMBO J 5: 1445-1454

Stachel SE, Zambryski PC (1986) *Agrobacterium tumefaciens* and the susceptible plant cell: a novel adaptation of extracellular recognition and DNA conjugation. Cell 47: 155-157

Stahl LE, Jacobs A, Binns AN (1998) The conjugal intermediate of plasmid RSF1010 inhibits *Agrobacterium tumefaciens* virulence and VirB-dependent export of VirE2. J Bacteriol 180: 3933-3939

Su S, Stephens BB, Alexandre G, Farrand SK (2006) Lon protease of the alpha-proteobacterium *Agrobacterium tumefaciens* is required for normal growth, cellular morphology and full virulence. Microbiology 152: 1197-1207

Sundberg CD, Ream W (1999) The *Agrobacterium tumefaciens* chaperone-like protein, VirE1, interacts with VirE2 at domains required for single-stranded DNA binding and cooperative interaction. J Bacteriol 181: 6850-6855

Tato I, Zunzunegui S, de la Cruz F, Cabezon E (2005) TrwB, the coupling protein involved in DNA transport during bacterial conjugation, is a DNA-dependent ATPase. Proc Natl Acad Sci USA 102: 8156-8161

Terradot L, Bayliss R, Oomen C, Leonard GA, Baron C, Waksman G (2005) Structures of two core subunits of the bacterial type IV secretion system, VirB8 from *Brucella suis* and ComB10 from *Helicobacter pylori*. Proc Natl Acad Sci USA 102: 4596-4601

Thanassi DG (2002) Ushers and secretins: channels for the secretion of folded proteins across the bacterial outer membrane. J Mol Microbiol Biotechnol 4: 11-20

Thompson DV, Melchers LS, Idler KB, Schilperoort RA, Hooykaas PJ (1988) Analysis of the complete nucleotide sequence of the *Agrobacterium tumefaciens virB* operon. Nucleic Acids Res 16: 4621-4636

Toro N, Datta A, Carmi OA, Young C, Prusti RK, Nester EW (1989) The *Agrobacterium tumefaciens virC1* gene product binds to overdrive, a T-DNA transfer enhancer. J Bacteriol 171: 6845-6849

Toro N, Datta A, Yanofsky M, Nester EW (1988) Role of the overdrive sequence in T-DNA border cleavage in *Agrobacterium*. Proc Natl Acad Sci USA 85: 8558-8562

Veluthambi K, Ream W, Gelvin SB (1988) Virulence genes, borders, and overdrive generate single-stranded T-DNA molecules from the A6 Ti plasmid of *Agrobacterium tumefaciens*. J Bacteriol 170: 1523-1532

Vergunst AC, Schrammeijer B, den Dulk-Ras A, de Vlaam CMT, Regensburg-Tuink TJ, Hooykaas PJJ (2000) VirB/D4-dependent protein translocation from *Agrobacterium* into plant cells. Science 290: 979-982

Vergunst AC, van Lier MCM, den Dulk-Ras A, Hooykaas PJJ (2003) Recognition of the *Agrobacterium* VirE2 translocation signal by the VirB/D4 transport system does not require VirE1. Plant Physiol 133: 978-988

Vergunst AC, van Lier MCM, den Dulk-Ras A, Stuve TAG, Ouwehand A, Hooykaas PJJ (2005) Positive charge is an important feature of the C-terminal transport signal of the VirB/D4-translocated proteins of *Agrobacterium*. Proc Natl Acad Sci USA 102: 832-837

Viala J, Chaput C, Boneca IG, Cardona A, Girardin SE, Moran AP, Athman R, Memet S, Huerre MR, Coyle AJ, DiStefano PS, Sansonetti PJ, Labigne A, Bertin J, Philpott DJ, Ferrero RL (2004) Nod1 responds to peptidoglycan delivered by the *Helicobacter pylori cag* pathogenicity island. Nat Immunol 5: 1166-1174

Vogel JP, Andrews HL, Wong SK, Isberg RR (1998) Conjugative transfer by the virulence system of *Legionella pneumophila*. Science 279: 873-876

Ward DV, Draper O, Zupan JR, Zambryski PC (2002) Peptide linkage mapping of the *Agrobacterium tumefaciens vir*-encoded type IV secretion system reveals protein subassemblies. Proc Natl Acad Sci USA 99: 11493-11500

Ward JE, Akiyoshi DE, Regier D, Datta A, Gordon MP, Nester EW (1988) Characterization of the virB operon from an *Agrobacterium tumefaciens* Ti plasmid. J Biol Chem 263: 5804-5814

Ward JE, Akiyoshi DE, Regier D, Datta A, Gordon MP, Nester EW (1990) Correction: characterization of the *virB* operon from *Agrobacterium tumefaciens* Ti plasmid. J Biol Chem 265: 4786

Ward JE, Jr., Dale EM, Binns AN (1991) Activity of the *Agrobacterium* T-DNA transfer machinery is affected by *virB* gene products. Proc Natl Acad Sci USA 88: 9350-9354

Waters VL (2001) Conjugation between bacterial and mammalian cells. Nat Genet 29: 375-376

Waters VL, Guiney DG (1993) Processes at the nick region link conjugation, T-DNA transfer and rolling circle replication. Mol Microbiol 9: 1123-1130

Weiss AA, Johnson FD, Burns DL (1993) Molecular characterization of an operon required for pertussis toxin secretion. Proc Natl Acad Sci USA 90: 2970-2974

Wessel M, Klusener S, Godeke J, Fritz C, Hacker S, Narberhaus F (2006) Virulence of *Agrobacterium tumefaciens* requires phosphatidylcholine in the bacterial membrane. Mol Microbiol 62: 906-915

Yeo HJ, Savvides SN, Herr AB, Lanka E, Waksman G (2000) Crystal structure of the hexameric traffic ATPase of the *Helicobacter pylori* type IV secretion system. Mol Cell 6: 1461-1472

Yeo HJ, Yuan Q, Beck MR, Baron C, Waksman G (2003) Structural and functional characterization of the VirB5 protein from the type IV secretion system encoded by the conjugative plasmid pKM101. Proc Natl Acad Sci USA 100: 15947-15952

Yuan Q, Carle A, Gao C, Sivanesan D, Aly KA, Hoppner C, Krall L, Domke N, Baron C (2005) Identification of the VirB4-VirB8-VirB5-VirB2 pilus assembly sequence of type IV secretion systems. J Biol Chem 280: 26349-26359

Zahrl D, Wagner M, Bischof K, Koraimann G (2006) Expression and assembly of a functional type IV secretion system elicit extracytoplasmic and cytoplasmic stress responses in *Escherichia coli*. J Bacteriol 188: 6611-6621

Zhao Z, Sagulenko E, Ding Z, Christie PJ (2001) Activities of *virE1* and the VirE1 secretion chaperone in export of the multifunctional VirE2 effector via an *Agrobacterium* Type IV secretion pathway. J Bacteriol 183: 3855-3865

Zhou XR, Christie PJ (1997) Suppression of mutant phenotypes of the *Agrobacterium tumefaciens* VirB11 ATPase by overproduction of VirB proteins. J Bacteriol 179: 5835-5842

Zhu J, Oger PM, Schrammeijer B, Hooykaas PJJ, Farrand SK, Winans SC (2000) The bases of crown gall tumorigenesis. J Bacteriol 182: 3885-3895

Zhu Y, Nam J, Humara JM, Mysore KS, Lee LY, Cao H, Valentine L, Li J, Kaiser AD, Kopecky AL, Hwang HH, Bhattacharjee S, Rao PK, Tzfira T, Rajagopal J, Yi H, Veena, Yadav BS, Crane YM, Lin K, Larcher Y, Gelvin MJ, Knue M, Ramos C, Zhao X, Davis SJ, Kim SI, Ranjith-Kumar CT, Choi YJ, Hallan VK, Chattopadhyay S, Sui X, Ziemienowicz A, Matthysse AG, Citovsky V, Hohn B, Gelvin SB (2003) Identification of Arabidopsis rat mutants. Plant Physiol 132: 494-505

Zorreguieta A, Geremia RA, Cavaignac S, Cangelosi GA, Nester EW, Ugalde RA (1988) Identification of the product of an *Agrobacterium tumefaciens* chromosomal virulence gene. Mol Plant Microbe Interact 1: 121-127

Chapter 10

INTRACELLULAR TRANSPORT OF *AGROBACTERIUM* T-DNA

Benoît Lacroix[1], Michael Elbaum[2], Vitaly Citovsky[1] and Tzvi Tzfira[3]

[1]Department of Biochemistry and Cell Biology, State University of New York, Stony Brook, NY 11794, USA; [2]Department of Materials and Interface, Weizmann Institute of Science, Rehovot 76100, Israel; [3]Department of Molecular, Cellular and Developmental Biology, The University of Michigan, Ann Arbor, MI 48109, USA

Abstract. To transfer genes to plants or other organisms, *Agrobacterium* exports its transferred DNA (T-DNA), along with several virulence proteins, into the host cell. The T-DNA must then be transported through the cytoplasm to the nuclear pore, pass through the nuclear pore complex, and finally move inside the nucleus toward a potential site of integration into the host genome. This T-DNA voyage inside the host cell results from a complex interplay between numerous bacterial and host factors, where host-cell machineries that allow macromolecular movements are employed by *Agrobacterium* to achieve the transfer and integration of T-DNA into the host genome.

1 INTRODUCTION

Numerous studies have shed new light on the interplay between bacterial and plant factors during DNA transfer and integration from *Agrobacterium tumefaciens* to its host genome, as reflected by several recent review articles on the different steps of the *Agrobacterium*-mediated genetic transformation of various host species (e.g., Zupan et al., 2000; e.g., Tzfira and Citovsky, 2002; Gelvin, 2003b; Tzfira et al., 2004a; Lacroix et al., 2006b; Tzfira and Citovsky, 2006 and other chapters in this volume). Nevertheless, there is still much to be discovered about the mechanisms governing the fate of the transferred DNA (T-DNA) following its translocation into the host cell, on its way to the nucleus, and to its point of integration, as well as the role that host cell factors play in mediating these processes. Within te host, *Agrobacterium* T-DNA travels as a nucleoprotein intermediate, the T-complex, composed of a single-stranded T-DNA copy (T-strand) associated with two virulence proteins, VirD2 and VirE2. A single molecule of VirD2 is bound to the 5' end of the T-strand, whereas multiple molecules of VirE2, the major protein component of the T-complex, coat the entire length of the T-strand (*Figure 10-1a*). The intracellular route taken by the T-complex can be roughly divided into three parts: (i) cytoplasmic transport, (ii) nuclear import and (iii) intranuclear transport. T-complexes must pass through the dense cytoplasmic structures, sneak through the narrow nuclear-pore complex and find their way in the confined nucleus volume to their point of integration. The T-complex is large on the scale of the sub-cellular structures, so its transport most likely requires interaction with host cell factors. As a pathogenic object it may exploit basic cellular mechanisms throughout its journey.

Some fundamental differences exist between the intracellular transport and nuclear import of the *Agrobacterium* T-complex and other DNA and nucleoprotein complexes such as viruses or artificial DNA molecules. In fact, many plant viruses do not undergo a nuclear stage and thus do not integrate into the host genome. Most of these viruses have developed special movement proteins which allow them, through interaction with various host factors and the host-cell cytoskeletal system, to move from cell to cell (Waigmann et al., 2004). On the other hand, viruses which replicate in the nucleus (reviewed in Whittaker and Helenius, 1998; Whittaker et al., 2000; Greber and Fassati, 2003; Whittaker, 2003; Krichevsky et al., 2006) have adopted the plant's nuclear-import mechanisms to deliver their genomes into the host-cell nucleus, in common with viruses affecting anima hosts. In direct genetic transformation, e.g. biolistic delivery, naked DNA has to reach the nucleus as well as to integrate into the host-cell genome. However, the methods used [involving mechanical (Taylor and Fauquet, 2002) or chemical disruption of biological membranes (Abel and Theologis, 1994; Mathur and Koncz, 1998)] probably affect the cell-membrane system and structure of the cytoplasm, which might provide an artifical route for DNA cytoplasmic transport and nuclear internalization. While we cannot rule out the possibility that T-DNA and naked DNA molecules use similar pathways for their subcellular transport, the existence of T-DNA as a nucleoprotein complex suggests a unique biological pathway for its transport to the integration site.

Cellular transport of the *Agrobacterium*'s T-complex thus implies several specific requirements to ensure a successful pathogenic infection. First, the size and three-dimensional structure of the DNA molecule must allow its movement in the cytoplasm and must be compatible with the size-exclusion limit of the nuclear pore. Second, the DNA segment must be protected from degradation in the host cytosol and nucleus. Third, interactions with the host nuclear-import machinery must occur to allow movement to and through the nuclear pore, and fourth, the T-complex must be directed to its point of integration. It also seems likely that T-DNA nuclear import is regulated, as a means of defense against the host organism, and that *Agrobacterium* species have evolved strategies to overcome plant-defense responses and to utilize the plant's machinery for their own benefit. Indeed, *Agrobacterium* has been shown to adopt and hijack many basic cellular processes and factors during the transformation of its host (for recent reviews see Lacroix et al., 2006a; Tzfira, 2006; for recent reviews see Tzfira and Citovsky, 2006). In the following sections of this chapter, we describe the current knowledge on these factors and mechanisms and propose a model for the intracellular transport of the *Agrobacterium* T-complex.

2 STRUCTURE AND FUNCTION OF THE T-COMPLEX

2.1 Structural requirements for T-complex subcellular transport

The need for T-DNA folding during its transport through the cell's compartments may be inferred from several constraints that limit the movement of large DNA molecules in an environment such as the cytoplasm of a eukaryotic cell, and thus their nuclear import. Three distinct barriers are generally recognized for the macromolecule's diffusion in the aqueous cellular compartment: fluid-phase viscosity, binding and crowding. In the cytoplasm, fluid-phase viscosity has been shown to be similar to that of water, but molecular crowding can reduce the mobility of large molecules, and binding may constitute a major impairment to DNA movement because DNA molecules are densely charged polyanions that may interact with cellular components (Verkman, 2002, and references therein). First, experiments realized in mammalian cells have shown that diffusion of naked DNA molecules (namely circular or linear plasmid DNA molecules) is extremely slow in the cytoplasm, and is negatively correlated with the size of the DNA segment (Leonetti et al., 1991; Lukacs et al., 2000). Specifically, over 250 bp, the mobility of DNA molecules is strongly reduced, which by extension suggests that naked DNA molecules of size typical to T-DNA (about 20 kb) cannot reach the nucleus by simple diffusion (Lukacs et al., 2000; Lechardeur and Lukacs, 2002; Dean et al., 2005). The cytoplasm is a crowded environment (Luby-Phelps, 2000), in which the passive diffusion of DNA molecules is impaired by interactions with cytoskeletal components, and perhaps more specifically with the actin network (Dauty and Verkman, 2005). Note that diffusion of chemically neutral dextran molecules of comparable size to 20 kb long DNA is not impaired, showing that not only the molecule's size but also its ability to chemically interact with other molecules in the cytoplasm affects the diffusion of DNA in the cytosol. Second, the T-DNA molecule is transferred in the form of a T-strand, namely a single-stranded DNA (ssDNA) corresponding

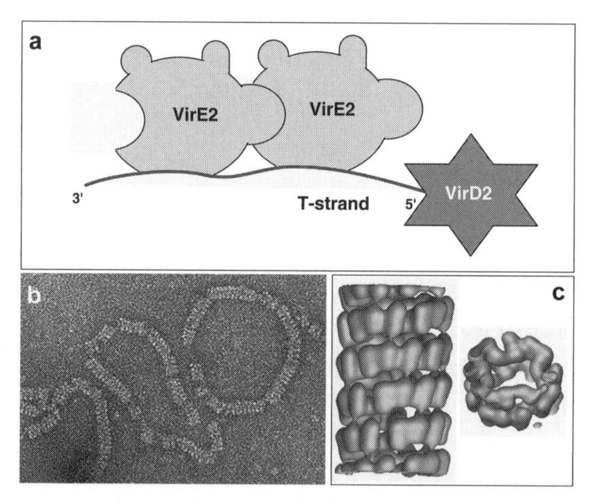

Figure 10-1. The T-complex structure. (a) The T-complex is composed of the T-strand, a single VirD2 which is bound to its 5' end, and numerous VirE2 molecules coating its entire length. (b) A complex of VirE2 proteins with ssDNA molecules, as seen by transmission electron microscopy. (c) A three-dimensional reconstruction of VirE2-ssDNA complexes showing the outer (left) and inner (right) structure of the solenoidal complex. Panels b and c reproduced with permission from (Abu-Arish et al., 2004).

to the coding strand of the T-DNA region (Lessl and Lanka, 1994). In the absence of packaging proteins, free T-strand molecules would most likely exist as polymeric random coils, forming a large structure that cannot be directed to the nuclear pore. Indeed, the typical size of a polymeric random coil is approximately the geometric mean of its extended length and its persistence length (Landau and EM, 1980; Briels, 1986), about 300 nm for a randomly coiled free ssDNA corresponding to a typical T-DNA size of 20 kb. Molecules of this size cannot move freely in the cytoplasm and are much larger than the nuclear-pore exclusion limit and than the nuclear pore itself. Moreover, recent studies have shown that remarkably long T-DNAs can be transferred and integrated into the host genome; indeed, by using binary vectors specifically designed to carry large inserts, it was demonstrated that T-DNA of up to 150 kb can be used for *Agrobacterium*-mediated transformation of tobacco (Hamilton et al., 1996) and tomato (Frary and Hamilton, 2001); such large molecules obviously need to be packaged in order to be transported in the cytoplasm toward the nucleus. Consequently, the T-strand must be organized in a specific spatial structure, allowing its movement in the cytoplasm and its entry into the nucleus; this structural organization must rely on interactions with packaging proteins. A multi-molecular complex composed of the T-strand and its associated bacterial and host proteins is likely to be the structure imported from the cytoplasm into the nucleus of the host cell. Nuclear import of naked DNA might also occur via interactions between regulatory sequences present on these DNA molecules and host transcription factors, which are synthesized in the cytoplasm before their import into the nucleus *Agrobacterium*-mediated transformation is not sequence-dependent, however, pointing to a direct role for the accompanying protein chaperones of the T-strand.

2.2 T-complex formation

Two bacterial virulence-induced proteins, namely the aforementioned VirD2 and VirE2, have been found to form the core of the T-complex, by their association with the T-strand (Ward and Barnes, 1988; Young and Nester, 1988; Citovsky et al., 1989; Sen et al., 1989). These two proteins are strictly required for the virulence of *Agrobacterium*, as shown by mutant studies (Stachel et al., 1985), and an increasing amount of data has allowed us to understand their multiple functions during the transfer of T-DNA and particularly during its nuclear import (Tzfira and Citovsky, 2002; Tzfira et al., 2005; Lacroix et al., 2006a).

The second bacterial protein implicated in the formation of the mature T-complex is VirE2. Along with other *Agrobacterium* effector proteins, VirE2 is most probably translocated to the plant cell independently of the VirD2-conjugated T-strand (Vergunst et al., 2000). Indeed, the virulence of a *virE2* mutant *Agrobacterium* strain can be complemented by co-infiltration with *Agrobacterium* strains containing *virE2* but not T-DNA (Otten et al., 1984), or by expression of VirE2 in the host-plant cells (Citovsky et al., 1992; Gelvin, 1998). Moreover, a functional genetic assay was employed to demonstrate the independent translocation of VirE2, and of other Vir proteins (namely VirD5, VirE3 and VirF), through the VirB/D4 channel (Vergunst et al., 2000; Schrammeijer et al., 2003; Vergunst et al., 2003; Lacroix et al., 2005; Vergunst et al., 2005). The formation of mature T-complex begins, in the host-cell cytoplasm, with the presumed association of the VirD2-conjugated T-strand with the VirE2 molecules. Indeed, VirE2 molecules bind cooperatively and nonspecifically to ssDNA with high affinity in common with most ssDNA-binding proteins (Christie et al., 1988; Citovsky et al., 1989; Sen et al., 1989). Inside the bacterial cell, the association between the T-strand and the VirE2 molecule is most likely prevented by VirE1, a VirE2 chaperone protein (Deng et al., 1999; Sundberg and Ream, 1999; Frenkiel-Krispin et al., 2006). Though initially thought to be implicated in the export of VirE2 from bacteria to host cells (Sundberg et al., 1996), more recent data indicate that VirE1 is not essential for this process (Vergunst et al., 2003). Like VirD2, VirE2 probably has multiple functions during T-DNA transfer and integration, besides its role in nuclear import. For example, VirE2 has been reported to form a channel-like structure in an artificial double-layer lipid membrane (Dumas et al., 2001), which suggests that this protein could also provide a route for the T-DNA to penetrate the host cell's periplasmic membrane. Interestingly, recent data indicate that the VirE1-VirE2 complex retains the ability to bind ssDNA with the same affinity as VirE2 alone (Duckely et al., 2005), and leads to the formation of a similar helicoidal structure. Moreover, the VirE1-VirE2 complex was also able to form channels in artificial lipid bilayer membranes (Duckely et al., 2005), suggesting that VirE1, if not exported to the plant-cell cytosol, might play a role in the association with VirE2 during the translocation of T-DNA.

2.3 The T-complex's three-dimensional structure

Association of the T-strand with both VirE2 and VirD2 thus results in the formation of a mature T-complex. To gain direct insight into the structural features of this macromolecular complex, the molecular assembly of a ssDNA from the bacteriophage M13 with purified VirE2 molecules was examined by scanning transmission electron microscopy and single-particle image-processing methods (Citovsky et al., 1997; Abu-Arish et al., 2004). These studies led to the discovery of a multi-molecular complex (*Figure 10-1b and c*) composed of a semi-rigid coiled "telephone cord"-like filament, with a hollow helical structure (Citovsky et al., 1997; Abu-Arish et al., 2004). The outer diameter is approximately 15 nm and in length the complex rises about 1 nm per 16 bases. On average, one turn of the coil contained 63.6 bases of DNA and bound to 3.4 molecules of VirE2; the helical pitch measured about 4.4 nm and the outer diameter 14 nm. For example, a typically sized 22-kb T-complex would have a total length of about 1.4 μm (Citovsky et al., 1997). Each molecule of VirE2 covers roughly 19 bases. Note that for a large T-DNA of 150 kb, 8000 molecules of VirE2 would be necessary to form the T-complex. Accordingly, additional copies of the *vir*E2 gene are required in the *Agrobacterium* strain to allow the successful transfer of this long T-DNA (Hamilton et al., 1996; Frary and Hamilton, 2001). The length of such a complex would reach about 10.5 μm, i.e. the same order of magnitude as the host cell itself, which raises the question of the need for a higher level of folding of the T-complex in order to ensure its subcellular transport. Nevertheless, the outer diameter of the VirE2-ssDNA assembly [approximately 15 nm, (Citovsky et al., 1997; Abu-Arish et al., 2004)] is compatible with the size-exclusion limit of receptor-mediated transport through the nuclear pore, which can reach up to 39 nm (Forbes, 1992; Pante and Kann, 2002; Suntharalingam and Wente, 2003; Fahrenkrog et al., 2004); however, it is much too large to pass the nuclear pore by free diffusion. Activation of the host-cell nuclear-import machinery is likely mediated by the VirD2 and VirE2 proteins. Consistently, T-DNA nuclear import was inhibited by non-hydrolyzable analogs of GTP, showing that this process is energy-dependent in a mannar similar to physiological import of nuclear proteins (Zupan et al., 1996).

2.4 Protection from host-cell nucleases

Besides packaging the T-strand in a way that allows its subcellular transport, protein coating provides the T-strand with protection against host cytosolic nucleases. Although the exact nature of the nucleases present in the cytosol of eukaryotic cells has not been fully elucidated, metabolic instability of nucleic acid in the cytosol is commonly observed when naked DNA or RNA is introduced into these cells (Lechardeur and Lukacs, 2002; Dean et al., 2005). It was shown, for example, that fluorescently labeled DNA fragments are degraded in the cytosol of mammalian cells or by cytosolic extracts (Lechardeur et al., 1999), likely via an as yet unidentified cytosolic, calcium-sensitive nuclease (Pollard et al., 2001). The role of the T-strand-coating proteins as a shield against nuclease activities was demonstrated by in-vitro nuclease-degradation assays of artificially reconstituted T-complexes. Indeed, a T-strand covalently linked to a VirD2 molecule was protected against the action of exonucleases (Durrenberger et al., 1989; Jasper et al., 1994), and a T-strand coated by VirE2 molecules was resistant to endonuclease activity (Christie et al., 1988; Sen et al., 1989). These results are also consistent with the observation that when integration of truncated T-DNA occurs, more deletions are generally observed on the LB (left border) side (3' end, unprotected) than on the RB (right border) side (5' end, covalently linked to a VirD2 molecule). Indeed, the percentage of deletions in integrated T-DNA was higher at the LB than at the RB, as shown in tobacco and Arabidopsis (Tinland, 1996), aspen (Kumar and Fladung, 2002), barley (Stahl et al., 2002), and rice (Kim et al., 2003; Afolabi et al., 2004). The importance of VirE2 in T-strand protection is also supported by experiments showing the instability of the T-strand of a VirE2 mutant strain of *Agrobacterium* inside the host cell, as compared to the wild-type strain (Yusibov et al., 1994).

3 CYTOPLASMIC TRANSPORT

As a consequence of its large size, cytoplasmic movement of the T-complex will be restricted by the dense structure of the cytoplasm (Luby-Phelps, 2000), and is most likely to be mediated by host-cytoskeleton-associated motors. Indeed, the nuclear import of various host (i.e. transcription factors) and pathogen (i.e. viral proteins and viral DNA-protein complexes) molecules (Greber and Way, 2006) has been reported to be mediated not only by interactions with the host's nuclear-import machinery, but also with the various molecular motors. Microscopic studies, for example, have demonstrated that nuclear import receptors co-localize with both actin and microtubule networks in plant cells (Smith and Raikhel, 1998), which suggests a role for the cell's cytoskeleton and possibly molecular motors in transporting these importins toward the nuclear-pore complex. In another example, the African swine fever virus protein p54 was reported to interact with a microtubulae-associated motor complex during its voyage to the cell's nucleus (Alonso et al., 2001). While current knowledge on the role of cytoskeletal elements and molecular motors in the intracellular transport of proteins and DNA-protein complexes has been derived mostly from analyzing the infectious pathways of mammalian viruses, recent data suggest that *Agrobacterium* may harness the host's intercellular transport system for the transport of its T-complex (Salman et al., 2005).

Using a combination of biochemical and biophysical techniques, Salman et al. (Salman et al., 2005) studied the movement of fluorescently labeled VirE2-ssDNA complexes in a reconstituted cytoskeletal network that contained microtubules, F-actin and associated motor proteins from *Xenopus* frog egg extract. Because native VirE2 molecules contain a nuclear-localization signal (NLS) that is not recognized by the animal cell's nuclear-import machineries (see below and Guralnick et al., 1996; Tzfira and Citovsky, 2001; Salman et al., 2005), the authors used a mutated form of VirE2, which was capable of being transported into animal cell nuclei (Guralnick et al., 1996; Salman et al., 2005). This 'animalized' form of VirE2 (anVirE2) differed from the native, 'plant-specific' VirE2 (plVirE2) by point mutations in one of the VirE2 NLS regions, which allowed the authors to compare the movement of anVirE2- and plVirE2-ssDNA complexes. Using automated single-particle-tracking methods coupled with statistical analysis, the authors discovered that anVirE2-ssDNA complexes are actively delivered along the reconstituted microtubule (but not actin) network (*Figure 10-2*), whereas plVirE2-ssDNA is incapable of active movement in the same system. Application of AMP-PNP—an ATP analog known to suppress the activity of kinesin-type motors by anchoring the motors and their cargo to the microtubule system (Lasek and Brady, 1985)—did not interfere with the active movement of anVirE2-ssDNA molecules. In contrast, the application of sodium orthovanadate effectively restricted their movement. Because sodium orthovanadate specifically inhibits dynein-motor activity (Shimizu, 1995), the authors suggested an active role for dynein motors in the intracellular transport of T-complexes. Furthermore, their results also indicated the requirement of an active NLS and pointed to a functional link between the presence of such a signal and the dynein-dependent transport along microtubules, which is similar to the NLS-dependent movement of many viral proteins and DNA-protein complexes (Greber and Way, 2006).

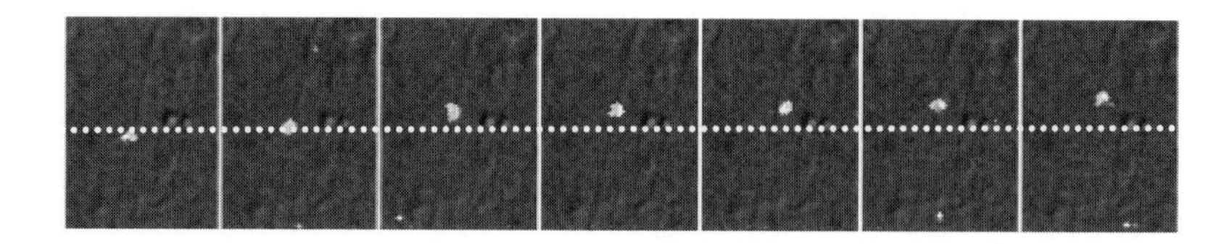

Figure 10-2. Movement of artificial T-complexes along the microtubule network. An example an anVirE2-ssDNA complex path along a microtubule. Reproduced with permission from (Salman et al., 2005).

The lack of movement of plVirE2-ssDNA complexes in animal systems (Salman et al., 2005) raises the question of how movement occurs in the plant host, and what type of motors *Agrobacterium* may use in plant cells. Preliminary evidence from a similar assay using tobaco BY-2 cell-free extracts points to a functional interaction between the plVirE2-ssDNA complex and active transport mediated by microtubule-associated motors.

We have recently isolated a component of a putative dynein-like plant motor, which may be involved in the transformation process (Tzfira, T., unpublished data). The dynein-like light chain (DLC3) protein shows high homology to dynein light chains from animals and human cells and interacted with another host protein, VIP1, which has been shown to be involved in the nuclear import of VirE2 and of the T-complex in plant cells by functioning as a mediator between VirE2 and the plant nuclear-import machinery (see below and Tzfira et al., 2001, 2002). DLC3 also colocalized with the plant microtubule system, and it is thus possible that it serves as a molecular link between VIP1-VirE2-T-DNA complexes and the microtubule network. Thus, whereas further studies are still required to identify additional components of such putative plant dynein-motors, and to determine the molecular mechanism of T-complex movement in plant cells, DLC3 may represent a new member of an unknown family of plant molecular motors (Lawrence et al., 2001; King, 2002; Scali et al., 2003) or may be part of the plant kinesin superfamily.

4 NUCLEAR IMPORT

The core T-complex, composed of ssT-DNA associated with a VirD2 molecule at its 5'end and coated with VirE2 molecules, is probably a stable structure during its travel inside the plant cell, considering the nature of the chemical interactions between the T-strand and its associated proteins [covalent with VirD2 (Herrera-Estrella et al., 1988; Ward and Barnes,

1988; Young and Nester, 1988), and noncovalent but cooperative and with strong affinity with VirE2 (Christie et al., 1988; Citovsky et al., 1989; Sen et al., 1989)], and consistent with the requirement for protection against nucleases during travel through the host cytosol. Its nuclear import is mediated by interactions with the host nuclear-import-machinery proteins, as well as with other plant and bacterial factors, which may bind to the coating proteins of the T-complex in a relatively more transient manner (for recent reviews see Tzfira et al., 2005; Lacroix et al., 2006b and other chapters in this volume).

4.1 Function of bacterial proteins in the nuclear import of T-complexes

VirD2 and VirE2 are likely to have different and complementary functions, as detailed further on, mediated by their interactions with various host and bacterial factors. The role of these two bacterial proteins was initially examined by introducing purified and labeled proteins and DNA molecules into plant cells. First, using microinjection in stamen hair cells of *Tradescentia virginiana*, Zupan et al. (Zupan et al., 1996) showed that fluorescently labeled ssDNA alone remains in the cytoplasm, while co-injection with VirE2 (without VirD2) leads to its transport into the nucleus. In contrast, double-stranded (ds) DNA remained cytoplasmic whether it was injected in the presence or absence of VirE2. The ability of VirE2 to mediate nuclear import of the T-strand in the absence of VirD2 is supported by several studies using VirD2 mutated in its NLS sequence. Indeed, *Agrobacterium* strains carrying this mutation show strongly attenuated virulence but always retain a residual ability to induce tumors (Shurvinton et al., 1992) or T-DNA gene expression (Narasimhulu et al., 1996). Moreover, an *Agrobacterium* strain with both VirE2 and VirD2 mutated in their NLS sequences was able to induce tumors in VirE2-expressing transgenic tobacco plants but not in wild-type plants (Gelvin, 1998). Second, using permeabilized and evacuolated tobacco protoplasts, Ziemienowicz et al. (Ziemienowicz et al., 2001) showed that a 25-bp single-stranded oligonucleotide linked to VirD2 is directed to the nucleus in the absence of VirE2, whereas longer ssDNA molecules (i.e. 250 or 1000 bp) require both VirD2 and VirE2 in order to be imported into the nucleus. In this system, an oligonucleotide bound to VirE2 molecules (without VirD2) remained in the cytoplasm, whereas VirE2 molecules alone were imported into the nucleus. The difference between the studies with respect to the function of VirE2 may reflect the difference in the system employed (microinjection of intact cells (Zupan et al., 1996) or permeabilized and evacuolated protoplasts (Ziemienowicz et al., 2001)).

In spite of minor inconsistencies between these different studies, a first model for the nuclear import of the T-strand can be proposed from these data. The molecular composition of the T-complex, i.e. its length coated by VirE2 molecules while a VirD2 molecule is covalently bound to its 5' end (Sheng and Citovsky, 1996), suggests that the polarity of this complex may play an important role in its nuclear import. Under natural conditions, both VirD2 and VirE2 likely contribute to the nuclear import of the T-DNA. Potentially, VirD2 is sufficient to target the T-DNA to the nuclear pore, while VirE2 is required for its passage through the pore. As the T-complex is typically much longer than the channel of the nuclear pore, the single VirD2 molecule will arrive in the nucleus at a relatively early stage. VirE2-binding would then present the T-DNA in a continuous structure compatible with its passage through the nuclear pore (Ziemienowicz et al., 2001) and provide an interaction with the host nuclear-import machinery to ensure complete delivery of a long T-DNA to the nucleus in a polar manner (Sheng and Citovsky, 1996; Citovsky et al., 1997; Tzfira et al., 2000).

4.2 Interactions of the T-complex with the host nuclear-import machinery

The VirD2 protein is imported into the host-cell nucleus via an evolutionarily conserved mechanism. Indeed, the nuclear localization of VirD2 fused to a reporter protein has been demonstrated not only in plant cells (Herrera-Estrella et al., 1990; Howard et al., 1992; Citovsky et al., 1994; Ziemienowicz et al., 2001), but also in yeast and animal cells (Guralnick et al., 1996; Relic et al., 1998; Ziemienowicz et al., 1999; Rhee et al., 2000). Moreover, the direct interaction between VirD2 and *Arabidopsis thaliana* karyopherin α—and by implication with the nuclear-import machinery of the host cell—was demonstrated by yeast two-hybrid and functional experiments (Ballas and Citovsky, 1997). Among the two distinct putative NLSs found in the VirD2 sequence, a bipartite NLS in its C-terminal part and a monopartite NLS in its N-terminal part (Herrera-Estrella et al., 1990; Howard et al., 1992), only the former was suggested to be essential for VirD2 and T-DNA nuclear import (Howard et al., 1992; Koukolikova-Nicola et al., 1993; Rossi et al., 1993; Mysore et al., 1998; Ziemienowicz et al., 2001). Indeed, mutations in the C-terminal NLS reduce *Agrobacterium* virulence, whereas mutations in the N-terminal NLS have no significant effect (Howard et al., 1992); consistent with this, mutations in the C-terminal, but not N-terminal NLS disrupt the nuclear localization of VirD2 in plant cells (Koukolikova-Nicola et al., 1993; Ziemienowicz et al., 2001). It is important to note that VirD2 also contains a secretion signal for export from the bacterium via interaction with VirD4 at the VirB channel. Both signals are rich in positively-charged amino acids, and there may possibly be some overlap in their functions.

Similarly, the VirE2 protein localizes in the nucleus of plant cells (Citovsky et al., 1992; Citovsky et al., 1994; Ziemienowicz et al., 2001); however, it remains in the cytoplasm in heterologous systems, such as yeast and mammalian cells (Guralnick et al., 1996; Rhee et al., 2000; Tzfira and Citovsky, 2001; Tzfira et al., 2001; Citovsky et al., 2004). Although VirE2 contains NLS-motif sequences, it did not interact directly with host karyopherin α in a yeast two-hybrid assay (Ballas and Citovsky, 1997), suggesting the intervention of another host factor to mediate the nuclear import of VirE2. Using yeast two-hybrid screening, VIP1 (VirE2 interacting protein 1) was identified as interacting with VirE2 (Tzfira et al., 2001). The VIP1 protein, containing a basic leucine zipper (bZIP) motif, was located in the eukaryotic cell nucleus, likely via a conserved mechanism as it contains a conventional NLS and interacts directly with karyopherin α (Tzfira et al., 2002; Citovsky et al., 2004). In non-plant systems, such as yeast and animal cells, the nuclear import of a green fluorescent protein (GFP)-VirE2 fusion protein was induced by co-expression of VIP1 (*Figure 10-3*). In plant cells, the reduction of VIP1 expression by antisense technology resulted in impaired nuclear targeting of GUS-VirE2, although GUS-VirD2 remained nuclear. Therefore, the antisense expression of VIP1 did not interfere with the general plant nuclear-import pathway, but rather specifically inhibited VirE2 nuclear import (Tzfira et al., 2001). Consistent with this, previous reports have shown that the absence of nuclear import of VirE2 in heterologous systems may be due to the nonfunctionality of its NLSs (Citovsky et al., 2004; Salman et al., 2005). The native VirE2 protein was not localized in the nucleus of animal cells; however, when this protein was slightly modified by reversing the order of two adjacent amino acids in either NLS, it was directed to the nucleus (Citovsky et al., 2004; Salman et al., 2005). Moreover, the rate of both transient and stable plant transformation is influenced by VIP1 activity: the efficiency of T-DNA transfer was strongly reduced in VIP1-antisense tobacco plants (Tzfira et al., 2001), while the opposite effect was observed in VIP1-overexpressing plants (Tzfira et al., 2002). VIP1 probably functions by forming a molecular adaptor between the T-complex and the karyopherin α-mediated nuclear-import machinery of the host cell. Indeed, the formation of ternary complexes was observed in vitro, comprised of VIP1, VirE2 and ssDNA (Tzfira et al., 2001) and VirE2, VIP1 and karyopherin α (Tzfira et al., 2001; Ward et al., 2002; Citovsky et al., 2004). VIP1, as well as other plant proteins, may thus represent a set of limiting host factors for *Agrobacterium*-mediated transformation. Their absence in a functional form may explain the recalcitrance of animal cells and of less susceptible to transformation plant species (e.g. Gelvin, 2003a).

Another translocated *Agrobacterium* protein, VirE3, is able to partially mimic VIP1 activity (Lacroix et al., 2005). GFP-tagged VirE3 exhibited nuclear localization in mammalian and plant cells mediated by two functional NLSs located in its N-terminal region. Moreover, VirE3 and VIP1 share similar properties: the VirE3 protein interacted with VirE2 and karyopherin α in a yeast two-hybrid assay (but not with VIP1), and was able to assist in the nuclear import of VirE2 in animal cells (*Figure 10-3*) and in VIP1-antisense plant cells. Moreover, in VIP1-antisense plants, VirE3 overexpression partially restored VirE2 nuclear import and susceptibility to *Agrobacterium* T-DNA transfer (Lacroix et al., 2005). *Agrobacterium* mutant analyses have shown that VirE3 is not essential for tumor formation on tobacco and *Kalanchoe* leaves *in vitro* (Winans et al., 1987; Kalogeraki et al., 2000). However, these two species are highly susceptible to *Agrobacterium*, and VirE3 may well play a role in *Agrobacterium*-mediated genetic transformation of other, species which lack an active form of VIP1 to act as a host-range factor (Hirooka and Kado, 1986).

Interestingly, unrelated strategies may be used by different strains of *Agrobacterium* to achieve transport of the T-strand inside the plant cell. Indeed, some strains of *A. rhizogenes* are able to infect plants, and thus to transfer and integrate DNA into the host genome, although they do not encode VirE1 or VirE2. In those strains, another protein—namely GALLS—has been suggested to fulfill a function similar to that of VirE2 (Hodges et al., 2004). Indeed, the pathogenicity of an *A. tumefaciens* strain mutated in the *vir*E1 and *vir*E2 genes was restored by mixed infection with a strain carrying the *GALLS* gene. Because GALLS and VirE2 sequences do not share any homology, the mechanism by which GALLS assists in T-complex nuclear import was suggested to be different from VirE2 activity (Hodges et al., 2004).

VirD2, the VIP1/VirE2 or VirE3/VirE2 complexes, and, by implication, the entire T-complex are imported into the host nucleus via a pathway, in which NLS-containing proteins are recognized by karyopherin α (Jans et al., 2000). In animals and yeast, this widely-conserved pathway depends on the nucleocytoplasmic transport receptor itself, karyopherin β (Görlich et al., 1995; Merkle, 2001). The latter protein mediates interaction with and passage through the nuclear pore, ferrying along NLS-containing proteins via the karyopherin α adapter (Görlich et al., 1995; Merkle, 2001). Nuclear import in plants may also occur via a karyopherin β-independent pathway (Hubner et al., 1999). On the other hand, an *Arabidopsis* mutant in a putative karyopherin β was resistant to *Agrobacterium* transformation (Zhu et al., 2003); moreover, *Arabidopsis* karyopherin α does not contain sequences known to be required for binding to the nuclear pore or to the regulatory GTPase Ran, although it carries the conserved karyopherin β-binding motif. This suggests that an as yet unidentified plant karyopherin β may play a role during T-complex nuclear import. Inside the nucleus, release of the imported protein is mediated by Ran. Accordingly, a nonhydrolyzable GTP analog should inhibit the nuclear import process. Indeed, GTPγS inhibits VirD2 and VirE2 nuclear import in plant cells (Zupan et al., 1996; Ziemienowicz et al., 2001).

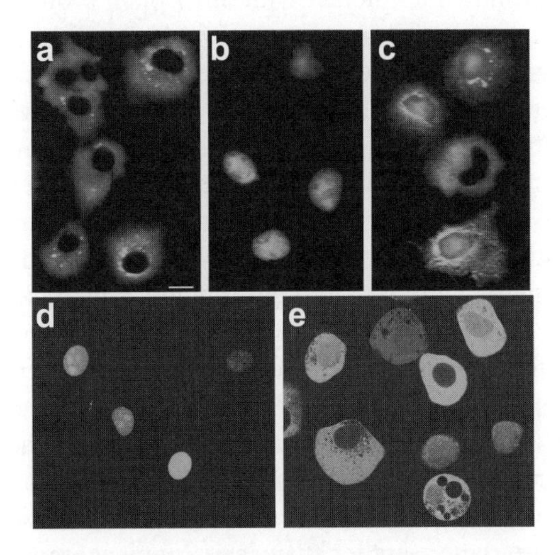

Figure 10-3. Host and bacterial proteins facilitate nuclear import of VirE2 in mammalian cells. (a) GFP-VirE2. (b) GFP-VIP1. (c) GFP-VirE2 + VIP1. (d) GFP-VirE3. (e) GFP-VirE2 + VirE3. Panels a, b, and c reproduced with permission from (Tzfira et al., 2001); panels d and e reproduced with permission from (Lacroix et al., 2005).

4.3 Regulation of T-DNA nuclear import

Plant factors may function indirectly during nuclear import of the T-complex by regulating that import. This may be the result of a defense reaction of the host cell against the presence of foreign DNA, or an adaptation of *Agrobacterium* to ensure efficient infection by utilizing such plant factors. The nuclear import of VirD2 may be regulated by various phosphorylation pathways (Bakó et al., 2003; Tao et al., 2004). Using a yeast two-hybrid assay, the C-terminal region of VirD2 was reported to interact with a protein from tomato, DIG3, a type 2C serine/threonine phosphatase (PP2C) (Tao et al., 2004). Overexpression of this protein inhibited the nuclear import of a GUS-VirD2 fusion in cultured tobacco cells and an *Arabidopsis* mutant in a PP2C gene (*abi1*) showed higher susceptibility to *Agrobacterium* infection. It was thus suggested that PP2C might negatively regulate VirD2 nuclear import, most probably by dephosphorylation of the VirD2 protein (Tao et al., 2004). This interaction of VirD2, and its phosphorylation by CAK2M [a plant ortholog of cyclin-dependent kinase-activating kinase (e.g., Bakó et al., 2003)], further support the notion that phosphorylation of VirD2 may also take place in the nucleus and may regulate the T-complex nuclear-import process.

VirD2 was also shown to interact with other plant proteins belonging to the family of plant cyclophilins, namely RocA, RocB and CypA (Deng et al., 1998). This study also revealed that cyclosporin A, which specifically inhibits interaction of cyclophilins with their target proteins, also inhibited T-DNA transfer during *Agrobacterium* infection (Deng et al., 1998). In plants, cyclophilins represent a large family of proteins (Romano et al., 2004); they are generally implicated in protein maturation, but with diverse cellular functions. Their role in T-DNA nuclear import and/or integration is still unknown, but these proteins could act by maintaining VirD2 in a conformation compatible with its nuclear import. Moreover, VIP2 was identified as another plant interactor of VirE2 ((Tzfira et al., 2000) and Anand, A., Krichevsky, A., Tzfira, T., Schornack, S., Lahaye, T., Citovsky, V., and Mysore, K.S., unpublished data), although whether this protein plays a role during nuclear import of the T-complex remains unknown. Unpublished data suggest that VIP2 is more likely to be involved in a later step of T-DNA integration, rather than in its nuclear import (Anand, A., Krichevsky, A., Tzfira, T., Schornack, S., Lahaye, T., Citovsky, V., and Mysore, K.S., unpublished data). Indeed, in *Nicotiana benthamiana* plants with VIP2 expression reduced by antisense technology, stable integration, but not transient T-DNA expression, is reduced. However, we cannot rule out that VIP2 may also be implicated in T-DNA import or its regulation, and that this role simply did not appear under these experimental conditions.

5 INTRANUCLEAR MOVEMENT OF THE T-COMPLEX

Little is known about movement of the T-complex inside the nucleus. In general, macromolecule trafficking in the nucleus is suggested to be under tight regulation (Phair and Misteli, 2000). For example, the transcription factors must reach their specific target DNA sequences (Zaidi et al., 2004; Zaidi et al., 2005). Although the site of T-DNA integration in the host genome is thought to be random (Alonso et al., 2003), the T-complex has to find a site in the nucleus where its interaction with the host chromatin is possible, and thus where integration may potentially occur. Several mechanisms may be invoked, although by no means exclusively, to explain this process. First, proteins of the T-complex may bind with host factors that themselves are involved in interactions with host genomic DNA, and direct the whole T-complex to a potential site of integration and/or play a role in the process of integration itself. In alfalfa cell nuclei, for example, VirD2 interacted with a plant ortholog of a cyclin-dependent kinase-acivating kinase, or CAK2M (Bakó et al., 2003). CAK2Ms are known to bind to and phosphorylate RNA polymerase II, which can recruit the TATA-box-binding protein. Moreover, VirD2 interacted with the TATA-box-binding protein in vitro and in *Agrobacterium*-transformed *Arabidopsis* cells (Bakó et al., 2003). Second, VIP1 is a multifunctional protein which probably also plays a role after nuclear import of the T-complex, perhaps by mediating the interaction between T-DNA and the host chromatin. Indeed, VIP1 has been shown to interact with histone H2A (Li et al., 2005; Loyter et al., 2005). A reverse genetic approach demonstrated that one domain in the N-terminal half of VIP1 is implicated in the interaction with karyopherin α and nuclear import of VirE2, while another domain in the C-terminal half of the protein is responsible for interaction with H2A and presumably plays a role in the T-DNA integration (Li et al., 2005). The importance of histone for *Agrobacterium*-mediated transformation has been previously shown: in particular, an *Arabidopsis* mutant in H2A was deficient in T-DNA integration but not transient expression (Mysore et al., 2000), and the susceptibility of *Arabidopsis* root cells to *Agrobacterium* transformation correlated with the level of expression of the *H2A-1* gene (Yi et al., 2002). Third, the promontory role of double-strand breaks (DSBs), and DSB-repair proteins, in T-DNA integration into the host genome was recently demonstrated (reviewed in Tzfira et al., 2004a). In this pathway, T-DNA is converted to a double-stranded intermediate before its integration, which most likely occurs by nonhomologous end-joining recombination (Puchta, 1998; Chilton and Que, 2003;

Tzfira et al., 2003; Tzfira et al., 2004a). DSBs then appear as preferential sites for the integration of T-DNA, and more generally of any foreign DNA, into the host genome. Moreover, DSB-repair proteins are efficiently recruited to the DSB sites (Mirzoeva and Petrini, 2001; Drouet et al., 2005), likely via intermediary histone modifications, such as H2A phosphorylation (Pilch et al., 2003; Shroff et al., 2004; Unal et al., 2004; van Attikum et al., 2004). It was thus suggested that DSB-repair proteins might be able to assist in directing the T-complex to a potential site of integration by interacting with components of that T-complex (Chilton and Que, 2003; Tzfira et al., 2003). Note that T-complex uncoating, as well as double-strand synthesis, are also required before T-DNA integration (see Tzfira et al., 2004b; Lacroix et al., 2006a), although it is not clear whether total or partial degradation of the coating proteins occurs before or after T-complex targeting to a potential site of integration in the host genome, or if these processes are coupled.

6 FROM THE CYTOPLASM TO THE CHROMATIN: A MODEL FOR T-COMPLEX IMPORT

The accumulated data on T-complex transport allows us to propose a working model for T-DNA transport inside the plant cell (*Figure 10-4*). The T-DNA's voyage begins with its entry into the plant-cell cytoplasm in the form of a ssDNA segment (T-strand), covalently linked with VirD2 at its 5' end. VirE2 molecules, translocated independently of the bacterial cell, will coat the T-strand's length and package the T-DNA in a semi-rigid helical structure with the fragile ssDNA wrapped and effectively sequestered inside the cylindrical protein shell (*Figure 10-4, step 1*). This becomes the core T-complex, also called mature T-complex. The multi-molecular assembly is believed to be a relatively stable structure with which several other bacterial and plant factors will interact transiently in order to mediate and regulate its transport toward the nuclear pore.

The VirD2 protein, via its direct interaction with importin α and perhaps other components of the nuclear-import machinery, is able to pilot the T-DNA to the nuclear pore (*Figure 10-4, step 2*). Due to the extremely large size of the T-complex, a single VirD2 molecule is probably not sufficient to mediate its movement through the cell cytoplasm. The coating VirE2 molecules, besides protecting the T-strand against cellular nucleases, enable this transport by interacting with various host-cell factors. VirE2 molecules interact with VIP1 (*Figure 10-4, step 2*), which in turn interacts with importin α and facilitates the transfer of T-DNA to the nucleus (*Figure 10-4, step 3*). It seems that molecular motors, e.g. kinesin or a putative plant dynein complex interacting with microtubules, are directly involved in T-DNA movement to the nuclear pore. Cytoplasmic streaming or another actomyosin-based transport process is also possible as an effector of T-complex movement. Interactions with diverse host cytoplasmic proteins, such as the VirD2 interactors PP2C and cyclophilins, may also regulate the movement of the T-complex to and through the nuclear pore. Whereas the VirD2 molecule probably allows docking of the T-complex at the nuclear pore, VirE2 molecules (in association with VIP1) successively mediate the passage of the whole T-DNA into the nucleus, via their interaction with components of the host's cellular import machinery (*Figure 10-4, step 4*).

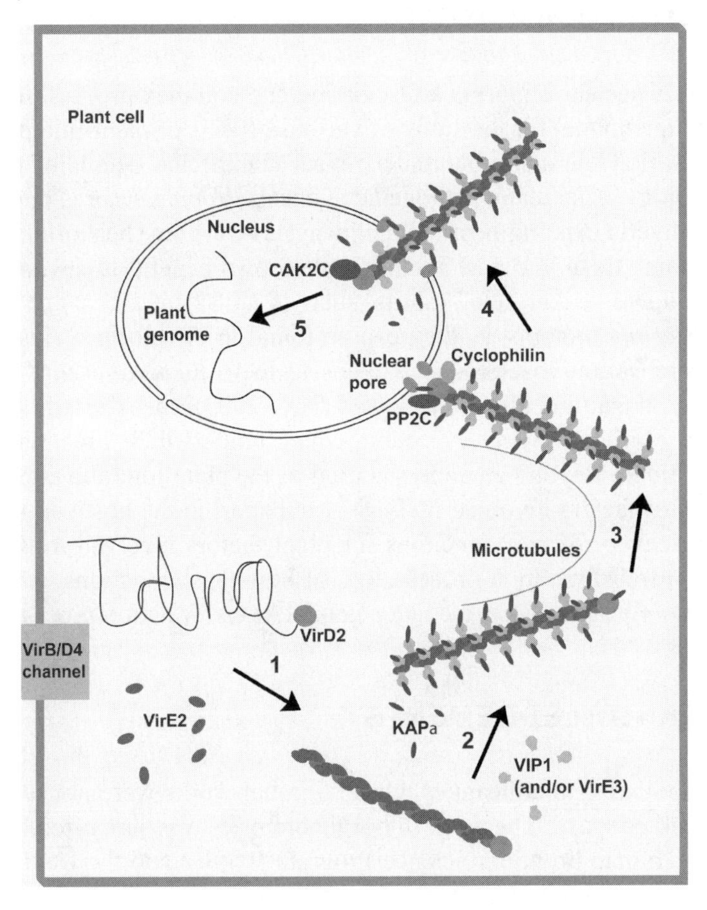

Figure 10-4. A model for T-complex cellular transport (see text for additional details).

Once in the nucleus, the T-complex is directed to its potential point of integration in the host chromatin (*Figure 10-4, step 5*). Several host factors are probably implicated in this process, although they have not yet been clearly identified. Potential host factors include: TATA-box-binding proteins that interact with VirD2, H2A histones or associated proteins that bind to VIP1, and proteins belonging to the DSB-repair machinery that binds to the double-stranded intermediate of the T-DNA before its integration by nonhomologous end-joining recombination. The T-complex must also be stripped of its associated proteins and undergo second-strand synthesis before its integration; however, the sequence of events that occurs inside the nucleus remains largely unknown.

7 FUTURE PROSPECTS

T-DNA nuclear import is a fascinating and complex process. It also offers a unique model for the study of various aspects of plant-microbe interactions, which can also potentially reveal unexpected details of the plant cell's biology. Like many pathogenic agents, *Agrobacterium* is opportunistic, and diverts existing host-cell pathways away from their original functions, turning them into dual agents for their own benefit. It is worth noting that functional "eukaryotic" motifs, such as NLS sequences, are found in *Agrobacterium* proteins, as they can be found in effector proteins translocated from various bacterial pathogens to their eukaryotic host. As suggested by Nagai and Roy (Nagai and Roy, 2003), these motifs probably originate from convergent evolution, which renders their functional annotation difficult. Beyond an understanding of the plant-microbe interactions, the study of factors involved in T-DNA transport in the host cell may lead to the discovery of new functions for plant factors involved in the *Agrobacterium* transformation process, and of how these functions can be utilized in the *Agrobacterium*-mediated genetic-transformation process.

8 ACKNOWLEDGEMENTS

We apologize to colleagues whose original works were not referred to due to lack of space. The work in our laboratories was supported by grants from the Human Frontiers Science Program (HFSP) and the US-Israel Binational Agricultural Research and Development Fund (BARD) to T.T. and M.E. and from the National Institutes of Health, National Science

Foundation, US-Israel Science Foundation (BSF), US Department of Agriculture, and BARD to V.C.

9 REFERENCES

Abel S, Theologis A (1994) Transient transformation of *Arabidopsis* leaf protoplasts: a versatile experimental system to study gene expression. Plant J 5: 421-427

Abu-Arish A, Frenkiel-Krispin D, Fricke T, Tzfira T, Citovsky V, Grayer Wolf S, Elbaum M (2004) Three-dimensional reconstruction of *Agrobacterium* VirE2 protein with single-stranded DNA. J Biol Chem 279: 25359-25363

Afolabi AS, Worland B, Snape JW, Vain P (2004) A large-scale study of rice plants transformed with different T-DNAs provides new insights into locus composition and T-DNA linkage configurations. Theor Appl Genet 109: 815-826

Alonso C, Miskin J, Hernaez B, Fernandez-Zapatero P, Soto L, Canto C, Rodriguez-Crespo I, Dixon L, Escribano JM (2001) African swine fever virus protein p54 interacts with the microtubular motor complex through direct binding to light-chain dynein. J Virol 75: 9819-9827

Alonso JM, Stepanova AN, Leisse TJ, Kim CJ, Chen H, Shinn P, Stevenson DK, Zimmerman J, Barajas P, Cheuk R, Gadrinab C, Heller C, Jeske A, Koesema E, Meyers CC, Parker H, Prednis L, Ansari Y, Choy N, Deen H, Geralt M, Hazari N, Hom E, Karnes M, Mulholland C, Ndubaku R, Schmidt I, Guzman P, Aguilar-Henonin L, Schmid M, Weigel D, Carter DE, Marchand T, Risseeuw E, Brogden D, Zeko A, Crosby WL, Berry CC, Ecker JR (2003) Genome-wide insertional mutagenesis of *Arabidopsis thaliana*. Science 301: 653-657

Bakó L, Umeda M, Tiburcio AF, Schell J, Koncz C (2003) The VirD2 pilot protein of *Agrobacterium*-transferred DNA interacts with the TATA box-binding protein and a nuclear protein kinase in plants. Proc Natl Acad Sci USA 100: 10108-10113

Ballas N, Citovsky V (1997) Nuclear localization signal binding protein from *Arabidopsis* mediates nuclear import of *Agrobacterium* VirD2 protein. Proc Natl Acad Sci USA 94: 10723-10728

Briels W (1986) The theory of polymer dynamics. Clarendon Press, Oxford

Chilton M-DM, Que Q (2003) Targeted integration of T-DNA into the tobacco genome at double-strand breaks: new insights on the mechanism of T-DNA integration. Plant Physiol 133: 956-965

Christie PJ, Ward JE, Winans SC, Nester EW (1988) The *Agrobacterium tumefaciens* virE2 gene product is a single-stranded-DNA-binding protein that associates with T-DNA. J Bacteriol 170: 2659-2667

Citovsky V, Guralnick B, Simon MN, Wall JS (1997) The molecular structure of *Agrobacterium* VirE2-single stranded DNA complexes involved in nuclear import. J Mol Biol 271: 718-727

Citovsky V, Kapelnikov A, Oliel S, Zakai N, Rojas MR, Gilbertson RL, Tzfira T, Loyter A (2004) Protein interactions involved in nuclear import of the *Agrobacterium* VirE2 protein *in vivo* and *in vitro*. J Biol Chem 279: 29528-29533

Citovsky V, Warnick D, Zambryski PC (1994) Nuclear import of *Agrobacterium* VirD2 and VirE2 proteins in maize and tobacco. Proc Natl Acad Sci USA 91: 3210-3214

Citovsky V, Wong ML, Zambryski PC (1989) Cooperative interaction of *Agrobacterium* VirE2 protein with single stranded DNA: implications for the T-DNA transfer process. Proc Natl Acad Sci USA 86: 1193-1197

Citovsky V, Zupan J, Warnick D, Zambryski PC (1992) Nuclear localization of *Agrobacterium* VirE2 protein in plant cells. Science 256: 1802-1805

Dauty E, Verkman AS (2005) Actin cytoskeleton as the principal determinant of size-dependent DNA mobility in cytoplasm: a new barrier for non-viral gene delivery. J Biol Chem 280: 7823-7828

Dean DA, Strong DD, Zimmer WE (2005) Nuclear entry of nonviral vectors. Gene Ther 12: 881-890

Deng W, Chen L, Peng WT, Liang X, Sekiguchi S, Gordon MP, Comai L, Nester EW (1999) VirE1 is a specific molecular chaperone for the exported single-stranded-DNA-binding protein VirE2 in *Agrobacterium*. Mol Microbiol 31: 1795-1807

Deng W, Chen L, Wood DW, Metcalfe T, Liang X, Gordon MP, Comai L, Nester EW (1998) *Agrobacterium* VirD2 protein interacts with plant host cyclophilins. Proc Natl Acad Sci USA 95: 7040-7045

Drouet J, Delteil C, Lefrancois J, Concannon P, Salles B, Calsou P (2005) DNA-dependent protein kinase and XRCC4-DNA ligase IV mobilization in the cell in response to DNA double strand breaks. J Biol Chem 280: 7060-7069

Duckely M, Oomen C, Axthelm F, Van Gelder P, Waksman G, Engel A (2005) The VirE1VirE2 complex of *Agrobacterium tumefaciens* interacts with single-stranded DNA and forms channels. Mol Microbiol 58: 1130-1142

Dumas F, Duckely M, Pelczar P, Van Gelder P, Hohn B (2001) An *Agrobacterium* VirE2 channel for transferred-DNA transport into plant cells. Proc Natl Acad Sci USA 98: 485-490

Durrenberger F, Crameri A, Hohn B, Koukolikova-Nicola Z (1989) Covalently bound VirD2 protein of *Agrobacterium tumefaciens* protects the T-DNA from exonucleolytic degradation. Proc Natl Acad Sci USA 86: 9154-9158

Fahrenkrog B, Koser J, Aebi U (2004) The nuclear pore complex: a jack of all trades? Trends Biochem Sci 29: 175-182

Forbes DJ (1992) Structure and function of the nuclear pore complex. Annu Rev Cell Biol 8: 495-527

Frary A, Hamilton CM (2001) Efficiency and stability of high molecular weight DNA transformation: an analysis in tomato. Transgenic Res 10: 121-132

Frenkiel-Krispin D, Grayer-Wolf S, Albeck S, Unger T, Peleg Y, Jacobovitch J, Michael Y, Daube S, Sharon M, Robinson CV, Svergun DI, Fass D, Tzfira T, Elbaum M (2006) Plant transformation by *Agrobacterium tumefaciens* regulation of ssDNA-VirE2 complex assembly by VirE1. 282: 3458-3464

Gelvin SB (2003a) Improving plant genetic engineering by manipulating the host. Trends Biotechnol 21: 95-98

Gelvin SB (1998) *Agrobacterium* VirE2 proteins can form a complex with T strands in the plant cytoplasm. J Bacteriol 180: 4300-4302

Gelvin SB (2003b) *Agrobacterium*-mediated plant transformation: the biology behind the "gene-jockeying" tool. Microbiol Mol Biol Rev 67: 16-37

Görlich D, Vogel F, Mills AD, Hartmann E, Laskey RA (1995) Distinct functions for the two importin subunits in nuclear protein import. Nature 377: 246-248

Greber UF, Fassati A (2003) Nuclear import of viral DNA genomes. Traffic 4: 136-143

Greber UF, Way M (2006) A superhighway to virus infection. Cell 124: 741-754

Guralnick B, Thomsen G, Citovsky V (1996) Transport of DNA into the nuclei of *Xenopus* oocytes by a modified VirE2 protein of *Agrobacterium*. Plant Cell 8: 363-373

Hamilton CM, Frary A, Lewis C, Tanksley SD (1996) Stable transfer of intact high molecular weight DNA into plant chromosomes. Proc Natl Acad Sci USA 93: 9975-9979

Herrera-Estrella A, Chen Z, Van Montagu M, Wang K (1988) VirD proteins of *Agrobacterium tumefaciens* are required for the formation of a covalent DNA protein complex at the 5' terminus of T-strand molecules. EMBO J 7: 4055-4062

Herrera-Estrella A, Van Montagu M, Wang K (1990) A bacterial peptide acting as a plant nuclear targeting signal: the amino-terminal portion of *Agrobacterium* VirD2 protein directs a beta-galactosidase fusion protein into tobacco nuclei. Proc Natl Acad Sci USA 87: 9534-9537

Hirooka T, Kado CI (1986) Location of the right boundary of the virulence region on *Agrobacterium tumefaciens* plasmid pTiC58 and a host specifying gene next to the boundary. J Bacteriol 168: 237-243

Hodges LD, Cuperus J, Ream W (2004) *Agrobacterium rhizogenes* GALLS protein substitutes for *Agrobacterium tumefaciens* single-stranded DNA-binding protein VirE2. J Bacteriol 186: 3065-3077

Howard E, Zupan J, Citovsky V, Zambryski PC (1992) The VirD2 protein of *A. tumefaciens* contains a C-terminal bipartite nuclear localization signal: implications for nuclear uptake of DNA in plant cells. Cell 68: 109-118

Hubner S, Smith HMS, Hu W, Chan CK, Rihs HP, Paschal BM, Raikhel NV, Jans DA (1999) Plant importin alpha binds nuclear localization sequences with high affinity and can mediate nuclear import independent of importin beta. J Biol Chem 274: 22610-22617

Jans DA, Xiao CY, Lam MH (2000) Nuclear targeting signal recognition: a key control point in nuclear transport? BioEssays 22: 532-544

Jasper F, Koncz C, Schell J, Steinbiss HH (1994) *Agrobacterium* T-strand production *in vitro*: sequence-specific cleavage and 5' protection of single-stranded DNA templates by purified VirD2 protein. Proc Natl Acad Sci USA 91: 694-698

Kalogeraki VS, Zhu J, Stryker JL, Winans SC (2000) The right end of the vir region of an octopine-type Ti plasmid contains four new members of the vir regulon that are not essential for pathogenesis. J Bacteriol 182: 1774-1778

Kim SR, Lee J, Jun SH, Park S, Kang HG, Kwon S, An G (2003) Transgene structures in T-DNA-inserted rice plants. Plant Mol Biol 52: 761-773

King SM (2002) Dyneins motor on in plants. Traffic 3: 930-931

Koukolikova-Nicola Z, Raineri D, Stephens K, Ramos C, Tinland B, Nester EW, Hohn B (1993) Genetic analysis of the virD operon of *Agrobacterium tumefaciens*: a search for functions involved in transport of T-DNA into the plant cell nucleus and in T-DNA integration. J Bacteriol 175: 723-731

Krichevsky A, Kozlovsky SV, Gafni Y, Citovsky V (2006) Nuclear import and export of plant virus proteins and genomes. Mol Plant Pathol 7: 131-146

Kumar S, Fladung M (2002) Transgene integratin in aspen: structures of integration sites and mechanism of T-DNA integration. Plant J 31: 543-551

Lacroix B, Li J, Tzfira T, Citovsky V (2006a) Will you let me use your nucleus? How *Agrobacterium* gets its T-DNA expressed in the host plant cell. Canadian J Physiol Pharmacol 84: 333-345

Lacroix B, Tzfira T, Vainstein A, Citovsky V (2006b) A case of promiscuity: *Agrobacterium*'s endless hunt for new partners. Trends Genet 22: 29-37

Lacroix B, Vaidya M, Tzfira T, Citovsky V (2005) The VirE3 protein of *Agrobacterium* mimics a host cell function required for plant genetic transformation. EMBO J 24: 428-437

Landau L, EM L (1980) Stastitical physics. Pergamon Press, Oxford

Lasek RJ, Brady ST (1985) Attachment of transported vesicles to microtubules in axoplasm is facilitated by AMP-PNP. Nature 316: 645-647

Lawrence CJ, Morris NR, Meagher RB, Dawe RK (2001) Dyneins have run their course in plant lineage. Traffic 2: 362-363

Lechardeur D, Lukacs GL (2002) Intracellular barriers to non-viral gene transfer. Curr Gene Ther 2: 183-194

Lechardeur D, Sohn KJ, Haardt M, Joshi PB, Monck M, Graham RW, Beatty B, Squire J, O'Brodovich H, Lukacs GL (1999) Metabolic instability of plasmid DNA in the cytosol: a potential barrier to gene transfer. Gene Ther 6: 482-497

Leonetti JP, Mechti N, Degols G, Gagnor C, Lebleu B (1991) Intracellular distribution of microinjected antisense oligonucleotides. Proc Natl Acad Sci USA 88: 2702-2706

Lessl M, Lanka E (1994) Common mechanisms in bacterial conjugation and Ti-mediated transfer to plant cells. Cell 77: 321-324

Li J, Krichevsky A, Vaidya M, Tzfira T, Citovsky V (2005) Uncoupling of the functions of the *Arabidopsis* VIP1 protein in transient and stable plant genetic transformation by Agrobacterium. Proc Nat Acad Sci USA 102: 5733-5738

Loyter A, Rosenbluh J, Zakai N, Li J, Kozlovsky SV, Tzfira T, Citovsky V (2005) The plant VIP1 protein — a molecular link between the *Agrobacterium* T-complex and the host cell chromatin? Plant Physiol 138: 1318-1321

Luby-Phelps K (2000) Cytoarchitecture and physical properties of cytoplasm: volume, viscosity, diffusion, intracellular surface area. Int Rev Cytol 192: 189-221

Lukacs GL, Haggie P, Seksek O, Lechardeur D, Freedman N, Verkman AS (2000) Size-dependent DNA mobility in cytoplasm and nucleus. J Biol Chem 275: 1625-1629

Mathur J, Koncz C (1998) PEG-mediated protoplast transformation with naked DNA. Methods Mol Biol 82: 267-276

Merkle T (2001) Nuclear import and export of proteins in plants: a tool for the regulation of signalling. Planta 213: 499-517

Mirzoeva OK, Petrini JH (2001) DNA damage-dependent nuclear dynamics of the Mre11 complex. Mol Cell Biol 21: 281-288

Mysore KS, Bassuner B, Deng XB, Darbinian NS, Motchoulski A, Ream LW, Gelvin SB (1998) Role of the *Agrobacterium tumefaciens* VirD2 protein in T-DNA transfer and integration. Mol Plant-Microbe Interact 11: 668-683

Mysore KS, Nam J, Gelvin SB (2000) An *Arabidopsis* histone H2A mutant is deficient in *Agrobacterium* T-DNA integration. Proc Natl Acad Sci USA 97: 948-953

Nagai H, Roy CR (2003) Show me the substrates: modulation of host cell function by type IV secretion systems. Cell Microbiol 5: 373-383

Narasimhulu SB, Deng X-B, Sarria R, Gelvin SB (1996) Early transcription of *Agrobacterium* T-DNA genes in tobacco and maize. Plant Cell 8: 873-886

Otten L, DeGreve H, Leemans J, Hain R, Hooykass P, Schell J (1984) Restoration of virulence of *vir* region mutants of *A. tumefaciens* strain B6S3 by coinfection with normal and mutant *Agrobacterium* strains. Mol Gen Genet 195: 159-163

Pante N, Kann M (2002) Nuclear pore complex is able to transport macromolecules with diameters of about 39 nm. Mol Biol Cell 13: 425-434

Phair RD, Misteli T (2000) High mobility of proteins in the mammalian cell nucleus. Nature 404: 604-609

Pilch DR, Sedelnikova OA, Redon C, Celeste A, Nussenzweig A, Bonner WM (2003) Characteristics of gamma-H2AX foci at DNA double-strand breaks sites. Biochem Cell Biol 81: 123-129

Pollard H, Toumaniantz G, Amos JL, Avet-Loiseau H, Guihard G, Behr JP, Escande D (2001) Ca2+-sensitive cytosolic nucleases prevent efficient delivery to the nucleus of injected plasmids. J Gene Med 3: 153-164

Puchta H (1998) Repair of genomic double-strand breaks in somatic plant cells by one-sided invasion of homologous sequences. Plant J 13: 77-78

Relic B, Andjelkovic M, Rossi L, Nagamine Y, Hohn B (1998) Interaction of the DNA modifying proteins VirD1 and VirD2 of *Agrobacterium tumefaciens*: analysis by subcellular localization in mammalian cells. Proc Natl Acad Sci USA 95: 9105-9110

Rhee Y, Gurel F, Gafni Y, Dingwall C, Citovsky V (2000) A genetic system for detection of protein nuclear import and export. Nat Biotechnol 18: 433-437

Romano PG, Horton P, Gray JE (2004) The *Arabidopsis* cyclophilin gene family. Plant Physiol 134: 1268-1282

Rossi L, Hohn B, Tinland B (1993) The VirD2 protein of *Agrobacterium tumefaciens* carries nuclear localization signals important for transfer of T-DNA to plant. Mol Gen Genet 239: 345-353

Salman H, Abu-Arish A, Oliel S, Loyter A, Klafter J, Granel R, Elbaum M (2005) Nuclear localization signal peptides induce molecular delivery along microtubules. Biophys J 89: 2134-2145

Scali M, Vignani R, Moscatelli A, Jellbauer S, Cresti M (2003) Molecular evidence for a cytoplasmic dynein heavy chain from Nicotiana tabacum L. Cell Biol Int 27: 261-262

Schrammeijer B, den Dulk-Ras A, Vergunst AC, Jurado Jacome E, Hooykaas PJJ (2003) Analysis of Vir protein translocation from *Agrobacterium tumefaciens* using *Saccharomyes cerevisiae* as a model: evidence for transport of a novel effector protein VirE3. Nucleic Acids Res 31: 860-868

Sen P, Pazour GJ, Anderson D, Das A (1989) Cooperative binding of *Agrobacterium tumefaciens* VirE2 protein to single-stranded DNA. J Bacteriol 171: 2573-2580

Sheng J, Citovsky V (1996) *Agrobacterium*-plant cell interaction: have virulence proteins – will travel. Plant Cell 8: 1699-1710

Shimizu T (1995) Inhibitors of the dynein ATPase and ciliary or flagellar motility. Methods Cell Biol 47: 497-501

Shroff R, Arbel-Eden A, Pilch D, Ira G, Bonner WM, Petrini JH, Haber JE, Lichten M (2004) Distribution and dynamics of chromatin modification induced by a defined DNA double-strand break. Curr Biol 14: 1703-1711

Shurvinton CE, Hodges L, Ream LW (1992) A nuclear localization signal and the C-terminal omega sequence in the *Agrobacterium tumefaciens* VirD2 endonuclease are important for tumor formation. Proc Natl Acad Sci USA 89: 11837-11841

Smith HMS, Raikhel NV (1998) Nuclear localization signal receptor importin alpha associates with the cytoskeleton. Plant Cell 10: 1791-1799

Stachel SE, An G, Flores C, Nester EW (1985) A Tn3 lacZ transposon for the random generation of beta-galactosidase gene fusions: application to the analysis of gene expression in *Agrobacterium*. EMBO J 4: 891-898

Stahl R, Horvath H, Van Fleet J, Voetz M, von Wettstein D, Wolf N (2002) T-DNA integration into the barley genome from single and double cassette vectors. Proc Natl Acad Sci USA 99: 2146-2151

Sundberg C, Meek L, Carrol K, Das A, Ream LW (1996) VirE1 protein mediates export of single-stranded DNA binding protein VirE2 from *Agrobacterium tumefaciens* into plant cells. J Bacteriol 178: 1207-1212

Sundberg CD, Ream LW (1999) The *Agrobacterium tumefaciens* chaperone-like protein, VirE1, interacts with VirE2 at domains required for single-stranded DNA binding and cooperative interaction. J Bacteriol 181: 6850-6855

Suntharalingam M, Wente SR (2003) Peering through the pore. Nuclear pore complex structure, assembly, and function. Dev Cell 4: 775-789

Tao Y, Rao PK, Bhattacharjee S, Gelvin SB (2004) Expression of plant protein phosphatase 2C interferes with nuclear import of the *Agrobacterium* T-complex protein VirD2. Proc Natl Acad Sci USA 101: 5164-5169

Taylor NJ, Fauquet CM (2002) Microparticle bombardment as a tool in plant science and agricultural biotechnology. DNA Cell Biol 21: 963-977

Tinland B (1996) The integration of T-DNA into plant genomes. Trends Plant Sci 1: 178-184

Tzfira T (2006) On tracks and locomotives: the long route of DNA to the nucleus. Trends Microbiol 14: 61-63

Tzfira T, Citovsky V (2001) Comparison between nuclear import of nopaline- and octopine-specific VirE2 protein of *Agrobacterium* in plant and animal cells. Mol Plant Pathol 2: 171-176

Tzfira T, Citovsky V (2002) Partners-in-infection: host proteins involved in the transformation of plant cells by *Agrobacterium*. Trends Cell Biol 12: 121-129

Tzfira T, Citovsky V (2006) *Agrobacterium*-mediated genetic transformation of plants: biology and biotechnology. Curr Opin Biotechnol 17: 147-154

Tzfira T, Frankmen L, Vaidya M, Citovsky V (2003) Site-specific integration of *Agrobacterium* T-DNA *via* double-stranded intermediates. Plant Physiol 133: 1011-1023

Tzfira T, Lacroix B, Citovsky V (2005) Nuclear Import of *Agrobacterium* T-DNA. In T Tzfira, V Citovsky, eds, Nuclear Import and Export in Plants and Animals. Eurekah.com and Kluwer Academic/Plenum, pp 83-99

Tzfira T, Li J, Lacroix B, Citovsky V (2004a) *Agrobacterium* T-DNA integration: molecules and models. Trends Genet 20: 375-383

Tzfira T, Rhee Y, Chen M-H, Citovsky V (2000) Nucleic acid transport in plant-microbe interactions: the molecules that walk through the walls. Annu Rev Microbiol 54: 187-219

Tzfira T, Vaidya M, Citovsky V (2001) VIP1, an Arabidopsis protein that interacts with *Agrobacterium* VirE2, is involved in VirE2 nuclear import and *Agrobacterium* infectivity. EMBO J 20: 3596-3607

Tzfira T, Vaidya M, Citovsky V (2002) Increasing plant susceptibility to *Agrobacterium* infection by overexpression of the *Arabidopsis* VIP1 gene. Proc Natl Acad Sci USA 99: 10435-10440

Tzfira T, Vaidya M, Citovsky V (2004b) Involvement of targeted proteolysis in plant genetic transformation by *Agrobacterium*. Nature 431: 87-92

Unal E, Arbel-Eden A, Sattler U, Shroff R, Lichten M, Haber JE, Koshland D (2004) DNA damage response pathway uses histone modification to assemble a double-strand break-specific cohesin domain. Mol Cell 16: 991-1002

van Attikum H, Fritsch O, Hohn B, Gasser SM (2004) Recruitment of the INO80 complex by H2A phosphorylation links ATP-dependent chromatin remodeling with DNA double-strand break repair. Cell 119: 777-788

Vergunst AC, Schrammeijer B, den Dulk-Ras A, de Vlaam CMT, Regensburg-Tuink TJ, Hooykaas PJJ (2000) VirB/D4-dependent protein translocation from *Agrobacterium* into plant cells. Science 290: 979-982

Vergunst AC, van Lier MCM, den Dulk-Ras A, Hooykaas PJJ (2003) Recognition of the *Agrobacterium* VirE2 translocation signal by the VirB/D4 transport system does not require VirE1. Plant Physiol 133: 978-988

Vergunst AC, van Lier MCM, den Dulk-Ras A, Stüve TA, Ouwehand A, Hooykaas PJJ (2005) Positive charge is an important feature of the C-terminal transport signal of the VirB/D4-translocated proteins of *Agrobacterium*. Proc Nat Acad Sci USA 102: 832-837

Verkman AS (2002) Solute and macromolecule diffusion in cellular aqueous compartments. Trends Biochem Sci 27: 27-33

Waigmann E, Ueki S, Trutnyeva K, Citovsky V (2004) The ins and outs of non-destructive cell-to-cell and systemic movement of plant viruses. Crit Rev Plant Sci 23: 195-250

Ward D, Zupan J, Zambryski PC (2002) *Agrobacterium* VirE2 gets the VIP1 treatment in plant nuclear import. Trends Plant Sci 7: 1-3

Ward E, Barnes W (1988) VirD2 protein of *Agrobacterium tumefaciens* very tightly linked to the 5' end of T-strand DNA. Science 242: 927-930

Whittaker GR (2003) Virus nuclear import. Adv Drug Deliv Rev 55: 733-747

Whittaker GR, Helenius A (1998) Nuclear import and export of viruses and virus genomes. Virology 246: 1-23

Whittaker GR, Kann M, Helenius A (2000) Viral entry into the nucleus. Annu Rev Cell Dev Biol 16: 627-651

Winans SC, Allenza P, Stachel SE, McBride KE, Nester EW (1987) Characterization of the virE operon of the *Agrobacterium* Ti plasmid pTiA6. Nucleic Acids Res 15: 825-837

Yi H, Mysore KS, Gelvin SB (2002) Expression of the *Arabidopsis* histone H2A-1 gene correlates with susceptibility to *Agrobacterium* transformation. Plant J 32: 285-298

Young C, Nester EW (1988) Association of the VirD2 protein with the 5' end of T-strands in *Agrobacterium tumefaciens*. J Bacteriol 170: 3367-3374

Yusibov VM, Steck TR, Gupta V, Gelvin SB (1994) Association of single-stranded transferred DNA from *Agrobacterium tumefaciens* with tobacco cells. Proc Natl Acad Sci USA 91: 2994-2998

Zaidi SK, Young DW, Choi JY, Pratap J, Javed A, Montecino M, Stein JL, Lian JB, van Wijnen AJ, Stein GS (2004) Intranuclear trafficking: organization and assembly of regulatory machinery for combinatorial biological control. J Biol Chem 279: 43363-43366

Zaidi SK, Young DW, Choi JY, Pratap J, Javed A, Montecino M, Stein JL, van Wijnen AJ, Lian JB, Stein GS (2005) The dynamic organization of gene-regulatory machinery in nuclear microenvironments. EMBO Rep 6: 128-133

Zhu Y, Nam J, Humara JM, Mysore K, Lee LY, Cao H, Valentine L, Li J, Kaiser A, Kopecky A, Hwang HH, Bhattacharjee S, Rao P, Tzfira T, Rajagopal J, Yi HC, Yadav VBS, Crane Y, Lin K, Larcher Y, Gelvin M, Knue M, Zhao X, Davis S, Kim SI, Kumar CTR, Choi YJ, Hallan V, Chattopadhyay S, Sui X, Ziemienowitz A, Matthysse AG, Citovsky V, Hohn B, Gelvin SB (2003) Identification of *Arabidopsis rat* mutants. Plant Physiol 132: 494-505

Ziemienowicz A, Görlich D, Lanka E, Hohn B, Rossi L (1999) Import of DNA into mammalian nuclei by proteins originating from a plant pathogenic bacterium. Proc Natl Acad Sci USA 96: 3729-3733

Ziemienowicz A, Merkle T, Schoumacher F, Hohn B, Rossi L (2001) Import of *Agrobacterium* T-DNA into plant nuclei: two distinct functions of VirD2 and VirE2 proteins. Plant Cell 13: 369-384

Zupan J, Citovsky V, Zambryski PC (1996) *Agrobacterium* VirE2 protein mediates nuclear uptake of ssDNA in plant cells. Proc Natl Acad Sci USA 93: 2392-2397

Zupan J, Muth TR, Draper O, Zambryski PC (2000) The transfer of DNA from *Agrobacterium tumefaciens* into plants: a feast of fundamental insights. Plant J 23: 11-28

Chapter 11

MECHANISMS OF T-DNA INTEGRATION

Alicja Ziemienowicz[1], Tzvi Tzfira[2] and Barbara Hohn[3]

[1]Department of Molecular Genetics, Faculty of Biochemistry, Biophysics and Biotechnology, Jagiellonian University, Gronostajowa 7, 30-387 Krakow, Poland; [2]Department of Molecular, Cellular and Developmental Biology, The University of Michigan, Ann Arbor, MI 48109, USA; [3]FMI for Biomedical Research, Maulbeerstrasse 66, CH-4058 Basel, Switzerland

Abstract. T-DNA integration is the final step of the transformation process. During this step, the T-DNA, which traveled as a single-stranded DNA molecule from the bacterial cell through the host-cell cytoplasm into the nucleus, must covalently attach itself to the host cell's double-stranded genomic DNA. To fulfil its destiny, the T-DNA needs to be directed to its point of integration in the host genome, to be stripped of some, if not all, of its bacterial and host escorting proteins, and to interact with and co-opt the host's DNA-repair proteins and machinery for its complementation into a double-stranded DNA molecule during its integration into the host genome. In the following chapter, we describe the current knowledge on the functions performed by the bacterial and host proteins, and the role that the host genome may play, during the integration process. We also present the dominant models used today to explain the complex mechanism of T-DNA integration in plant cells.

1 INTRODUCTION

T-DNA integration is an exciting process, from two vantage points: first, a prokaryotic segment of DNA, originating from a bacterium, becomes covalently linked to eukaryotic (plant) genomic DNA. This interkingdom marriage of DNA molecules is unique in nature and therefore of great importance for basic and applied science (reviewed in Tzfira and Citovsky, 2002; Gelvin, 2003). Second, these integration events, documented for many plants (and some non-plant species, see Michielse et al., 2005; Lacroix et al., 2006 and chapter 18), are of enormous applied relevance since most transgenic plants in the field at present have been generated by *Agrobacterium tumefaciens*-mediated transformation. This is due to the fact that transgenesis *via* this process yields plants with mostly reliable transgene expression, due to relatively "clean" integration events (for a recent review on plant transformation see chapter 3 and Gelvin, 2003). Nevertheless, several questions concerning bacterium-plant interactions still need to be answered before we can unveil the mechanisms of T-DNA integration in plant cells. These questions include:

- How is the T-DNA excised from its precursor molecule, the Ti plasmid, within the bacterium?
- Which virulence proteins accompany the single-stranded (ss) T-DNA (the T-strand) into the plant and what roles do these proteins play inside the plant cell?
- Which plant proteins aid in the process of T-DNA delivery into plant cells and in intracellular T-DNA trafficking?
- Since T-DNA is delivered into plant cells in single-stranded form and the final product of the integration event is a continuously double-stranded (ds) molecule, a question arises as to the timing of the conversion of T-DNA to a double-stranded form: does it occur before or during integration?
- An especially challenging question concerns the determination of the genomic position of T-DNA integration. What is the precondition for a genomic site to be "chosen" as a new home for the T-DNA? The existence of a break in the DNA? The availability of chromatin with an open conformation and reduced nucleosome content? If so, are chromatin-remodelling mechanisms involved? Is the integration process linked to DNA replication and/or transcription? Related to this, where with respect to the genes and intergenic regions does T-DNA integrate? Does

the T-DNA, or rather one of its accompanying proteins, have an influence on target-site selection?
- Recently, new knowledge has been gained on gene-silencing processes, raising the question of whether transgenes become silenced at some stage after transformation and/or are already silenced upon transgenesis.

Several reviews cover the various aspects of T-DNA integration and their detailed study is highly recommended (Gelvin, 2000; Wu and Hohn, 2003; Tzfira et al., 2004a; Citovsky et al., 2007). In addition, several chapters in this book cover some of the above-listed questions. In this chapter, we concentrate on describing the role of bacterial and host proteins in the integration process and the possible mechanisms governing the integration of T-DNA molecules into the host-cell genome.

2 THE T-DNA MOLECULE

T-DNA is a ssDNA molecule which is excised from the *Agrobacterium* Ti plasmid, where it was originally delimited by two direct 25-bp repeats, termed left and right T-DNA borders (reviewed by Zambryski, 1992). Induction of the *Agrobacterium*'s virulence machinery by specific host signals leads to expression of the VirD1 and VirD2 proteins, which are responsible for nicking both borders in the bottom strand, thereby releasing the T-strand, *i.e.* the transported ssT-DNA molecule (for more details see chapter 8). The T-strand, with one VirD2 molecule covalently attached to its 5' end, is then exported, together with several other virulence proteins, through the bacterial type IV secretion system (for further details see chapter 9) where it is most likely coated with many VirE2 molecules, becoming the transported form of the T-DNA, the transport complex (T-complex). This complex is then imported into the host-cell nucleus (for further details see chapter 10). Once inside the nucleus, the substrate molecule(s) destined for integration can potentially be (i) a naked ssDNA molecule with a single VirD2 molecule attached to its 5' end, (ii) a VirE2-coated ssDNA molecule with a single VirD2 molecule attached to its 5' end, and/or (iii) a dsDNA molecule which may or may not be attached to a VirD2 molecule but is not covered by the single-stranded binding protein VirE2. In the following sections, we discuss the possible roles of bacterial and host proteins in T-DNA integration and propose mechanisms for this process.

3 PROTEINS INVOLVED IN T-DNA INTEGRATION

In higher eukaryotic organisms, such as plants, illegitimate recombination is the predominant mechanism of integration for naked DNA (Paszkowski et al., 1988). Likewise, T-DNA molecules are integrated into the plant genome by nonhomologous recombination (NHR) (Gheysen et al., 1991; Mayerhofer et al., 1991). As T-DNA in the plant cell is accompanied by the bacterial proteins VirD2 and VirE2, it was not clear whether bacterial and/or plant factors mediate T-DNA integration. The VirD2 protein was suggested to function in this process as an integrase or ligase. However, this was was never confirmed by the experimental data. Therefore, VirD2 is likely to act in the recruitment of plant factors, such as DNA ligases, cyclins, etc., to the site of integration and in directing the T-DNA complex to a potential integration site by interacting with host transcription factors. The second bacterial protein forming a complex with T-DNA, VirE2, may protect T-DNA from nucleolytic degradation and, by interacting with host factors such as VIP1, may form a link between the T-DNA complex and plant chromatin. T-DNA integration is a complex process that may occur *via* different mechanisms depending on the actual condition of the host genome, especially locally, and thus likely requires the engagement of different sets of factors.

3.1 The role of VirD2 in the integration process

The VirD2 protein is suggested to function in T-DNA integration since it is covalently attached to the 5' end of T-DNA, pilots the T-DNA to the plant nucleus, and likely stays attached to it until the actual integration step. Two hypotheses of VirD2's function during T-DNA integration have been proposed: (i) VirD2 acts as an integrase, and (ii) VirD2 acts as a ligase. Analysis of the amino-acid sequence of VirD2 revealed the presence of an H-R-Y motif that is typical of bacteriophage λ integrase and other site-specific recombinases. An R-to-G mutation introduced into this H-R-Y motif resulted in a loss of precision of T-DNA integration, without any change in its efficiency (Tinland et al., 1995). The unchanged efficiency argues against VirD2's function as an integrase, whereas the loss of integration precision (defined as a lack of conservation of the 5'-end nucleotide attached to VirD2 in the integrated T-DNA) suggests the importance of VirD2 in the T-DNA integration process. The second hypothesis was based on results of *in vitro* studies of VirD2-mediated cleavage of the right border sequence. VirD2 was found able not only to cleave ssDNA at the border sequence but also to ligate cleaved ssDNA to the 3' preformed end of another ssDNA molecule (Pansegrau et al., 1993), suggesting a ligase function for VirD2 in T-DNA integration. However, both cleavage and joining reactions were sequence-specific, while *in vivo*, T-DNA integration shows a limited requirement for sequence homology. Nevertheless, this hypothesis was strongly believed, and therefore, was subjected to extensive examination. The potential function of VirD2 in ligating the 5' end of the T-DNA to the 3' end of plant DNA was analyzed *in vitro* (Ziemienowicz et al., 2000) and VirD2 was found unable to perform T-DNA ligation. This result suggested that other factors, most probably from the plant, are involved in T-DNA ligation/integration. Indeed, such activity was found in plant extracts from tobacco BY-2 cells and pea axes (Ziemienowicz et al., 2000). Moreover, more recent data indicate that *Arabidopsis thaliana* type I DNA ligase can function as a ligase for T-DNA *in vitro* (Wu, 2002).

Although VirD2 was shown not to act as a ligase for T-DNA, this does not preclude its potential function in other steps of T-DNA integration, for example, in recruiting plant enzymes involved in DNA repair or recombination to the site of the integration and/or in interacting with some structural chromatin proteins. Indeed, VirD2 has been shown to interact with a number of plant proteins that may be involved (directly or indirectly) in T-DNA integration. Similar interactions have been previously described for proteins facilitating the nuclear import of VirD2 (and most likely of the T-DNA complex), including AtKAPα (Ballas and Citovsky, 1997). Additionally, VirD2 has been shown to interact in the nuclei of alfalfa cells with CAK2Ms and with TATA-box-binding protein (Bako et al., 2003). CAK2Ms is a conserved plant orthologue of cyclin-dependent kinase-activating kinases. It binds to and phosphorylates the C-terminal regulatory domain of RNA polymerase II's largest subunit, which recruits the TATA-box-binding protein. VirD2 not only interacted with both proteins, but also became phosphorylated by CAK2Ms kinase, indicating that it is recognized by plant nuclear factors. Other studies revealed interaction of VirD2 with plant cyclophilins (CyPs) (Deng et al., 1998). This interaction was disrupted by cyclosporin A, which also inhibited *Agrobacterium*-mediated transformation of *Arabidopsis* and tobacco plants, strongly suggesting that the VirD2-CyP interaction plays a role in T-DNA transfer. Interestingly, CyPs exert their functions in different cellular processes, leading to speculation as to their possible role in T-DNA transfer. The chaperone activities of some CyPs suggest that they may play a role in maintaining VirD2 in a functional and transfer-competent state. The CyPs' subcellular localization, to both the nucleus and the cytoplasm, raises the possibility that they may function in the T-complex's transfer in these compartments. In addition, since several CyPs have been shown to possess nuclease activity (e.g. Montague et al., 1997), we can also suggest that they function during T-DNA integration into the plant genome.

It is not clear how VirD2 is removed from T-DNA upon integration. One can speculate that the mechanism is similar to that of Spo11 in yeast (Neale et al., 2005). Spo11 initiates meiotic recombination by forming DNA double-strand breaks (DSBs) with protein covalently attached to the 5' termini. Spo11 must be removed from the DSB before the latter is repaired. In yeast, DSBs have been shown to be processed by endonucleolytic cleavage which releases Spo11 attached to an oligonucleotide with free 3' OH. However, such processing of the T-DNA-VirD2 complex would lead to a loss of nucleotides from the 5' terminus, which in turn would not allow for precise T-DNA integration.

Interestingly, experiments involving the transformation of HeLa cells with *Agrobacterium* T-DNA complex reconstructed *in vitro* showed that VirD2 and VirE2 are the only virulence proteins absolutely required for precise T-DNA integration (Pelczar et al., 2004).

3.2 The role of VirE2 in the integration process

The ssDNA-binding protein VirE2 coats T-DNA inside the plant cell. VirE2 binds to T-DNA cooperatively and protects it from degradation. *In vivo* transformation of tobacco seedlings with an *Agrobacterium* strain deficient in VirE2 production showed that in the absence of VirE2 protein, mainly truncated versions of T-DNA are integrated into the plant genome (Rossi et al., 1996). The molecular structure of VirE2-ssDNA complexes revealed by scanning transmission electron microscopy (STEM) showed a solenoidal organization (Citovsky et al., 1997). More recent electron microscopy analyses accompanied by single-particle image-processing methods have provided even more precise measurements of the VirE2-ssDNA complex. The helical structure of the complex was found to form a hollowed solenoid shape with a putative ssDNA-binding site near the inner diameter of the structure (Abu-Arish et al., 2004). Such a structure is expected to sequester the DNA from cytosolic nucleases and to enable interaction of VirE2 with host factors, facilitating, for example, nuclear import of the T-DNA complex. On the other hand, according to the results of atomic force microscopy (AFM) analyses of the VirE2-ssT-DNA complex, it is composed of the T-DNA and protein globules attached to the DNA at separate sites (Volokhina and Chumakov, 2007). However, in this case, VirE2 was also able to protect the ssDNA *in vitro* from degradation by S1 nuclease. Interestingly, another ssDNA-binding protein, *Escherichia coli* SSB, was not efficient in this reaction. This finding is in agreement with previous observations that SSB is unable to replace VirE2 in its function in the nuclear import of T-DNA (Ziemienowicz et al., 2001).

Protection from nucleolytic degradation probably represents only an indirect function of VirE2 in T-DNA integration. However, the protein interacting with VirE2, VIP1, may aid integration more directly, possibly by creating a link to the histone constituents of the host chromatin. *At*VIP1 protein has been shown to interact with the plant histone H2A *in planta* (Li et al., 2005a), as well as with purified *Xenopus* core histones H2A, H2B, H3 and H4 *in vitro* (Loyter et al., 2005).

Upon T-DNA integration, VirE2 protein must be removed from the DNA, by being stripped off and/or *via* degradation-mediated processes. It seems very likely that VirE2 is replaced in the plant cell by a host ssDNA-binding protein. Such displacement ability has already been shown for the P30 protein of Tobacco mosaic virus (TMV) (Citovsky et al., 1997), although no plant protein exerting similar activity has been found to date. Replacement of VirE2 by host SSB protein may play a crucial role in the recruitment of other host factors involved in T-DNA integration. Although displacement of VirE2 by P30 *in vitro* did not require additional factors, this process could be facilitated by protein degradation. VirF, the bacterial virulence protein that is exported to plant cells upon *Agrobacterium*-mediated transformation, was shown to be involved in the targeted proteolysis of VIP1 and VirE2 (Tzfira et al., 2004b). VirF contains an F-box motif that binds the plant homologue of yeast Skp1, ASK1, a component of a Skp1-Cdc53-cullin-F-box (SCF) complex involved in protein degradation. In addition, mutation in an F-box-encoding *Arabidopsis* gene resulted in decreased susceptibility of the plant roots to *Agrobacterium*-mediated transformation (Zhu et al., 2003), thus confirming the requirement for protein degradation mediated by host factors. Degraded VirE2 may then be replaced by a single-stranded host protein or, if the integration process occurs quickly enough, the replacement may not be necessary at all, since the integrated T-DNA will be protected by the chromatin structure.

3.3 The role of host proteins in the integration process

T-DNA integration is one of the last steps in the transformation process and it relies heavily on the functions of host proteins. Host factors are required to convert ssT-DNA molecules into double-stranded form, to provide breaks in the plant DNA that may serve as T-DNA integration sites, and to incorporate the T-DNA molecules into plant DNA *via* ligation. Analyses of T-DNA integration in yeast and plant cells have revealed that the mechanism of, and factors involved in this process fully depend on the DNA-recombination/repair mechanism functioning in these cells. In yeast, where homologous recombination (HR) is the predominant DNA-rearrangement mechanism, the same pathway is employed for the integration of T-DNA molecules sharing homology with the host genome. Lack of homology directs T-DNA integration to the NHR (nonhomologous recombination) pathway. Each pathway depends on its own set of factors, Rad52 and Ku70 being the key regulators directing T-DNA integration to HR or NHR pathways, respectively. In plants, the NHR pathway (which is usually referred to as nonhomologous end joining, or NHEJ) is employed for T-DNA integration, regardless of the presence of homology between T-DNA and plant DNA sequences. As NHEJ also represents the DNA-repair mechanism, factors involved in DNA repair are expected to be required for T-DNA integration as well. However, contradictory data have been reported concerning the requirement for NHEJ factors such as Ku70/80 heterodimers and type IV DNA ligase, while the involvement of other DNA repair/replication factors, such as DNA polymerase, has not yet been proven. In addition to plant proteins functioning directly during T-DNA integration *via* recombination/repair, other factors may also influence this process. The structure of chromatin, affecting the accessibility of the host genome for the integration of T-DNA molecules, is definitely crucial. Indeed, genes encoding chromatin structural proteins (histones), as well as proteins that modify histones, have been found to be involved in T-DNA integration. However, although our knowledge of the factors involved in T-DNA integration has increased in the last decade, many aspects of this process and the engagement of as-yet unidentified factors still remain obscure.

3.3.1 A lesson learnt from yeast

Although plants are the natural host for *Agrobacterium*, T-DNA transfer can also occur, at least under laboratory conditions, in yeast and fungi (reviewed inLacroix et al., 2006). When the T-DNA is homologous to the yeast genome, its integration occurs *via* HR (Bundock et al., 1995); in the absence of DNA homology, integration occurs *via* NHJ, a process similar to NHEJ in plants (Bundock and Hooykaas, 1996). However, as already mentioned, in plants, even if the T-DNA shares extensive homology with the plant genome, it still integrates mainly by NHR (Offringa et al., 1990). These findings indicate that the process of T-DNA integration into the host genome is predominantly determined by host factors.

Host proteins involved in T-DNA integration into the yeast genome *via* NHR include factors such as Ku70, Rad50, Mre11, Xrs2, Lig4 and Sir4 (van Attikum et al., 2001). These proteins have been previously described to have distinct functions in the repair of genomic DSBs by NHR (Tsukamoto and Ikeda, 1998; Lewis and Resnick, 2000). These findings imply that DSB repair by NHJ in yeast and NHEJ in plants provides a pathway for T-DNA integration. Indeed, previous studies have already shown that T-DNA can be captured during DSB repair in plants (Salomon and Puchta, 1998). However, since the right ends of the T-DNAs are all truncated, this may not represent the most common form of T-DNA integration. Nevertheless, T-DNA integration in yeast is dependent on the NHEJ enzymes. In addition, a minor pathway for T-DNA integration was discovered in yeast cells—integration at the telomeric regions, which was even enhanced in the absence of Rad50, Mre11 or Xrs2, but not operational in the absence of Ku70 (van Attikum et al., 2001). Since Rad50, Mre11 and Xrs2 proteins play a role, albeit a minor one, in telomeric silencing (Boulton and Jackson, 1998), the reduced silencing in the telomeric region could make this part of the chromosome more accessible to T-DNA, thereby facilitating T-DNA integration in (sub)telomeric regions.

Integration of *Agrobacterium* T-DNA *via* HR has been shown to require the recombination/repair proteins Rad51 and Rad52, but not Rad50, Mre11, Xrs2, Ku70 or Lig4 (van Attikum and Hooykaas, 2003). In a *rad51* mutant, residual integration occurred predominantly by HR, whereas in the *rad52* mutant, integration occurred exclusively by NHR, indicating that Rad52 is the key regulator of T-DNA integration *via* HR. Similarly, T-DNA integration was shown to be abolished in a *ku70* mutant (van Attikum et al., 2001), implying that Ku70 is the key regulator of T-DNA integration *via* NHR. These two regulators channel integration into either the HR or NHR pathways. Not surprisingly, double mutation of the *ku70* and *rad52* genes resulted in complete abolishment of T-DNA integration (van Attikum and Hooykaas, 2003).

3.3.2 Plant proteins

A few years ago, a novel test for the identification of plant factors involved in T-DNA integration was employed. This test was based on the laborious screening of T-DNA tagged (or antisense) *A. thaliana* lines to find mutants that are resistant to *Agrobacterium*-mediated transformation (*rat* mutants). Over 120 *Arabidopsis* plant mutant lines showing the *rat* phenotype were isolated and the affected genes identified (Zhu et al., 2003). These genes can be classified into several categories, including (i) chromatin structure and remodeling genes, (ii) nuclear-targeting genes, (iii) cytoskeleton genes, (iv) cell-wall structural and metabolism genes and (iv) other *rat* genes likely involved in signal transduction, gene expression and protein function. This method allowed identification of the first plant factor involved in T-DNA integration, namely histone H2A (Mysore et al., 2000b), thus confirming the involvement of structural proteins in this process.

A *rat5* mutant containing a T-DNA insertion in the 3' UTR of the histone *H2A-1* gene is deficient in its ability to integrate T-DNA into the plant genome (Mysore et al., 2000b). Interestingly, although roots of the *rat5* mutant were not susceptible to transformation by *Agrobacterium,* transformation by the flower vacuum-infiltration method was as efficient as that of wild-type plants (Mysore et al., 2000a). This result, together with those of other groups, indicated that specific host cells and stages of the cell cycle may be important targets for transformation. However, exposure of root segments to phytohormones or wounding resulted in increases in both the expression of histone *H2A-1* and transformation, whereas the response of a cyclin gene (*cyc1At*) which is important for cell division (Ferreira et al., 1994) was not affected by these treatments (Yi et al., 2002). It was thus proposed that *H2A-1* gene expression is not strictly linked to the S-phase of the mitotic cell cycle, but that nevertheless, expression of this gene is both a marker for, and a predictor of, plant cells most susceptible to *Agrobacterium* transformation. Recently, the expression levels and patterns of several *Arabidopsis HTA* genes, encoding H2A histones, have been investigated and tested for their ability to complement the *rat5* phenotype. The multiple histone *HTA* genes were shown to be able to compensate for loss of *HTA1* (*H2A-1*) gene activity when overexpressed from a strong promoter (Yi et al., 2006). However, only the *HTA1* gene could phenotypically complement *rat5*-mutant plants when overexpressed from their native promoters.

In addition to chromatin structure and remodeling functions, other host factors/pathways may also play a crucial role in T-DNA integration. These include, first and foremost, DNA repair and recombination. As T-DNA is integrated into the plant genome *via* illegitimate recombination—a mechanism which does not require extensive homology between the recombining DNA molecules, it is suggested that factors involved in DSB repair *via* NHEJ (NHR) are also involved in T-DNA integration. Indeed, DSBs may serve as target sites for T-DNA integration, as shown by inducing the breaks with the rare cutting restriction enzyme I-*Sce*I (Salomon and Puchta, 1998). In yeast cells, DSBs are repaired predominantly *via* HR by the factors Rad51 and Rad52. Nevertheless, the NHEJ pathway, which is dependent on the factors Ku70, Ku80, Rad50, Mre11, Xrs2, Lif1, Nej1, Lig4 and Sir4, can also be used (Haber, 2000). Both pathways can be employed to integrate T-DNA into the yeast genome: HR represents the principal mechanism, but lack of homology between the T-DNA and the yeast genome triggers the NHR mechanism (van Attikum et al., 2001; van Attikum and Hooykaas, 2003). In higher eukaryotes, NHR is the dominant pathway for DSB repair. In mammalian cells, DNA breaks are recognized by the Ku70/80 heterodimer, which then recruits other factors, such as DNA-PK_{cs} protein, Artemis, XRCC4 and DNA ligase IV (Weterings and van Gent, 2004). In plants, similar factors have been shown to mediate NHEJ: Ku70, Ku80, the Mre11/Rad50/Nbs1 complex (MRN), XRCC4 and Lig4 (reviewed in Bray and West, 2005). Alternatively, DSBs can be also repaired *via* HR employing either the DSBR (DNA double-strand break repair) or SDSA (synthesis-dependent single-strand annealing) models with involvement of the factors RPA, Rad51 and likely Rad54/Rad57 (reviewed in Bray and West, 2005).

From the above list of proteins, the function of only a few factors, including Ku80 and DNA ligases, in their potential involvement in T-DNA integration has been studied to date. In general, mutation of *Arabidopsis Ku* genes results in hypersensitivity of the mutants to DNA-damaging agents (γ-irradiation, bleomycin and MMS), but these plants exhibit no growth or developmental defects under standard growth conditions (Riha et al., 2002; West et al., 2002). This is in contrast to animal systems, where homozygous *Ku80*-deficient human cells undergo apoptosis. In addition, lack of Ku70 results in dramatic deregulation of telomere-length control, as the mutants possess expanded telomeres (Bundock et al., 2002). Interestingly, a novel role for mammalian Ku70 protein has been recently proposed: it may function as a receptor for the pathogenic bacterium *Rickettsia conorii*, which undergoes ubiquitination upon bacterial infection (Martinez et al., 2005). Whether Ku70/80 proteins play a role in *Agrobacterium*-mediated plant transformation, and in particular in T-DNA integration, is still a topic of extensive debate, mainly due to contradictions in the data obtained so far in different laboratories. When T-DNA integration was analyzed in somatic cells, defectiveness of *Arabidopsis* insertional mutants was observed in the *Ku80* gene (Li et al., 2005b). In addition, formation of complexes between Ku80 and dsT-DNA molecules in

Agrobacterium-infected plants has been shown by co-immunoprecipitation. Involvement of Ku80 in the transformation of germ-line cells (floral-dip method) is, however, not so clear-cut, since the loss of Ku80 in germline cells has a negligible effect on the transformation frequency (Friesner and Britt, 2003; Gallego et al., 2003).

An important step in T-DNA integration is ligation of the T-DNA to the plant cell DNA at the site of integration. Since VirD2 protein was shown not to be able to perform ligation of T-DNA to plant DNA sequences *in vitro*, this function must be exerted by plant enzymes (Ziemienowicz et al., 2000). Indeed, such activity has been found in extracts from tobacco BY2 cells and pea meristems. This reaction was dependent on ATP being hydrolyzable to AMP and sensitive to dTTP. These results confirm the involvement of plant DNA ligases in T-DNA ligation *in vitro*, since ATP is known to be a cofactor of eukaryotic and viral DNA ligases, serving as a source of AMP groups for adenylation of the enzyme and DNA substrate (reviewed in Timson et al., 2000), whereas dTTP has been previously shown to inhibit the activity of plant DNA ligases (Daniel and Bryant, 1985). *In vitro*, the T-DNA ligation reaction seems to not be specific for a particular enzyme since all DNA ligases tested to date (including bacterial enzymes as well, such as *E. coli* DNA ligase, *Taq* DNA ligase and T4 DNA ligase) have been able to perform this reaction. It is most likely that the involvement of a particular type of DNA ligase during T-DNA ligation is determined by the DNA-repair pathway *via* which the T-DNA integration is actually occurring.

In mammalian cells, type I DNA ligase was shown to be involved in DNA replication (joining of Okazaki fragments) as well as in long patch BER (base-excision repair) and NER (nucleotide-excision repair) pathways, whereas the short patch BER pathway requires type IIIα DNA ligase which joins single-strand breaks in the DNA (reviewed in Sancar et al., 2004). DNA ligase IIIβ is involved in joining DNA during meiotic recombination. In plants, type I DNA ligase was shown indispensable for cell function and survival (mutations in the *AtLig1* gene are lethal), whereas no clear homologues of DNA ligase III have been found in plants. Type IV DNA ligase has been shown to be involved in DSB repair during NHR by joining of nonhomologous ends (NHEJ) in yeast, mammalian and plant cells (Teo and Jackson, 1997; Sancar et al., 2004; Bray and West, 2005). Every mechanism of DNA repair which involves joining of DNA may potentially be employed for T-DNA ligation. *Arabidopsis* DNA ligase I was the first plant ligase tested for its ability to ligate T-DNA. Since mutation in DNA ligase I is lethal in homozygous mutant plants (Babiychuk et al.,

1998), such mutants could not be used to test for their resistance (or susceptibility) to *Agrobacterium*-mediated transformation. Therefore, data suggesting the potential involvement of type I DNA ligase in T-DNA ligation derive from *in vitro* experiments. *Arabidopsis* type I DNA ligase (*At*-Lig1)-VirD2 was shown to ligate model T-DNAs to model chromosomal DNA *in vitro* (Wu, 2002). Moreover, VirD2 protein was demonstrated to interact with *At*Lig1 *in vitro* and to stimulate, in free form or attached to ssDNA, adenylation and enzymatic activity of the ligase (Wu Y.-Q. and Hohn B., unpublished data). The second enzyme that was suggested to be involved in T-DNA ligation was type IV DNA ligase, mainly due to its engagement in DSB repair, the mechanism shown to capture the integrating T-DNA. Contradictory data have, however, been obtained in experiments testing *At*Lig4 mutant plants for their recalcitrance/susceptibility to *Agrobacterium* T-DNA transfer. T-DNA insertion rates were decreased in the mutant plants transformed by *Agrobacterium* using the floral-dip method, suggesting that *At*Lig4 is required in the transformation of germ-line cells (Friesner and Britt, 2003). In contrast, *At*Lig4 was shown to be dispensable for T-DNA integration (van Attikum et al., 2003). Very intriguing is the fact that impaired T-DNA integration in *AtLig4*-mutant plants was demonstrated using both somatic-cell transformation (*rat* test) and germ-line transformation (flower dip) assays. In this case as well, two explanations may be proposed for this discrepancy: the nature of the flower-dip transformation method and the use of different allelic mutants. The latter possibility does not seem likely since the same source of mutant lines was indicated by both research groups.

Another approach to investigating the involvement of *At*Lig1 and *At*-Lig4 in T-DNA ligation was to transiently overexpress wild-type ligases and mutants in the active site and to analyze their effect on T-DNA integration. Overexpression of wild-type enzymes led to a slightly increased number of transformants whereas overexpression of the mutant versions led to a lower efficiency of transformation than that in control experiments with empty vectors (Valentine et al., unpublished). This may be an indication that binding of VirD2/T-DNA to inactive ligases interferes or competes with the binding of the complex to functional enzymes. In addition, *Arabidopsis* lines stably overexpressing the wild-type *At*Lig4 gene were more susceptible to *Agrobacterium*-mediated transformation (Valentine et al., unpublished). These data indicate that both DNA ligases may be involved in T-DNA ligation to the plant genome, the choice lying in one or several of factors, including the availabilty of single- or double-strand breaks, enzymes, or additional components such as chromatin-remodeling machineries or histones. Replication repair and transcription-coupled repair (leading to BER or NER, respectively) implicate the use of type I DNA ligase, whereas the NHEJ repair pathway uses type IV enzyme. However, since the T-DNA molecules that were integrated into the plant genome *via* the induced NHEJ/DSB repair pathway were truncated at their extremities (Salomon and Puchta, 1998), it is possible that DSB repair by NHEJ does not represent the most common form of T-DNA integration.

In addition to a DNA ligase responsible for joining T-DNA with plant DNA, there must be a host DNA polymerase(s) that converts the ssT-DNA to a double-stranded form, either before or during the integration event; however, such an activity has yet to be shown. One can speculate that type δ and/or ε DNA polymerases might be good candidates due to their involvement in many DNA-repair processes, at least in mammalian cells (for example in DSB repair via HR, Bebenek and Kunkel, 2004), although the involvement of other types of DNA polymerases (λ or μ, involved in NHEJ in mammalian cells) cannot be excluded. Moreover, other DNA-repair/recombination factors are expected to be involved in T-DNA integration. Identification of the UE-1 *Arabidopsis* ecotype showing high transient but low stable (post-integration) expression of T-DNA-encoded reporter gene supports this hypothesis (Nam et al., 1997). Since UE-1 plants are slightly radiation-sensitive, their recalcitrance to T-DNA integration may result from deficiencies in DNA repair and/or recombination.

4 GENOMIC ASPECTS OF T-DNA INTEGRATION/TARGET-SITE SELECTION

Upon their arrival to the host chromosome, T-DNA molecules may encounter highly complex genomic DNA structures. Different areas of the host genome may be engaged in transcription activities while others may be silenced and thus tightly packed. Various parts of the host genome may be relatively stable while others, more prone to damage and breaks. The latter may thus be repaired and/or maintained by the host DNA-repair machinery. Such factors, as well as others, can not only influence the access of T-DNA molecules and their escorting proteins to points of integration but can also determine their fate, and the rate and mode of integration.

Early work on tobacco and *A. thaliana* led to the conclusion that T-DNA preferentially integrates into (potentially) transcriptionally active regions of the genome (Tinland and Hohn, 1995; Tinland, 1996). The evidence was based, in part, on the fact that T-DNAs containing a selectable gene and a promoterless screenable or selectable gene yielded fusions with resident promoters at relatively high frequency (Koncz et al., 1989; Herman et al., 1990; Kertbundit et al., 1991). Similar fusion frequencies of about 30% were found for tobacco and *Arabidopsis* species with very different genome complexities but similar gene contents; this suggested that it was the genes, active or nonactive at the time of transformation, that were the targets for T-DNA integration. However, only an analysis of large T-DNA integration libraries, in conjunction with the availability of the sequence of *A. thaliana*, enabled a generalized conclusion on the rules by which T-DNA chooses its integration sites in this model plant (Brunaud et al., 2002; Sessions et al., 2002; Szabados et al., 2002; Alonso et al., 2003; Forsbach et al., 2003; Rosso et al., 2003; Schneeberger et al., 2005) or in rice (An et al., 2003; Chen et al., 2003; Sallaud et al., 2004; Zhang et al., 2007). It turned out that the distribution, originally claimed to be uniform, is nonrandom at both the gene and chromosome levels, although the latter conclusion may need to be modified somewhat.

4.1 T-DNA integration at the gene level

Analysis of very large T-DNA insertion libraries led to the conclusion that the distribution of T-DNA insertions is closely correlated with gene density on all five chromosomes of *A. thaliana* (Brunaud et al., 2002; Sessions et al., 2002; Szabados et al., 2002; Alonso et al., 2003; Forsbach et al., 2003; Rosso et al., 2003; Schneeberger et al., 2005). Recent compilations of data from several insertion libraries have indicated that the insertion frequencies are highest at the sites of transcription initiation and termination. These frequencies were calculated to be far higher than those for insertions at average locations within a gene (Schneeberger et al., 2005; Li et al., 2006). In contrast, the structural part of the genes, the so-called "genebodies", as well as the intergenic regions, exhibit a lower-than-average percentage of insertions. This distribution is in sharp contrast to that of T-DNA insertions around the mostly nontranscribed pseudogenes: insertion frequency peaks are lacking around their open reading frames (Li et al., 2006). Interestingly, T-DNA insertions into the rice genome were also found biased in both the 5' and 3' regulatory regions of genes, outside the coding sequences (Chen et al., 2003; Zhang et al., 2007). This clearly points to transcriptional initiation and termination regions on the DNA as preferred targets for T-DNA integration. Preference for these locations, however, has been suggested to not be correlated with the level of transcription (Alonso et al., 2003). This was inferred from a comparison of T-DNA-integration density and the levels of genome-wide expression using data from expressed sequence tags and microarray analysis. It may thus be concluded that an open configuration of chromatin, which is probably required for transcription, is sufficient to allow T-DNA to enter, independent of the density of transcriptional rounds. In another study, the increased insertion frequencies in 5' upstream regions were found to be positively correlated with gene expression (Schneeberger et al., 2005). However, direct assessment of transcription in female gametophytic tissue, the target of T-DNA integration in the commonly used flower-dip procedure, is not easy; therefore the precise role of transcriptional activity in T-DNA integration may have to be reevaluated.

A fascinating link between gene activity and intranuclear localization of active genes was recently established for *Saccharomyces cerevisiae*. Since the T-DNA-containing complex has to enter the nucleus *via* the nuclear pore complex (NPC, see chapter 10), physical proximity of the complex and active chromatin may provide an attractive hypothesis to explain integration preferences into active genes. The T-DNA complex, upon release from the NPC, may find accessible genes that are already closely associated with the NPC. The physical link between components of the basket (i.e. inner) part of the NPC and active genes has been exclusively analyzed in yeast (Casolari et al., 2004; Rodriguez-Navarro et al., 2004; Cabal et al., 2006; Schmid et al., 2006; Taddei et al., 2006; Luthra et al., 2007). These data corroborate the "gene-gating" hypothesis, according to which "all transcripts of a given gated gene would leave the nucleus by way of that pore complex to which the gene is gated" (Blobel, 1985). Extension of this interesting concept to higher eukaryotic organisms was anticipated to represent an evolutionarily conserved mechanism for gene regulation (Luthra et al., 2007).

An interesting distribution of DNA methylation in *A. thaliana*, usually regarded as having a negative influence on gene expression, was recently described (Zhang et al., 2006; Zilberman et al., 2007). While most genes were unmethylated, a substantial fraction (20 to 30% of them) was found methylated, usually preferentially at the transcribed part of the gene, i.e. its "body". These genes were found, for the most part, to be highly expressed and constitutively active. Transposons were normally methylated while some special genes were preferentially methylated at their promoters. Occupation of transcriptional entry or exit positions by the huge transcription machinery may promote access to foreign DNA such as T-DNA, whereas methylation of the structural part of the gene, exons and introns, may inhibit unorthodox transcription initiation, as suggested by Zilberman et al.

(2007), but may also keep these regions from T-DNA access. However, it remains to be determined whether this nonrandom distribution of methylation corresponds to the nonrandom distribution of T-DNA insertions within genes and if so, what the underlying mechanism is. In addition, it should be remembered that only a fraction of the *Arabidopsis* genes possess this special methylation distribution. Further detailed analysis is required to resolve the issue of methylation-dependent chromatin organization and accessibility for T-DNA integration.

The T-DNA preintegration complex consists of the ssT-DNA, the virulence proteins VirD2, VirE2 and others and the plant-derived proteins interacting with these virulence proteins as previously described. VirD2 and VirE3 seem to be of special relevance to target-site selection. VirD2, which is covalently attached to the 5' terminus of the ssT-DNA, has been found to interact in plants with a nuclear protein kinase (Bako et al., 2003): VirD2 is phosphorylated by the kinase CAK2M which also binds to and phosphorylates the C-terminal regulatory domain of RNA polymerase II, which can recruit the TATA-box-binding protein. VirD2 has also been found in tight association with the TATA-box-binding protein *in planta* (Bako et al., 2003). These findings lend credence to the suggestion that T-DNA, with the help of its ingenious protein VirD2, interacts directly with the transcription machinery and aids in, if not enables, T-DNA integration. However, the exact mechanism still needs to be worked out: one can imagine that the huge transcription complex, by removing nucleosomes, promotes access. Alternatively, or additionally, transcription may, in rare cases, lead to breaks in the DNA that can be used by the T-DNA for entry into the genomic DNA. Transcription-coupled repair, a process which removes DNA lesions from transcribed genes, is an exciting possible mechanism for T-DNA integration.

The other virulence protein with a possible function in T-DNA integration is VirE3. This virulence protein, while not absolutely essential, has been shown to be transported to the plant nucleus, to possess transcription-induction activity, at least in yeast, and to interact with a plant-specific general transcription factor belonging to the TFIIB family (Garcia-Rodriguez et al., 2006). Although it is not clear if VirE3 translocates to the plant cell in conjunction with the T-DNA, the possible interaction of this protein with the host transcription machinery may be an attractive additional instrument in the integration of the pathogenic DNA.

From the point of view of the bacterium, it is a "clever" idea for the T-DNA to integrate into transcriptional start and stop sites rather than into the structural parts of genes; as discussed in Li et al. (2006), this strategy guarantees the expression of T-DNA-derived genes, while the plant genes' functions are left undisturbed. From the point of view of the scientist who aims to isolate plant mutants, this is not the optimal strategy. However, given the number of very large T-DNA insertion libraries available, the goal of achieving saturating levels is becoming more realistic (see below).

4.2 T-DNA integration at the chromosome level

It is a matter of considerable interest for both basic and applied studies on T-DNA integration and plant-mutant isolation to find out where in the genome the T-DNA inserts. Aspects of gene activity, accessibility of chromatin and availability of DNA breaks in the plant genome play a role, as does the abundance of repair proteins.

T-DNA insertions are generally random with respect to plant genomes. This means that in the analyzed cases, integration of the bacterially derived DNA occurs in all chromosomes; this has been tested by *in-situ* hybridization in *Crepis capillaries* (Ambros et al., 1986), *Petunia hybrida* (Wallroth et al., 1986) and tomato (Chyi et al., 1986). However, with the advent of the *A. thaliana* genome sequence as well as the draft sequence of rice (Goff et al., 2002; Yu et al., 2002), and the availability of large T-DNA insertion collections, the picture has become more complex: in general, the frequency of insertion of the T-DNA units follows the patterns of gene densities over the chromosomes rather precisely. In all five chromosomes of *A. thaliana*, T-DNAs land in gene-rich regions whereas the centromeric, paracentromeric and telomeric sequences attract far fewer T-DNA insertions (Brunaud et al., 2002; Sessions et al., 2002; Szabados et al., 2002; Alonso et al., 2003; Rosso et al., 2003). Analysis of T-DNA-insertion frequencies in the cereal rice, having much greater genomic complexity, resulted in similar conclusions: all 12 chromosomes turned out to be targets for integration; the distribution of independent T-DNA insertions along the chromosomes did not differ from that of the predicted coding sequences which are clustered in the subtelomeric regions. Around the centromeric regions which are rich in repetitive elements, T-DNA insertions were more scarce (An et al., 2003; Chen et al., 2003; Sallaud et al., 2004; Zhang et al., 2007). Thus the rules dictating T-DNA integration seem to be similar for *Arabidopsis* and rice, two species with very different genome complexities and very different susceptibilities to *Agrobacterium tumefaciens*-mediated transformation and tumorigenesis. The mechanism of transformation and target-site selection seems to be dictated by the T-DNA complex and the functions of auxiliary proteins. A similar argument can be raised for target-site selection at the gene level (see earlier). In line with these arguments is the relatively high AT content flanking T-DNA insertion sites, as found for several species. This has been suggested to enrich the flexibility of the target site and its ability to be bent, thereby promoting access to T-DNA and repair machineries (Brunaud et al., 2002). Analyses of the predicted bendability of DNA and its correlation with the probability of acting as an acceptor for T-DNA integration indicated elevated bendability around the integration sites in *Arabidopsis* (Schneeberger et al., 2005) and rice (Zhang et al., 2007). A fascinating "detail" in the rice study was a sudden drop in bendability at the very insertion sites. These data were extended to a more general view by analysis of several other rice insertion libraries (Zhang et al., 2007). Thus, the bendability peak around the preinsertion site seems to be a common aspect of chromatin recognition by the T-DNA integration complex. It will be interesting to explore the molecular and structural bases for this target-site preference.

On the other hand, CACTA transposons in *Arabidopsis* and *Tos17* retrotransposons in rice choose target insertion sites by using "rules" similar to those used by T-DNA: CACTA transposons, activated by the presence of the DNA-methylation mutation *ddm*, are not found enriched in heterochromatic regions; this is in sharp contrast to natural populations of *Arabidopsis* in which inert CACTA elements tend to occupy the rather transcriptionally inactive pericentromeric regions of the chromosomes (Kato et al., 2004). Insertion preference of the rice retrotransposon *Tos17* mimicked that of T-DNA in that the density of integration events strictly followed that of gene density (Miyao et al., 2003).

An intriguing question is pertinent to these arguments: are the analyses biased because only actively expressing T-DNA inserts are being studied? In other words, are we looking at a subpopulation of transformants recovered only on the basis of strong, selectable expression of a T-DNA-derived marker gene? It could indeed be expected that a T-DNA that landed in a transcriptionally inert region of the chromosome would not be able to express its marker gene and therefore would escape detection. Two complementary approaches have been taken to answer this question experimentally, although extensive analyses have yet to be performed: in one approach T-DNA-derived transformants were regenerated in the absence of selection for expression of a T-DNA marker. In the other approach, transformed plants were produced in which (a part of) the mechanism thought to be involved in (trans)gene silencing was inactivated.

Francis and Spiker analyzed *Arabidopsis* plants recovered from a transformation experiment in which selection was omitted (Francis and Spiker, 2005). PCR-positive plants were compared to those originating from selection, performed in a parallel experiment. Comparative analysis of these two populations revealed that about 30% of the plants recovered in the absence of selection could not have been retrieved using selection. A fascinating result was that a large fraction of the nonselected transformants mapped to positions in the *A. thaliana* genome that were underrepresented as targets for T-DNA-integration events recovered *via* selection. It was concluded that the selection bias could account for at least part of the observed nonrandom integration of T-DNA, at least in the *Arabidopsis* genome. However, only two silenced lines could be shown to contain T-DNA inserts in heterochromatic regions, thus precluding generalizations. In very recent work (Kim et al., 2007), an *Arabidopsis* tissue-culture line was inoculated with *Agrobacterium* and after a few days of incubation, DNA was isolated and T-DNA/plant DNA junction sequences were amplified, sequenced and mapped to various chromosomal locations. About 10% of the T-DNA insertions from this library mapped to repeated DNA regions in centromeric and telomeric sequences in the *Arabidopsis* genome. This was compared to a similar library isolated under selective conditions in which only 4.6% of the T-DNA insertions mapped to these repeated regions (Kim et al., 2007). Since in this experimental regime, transformed plants could not be recovered, the state of expression of the selectable gene could not be tested. Nevertheless, the study indicated that *bona fide* T-DNA integration can be discovered in areas of the genome that were previously underrepresented in T-DNA collections.

Additional useful information has been derived from transformation experiments in citrus plants. Although genomic information on this organism is largely lacking, transformation experiments avoiding selection steps during plant regeneration yielded transformants at a frequency that was about 30% higher than that of events recovered through selection (Dominguez et al., 2002). Although individual plants regenerated in the absence of selection were not challenged by selective pressure, the resultant data strongly speak for the recovery of an appreciable amount of transformed plants that would not have survived selection. One would expect that transformed lines containing T-DNAs with direct or indirect repeats would be the predominant fraction of the silenced transformants. However, no such bias was found. This certainly indicates that gene silencing of transgenes is an important phenomenon but also that we need to learn more about the reasons for its occurrence. Silent transformants need to be analyzed on the basis of their position in the genome, an undertaking which is not yet possible in citrus.

Pertinent to this discussion are the results of an analysis of *Arabidopsis* transformants carrying single-copy transgenes with low levels of expression: RNA silencing (post-transcriptional gene silencing) was triggered if the transcript level of the gene in question had surpassed a gene-specific level. This silencing activity was suggested to account for the strong variations in transgene expression found among the transformants (Schubert et al., 2004).

If silencing accounts for the reduction in transgene expression, the use of silencing mutants, of which there exist many that interfere at different levels of the silencing pathway, is expected to abolish this effect. This was indeed the case in analyses involving *Arabidopsis* post-transcriptional gene-silencing mutants (Elmayan et al., 1998; Butaye et al., 2004; De Bolle et al., 2006). The average expression levels of transgenes in these mutants were markedly higher than in wild-type plants transformed with the same genes. However, analysis of a possible correlation between the transgenes' positions within the *Arabidopsis* genome and the effect of silencing inhibition was missing, as was a comparison between individual transformants in wild-type and silencing-mutant backgrounds. Nevertheless, the general impression remains that T-DNA, if its expression is selected for, is preferentially inserted in transcriptionally active regions, if gene silencing is not inhibited. It should be noted, however, that *A. tumefaciens*-mediated transformation itself involves the host's silencing machinery, and conclusions concerning transgene silencing will have to take this into account (Dunoyer et al., 2006).

4.3 The chromatin connection

As is evident from the last section dealing with integration preferences, T-DNA integrates into chromatin and not naked DNA. A number of interesting questions ensue: How does the bacterial DNA recognize chromatin? How does it recognize a break? Are the nucleosomes moved away to afford the foreign DNA entrance? Is chromatin remodeling involved?

Several indications of specific recognition between chromatin and components of the T-DNA complex have been described. As mentioned in section 4.1, the bacterial virulence protein VirD2 has been found associated with the TATA-box-binding protein *in vitro* and in transformed *Arabidopsis* plants (Bako et al., 2003). Since the TATA-binding protein acts not only in transcription but also in transcription-coupled repair, there are potentially two pathways that the T-DNA can hijack to reach its goal, nuclear integration. The VirE3 protein, which also accompanies the T-DNA to the nucleus, uses a similar trick: it interacts with several plant components, including a transcription factor IIB-related protein (Garcia-Rodriguez et al., 2006). Although this interaction, shown in yeast two-hybrid analysis, needs to be confirmed inside the plant nucleus, it could be another illustration of the sophisticated way in which the bacterial DNA makes use of essential plant activities. Involvement of the *Arabidopsis* Ku80 protein, known to be active in repair in the NHEJ process and shown to be of special importance in T-DNA integration, may be more direct, by bridging T-DNA and plant DNA at DSB sites (Li et al., 2005b).

A promising finding involves the detection of an interaction between the VirE2-interacting protein *At*VIP1 and histones (Li et al., 2005a; Loyter et al., 2005) and the reduced transcript levels of several histone genes in VirE2-interacting protein *At*VIP2 mutant *Arabidopsis* plants (Anand et al., 2007). Specific binding to all four histones was demonstrated *in vitro* and binding to H2A was shown in tobacco nuclei using the bimolecular fluorescence complementation assay (Li et al., 2005a; Loyter et al., 2005). Although the binding of *At*VIP1 to nucleosomes, and not just histones, still needs to be demonstrated, the results nevertheless allow the speculation that the T-DNA complex, *via* its protein components VirE2 and VIP1, is recruited to chromatin. Furthermore, VIP2 was found to be essential for T-DNA integration in plants. That VIP2 interacts with both VirE2 and VIP1 and that several histone genes showed reduced expression levels in *Atvip2* plants, suggest a possible link between the T-DNA-VirE2-VIP1-VIP2 complex and the plant chromatin (Anand et al., 2007). Again, plant-derived proteins and bacterial proteins collaborate to achieve proper subcellular localization of the T-DNA.

These experiments may help understand the role histone H2A has been shown to play in *Agrobacterium*-mediated transformation (Mysore et al., 2000b). *Arabidopsis* plants lacking a functional histone H2A1 protein were severely affected in stable transformation by *Agrobacterium* (Mysore et al., 2000b). Although H2A1 is a member of a 13-gene histone H2A family in *Arabidopsis*, other members could not fully compensate for the H2A defect. Overexpression of *Arabidopsis* H2A as well as of other histones increased the plant's sensitivity to *Agrobacterium*-mediated transformation (Yi et al., 2006). Using RNAi, expression inhibition of a series of other chromatin components also yielded strains with reduced competence for transformation (Gelvin and Kim 2007). Since a clear picture is not yet in place and mechanistic aspects are missing, the involvement of chromatin in T-DNA integration remains a challenging subject for the future (reviewed in Gelvin and Kim, 2007).

As will be discussed further on, retrovirus integration into mammalian chromatin is a fascinating topic, to which T-DNA integration can be compared. One obvious difference is the availability of a dedicated integrase to the mammalian human immunodeficiency virus (HIV). This enzyme has been documented to adhere to chromatin with the help of a tether, the transcriptional coactivator LEDGF/p75 (Ciuffi et al., 2005 and references therein). It remains to be determined whether any of the aforementioned proteins, being part of or interacting with the T-DNA complex, have functions reminiscent of the mammalian tether linking the preintegration complex to chromatin.

T-DNA, on its way to the chromosomal DNA wrapped in chromatin, is expected to recruit activities that will remove or remodel the chromatin. In the absence of chromatin assembly factor (CAF), the accessibility of chromosomal DNA to T-DNA has been suggested to increase. Indeed, increased T-DNA integration efficiency was observed in mutants in which one of the three subunits of *Arabidopsis* CAF had been inactivated (Endo et al., 2006). CAF, of critical importance in DNA replication and nucleotide-excision repair, can thus be regarded as controlling T-DNA integration by protecting nuclear DNA.

CAF has not only been shown to control T-DNA integration but also somatic HR (Endo et al., 2006; Kirik et al., 2006). In contrast, the requirement for a subunit of the chromatin remodeling complex, INO80, is specific to HR; the efficiency of T-DNA integration is unchanged in *ino80* mutants (Fritsch et al., 2004). The HR-specific activity of this complex may be one example of a series of unexplored differences between HR and NHEJ, of which T-DNA is a special case.

4.4 Who makes the cut?

DNA recombination is dependent on the availability of cuts in the DNA. Consequently, T-DNA integration is dependent on breaks in the plant's DNA. As has been clearly demonstrated, VirD2 protein, attached to the 5' terminus of T-DNA, cannot promote ligation of the complex to nonspecific DNA, at least not *in vitro* (Ziemienowicz et al., 2000). Since by itself, the T-DNA complex most likely cannot initiate cuts in the plant DNA, site-specific integration of T-DNA into plant DNA is expected to be very rare, and this has been demonstrated in many cases. The question then arises as to which activity might be responsible for introducing a break in the plant DNA that can be used for insertion of DNA (the T-DNA point of view) or for DNA repair (the plant point of view).

There are several activities which may introduce breaks into plant DNA: there are artificial means of introducing breaks and there are hypothesized ones. Several genomic functions, including DNA replication, DNA repair, recombination, transcription and flexibility in chromosome architecture require local and temporary unwinding of DNA. Chromosomal ssDNA is most likely to be prone to mechanically or enzymatically introduced breaks, which the T-DNA may exploit to gain entry. DNA topoisomerase II is involved in most, if not all of these processes for controlling DNA topology. *At*TopoII sites mapped on the chromosomes of *A. thaliana* have been found to be strongly associated with T-DNA-integration sites (Makarevitch and Somers, 2006). Unfortunately, the importance of this finding cannot be tested genetically because *topoII* mutants are lethal.

Other structural features that may influence the attractiveness of chromosomal sites for T-DNA integration seem to be palindromic sequences in the genomic DNA of the host plant. Indeed, such sequences have been found in very close proximity to T-DNA insertion sites (Muller et al., 2007). These palindromic elements, all of similar sizes, have been suggested to have a high recombinogenic potential, permitting T-DNA integration. Several mechanisms have been suggested to explain these data, including increased susceptibility of palindromic sequences to breakage or endonucleolytic cleavage. Interestingly, weakly conserved palindromic sequences have also been discovered at the retroviruses' position of integration (see section 4.5).

Over the years, scientists have attempted to influence the attractiveness of plant DNA for T-DNA integration. This was accomplished by artificially introducing breaks in the plant DNA. In a number of studies, cleavage sites for rare-cutting restriction sites were introduced by *Agrobacterium*-mediated transgenesis into tobacco plants (Salomon and Puchta, 1998; Chilton and Que, 2003; Tzfira et al., 2003). The gene for the respective restriction enzyme, as well as a T-DNA whose fate was to be monitored, were introduced by coinoculation into the test plants and stable transformants were selected. T-DNA integration events into the cleaved restriction sites were indeed recovered, proving that DSBs had been efficiently used. Further development of this strategy for real gene targeting is being attempted in several laboratories, involving target-site-specific zinc-finger nucleases.

As already mentioned, in the absence of one of the components of CAF, the incidence of DSBs is slightly increased. In parallel, efficiency of T-DNA integration is enhanced (Endo et al., 2006). This may be due not only to improved accessibility of nuclear DNA to the T-DNA, but also to the larger number of breaks in the plant DNA.

Could *Agrobacterium* contact with plant cells induce the formation of DSBs? The answer to this question is not known, but an interesting finding may at least validate the question: *E. coli* strains containing a pathogenicity island coding for giant non-ribosomal peptide and polyketide synthases have been shown to induce DSBs in mammalian host cells (Nougayrede et al., 2006).

4.5 Target-site selection—a peek "over the fence"

Integration into mammalian chromatin is a required step for the life cycle of retroviruses. A DNA copy of the viral RNA, in the form of a preintegration complex, is integrated. Integration-site selection is not DNA-sequence specific. However, in all cases studied, palindromic sequences have been detected at the position of integration (quoted in Lewinski et al., 2006). Thus, this must be a general feature for the integration of pathogenic DNA into higher eukaryotic DNA, since in a very recent study on T-DNA integration, about 50% of the analyzed cases involved the detection of palindromic sequences at the position of integration (Muller et al., 2007). Could this mean that palindromes are not wrapped in chromatin and are therefore accessible?

Although palindromes seem to be a common feature, target-site selection among several retroviruses exhibits marked differences. HIV integrates preferentially into active transcription units whereas murine leukemia virus (MLV) targets transcription start sites and avian sarcoma-leukosis virus (ASLV) shows only a weak preference for active genes and no interest in transcription start sites (Mitchell et al., 2004). Dedicated integrases are responsible for the enzymology, but target-site selection is probably due to their interaction with cellular factors. HIV integrase binds to INI1, a subunit of a chromatin-remodeling complex that alters nucleosome interactions with DNA and stimulates viral integration (Kalpana et al., 1994). The cellular protein LEDGF/p75 has been shown to bind both HIV integrase and chromosomal DNA (Ciuffi et al., 2005). In cells depleted of this protein, integration was found less frequently in transcription units. Viral DNA also interacts with chromatin *via* emerin, an integral inner-nuclear-envelope protein; in its absence, viral DNA remains largely an episome (Jacque and Stevenson, 2006). Emerin, by bridging the interface between the inner nuclear envelope and chromatin, is suggested to establish a link between viral DNA and chromatin. A central player, however, in target-site selection is integrase itself, since in domain-swapping experiments with the MLV integrase, HIV changed its integration behavior to that of MLV. Thus, the virally encoded protein, in concert with cellular functions, determines its integration target.

Certainly T-DNA integration has some parallels: the T-DNA preintegration complex, consisting of the DNA and proteins imported from the bacterium, is escorted by cellular proteins on its way to the nucleus. Some of these are known, but others undoubtedly still await discovery.

5 MODELS FOR T-DNA INTEGRATION

Molecular modeling of biological processes requires a characterization of the molecules involved in the process and an understanding of the outcome. In the previous sections, we explored the functions of bacterial and host proteins in the integration process and analyzed the role that host genomes must be playing in accommodating the invading T-DNA molecule. Nevertheless, we are still missing information on the final structure of the most important component—the T-DNA—during the integration process. In other words, we still do not know whether T-DNAs integrate as a single- or double-stranded molecule or whether both forms can serve as substrates for a successful integration event. Thus, it is certainly possible that T-DNA molecules integrate into plant-cell genomes by not one, but several mechanisms. In the following sections, we present the dominant models that exist today, and point out each model's weaknesses and strengths.

5.1 The single- and double-stranded T-DNA integration models

One of the earliest models describing the mechanism of T-DNA integration into plant cells relied on analyzing the products of the integration process, i.e. analyzing the junctions and structures of the integrating T-DNA molecules (Gheysen et al., 1991; Mayerhofer et al., 1991). Supported by numerous reports years later (e.g. Chen et al., 2003; Li et al., 2006) (Kumar and Fladung, 2002; Alonso et al., 2003; Sha et al., 2004; Schneeberger et al., 2005), these pioneering studies demonstrated that T-DNA integration was not site-specific and that T-DNA molecules integrated at random locations throughout plant genomes. It should be noted that while it is still well accepted that T-DNA integration is indeed a random event, several large-scale T-DNA integration studies (Alonso et al., 2003; Rosso et al., 2003) argue that intergenic regions are more susceptible to T-DNA integration; nevertheless, even within these regions, T-DNA integration at the genome level is still random.

Together, Mayerhofer et al. (1991) and Gheysen et al. (1991) analyzed a total of 13 T-DNA inserts in *Arabidopsis* and two in tobacco plants and discovered that the integrating T-DNA molecules lose part of their sequences during the integration process, as evidenced by sequencing their integration junctions with the plant genomes. Sequencing also revealed that small deletions occur in the host genome at the integration site (Gheysen et al., 1991; Mayerhofer et al., 1991). Interestingly, while integration of the T-DNA's 3' end usually resulted in deletion of several nucleotides from the T-DNA molecule, integration of its 5' end was somewhat more precise and in some cases, the entire 5' end of the T-DNA molecule integrated into the plant genome (Gheysen et al., 1991; Mayerhofer et al., 1991). Further analysis of the integration sites revealed a certain homology between the T-DNA's ends and the preintegration site. This led the authors to suggest that T-DNA integration is driven not only by the function of the *Agrobacterium*'s VirD2 protein, but by homology between the T-DNA and the host genome. More importantly, because the homology between the T-DNA and the integration site was higher at the T-DNA's 3' end than at its 5' end, and because integration at the 5' end was more precise, the authors proposed a model in which a specific role for each side of the T-DNA molecule could be assigned. These two early models (Mayerhofer et al., 1991) for T-DNA integration—named the 'double-strand-break repair' (DSBR) and 'single-strand-gap repair' (SSGR) integration models, are illustrated in *Figures 11-1* and *11-2*, respectively.

According to the DSBR integration model (*Figure 11-1*), integration begins with the creation of a DSB in the host DNA and requires a dsT-DNA molecule (*Figure 11-1*, step 1). The broken host DNA, which may lose several nucleotides due to the activity of host exonucleases, unwinds and allows the dsT-DNA's free ends to anneal to it (*Figure 11-1*, step 2). Host exonucleases (and/or endonucleases) are also responsible for trimming the ssT-DNA overhang, which allows for the final repair and ligation of the dsT-DNA into the host genome (*Figure 11-1*, step 3).

In the SSGR integration model (*Figure 11-2*), the prerequisite for initiation of integration is not a complete DSB, but rather, a single nick, which is later converted to a gap by a 5'—>3' endonuclease (*Figure 11-2*,

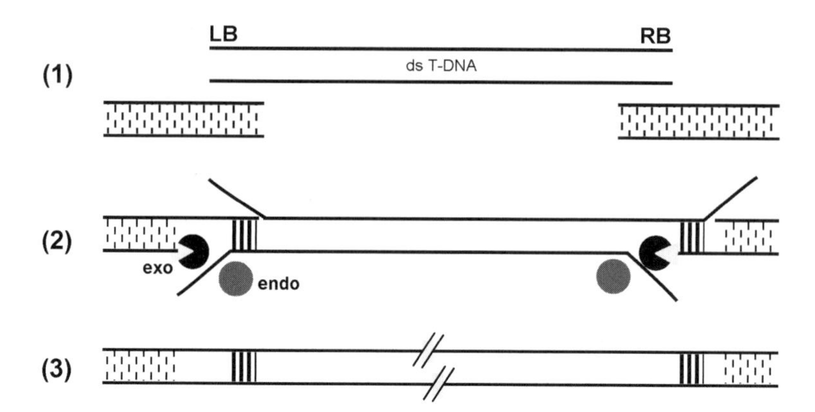

Figure 11-1. Double-strand-break repair (DSBR) model for T-DNA integration. (1) Conversion of the T-DNA molecule to a double-stranded (ds) intermediate and the production of a genomic double-strand break (DSB) initiate the integration process. (2) The DSB and the T-DNA ends are processed by exonucleases and microhomology between the resulting single-stranded (ss) ends, and the ss ends of the DSB fix the dsT-DNA in place. The ss overhangs are then removed by exonucleases (exo) and/or endonucleases (endo). (3) The ends are finally repaired and ligated. Adapted from Mayerhofer et al. (1991) and Tzfira et al. (2004) with permission.

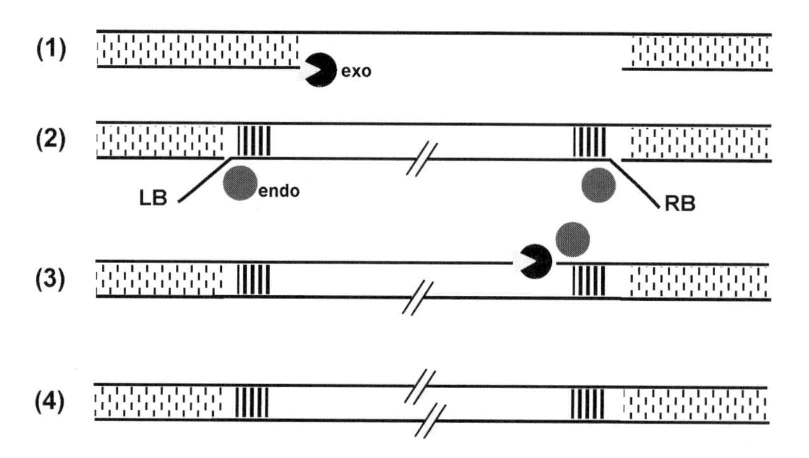

Figure 11-2. Single-strand-gap repair (SSGR) model for T-DNA integration. (1) A nick in the host genome is converted to a single-stranded (ss) gap by exonucleases (exo). (2) Microhomology between the ssT-DNA's 5' and 3' ends and the target DNA gap fix the T-DNA in place and the T-strand overhangs are trimmed by endonucleases (endo). (3) The T-strand ligates to the target DNA and a second nick is produced and extended to a gap by the combined action of endo- and exonucleases. (4) The gap is repaired and the T-strand is complemented to a double strand. Adapted from Mayerhofer et al. (1991) and Tzfira et al. (2004) with permission.

step 1). Here too, annealing is required between the T-strand and host DNA, and the T-DNA's overhangs are trimmed by the host nucleases (*Figure 11-2*, step 2). Next, the ssT-strand ligates to the target DNA (a function which was originally proposed to be carried out by VirD2) and a second nick is introduced (this time in the upper strand of the target DNA) and extended to a gap by host exonucleases (*Figure 11-2*, step 3). T-DNA integration is complete when the T-strand is complemented and its 3' end is also ligated to the target DNA (*Figure 11-2*, step 4).

These two models are fundamentally different. While the DSBR integration model requires that the T-DNA be converted to double-stranded form prior to its integration, the SSGR integration model assigns a specific role for the ssT-DNA molecule. Several observations have led to the establishment of SSGR as the dominant integration model and to the notion that ssT-DNA molecules are preferred substrates for integration. First, it is a known fact that T-DNA is transferred into the host cell as a single-stranded molecule and it is thus natural to presume that T-DNA integration occurs *via* a single-stranded intermediate. Second, experimental evidence has shown that the transformation frequency of artificial DNA molecules is higher for ssDNA than for dsDNA (Rodenburg et al., 1989) and again, it is therefore natural to assume that T-DNAs integrate as single-stranded molecules. Third, the relative integration accuracy of the T-DNA 5' end suggests that VirD2, which binds to a single-stranded, but not a double-stranded molecule, may protect the ssT-DNA during the integration process. Fourth, integration of T-DNA molecules with an uninterrupted left border, where priming of the complemented strand did not occur at the first base, suggests that such molecules integrated as a single-stranded, but not a double-stranded molecule.

Indeed, it was later suggested that VirD2 is directly involved in the integration process, not only in protecting the T-DNA, but also in directing it to nicks in the plant DNA and in ligating the T-strand's 5' end to the target DNA (Pansegrau et al., 1993). However, this notion was then shown to be incorrect (Ziemienowicz et al., 2000).

5.2 The microhomology-based T-strand integration model

The proposed function of VirD2 as a plant DNA ligase (Pansegrau et al., 1993) and its role in T-DNA integration precision (Tinland et al., 1995) helped in establishing the SSGR integration model and the notion that T-DNA integrates as a single-stranded molecule. Various reports further strengthened the role of VirD2 in the integration process, assigning it a DNA ligase-like activity *in planta* (Pansegrau et al., 1993). These included the observation that VirD2 is able to re-join the products of its nuclease activity *in vivo* (Pansegrau et al., 1993) and that certain mutations in the VirD2 integrase motif affect the integration process, rendering it less precise, albeit no less efficient (Tinland et al., 1995). This loss of precision was reflected by the loss of part of the 5'-end of the T-DNA sequence at the integration sites and led the authors to suggest a revised model for T-DNA integration (Tinland and Hohn, 1995). In their model (*Figure 11-3*), Tinland et al. (Tinland and Hohn, 1995) revisited the homology between the T-DNA and the host genome (Gheysen et al., 1991; Mayerhofer et al., 1991), narrowing it down to microhomologies which could be observed between T-DNA borders and preintegration genomic sites. T-DNA integration was proposed to be driven by microhomology-dependent annealing of the T-strand's 3' end, or its adjacent sequences, to the unwound, but not necessarily nicked, host DNA region (*Figure 11-3*, step 1). Endoucleases then trim the 3'-end overhang of the T-strand and also produce the first nick, in the bottom strand of the target DNA (*Figure 11-3*, step 2). Once the 5' end of the T-strand anneals to the target DNA, a second nick, this time in the top strand of the target DNA, is produced (*Figure 11-3*, step 3), and the VirD2 (or as later proposed, a plant DNA ligase) ligates the T-strand to the plant DNA (*Figure 11-3*, step 4). Finally, complementation of the T-strand to a double-stranded molecule completes the integration (*Figure 11-3*, step 4).

This microhomology-based T-DNA integration model relies on the role of VirD2 as a DNA ligase *in planta* (Tinland and Hohn, 1995). Nevertheless, this notion still remains controversial as a recent study has clearly shown that VirD2 does not possess DNA-ligase activity (Ziemienowicz et al., 2000). Using an *in-vitro* integration assay, it was shown that plant extracts or prokaryotic DNA ligase, but not VirD2, could facilitate the ligation of ssDNA molecules to various targets. It was thus suggested that a plant ligase, but not VirD2, is the functional enzyme during T-DNA integration. This observation does not profoundly affect the microhomology-based model, but it revises it slightly in that it provides a broader role for VirD2 in the integration process: VirD2 may still be needed to recruit plant ligase to the integration site, or perhaps other plant factors may be required for VirD2 to act as a ligase.

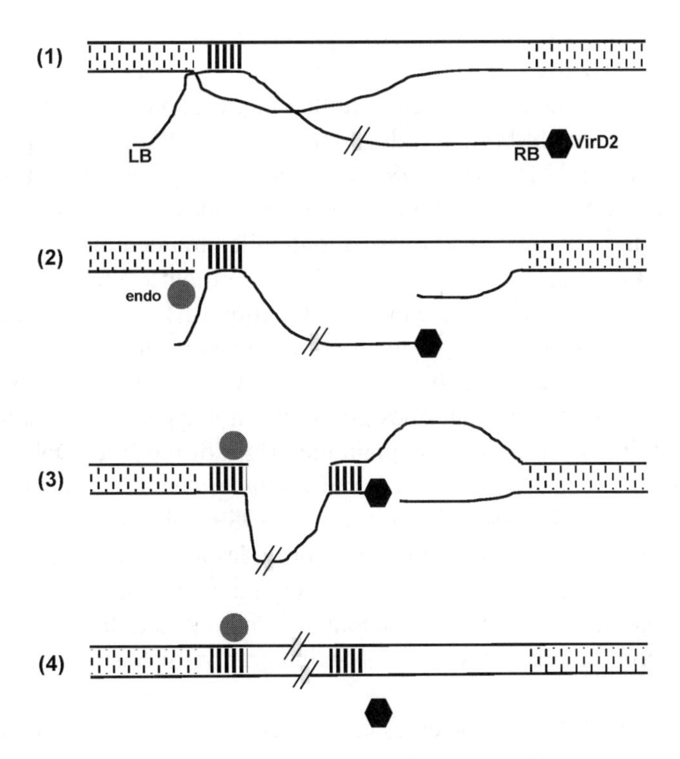

Figure 11-3. Microhomology-dependent model for T-DNA integration. (1) Microhomology-based annealing of the T-strand's 3'-end, or its adjacent sequences, to the target DNA. (2) Trimming of the T-strand's 3' end overhang and nicking of the target DNA's bottom strand by endonucleases (endo). (3) Additional microhomologies between the T-strand's 5' end and the target DNA immobilize the T-strand and the target DNA's upper stand is nicked. (4) Complementation of the T-strand to dsDNA and its ligation, possibly with the assistance of VirD2 and/or other plant factors, to the host DNA. Adapted from Tinland and Hohn (1995) and Tzfira et al. (2004) with permission.

5.3 A model for double strand T-DNA integration into double strand breaks

The microhomology-based T-strand integration model (Tinland and Hohn, 1995) provides a good explanation for the integration of a single-copy T-DNA, and variations on this model could also explain some of the different complex integration patterns often observed in transgenic plants (see chapter 12). Nevertheless, this model cannot provide a simple explanation for the integration of two T-DNA molecules arranged in the same orientation relative to one another (e.g., Journin et al., 1989; De Neve et al., 1997; Krizkova and Hrouda, 1998; De Buck et al., 1999). Sequence analysis of such events revealed that there is sometimes precise fusion between two right-border ends (Journin et al., 1989; De Neve et al., 1997; Krizkova and Hrouda, 1998; De Buck et al., 1999). Because ssT-DNA molecules cannot recombine at their right borders, it was suggested that these molecules must be converted to double-stranded molecules prior to their integration (De Neve et al., 1997). According to this model, integration of dsT-DNA molecules and recombination of T-DNA molecules to each other can occur by a simple ligation mechanism. The model also suggests that VirD2 remains attached to the T-DNA, even after its conversion to a double-stranded form, but does not assign it a specific role as a plant ligase. Furthermore, it does not explain the origin of the filler DNA sometimes found between repeating copies of co-integrated T-DNA molecules (e.g., Bakkeren et al., 1989; Journin et al., 1989; Gheysen et al., 1991; Mayerhofer et al., 1991). Nevertheless, it provides a solid basis for the notion that T-DNA molecules can indeed integrate as double-stranded intermediates. (For further discussion about this model and for an in-depth analysis of complex patterns of T-DNA integration in host genomes, see chapter 12.)

Integration of T-DNA molecules can sometimes be accompanied by the insertion of filler DNA at the T-DNA/host genome junction sites and between integrated T-DNA molecules. Filler T-DNA can also be observed during integration of double-stranded plasmid DNA into the host genome (Gorbunova and Levy, 1997). Filler DNA is usually made up of scrambled sequences derived from the invading DNA molecule or the host genome, and it is often linked to the repair of DSBs. Indeed, it has been shown that DSB repair (artificially induced using a rare-cutting restriction enzyme), which typically results in small DNA insertions and deletions at the break site, can also lead to the incorporation of T-DNA molecules into these sites (Salomon and Puchta, 1998). These observations led to the proposal that DSBs play a significant role in T-DNA integration, and that the number of breaks in the host genome might be the limiting factor for this integration (Salomon and Puchta, 1998). Nevertheless, whether T-DNA molecules integrate into the DSBs as single-stranded molecules *via* a synthesis-dependent annealing (SDA) mechanism or whether they integrate as double-stranded intermediates by a simple ligation step was not determined. It was later shown that combining the induction of genomic DSBs by rare-cutting restriction enzymes with the induction of similar breaks on the T-DNA molecules results in precise ligation of truncated T-DNA molecules into the targeted genome (Chilton and Que, 2003; Tzfira et al., 2003). Because these enzymes can only digest double-stranded molecules, it was obvious that the T-DNA molecules had been at least partially converted to a double-stranded form before their integration into the host genome (Chilton and Que, 2003; Tzfira et al., 2003).

That X-ray irradiation, known to cause DSBs (Leskov et al., 2001), enhances transgene integration (Kohler et al., 1989) and that KU80, a dsDNA-repair protein, was found important for T-DNA integration in *Arabidopsis* plants (Li et al., 2005b), further support the role of DSBs in T-DNA integration. Thus, a model for the functions of DSBs, dsT-DNA intermediates, and DSB-repair proteins can be formulated (*Figure 11-4*).

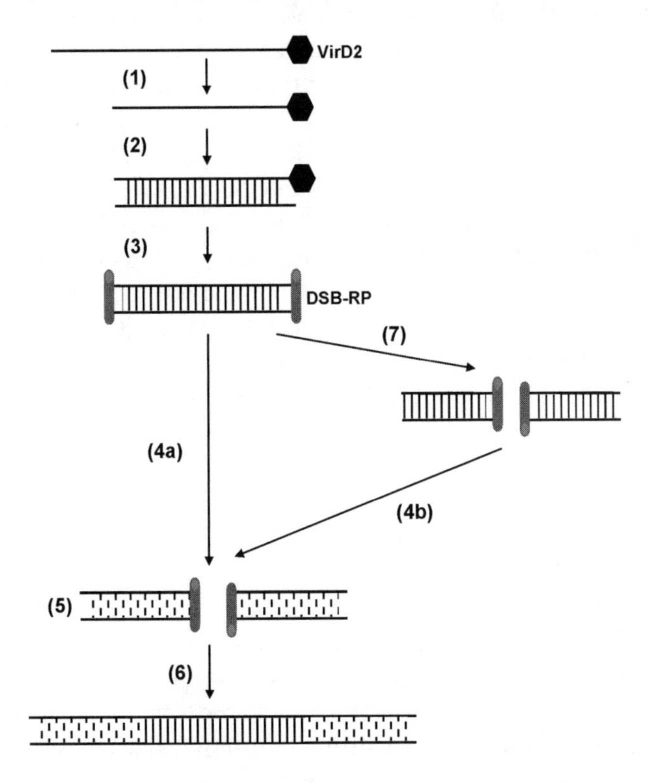

Figure 11-4. A model for double-stranded (ds) T-DNA integration into double-strand breaks (DSBs). (1) Exonucleases may degrade part of the T-strand's unprotected 3' end. (2) The T-strand is converted into ds form. (3) VirD2 is replaced by DNA-repair proteins (DSB-RP). (4a and 4b) dsT-DNA is directed into DSBs. (5) Occurrence of genomic DSBs and stabilization of DSBs by the host's DNA-repair proteins. (6) Ligation of dsT-DNA into DSBs. (7) Ligation of dsT-DNA to each other prior to their ligation into genomic DSBs. Tzfira et al. (2004) with permission.

According to this model, the T-strand can degrade slightly at its unprotected 3' end, but not at its VirD2-protected 5' end, on the way to the nucleus (*Figure 11-4*, step 1). The T-strand is then converted to a double-stranded molecule inside the nucleus (*Figure 11-4*, step 2) and serves as the actual substrate for the integration process. Replacement of VirD2 by host DNA-repair proteins (e.g., KU80) (*Figure 11-4*, step 3) stabilizes and directs the dsDNA molecule to repair *via* the plant nonhomologous DNA-repair mechanism (*Figure 11-4*, step 4a). Meanwhile, DNA maintenance reactions and/or transcription lead to DNA unpacking, possibly through genome modifications, and to the formation of DSBs (*Figure 11-4*, step 5); this DNA is also stabilized and directed to repair by the host DNA-repair machinery. How the new dsT-DNA molecule ligates into the DSBs is still unclear (*Figure 11-4*, step 6), but the process is likely to result in the addition or removal of a few base pairs from the host genome and the T-DNA molecule (*Figure 11-4*, step 6). This model can also explain the co-integration of multiple T-DNA molecules into the same site since several dsDNA molecules can become ligated together (*Figure 11-4*, step 7) prior to their integration into the break site (*Figure 11-4*, step 4b).

Thus, the integration of dsT-DNA molecules into genomic DSBs may represent another mode of T-DNA entry into the host genome (Chilton and Que, 2003; Tzfira et al., 2003). Whether this mode is a major or minor pathway is still a matter of debate and further research is required to identify all the host and bacterial factors functioning in this process.

6 FUTURE DIRECTIONS

Ever since 1977, when T-DNA integration into plant DNA was first documented, the rules followed by plant and agrobacterial DNA and proteins to accomplish their task of integration have been busily researched and debated. Whereas the roles of virulence proteins are easier to analyze—at least we know what they are—the functions of plant proteins in this process are more difficult to study; educated guesses on the proteins involved and their analysis have led to the elucidation of their important contributions to integration. However, only a screen for plants impaired in T-DNA integration will yield information on potentially unsuspected players in the field. Such screens are being performed, with the caveat that *Agrobacterium* selfishly hijacks essential plant activities for its own goals. Consequently such mutants cannot be isolated as homozygous plants.

An elucidation of the positions of T-DNA integration into the plant genome will remain a fascinating aspect of this research. As we improve our knowledge of the plant genome, its structure and its function, the analysis of T-DNA integration sites serves as a means to further explore it. Conversely, the principles by which T-DNA integration is accomplished provide information on the specialized roles of the proteins attached to the T-DNA. In addition, it will be imperative to determine what kind of breaks in the genomic DNA attract T-DNA and what kind of repair activities are being recruited to the sites of repair. As should be obvious from this review, T-DNA most probably integrates in more than just one mode—most likely as single-stranded and double-stranded versions, accompanied by different proteins in each case. As an extension, one can ask: Which kind of breaks in the genomic DNA attract which kind of T-DNA, using which kind of repair activity? What about plant chromatin? Does a repressive, tightly packed chromatin influence the availability of genomic DNA for integration? Further work is clearly needed to enable improved insight into the choice of site and the mechanism of T-DNA integration.

7 ACKNOWLEDGEMENTS

The work in our labs is supported by grants from MNiSzW to A.Z, from the HFSP and BRDC to T.T. and from Novartis Research Foundation to B.H. The Authors wish to thank Drs. Stan B. Gelvin and Bernd Weisshaar for the stimulating discussions about T-DNA integration.

8 REFERENCES

Abu-Arish A, Frenkiel-Krispin D, Fricke T, Tzfira T, Citovsky V, Grayer Wolf S, Elbaum M (2004) Three-dimensional reconstruction of *Agrobacterium* VirE2 protein with single-stranded DNA. J Biol Chem 279: 25359-25363

Alonso JM, Stepanova AN, Leisse TJ, Kim CJ, Chen H, Shinn P, Stevenson DK, Zimmerman J, Barajas P, Cheuk R, Gadrinab C, Heller C, Jeske A, Koesema E, Meyers CC, Parker H, Prednis L, Ansari Y, Choy N, Deen H, Geralt M, Hazari N, Hom E, Karnes M, Mulholland C, Ndubaku R, Schmidt I, Guzman P, Aguilar-Henonin L, Schmid M, Weigel D, Carter DE, Marchand T, Risseeuw E, Brogden D, Zeko A, Crosby WL, Berry CC, Ecker JR (2003) Genome-wide insertional mutagenesis of *Arabidopsis thaliana*. Science 301: 653-657

Ambros PF, Matzke AJ, Matzke MA (1986) Localization of *Agrobacterium rhizogenes* T-DNA in plant chromosomes by *in situ* hybridization. EMBO J 5: 2073-2077

An S, Park S, Jeong DH, Lee DY, Kang HG, Yu JH, Hur J, Kim SR, Kim YH, Lee M, Han S, Kim SJ, Yang J, Kim E, Wi SJ, Chung HS, Hong JP, Choe V, Lee HK, Choi JH, Nam J, Park PB, Park KY, Kim WT, Choe S, Lee CB, An G (2003) Generation and analysis of end sequence database for T-DNA tagging lines in rice. Plant Physiol 133: 2040-2047

Anand A, Krichevsky A, Schornack S, Lahaye T, Tzfira T, Tang Y, Citovsky V, Mysore KS (2007) Arabidopsis VirE2 interacting protein2 is required for *Agrobacterium* T-DNA Integration in Plants. Plant Cell 19: 1695-1708

Babiychuk E, Cottrill PB, Storozhenko S, Fuangthong M, Chen Y, O'Farrell MK, Van Montagu M, Inze D, Kushnir S (1998) Higher plants possess two structurally different poly(ADP-ribose) polymerases. Plant J 15: 635-645

Bakkeren G, Koukolikova-Nicola Z, Grimsley N, Hohn B (1989) Recovery of *Agrobacterium tumefaciens* T-DNA molecules from whole plants early after transfer. Cell 57: 847-857

Bako L, Umeda M, Tiburcio AF, Schell J, Koncz C (2003) The VirD2 pilot protein of *Agrobacterium*-transferred DNA interacts with the TATA box-binding protein and a nuclear protein kinase in plants. Proc Natl Acad Sci USA 100: 10108-10113

Ballas N, Citovsky V (1997) Nuclear localization signal binding protein from *Arabidopsis* mediates nuclear import of *Agrobacterium* VirD2 protein. Proc Natl Acad Sci USA 94: 10723-10728

Bebenek K, Kunkel TA (2004) Functions of DNA polymerases. Adv Protein Chem 69: 137-165

Blobel G (1985) Gene gating: a hypothesis. Proc Natl Acad Sci USA 82: 8527-8529

Boulton SJ, Jackson SP (1998) Components of the Ku-dependent non-homologous end-joining pathway are involved in telomeric length maintenance and telomeric silencing. EMBO J 17: 1819-1828

Bray CM, West CE (2005) DNA repair mechanisms in plants: crucial sensors and effectors for the maintenance of genome integrity. New Phytol 168: 511-528

Brunaud V, Balzergue S, Dubreucq B, Aubourg S, Samson F, Chauvin S, Bechtold N, Cruaud C, DeRose R, Pelletier G, Lepiniec L, Caboche M, Lecharny A (2002) T-DNA integration into the *Arabidopsis* genome depends on sequences of pre-insertion sites. EMBO Rep 3: 1152-1157

Bundock P, den Dulk-Ras A, Beijersbergen A, Hooykaas PJJ (1995) Trans-kingdom T-DNA transfer from *Agrobacterium tumefaciens* to *Saccharomyces cerevisiae*. EMBO J 14: 3206-3214

Bundock P, Hooykaas PJJ (1996) Integration of *Agrobacterium tumefaciens* T-DNA in the *Saccharomyces cerevisiae* genome by illegitimate recombination. Proc Natl Acad Sci USA 93: 15272-15275

Bundock P, van Attikum H, den Dulk-Ras A, Hooykaas PJ (2002) Insertional mutagenesis in yeasts using T-DNA from *Agrobacterium tumefaciens*. Yeast 19: 529-536

Butaye KM, Goderis IJ, Wouters PF, Pues JM, Delaure SL, Broekaert WF, Depicker A, Cammue BP, De Bolle MF (2004) Stable high-level transgene expression in *Arabidopsis thaliana* using gene silencing mutants and matrix attachment regions. Plant J 39: 440-449

Cabal GG, Genovesio A, Rodriguez-Navarro S, Zimmer C, Gadal O, Lesne A, Buc H, Feuerbach-Fournier F, Olivo-Marin JC, Hurt EC, Nehrbass U (2006) SAGA interacting factors confine sub-diffusion of transcribed genes to the nuclear envelope. Nature 441: 770-773

Casolari JM, Brown CR, Komili S, West J, Hieronymus H, Silver PA (2004) Genome-wide localization of the nuclear transport machinery couples transcriptional status and nuclear organization. Cell 117: 427-439

Chen S, Jin W, Wang M, Zhang F, Zhou J, Jia Q, Wu Y, Liu F, Wu P (2003) Distribution and characterization of over 1000 T-DNA tags in rice genome. Plant J 36: 105-113

Chilton M-D, Que Q (2003) Targeted integration of T-DNA into the tobacco genome at double-stranded breaks: new insights on the mechanism of T-DNA integration. Plant Physiol 133: 956-965

Chyi YS, Jorgensen RA, Goldstein D, Tanksley SD, Loaiza-Figueroa F (1986) Locations and stability of *Agrobacterium*-mediated T-DNA insertions in the *Lycopersicon* genome. Mol Gen Genet 204: 64-69

Citovsky V, Guralnick B, Simon MN, Wall JS (1997) The molecular structure of *Agrobacterium* VirE2-single stranded DNA complexes involved in nuclear import. J Mol Biol 271: 718-727

Citovsky V, Kozlovsky SV, Lacroix B, Zaltsman A, Dafny-Yelin M, Vyas S, Tovkach A, Tzfira T (2007) Biological systems of the host cell involved in *Agrobacterium* infection. Cell Microbiol 9: 9-20

Ciuffi A, Llano M, Poeschla E, Hoffmann C, Leipzig J, Shinn P, Ecker JR, Bushman F (2005) A role for LEDGF/p75 in targeting HIV DNA integration. Nat Med 11: 1287-1289

Daniel PP, Bryant JA (1985) DNA ligase in pea (*Pisum sativum* L.) seedlings: changes in activity during germination and effects of deoxyribonucleotides. J Exp Bot 39: 481-486

De Bolle MF, Butaye KM, Goderis IJ, Wouters PF, Jacobs A, Delaure SL, Depicker A, Cammue BP (2006) The influence of matrix attachment regions on transgene expression in *Arabidopsis thaliana* wild type and gene silencing mutants. Plant Mol Biol 63: 533-543

De Buck S, Jacobs A, Van Montagu M, Depicker A (1999) The DNA sequences of T-DNA junctions suggest that complex T-DNA loci are formed by a recombination process resembling T-DNA integration. Plant J 20: 295-304

De Neve M, De Buck S, Jacobs A, Van Montagu M, Depicker A (1997) T-DNA integration patterns in co-transformed plant cells suggest that T-DNA repeats originate from co-integration of separate T-DNAs. Plant J 11: 15-29

Deng W, Chen L, Wood DW, Metcalfe T, Liang X, Gordon MP, Comai L, Nester EW (1998) *Agrobacterium* VirD2 protein interacts with plant host cyclophilins. Proc Natl Acad Sci USA 95: 7040-7045

Dominguez A, Fagoaga C, Navarro L, Moreno P, Pena L (2002) Regeneration of transgenic citrus plants under non selective conditions results in high-frequency recovery of plants with silenced transgenes. Mol Genet Genomics 267: 544-556

Dunoyer P, Himber C, Voinnet O (2006) Induction, suppression and requirement of RNA silencing pathways in virulent *Agrobacterium tumefaciens* infections. Nat Genet 38: 258-263

Elmayan T, Balzergue S, Beon F, Bourdon V, Daubremet J, Guenet Y, Mourrain P, Palauqui JC, Vernhettes S, Vialle T, Wostrikoff K, Vaucheret H (1998) *Arabidopsis* mutants impaired in cosuppression. Plant Cell 10: 1747-1758

Endo M, Ishikawa Y, Osakabe K, Nakayama S, Kaya H, Araki T, Shibahara K, Abe K, Ichikawa H, Valentine L, Hohn B, Toki S (2006) Increased frequency of homologous recombination and T-DNA integration in *Arabidopsis* CAF-1 mutants. EMBO J 25: 5579-5590

Ferreira PC, Hemerly AS, Engler JD, van Montagu M, Engler G, Inze D (1994) Developmental expression of the *Arabidopsis* cyclin gene cyc1At. Plant Cell 6: 1763-1774

Forsbach A, Schubert D, Lechtenberg B, Gils M, Schmidt R (2003) A comprehensive characterization of single-copy T-DNA insertions in the *Arabidopsis thaliana* genome. Plant Mol Biol 52: 161-176

Francis KE, Spiker S (2005) Identification of *Arabidopsis thaliana* transformants without selection reveals a high occurrence of silenced T-DNA integrations. Plant J 41: 464-477

Friesner J, Britt AB (2003) *Ku80*- and *DNA ligase IV*-deficient plants are sensitive to ionizing radiation and defective in T-DNA integration. Plant J 34: 427-440

Fritsch O, Benvenuto G, Bowler C, Molinier J, Hohn B (2004) The INO80 protein controls homologous recombination in *Arabidopsis thaliana*. Mol Cell 16: 479-485

Gallego ME, Bleuyard JY, Daoudal-Cotterell S, Jallut N, White CI (2003) Ku80 plays a role in non-homologous recombination but is not required for T-DNA integration in *Arabidopsis*. Plant J 35: 557-565

Garcia-Rodriguez FM, Schrammeijer B, Hooykaas PJ (2006) The *Agrobacterium* VirE3 effector protein: a potential plant transcriptional activator. Nucleic Acids Res 34: 6496-6504

Gelvin SB (2000) *Agrobacterium* and plant genes involved in T-DNA transfer and integration. Annu Rev Plant Physiol Plant Mol Biol 51: 223-256

Gelvin SB (2003) *Agrobacterium*-mediated plant transformation: the biology behind the "gene-jockeying" tool. Microbiol Mol Biol Rev 67: 16-37

Gelvin SB, Kim SI (2007) Effect of chromatin upon *Agrobacterium* T-DNA integration and transgene expression. Biochim Biophys Acta 1769: 409-420

Gheysen G, Villarroel R, Van Montagu M (1991) Illegitimate recombination in plants: a model for T-DNA integration. Genes Dev 5: 287-297

Goff SA, Ricke D, Lan TH, Presting G, Wang R, Dunn M, Glazebrook J, Sessions A, Oeller P, Varma H, Hadley D, Hutchison D, Martin C, Katagiri F, Lange BM, Moughamer T, Xia Y, Budworth P, Zhong J, Miguel T, Paszkowski U, Zhang S, Colbert M, Sun WL, Chen L, Cooper B, Park S, Wood TC, Mao L, Quail P, Wing R, Dean R, Yu Y, Zharkikh A, Shen R, Sahasrabudhe S, Thomas A, Cannings R, Gutin A, Pruss D, Reid J, Tavtigian S, Mitchell J, Eldredge G, Scholl T, Miller RM, Bhatnagar S, Adey N, Rubano T, Tusneem N, Robinson R, Feldhaus J, Macalma T, Oliphant A, Briggs S (2002) A draft sequence of the rice genome (*Oryza sativa* L. ssp. japonica). Science 296: 92-100

Gorbunova V, Levy AA (1997) Non-homologous DNA end joining in plant cells is associated with deletions and filler DNA insertions. Nucleic Acids Res 25: 4650-4657

Haber JE (2000) Lucky breaks: analysis of recombination in *Saccharomyces*. Mutat Res 451: 53-69

Herman L, Jacobs A, Van Montagu M, Depicker A (1990) Plant chromosome/marker gene fusion assay for study of normal and truncated T-DNA integration events. Mol Gen Genet 224: 248-256

Jacque JM, Stevenson M (2006) The inner-nuclear-envelope protein emerin regulates HIV-1 infectivity. Nature 441: 641-645

Journin L, Bouchezs D, Drong RF, Tepfer D, Slightom JL (1989) Analysis of TR-DNA/plant junctions in the genome of *Convolvulus arvensis* clone transformed by *Agrobacterium rhizogenes* strain A4. Plant Mol Biol 12: 72-85

Kalpana GV, Marmon S, Wang W, Crabtree GR, Goff SP (1994) Binding and stimulation of HIV-1 integrase by a human homolog of yeast transcription factor SNF5. Science 266: 2002-2006

Kato M, Takashima K, Kakutani T (2004) Epigenetic control of CACTA transposon mobility in *Arabidopsis thaliana*. Genetics 168: 961-969

Kertbundit S, De Greve H, Deboeck F, Van Montagu M, Hernalsteens JP (1991) *In vivo* random beta-glucuronidase gene fusions in *Arabidopsis thaliana*. Proc Natl Acad Sci USA 88: 5212-5216

Kim S-I, Veena, Gelvin SB (2007) Genome-wide analysis of *Agrobacterium* T-DNA integration sites in the *Arabidopsis* genome generated under non-selective conditions. Plant J 51: 779-791

Kirik A, Pecinka A, Wendeler E, Reiss B (2006) The chromatin assembly factor subunit FASCIATA1 is involved in homologous recombination in plants. Plant Cell 18: 2431-2442

Kohler F, Cardon G, Pohlman M, Gill R, Schieder O (1989) Enhancement of transformation rates in higher plants by low-dose irradiation: are DNA repair systems involved in the incorporation of exogenous DNA into the plant genome? Plant Mol Biol 12: 189-199

Koncz C, Martini N, Mayerhofer R, Koncz-Kalman Z, Korber H, Redei GP, Schell J (1989) High-frequency T-DNA-mediated gene tagging in plants. Proc Natl Acad Sci USA 86: 8467-8471

Krizkova L, Hrouda M (1998) Direct repeats of T-DNA integrated in tobacco chromosome: characterization of junction regions. Plant J 16: 673-680

Kumar S, Fladung M (2002) Transgene integration in aspen: structures of integration sites and mechanism of T-DNA integration. Plant J 31: 543-551

Lacroix B, Tzfira T, Vainstein A, Citovsky V (2006) A case of promiscuity: *Agrobacterium*'s endless hunt for new partners. Trends Genet 22: 29-37

Leskov KS, Criswell T, Antonio S, Li J, Yang CR, Kinsella TJ, Boothman DA (2001) When X-ray-inducible proteins meet DNA double strand break repair. Semin Radiat Oncol 11: 352-372

Lewinski MK, Yamashita M, Emerman M, Ciuffi A, Marshall H, Crawford G, Collins F, Shinn P, Leipzig J, Hannenhalli S, Berry CC, Ecker JR, Bushman FD (2006) Retroviral DNA integration: viral and cellular determinants of target-site selection. PLoS Pathog 2: e60

Lewis LK, Resnick MA (2000) Tying up loose ends: nonhomologous end-joining in *Saccharomyces cerevisiae*. Mutat Res 451: 71-89

Li J, Krichevsky A, Vaidya M, Tzfira T, Citovsky V (2005a) Uncoupling of the functions of the *Arabidopsis* VIP1 protein in transient and stable plant genetic transformation by *Agrobacterium*. Proc Natl Acad Sci USA 102: 5733-5738

Li J, Vaidya M, White C, Vainstein A, Citovsky V, Tzfira T (2005b) Involvement of KU80 in T-DNA integration in plant cells. Proc Natl Acad Sci USA 102: 19231-19236

Li Y, Rosso MG, Ulker B, Weisshaar B (2006) Analysis of T-DNA insertion site distribution patterns in *Arabidopsis thaliana* reveals special features of genes without insertions. Genomics 87: 645-652

Loyter A, Rosenbluh J, Zakai N, Li J, Kozlovsky SV, Tzfira T, Citovsky V (2005) The plant VirE2 interacting protein 1. A molecular link between the *Agrobacterium* T-complex and the host cell chromatin? Plant Physiol 138: 1318-1321

Luthra R, Kerr SC, Harreman MT, Apponi LH, Fasken MB, Ramineni S, Chaurasia S, Valentini SR, Corbett AH (2007) Actively transcribed GAL genes can be physically linked to the nuclear pore by the SAGA chromatin modifying complex. J Biol Chem 282: 3042-3049

Makarevitch I, Somers DA (2006) Association of *Arabidopsis* topoisomerase IIA cleavage sites with functional genomic elements and T-DNA loci. Plant J 48: 697-709

Martinez JJ, Seveau S, Veiga E, Matsuyama S, Cossart P (2005) Ku70, a component of DNA-dependent protein kinase, is a mammalian receptor for *Rickettsia conorii*. Cell 123: 1013-1023

Mayerhofer R, Koncz-Kalman Z, Nawrath C, Bakkeren G, Crameri A, Angelis K, Redei GP, Schell J, Hohn B, Koncz C (1991) T-DNA integration: a mode of illegitimate recombination in plants. EMBO J 10: 697-704

Michielse CB, Hooykaas PJ, van den Hondel CA, Ram AF (2005) *Agrobacterium*-mediated transformation as a tool for functional genomics in fungi. Curr Genet 48: 1-17

Mitchell RS, Beitzel BF, Schroder AR, Shinn P, Chen H, Berry CC, Ecker JR, Bushman FD (2004) Retroviral DNA integration: ASLV, HIV, and MLV show distinct target site preferences. PLoS Biol 2: E234

Miyao A, Tanaka K, Murata K, Sawaki H, Takeda S, Abe K, Shinozuka Y, Onosato K, Hirochika H (2003) Target site specificity of the Tos17 retrotransposon shows a preference for insertion within genes and against insertion in retrotransposon-rich regions of the genome. Plant Cell 15: 1771-1780

Montague JW, Hughes FM, Jr., Cidlowski JA (1997) Native recombinant cyclophilins A, B, and C degrade DNA independently of peptidylprolyl cis-trans-isomerase activity. Potential roles of cyclophilins in apoptosis. J Biol Chem 272: 6677-6684

Muller AE, Atkinson RG, Sandoval RB, Jorgensen RA (2007) Microhomologies between T-DNA ends and target sites often occur in inverted orientation and may be responsible for the high frequency of T-DNA-associated inversions. Plant Cell Rep 26: 617-630

Mysore KS, Kumar CTR, Gelvin SB (2000a) *Arabidopsis* ecotypes and mutants that are recalcitrant to *Agrobacterium* root transformation are susceptible to germ-line transformation. Plant J 21: 9-16

Mysore KS, Nam J, Gelvin SB (2000b) An *Arabidopsis* histone H2A mutant is deficient in *Agrobacterium* T-DNA integration. Proc Natl Acad Sci USA 97: 948-953

Nam J, Matthysse AG, Gelvin SB (1997) Differences in susceptibility of *Arabidopsis* ecotypes to crown gall disease may result from a deficiency in T-DNA integration. Plant Cell 9: 317-333

Neale MJ, Pan J, Keeney S (2005) Endonucleolytic processing of covalent protein-linked DNA double-strand breaks. Nature 436: 1053-1057

Nougayrede JP, Homburg S, Taieb F, Boury M, Brzuszkiewicz E, Gottschalk G, Buchrieser C, Hacker J, Dobrindt U, Oswald E (2006) *Escherichia coli* induces DNA double-strand breaks in eukaryotic cells. Science 313: 848-851

Offringa R, de Groot MJA, Haagsman HJ, Does MP, van den Elzen PJM, Hooykaas PJJ (1990) Extrachromosomal homologous recombination and gene targeting in plant cells after *Agrobacterium* mediated transformation. EMBO J 9: 3077-3084

Pansegrau W, Schoumacher F, Hohn B, Lanka E (1993) Site-specific cleavage and joining of single-stranded DNA by VirD2 protein of *Agrobacterium tumefaciens* Ti plasmids: analogy to bacterial conjugation. Proc Natl Acad Sci USA 90: 11538-11542

Paszkowski J, Baur M, Bogucki A, Potrykus I (1988) Gene targeting in plants. EMBO J 7: 4021-4026

Pelczar P, Kalck V, Gomez D, Hohn B (2004) *Agrobacterium* proteins VirD2 and VirE2 mediate precise integration of synthetic T-DNA complexes in mammalian cells. EMBO Rep 5: 632-637

Riha K, Watson JM, Parkey J, Shippen DE (2002) Telomere length deregulation and enhanced sensitivity to genotoxic stress in *Arabidopsis* mutants deficient in Ku70. EMBO J 21: 2819-2826

Rodenburg KW, de Groot MJ, Schilperoort RA, Hooykaas PJ (1989) Single-stranded DNA used as an efficient new vehicle for transformation of plant protoplasts. Plant Mol Biol 13: 711-719

Rodriguez-Navarro S, Fischer T, Luo MJ, Antunez O, Brettschneider S, Lechner J, Perez-Ortin JE, Reed R, Hurt E (2004) Sus1, a functional component of the SAGA histone acetylase complex and the nuclear pore-associated mRNA export machinery. Cell 116: 75-86

Rossi L, Hohn B, Tinland B (1996) Integration of complete transferred DNA units is dependent on the activity of virulence E2 protein of *Agrobacterium tumefaciens*. Proc Natl Acad Sci USA 93: 126-130

Rosso MG, Li Y, Strizhov N, Reiss B, Dekker K, Weisshaar B (2003) An *Arabidopsis thaliana* T-DNA mutagenized population (GABI-Kat) for flanking sequence tag-based reverse genetics. Plant Mol Biol 53: 247-259

Sallaud C, Gay C, Larmande P, Bes M, Piffanelli P, Piegu B, Droc G, Regad F, Bourgeois E, Meynard D, Perin C, Sabau X, Ghesquiere A, Glaszmann JC, Delseny M, Guiderdoni E (2004) High throughput T-DNA insertion mutagenesis in rice: a first step towards *in silico* reverse genetics. Plant J 39: 450-464

Salomon S, Puchta H (1998) Capture of genomic and T-DNA sequences during double-strand break repair in somatic plant cells. EMBO J 17: 6086-6095

Sancar A, Lindsey-Boltz LA, Unsal-Kacmaz K, Linn S (2004) Molecular mechanisms of mammalian DNA repair and the DNA damage checkpoints. Annu Rev Biochem 73: 39-85

Schmid M, Arib G, Laemmli C, Nishikawa J, Durussel T, Laemmli UK (2006) Nup-PI: the nucleopore-promoter interaction of genes in yeast. Mol Cell 21: 379-391

Schneeberger RG, Zhang K, Tatarinova T, Troukhan M, Kwok SF, Drais J, Klinger K, Orejudos F, Macy K, Bhakta A, Burns J, Subramanian G, Donson J, Flavell R, Feldmann KA (2005) *Agrobacterium* T-DNA integration in *Arabidopsis* is correlated with DNA sequence compositions that occur frequently in gene promoter regions. Funct Integr Genomics 5: 240-253

Schubert D, Lechtenberg B, Forsbach A, Gils M, Bahadur S, Schmidt R (2004) Silencing in *Arabidopsis* T-DNA transformants: the predominant role of a gene-specific RNA sensing mechanism versus position effects. Plant Cell 16: 2561-2572

Sessions A, Burke E, Presting G, Aux G, McElver J, Patton D, Dietrich B, Ho P, Bacwaden J, Ko C, Clarke JD, Cotton D, Bullis D, Snell J, Miguel T, Hutchison D, Kimmerly B, Mitzel T, Katagiri F, Glazebrook J, Law M, Goff SA (2002) A high-throughput *Arabidopsis* reverse genetics system. Plant Cell 14: 2985-2994

Sha Y, Li S, Pei Z, Luo L, Tian Y, He C (2004) Generation and flanking sequence analysis of a rice T-DNA tagged population. Theor Appl Genet 108: 306-314

Szabados L, Kovacs I, Oberschall A, Abraham E, Kerekes I, Zsigmond L, Nagy R, Alvarado M, Krasovskaja I, Gal M, Berente A, Redei GP, Haim AB, Koncz C (2002) Distribution of 1000 sequenced T-DNA tags in the *Arabidopsis* genome. Plant J 32: 233-242

Taddei A, Van Houwe G, Hediger F, Kalck V, Cubizolles F, Schober H, Gasser SM (2006) Nuclear pore association confers optimal expression levels for an inducible yeast gene. Nature 441: 774-778

Teo SH, Jackson SP (1997) Identification of *Saccharomyces cerevisiae* DNA ligase IV: involvement in DNA double-strand break repair. EMBO J 16: 4788-4795

Timson DJ, Singleton MR, Wigley DB (2000) DNA ligases in the repair and replication of DNA. Mutat Res 460: 301-318

Tinland B (1996) The integration of T-DNA into plant genomes. Trends Plant Sci 1: 178-184

Tinland B, Hohn B (1995) Recombination between prokaryotic and eukaryotic DNA: integration of *Agrobacterium tumefaciens* T-DNA into the plant genome. Genet Eng 17: 209-229

Tinland B, Schoumacher F, Gloeckler V, Bravo-Angel AM, Hohn B (1995) The *Agrobacterium tumefaciens* virulence D2 protein is responsible for precise integration of T-DNA into the plant genome. EMBO J 14: 3585-3595

Tsukamoto Y, Ikeda H (1998) Double-strand break repair mediated by DNA end-joining. Genes Cells 3: 135-144

Tzfira T, Citovsky V (2002) Partners-in-infection: host proteins involved in the transformation of plant cells by *Agrobacterium*. Trends Cell Biol 12: 121-129

Tzfira T, Frankman LR, Vaidya M, Citovsky V (2003) Site-specific integration of *Agrobacterium tumefaciens* T-DNA *via* double-stranded intermediates. Plant Physiol 133: 1011-1023

Tzfira T, Li J, Lacroix B, Citovsky V (2004a) *Agrobacterium* T-DNA integration: molecules and models. Trends Genet 20: 375-383

Tzfira T, Vaidya M, Citovsky V (2004b) Involvement of targeted proteolysis in plant genetic transformation by *Agrobacterium*. Nature 431: 87-92

van Attikum H, Bundock P, Hooykaas PJJ (2001) Non-homologous end-joining proteins are required for *Agrobacterium* T-DNA integration. EMBO J 20: 6550-6558

van Attikum H, Bundock P, Overmeer RM, Lee LY, Gelvin SB, Hooykaas PJJ (2003) The *Arabidopsis AtLIG4* gene is required for the repair of DNA damage, but not for the integration of *Agrobacterium* T-DNA. Nucleic Acids Res 31: 4247-4255

van Attikum H, Hooykaas PJJ (2003) Genetic requirements for the targeted integration of *Agrobacterium* T-DNA in *Saccharomyces cerevisiae*. Nucleic Acids Res 31: 826-832

Volokhina I, Chumakov M (2007) Study of the VirE2-ssT-DNA complex formation by scanning probe microscopy and gel electrophoresis- T-complex visualization. Microsc Microanal 13: 51-54

Wallroth M, Gerats AGM, Rogers SG, Fraley RT, Horsch RB (1986) Chromosomal localization of foreign genes in *Petunia hybrida*. Mol Gen Genet 202: 6-15

West CE, Waterworth WM, Story GW, Sunderland PA, Jiang Q, Bray CM (2002) Disruption of the *Arabidopsis AtKu80* gene demonstrates an essential role for

AtKu80 protein in efficient repair of DNA double-strand breaks *in vivo*. Plant J 31: 517-528

Weterings E, van Gent DC (2004) The mechanism of non-homologous end-joining: a synopsis of synapsis. DNA Repair (Amst) 3: 1425-1435

Wu Y-Q (2002) Protein-protein interaction between VirD2 and DNA ligase: an essential step of Agrobacterium tumefaciens T-DNA integration. Ph.D. thesis. University of Basel

Wu Y-Q, Hohn B (2003) Cellular transfer and Chromosomal integration of T-DNA during *Agrobacterium tumefaciens*-mediated plant transformation. *In* G Stacey, NT Keen, eds, Plant-Microbe Interactions Vol 6, pp 1-18

Yi H, Mysore KS, Gelvin SB (2002) Expression of the *Arabidopsis* histone *H2A-1* gene correlates with susceptibility to *Agrobacterium* transformation. Plant J 32: 285-298

Yi H, Sardesai N, Fujinuma T, Chan CW, Veena, Gelvin SB (2006) Constitutive expression exposes functional redundancy between the *Arabidopsis* histone H2A gene HTA1 and other H2A gene family members. Plant Cell 18: 1575-1589

Yu J, Hu S, Wang J, Wong GK, Li S, Liu B, Deng Y, Dai L, Zhou Y, Zhang X, Cao M, Liu J, Sun J, Tang J, Chen Y, Huang X, Lin W, Ye C, Tong W, Cong L, Geng J, Han Y, Li L, Li W, Hu G, Li J, Liu Z, Qi Q, Li T, Wang X, Lu H, Wu T, Zhu M, Ni P, Han H, Dong W, Ren X, Feng X, Cui P, Li X, Wang H, Xu X, Zhai W, Xu Z, Zhang J, He S, Xu J, Zhang K, Zheng X, Dong J, Zeng W, Tao L, Ye J, Tan J, Chen X, He J, Liu D, Tian W, Tian C, Xia H, Bao Q, Li G, Gao H, Cao T, Zhao W, Li P, Chen W, Zhang Y, Hu J, Liu S, Yang J, Zhang G, Xiong Y, Li Z, Mao L, Zhou C, Zhu Z, Chen R, Hao B, Zheng W, Chen S, Guo W, Tao M, Zhu L, Yuan L, Yang H (2002) A draft sequence of the rice genome (*Oryza sativa* L. ssp. indica). Science 296: 79-92

Zambryski PC (1992) Chronicles from the *Agrobacterium*-plant cell DNA transfer story. Annu Rev Plant Physiol Plant Mol Biol 43: 465-490

Zhang J, Guo D, Chang Y, You C, Li X, Dai X, Weng Q, Chen G, Liu H, Han B, Zhang Q, Wu C (2007) Non-random distribution of T-DNA insertions at various levels of the genome hierarchy as revealed by analyzing 13 804 T-DNA flanking sequences from an enhancer-trap mutant library. Plant J 49: 947-959

Zhang X, Yazaki J, Sundaresan A, Cokus S, Chan SW, Chen H, Henderson IR, Shinn P, Pellegrini M, Jacobsen SE, Ecker JR (2006) Genome-wide high-resolution mapping and functional analysis of DNA methylation in *Arabidopsis*. Cell 126: 1189-1201

Zhu Y, Nam J, Humara JM, Mysore KS, Lee LY, Cao H, Valentine L, Li J, Kaiser AD, Kopecky AL, Hwang HH, Bhattacharjee S, Rao PK, Tzfira T, Rajagopal

J, Yi H, Veena, Yadav BS, Crane YM, Lin K, Larcher Y, Gelvin MJ, Knue M, Ramos C, Zhao X, Davis SJ, Kim SI, Ranjith-Kumar CT, Choi YJ, Hallan VK, Chattopadhyay S, Sui X, Ziemienowicz A, Matthysse AG, Citovsky V, Hohn B, Gelvin SB (2003) Identification of *Arabidopsis rat* mutants. Plant Physiol 132: 494-505

Ziemienowicz A, Merkle T, Schoumacher F, Hohn B, Rossi L (2001) Import of *Agrobacterium* T-DNA into plant nuclei: Two distinct functions of VirD2 and VirE2 proteins. Plant Cell 13: 369-384

Ziemienowicz A, Tinland B, Bryant J, Gloeckler V, Hohn B (2000) Plant enzymes but not *Agrobacterium* VirD2 mediate T-DNA ligation *in vitro*. Mol Cell Biol 20: 6317-6322

Zilberman D, Gehring M, Tran RK, Ballinger T, Henikoff S (2007) Genome-wide analysis of *Arabidopsis thaliana* DNA methylation uncovers an interdependence between methylation and transcription. Nat Genet 39: 61-69

Chapter 12

AGROBACTERIUM TUMEFACIENS-MEDIATED TRANSFORMATION: PATTERNS OF T-DNA INTEGRATION INTO THE HOST GENOME

Pieter Windels[1,2], Sylvie De Buck[3] and Ann Depicker[3]

[1]Pioneer Hi-Bred International, Avenue des Arts 44, B-1040 Brussels, Belgium; [2]Department of Plant Genetics and Breeding, Centre for Agricultural Research, Caritasstraat 21, B-9090 Melle, Belgium; [3]Department of Plant Systems Biology, Flanders Interuniversity Institute for Biotechnology (VIB), Ghent University, Technologiepark 927, B-9052 Gent, Belgium

Abstract. The *Agrobacterium tumefaciens*-mediated transformation system is widely used to introduce genes into plants and is based on the conjugative transfer of the T-DNA to the plant nucleus. In this process, T-DNA formation, T-DNA transfer, and T-DNA integration via illegitimate recombination can be distinguished. In addition to some transformants with one T-DNA copy, transformants with multicopy T-DNA loci are also often found. In these multicopy loci, the T-DNAs often occur as inverted repeats about the right or left border. The T-DNA plant junctions frequently contain insertions of filler DNA, short regions of microhomology, small deletions of both T-DNA ends and target sequences, and integration of vector backbone sequences. To date, extensive scientific research has paved the way for a better understanding of the bacterial and plant host-driven molecular mechanisms that underlie the different steps in the *Agrobacterium*-mediated plant cell transformation process. The aim of this chapter is to discuss the final stage and outcome of the T-DNA transformation process, *i.e.* to focus on the molecular mechanism that integrates the T-DNA and, in addition, to describe the various patterns documented in the literature.

1 INTRODUCTION

The scientific story behind the interaction between the Gram-negative soil bacterium *Agrobacterium tumefaciens* and the plant cell starts back in 1907, when Smith and Townsend (1907) reported that *A. tumefaciens* was the causative agent of the crown gall disease that is characterized by the formation of tumors on infected plant tissues. Early pioneering work by Braun (1958) led to the insight that crown gall tumors could be excised from the plant and could be cultured *in vitro* without the need for exogenously supplied plant growth hormones. Braun proposed that the agrobacteria transferred "something" into the plant cells, which he referred to as a tumor-inducing principle. Later on, the observations that agrobacteria harbor a large tumor-inducing (Ti) plasmid that is present as an extrachromosomal unit in the bacterial cell and that part of this plasmid DNA can become stably incorporated into the nuclear genome of the susceptible plant cell, lay the foundations for studying the interaction between the bacterial cell and the plant cell (Zaenen et al., 1974; Chilton et al., 1977).

The portion of the Ti plasmid that is transferred to the plant cell was later identified as the transferred DNA or T-DNA that is delineated by two 25-bp imperfect repeat sequences, designated the left and right T-DNA borders (Zambryski et al., 1982; Van Haaren et al., 1987). These left and right T-DNA borders are recognized as the boundaries of the T-DNA by the bacterial DNA transfer protein machinery and usually only the region in-between both borders is transferred to the plant cell. In this manner, a series of successive steps, including chemotactic sensing, bacterial attachment to the plant cell, T-DNA formation, T-DNA transfer and nuclear import of the T-DNA, eventually lead to stable integration of the T-DNA into the plant host genome (Zupan et al., 2000; Escobar and Dandekar, 2003; Gelvin, 2003; Valentine, 2003; Komari et al., 2004; Tzfira et al., 2004a).

2 T-DNA INTEGRATION MECHANISM: SUCCESSIVE STEPS LEADING TO STABLE INTEGRATION OF THE T-DNA INTO THE PLANT HOST GENOME

The perception of and the movement toward the plant cell is the first step in the T-DNA transformation process by *A. tumefaciens*. This early step in *Agrobacterium*-mediated plant cell transformation is followed by the attachment of the infecting agrobacteria to the plant cell, forming a biofilm of bacteria on the surface of the plant tissue. The sensing of the wounded plant cell induces a well-studied two-component regulatory system in the agrobacteria: the virulence VirA-VirG system. The activated VirG allows induction of the VirD operon, containing the *virD1* and *virD2* genes that encode a site-specific endonuclease and a relaxase, respectively. Together, the VirD1 and VirD2 proteins recognize and produce a single-stranded nick in the lower strand of the 25-bp double-stranded right and left T-DNA border sequence between the 3rd and 4th base of the repeat (Scheiffele et al., 1995; Relić et al., 1998). After nicking the right T-DNA border sequence, a single-stranded T-strand is set free by a unidirectional process that proceeds toward the left T-DNA border (Albright et al., 1987). Upon nicking, the VirD2 protein remains covalently attached to the 5' end of the processed T-strand (Herrera-Estrella et al., 1988). As a consequence, a single-stranded protein/nucleic acid complex is transferred to the plant cell (Tinland et al., 1994). The VirD2 protein has been implicated in different functions during transfer of the T-strand, such as protection against exonucleolytic degradation (Dürrenberger et al., 1989), nuclear localization by means of identified nuclear localization signals (NLS) at the amino and carboxyl termini (Herrera-Estrella et al., 1990; Howard et al., 1992; Tinland et al., 1992), and precise integration and ligation to the plant DNA (Tinland et al., 1995). However, with regard to the latter function, it needs to be mentioned that not VirD2, but specific plant enzymes might be involved in ligation of the T-DNA to the plant DNA (Ziemienowicz et al., 2000). After transfer to the plant cell, the T-DNA is coated along its length with VirE2 proteins (Tzfira et al., 2000) and both the VirD2 and the VirE2 direct the nuclear uptake of the T-complex (Citovsky et al., 1992; Shurvinton et al., 1992).

Once the T-complex enters the plant cell nucleus, it is further processed to integrate into the plant host genome by targeted proteolysis of T-complex-associated virulence proteins, for instance (Tzfira et al., 2004b). By performing extrachromosomal recombination assays, the T-strand has been shown to be imported into the plant cell nucleus as a single-stranded nucleoprotein complex (Tinland et al., 1994). This raised the question as to the exact nature of the substrate for T-DNA integration, *i.e.* whether the T-complex becomes double-stranded prior or during T-DNA integration. First, the assumption that T-strands can be made double stranded prior to T-DNA integration came from the observations that already some hours after co-cultivation T-DNA transient expression can be detected in plant cells (Janssen and Gardner, 1990) and also from the results of recombination assays (Offringa et al., 1990). In addition, De Neve et al. (1997) and De Buck et al. (1999) recognized that the frequently observed linkages of two different T-DNAs about the right border in transformed plants can only arise when second-strand synthesis results in a double-stranded T-DNA substrate prior to linkage. The double-stranded nature of the T-DNA prior to integration was further substantiated by the observation that a fragment can be released from the T-DNA by restriction enzymes (Chilton and Que, 2003). Therefore, the model has been put forward in which primases could synthesize primers for the 3' left end of the T-strand to drive second-strand synthesis by DNA polymerases present in repair complexes in the nucleus (De Neve et al., 1997). On the other hand, the T-complex has been hypothesized to be made double stranded during T-DNA integration by interaction of the single-stranded T-complex with the plant target site. After this initial interaction, the T-strand could be used as template for second-strand synthesis starting from a free 3' end residing in the plant DNA (Gheysen et al., 1991; Mayerhofer et al., 1991; Brunaud et al., 2002; Kumar and Fladung, 2002). Alternatively, the T-strand could also be made double stranded when a single-stranded T-strand commences integration at the right border, while simultaneously the 3' end loops back and anneals to itself (Meza et al., 2002). In this manner, a 3'-OH end is provided that can prime second-strand synthesis. In conclusion, and since experimental evidence has been provided for all the above hypotheses, these mechanisms might be not mutually exclusive.

2.1 T-DNA integration can be a serious bottleneck to obtaining transgenic plants with a high efficiency

It is still unclear whether one or many T-DNA copies enter one plant cell and what percentage of the transferred T-DNAs becomes subsequently integrated. Several studies indicated that important restrictions for stable integration are situated at the level of either plant cell accessibility, T-DNA transfer to the nucleus or T-DNA integration (De Block and Debrouwer, 1991; De Buck et al., 1998; Gheysen et al., 1998).

Many environmental factors as well as bacterial and plant genes that affect T-DNA transfer from *Agrobacterium* to plants have been identified (Dillen et al., 1997; Tzfira and Citovsky, 2002; Komari et al., 2004; Kumar and Rajam, 2005), forming the basis for improving transformation procedures. For example, Zambre et al. (2003), recently identified light as an important factor that strongly promotes gene transfer from *Agrobacterium* to cells of *Arabidopsis thaliana* and *Phaseolus acutifolius*. Light conditions may affect physiological factors, such as plant hormone levels and cell proliferation that, in turn, could influence the competence for *Agrobacterium* attachment and subsequently T-DNA transfer (de Kathen and Jacobsen, 1995; Villemont et al., 1997; Chateau et al., 2000). At the same time, these variables will affect tissue viability and regeneration potential, two parameters that are also very important in the recovery of transgenic plants. As a matter of fact, regenerating cells must not only be accessible for the agrobacteria, but should also be competent to receive the T-DNA and to integrate it into the chromosomal DNA (Gheysen et al., 1998).

In most systems, selection for an efficient transformation system is essential, because the fraction of stably transformed cells is usually small (De Buck et al., 1998; Gheysen et al., 1998). Both T-DNA transfer and T-DNA integration limit the transformation and cotransformation frequencies, on which plant cell competence for transformation is based on (De Buck et al., 2000a). Indeed, without selection for transformation competence, none of the 84 regenerants, obtained after *Arabidopsis* root transformation, were stably transformed. However, in four of these regenerants, the T-DNA was transiently expressed, which indicated that an effective *Agrobacterium*-plant cell interaction had been established, but without T-DNA integration. Therefore, although the transformation frequency was below 0.5% during *Agrobacterium*-mediated root transformation, the T-DNA transfer frequency was 10-fold or more above this frequency (De Buck et al., 2000a). Nevertheless, the stable transformation frequency of co-cultivated tobacco *(Nicotiana tabacum)* protoplasts that regenerate into shoots was found to be relatively high: 18.5% of the tobacco shoots, regenerated on non-selective medium, were transformed (De Buck et al., 1998), indicating that a large proportion of the regenerating tobacco protoplasts were competent for transformation (Depicker et al., 1985; De Buck et al., 1998). The difference in competence for stable transformation can be based on several factors, such as the plant genotype, differentiation, physiology of the plant cells (De Buck et al., 1998), and, in view of the recent results, might depend on the availability of double-stranded breaks (DSBs; see below).

On the other hand, when the integration of one T-DNA had been selected, approximately 50% of these transformants (24/56) expressed transiently a second nonselected T-DNA, and 25% (11/56) had the second T-DNA also integrated into the *Arabidopsis* genome (cotransformation frequency 25%; De Buck et al., 2000a). These results suggest that transformation frequencies in root explants are especially hampered by T-DNA integration. Similarly, also in apple (*Malus domestica*), maize (*Zea mays*), and *Phaseolus* sp., factors others than T-DNA transfer are rate limiting (Narasimhulu et al., 1996; Dillen et al., 1997; Maximova et al., 1998). Additionally, agroinfection studies indicated that although T-DNA transfer to monocotyledonous cells occur, the block of transformation by *Agrobacterium* may lie in a lack of competence for T-DNA transport to the nucleus or its integration (Grimsley et al., 1987, 1988).

Cells that can stabilize one T-DNA have a higher chance to stabilize a second incoming T-DNA than cells that do not. It is possible that the increased efficiency of stabilization of a second T-DNA results from properties of the T-DNA transfer mechanism and the activity of certain DNA repair enzymes (Nam et al., 1997; De Buck et al., 1998). However, we favor the possibility that the availability of an integration site corresponding to a DSB (see below) might explain the high integration frequency of a second non-selected T-DNA.

2.2 T-DNA integration: Involvement of bacterial and plant host factors

Eventually, nuclear uptake of the T-complex can result in stable T-DNA integration. Because together with the nucleoprotein T-complex also other virulence proteins (e.g. VirF, VirE3, and VirD5) are transferred to the plant cell (Vergunst et al., 2005), bacterial virulence factors are probably involved in T-DNA integration. As described above, it is well-established that the VirD2 protein could influence nuclear uptake (Ballas and Citovsky, 1997; Deng et al., 1998) as well as the precision and the efficiency of T-DNA integration (Shurvinton et al., 1992; Tinland et al., 1995; Mysore et al., 1998). In addition, the VirD2 protein can be bound by nuclear TATA-box binding proteins and interact with nuclear cyclin-dependent kinase-activating kinases (Bakó et al., 2003). Because these TATA-box binding proteins can attach to lesions in the DNA to initiate transcription-coupled repair (Vichi et al., 1997; Coin et al., 1998), this finding implicates that the VirD2 protein could be a link between T-DNA integration and transcription-coupled repair (Bakó et al., 2003). Furthermore, two lines of evidence indicate that the VirE2 participates in T-DNA integration. First, because the VirE2 protein can bind a VirE2-interacting protein 1 (VIP1; Tzfira et al., 2002) that interacts with histones *in vitro* and *in planta*, this bound could act as a molecular link between the T-complex and the histone constituents of the host chromatin (Li et al., 2005; Loyter et al., 2005). Second, in mutants of VirE2, the precision of T-DNA integration, especially at the 3' end, is diminished, presumably because of the exonucleolytic degradation of the T-complex ends in the absence of the VirE2 protein (Rossi et al., 1996).

That not only bacterial virulence factors are involved in T-DNA integration but also plant host factors, was first demonstrated by Sonti et al. (1995), who characterized two radiation-sensitive *A. thaliana* mutants, *uvh1* and *rad5*, that were susceptible to transient transformation but were recalcitrant to stable transformation. Also, radiosensitive *Arabidopsis* ecotypes were shown to be recalcitrant for *Agrobacterium* infection (Nam et al., 1997). Furthermore, Nam et al. (1999) used a T-DNA insertion library and identified five *rat* mutants, *rat5*, *rat17*, *rat18*, *rat20*, and *rat22*, all deficient in T-DNA integration. Two of these mutants have been characterized in more detail. The *rat17* mutant harbors a disrupted gene for a *myb*-like transcription factor (Gelvin, 2000), whereas the *rat5* mutant carries the T-DNA in a histone *H2A* gene (Mysore et al., 2000). This histone H2A might affect the chromatin structure at the T-DNA integration site and influence efficiency of T-DNA integration (Mysore et al., 2000). In addition, the plant's repair machinery to repair DSBs might be involved in T-DNA integration (see below).

That the T-DNA integration process is a complex interplay between bacterial virulence and host-dependent factors is also nicely illustrated by the fact that the *A. thaliana* ecotype and the tissue that is transformed will influence T-DNA integration efficiency. For instance, the *rat5* mutants described above are recalcitrant to stable root transformation, but are easily transformed by flower vacuum infiltration (Mysore et al., 2000).

Other plant genes involved in *Agrobacterium* transformation have been identified by using cDNA-AFLP (Ditt et al., 2001; Gelvin, 2003) and DNA macroarrays (Veena et al., 2003). Analysis of the genes that are differentially expressed upon *Agrobacterium* infection indicate that, in general, plant genes necessary for T-DNA transformation (e.g. histones and ribosomal proteins) are induced, while host defense responses are repressed simultaneously.

2.3 The molecular mechanism that drives T-DNA integration: Illegitimate recombination

Despite the identification of several *A. tumefaciens*-specific virulence proteins as well as plant host proteins involved in the T-DNA transformation process, little is known about the exact molecular mechanism of T-DNA integration. Sequence analysis of isolated T-DNA inserts provided the first clues about the T-DNA integration mechanism, whereas sequence analysis of the boundaries of a set of 13 T-DNA inserts in *Arabidopsis* and two in tobacco revealed that all T-DNA/plant DNA junctions exhibit some striking and common characteristics (Gheysen et al., 1991; Mayerhofer et al., 1991). Overall, these plant DNA/T-DNA junctions involve (i) intact right T-DNA borders and processed left T-DNA borders, (ii) small regions of homology, referred to as microhomology between the interacting T-DNA end and the plant genomic target site, and (iii) insertion of small sequences at the plant DNA/T-DNA junctions, referred to as filler DNAs. These findings have been substantiated by studies reporting on the boundaries of a large set of plant DNA/T-DNA junctions (Fladung, 1999; Zheng et al., 2001; Brunaud et al., 2002; Krysan et al., 2002; Kumar and Fladung, 2002; Meza et al., 2002; Windels et al., 2003). The characteristics observed for these plant DNA/T-DNA junctions, together with the absence of the requirement for large stretches of sequence homology are all in line with the original conclusion that T-DNA integration is mediated by the plant's illegitimate recombination machinery (Gheysen et al., 1991).

2.4 T-DNA integration: Single-stranded gap repair (SSGR) vs. double-stranded break repair (DSBR) models

Although it is generally accepted that T-DNA integration uses the plant's illegitimate recombination apparatus, several slightly different models can be used to explain the actual first molecular steps that lead to integration of the incoming T-DNA. As discussed above, it is currently not completely clear whether a single-stranded T-complex or a partial double-stranded T-complex is the preferred substrate for T-DNA integration. Therefore, some of the models use single-stranded T-strands as substrate for T-DNA integration, while others propose double-stranded T-DNAs with free, single-stranded T-DNA ends to be the substrate for T-DNA integration.

Based on sequence characterization of a set of plant DNA/T-DNA junctions, Gheysen et al. (1991) and Mayerhofer et al. (1991) proposed a model for T-DNA integration. We will discuss the former, referred to as the single-stranded gap repair (SSGR) model (*Figure 12-1*) and then, the latter referred to as the double-stranded break repair (DSBR) model (*Figure 12-2*).

2.4.1 SSGR model

The SSGR model for T-DNA integration was proposed for T-DNAs that have either single-stranded or double-stranded T-DNA ends with the body of the T-DNA. According to the SSGR model, T-DNA integration requires the presence of a nick at the T-DNA target site, found, for instance, at a replication fork or in regions subjected to excision repair. Due to 5'-to-3'-exonuclease activity, this single-stranded nick or site can be enlarged and the T-DNA ends can interact with the single-stranded region of the target site by means of microhomology pairing (*Figure 12-1a*). The non-complementary overhangs of the T-DNA ends are removed and the T-DNA is ligated to the plant target ends. Removal of the non-complementary overhangs might result in the frequently observed truncated left and/or right borders. Eventually, the heteroduplex is resolved by nicking the opposite strand of the target DNA, which might be the reason why T-DNA integration might go together with a small deletion of the plant DNA at the target site. In case the incoming T-DNA is a single-stranded molecule, the gap is repaired by using the T-strand as template. The complementary strand of the T-strand is copied into the plant genome to start synthesis at a free 3' plant DNA end. In case a double-stranded T-strand is the substrate for integration, the ends of the double-stranded T-DNA are simply ligated to the free plant DNA ends. Gheysen et al. (1991) also proposed a model for filler formation, by hypothesizing that the observed filler was either formed through ligation of preformed sequence blocks or by a slip/mispair/repair process, the latter resulting in a duplicated region. Polymerase slipping and extensive template switching explained more complex filler. Template switching might even involve folding back of the T-DNA ends to find microhomology. Based on the findings that left border T-DNA/plant DNA junctions were frequently imprecise and processed, whereas right border T-DNA/plant DNA junctions were more frequently conserved, the SSGR model for T-DNA integration was further refined (*Figure 12-1b*). Tinland (1996) proposed that T-DNA integration started at a D-loop formed in a double-stranded target DNA by invasion of the target DNA at the 3' end of the incoming T-strand, hence the name "LB first" model. As such, the 3' end of the T-DNA invades a

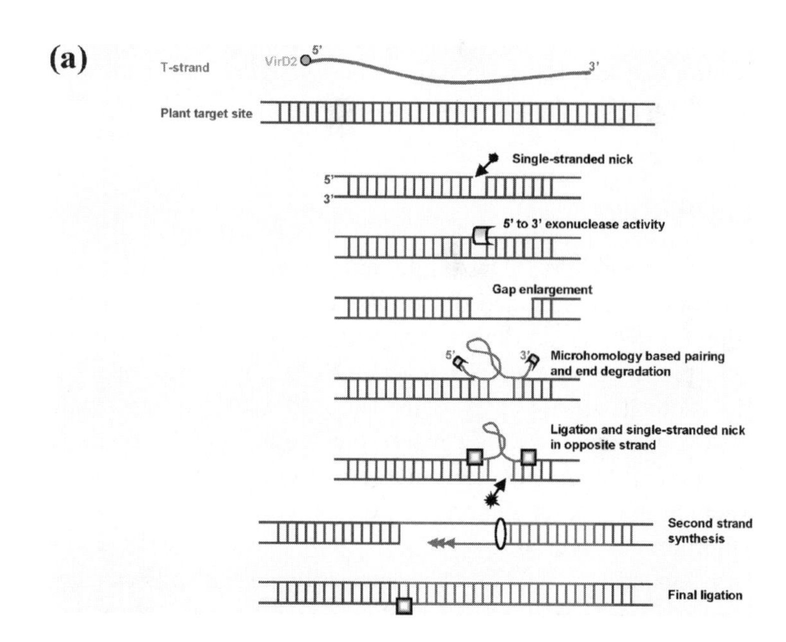
(a)
T-strand
VirD2
5'
3'
Plant target site
Single-stranded nick
5'
3'
5' to 3' exonuclease activity
Gap enlargement
5'
3'
Microhomology based pairing and end degradation
Ligation and single-stranded nick in opposite strand
Second strand synthesis
Final ligation

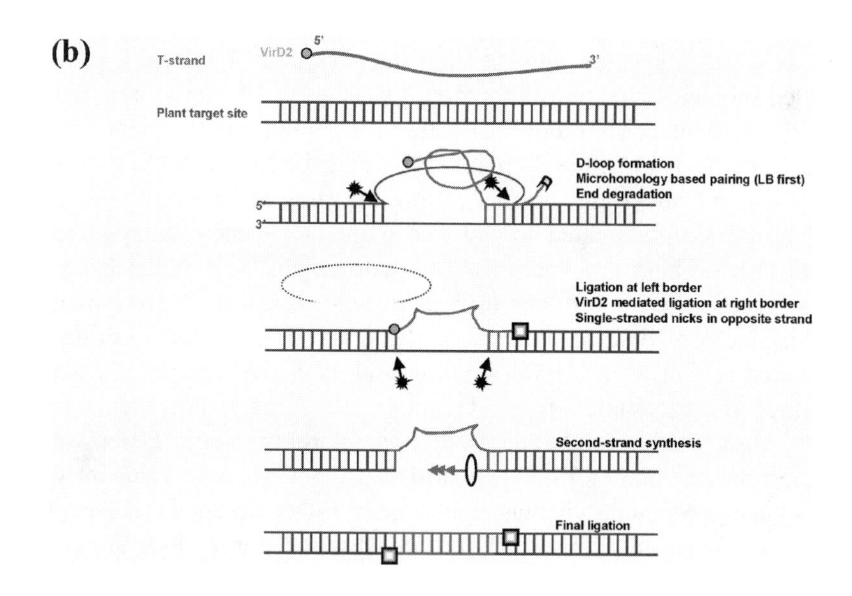
(b)
T-strand
VirD2
5'
3'
Plant target site
D-loop formation
Microhomology based pairing (LB first)
End degradation
5'
3'
Ligation at left border
VirD2 mediated ligation at right border
Single-stranded nicks in opposite strand
Second-strand synthesis
Final ligation

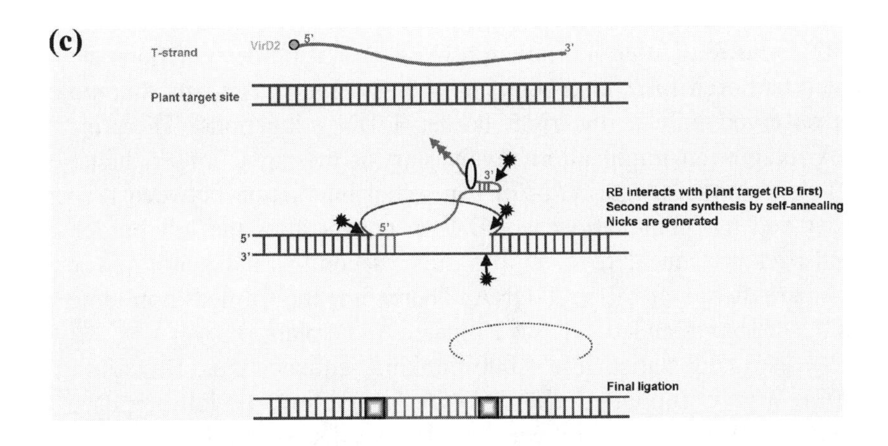

Figure 12-1. Single-stranded gap repair (SSGR) as a model for T-DNA integration. (a) Schematic representation of the single-stranded gap repair model for T-DNA integration as proposed by Gheysen et al. (1991). T-DNA integration is initiated at a single-stranded nick (arrowed asterisk). Upon gap enlargement, due to exonucleolytic degradation, a T-DNA with single-stranded ends and a body that is either single- or double-stranded interacts with the plant DNA target. This initial interaction is stabilized by means of microhomology pairing. Subsequently, the T-DNA ends become ligated to the plant target by the coordinated action of DNA ligases (purple square). A single-stranded nick is induced in the opposite strand of the plant target and polymerase activity copies the T-DNA into the plant genome. (b) T-DNA integration by SSGR as proposed by Tinland (1996). This model is based on the fact that T-DNA integration commences at the left border. The orange circle represents the VirD2 protein, whereas the yellow oval represents a DNA polymerase. (c) Schematic representation of the SSGR "RB first" model for T-DNA integration as proposed by Meza et al. (2002). This model is based on microhomologies between T-DNA and genomic DNA near the T-DNA right border. After the 3' end of the single-stranded T-DNA loops back and anneals to it self, the T-DNA that is still single stranded near the 5' end, invades the genomic DNA and finds a short stretch of microhomology. Nicks will occur in the genomic DNA and the double-stranded 3' end loop will be degraded resulting in a left end deletion. VirD2 and the 5' end of the T-DNA upstream of the microhomology are removed, single-stranded gaps are repaired, and the double-stranded break between the T-DNA left end and the genomic DNA as well, thereby generating filler DNA.

target site at a locally denatured region. Again this first interaction is based on microhomology pairing. Subsequently, the displaced plant DNA strand is removed by endonucleolytic digestion and the 5' end of the T-strand binds to a region of microcomplementarity. This binding might be mediated by the energy-rich covalent bond between the 5' end and the VirD2 protein. Thereafter, the opposite strand of the plant target is nicked and removed. As such, a reactive plant DNA 3' end is formed that is used as a primer to copy the T-strand into the plant genome (*Figure 12-1b*).

A variation on the "LB first" model has been proposed by Meza et al. (2002), who found that more than half of the left border junctions characterized harbored filler insertions and that, in some cases, microhomology was observed only at the right border T-DNA junctions. Therefore, T-DNA integration might alternatively start at the right border, hence the "RB first" model (*Figure 12-1c*). An initial interaction between the right border and the plant target at a D-loop occurs after the left border end loops back to anneal to itself. As such, second-strand synthesis occurs from the left border of the T-DNA. Thereafter, the formed loop is nicked and the double-stranded T-DNA is ligated to the plant DNA.

By analyzing statistically 9,000 flanking sequence tags, Brunaud et al. (2002) largely supported the T-DNA integration model proposed by Tinland (1996). The novelty was that the initial step in T-DNA integration is seemingly driven by the requirement of the vicinity of a T-rich region at the 3' end of the invading T-DNA. In addition, microcomplementarity between the 5' T-DNA end and the plant target seems to involve a G-base and another, random nucleotide upstream of it (Brunaud et al., 2002).

2.4.2 DSBR model

Based on their findings, Mayerhofer et al. (1991) proposed that T-DNA integration could either occur by means of a SSGR mechanism for single-stranded T-strand substrates (see above) or by means of a pathway that involves DSBs when double-stranded T-DNAs are the substrate for integration (*Figure 12-2*). This latter pathway for T-DNA integration requires an initial DSB lesion. Prior to integration of the T-DNA, the free plant DNA ends are processed by 5'-to-3' exonuclease activity, generating free 3' sticky overhangs. Subsequently, the ends of the double-stranded T-DNA interact with the free plant DNA ends. Three alternative pathways have been proposed by Windels et al. (2003) while investigating the relationship between DSBR and T-DNA integration (*Figure 12-2*). In a first pathway, which was observed in a minority of T-DNA/plant DNA junctions (10%), the double-stranded T-DNA is simply ligated into the plant DSB, resulting in a junction without microhomology or filler DNA insertions. A second pathway takes into account the possible integration of a double-stranded T-DNA by two consecutive single-strand annealing (SSA)-like reactions. Here, the left and right T-DNA borders interact with both plant DNA ends of the DSB, an interaction mediated by the presence of small stretches of sequence microhomology. This initial pairing is followed by removal of non-complementary overhangs by exo- or endonucleases. Finally, the interacting ends are ligated to each other. This second pathway was observed in approximately 50% of all analyzed junctions (Windels et al., 2003). A third pathway, present in the remaining 40% of the DNA/T-DNA junctions, uses synthesis-dependent strand annealing (SDSA) mechanisms to explain the integration of double-stranded T-DNAs at DSBs in plant genomes (Salomon and Puchta, 1998; Windels et al., 2003). The initial interaction between the invading T-DNA end and the plant target is not stabilized immediately but a free 3' end protruding T-DNA end either from the left or right T-DNA end lands and screens for microcomplementarity (Windels et al., 2003; *Figures 12-2* and *12-3*). Upon landing, the free 3' end is taken as primer for simultaneous template-based DNA synthesis, and filler DNA is formed. The resulting junctions contain up to 51-bp-long filler sequences between plant DNA and T-DNA contiguous sequences. These filler segments were built from several short sequence motifs, identical to sequence blocks that are present in the T-DNA ends and/or the plant DNA close to the integration site (*Figure 12-3*). Mutual microhomologies among the sequence motifs that constitute a filler segment were frequently observed. This frequency and the composition of filler sequences during T-DNA integration is the most conspicuous difference with DSB repair in *Arabidopsis*, because no filler junctions were found at repaired DSB junctions (Kirik et al., 2000). Additionally, whereas so-called non-homologous end-joining (NHEJ) repairs in *Arabidopsis* (see below) are characterized by the occurrence of large deletions, T-DNA insertions are normally associated with much smaller ones (Kirik et al., 2000; Windels et al., 2003).

2.5 T-DNA integration: involvement of DSBR *via* non-homologous end-joining (NHEJ)

To discuss in more detail the interaction between DSBs, the plant host DSBR machinery, and T-DNA integration, we will give a general overview of DSBR in plant cells and will summarize the evidence for the involvement of DSBR in T-DNA integration. The maintenance of the information and structural organization of genomes is a prerequisite for the expression as well as for the transfer of this genetic information to next generations. DSBs are only one of numerous outcomes of DNA damage and can occur through a variety of different processes (Britt, 1999). It is self-evident that DSBs are a major threat to unicellular and multicellular organisms if left unrepaired. As a consequence, prokaryotic as well as eukaryotic organisms dispose of mechanisms to deal with these kind of lesions. The fact that at a DSB, the damage occurs in both strands of the

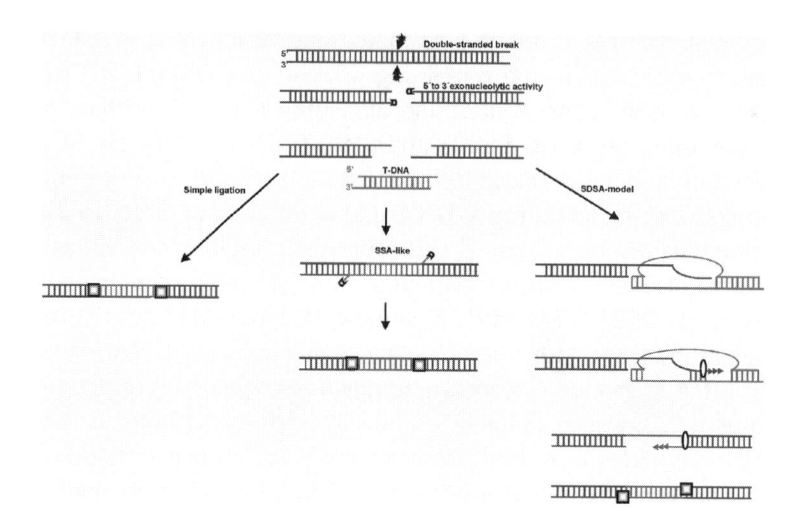

Figure 12-2. Double-stranded break repair (DSBR) as a pathway for T-DNA integration. A double-stranded break (DSB), generated at the plant DNA target site is required for T-DNA integration. After 5' to 3' exonucleolytic degradation of the 5' termini of the DSB, 3' sticky ends are available. A double-stranded DNA is the preferred substrate for integration. Subsequently, three different pathways are represented for resolving a DSB by an incoming T-DNA: simple ligation, a single-strand annealing (SSA)-like pathway, and the SDSA are shown. Simple ligation involves the ligation of the T-DNA ends to the plant DNA ends without deletion of the termini involved and without insertion of filler sequences. An SSA-like mechanism combines microhomology pairing between plant DNA and T-DNA ends, followed by resection of non-complementary overhangs and end ligation. Finally, the most complex mechanism involves SDSA. In these junctions, filler insertions are observed at T-DNA/plant DNA junctions. They originate when the initial interaction between the invading T-DNA end and the plant target is not stabilized yet, but landing and taking off of the T-DNA is repeated regularly. Upon landing, the free 3' end is taken as primer for simultaneous template-based DNA synthesis. Different regions, indicated by green numbers, such as surroundings of the T-DNA target site and the T-DNA plasmid sequence can be invaded, resulting in patchwork-like filler sequence. A reactive plant DNA end invades the double-stranded T-DNA and the T-DNA is copied into the plant genome. Finally, T-DNA and plant DNA are ligated by DNA ligase activity. DNA ligases, purple squares; DNA polymerase, yellow oval (see text, for more details).

DNA molecule implies that no longer an original template is present that can be used to repair the break.

Principally, two pathways are available for the restoration of DSBs. Depending on the extent to which sequence homology between the interacting DNA molecules is needed, a distinction is made between DSBR via homologous recombination and DSBR without the need for extensive sequence homology, the so-called NHEJ. DSBR involves sealing of two

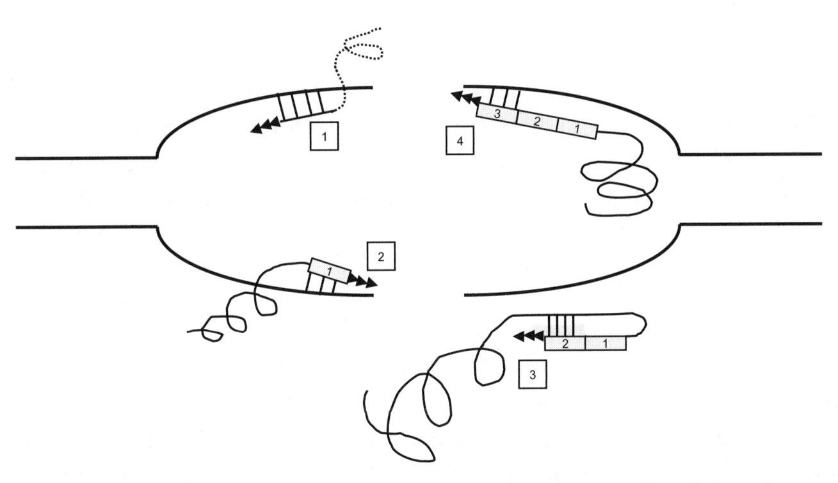

Figure12-3. Model for filler DNA formation at T-DNA junctions. The origin and composition of filler DNA observed at T-DNA junctions suggests that T-DNA-associated filler results from an unstable, initial interaction between the invading T-DNA end and the plant target site. Free 3' protruding T-DNA ends, available as termini of single-stranded T-DNA or available because of breathing of double-stranded T-DNA ends, interact with the host genome. Therefore, the proposed model holds true for single-stranded as well as double-stranded T-DNAs. Initially, the T-DNA end lands at regions of microcomplementarity (step 1), followed by polymerase activity. Because this initial interaction is not yet stabilized by host-dependent non-homologous end joining (NHEJ)-associated protein complexes, the T-DNA takes off and lands regularly. As such, different regions, such as the surroundings of the T-DNA target site and the T-DNA plasmid sequence, are invaded (steps 2, 3, and 4). This landing and take off principle explain the complex filler DNA consisting of several small repeats, each with mutual microhomologies. Eventually, the interaction becomes stabilized. Exonuclease activity might be associated with this final interaction. Finally, the junction is sealed by means of a SSA-like mechanism. The incoming T-DNA is drawn as a single strand for the sake of simplicity.

DNA ends, and here the lesion is repaired by an illegitimate recombination event. Because no extensive sequence homology between the interacting partners is needed by NHEJ, almost all molecules can interact with each other, ultimately forming chromosomal inversions, deletions, and translocations. As a consequence, DSBR via NHEJ is often referred to as an error-prone pathway of DNA repair. It seems that which pathway is followed for DBSR depends more on the organism studied than on the degree of sequence homology available. Homology-based DSBR is mainly observed in organisms with relatively compact genomes (bacteria and yeast), whereas the error-prone NHEJ pathway is predominantly present for the repair of DSBs in plant cells and mammalian cells (Pâques and Haber, 1999). Indeed, even when large stretches of sequence homology are provided, DSB repair in mammals and plants will follow the NHEJ pathway.

The molecular mechanism underlying NHEJ-mediated repair has been mainly studied in yeast and mammalian cells (for excellent reviews on NHEJ, see Critchlow and Jackson, 1998; Tsukamoto and Ikeda, 1998; Haber, 2000). The first notion on the type of proteins necessary in NHEJ-mediated repair came from complementation studies in mammalian cells that were hypersensitive to ionizing radiation (Jackson and Jeggo, 1995). A DNA-dependent protein serine/threonine kinase, the DNA-PK, was identified. This multiprotein complex consists of a catalytic subunit, the DNA-PK_{cs}, and a heterodimeric Ku protein, which comprises units of 70 kDa and 80 kDa, hence their name Ku70 and Ku80, respectively (Dynan and Yoo, 1998). The molecular function of the DNA-PK_{cs} and the heterodimeric Ku proteins is multiple. The catalytic subunit might be important for the phosphorylation of DNA-bound proteins and in regulating their function. Secondly, the DNA-PK_{cs} is homologous to proteins that have been identified as regulators in DNA damage cell cycle checkpoints. Alternatively, or in combination, its function might also be required for the alteration of the chromatin structure in the immediate surroundings of the DSB. Restructuring of the nucleosomal DNA is most probably mediated by an interaction between the silencing protein Sir4p and the Ku70 subunit (Tsukamoto et al., 1997). SirP4 interacts with the N-terminal tails of the histone proteins and the interaction might induce the DNA to condense and form a heterochromatin-like structure (Grunstein, 1997). This condensation of the DNA might be necessary to isolate physically the DNA at a DSB from the DNA that undergoes active transcription and replication. In addition, the Ku heterodimeric protein has also been implicated in several additional steps of the NHEJ repair process, such as: (i) to facilitate NHEJ itself, (ii) to juxtapose two DNA ends, (iii) to recruit other proteins, for instance DNA ligase IV, necessary in DNA repair, (iv) to preserve the free DNA ends from extensive degradation and (v) to promote efficient and accurate illegitimate end joining. Mutations in the yeast Ku homolog decrease the efficiency of NHEJ repair. Interestingly, the residual repaired junctions seem to exhibit the features of an error-prone DSB repair, meaning that most of the rejoined junctions in a Ku-deficient background are imprecisely repaired (Boulton and Jackson, 1996; Milne et al., 1996). Furthermore, Ku seems to greatly facilitate the efficient joining of blunt DNA ends and ends with only one or two bases complementary overhangs, whereas the effect of Ku is much smaller when four bases of complementarity are present. Taken together, these latter observations indicate that Ku is important for stabilizing two DNA ends in the absence of extensive sequence complementarity. When no Ku is present, these DNA ends will search for microcomplementarity leading to a higher degree of imprecise joints. Recognition of a DSB by the Ku heterodimer and subsequent recruitment of the DNA-PK_{cs} seem to be the first steps to an efficient joining reaction of two non-homologous DNA ends. Another important protein is the XRCC4 in mammalian cells and its yeast homolog Lif1. In mammalian cells, XRCC4 has a weak affinity for the DNA-PK and, *in vitro*, has been shown to strongly interact with the DNA ligase IV and to activate the ligase activity. Therefore, XRCC4 might be necessary for ligase activity and be the docking protein for the interaction between the DNA-PK_{cs} and the DNA ligase IV. A third player during NHEJ repair is the Rad50/Xrs2/Mre11 complex, which has a role in DNA end processing (Boulton and Jackson, 1996; Goedecke et al., 1999).

For many of the molecular components involved in NHEJ in yeast and in mammals, putative homologs in plants have been identified: a DNA ligase IV, an XRCC4 homolog in *Arabidopsis* (West et al., 2000) and Ku70 and Ku80 homologs as well. Transcription of the AtLIG4 has been shown to be induced after γ-irradiation and the Ku70/Ku80 heterodimer to be required for efficient DSBR DNA *in vivo* in *Arabidopsis* (Tamura et al., 2002; West et al., 2002). From a historical viewpoint, NHEJ in plants has been mainly studied by footprint analysis of transposons (Rinehart et al., 1997) and by analyzing T-DNA integration events (Gheysen et al., 1991; Mayerhofer et al., 1991). More recently, more direct ways of analyzing NHEJ in plants have been developed. For instance, DSBs are induced in one way or another and the repaired junctions are sequenced. Gorbunova and Levy (1997) used linearized plasmids to transform tobacco cells. After transformation, circularized plasmids were recovered and the novel joints were sequenced. Similarly, Salomon and Puchta (1998) analyzed repaired DSBs induced at transgenic I-SceI recognition sites in tobacco. In summary, these studies revealed that (i) end joining by simple ligation, with no sequence alteration is rare; (ii) end joining is usually associated with deletions; (iii) rejoining frequently occurs at short repeats and (iv) frequently, large insertions of filler DNA are present at the repaired break (Gorbunova and Levy, 1999). Taken together, NHEJ in plants seems more error-prone than in other species. In addition, species-specific differences in NHEJ repair might occur in plants. In *Arabidopsis*, deletions were on average larger than in tobacco and not associated with insertions (Kirik et al., 2000; Orel and Puchta, 2003). These differences in DSBR via NHEJ have been correlated with the genome size of both species and with the fact that free DNA ends are more stable in tobacco than in *Arabidopsis*.

The findings that repair activities in the plant nucleus might relate to the process of integration of exogenous DNA dates back to the study of Köhler et al. (1989), who enhanced significantly the transformation rate of the plants treated with low-dose irradiation. That DSBs in the host genome can be the substrate for T-DNA integration was evidenced directly by the frequent capture of T-DNAs at induced DSBs in plant genomes (Salomon and Puchta, 1998). That the NHEJ machinery of an organism might be involved in T-DNA integration was first demonstrated in *Saccharomyces cerevisiae* by Bundock and Hooykaas (1996), who found that *Agrobacterium* T-DNA integrated into the yeast genome by an illegitimate recombination event. This ability of *Agrobacterium* was used to study the role of host proteins in the integration of T-DNA by NHEJ (van Attikum et al., 2001). In yeast, NHEJ proteins were found to be absolutely required for *Agrobacterium* T-DNA integration, implicating the requirement of the Ku heterodimer in *Agrobacterium*-mediated T-DNA transformation. In analogy, in *Arabidopsis*, plants deficient in Ku80 (Gallego et al., 2003) and DNA ligase IV (van Attikum et al., 2003) exhibited decreased T-DNA integration efficiency (Friesner and Britt, 2003) and were hypersensitive to the DNA-damaging agent methyl-methane-sulfonate without defect in the efficiency of T-DNA transformation, in contrast to NHEJ repair-deficient with normal T-DNA integration frequencies.

3 PATTERNS OF T-DNA INTEGRATION INTO THE HOST GENOME

3.1 Distribution of T-DNA inserts

When T-DNA integration patterns in the plant host genome are studied, an obvious first question is whether T-DNA integration is a random process or not. This question relates back to the notion that genomic regions can have different chromatin states that could either promote or impede T-DNA integration.

From a historical viewpoint, *in situ* hybridization and genetic analyses were the first tools used to locate T-DNA inserts on the plant genome. Analysis of a small number of T-DNA loci suggested that the T-DNA integrates at random loci into the host plant genome (Chyi et al., 1986;

Wallroth et al., 1986; Azpiroz-Leehan and Feldmann, 1997). However, more detailed research and novel approaches indicated that T-DNA integration could also be non-random. For instance, by using promoterless marker genes located near the border regions of the T-DNA, T-DNAs might preferentially integrate into transcriptionally active regions in the *Arabidopsis* and tobacco genome (Koncz et al., 1989; Herman et al., 1990). Besides a preference for the integration into transcriptionally active and gene-rich regions, the use of fluorescent *in situ* hybridization analysis showed that T-DNAs might be targeted toward the distal regions of the chromosome arms (Wang et al., 1995; ten Hoopen et al., 1996; Dong et al., 2001; Khrustaleva and Kik, 2001). However, a drawback was that many studies used *Arabidopsis* that is known for its very small genome with relatively few heterochromatic genome regions. Therefore, the observed preference for transcriptionally active gene-rich genomic loci could be explained as an artefact of the study design. However, more direct evidence to support the hypothesis that gene-rich regions are the preferred targets for T-DNA integration was provided by Barakat et al. (2000), who found that in *Arabidopsis*, whose coding sequences are evenly spread along the genome, T-DNA integrates essentially everywhere in the genome. On the contrary, in rice (*Oryza sativa*) that has a genome that comprises large gene-rich regions combined with vast gene-empty repeated sequences, T-DNA integrates only into the gene-rich and transcriptionally active regions of the genome. This bias for T-DNA integration into gene-rich, transcriptionally active regions has been explained as the result of the better accessibility of these regions because of their specific chromatin structures. The non-genic or heterochromatic regions of the genome could then physically prevent T-DNA integration (Herman et al., 1990; Topping et al., 1995). Furthermore, the T-DNA integration and the chromatin state of the host genome locus might possibly be correlated, because the Vip1 protein of *Arabidopsis*, which interacts with the T-DNA-associated VirE2 protein during T-DNA transformation, attaches to plant chromosomes via histones (Loyter et al., 2005).

Recently, in-depth analyses of T-DNA distribution patterns by means of genome-wide T-DNA insertional mutagenesis has shed more light on this topic. By analyzing the integration locus of 1000 (Szabados et al., 2002), 850 (Ortega et al., 2002), 583 (Krysan et al., 2002), and 112 T-DNA inserts (Forsbach et al., 2003) in the *Arabidopsis* genome and over 1000 T-DNA tags in the rice genome (Chen et al., 2003; Sha et al., 2004), it appears that the distribution pattern of a large number of T-DNAs in the nuclear genome of *Arabidopsis* and rice (Alonso et al., 2003; Sallaud et al.,

2004) along the chromosome is not random. The frequency of T-DNA insertion is significant and higher in gene-dense chromosomal regions and only a small number of T-DNA inserts are found in centromeric, telomeric, or rDNA repeats. When integration into "genic" (defined as the genomic sequence between the start and stop codons of a gene) and "intergenic" regions are compared, a small bias toward recovering T-DNA integrations within intergenic regions was reported in some, but not all studies (Krysan et al., 2002; Forsbach et al., 2003). Integration within the genic region is either evenly spread (Szabados et al., 2002) or reduced in exons (Krysan et al., 2002; Chen et al., 2003) or introns as well (Forsbach et al., 2003). In addition, the data provided by Szabados et al. (2002) and Chen et al. (2003) suggest that 5' and 3' regulatory regions might be more accessible for T-DNA integration. However, it should be noted that most of the above described lines were obtained after selecting or screening for transgene expression, implying that transformants with integrations into genomic regions that suppress transcription of the selection marker will not be identified. Francis and Spiker (2005) determined that approximately 30% of the transformation events might result in non-expressing transgenes that would preclude identification by selection and thus that selection bias can account for at least some of the observed non-random integration of T-DNA into the *Arabidopsis* genome.

3.2 T-DNA integration results in a transgene locus that is either simple or complex

After excision of the T-strand from the Ti plasmid, a VirD2-capped single-stranded T-region harboring residual left and right border sequences is formed that becomes stably integrated into the host genome. Usually, the integration pattern observed for *Agrobacterium*-mediated T-DNA transformation is less complex and more precise than those observed after direct gene transformation (Kohli et al., 2003). The process of T-DNA integration is a precise process usually resulting in a significant number of simple and single-copy T-DNA inserts. However, very often complex integration patterns, in combination with T-DNA truncations and host genome rearrangements, are observed (*Figure 12-4*). Hereafter, truncation of T-DNA ends, integration of complex and multiple T-DNAs at the same host genomic locus, integration of vector backbone sequences, and T-DNA induced rearrangements of the host plant genome will be discussed.

3.3 T-DNA integration can result in truncated T-DNA inserts

Besides simple and intact T-DNA inserts (Koncz and Schell, 1986), integrated T-DNAs are found with truncations (deletions) at one or both ends of the T-DNA (Deroles and Gardner, 1988; Brunaud et al., 2002; Windels et al., 2003) or with an internal break in the T-DNA (Herman et al., 1990; Meza et al., 2002; Lechtenberg et al., 2003) (see *Figure 12-4*). Truncated T-DNAs might arise from the recognition of "pseudoborders" during T-strand synthesis in the bacterium (Van Lijsebettens et al., 1986). In order to verify this statement, Herman et al. (1990) analyzed the activation frequency of a promoterless gene located internally in the T-DNA. The activation frequency of 1% remained constant for T-DNAs with different lengths and different sequences upstream of the activated gene, implicating the improbable involvement of pseudoborders (Herman et al., 1990). Furthermore, sequencing several aberrant T-DNA/plant junctions did not reveal homology with the border consensus sequence (Gheysen et al., 1990). Therefore, because pseudoborder sequences seem responsible for the generation of aberrant T-DNAs only at a low frequency, other mechanisms resulting in random distribution of T-DNA truncation points must occur, such as (i) the use of random nicks for the synthesis of shortened T-strands, (ii) digestion of the T-DNA ends prior to integration, (iii) breakage of the T-DNA during the transfer process, (iv) rearrangements generated during integration, and (v) endonucleolytic digestion of the T-DNA after annealing of the central region of the T-DNA to the plant target site (Herman et al., 1990; Laufs et al., 1999).

Concerning truncation of the T-DNA ends, the left border might be more prone to truncation than the right border sequence, implying a protective role for the 5' end attached VirD2 protein, but this suggestion was based on a small number of T-DNA/plant DNA sequences (Dürrenberger et al., 1989; Tinland, 1996). Indeed, a large number of sequenced T-DNA/plant DNA junctions in *Arabidopsis* indicate that both border regions are equally prone to processing, although the deletion size at the left border junctions is usually larger than that at the right border junctions (Brunaud et al., 2002; Krysan et al., 2002; Meza et al., 2002; Kim et al., 2003; Windels et al., 2003). These truncations can be formed in a number of manners. On the one hand, they could be the result of T-DNA end degradation by nucleolytic attack during T-DNA transfer. The presence of VirD2 and VirE2 proteins has been implicated in conservation of the integrity of the T-strand. On the other hand, deletions at the T-DNA ends might be explained by the subtle mechanisms that play during T-DNA integration. During the pairing of the T-DNA ends with the plant target site, non-complementary overhanging ends are removed by exonucleolytic degradation, which might account for the frequently observed small truncations at the T-DNA ends. In addition, the context of the border region of the Ti plasmid, *i.e.* natural versus synthetic, might influence the degree of T-DNA end processing (De Buck et al., 2000b; Meza et al., 2002).

3.4 T-DNA integration can result in multicopy T-DNA loci

Upon T-DNA transfer, more than one T-DNA copy can integrate into the plant host genome (Koncz et al., 1989) (*Figure 12-4*). These multiple copies of the T-DNA can be clustered in the same genomic locus or they can be present at different loci. If multiple copies of the T-DNA integrate at the same genomic locus, they are frequently found as T-DNA copies that are organized in direct or inverted orientation or as a combination of both types (Jorgensen et al., 1987). These kinds of complex T-DNA integrations have been described in several plant species, such as *Arabidopsis* (De Neve et al., 1997), tobacco (De Neve et al., 1997; Krizkova and Hrouda, 1998), petunia (*Petunia hybrida*; Cluster et al., 1996), potato (*Solanum tuberosum*; Wolters et al., 1998), aspen (*Populus tremuloides*; Kumar and Fladung, 2000), and rice (Kim et al., 2003; Sallaud et al., 2003). Because these multimeric T-DNA forms have never been observed in the bacterial cell, these structures are assumed to arise in the plant cell prior (De Neve et al., 1997; De Buck et al., 1999) or during T-DNA integration (Krizkova and Hrouda, 1998; Kumar and Fladung, 2000).

Different mechanisms for the formation of these T-DNA repeats have been postulated. The replication model hypothesizes that multiple T-DNAs originate from a single T-DNA by replication and repair before or during insertion into the plant genome (Jorgensen et al., 1987). The main argument favoring this model is that T-DNAs involved in direct or inverted repeats can have analogous breakpoints in the restriction analysis (Van Lijsebettens et al., 1986; Jorgensen et al., 1987). However, when plant cells co-transformed with different T-DNAs originating from separated *Agrobacterium* strains are analyzed, the co-transferred T-DNAs seem frequently linked to each other (De Block and Debrouwer, 1991; De Neve et al., 1997; De Buck et al., 1999). Therefore, ligation of separate T-DNAs is a proven mechanism in the formation of the frequently observed multiple T-DNA inserts (De Neve et al., 1997). The fact that transformants were found that harbored four T-DNAs precisely linked to each other led to the idea that T-DNA repeats are formed by extrachromosomal ligation of separate T-DNAs prior to integration (De Neve et al., 1997).

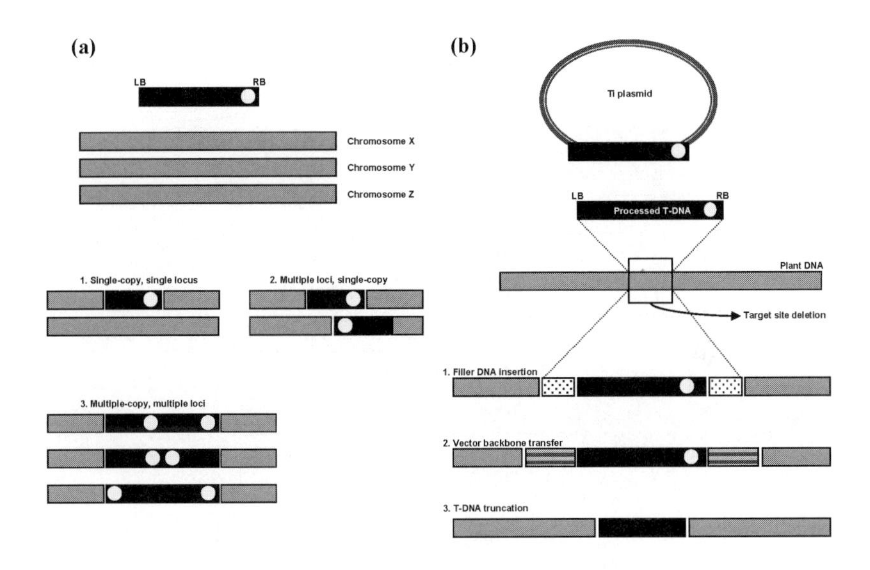

Figure 12-4. T-DNA integration patterns: possible outcome of the T-DNA integration process. (a) During T DNA transformation, more than one copy of the T-DNA can become stably integrated into the host nuclear genome. Three different possibilities are represented: a single copy at a single insertion locus (a1), multiple single copies at different genomic loci (a2), or several T-DNA inserts targeted to a single genomic locus (a3). In the latter case, a number of variations have been observed: tandem repeat organization (upper), inverted repeat organization over the right border (middle), and inverted orientation over the left border (bottom). More complex integration patterns, consisting of a mixture of all the above have also been reported (not shown in figure). (b) During T-DNA integration, several aberrations can occur: a target site deletion at the plant host locus, insertion of filler sequences at the T-DNA border regions, co-transfer of vector backbone sequences, and truncation or deletion of the T-DNA borders. Again, the complexity of an actual T-DNA integration pattern could be a mixture of all of the above. LB, left border; RB, right border.

Complex T-DNA integration loci were further characterized by sequencing the junctions between two T-DNAs. By cotransforming *Arabidopsis* root cells with two different T-DNAs with distinct T-DNA ends and screening for transgenic loci with two different T-DNAs, it is possible to sequence left-left, right-right, and right-left T-DNA border junctions (*Figure 12-5*). Precise end-to-end fusions were found between two right border ends, whereas imprecise fusions and filler DNA were detected in T-DNA linkages containing a left border end. This observation implied that end-to-end ligation of double-stranded T-DNAs occurred especially between right T-DNA ends, whereas illegitimate recombination on the basis of microhomology, deletions, repair activities, and insertions

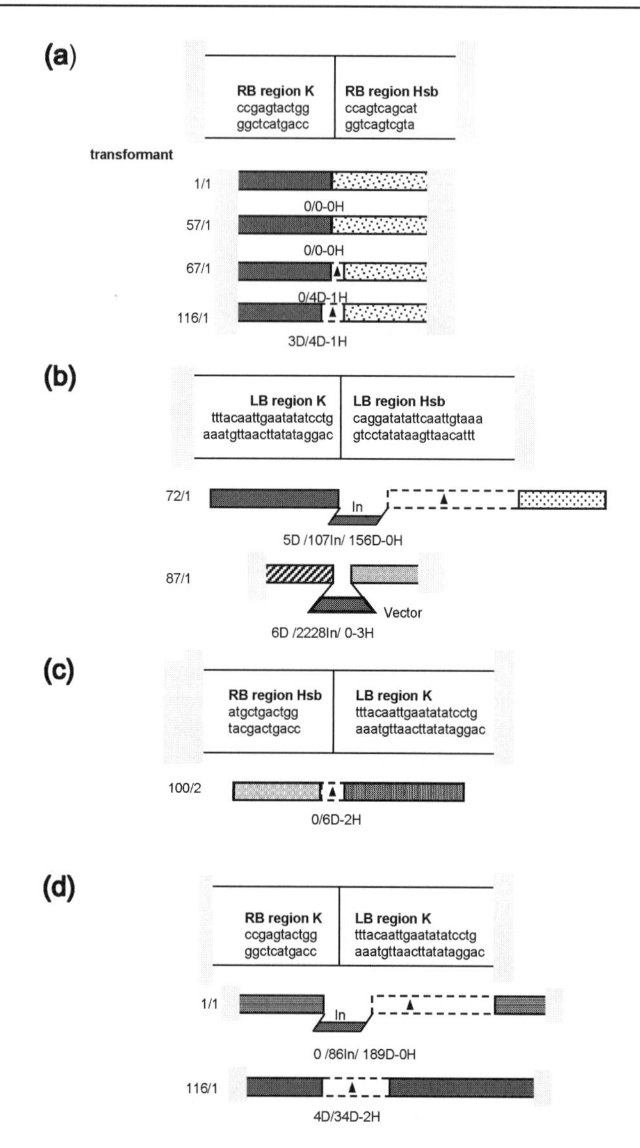

Figure 12-5. Structure of the sequenced junctions between linked T-DNAs. Sequenced junctions between an Hsb and K T DNA in an inverted repeat about the right border (a), an inverted repeat about the left border (b), a tandem configuration (c), and between two directly repeated K T-DNAs (d). At the top, the sequence of the left or right T-DNA borders is given. For each type of junction, a schematic overview is presented of the structure and the transformants in which they occur. The Hsb T-DNA and K T-DNA are represented by dotted and hatched bars, respectively. Abbreviations: D, the number of bp deleted relative to the nick position in the right and left border repeats; H, number of homologous bp between the two T-DNA ends at the cross-over position; In, number of bp inserted that are not colinear with either of the two T-DNAs; vector, vector backbone sequences that lay outside of the right and left border repeats.

of filler DNA was more frequently observed in the formation of left border T-DNA junctions (De Buck et al., 1999). No filler DNA was observed in the inverted configurations about the right borders, and in the other configurations, ligation could be accompanied by repair giving rise to deletions or filler DNA at the junction of the linked T-DNAs (*Figure 12-5*). Additionally, none of the sequenced T-DNA junctions contained plant DNA, favoring the idea that T-DNA recombination and ligation occurred before integration (De Buck et al., 1999). However, in other studies small stretches of plant genomic sequences are sometimes found in between different T-DNAs that constitute a multimeric T-DNA repeat (Krizkova and Hrouda, 1998). Therefore, multimeric T-DNA repeats have been suggested to be formed during T-DNA integration. A model in which T-DNA repeats arise because of co-integration of several intermediates into one target site, which is a "hot spot" for integration has been proposed (Krizkova and Hrouda, 1998), but cannot explain the formation of inverted repeats about the right border.

The fact that independently transferred T-DNA complexes integrate at the same locus into the plant genome implies their sufficient proximity and compartmentalization. Probably, association with proteins involved in DNA repair might be a driving force (De Neve et al., 1997; De Buck et al., 1999). Moreover, VirD2 molecules that are attached to the right borders of both T-DNAs can interact and bring the T-DNAs together in a head-to-head orientation before their ligation by host factors. The ability of VirD2 molecules to interact with each other (Relić et al., 1998) as well as with a cellular DNA-binding protein (Bakó et al., 2003) further supports the role of VirD2 during the formation and ligation of T-DNA repeat structures. Nevertheless, not the VirD2 protein but the plant enzymes have been revealed to mediate T-DNA ligation *in vitro* (Ziemienowicz et al., 2000), without excluding the potential function of VirD2 in other steps of T-DNA integration, such as recruitment of plant enzymes involved in DNA repair and/or interaction with some structural chromatin proteins.

3.5 Transformation conditions may influence the number of integrated T-DNAs

Whereas in some studies single T-DNA copy inserts are found predominantly (Deroles and Gardner, 1988), these simple insertion patterns are less frequent than complex T-DNA loci with multiple T-DNA copies oriented in direct or inverted repeats in other transformation experiments (Jorgensen et al., 1987; De Block and Debrouwer, 1991; Grevelding et al.,

1993; De Neve et al., 1997; De Buck et al., 2004). Which factors influence the number of integrated T-DNAs are still undetermined. Several possibilities have been put forward, but neither of them has been confirmed and contradictory results have been reported.

The high frequency of inverted repeats has been attributed to the use of nopaline-type Vir functions (Jorgensen et al., 1987). However, they were also found with octopine Vir proteins (De Neve et al., 1997). Several studies formulated also contradictory conclusions about the fact that the T-DNA copy number would be determined by the type of plant material and protocol used during transformation (Grevelding et al., 1993; De Neve et al., 1997; De Buck et al., 2004). Whereas Grevelding et al. (1993) reported that *Arabidopsis* leaf transformation resulted in a higher percentage of multicopy T-DNA transformants than root transformation, the type of explant (root or leaf) had seemingly no major influence on this frequency according to De Neve et al. (1997). The number of multimeric T-DNAs might correlate with the number and the physiology of infecting bacteria affecting the number of T-DNA copies delivered into the plant cell (De Neve et al., 1997). The physiology and competence of the plant tissue to be transformed clearly influence the number and pattern of inserted T-DNAs. The presence of truncated T-DNAs has been correlated with plant cells under high recombination or repair activity. This hypothesis was evidenced by the correlation of selection for truncation of the T-DNA and a statistically significant increase in the number of integrated T-DNA copies (Gheysen et al., 1990). The frequency of single-copy transformants is much higher after *Arabidopsis* root transformation than after floral dip transformation (De Buck et al., 2004).

3.6 Integration of vector backbone sequences

Nicking the right and left border of the T-strand during processing is not always exact as reported by Martineau et al. (1994), who drew the attention to the fact that 30% of a population of transgenic plants harbored the vector backbone DNA sequences (*Figure 12-4*). This backbone transfer could either happen linked to the T-DNA in a colinear fashion or completely independently from the T-DNA (Kononov et al., 1997; De Buck et al., 2000b; Kim et al., 2003). Several models have been proposed for this backbone transfer. Transfer could erroneously start at the left border, proceeding through the binary vector, toward the right border, and finish only when the left border is encountered for the second time (Ramanathan and Veluthambi, 1995; van der Graaff et al., 1996). Alternatively, backbone transfer could be the result of initiation of transfer at the right border and readthrough over the left border (Kononov et al., 1997). Readthrough could then continue along the complete vector backbone and the T-DNA and finish when the left border is met for the second time (De Buck et al., 2000b). The frequency of these backbone sequences might rise up to 75% or 80% from the transgenic population (Kononov et al., 1997; Wenck et al., 1997; Vain et al., 2003). In *Arabidopsis*, 6 out of 99 (Forsbach et al., 2003) and 5 out of 37 single-copy T-DNAs (Meza et al., 2002) harbored these backbone sequences, suggesting that single-copy T-DNA transformants are negatively correlated with the occurrence of vector backbone sequences. To reduce integration of vector backbone sequences, Hanson et al. (1999) developed a method to enrich for transformants with only the T-DNA sequences. By incorporating a lethal gene into the non-T-DNA portion of the vector, the number of tobacco plants containing vector backbone sequences in the transgenic population was highly reduced (Hanson et al., 1999), but transformants with only limited amounts of vector backbone sequences could still be detected (Eamens et al., 2004). Recently, the presence of four copies of the left border repeat have been shown to positively prevent readthrough at the left border in rice transformants (Kuraya et al., 2004) but efficient T-DNA synthesis termination at multiple left border repeats cannot be generalized (Podevin et al., 2006).

3.7 Rearrangements of the host genomic locus as a result of T-DNA integration

In addition to rearrangements that are linked to the structural organization of the integrated T-DNA itself, *Agrobacterium*-mediated transformation might also induce small rearrangements that alter the host genomic locus into which the T-DNA becomes integrated. It has been generally accepted that the T-DNA becomes integrated into the plant genome by means of the illegitimate recombination machinery of the plant cell (Gheysen et al., 1991; Mayerhofer et al., 1991). Inherent to the process of illegitimate recombination is the occurrence of a deletion at the plant genomic target site and the possible insertion of so-called filler sequences at the boundaries between the plant DNA and the inserted T-DNA. In general, target site deletions are small. Most of the target site deletions reported in the literature are smaller than 75 bp (Gheysen et al., 1991; Mayerhofer et al., 1991; Krysan et al., 2002; Meza et al., 2002; Kim et al., 2003; Windels et al., 2003; De Buck et al., 2004), but large target site deletions, in the kilobase range, have been reported as well (Kertbundit et al.,

1998; Kaya et al., 2000). Filler sequences are scrambled DNA segments that originate from the plant DNA and/or the T-DNA. The presence of filler sequences at recombinant plant DNA/T-DNA junctions has been reported with variable frequencies and, in general, the size of the filler insertions is less than 100 bp (Gheysen et al., 1991; Mayerhofer et al., 1991; Meza et al., 2002; Forsbach et al., 2003; Kim et al., 2003; Windels et al., 2003). Taken together, the process of illegitimate recombination usually introduces only small target site rearrangements.

However, sometimes the joining reaction between the incoming T-DNA and the plant DNA can profoundly change the plant target site after T-DNA integration. Gheysen et al. (1987) were the first to analyze a plant target in tobacco before and after T-DNA integration and found that several target site rearrangements coincided with T-DNA integration. The most dramatic rearrangement was a 158-bp duplication of the plant target in combination with a 27-bp deletion. Similar results have been observed by Ohba et al. (1995) and Forsbach et al. (2003). Rearrangements of the plant genome because of T-DNA integration are not restricted to the immediate vicinity of the plant target site, but more extensive chromosomal rearrangements have been reported (Castle et al., 1993; Nacry et al., 1998; Laufs et al., 1999; Tax and Vernon, 2001; Forsbach et al., 2003). In a transgenic *Arabidopsis* line that harbors two T-DNA loci, the first locus constitutes a tandem repeat of two T-DNAs, while the second locus is a truncated single T-DNA insertion consisting of a left T-DNA end (Nacry et al., 1998). Extensive molecular characterization of this transgenic line reveals that the integration of these T-DNAs is accompanied by a reciprocal translocation, a large inversion, and a 1.4-kb deletion. Similarly, another *Arabidopsis* transformant was found to harbor a single T-DNA in combination with a T-DNA with a truncated right border (Laufs et al., 1999). Comparison of the T-DNA target site before and after transformation revealed that due to the integration of the two T-DNAs, the genomic sequence in between both T-DNAs was inverted, resulting in a 26-cM paracentromeric inversion. A T-DNA-associated dupli-cation/translocation in *Arabidopsis* has been described as well (Tax and Vernon, 2001). Based on these data, models have been put forward to explain T-DNA integration-associated chromosomal rearrangements. The possibility that two independent T-DNAs are inserted into the genome and then recombine, resulting in the observed chromosomal rearrangements, seems not likely. The final outcome of the chromosomal rearrangement would require two independent recombination events what is also improbable (Nacry et al., 1998). Therefore, T-DNA-associated chromosomal rearrangements might occur as a consequence of aberrant break and repair functions. Essentially, two different models account for the paracentromeric inversion observed (Laufs et al., 1999): DSBs are assumed to occur prior to T-DNA integration. If plant ends of these two DSBs are exchanged prior to T-DNA integration or if one incoming T-DNA interacts with two different genomic DSBs, then a chromosomal inversion is the result. Another intriguing aspect of T-DNA-induced chromosomal mutations is the occurrence of DNA sequence rearrangements at genomic loci without a T-DNA integrated at that locus (Forsbach et al., 2003). For instance, Azpiroz-Leehan and Feldmann (1997) found that in a T-DNA-mutagenized population approximately 65% of the mutant phenotypes were not linked to a T-DNA insertion, suggesting that abortive T-DNA integration could induce deletions, additions, and base substitutions (Negruk et al., 1996). Another possibility for the frequently occurring mutations in transformants that are unlinked to the T-DNA insert might be the generation of the transformed population based on cells with a higher number of DSBs than non-transformed cells. Hence, in these cells DSBR might explain the high mutation frequency in transformed plants.

4 ACKNOWLEDGMENTS

The authors thank Martine De Cock for help with the manuscript. This work was supported by a grant from the European Union BIOTECH program (QLRT-2000-00078) and by a STREP project under the 6th framework program of the European Union "GENINTEG" (LSHG-CT2003-503303) with additional co-financing from the Flemish community.

5 REFERENCES

Albright LM, Yanofsky MF, Leroux B, Ma DQ, Nester EW (1987) Processing of the T-DNA of *Agrobacterium tumefaciens* generates border nicks and linear, single-stranded T-DNA. J Bacteriol 169: 1046-1055

Alonso JM, Stepanova AN, Leisse TJ, Kim CJ, Chen H, Shinn P, Stevenson DK, Zimmerman J, Barajas P, Cheuk R, Gadrinab C, Heller C, Jeske A, Koesema E, Meyers CC, Parker H, Prednis L, Ansari Y, Choy N, Deen H, Geralt M, Hazari N, Hom E, Karnes M, Mulholland C, Ndubaku R, Schmidt I, Guzman P, Aguilar-Henonin L, Schmid M, Weigel D, Carter DE, Marchand T, Risseeuw E,

Brogden D, Zeko A, Crosby WL, Berry CC, Ecker JR (2003) Genome-wide insertional mutagenesis of *Arabidopsis thaliana*. Science 301: 653-657

Azpiroz-Leehan R, Feldmann KA (1997) T-DNA insertion mutagenesis in *Arabidopsis*: going back and forth. Trends Genet 13: 152-156

Bakó L, Umeda M, Tiburcio AF, Schell J, Koncz C (2003) The VirD2 pilot protein of *Agrobacterium*-transferred DNA interacts with the TATA box-binding protein and a nuclear protein kinase in plants. Proc Natl Acad Sci USA 100: 10108-10113

Ballas N, Citovsky V (1997) Nuclear localization signal binding protein from *Arabidopsis* mediates nuclear import of *Agrobacterium* VirD2 protein. Proc Natl Acad Sci USA 94: 10723-10728

Barakat A, Gallois P, Raynal M, Mestre-Ortega D, Sallaud C, Guiderdoni E, Delseny M, Bernardi G (2000) The distribution of T-DNA in the genomes of transgenic *Arabidopsis* and rice. FEBS Lett 471: 161-164

Boulton SJ, Jackson SP (1996) *Saccharomyces cerevisiae* Ku70 potentiates illegitimate DNA double-strand break repair and serves as a barrier to error-prone DNA repair pathways. EMBO J 15: 5093-5103

Braun AC (1958) A physiological basis for autonomous growth of the crown-gall tumor cell. Proc Natl Acad USA 44: 344-349

Britt AB (1999) Molecular genetics of DNA repair in higher plants. Trends Plant Sci 4: 20-25

Brunaud V, Balzergue S, Dubreucq B, Aubourg S, Samson F, Chauvin S, Bechtold N, Cruaud C, DeRose R, Pelletier G, Lepiniec L, Caboche M, Lecharny A (2002) T-DNA integration into the *Arabidopsis* genome depends on sequences of pre-insertion sites. EMBO Rep 3: 1152-1157

Bundock P, Hooykaas PJJ (1996) Integration of *Agrobacterium tumefaciens* T-DNA in the *Saccharomyces cerevisiae* genome by illegitimate recombination. Proc Natl Acad Sci USA 93: 15272-15275

Castle LA, Errampalli D, Atherton TL, Franzmann LH, Yoon ES, Meinke DW (1993) Genetic and molecular characterization of embryonic mutants identified following seed transformation in *Arabidopsis*. Mol Gen Genet 241: 504-514

Chateau S, Sangwan RS, Sangwan-Norreel BS (2000) Competence of *Arabidopsis thaliana* genotypes and mutants for *Agrobacterium tumefaciens*-mediated gene transfer: role of phytohormones. J Exp Bot 51: 1961-1968

Chen S, Jin W, Wang M, Zhang F, Zhou J, Jia Q, Wu Y, Liu F, Wu P (2003) Distribution and characterization of over 1000 T-DNA tags in rice genome. Plant J 36: 105-113

Chilton M-D, Drummond MH, Merio DJ, Sciaky D, Montoya AL, Gordon MP, Nester EW (1977) Stable incorporation of plasmid DNA into higher plant cells: the molecular basis of crown gall tumorigenesis. Cell 11: 263-271

Chilton M-D, Que Q (2003) Targeted integration of T-DNA into the tobacco genome at double-strand breaks: new insights on the mechanism of T-DNA integration. Plant Physiol 133: 956-965

Chyi YS, Jorgensen RA, Goldstein D, Tanksley SD, Loaiza-Figueroa F (1986) Locations and stability of *Agrobacterium*-mediated T-DNA insertions in the *Lycopersicon* genome. Mol Gen Genet 204: 64-69

Citovsky V, Zupan J, Warnick D, Zambryski PC (1992) Nuclear localization of *Agrobacterium* VirE2 protein in plant cells. Science 256: 1802-1805

Cluster PD, O'Dell M, Metzlaff M, Flavell RB (1996) Details of T-DNA structural organization from a transgenic Petunia population exhibiting co-suppression. Plant Mol Biol 32: 1197-1203

Coin F, Frit P, Viollet B, Salles B, Egly JM (1998) TATA binding protein discriminates between different lesions on DNA, resulting in a transcription decrease. Mol Cell Biol 18: 3907-3914

Critchlow SE, Jackson SP (1998) DNA end-joining: from yeast to man. Trends Biochem Sci 23: 394-398

De Block M, Debrouwer D (1991) Two T-DNA's co-transformed into *Brassica napus* by a double *Agrobacterium tumefaciens* infection are mainly integrated at the same locus. Theor Appl Genet 82: 257-263

De Buck S, De Wilde C, Van Montagu M, Depicker A (2000a) Determination of the T-DNA transfer and the T-DNA integration frequencies upon cocultivation of *Arabidopsis thaliana* root explants. Mol Plant-Microbe Interact 13: 658-665

De Buck S, De Wilde C, Van Montagu M, Depicker A (2000b) T DNA vector backbone sequences are frequently integrated into the genome of transgenic plants obtained by *Agrobacterium*-mediated transformation. Mol Breed 6: 459-468

De Buck S, Jacobs A, Van Montagu M, Depicker A (1998) *Agrobacterium tumefaciens* transformation and cotransformation frequencies of *Arabidopsis thaliana* root explants and tobacco protoplasts. Mol Plant-Microbe Interact 11: 449-457

De Buck S, Jacobs A, Van Montagu M, Depicker A (1999) The DNA sequences of T-DNA junctions suggest that complex T-DNA loci are formed by a recombination process resembling T-DNA integration. Plant J 20: 295-304

De Buck S, Windels P, De Loose M, Depicker A (2004) Single-copy T-DNAs integrated at different positions in the *Arabidopsis* genome display uniform and comparable β-glucuronidase accumulation levels. Cell Mol Life Sci 61: 2632-2645

de Kathen A, Jacobsen HJ (1995) Cell competence for *Agrobacterium*-mediated DNA transfer in *Pisum sativum* L. Transgenic Res 4: 184-191

De Neve M, De Buck S, Jacobs A, Van Montagu M, Depicker A (1997) T-DNA integration patterns in co-transformed plant cells suggest that T-DNA repeats originate from co-integration of separate T-DNAs. Plant J 11: 15-29

Deng W, Chen L, Wood DW, Metcalfe T, Liang X, Gordon MP, Comai L, Nester EW (1998) *Agrobacterium* VirD2 protein interacts with plant host cyclophilins. Proc Natl Acad Sci USA 95: 7040-7045

Depicker A, Herman L, Jacobs A, Schell J, Van Montagu M (1985) Frequencies of simultaneous transformation with different T-DNAs and their relevance to the *Agrobacterium*/plant cell interaction. Mol Gen Genet 201: 477-484

Deroles SC, Gardner RC (1988) Analysis of the T DNA structure in a large number of transgenic petunias generated by *Agrobacterium*-mediated transformation. Plant Mol Biol 11: 365-377

Dillen W, De Clercq J, Kapila J, Zambre M, Van Montagu M, Angenon G (1997) The effect of temperature on *Agrobacterium tumefaciens*-mediated gene transfer to plants. Plant J 12: 1459-1463

Ditt RF, Nester EW, Comai L (2001) Plant gene expression response to *Agrobacterium tumefaciens*. Proc Natl Acad Sci USA: 10954-10959

Dong J, Kharb P, Teng W, Hall TC (2001) Characterization of rice transformed via an *Agrobacterium*-mediated inflorescence approach. Mol Breed 7: 187-194

Dürrenberger F, Crameri A, Hohn B, Koukolíková-Nicola Z (1989) Covalently bound VirD2 protein of *Agrobacterium tumefaciens* protects the T-DNA from exonucleolytic degradation. Proc Natl Acad Sci USA 86: 9154-9158

Dynan WS, Yoo S (1998) Interaction of Ku protein and DNA-dependent protein kinase catalytic subunit with nucleic acids. Nucleic Acids Res 26: 1551-1559

Eamens AL, Blanchard CL, Dennis ES, Upadhyaya NM (2004) A bidirectional gene trap construct suitable for T-DNA and *Ds*-mediated insertional mutagenesis in rice (*Oryza sativa* L.). Plant Biotechnol J 2: 367-380

Escobar MA, Dandekar AM (2003) *Agrobacterium tumefaciens* as an agent of disease. Trends Plant Sci 8: 380-386

Fladung M (1999) Gene stability in transgenic aspen (*Populus*). I. Flanking DNA sequences and T-DNA structure. Mol Gen Genet 260: 574-581

Forsbach A, Schubert D, Lechtenberg B, Gils M, Schmidt R (2003) A comprehensive characterization of single-copy T-DNA insertions in the *Arabidopsis thaliana* genome. Plant Mol Biol 52: 161-176

Francis KE, Spiker S (2005) Identification of *Arabidopsis thaliana* transformants without selection reveals a high occurrence of silenced T-DNA integrations. Plant J 41: 464-477

Friesner J, Britt AB (2003) *Ku80*- and *DNA ligase IV*-deficient plants are sensitive to ionizing radiation and defective in T-DNA integration. Plant J 34: 427-440

Gallego ME, Bleuyard JY, Daoudal-Cotterell S, Jallut N, White CI (2003) Ku80 plays a role in non-homologous recombination but is not required for T-DNA integration in *Arabidopsis*. Plant J 35: 557-565

Gelvin SB (2000) *Agrobacterium* and plant genes involved in T-DNA transfer and integration. Annu Rev Plant Physiol Plant Mol Biol 51: 223-256

Gelvin SB (2003) *Agrobacterium*-mediated plant transformation: the biology behind the "gene-jockeying" tool. Microbiol Mol Biol Rev 67: 16-37

Gheysen G, Angenon G, Van Montagu M (1998) *Agrobacterium*-mediated plant transformation: a scientifically intriguing story with significant applications. *In* K Lindsey, ed, Transgenic Plant Research. Harwood Academic Publishers, Amsterdam, pp 1-33

Gheysen G, Herman L, Breyne P, Gielen J, Van Montagu M, Depicker A (1990) Cloning and sequence analysis of truncated T-DNA inserts from *Nicotiana tabacum*. Gene 94: 155-163

Gheysen G, Van Montagu M, Zambryski P (1987) Integration of *Agrobacterium tumefaciens* transfer DNA (T-DNA) involves rearrangements of target plant DNA. Proc Natl Acad Sci USA 84: 6169-6173

Gheysen G, Villarroel R, Van Montagu M (1991) Illegitimate recombination in plants: a model for T-DNA integration. Genes Dev 5: 287-297

Goedecke W, Eijpe M, Offenberg HH, van Aalderen M, Heyting C (1999) Mre11 and Ku70 interact in somatic cells, but are differentially expressed in early meiosis. Nat Genet 23: 194-198

Gorbunova V, Levy AA (1997) Non-homologous DNA end joining in plant cells is associated with deletions and filler DNA insertions. Nucleic Acids Res 25: 4650-4657.

Gorbunova VV, Levy AA (1999) How plants make ends meet: DNA double-strand break repair. Trends Plant Sci 4: 263-269

Grevelding C, Fantes V, Kemper E, Schell J, Masterson R (1993) Single-copy T-DNA insertions in *Arabidopsis* are the predominant form of integration in root-derived transgenics, whereas multiple insertions are found in leaf discs. Plant Mol Biol 23: 847-860

Grimsley N, Hohn T, Davies JW, Hohn B (1987) *Agrobacterium*-mediated delivery of infectious maize streak virus into maize plants. Nature 325: 177-179

Grimsley NH, Ramos C, Hein T, Hohn B (1988) Meristematic tissues of maize plants are most susceptible to agroinfection with maize streak virus. Bio/Technology 6: 185-189

Grunstein M (1997) Histone acetylation in chromatin structure and transcription. Nature 389: 349-352

Haber JE (2000) Lucky breaks: analysis of recombination in *Saccharomyces*. Mutat Res 451: 53-69

Hanson B, Engler D, Moy Y, Newman B, Ralston E, Gutterson N (1999) A simple method to enrich an *Agrobacterium*-transformed population for plants containing only T-DNA sequences. Plant J 19: 727-734

Herman L, Jacobs A, Van Montagu M, Depicker A (1990) Plant chromosome/marker gene fusion assay for study of normal and truncated T-DNA integration events. Mol Gen Genet 224: 248-256

Herrera-Estrella A, Chen Z, Van Montagu M, Wang K (1988) VirD proteins of *Agrobacterium tumefaciens* are required for the formation of a covalent DNA protein complex at the 5' terminus of T-strand molecules. EMBO J 7: 4055-4062

Herrera-Estrella A, Van Montagu M, Wang K (1990) A bacterial peptide acting as a plant nuclear targeting signal: the amino-terminal portion of *Agrobacterium* VirD2 protein directs a β-galactosidase fusion protein into tobacco nuclei. Proc Natl Acad Sci USA 87: 9534-9537

Howard EA, Zupan JR, Citovsky V, Zambryski PC (1992) The VirD2 protein of *A. tumefaciens* contains a C-terminal bipartite nuclear localization signal: implications for nuclear uptake of DNA in plant cells. Cell 68: 109-118

Jackson SP, Jeggo PA (1995) DNA double-strand break repair and V(D)J recombination: involvement of DNA-PK. Trends Biochem Sci 20: 412-415

Janssen B-J, Gardner RC (1990) Localized transient expression of GUS in leaf discs following cocultivation with *Agrobacterium*. Plant Mol Biol 14: 61-72

Jorgensen R, Snyder C, Jones JDG (1987) T-DNA is organized predominantly in inverted repeat structures in plants transformed with *Agrobacterium tumefaciens* C58 derivatives. Mol Gen Genet 207: 471-477

Kaya H, Sato S, Tabata S, Kobayashi Y, Iwabuchi M, Araki T (2000) *hosoba toge toge*, a syndrome caused by a large chromosomal deletion associated with a T-DNA insertion in *Arabidopsis*. Plant Cell Physiol 41: 1055-1066

Kertbundit S, Linacero R, Rouzé P, Galis I, Macas J, Deboeck F, Renckens S, Hernalsteens J-P, De Greve H (1998) Analysis of T-DNA-mediated translational ß-glucuronidase gene fusions. Plant Mol Biol 36: 205-217

Khrustaleva LI, Kik C (2001) Localization of single-copy T-DNA insertion in transgenic shallots (*Allium cepa*) by using ultra-sensitive FISH with tyramide signal amplification. Plant J 25: 699-707

Kim S-R, Lee J, Jun S-H, Park S, Kang H-G, Kwon S, An G (2003) Transgene structures in T-DNA-inserted rice plants. Plant Mol Biol 52: 761-773

Kirik A, Salomon S, Puchta H (2000) Species-specific double-strand break repair and genome evolution in plants. EMBO J 19: 5562-5566

Köhler F, Cardon G, Pöhlman M, Gill R, Schieder O (1989) Enhancement of transformation rates in higher plants by low-dose irradiation: are DNA repair systems involved in the incorporation of exogenous DNA into the plant genome? Plant Mol Biol 12: 189-199

Kohli A, Twyman RM, Abranches R, Wegel E, Stoger E, Christou P (2003) Transgene integration, organization and interaction in plants. Plant Mol Biol 52: 247-258

Komari T, Ishida Y, Hiei Y (2004) Plant transformation technology: *Agrobacterium*-mediated transformation. *In* P Christou, H Klee, eds, Handbook of Plant Biotechnology, Vol 1. John Wiley and Sons, Chichester, pp. 233-261

Koncz C, Martini N, Mayerhofer R, Koncz-Kalman Z, Körber H, Redei GP, Schell J (1989) High-frequency T-DNA-mediated gene tagging in plants. Proc Natl Acad Sci USA 86: 8467-8471

Koncz C, Schell J (1986) The promoter of T_L-DNA gene *5* controls the tissue-specific expression of chimaeric genes carried by a novel type of *Agrobacterium* binary vector. Mol Gen Genet 204: 383-396

Kononov ME, Bassuner B, Gelvin SB (1997) Integration of T-DNA binary vector 'backbone' sequences into the tobacco genome: evidence for multiple complex patterns of integration. Plant J 11: 945-957

Krizkova L, Hrouda M (1998) Direct repeats of T-DNA integrated in tobacco chromosome: characterization of junction regions. Plant J 16: 673-680

Krysan PJ, Young JC, Jester PJ, Monson S, Copenhaver G, Preuss D, Sussman MR (2002) Characterization of T-DNA insertion sites in *Arabidopsis thaliana* and the implications for saturation mutagenesis. OMICS 6: 163-174

Kumar S, Fladung M (2000) Transgene repeats in aspen: molecular characterisation suggests simultaneous integration of independent T-DNAs into receptive hotspots in the host genome. Mol Gen Genet 264: 20-28

Kumar S, Fladung M (2002) Transgene integration in aspen: structures of integration sites and mechanism of T-DNA integration. Plant J 31: 543-551

Kumar SV, Rajam MV (2005) Polyamines enhance *Agrobacterium tumefaciens vir* gene induction and T-DNA transfer. Plant Sci 168: 475-480

Kuraya Y, Ohta S, Fukuda M, Hiei Y, Murai N, Hamada K, Ueki J, Imaseki H, Komari T (2004) Suppression of transfer of non-T-DNA vector backbone' sequences by multiple left border repeats for transformation of higher plant mediated by *Agrobacterium tumefaciens*. Mol Breed 14: 309-320

Laufs P, Autran D, Traas J (1999) A chromosomal paracentric inversion associated with T-DNA integration in *Arabidopsis*. Plant J 18: 131-139

Lechtenberg B, Schubert D, Forsbach A, Gils M, Schmidt R (2003) Neither inverted repeat T-DNA configurations nor arrangements of tandemly repeated transgenes are sufficient to trigger transgene silencing. Plant J 34: 507-517

Li J, Krichevsky A, Vaidya M, Tzfira T, Citovsky V (2005) Uncoupling of the functions of the *Arabidopsis* VIP1 protein in transient and stable plant genetic transformation by *Agrobacterium*. Proc Natl Acad Sci USA 102: 5733-5738

Loyter A, Rosenbluh J, Zakai N, Li J, Kozlovsky SV, Tzfira T, Citovsky V (2005) The plant VirE2 interacting protein 1. A molecular link between the *Agrobacterium* T-complex and the host cell chromatin? Plant Physiol 138: 1318-1321

Martineau B, Voelker TA, Sanders RA (1994) On defining T-DNA. Plant Cell 6: 1032-1033

Maximova SN, Dandekar AM, Guiltinan MJ (1998) Investigation of *Agrobacterium*-mediated transformation of apple using green fluorescent protein: high transient expression and low stable transformation suggest that factors other than T-DNA transfer are rate-limiting. Plant Mol Biol 37: 549-559

Mayerhofer R, Koncz-Kalman Z, Nawrath C, Bakkeren G, Crameri A, Angelis K, Redei GP, Schell J, Hohn B, Koncz C (1991) T-DNA integration: a mode of illegitimate recombination in plants. EMBO J 10: 697-704

Meza TJ, Stangeland B, Mercy IS, Skårn M, Nymoen DA, Berg A, Butenko MA, Håkelien AM, Haslekås C, Meza-Zepeda LA, Aalen RB (2002) Analyses of single-copy *Arabidopsis* T-DNA-transformed lines show that the presence of vector backbone sequences, short inverted repeats and DNA methylation is not sufficient or necessary for the induction of transgene silencing. Nucleic Acids Res 30: 4556-4566

Milne GT, Jin S, Shannon KB, Weaver DT (1996) Mutations in two Ku homologs define a DNA end-joining repair pathway in *Saccharomyces cerevisiae*. Mol Cell Biol 16: 4189-4198

Mysore KS, Bassuner B, Deng X-B, Darbinian NS, Motchoulski A, Ream LW, Gelvin SB (1998) Role of the *Agrobacterium tumefaciens* VirD2 protein in T-DNA transfer and integration. Mol Plant-Microbe Interact 11: 668-683

Mysore KS, Nam J, Gelvin SB (2000) An *Arabidopsis* histone H2A mutant is deficient in *Agrobacterium* T-DNA integration. Proc Natl Acad Sci USA 97: 948-953

Nacry P, Camilleri C, Courtial B, Caboche M, Bouchez D (1998) Major chromosomal rearrangements induced by T-DNA transformation in Arabidopsis. Genetics 149: 641-650

Nam J, Matthysse AG, Gelvin SB (1997) Differences in susceptibility of Arabidopsis ecotypes to crown gall disease may result from a deficiency in T-DNA integration. Plant Cell 9: 317-333

Nam J, Mysore KS, Zheng C, Knue MK, Matthysse AG, Gelvin SB (1999) Identification of T-DNA tagged *Arabidopsis* mutants that are resistant to transformation by *Agrobacterium*. Mol Gen Genet 261: 429-438

Narasimhulu SB, Deng X-B, Sarria R, Gelvin SB (1996) Early transcription of Agrobacterium T-DNA genes in tobacco and maize. Plant Cell 8: 873-886

Negruk V, Eisner G, Lemieux B (1996) Addition-deletion mutations in transgenic *Arabidopsis thaliana* generated by the seed co-cultivation method. Genome 39: 1117-1122

Offringa R, de Groot MJA, Haagsman HJ, Does MP, van den Elzen PJM, Hooykaas PJJ (1990) Extrachromosomal homologous recombination and gene targeting in plant cells after *Agrobacterium* mediated transformation. EMBO J 9: 3077-3084

Ohba T, Yoshioka Y, Machida C, Machida Y (1995) DNA rearrangement associated with the integration of T-DNA in tobacco: an example for multiple duplications of DNA around the integration target. Plant J 7: 157-164

Orel N, Puchta H (2003) Differences in the processing of DNA ends in *Arabidopsis thaliana* and tobacco: possible implications for genome evolution. Plant Mol Biol 51: 523-531

Ortega D, Raynal M, Laudié M, Llauro C, Cooke R, Devic M, Genestier S, Picard G, Abad P, Contard P, Sarrobert C, Nussaume L, Bechtold N, Horlow C, Pelletier G, Delseny M (2002) Flanking sequence tags in *Arabidopsis thaliana* T-DNA insertion lines: a pilot study. C R Biol 325: 773-780

Pâques F, Haber JE (1999) Multiple pathways of recombination induced by double-strand breaks in *Saccharomyces cerevisiae*. Microbiol Mol Biol Rev 63: 349-404

Podevin N, De Buck S, De Wilde C, Depicker A (2006) Insight into recognition of the T-DNA border repeats as termination sites for T-strand synthesis by *Agrobacterium tumefaciens*. Transgenic Res, in press

Ramanathan V, Veluthambi K (1995) Transfer of non-T-DNA portions of the *Agrobacterium tumefaciens* Ti plasmid pTiA6 from the left terminus of T_L-DNA. Plant Mol Biol 28: 1149-1154

Relić B, Andjelković M, Rossi L, Nagamine Y, Hohn B (1998) Interaction of the DNA modifying proteins VirD1 and VirD2 of *Agrobacterium tumefaciens*: analysis by subcellular localization in mammalian cells. Proc Natl Acad Sci USA 95: 9105-9110

Rinehart TA, Dean C, Weil CF (1997) Comparative analysis of non-random DNA repair following *Ac* transposon excision in maize and *Arabidopsis*. Plant J 12: 1419-1427

Rossi L, Hohn B, Tinland B (1996) Integration of complete transferred DNA units is dependent on the activity of virulence E2 protein of *Agrobacterium tumefaciens*. Proc Natl Acad Sci USA 93: 126-130

Sallaud C, Gay C, Larmande P, Bès M, Piffanelli P, Piégu B, Droc G, Regad F, Bourgeois E, Meynard D, Périn C, Sabau X, Ghesquière A, Glaszmann JC, Delseny M, Guiderdoni E (2004) High throughput T-DNA insertion mutagenesis in rice: a first step towards *in silico* reverse genetics. Plant J 39: 450-464

Sallaud C, Meynard D, van Boxtel J, Gay C, Bès M, Brizard JP, Larmande P, Ortega D, Raynal M, Portefaix M, Ouwerkerk PBF, Rueb S, Delseny M, Guiderdoni E (2003) Highly efficient production and characterization of T-DNA plants for rice (*Oryza sativa* L.) functional genomics. Theor Appl Genet 106: 1396-1408

Salomon S, Puchta H (1998) Capture of genomic and T-DNA sequences during double-strand break repair in somatic plant cells. EMBO J 17: 6086-6095

Scheiffele P, Pansegrau W, Lanka E (1995) Initiation of *Agrobacterium tumefaciens* T-DNA processing. Purified proteins VirD1 and VirD2 catalyze site- and strand-specific cleavage of superhelical T-border DNA *in vitro*. J Biol Chem 270: 1269-1276

Sha Y, Li S, Pei Z, Luo L, Tian Y, He C (2004) Generation and flanking sequence analysis of a rice T-DNA tagged population. Theor Appl Genet 108: 306-314

Shurvinton CE, Hodges L, Ream LW (1992) A nuclear localization signal and the C-terminal omega sequence in the *Agrobacterium tumefaciens* VirD2 endonuclease are important for tumor formation. Proc Natl Acad Sci USA 89: 11837-11841

Smith EF, Townsend CO (1907) A plant tumor of bacterial origin. Science 25: 671-673

Sonti RV, Chiurazzi M, Wong D, Davies CS, Harlow GR, Mount DW, Signer ER (1995) *Arabidopsis* mutants deficient in T-DNA integration. Proc Natl Acad Sci USA 92: 11786-11790

Szabados L, Kovács I, Oberschall A, Ábrahám E, Kerekes I, Zsigmond L, Nagy R, Alvarado M, Krasovskaja I, Gál M, Berente A, Rédei GP, Haim AB, Koncz C (2002) Distribution of 1000 sequenced T-DNA tags in the *Arabidopsis* genome. Plant J 32: 233-242

Tamura K, Adachi Y, Chiba K, Oguchi K, Takahashi H (2002) Identification of Ku70 and Ku80 homologues in *Arabidopsis thaliana*: evidence for a role in the repair of DNA double-strand breaks. Plant J 29: 771-781

Tax FE, Vernon DM (2001) T-DNA-associated duplication/translocations in Arabidopsis. Implications for mutant analysis and functional genomics. Plant Physiol 126: 1527-1538

ten Hoopen R, Robbins TP, Fransz PF, Montijn BM, Oud O, Gerats AGM, Nanninga N (1996) Localization of T-DNA insertions in Petunia by fluorescence in-situ hybridization: physical evidence for suppression of recombination. Plant Cell 8: 823-830

Tinland B (1996) The integration of T-DNA into plant genomes. Trends Plant Sci 1: 178-184

Tinland B, Hohn B, Puchta H (1994) *Agrobacterium tumefaciens* transfers single-stranded transferred DNA (T-DNA) into the plant cell nucleus. Proc Natl Acad Sci USA 91: 8000-8004

Tinland B, Koukolíková-Nicola Z, Hall MN, Hohn B (1992) The T-DNA-linked VirD2 protein contains two distinct nuclear localization signals. Proc Natl Acad Sci USA 89: 7442-7446

Tinland B, Schoumacher F, Gloeckler V, Bravo-Angel AM, Hohn B (1995) The *Agrobacterium tumefaciens* virulence D2 protein is responsible for precise integration of T-DNA into the plant genome. EMBO J 14: 3585-3595

Topping JF, Wei W, Clarke MC, Muskett P, Lindsey K (1995) *Agrobacterium*-mediated transformation of *Arabidopsis thaliana*: application in T DNA tagging. *In* H Jones, ed, Plant Gene Transfer and Expression Protocols, Methods in Molecular Biology, Vol 49. Humana Press, Totowa, pp 63-76

Tsukamoto Y, Ikeda H (1998) Double-strand break repair mediated by DNA end-joining. Genes Cells 3: 135-144

Tsukamoto Y, Kato J-I, Ikeda H (1997) Silencing factors participate in DNA repair and recombination in *Saccharomyces cerevisiae*. Nature 388: 900-903

Tzfira T, Citovsky V (2002) Partners-in-infection: host proteins involved in the transformation of plant cells by *Agrobacterium*. Trends Cell Biol 12: 121-129

Tzfira T, Li J, Lacroix B, Citovsky V (2004a) *Agrobacterium* T-DNA integration: molecules and models. Trends Genet 20: 375-383

Tzfira T, Rhee Y, Chen M-H, Kunik T, Citovsky V (2000) Nucleic acid transport in plant-microbe interactions: the molecules that walk through the walls. Annu Rev Microbiol 54: 187-219

Tzfira T, Vaidya M, Citovsky V (2002) Increasing plant susceptibility to *Agrobacterium* infection by overexpression of the *Arabidopsis* nuclear protein VIP1. Proc Natl Acad Sci USA 99: 10435-10440

Tzfira T, Vaidya M, Citovsky V (2004b) Involvement of targeted proteolysis in plant genetic transformation by *Agrobacterium*. Nature 431: 87-92

Vain P, James VA, Worland B, Snape JW (2003) Transgene behaviour across two generations in a large random population of transgenic rice plants produced by particle bombardment. Theor Appl Genet 105: 878-889

Valentine L (2003) *Agrobacterium tumefaciens*: the David and Goliath of modern genetics. Plant Physiol 133: 948-955

van Attikum H, Bundock P, Hooykaas PJJ (2001) Non-homologous end-joining proteins are required for *Agrobacterium* T-DNA integration. EMBO J 20: 6550-6558

van Attikum H, Bundock P, Overmeer RM, Lee LY, Gelvin SB, Hooykaas PJJ (2003) The *Arabidopsis AtLIG4* gene is required for the repair of DNA damage, but not for the integration of *Agrobacterium* T-DNA. Nucleic Acids Res 31: 4247-4255

van der Graaff E, den Dulk-Ras A, Hooykaas PJJ (1996) Deviating T-DNA transfer from *Agrobacterium tumefaciens* to plants. Plant Mol Biol 31: 677-681

Van Haaren MJJ, Pronk JT, Schilperoort RA, Hooykaas PJJ (1987) Functional analysis of the *Agrobacterium tumefaciens* octopine Ti-plasmid left and right T-region border fragments. Plant Mol Biol 8: 95-104

Van Lijsebettens M, Inzé D, Schell J, Van Montagu M (1986) Transformed cell clones as a tool to study T DNA integration mediated by *Agrobacterium tumefaciens*. J Mol Biol 188: 129-145

Veena, Jiang H, Doerge RW, Gelvin SB (2003) Transfer of T-DNA and Vir proteins to plant cells by *Agrobacterium tumefaciens* induces expression of host genes involved in mediating transformation and suppresses host defense gene expression. Plant J 35: 219-236

Vergunst AC, van Lier MCM, den Dulk-Ras A, Grosse Stüve TA, Ouwehand A, Hooykaas PJJ (2005) Positive charge is an important feature of the C-terminal transport signal of the VirB/D4-translocated proteins of *Agrobacterium*. Proc Natl Acad Sci USA 102: 832-837

Vichi P, Coin F, Renaud J-P, Vermeulen W, Hoeijmakers JHJ, Moras D, Egly J-M (1997) Cisplatin- and UV-damaged DNA lure the basal transcription factor TFIID/TBP. EMBO J 16: 7444-7456

Villemont E, Dubois F, Sangwan RS, Vasseur G, Bourgeois Y, Sangwan-Norreel BS (1997) Role of the host cell cycle in the *Agrobacterium*-mediated genetic transformation of *Petunia*: evidence of an S-phase control mechanism for T-DNA transfer. Planta 201: 160-172

Wallroth M, Gerats AGM, Rogers SG, Fraley RT, Horsch RB (1986) Chromosomal localization of foreign genes in *Petunia hybrida*. Mol Gen Genet 202: 6-15

Wang J, Lewis ME, Whallon JH, Sink KC (1995) Chromosomal mapping of T-DNA inserts in transgenic *Petunia* by in situ hybridization. Transgenic Res 4: 241-246

Wenck A, Czakó M, Kanevski I, Márton L (1997) Frequent collinear long transfer of DNA inclusive of the whole binary vector during *Agrobacterium*-mediated transformation. Plant Mol Biol 34: 913-922

West CE, Waterworth WM, Jiang Q, Bray CM (2000) *Arabidopsis* DNA ligase IV is induced by γ-irradiation and interacts with an *Arabidopsis* homologue of the double strand break repair protein XRCC4. Plant J 24: 67-78

West CE, Waterworth WM, Story GW, Sunderland PA, Jiang Q, Bray CM (2002) Disruption of the *Arabidopsis AtKu80* gene demonstrates an essential role for AtKu80 protein in efficient repair of DNA double-strand breaks *in vivo*. Plant J 31: 517-528

Windels P, De Buck S, Van Bockstaele E, De Loose M, Depicker A (2003) T-DNA integration in Arabidopsis chromosomes. Presence and origin of filler DNA sequences. Plant Physiol 133: 2061-2068

Wolters A-MA, Trindade LM, Jacobsen E, Visser RGF (1998) Fluorescence *in situ* hybridization on extended DNA fibres as a tool to analyse complex T-DNA loci in potato. Plant J 13: 837-847

Zaenen I, Van Larebeke N, Teuchy H, Van Montagu M, Schell J (1974) Super-coiled circular DNA in crown gall inducing *Agrobacterium* strains. J Mol Biol 86: 109-127

Zambre M, Terryn N, De Clercq J, De Buck S, Dillen W, Van Montagu M, Van Der Straeten D, Angenon G (2003) Light strongly promotes gene transfer from *Agrobacterium tumefaciens* to plant cells. Planta 216: 580-586

Zambryski PC, Depicker A, Kruger K, Goodman HM (1982) Tumor induction by *Agrobacterium tumefaciens*: analysis of the boundaries of T-DNA. J Mol Appl Genet 1: 361-370

Zheng SJ, Henken B, Sofiari E, Jacobsen E, Krens FA, Kik C (2001) Molecular characterization of transgenic shallots (*Allium cepa* L.) by adaptor ligation PCR (AL-PCR) and sequencing of genomic DNA flanking T-DNA borders. Transgenic Res 10: 237-245

Ziemienowicz A, Tinland B, Bryant J, Gloeckler V, Hohn B (2000) Plant enzymes but not *Agrobacterium* VirD2 mediate T-DNA ligation in vitro. Mol Cell Biol 20: 6317-6322

Zupan J, Muth TR, Draper O, Zambryski PC (2000) The transfer of DNA from *Agrobacterium tumefaciens* into plants: a feast of fundamental insights. Plant J 23: 11-28

Chapter 13

FUNCTION OF HOST PROTEINS IN THE *AGROBACTERIUM*-MEDIATED PLANT TRANSFORMATION PROCESS

Stanton B. Gelvin

Department of Biological Sciences, Purdue University, West Lafayette, IN 47907 USA

Abstract. Genetic transformation results from a complex interaction between *Agrobacterium* and host plant cells. Many decades of genetic, biochemical, and molecular analyses have revealed in detail those events taking place within the bacterium that contribute to T-DNA and Virulence protein transfer. However, we understand much less about the plant contribution to the transformation process. Plant species, and even varieties/ecotypes, differ markedly in their susceptibility to *Agrobacterium*. A genetic component underlies these differences, permitting scientists to identify specific host genes and proteins mediating transformation. In this chapter, I review what is known about the plant contribution to transformation, and the tools which scientists are using to reveal the mechanisms by which host genes and proteins function in various steps of the transformation process.

1 INTRODUCTION

More than three decades of extensive genetic, biochemical, and molecular analyses have resulted in a reasonably complete understanding of the process of plant genetic transformation from the perspective of *Agrobacterium*. We now understand in relative detail signaling events resulting in virulence (*vir*) gene induction, T-DNA processing, and T-DNA and Vir protein transport through the Type IV secretion apparatus (for recent reviews, see Christie and Vogel, 2000; Gelvin, 2000; Tzfira et al., 2000; Zupan et al., 2000; Tzfira and Citovsky, 2002; Gelvin, 2003; Tzfira and Citovsky, 2003; Valentine, 2003; Brencic and Winans, 2005; Lacroix et al., 2006, and other chapters in this volume). However, our knowledge of the host contribution to the transformation process has lagged. It is clear that although *Agrobacterium* has an enormous host range encompassing species of numerous phylogenetic kingdoms, differences in susceptibility to *Agrobacterium*-mediated transformation abound among plant species. Physiological or environmental effects may account for some of these disparities, but a genetic basis also underlies host susceptibility or resistance. For example, a recent survey of approximately 40 *Arabidopsis thaliana* ecotypes indicated vastly disparate responses to root transformation (Nam et al., 1997). The mechanisms accounting for resistance among these ecotypes varied from lack of bacterial attachment and biofilm formation to lack of T-DNA integration into the plant genome. In addition, various plant tissues or organs can respond differently to infection by a particular *Agrobacterium* strain (Grevelding et al., 1993; Mysore et al., 2000b).

The existence of a genetic basis underlying host susceptibility to *Agrobacterium*-mediated transformation has allowed us to conduct genetic screens for specific plant genes that contribute to the transformation process. "Forward" and "reverse" genetic approaches have thus been developed to probe the plant genome for host "transformation" genes. These screens have resulted the identification of more than 125 *Arabidopsis* genes involved in transformation (Zhu et al., 2003b). Cell and molecular biology methodologies have also been combined with bioinformatics approaches to investigate plant genes responding to *Agrobacterium* infection and which, thus, may play a role in transformation (Ditt et al., 2001; Veena et al., 2003; Ditt et al., 2005).

In this chapter, I shall review some of the methodologies that we and others have used to identify host genes that mediate plant genetic transformation by *Agrobacterium*. I shall also describe how we believe these genes function to help effect transformation. Finally, I shall illustrate how manipulation of some of these genes may be used to improve the transformation of recalcitrant plant species.

2 A GENETIC BASIS EXISTS FOR HOST SUSCEPTIBILITY TO *AGROBACTERIUM*-MEDIATED TRANSFORMATION

It has long been recognized that differences exist among plant species with regard to susceptibility to crown gall or hairy root disease caused by oncogenic strains of *Agrobacterium tumefaciens* or *A. rhizogenes* (Owens and Cress, 1984; Szegedi and Kozma, 1984; Smarrelli et al., 1986; Robbs et al., 1991; Bailey et al., 1994; Mauro et al., 1995; Bliss et al., 1999; Sparrow et al., 2004). Such differences may simply have reflected the particular *Agrobacterium* strain or assay condition used to determine susceptibility. For example, the strain *A. tumefaciens* Bo542 cannot incite tumors on the legumes alfalfa and soybean, whereas a similar strain (A281, which contains pTiBo542 in a different bacterial chromosomal background) very efficiently generates tumors on these host species (Hood et al., 1987). Monocotyledonous plants were initially considered non-hosts for *Agrobacterium*, primarily because they could not be shown to support the growth of crown gall tumors. However, tumor production was eventually detected on the monocot species *Asparagus officinalis* and *Dioscorea bulbifera* (Bytebier et al., 1987; Schafer et al., 1987), and genetic transformation (although not Crown Gall tumorigenesis) of rice, maize, barley, wheat, and onion has been reported and, for some species, is now routine (Dommisse et al., 1990; Chan et al., 1992; Conner and Dommisse, 1992; Hiei et al., 1994; Dong et al., 1996; Ishida et al., 1996; Rashid et al., 1996; Cheng et al., 1997; Tingay et al., 1997; Toki, 1997; Lu et al., 2001; An et al., 2003; Chen et al., 2003; Sallaud et al., 2004; An et al., 2005).

As mentioned above, the first extensive investigation of a genetic basis for plant susceptibility to *Agrobacterium*-mediated transformation was a study conducted by Nam et al. (1997). These authors determined root transformation-susceptibility of approximately 40 different ecotypes of *Arabidopsis thaliana*. By evaluating genetic crosses between a highly susceptible (Aa-0) and highly recalcitrant (UE-1) ecotype, they concluded that a genetic basis for susceptibility exists, and that in this cross, susceptibility segregated as a single major genetic locus. Recalcitrance in UE-1 was traced to a deficiency in the process of T-DNA integration.

It is possible to identify host "transformation competence" genes using bulk segregant analysis followed by positional cloning. However, our laboratory chose another approach. We screened ~20,000 *Arabidopsis* T-DNA insertion lines for mutants that were resistant to *Agrobacterium* transformation (*rat* mutants). Although many of the mutagenized *Arabidopsis* lines initially showed a rat phenotype when their roots were infected with

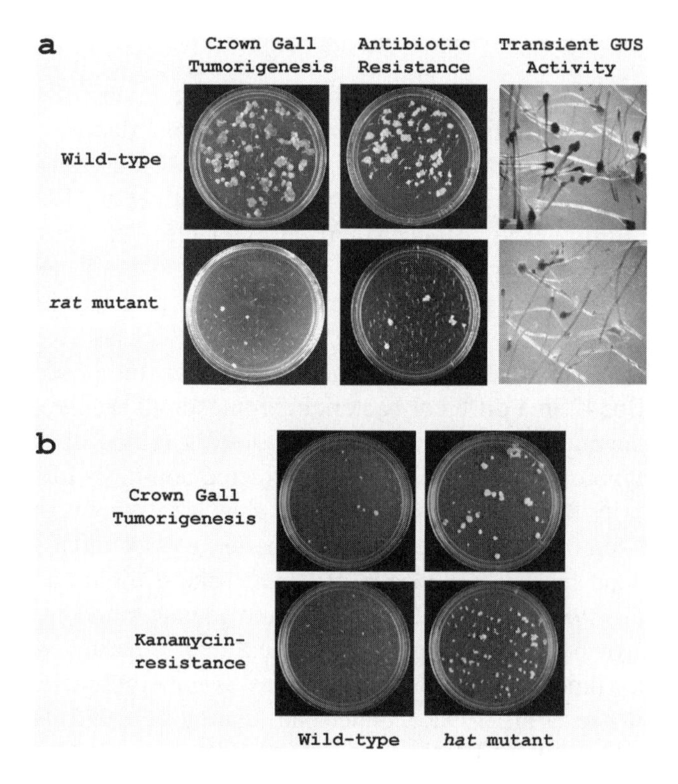

Figure 13-1. Identification of *Arabidopsis rat* (a) and *hat* (b) mutants. (a) For identification of *rat* mutants, axenic root segments of wild-type or mutant *Arabidopsis* plants were inoculated with various *Agrobacterium* strains (10^8 cells/ml) and examined either one month (crown gall tumorigenesis and antibiotic resistance) or six days later (for transient GUS activity). (b) For identification of *hat* mutants, axenic root segments of *Arabidopsis* wild-type or mutant plants containing a T-DNA "activation tag" were inoculated with a low concentration (10^5-10^6 cells/ml) of *A. tumefaciens* cells and examined one month (crown gall tumorigenesis and antibiotic resistance) later. For Crown Gall tumorigenesis assays, the root segments were inoculated with *A. tumefaciens* A208, a tumorigenic strain, and incubated on MS medium lacking phytohormones. For antibiotic resistance, the segments were inoculated with *A. tumefaciens* At849 (containing plant-active *nptII* and *gusA*-intron genes) and incubated on callus inducing medium (CIM) containing 50 mg/l kanamycin. For transient GUS activity, the root segments were incubated on CIM for six days, and then stained with X-gluc.

the tumorigenic strain *A. tumefaciens* A208, subsequent re-testing of these lines indicated that most (~90%) were not *rat* mutants. We were, however, able to recover approximately 100 T-DNA insertion mutants that consistently displayed a rat phenotype (Nam et al., 1999; Zhu et al., 2003b). The rat phenotype was confirmed in these mutant lines by repeated rounds of infection of progeny plants using several *Agrobacterium* strains and scoring for crown gall tumorigenesis, antibiotic/herbicide resistance, or transient β-glucuronidase (GUS) activity directed by a *gusA*-intron transgene (*Figure 13-1a* and *Table 13-1*). The high initial rate of "false positive" *rat* mutant identification serves as a reminder that plant transformation-competence may often be determined by physiological or environmental factors, and care must be taken in interpreting the results of screens for transformation competence or recalcitrance.

Table 13-1. Classes of *Arabidopsis rat* mutants

Step in transformation	**Defect in transient or stable transformation**	**Examples of *rat* genes**
Bacterial attachment or biofilm formation	Transient and stable	Arabinogalactan proteins, cellulose synthase-like proteins
T-DNA and Virulence protein transfer	Transient and stable	BTI proteins
Cytoplasmic trafficking	Transient and stable	Actin, kinesin
Nuclear targeting and import	Transient and stable	Importin α, transportin, VIP1
T-DNA integration	Stable	Histones, histone deacetylases, histone acetyltransferases, nucleosome assembly factors

In accord with the results of others who investigated tissue-specific responses to *Agrobacterium*-mediated plant transformation (Akama, 1992; Yi et al., 2002), we noted that the rat phenotype, as initially characterized by recalcitrance to root transformation, frequently did not extend to *Arabidopsis* flower dip transformation (Mysore et al., 2000a). In flower dip transformation (or the closely related flower vacuum infiltration method; Bechtold et al., 1993), emerging flowers of *Arabidopsis* plants are immersed in a solution of *Agrobacterium* cells (Clough and Bent, 1998; Bent, 2000). Seeds are collected from the siliques of these dipped flowers and are plated onto medium to select for plants expressing an antibiotic/herbicide resistance marker. Several reports have indicated that the female gametophyte is the target for flower dip transformation (Ye et al., 1999; Bechtold et al., 2000; Desfeux et al., 2000; Bechtold et al., 2003). We tested a number of *Arabidopsis rat* mutants and recalcitrant ecotypes for susceptibility to flower dip transformation. Although all the tested mutants and ecotypes were highly recalcitrant to somatic (root) transformation, only the *rad5* mutant (Nam et al., 1998) was resistant to flower dip transformation. All other *rat* mutants were highly susceptible to flower dip transformation (Mysore et al., 2000a). Furthermore, different cell types within plants show markedly different susceptibility to *Agrobacterium*-mediated transformation (Sangwan et al., 1992; Geier and Sangwan, 1996; Yi et al., 2002). In some instances, tissue- or cell-specific expression of particular host genes correlated with transformation competence (Yi et al., 2002; Zhu et al., 2003a; Yi et al., 2006). Thus, one needs to take into consideration the host tissue being assayed for transformation competence when evaluating host susceptibility to *Agrobacterium*-mediated transformation.

In an effort to identify plant genes that when over-expressed would result in increased transformation-susceptibility, our laboratory has recently initiated a genetic screen of *Arabidopsis* T-DNA activation-tagged lines that are hyper-susceptible to *Agrobacterium* transformation (*hat* mutants). Using a low concentration of bacterial cells (10^5-10^6/ml, compared to 10^8/ml as used for *rat* mutant screening) as an inoculum, we have identified eight *hat* mutants (S.B. Gelvin, unpublished; *Figure 13-1b*). Current efforts include identification of *hat* genes and verification of their roles in transformation.

3 THE PLANT RESPONSE TO *AGROBACTERIUM*: STEPS IN THE TRANSFORMATION PROCESS, AND PLANT GENES/PROTEINS INVOLVED IN EACH OF THESE STEPS

Agrobacterium-mediated plant transformation is a complex process involving numerous steps. From the plant's perspective, these include bacterial attachment/biofilm formation, T-DNA and virulence (vir) protein transfer, cytoplasmic trafficking and nuclear targeting of the T-complex, "stripping" of proteins from the T-DNA, T-DNA integration, and T-DNA gene expression. Interruption of any of these steps can result in disruption of the transformation process. I shall review what is known about plant proteins involved in each of these steps.

3.1 Bacterial attachment and biofilm formation

In order for efficient transformation to occur, *Agrobacterium* must attach to the surface of wounded plant tissues. Bacterial attachment is a complex process involving bacterial exopolysaccharides and, most likely, plant cell wall proteins. All bacterial mutants which are "non-attaching" are attenuated in virulence (Matthysse, 1987; Crews et al., 1990). Efficient transformation generally involves wounding of plant tissue. However, several reports have indicated that "non-wounded" plant tissue can also be infected, albeit with low efficiency (Escudero and Hohn, 1997; Brencic et al., 2005).

A major problem in studying bacterial attachment is in defining attachment. *Agrobacterium* makes and secretes numerous exopolysaccharides, including a cyclic 1,2-β-D-glucan and cellulose (Matthysse et al., 1981; Deasey and Matthysse, 1984; Puvanesarajah et al., 1985; Zorreguieta and Ugalde, 1986; Robertson et al., 1988; Cangelosi et al., 1989; de Iannino and Ugalde, 1989; Hawes and Pueppke, 1989; Kamoun et al., 1989; Reuhs et al., 1997; O'Connell and Handelsman, 1999). This extracellular matrix "entraps" thousands of bacteria near the cell surface. However, most bacteria do not bind directly to the plant cell surface; rather, the bacteria form a biofilm (Matthysse et al., 2005). Biofilm formation is required for efficient transformation to occur, and bacterial mutants deficient in biofilm formation are either avirulent or highly attenuated in virulence (Douglas et al., 1982; Douglas et al., 1985; Cangelosi et al., 1987; Thomashow et al., 1987; Ramey et al., 2004).

Because of the importance of biofilm formation in plant transformation, it is at times difficult to determine the role of specific plant genes and proteins in bacterial attachment directly to the plant cell. I shall call this "productive attachment", defined as the attachment of bacteria to plant cells that directly results in T-DNA and virulence protein transfer. Early reports suggested a proteinaceous material on the plant surface is required for bacterial attachment (Neff and Binns, 1985; Gurlitz et al., 1987), and several reports suggested that specific plant proteins mediated bacterial attachment (Wagner and Matthysse, 1992; Swart et al., 1994). However, these observations were never confirmed or followed up.

Approximately 10 years ago, our laboratory initiated genetic studies to identify *Arabidopsis* ecotypes and mutants that were resistant to *Agrobacterium*-mediated transformation. Among the recalcitrant ecotypes identified were several, including Bl-1 and Petergof, which would not support bacterial attachment (Nam et al., 1997). *Arabidopsis* ecotype Ws is highly susceptible to *Agrobacterium*-mediated transformation, and an examination of T-DNA insertion mutants of this ecotype revealed three which were deficient, to various degrees, in bacterial attachment and biofilm formation (Nam et al., 1999). The mutant most severely deficient in bacterial attachment under all conditions examined was *rat1*. *Rat1* encodes the arabinogalactan protein AtAGP17 (Gaspar et al., 2004). AtAGP17 contains a basic amino acid motif in the carboxy-terminal region, and a GPI anchor. The protein is expressed at very low levels in *Arabidopsis* roots, but introduction of either a cDNA or a genomic clone encoding AtAGP17 into the *rat1* mutant could complement the rat phenotype (Gaspar et al., 2004). Recent data from our laboratory indicates that the rat phenotype of *Arabidopsis* ecotype Bl-1 (which is deficient in supporting *Agrobacterium* biofilm formation on its roots) can be reversed by introduction of the *AtAGP17* gene (E. Wilkinson, T. Muth and S.B. Gelvin, unpublished and *Figure 13-2*). Thus, at least one arabinogalactan protein gene is required for bacterial attachment, biofilm formation, and transformation.

A second *Arabidopsis* gene, encoding the cellulose synthase-like protein AtCslA9, is also required for bacterial biofilm formation on *Arabidopsis* roots. A T-DNA insertion in the 3' untranslated region of this gene results in the *rat4* mutant (Zhu et al., 2003a). This gene is expressed in the hypocotyls of young *Arabidopsis* plants and the elongation zone of mature roots (Zhu et al., 2003a), the region of the root that we had previously identified as most susceptible to *Agrobacterium*-mediated transformation (Yi et al., 2002). Although the chemical composition of cell walls extracted from total *rat4* plants was similar to that of wild-type plants, Fourier transform infrared spectroscopy of the elongation zone of *rat4* roots showed a great enrichment for cellulose (M. McCann and S.B. Gelvin, unpublished). Thus, a mutant which altered the polysaccharide composition of plant cell walls could affect *Agrobacterium* attachment, the first step in transformation.

A third *Arabidopsis* mutant, *rat3*, also altered *Agrobacterium* biofilm formation on plant roots. *Rat3* encodes a protein with little homology to other proteins in databases. However, the *rat3* mutant could be complemented with a *Rat3* genomic clone, indicating that the rat phenotype of this mutant was caused by the T-DNA insertion in the *Rat3* gene (Y. Zhu and S.B. Gelvin, unpublished).

3.1.1 Enhancement of plant defense signaling can result in decreased Agrobacterium biofilm formation

While investigating changes in plant gene expression soon after inoculation with *Agrobacterium*, we noted that many plant defense genes are rapidly induced (Veena et al., 2003; Veena and S.B. Gelvin, unpublished). We therefore investigated the effects of *Arabidopsis* defense signaling mutants, and chemical elicitors of defense responses, on *Agrobacterium*-mediated transformation. The *cep1* (constitutive expression of PR genes) mutant of *Arabidopsis* shows enhanced resistance to several bacterial and fungal pathogens (Silva et al., 1999). We therefore investigated the susceptibility of *cep1* to *Agrobacterium*-mediated transformation. *cep1* mutant roots display a strong rat phenotype and do not support *Agrobacterium* biofilm formation (Veena and Gelvin, unpublished). Furthermore, chemical elicitation of defense responses by salicylic acid, BTH, methyljasmonate, or ethephon also conferred a rat phenotype upon *Arabidopsis* roots. As seen with the *cep1* mutant, chemically elicited *Arabidopsis* roots did not support *Agrobacterium* biofilm formation (Veena and S.B. Gelvin, unpublished). Thus, expression of plant defense responses, either by chemical elicitation or by mutation, could affect *Agrobacterium* attachment and transformation efficiency.

3.1.2 T-DNA and virulence protein transfer: A putative receptor for the Agrobacterium T-pilus

Agrobacterium has an extremely broad host range (De Cleene and De Ley, 1976; Anderson and Moore, 1979; van Wordragen and Dons, 1992; Pena and Seguin, 2001). In addition to dicot and monocot plant species, *Agrobacterium* can transform gymnosperms (Morris and Morris, 1990; Stomp et al., 1990; McAfee et al., 1993; Levee et al., 1999; Wenck et al., 1999). "Hearsay" evidence suggests that *Agrobacterium* can infect ferns and algal species. *Agrobacterium* can also transform fungal species (Bundock et al., 1995; Bundock and Hooykaas, 1996; Piers et al., 1996; de Groot et al., 1998; Abuodeh et al., 2000; Bundock et al., 2002; Schrammeijer et al., 2003; van Attikum and Hooykaas, 2003; Michielse et al., 2004). *Agrobacterium* has also been reported to transform mammalian cells (Kunik et al., 2001). Given the broad range of susceptible host species, it is likely that there is either a "common" receptor for *Agrobacterium* attachment, or there is no receptor at all.

Upon *vir* gene induction, *Agrobacterium* generates a T-pilus as part of a Type IV secretion system (Kado, 1994; Baron and Zambryski, 1996;

Fullner et al., 1996). The major pilin is a processed and cyclized protein encoded by *virB2* (Lai and Kado, 1998; Eisenbrandt et al., 1999), although other VirB-encoded proteins (VirB5 and VirB7) form minor T-pilus components (Schmidt-Eisenlohr et al., 1999; Sagulenko et al., 2001). Although pili have been postulated to form the conduit through which DNA and proteins are transferred to recipient organisms during conjugation, recent results suggest that T-DNA transfer can occur in the absence of T-pili (but not in the absence of VirB2 protein; Jakubowski et al., 2005). The T-pilus may therefore serve as a "grappling hook" to bring *Agrobacterium* and the recipient cell into close enough proximity for conjugation to occur. We therefore searched for a plant protein which would specifically bind to VirB2 in the hopes that such a protein would mediate *Agrobacterium* "productive" attachment and T-DNA transfer.

We screened an *Arabidopsis* cDNA library for clones encoding proteins that would interact with the processed (but not cyclized) form of VirB2 in yeast. Amongst the approximately three million colonies screened, we recurrently identified four cDNAs. Three of these encoded highly related proteins, which we termed BTI (VirB2 interacting) proteins, and the fourth encoded a Rab8 GTPase (Hwang and Gelvin, 2004). The BTI proteins contain a "reticulon domain", found in many eukaryotic but not prokaryotic proteins. Rab8 GTPases have been implicated in retrograde trafficking of proteins from the ER and Golgi to the cellular membrane. Inhibition of expression of the genes encoding these proteins in *Arabidopsis*, using T-DNA "knockout" insertion mutations, anti-sense RNA, and RNAi, severely attenuated *Agrobacterium*-mediated transformation, whereas over-expression of BTI1 in transgenic *Arabidopsis* increased transformation efficiency. The BTI1 protein is transiently induced in *Arabidopsis* suspension cells upon infection by *Agrobacterium*. Importantly, YFP fusions to BTI1, 2, and 3 proteins localized to the cellular periphery (but not the cell wall) in transgenic *Arabidopsis* plants, whereas the corresponding YFP fusion to the Rab8 GTPase localized to the cytoplasm. Because there is no evidence that the T-pilus "penetrates" into the plant cell (as do the "needles" of Type III secretion systems), any protein interacting with the T-pilus should localize to the surface of plant cells. Finally, pre-incubation of acetosyringone-induced *Agrobacterium* cells with recombinant BTI1 protein inhibited subsequent transformation of plant cells, suggesting that the BTI1 protein had "coated" the T-pilus and functioned as a competitor to BTI proteins on the plant surface (Hwang and Gelvin, 2004). Future experiments are aimed at understanding the mechanism by which the BTI and Rab8 proteins function in the transformation process.

3.2 T-DNA cytoplasmic trafficking and nuclear targeting

Once the T-DNA has entered the plant cell, it must traffic through the cytoplasm and enter the nucleus. Because any DNA sequence inserted between T-DNA borders can ultimately integrate into host chromosomes, information regarding cytoplasmic trafficking must necessarily reside not in the T-DNA sequence itself, but rather in proteins (both *Agrobacterium*- and host-encoded) that interact with the T-DNA.

The VirD2 endonuclease which processes the T-DNA region from the Ti-plasmid covalently links to the 5' end of the T-strand (a single-stranded T-DNA molecule) through a phospho-tyrosine bond (Herrera-Estrella et al., 1988; Ward and Barnes, 1988; Young and Nester, 1988; Durrenberger et al., 1989; Howard et al., 1989; Vogel and Das, 1992). Another protein, VirE2, has been proposed to interact with the single-stranded T-DNA (Gietl et al., 1987; Christie et al., 1988; Citovsky et al., 1988; Das, 1988; Citovsky et al., 1989; Sen et al., 1989). The complex formed by the T-strand covalently linked to a single molecule of VirD2 and "coated" by multiple molecules of VirE2 has been termed the T-complex. Although the formation of the T-complex was originally postulated to occur within *Agrobacterium* (Christie et al., 1988; Howard and Citovsky, 1990), it is now clear that VirE2 is separately transported to plant cells through the Type IV secretion system, and that this putative complex is formed outside the bacterial cytoplasm, possibly in the plant cell (Otten et al., 1984; Citovsky et al., 1992; Binns et al., 1995; Gelvin, 1998; Lee et al., 1999; Cascales and Christie, 2004). Both VirD2 and VirE2 contain nuclear localization signal (NLS) sequences which can target reporter protein fusions to the nucleus in plant, yeast, and (in some instances) mammalian cells (Herrera-Estrella et al., 1990; Citovsky et al., 1992; Howard et al., 1992; Tinland et al., 1992; Koukolikova-Nicola and Hohn, 1993; Koukolikova-Nicola et al., 1993; Rossi et al., 1993; Citovsky et al., 1994; Guralnick et al., 1996; Relic et al., 1998; Rhee et al., 2000). These NLS sequences have been proposed to mediate nuclear targeting of the T-strand (Zupan et al., 1996; Citovsky et al., 1997; Mysore et al., 1998; Ziemienowicz et al., 1999; Ziemienowicz et al., 2001).

Nuclear targeting of many karyophilic proteins is mediated by the importin α/β (karyopherin) pathway. Importin α interacts with many "classical" NLS sequences, such as those found in VirD2 and VirE2, and serves as an "adaptor" molecule for further interaction with importin β. The importin α/β-cargo protein is shuttled into the nucleus, an event which is coupled with GTP hydrolysis in the cytoplasm by the small GTPase Ran

(Macara, 2001; Merkle, 2004). Using a yeast two-hybrid system, Ballas and Citovsky (1997) first showed that VirD2 could interact with the *Arabidopsis* importin α protein AtKapα (now known as AtImpa-1). They also showed that AtKapα promoted nuclear import of fluorescently labeled

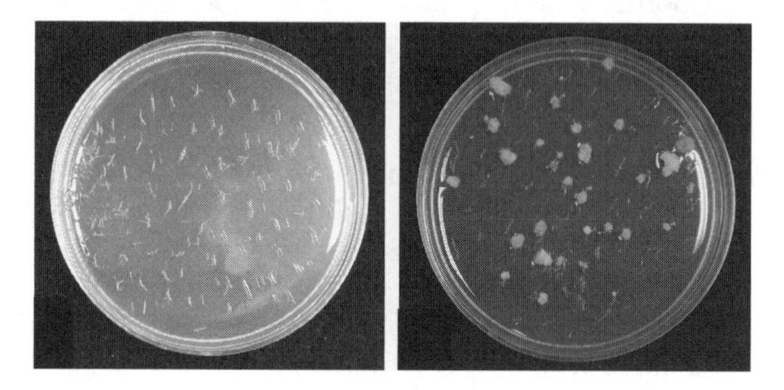

Figure 13-2. Over-expression of the AtAGP17 gene increases the re-transformation frequency of *Arabidopsis*. Axenic root segments from wild-type plants (ecotype Bl-1) (left) or Bl-1 transgenic plants harboring additional copies of the AtAGP17 gene (right) were inoculated with the tumorigenic strain *A. tumefaciens* A208. The segments were incubated on MS medium lacking phytohormones and photographed after one month.

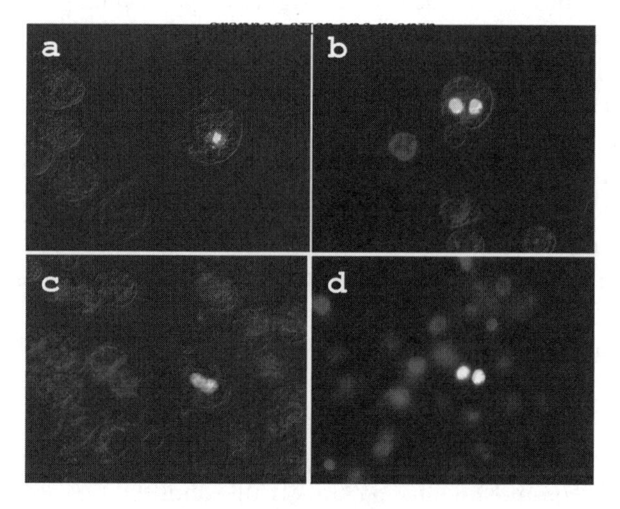

Figure 13-3. Bimolecular fluorescence complementation reveals that VirD2 interacts with several different importin alpha proteins in plant cells. Tobacco BY-2 protoplasts were co-electroporated with constructions containing nYFP-VirD2 and one of four importin alpha-cYFP proteins: AtKAPα (a), AtImpa-4 (b), AtImpa-7 (c) or AtImpa-9 (d). After 24 hours, the cells were visualized using an epifluorescence microscope. For interaction with At-KAPα, AtImpa-4, and AtImpa-7, overlay images are shown with the YFP signal (yellow) in the nucleus and the bright-field image of the cells pseudo-colored in blue. For the interaction with AtImpa-9, an overlay image is shown with the YFP signal (yellow) in the nucleus and with nuclei stained blue with Hoechst 33242.

VirD2 in permeabilized yeast cells. Báko et al. (2003) also noted that VirD2 could interact in yeast with several *Arabidopsis* importin α proteins, including AtKapα.

In *Arabidopsis*, the importin α family is made up of nine closely related proteins (Merkle, 2004; S. Bhattacharjee and S.B. Gelvin, unpublished). We have cloned cDNAs for most of these proteins, and have shown that both in yeast and *in vitro*, VirD2 interacts with most of them (S. Bhattacharjee and S.B. Gelvin, unpublished). Using bimolecular fluorescence complementation (BiFC), we have also shown that VirD2 interacts with several members of the importin α family in plants, and that interaction occurs in the nucleus (L.-Y. Lee and S.B. Gelvin, unpublished and *Figure 13-3*). Genetic analysis indicated that AtImpa-4, not AtImpa-1 (AtKapα) is essential for *Agrobacterium*-mediated transformation: T-DNA insertions in all *Arabidopsis* importin α genes tested, except for *AtImpa-4*, had no effect on transformation. However, T-DNA disruption of *AtImpa*-4 resulted in a rat phenotype (S. Bhattacharjee and S.B. Gelvin, unpublished and *Figure 13-4*). The rat phenotype of the Impa-4 mutant could be complemented by an importin α cDNA, indicating that disruption of the *AtImpa-4* gene was responsible for the rat phenotype (S. Bhattacharjee, H. Oltmanns and S.B. Gelvin, unpublished). Interestingly, expression of a VirD2-YFP fusion protein in *AtImpa-4* mutant plants still resulted in nuclear localization of VirD2, indicating that the rat phenotype of these mutant plants did not result from mis-targeting of VirD2 protein (S. Bhattacharjee and S.B. Gelvin, unpublished).

The role of VirE2 in nuclear targeting of the T-complex remains controversial. Gelvin (1998) showed that in the absence of all known NLS sequences in VirD2, VirE2-expressing transgenic plants supported transformation. These results suggested that other nuclear targeting sequences (such as those in VirE2) may compensate for the lack of nuclear targeting by VirD2. Zupan et al. (1996) showed that VirE2 by itself could mediate nuclear targeting of labeled single-stranded DNA introduced into plant cells. However, Ziemienowicz et al. (2001) showed that both VirD2 and VirE2 were required for nuclear import of “long” T-DNA molecules assembled *in vitro* and introduced into permeabilized tobacco cells. Thus, the relative roles of VirD2 and VirE2 in T-DNA nuclear targeting remain unknown.

Perhaps the most currently controversial aspect of T-DNA nuclear targeting involves the trafficking of VirE2 in plant cells. Citovsky et al. (1992) showed that a GUS-VirE2 fusion protein expressed in plant cells localized predominantly to the nucleus, although careful inspection of the data indicate some cytoplasmic GUS activity. Ziemienowicz et al. (2001) also showed that fluorescently tagged VirE2 protein localized to the nucleus of permeabilized tobacco cells. However, Dumas et al. (2001) and Duckely et al. (2005) demonstrated that VirE2 could form voltage-gated channels which would allow the passage of single-stranded DNA molecules through artificial membranes. Duckely et al. (2003) further showed cytoplasmic, not nuclear localization of VirE2 in plant cells. These authors have suggested that VirE2 may be exported from *Agrobacterium* to the plant where it remains in the plasma membrane, "waiting" to interact with the incoming T-strand.

We have recently shown that a VirE2-YFP fusion protein localizes completely to the cytoplasm of tobacco and *Arabidopsis* cells (S. Bhattacharjee and S.B. Gelvin, unpublished), and that VirE2 interacts with itself in the cytoplasm, not the nucleus, of plant cells (L.-Y. Lee and S.B. Gelvin, unpublished and *Figure 13-5*). The YFP (and half-YFP) fusions that we used were to the C-terminus of VirE2, and "extra-cellular complementation" experiments (Otten et al., 1984) indicated that the tagged VirE2 proteins were functionally active to effect transformation. However, N-terminal fusions to VirE2, as were used by Citovsky et al. (1992), were not functionally active (S. Bhattacharjee and S.B. Gelvin, unpublished). Cytoplasmic (or plasma membrane) localization of VirE2 makes sense in that rapid nuclear import of VirE2 may preclude its interaction with the T-strand. For all these experiments, it should be noted that over-expression of VirE2 in transgenic plant cells may cause self-aggregation and mis-localization. The extra-cellular complementation experiments of Bhattacharjee indicate, however, that at least some VirE2 molecules remain functional and can participate in processes resulting in transformation.

What plant proteins, then, are involved in the ultimate nuclear localization of VirE2? Initial experiments by Ballas and Citovsky (1997) indicated that VirE2 did not interact with AtKapα in yeast. This group therefore conducted a yeast two-hybrid screen for *Arabidopsis* proteins which would interact with VirE2, and identified two proteins, VIP1 and VIP2 (Tzfira et al., 2001). VIP1, a basic leucine zipper (bZIP) protein which contains a "classical" NLS, could mediate VirE2 nuclear import in yeast and mammalian cells (Tzfira et al., 2001), and at least in yeast this import was dependent on the importin α/β pathway (Tzfira et al., 2002). Involvement of VIP1 in the transformation process was further indicated by the observation that tobacco plants expressing a VIP1 anti-sense gene showed reduced transformation, and that over-expression of VIP1 increased the transformation efficiency of plant cells (Tzfira et al., 2001, 2002). The authors speculated that VIP1, which by itself could not interact with single-stranded DNA, could serve as an "adaptor" molecule by binding both to VirE2 and to importin α, thus mediating nuclear import of the T-complex (Tzfira et al., 2001, 2002).

Recent experiments in our laboratory, however, indicate that VirE2 can interact with several *Arabidopsis* importin α isoforms *in vitro*, in yeast, and in plants (S. Bhattacharjee, L.-Y. Lee and S.B. Gelvin, unpublished). Somewhat surprisingly, interaction of VirE2 with importin α occurs in the plant cytoplasm and, unlike the situation with VirD2, does not generally result in nuclear translocation of the interacting proteins (some nuclear localization of VirE2 may occur when AtImpa-4 is used as the interacting partner; however, nuclear localization of VirE2 was never observed when other isoforms of importin α were used in BiFC; *Figure 13-5*). We hypothesize that when VirE2 enters the plant cell, it remains in the cytoplasm until a complex is formed with importin α and the incoming T-strand. We have shown that *in vitro*, such complexes can be formed (S. Bhattacharjee and S.B. Gelvin, unpublished). We are currently conducting experiments to test this hypothesis and the role that VIP1 may play in this process.

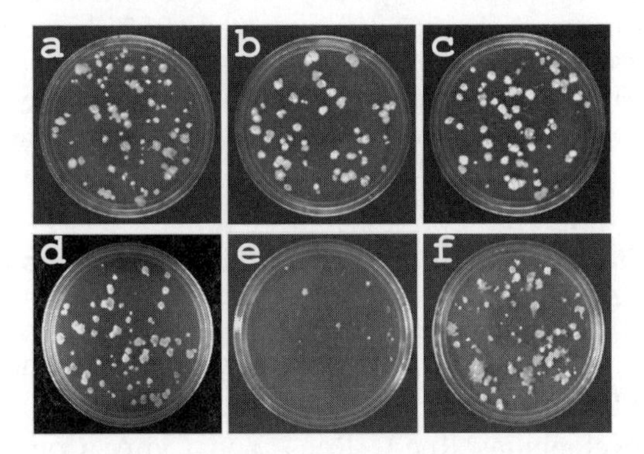

Figure 13-4. Mutation of the *AtImpa-4* gene, but not other importin *a* genes, results in a *rat* phenotype. Axenic root segments of wild-type (a) plants and T-DNA insertions into *At-Kapα* (b), *AtImpa-2* (c), *AtImpa-3* (d) and *AtImpa-4* (e) were inoculated with the tumorigenic strain *A. tumefaciens* A208. Note that disruption of only *AtImpa-4* resulted in a rat phenotype. The right panel (f) shows that complementation of the *AtImpa-4* mutant with an *AtImpa-4* cDNA, under the control of a CaMV 35S promoter, restores transformation susceptibility, indicating that the rat phenotype of the *AtImpa-4* mutant results from disruption of the *AtImpa-4* gene.

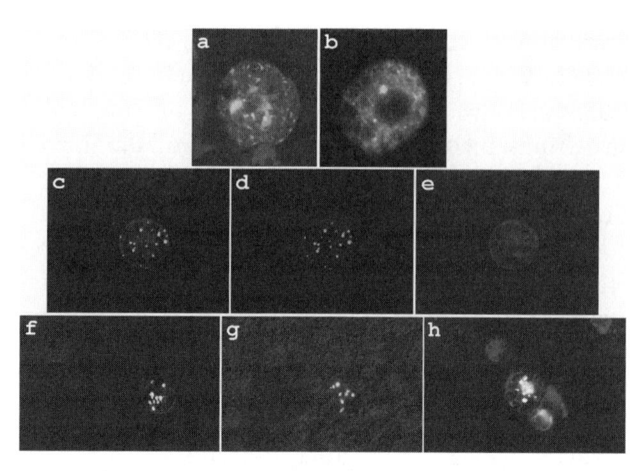

Figure 13-5. Localization and interactions of VirE2. (a, b) VirE2 localizes to the cytoplasm, not the nucleus, of tobacco cells. (c-h) VirE2 interacts with several different importin α proteins in plant cells using the bimolecular fluorescence complementation assay. Tobacco BY-2 protoplasts were electroporated with a construct expressing VirE2-YFP (a), VirE2-cYFP and VirE2-nYFP (b) VirE2-nYFP and AtImpα-4-cYFP (c-e), or VirE2-nYFP with AtKAPα-cYFP (f), AtImpα-7-cYFP (g) or AtImpα-9-cYFP (h). After 24 hours, the cells were visualized using an epifluorescence microscope. Panel a shows an overlay image of YFP fluorescence (yellow) with nuclei stained blue with Hoechst 33242 and panel b shows an overlay image of YFP fluorescence (yellow) with a bright-field image pseudo-colored in blue. The center set of panels show the YFP image (d) the RFP image of the entire cell (e), and the overlay images (c). For interaction with AtKAPα (f), AtImpα-7 (g), and AtImpα-9(h), overlay images are shown with the YFP signal (yellow) in the cytoplasm and the bright-field image of the cells pseudo-colored in blue.

3.2.1 Interaction of the T-complex with other proteins in the plant cytoplasm

Using yeast two-hybrid screens, several groups have shown that VirD2 interacts with additional plant proteins. Deng et al. (1998) showed interaction with the cyclophilins RocA, Roc4, and CypA. Consistent with this observation, Cyclosporin A, a cyclophilin inhibitor, reduced plant transformation. Although the precise role of cyclophilins in *Agrobacterium*-mediated transformation remains unknown, these authors suggested that cyclophilins may be involved in maintaining an active conformation of VirD2 during the processes of T-complex nuclear localization or T-DNA integration. Interaction of VirD2 with cyclophilins was also noted by Báko et al. (2003).

Tao et al. (2004) demonstrated interaction of VirD2 with a tomato Type 2C protein phosphatase (PP2C) in yeast. Expression of a GUS-VirD2-NLS protein in tobacco protoplasts resulted in nuclear localization of the fusion protein. However, co-expression of the tomato PP2C shifted the localization of GUS activity predominantly to the cytoplasm. A C-terminal fragment of VirD2 is phosphorylated in plants, most likely at a serine residue two amino acids preceding the bipartite NLS (Tao, 1998). Phosphorylation of serine residues near NLS sequences of proteins frequently alters their intracellular localization (Vandromme et al., 1996; Shibasaki et al., 1996, 1997), and mutation of the VirD2 serine residue preceding the NSL altered nuclear localization of a GUS-VirD2 NLS fusion protein (Tao et al., 2004). Furthermore, the *Arabidopsis abi1* mutant, a PP2C mutant, shows increased transformation-susceptibility to *Agrobacterium* (*hat* phenotype; Tao, 1998). These results suggest that a plant protein phosphatase may serve as a negative regulator of transformation, perhaps by altering the phosphorylation status of VirD2 and its subsequent ability to target the nucleus.

Báko et al. (2003) demonstrated interaction of VirD2 with the protein kinase CAK2M, a cyclin-dependent kinase-activating kinase, as well as a TATA-box binding factor. However, this kinase is found in the nucleus, and the authors suggested that it may be involved in targeting the T-complex to transcriptionally active regions of chromatin for integration, rather than targeting of the T-complex to the nucleus.

3.2.2 Does the T-complex utilize the plant cytoskeleton for intracellular trafficking?

Many pathogens utilize the cytoskeleton for trafficking and/or assembly of, e.g., viral components (reviewed in Gouin et al., 2005). In addition, importin α proteins, known to be involved in *Agrobacterium*-mediated transformation (Ballas and Citovsky, 1997; S. Bhattacharjee and S.B. Gelvin, unpublished), often associate with the cytoskeleton (Smith and Raikhel, 1998). We therefore investigated the possible involvement of the plant cytoskeleton in cytoplasmic trafficking of T-complex components (Rao, 2002).

Recombinant VirD2, and possibly VirE2, proteins interact *in vitro* with pre-polymerized f-actin microfilaments, but not with pre-polymerized microtubules. The *Arabidopsis* mutants *act2* and *act7*, which do not make root-expressed forms of actin, are resistant to *Agrobacterium*-mediated root transformation, whereas mutation of a pollen-expressed actin gene (*act12* mutant) does not result in alteration of root transformation efficiency. Finally, incubation of tobacco BY-2 cells with pharmacological agents that inhibit actin microfilament structure or function reversibly reduce *Agrobacterium*-mediated transformation frequency; inhibitors of microtubule structure do not (P. Rao, M. Duckely, B. Hohn and S.B. Gelvin, unpublished). Taken together, these results suggest a role for the actin cytoskeleton in *Agrobacterium*-mediated transformation.

Recently, Salman et al. (2005) followed the movement of VirE2-ssDNA complexes in extracts from *Xenopus laevis* oocytes. These authors demonstrated directed, rather than random, movement of these molecules if VirE2 had been "animalized" by changes in amino acid sequence that would allow nuclear targeting in animal, rather than in plant cells (Guralnick et al., 1996). This "directed" movement could be disrupted either by using the "plant" (wild-type) form of VirE2, or by inhibitors, such as nocodazole, directed against microtubules. In addition, inhibitor studies indicated that movement was dependent on dynein but not kinesin. Salman et al. suggested that the T-complex may track the microtubule cytoskeleton to reach the nucleus. However, these experiments need to be performed in plants. It is interesting to note that nuclear targeting of VirE2-ssDNA complexes in these experiments occurred in animal systems in the absence of VIP1 protein, calling into question the importance of VIP1 in T-DNA trafficking, at least in animal cells. During the natural process of infection of cells by *Agrobacterium*, VirE3 protein may fulfill the role of VIP1 in those plant species lacking VIP1 activity (Lacroix et al., 2005).

3.3 "Uncoating" the T-strand in the nucleus

Once in the nucleus, the T-strand must presumably have proteins such as VirE2 stripped off either prior to or during the course of T-DNA integration into the plant genome. Recent experiments implicate the protein VirF in this process. VirF is a "host range factor" implicated in transformation of some plant species, but dispensable for infection of others (Melchers et al., 1990; Regensburg-Tuink and Hooykaas, 1993). In yeast, VirF interacts with Skp1 protein, a component of the SCF complex involved in targeted proteolysis of other proteins via the ubiquitin and 26S proteosome pathway (Schrammeijer et al., 2001). VirF contains a "F box", a peptide motif involved in "selecting" particular proteins for ubiquitination and, subsequently, proteolysis. VirF localizes to plant nuclei and, in yeast, interacts with VIP1. Tzfira et al. (2004b) showed that in yeast, VirF could mediate proteolysis of VirE2 and VIP1, and that this process could be prevented in the presence of an inhibitor directed against 26S proteosome function. The authors suggested that in most host species, a plant F box protein is responsible for targeting VirE2 and VIP1 for degradation, whereas in those plants lacking a F box protein with this particular specificity, VirF takes on this function.

3.4 Proteins involved in T-DNA integration

The mechanism of T-DNA integration into the plant genome remains obscure. Several models have been presented, and are reviewed in Tzfira et al. (2004a) and other chapters in this volume. One proposed model is the single-strand T-DNA "strand invasion" or single-strand "gap repair" model. According to this paradigm, the invading T-strand "opens up" the host DNA at sites of micro-homology and ligates to a nick site in one host DNA strand. A nick is made on the other host DNA strand, and repair replication generates the complementary T-DNA strand. Indeed, micro-homologies of T-DNA target sites and regions within the T-strand have been noted in many, but not all, instances (Tinland and Hohn, 1995; Tinland, 1996). This mechanism nicely explains integration of single copies of T-DNA, but does not adequately account for integration of multiple linked copies of T-DNA, especially when they are in inverted repeat (head-to-head or tail-to-tail) configuration.

A second model postulates that T-DNA becomes double-stranded prior to integration into double-strand breaks in the host DNA (the "double-strand break repair" model). Double-strand T-DNA molecules clearly exist in the plant nucleus prior to integration and can be transcribed (Narasimhulu et al., 1996; Mysore et al., 1998). In addition, creation of double strand breaks in plant DNA using rare-cutting "homing endonucleases" results in increased frequency of site-specific T-DNA integration (Salomon and Puchta, 1998; Chilton and Que, 2003; Tzfira et al., 2003). The double-strand break repair model can easily account for integration of multiple T-DNA molecules at the same site (De Block and Debrouwer, 1991; De Neve et al., 1997; De Buck et al., 1999). However, it does not fully explain "filler" DNA sequences frequently found between integrated T-DNA sequences (Gheysen et al., 1991; Mayerhofer et al., 1991) in some plant species such as tobacco and maize but less frequently found in *Arabidopsis* (Kirik et al., 2000; Windels et al., 2003).

However, both of these models have been derived from characterization of plant DNA/T-DNA junction sequences and attempts to reconstruct what must have happened to generate these junctions. One of the difficulties in understanding T-DNA integration is the current lack of an *in vitro* T-DNA system which can be used to follow the actual process of integration. An alternative approach has been to use genetic screens to identify host proteins involved in T-DNA integration, and cell biology/biochemical approaches to identify host proteins which interact with T-complex proteins.

In the course of screening *Arabidopsis* T-DNA insertion mutants that are resistant to *Agrobacterium* transformation, we have noted several *rat* mutants which remain susceptible to transient transformation but resistant to stable transformation. Because transient expression of transgenes requires nuclear translocation and conversion of the T-strand to a double-stranded transcription-competent form, these mutants are potential T-DNA integration mutants (Nam et al., 1999; Zhu et al., 2003b). Alternatively, these mutants could integrate T-DNA but not express the encoded transgenes (silencing mutants).

In order to distinguish between these latter possibilities, we have developed a biochemical assay that directly measures T-DNA integration into high molecular weight plant DNA without recourse to T-DNA expression analysis (Mysore et al., 2000b). Wild-type and mutant *Arabidopsis* root segments are infected with a non-tumorigenic (disarmed) *Agrobacterium* strain harboring a T-DNA binary vector containing a *gusA*-intron transgene. Two days after inoculation, the root segments are moved to solidified callus inducing medium (CIM) containing antibiotics to kill the bacteria. Calli are permitted to grow for several weeks in the absence of selection for any T-DNA-encoded gene. Calli are pooled and moved to liquid CIM where they continue to grow in the absence of selection. After

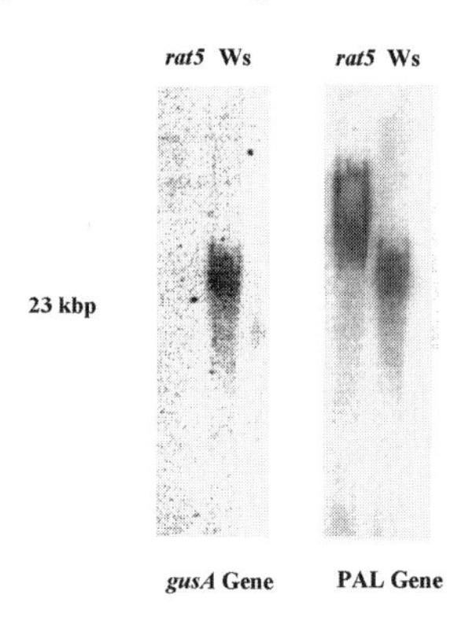

Figure 13-6. Biochemical assay for T-DNA integration. Axenic root segments of a wild-type plant (right lane of each panel, ecotype Ws) or a *rat* mutant plant (left lane of each panel) were inoculated with *A. tumefaciens* AtA49 (containing a *gusA*-intron gene within the T-DNA of a the binary vector pBISN1). After two days, the root segments were transferred to CIM medium containing timentin (to kill the bacteria) and phytohormones and incubated for ~2 months. High molecular weight DNA was extracted and subjected to electrophoresis through 0.7% agarose gels. The DNA was blotted onto nitrocellulose and hybridized with a *gusA* gene probe (left panel) or a phenylalanine ammonia lyase (*PAL*) gene probe (control, right panel). Note that little T-DNA integrated into the DNA of the *rat* mutant plant.

several months (during which several different antibiotics are rotated through the medium to make sure that all bacteria have been killed), high molecular weight DNA is isolated form the calli and resolved by electrophoresis (without restriction endonuclease digestion) through agarose gels. The DNA is blotted onto membranes and hybridized with a *gusA* probe. Control hybridizations are conducted to assure that bacterial DNA does not contaminate the plant DNA. Comparison of the hybridization signal of the *gusA* probe between wild-type and mutant DNA indicates the extent to which the mutant *Arabidopsis* line is able to integrate T-DNA (*Figure 13-6*).

Using this assay, Mysore et al. (2000b) demonstrated biochemically that the *rat5* mutant was a T-DNA integration-deficient mutant. *Rat5* encodes one member of a 13-member histone H2A (*HTA*) gene family, *HTA1*. Complementation of *rat5* with either a genomic (Mysore et al., 2000b) or cDNA (Yi et al., 2006) clone restored transformation competence. HTA1 is a likely a replacement histone. The gene is expressed in many different cell types at a low level, including cells of the root elongation zone which are not undergoing mitotic cell cycling (Yi et al., 2002; Yi et al., 2006). *HTA1* is the only tested histone H2A gene which is induced by *Agrobacterium* inoculation (Yi et al., 2006).

With *HTA1* expression so low in various cells, why can't the other *HTA* genes (the transcripts of some of which accumulate up to 1000-fold more abundantly than do *HTA1* transcripts) compensate for loss of expression of this particular histone? Are the other histone H2A proteins not functionally redundant with HTA1? To investigate this, Yi et al. (2006) individually expressed cDNAs of other *HTA* genes under constitutive promoter control in the *rat5* mutant. All tested *HTA* cDNAs could phenotypically complement the rat phenotype. Yi et al. (2006) also introduced genomic clones of *HTA* genes into the *rat5* mutant. In this case, however, only the *HTA1* gene could complement the rat phenotype; other *HTA* genes could not. These experiments indicate that all histone H2A proteins are functionally redundant with respect to *Agrobacterium*-mediated transformation. However, in order to show this redundancy, they needed to be "mis-expressed" at high levels in all cell types. The promoters of *HTA* genes other than *HTA1* would not permit such high level expression in root cell types known to correlate with cells most susceptible to *Agrobacterium*-mediated transformation (Yi et al., 2002; Yi et al., 2006).

Is *HTA1* the only histone gene required for transformation? We conducted both forward and reverse genetic screens of *Arabidopsis* T-DNA insertion libraries. There are 46 "core" histone genes (histone H2A, H2B, H3, and H4) in *Arabidopsis*. Using the available databases, we were able to identify very few T-DNA insertions into histone gene exons (indeed, the *rat5* mutant contains a T-DNA insertion into the 3' untranslated region of the *HTA1* gene; Mysore et al., 2000b). Screening of many T-DNA insertion mutants revealed only one other mutant which had a moderately strong rat phenotype. This mutant contained a T-DNA insertion between the two closely-positioned histone H3 genes *HTR5* and *HTR4* (Zhu et al., 2003b; Y. Zhu and S.B. Gelvin, unpublished). Complementation of this mutant with a genomic clone containing both *HTR5* and *HTR4*, or *HTR5* alone could complement the rat phenotype. However, a genomic clone containing only the *HTR4* gene could not (Y. Zhu and S.B. Gelvin, unpublished). *HTR5* and *HTR4* encode histone H3 proteins with the same amino acid sequence; therefore, complementation of the rat phenotype of this mutant with cDNAs of either of these two genes was successful (Y. Zhu and S.B. Gelvin, unpublished).

With at least two histone genes playing a role in transformation, is there a direct link between histones and T-complex proteins? Li et al. (2005a) and Loyter et al. (2005) noted that histones could interact with VIP1 *in vitro* and in plant cells. These authors suggested that VIP1 may target T-DNA to chromatin via interaction with histones. The almost universal presence of histones throughout the genome could explain both targeting of T-DNA to chromatin and the random integration of T-DNA into the genome.

3.4.1 Role of "recombination" proteins in T-DNA integration

Because T-DNA integration does not require extensive target site homology, this process has been proposed to occur by "illegitimate" recombination, or "non-homologous end-joining (NHEJ). In yeast, many proteins involved in NHEJ are known. van Attikum et al. (2001) demonstrated that the proteins Yku70, Rad50, Mre11, Xrs2, Lig4, and Sir4 are required for T-DNA integration in the absence of homology between T-DNA and yeast genome target sites. In the absence of Ku70, T-DNA integration could only occur via homologous recombination, a process also requiring the yeast proteins Rad51 and Rad52 (van Attikum and Hooykaas, 2003).

In plants, however, the mechanism of T-DNA via the "classical" NHEJ pathway is not clear. Integration in *Arabidopsis* roots does not require DNA ligase IV, a protein required for NHEJ in other organisms (van Attikum et al., 2003). Because there is only one annotated DNA ligase IV gene in *Arabidopsis*, either one of the other DNA ligases is involved in T-DNA integration, or integration by NHEJ may not use enzymes required in other systems. Friesner and Britt (2003), however, reported that an *Arabidopsis* mutant in DNA ligase IV had a slightly lower transformation efficiency using a flower dip protocol.

The role of Ku80 in *Arabidopsis* transformation also is unclear. Gallego et al. (2003) reported that this protein was dispensable for flower dip transformation, whereas Friesner and Britt (2003) indicated that it was required for efficient transformation by this method. Recently, Li et al. (2005b) showed that Ku80 was required for root transformation, and that over-expression of AtKu80 could enhance the transformation frequency of plant cells. A role for VIP1 in T-DNA integration has also been proposed: Li et al. (2005a) showed that an *Arabidopsis* mutant encoding a truncated VIP1 protein that was capable of nuclear import of VirE2 was deficient in T-DNA integration.

3.4.2 Role of chromatin proteins in Agrobacterium-mediated transformation

In addition to histones, other chromatin proteins may be involved in transformation, either at the steps of T-DNA integration or transgene expression. For example, a T-DNA insertion into the *Arabidopsis HDA19* gene (formerly known as *HDA1*) resulted in a rat phenotype (Tian et al., 2003; Zhu et al., 2003b).

Recently, our laboratory conducted an extensive screen of *Arabidopsis* lines carrying RNAi constructs individually targeting more than 100 different chromatin genes. At least three independent lines for each targeted gene were examined, and both stable and transient root transformation assays were conducted (Y.M. Crane and S.B. Gelvin, unpublished). A few RNAi constructions, such as those targeting the genes *NFC1* and *SDG15*, resulted in abnormal plant development, and therefore were not considered further for investigation. However, two RNAi constructions resulted in normal plants that displayed a strong rat phenotype. These constructions were targeted against the histone deacetylase gene *HDT2* and the gene *SGA1*. Analysis of DNA from non-selected calli derived from these plants after infection by *Agrobacterium* indicated that T-DNA had not efficiently integrated into high molecular weight plant DNA. For these experiments, it was important to conduct direct biochemical assays for T-DNA integration as described above because many of these chromatin genes are known to be involved in gene silencing. RNAi constructions directed against DNA methylation genes, general transcription factors, nucleosomal assembly factors, histone acetyltransferases, and other histone deacetylases also demonstrated a more moderate or weak rat phenotype (Y.M. Crane and S.B. Gelvin, unpublished).

3.4.3 Over-expression of some "rat" genes may alter transgene expression

The histone H2A gene *HTA1* encoded by the *Rat5* locus is involved *Agrobacterium*-mediated transformation: mutation of this gene results in a deficiency in T-DNA integration (Mysore et al., 2000b). Because over-expression of *HTA1* increases the transformation frequency of plants, it would be logical to conclude that the mechanism by which this increase occurs is by increasing T-DNA integration. However, several experiments indicated that this mechanism may not explain the higher transformation frequency of *HTA1* over-expressing plants. *HTA1* over-expression increased the transient transformation frequency of plants using "wild-type" (but disarmed) *Agrobacterium* strains, a process not requiring T-DNA integration. In addition, transient transformation by an *Agrobacterium* strain that contained a mutant VirD2 protein that could not effect T-DNA integration (Mysore et al., 1998) was also increased by over-expression of *HTA1* (S. Johnson and S.B. Gelvin, unpublished). We therefore examined whether over-expression of *HTA1* could enhance transgene expression following gene introduction by methods other than *Agrobacterium*-mediated

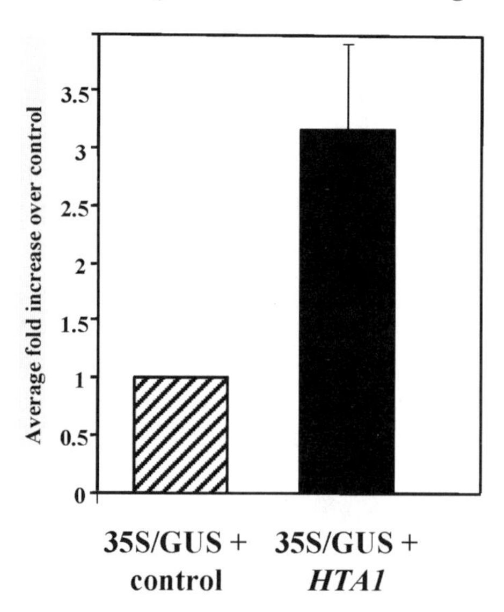

Figure 13-7. Expression of a histone H2A-1 (*HTA1*) cDNA increases transient GUS expression of a *gusA* gene in plant cells. Tobacco BY-2 protoplasts were co-electroporated with constructions expressing a gusA gene (under the control of a CaMV 35S promoter) and either a *HTA1* cDNA (under the control of a CaMV 35S promoter; right) or an "empty vector" (left). After 48 h, the percentage of cells staining blue with X-gluc was determined. The figure shows the fold-increase of the experimental cells over the control cells.

transformation. We co-electroporated tobacco BY-2 cells with a plant-active *gusA* gene and either a *HTA1*-expressing clone or an "empty vector" construction. When co-electroporated with the *HTA1* gene, we consistently saw a 3- to 4-fold increase in the percentage of cells expressing GUS activity (*Figure 13-7*). Because in these experiments only transient expression was measured, and because the transgenes were introduced by a method not involving *Agrobacterium*, we conclude that *HTA1* over-expression increases transgene expression. Thus, *HTA1* over-expression may allow one to recover more transgenic events because of increased selectable marker and/or reporter gene expression.

Is *HTA1* the only *Arabidopsis* histone gene that, when over-expressed, increases *Agrobacterium*-mediated transformation? The *Arabidopsis* genome contains 46 core histone genes that encode 33 different histone proteins. We over-expressed 20 different representative histone cDNAs in *Arabidopsis* and tested the roots of these transgenic lines for transformation-competence using dilute inocula of *Agrobacterium* (J. Spantzel, Y. Zhu, S. Bhullar and S.B. Gelvin, unpublished). We examined a minimum of 50 independent lines for each gene construction, and quantified the number of lines which enhanced transformation by at least two-fold. All tested histone H2A (*HTA*) cDNAs increased transformation, as did the histone H4 (*HFO*) cDNA (all *Arabidopsis HFO* genes encode the same protein). None of the tested histone H2B (*HTB*) or H3 (*HTR*) genes had a substantial effect on transformation frequency. Thus, over-expression of multiple members of particular classes of histone genes could increase the frequency of *Agrobacterium*-mediated transformation.

4 CONCLUSIONS

The process of T-DNA and Virulence protein uptake into the host, its trafficking through the host cytoplasm into the nucleus, and the eventual integration of T-DNA into the host genome is a complex process utilizing numerous host proteins and sub-organellar structures. Our efforts to date indicate that there are likely several hundred *Arabidopsis* genes which encode proteins that, directly or indirectly, affect the transformation process. These proteins are, obviously, not found in the *Arabidopsis* genome in case the cells were infected by *Agrobacterium*. Rather, this extraordinary bacterium has learned how to "pirate" the host's normal cell biology machinery for its own advantage. An understanding of T-DNA and Virulence protein trafficking through the plant cell will not only indicate how *Agrobacterium* manipulates its host. It will also help explain how plants conduct their normal biological processes.

5 ACKNOWLEDGMENTS

Work in this author's laboratory is supported by grants from the National Science Foundation, the US Department of Agriculture, the US Department of Energy, the Biotechnology Research and Development Corporation, and the Corporation for Plant Biotechnology Research.

6 REFERENCES

Abuodeh RO, Orbach MJ, Mandel MA, Das A, Galgiani JN (2000) Genetic transformation of *Coccidioides immitis* facilitated by *Agrobacterium tumefaciens*. J Infect Dis 181: 2106-2110

Akama K, Shiraishi H, Ohta S, Nakamura K, Okada K, Shimura Y (1992) Efficient transformation of *Arabidopsis thaliana*: Comparison of the efficiencies with various organs, plant ecotypes and *Agrobacterium* strains. Plant Cell Rep 12: 7-11

An G, Lee S, Kim S-H, Kim S-R (2005) Molecular genetics using T-DNA in rice. Plant Cell Physiol 46: 14-22

An S, Park S, Jeong DH, Lee DY, Kang HG, Yu JH, Hur J, Kim SR, Kim YH, Lee M, Han S, Kim SJ, Yang J, Kim E, Wi SJ, Chung HS, Hong JP, Choe V, Lee HK, Choi JH, Nam J, Park PB, Park KY, Kim WT, Choe S, Lee CB, An G (2003) Generation and analysis of end sequence database for T-DNA tagging lines in rice. Plant Physiol 133: 2040-2047

Anderson AR, Moore LW (1979) Host specificity in the genus *Agrobacterium*. Phytopathology 69: 320-323

Bailey MA, Boerma HR, Parrott WA (1994) Inheritance of *Agrobacterium tumefaciens*-induced tumorigenesis of soybean. Crop Sci 34: 514-519

Bakó L, Umeda M, Tiburcio AF, Schell J, Koncz C (2003) The VirD2 pilot protein of *Agrobacterium*-transferred DNA interacts with the TATA box-binding protein and a nuclear protein kinase in plants. Proc Natl Acad Sci USA 100: 10108-10113

Ballas N, Citovsky V (1997) Nuclear localization signal binding protein from *Arabidopsis* mediates nuclear import of *Agrobacterium* VirD2 protein. Proc Natl Acad Sci USA 94: 10723-10728

Baron C, Zambryski PC (1996) Plant transformation: a pilus in *Agrobacterium* T-DNA transfer. Curr Biol 6: 1567-1569

Bechtold N, Ellis J, Pelletier G (1993) *In planta Agrobacterium* mediated gene transfer by infiltration of adult *Arabidopsis thaliana* plants. C R Acad Sci 316: 1194-1199

Bechtold N, Jaudeau B, Jolivet S, Maba B, Vezon D, Voisin R, Pelletier G (2000) The maternal chromosome set is the target of the T-DNA in the *in planta* transformation of *Arabidopsis thaliana*. Genetics 155: 1875-1887

Bechtold N, Jolivet S, Voisin R, Pelletier G (2003) The endosperm and the embryo of *Arabidopsis thaliana* are independently transformed through infiltration by *Agrobacterium tumefaciens*. Transgenic Res 12: 509-517

Bent AF (2000) *Arabidopsis in planta* transformation. Uses, mechanisms, and prospects for transformation of other species. Plant Physiol 124: 1540-1547

Binns AN, Beaupre CE, Dale EM (1995) Inhibition of VirB-mediated transfer of diverse substrates from *Agrobacterium tumefaciens* by the IncQ plasmid RSF1010. J Bacteriol 177: 4890-4899

Bliss FA, Almehdi AA, Dandekar AM, Schuerman PL, Bellaloui N (1999) Crown gall resistance in accessions of 20 *Prunus* species. Hort Sci 34: 326-330

Brencic A, Angert ER, Winans SC (2005) Unwounded plants elicit *Agrobacterium vir* gene induction and T-DNA transfer: transformed plant cells produce opines yet are tumour free. Mol Microbiol 57:1522-1531

Brencic A, Winans SC (2005) Detection of and response to signals involved in host-microbe interactions by plant-associated bacteria. Microbiol Mol Biol Rev 69: 155-194

Bundock P, den Dulk-Ras A, Beijersbergen A, Hooykaas PJJ (1995) Trans-kingdom T-DNA transfer from *Agrobacterium tumefaciens* to *Saccharomyces cerevisiae*. EMBO J 14: 3206-3214

Bundock P, Hooykaas PJJ (1996) Integration of *Agrobacterium tumefaciens* T-DNA in the Saccharomyces cerevisiae genome by illegitimate recombination. Proc Natl Acad Sci USA 93: 15272-15275

Bundock P, van Attikum H, den Dulk-Ras A, Hooykaas PJJ (2002) Insertional mutagenesis in yeasts using T-DNA from *Agrobacterium tumefaciens*. Yeast 19: 529-536

Bytebier B, Deboeck F, De Greve H, Van Montagu M, Hernalsteens JP (1987) T-DNA organization in tumor cultures and transgenic plants of the monocotyledon *Asparagus officinalis*. Proc Natl Acad Sci USA 84: 5345-5349

Cangelosi GA, Hung L, Puvanesarajah V, Stacey G, Ozga DA, Leigh JA, Nester EW (1987) Common loci for *Agrobacterium tumefaciens* and *Rhizobium meliloti* exopolysaccharide synthesis and their roles in plant interactions. J Bacteriol 169: 2086-2091

Cangelosi GA, Martinetti G, Leigh JA, Lee CC, Theines C, Nester EW (1989) Role for [corrected] *Agrobacterium tumefaciens* ChvA protein in export of beta-1,2-glucan. J Bacteriol 171: 1609-1615

Cascales E, Christie PJ (2004) Definition of a bacterial type IV secretion pathway for a DNA substrate. Science 304: 1170-1173

Chan M-T, Lee T-M, Chang H-H (1992) Transformation of Indica rice (*Oryza sativa* L.) mediated by *Agrobacterium tumefaciens*. Plant Cell Physiol 33: 577

Chen S, Jin W, Wang M, Zhang F, Zhou J, Jia Q, Wu Y, Liu F, Wu P (2003) Distribution and characterization of over 1000 T-DNA tags in rice genome. Plant J 36: 105-113

Cheng M, Fry JE, Pang S, Zhou H, Hironaka CM, Duncan DR, Conner TW, Wan Y (1997) Genetic transformation of wheat mediated by *Agrobacterium tumefaciens*. Plant Physiol 115: 971-980

Chilton M-D, Que Q (2003) Targeted integration of T-DNA into the tobacco genome at double-stranded breaks: new insights on the mechanism of T-DNA integration. Plant Physiol 133: 956-965

Christie PJ, Vogel JP (2000) Bacterial type IV secretion: conjugation systems adapted to deliver effector molecules to host cells. Trends Microbiol 8: 354-360

Christie PJ, Ward JE, Winans SC, Nester EW (1988) The *Agrobacterium tumefaciens virE2* gene product is a single-stranded-DNA-binding protein that associates with T-DNA. J Bacteriol 170: 2659-2667

Citovsky V, De Vos G, Zambryski P (1988) Single-stranded DNA binding protein encoded by the *virE* locus of *Agrobacterium tumefaciens*. Science 240: 501-504

Citovsky V, Guralnick B, Simon MN, Wall JS (1997) The molecular structure of *Agrobacterium* VirE2-single stranded DNA complexes involved in nuclear import. J Mol Biol 271: 718-727

Citovsky V, Warnick D, Zambryski P (1994) Nuclear import of *Agrobacterium* VirD2 and VirE2 proteins in maize and tobacco. Proc Natl Acad Sci USA 91: 3210-3214

Citovsky V, Wong ML, Zambryski P (1989) Cooperative interaction of *Agrobacterium* VirE2 protein with single stranded DNA: implications for the T-DNA transfer process. Proc Natl Acad Sci USA 86: 1193-1197

Citovsky V, Zupan J, Warnick D, Zambryski P (1992) Nuclear localization of *Agrobacterium* VirE2 protein in plant cells. Science 256: 1802-1805

Clough SJ, Bent AF (1998) Floral dip: a simplified method for *Agrobacterium*-mediated transformation of *Arabidopsis thaliana*. Plant J 16: 735-743

Conner AJ, Dommisse EM (1992) Monocotyledonous plants as hosts for *Agrobacterium*. Int J Plant Sci 153: 550-555

Crews JL, Colby S, Matthysse AG (1990) *Agrobacterium rhizogenes* mutants that fail to bind to plant cells. J Bacteriol 172: 6182-6188

Das A (1988) *Agrobacterium tumefaciens virE* operon encodes a single-stranded DNA-binding protein. Proc Natl Acad Sci USA 85: 2909-2913

De Block M, Debrouwer D (1991) Two T-DNA's co-transformed into *Brassica napus* by a double *Agrobacterium tumefaciens* infection are mainly integrated at the same locus. Theor Appl Genet 82: 257-263

De Buck S, Jacobs A, Van Montagu M, Depicker A (1999) The DNA sequences of T-DNA junctions suggest that complex T-DNA loci are formed by a recombination process resembling T-DNA integration. Plant J 20: 295-304

De Cleene M, De Ley J (1976) The host range of crown gall. Bot Rev 42: 389-466

de Groot MJ, Bundock P, Hooykaas PJJ, Beijersbergen AG (1998) *Agrobacterium tumefaciens*-mediated transformation of filamentous fungi. Nat Biotechnol 16: 839-842

de Iannino NI, Ugalde RA (1989) Biochemical characterization of avirulent *Agrobacterium tumefaciens chvA* mutants: synthesis and excretion of β-(1-2)glucan. J Bacteriol 171: 2842-2849

De Neve M, De Buck S, Jacobs A, Van Montagu M, Depicker A (1997) T-DNA integration patterns in co-transformed plant cells suggest that T-DNA repeats originate from co-integration of separate T-DNAs. Plant J 11: 15-29

Deasey MC, Matthysse AG (1984) Interactions of wild-type and a cellulose-minus mutant of *Agrobacterium tumefaciens* with tobacco mesophyll and tobacco tissue culture cells. Phytopathol 74: 991-994

Deng W, Chen L, Wood DW, Metcalfe T, Liang X, Gordon MP, Comai L, Nester EW (1998) *Agrobacterium* VirD2 protein interacts with plant host cyclophilins. Proc Natl Acad Sci USA 95: 7040-7045

Desfeux C, Clough SJ, Bent AF (2000) Female reproductive tissues are the primary target of *Agrobacterium*-mediated transformation by the *Arabidopsis* floral-dip method. Plant Physiol 123: 895-904

Ditt RF, Nester E, Comai L (2005) The plant cell defense and *Agrobacterium tumefaciens*. FEMS Microbiol Lett 247: 207-213

Ditt RF, Nester EW, Comai L (2001) Plant gene expression response to *Agrobacterium tumefaciens*. Proc Natl Acad Sci USA: 10954-10959

Dommisse EM, Leung DWM, Shaw ML, Conner AJ (1990) Onion is a monocotyledonous host for *Agrobacterium*. Plant Sci 69: 249-257

Dong J, Teng W, Buchholz WG, Hall TC (1996) *Agrobacterium*-mediated transformation of Javanica rice. Mol Breed 2: 267-276

Douglas CJ, Halperin W, Nester EW (1982) *Agrobacterium tumefaciens* mutants affected in attachment to plant cell. J Bacteriol 152: 1265-1275

Douglas CJ, Staneloni RJ, Rubin RA, Nester EW (1985) Identification and genetic analysis of an *Agrobacterium tumefaciens* chromsomal virulence region. J Bacteriol 161: 850-860

Duckely M, Hohn B (2003) The VirE2 protein of *Agrobacterium tumefaciens*: the Yin and Yang of T-DNA transfer. FEMS Microbiol Lett 223: 1-6

Duckely M, Oomen C, Axthelm F, Van Gelder P, Waksman G, Engel A (2005) The VirE1VirE2 complex of *Agrobacterium tumefaciens* interacts with single-stranded DNA and forms channels. Mol Microbiol 58: 1130-1142

Dumas F, Duckely M, Pelczar P, Van Gelder P, Hohn B (2001) An *Agrobacterium* VirE2 channel for transferred-DNA transport into plant cells. Proc Natl Acad Sci USA 98: 485-490

Durrenberger F, Crameri A, Hohn B, Koukolikova-Nicola Z (1989) Covalently bound VirD2 protein of *Agrobacterium tumefaciens* protects the T-DNA from exonucleolytic degradation. Proc Natl Acad Sci USA 86: 9154-9158

Eisenbrandt R, Kalkum M, Lai EM, Lurz R, Kado CI, Lanka E (1999) Conjugative pili of IncP plasmids, and the Ti plasmid T pilus are composed of cyclic subunits. J Biol Chem 274: 22548-22555

Escudero J, Hohn B (1997) Transfer and integration of T-DNA without cell injury in the host plant. Plant Cell 9: 2135-2142

Friesner J, Britt AB (2003) Ku80- and DNA ligase IV-deficient plants are sensitive to ionizing radiation and defective in T-DNA integration. Plant J 34: 427-440

Fullner KJ, Lara JC, Nester EW (1996) Pilus assembly by *Agrobacterium* T-DNA transfer genes. Science 273: 1107-1109

Gallego ME, Bleuyard JY, Daoudal-Cotterell S, Jallut N, White CI (2003) Ku80 plays a role in non-homologous recombination but is not required for T-DNA integration in *Arabidopsis*. Plant J 35: 557-565

Gaspar YM, Nam J, Schultz CJ, Lee LY, Gilson PR, Gelvin SB, Bacic A (2004) Characterization of the *Arabidopsis* lysine-rich arabinogalactan-protein AtAGP17 mutant (*rat1*) that results in a decreased efficiency of *Agrobacterium* transformation. Plant Physiol 135: 2162-2171

Geier T, Sangwan RS (1996) Histology and chimeral segreation reveal cell-specific differences in the competence for shoot regeneration and Agrobacerium-mediated transformation in *Kohleria* internode explants. Plant Cell Rep 15: 386-390

Gelvin SB (1998) *Agrobacterium* VirE2 proteins can form a complex with T strands in the plant cytoplasm. J Bacteriol 180: 4300-4302

Gelvin SB (2000) *Agrobacterium* and plant genes involved in T-DNA transfer and integration. Annu Rev Plant Physiol Plant Mol Biol 51: 223-256

Gelvin SB (2003) *Agrobacterium*-mediated plant transformation: the biology behind the "gene-jockeying" tool. Microbiol Mol Biol Rev 67: 16-37

Gheysen G, Villarroel R, Van Montagu M (1991) Illegitimate recombination in plants: a model for T-DNA integration. Genes Dev 5: 287-297

Gietl C, Koukolikova-Nicola Z, Hohn B (1987) Mobilization of T-DNA from *Agrobacterium* to plant cells involves a protein that binds single-stranded DNA. Proc Natl Acad Sci USA 84: 9006-9010

Gouin E, Welch MD, Cossart P (2005) Actin-based motility of intracellular pathogens. Curr Opin Microbiol 8: 35-45

Grevelding C, Fantes V, Kemper E, Schell J, Masterson R (1993) Single-copy T-DNA insertions in *Arabidopsis* are the predominant form of integration in root-derived transgenics, whereas multiple insertions are found in leaf discs. Plant Mol Biol 23: 847-860

Guralnick B, Thomsen G, Citovsky V (1996) Transport of DNA into the nuclei of *Xenopus* oocytes by a modified VirE2 protein of *Agrobacterium*. Plant Cell 8: 363-373

Gurlitz RHG, Lamb PW, Matthysse AG (1987) Involvement of carrot cell surface proteins in attachment of *Agrobacterium tumefaciens*. Plant Physiol 83: 564-568

Hawes MC, Pueppke SG (1989) Variation in binding and virulence of *Agrobacterium tumefaciens* chromosomal virulence (*chv*) mutant bacteria on different plant species. Plant Physiol 91: 113-118

Herrera-Estrella A, Chen Z-M, Van Montagu M, Wang K (1988) VirD proteins of *Agrobacterium tumefaciens* are required for the formation of a covalent DNA protein complex at the 5' terminus of T-strand molecules. EMBO J 7: 4055-4062

Herrera-Estrella A, Van Montagu M, Wang K (1990) A bacterial peptide acting as a plant nuclear targeting signal: the amino-terminal portion of *Agrobacterium* VirD2 protein directs a beta-galactosidase fusion protein into tobacco nuclei. Proc Natl Acad Sci USA 87: 9534-9537

Hiei Y, Ohta S, Komari T, Kumashiro T (1994) Efficient transformation of rice (*Oryza sativa* L.) mediated by *Agrobacterium* and sequence analysis of the boundaries of the T-DNA. Plant J 6: 271-282

Hood EE, Fraley RT, Chilton M-D (1987) Virulence of *Agrobacterium tumefaciens* strain A281 on legumes. Plant Physiol 83: 529-534

Howard EA, Citovsky V (1990) The emerging structure of the *Agrobacterium* T-DNA transfer complex. BioEssays 12: 103-108

Howard EA, Winsor BA, De Vos G, Zambryski P (1989) Activation of the T-DNA transfer process in *Agrobacterium* results in the generation of a T-strand protein complex: tight association of VirD2 with the 5' ends of T-strands. Proc Natl Acad Sci USA 86: 4017-4021

Howard EA, Zupan JR, Citovsky V, Zambryski P (1992) The VirD2 protein of *A. tumefaciens* contains a C-terminal bipartite nuclear localization signal: implications for nuclear uptake of DNA in plant cells. Cell 68: 109-118

Hwang HH, Gelvin SB (2004) Plant proteins that interact with VirB2, the *Agrobacterium tumefaciens* pilin protein, mediate plant transformation. Plant Cell 16: 3148-3167

Ishida Y, Saito H, Ohta S, Hiei Y, Komari T, Kumashiro T (1996) High efficiency transformation of maize (*Zea mays* L.) mediated by *Agrobacterium tumefaciens*. Nat Biotechnol 14: 745-750

Jakubowski SJ, Cascales E, Krishnamoorthy V, Christie PJ (2005) *Agrobacterium tumefaciens* VirB9, an outer-membrane-associated component of a type IV secretion system, regulates substrate selection and T-pilus biogenesis. J Bacteriol 187: 3486-3495

Kado CI (1994) Promiscuous DNA transfer system of *Agrobacterium tumefaciens*: role of the *virB* operon in sex pilus assembly and synthesis. Mol Microbiol 12: 17-22

Kamoun S, Cooley MB, Rogowsky PM, Kado CI (1989) Two chromosomal loci involved in production of exopolysaccharide in *Agrobacterium tumefaciens*. J Bacteriol 171: 1755-1759

Kirik A, Salomon S, Puchta H (2000) Species-specific double-strand break repair and genome evolution in plants. Embo J 19: 5562-5566

Koukolikova-Nicola Z, Hohn B (1993) How does the T-DNA of *Agrobacterium tumefaciens* find its way into the plant cell nucleus? Biochimie 75: 635-638

Koukolikova-Nicola Z, Raineri D, Stephens K, Ramos C, Tinland B, Nester EW, Hohn B (1993) Genetic analysis of the *virD* operon of *Agrobacterium tumefaciens*: a search for functions involved in transport of T-DNA into the plant cell nucleus and in T-DNA integration. J Bacteriol 175: 723-731

Kunik T, Tzfira T, Kapulnik Y, Gafni Y, Dingwall C, Citovsky V (2001) Genetic transformation of HeLa cells by *Agrobacterium*. Proc Natl Acad Sci USA 98: 1871-1876

Lacroix B, Tzfira T, Vainstein A, Citovsky V (2006) A case of promiscuity: *Agrobacterium*'s endless hunt for new partners. Trends Genet 22: 29-37

Lacroix B, Vaidya M, Tzfira T, Citovsky V (2005) The VirE3 protein of *Agrobacterium* mimics a host cell function required for plant genetic transformation. EMBO J 24: 428-437

Lai EM, Kado CI (1998) Processed VirB2 is the major subunit of the promiscuous pilus of *Agrobacterium tumefaciens*. J Bacteriol 180: 2711-2717

Lee LY, Gelvin SB, Kado CI (1999) pSa causes oncogenic suppression of *Agrobacterium* by inhibiting VirE2 protein export. J Bacteriol 181: 186-196

Levee V, Garin E, Klimaszewska K, Seguin A (1999) Stable genetic transformation of white pine (*Pinus strobus* L.) after cocultivation of embryogenic tisues with *Agrobacterium tumefaciens*. Mol Breed 5: 429-440

Li J, Krichevsky A, Vaidya M, Tzfira T, Citovsky V (2005a) Uncoupling of the functions of the *Arabidopsis* VIP1 protein in transient and stable plant genetic transformation by *Agrobacterium*. Proc Natl Acad Sci USA 102: 5733-5738

Li J, Vaidya M, White C, Vainstein A, Citovsky V, Tzfira T (2005b) Involvement of KU80 in T-DNA integration in plant cells. Proc Natl Acad Sci USA 102: 19231-19236

Loyter A, Rosenbluh J, Zakai N, Li J, Kozlovsky SV, Tzfira T, Citovsky V (2005) The plant VirE2 interacting protein 1. A molecular link between the *Agrobacterium* T-complex and the host cell chromatin? Plant Physiol 138: 1318-1321

Lu H-J, Zhou X-R, Gong Z-X, Upadhyaya NM (2001) Generation of selectable marker-free transgenic rice using double right-border (DRB) vectors. Aust J Plant Physiol 28: 241-248

Macara IG (2001). Transport into and out of the nucleus. Microbiol Mol Biol Rev 65: 570-594

Matthysse AG (1987) Characterization of nonattaching mutants of *Agrobacterium tumefaciens*. J Bacteriol 169: 313-323

Matthysse AG, Holmes KV, Gurlitz RH (1981) Elaboration of cellulose fibrils by *Agrobacterium tumefaciens* during attachment to carrot cells. J Bacteriol 145: 583-595

Matthysse AG, Marry M, Krall L, Kaye M, Ramey BE, Fuqua C, White AR (2005) The effect of cellulose overproduction on binding and biofilm formation on roots by *Agrobacterium tumefaciens*. Mol Plant Microbe Interact 18: 1002-1010

Mauro AO, Pfeiffer TW, Collins GB (1995) Inheritance of soybean susceptibility to *Agrobacterium tumefaciens* and its relationship to transformation. Crop Sci 35: 1152-1156

Mayerhofer R, Koncz-Kalman Z, Nawrath C, Bakkeren G, Crameri A, Angelis K, Redei GP, Schell J, Hohn B, Koncz C (1991) T-DNA integration: a mode of illegitimate recombination in plants. EMBO J 10: 697-704

McAfee BJ, White EE, Pelcher LE, Lapp MS (1993) Root induction in pine (*Pinus*) and larch (*Larix*) spp. using *Agrobacterium rhizogenes*. Plant Cell Tissue Organ Cultur 34: 53-62

Melchers LS, Maroney MJ, den Dulk-Ras A, Thompson DV, van Vuuren HAJ, Schilperoort RA, Hooykaas PJJ (1990) Octopine and nopaline strains of

Agrobacterium tumefaciens differ in virulence; molecular characterization of the virF locus. Plant Mol Biol 14: 249-259

Merkle T (2004) Nucleo-cytoplasmic partitioning of proteins in plants: Implications for the regulation of environmental and developmental signaling. Curr Genet 44: 231-260

Michielse CB, Ram AFJ, Hooykaas PJJ, van den Hondel CAMJJ (2004) *Agrobacterium*-mediated transformation of *Aspergillus awamori* in the absence of full-length VirD2, VirC2, or VirE2 leads to insertion of aberrant T-DNA structures. J Bacteriol 186: 2038-2045

Morris JW, Morris RO (1990) Identification of an *Agrobacterium tumefaciens* virulence gene inducer from the pinaceous gymnosperm *Pseudotsuga menziesii*. Proc Natl Acad Sci USA 87: 3614-3618

Mysore KS, Bassuner B, Deng X-B, Darbinian NS, Motchoulski A, Ream W, Gelvin SB (1998) Role of the *Agrobacterium tumefaciens* VirD2 protein in T-DNA transfer and integration. Mol Plant-Microbe Interact 11: 668-683

Mysore KS, Kumar CTR, Gelvin SB (2000a) Arabidopsis ecotypes and mutants that are recalcitrant to *Agrobacterium* root transformation are susceptible to germ-line transformation. Plant J 21: 9-16

Mysore KS, Nam J, Gelvin SB (2000b) An *Arabidopsis* histone H2A mutant is deficient in *Agrobacterium* T-DNA integration. Proc Natl Acad Sci USA 97: 948-953

Nam J, Matthysse AG, Gelvin SB (1997) Differences in susceptibility of *Arabidopsis* ecotypes to crown gall disease may result from a deficiency in T-DNA integration. Plant Cell 9: 317-333

Nam J, Mysore KS, Gelvin SB (1998) *Agrobacterium tumefaciens* transformation of the radiation hypersensitive *Arabidopsis thaliana* mutants uvh1 and rad5. Mol Plant-Microbe Interact 11: 1136-1141

Nam J, Mysore KS, Zheng C, Knue MK, Matthysse AG, Gelvin SB (1999) Identification of T-DNA tagged *Arabidopsis* mutants that are resistant to transformation by *Agrobacterium*. Mol Gen Genet 261: 429-438

Narasimhulu SB, Deng X-B, Sarria R, Gelvin SB (1996) Early transcription of *Agrobacterium* T-DNA genes in tobacco and maize. Plant Cell 8: 873-886

Neff NT, Binns AN (1985) *Agrobacterium tumefaciens* interaction with suspension-cultured tomato cells. Plant Physiol 77: 35-42

O'Connell KP, Handelsman J (1999) chvA locus may be involved in export of neutral cyclic beta-1,2 linked D-glucan from *Agrobacterium tumefaciens*. Mol Plant Microbe Interact 2: 11-16

Otten L, DeGreve H, Leemans J, Hain R, Hooykass P, Schell J (1984) Restoration of virulence of *vir region* mutants of *Agrobacterium tumefaciens* strain B6S3 by coinfection with normal and mutant *Agrobacterium* strains. Mol Gen Genet 195: 159-163

Owens LD, Cress DE (1984) Genotypic variability of soybean response to *Agrobacterium* strains harboring the Ti or Ri plasmids. Plant Physiol 77: 87-94

Pena L, Seguin A (2001) Recent advances in the genetic transformation of trees. Trends Biotechnol 19: 500-506

Piers KL, Heath JD, Liang X, Stephens KM, Nester EW (1996) *Agrobacterium tumefaciens*-mediated transformation of yeast. Proc Natl Acad Sci USA 93: 1613-1618

Puvanesarajah V, Schell FM, Stacey G, Douglas CJ, Nester EW (1985) Role for 2-linked-β-D-glucan in the virulence of *Agrobacterium tumefaciens*. J Bacteriol 164: 102-106

Ramey BE, Matthysse AG, Fuqua C (2004) The FNR-type transcriptional regulator SinR controls maturation of *Agrobacterium tumefaciens* biofilms. Mol Microbiol 52: 1495-1511

Rao P (2002) Involvement of the plant cytoskeleton in *Agrobacterium* transformation. PhD Thesis, Purdue University

Rashid H, Yokoi S, Toriyama K, Hinata K (1996) Transgenic plant production mediated by *Agrobacterium* in Indica rice. Plant Cell Rep 15: 727-730

Regensburg-Tuink AJ, Hooykaas PJJ (1993) Transgenic N. glauca plants expressing bacterial virulence gene *virF* are converted into hosts for nopaline strains of *A. tumefaciens*. Nature 363: 69-71

Relic B, Andjelkovic M, Rossi L, Nagamine Y, Hohn B (1998) Interaction of the DNA modifying proteins VirD1 and VirD2 of *Agrobacterium tumefaciens*: analysis by subcellular localization in mammalian cells. Proc Natl Acad Sci USA 95: 9105-9110

Reuhs BL, Kim JS, Matthysse AG (1997) Attachment of *Agrobacterium tumefaciens* to carrot cells and *Arabidopsis* wound sites is correlated with the presence of a cell-associated, acidic polysaccharide. J Bacteriol 179: 5372-5379

Rhee Y, Gurel F, Gafni Y, Dingwall C, Citovsky V (2000) A genetic system for detection of protein nuclear import and export. Nat Biotechnol 18: 433-437

Robbs SL, Hawes MC, Lin H-J, Pueppke SG, Smith LY (1991) Inheritance of resistance to crown gall in *Pisum sativum*. Plant Physiol 95: 52-57

Robertson JL, Holliday T, Matthysse AG (1988) Mapping of *Agrobacterium tumefaciens* chromosomal genes affecting cellulose synthesis and bacterial attachment to host cells. J Bacteriol 170: 1408-1411

Rossi L, Hohn B, Tinland B (1993) The VirD2 protein of *Agrobacterium tumefaciens* carries nuclear localization signals important for transfer of T-DNA to plant. Mol Gen Genet 239: 345-353

Sagulenko V, Sagulenko E, Jakubowski S, Spudich E, Christie PJ (2001) VirB7 lipoprotein is exocellular and associates with the *Agrobacterium tumefaciens* T pilus. J Bacteriol 183: 3642-3651

Sallaud C, Gay C, Larmande P, Bes M, Piffanelli P, Piegu B, Droc G, Regad F, Bourgeois E, Meynard D, Perin C, Sabau X, Ghesquiere A, Glaszmann JC, Delseny M, Guiderdoni E (2004) High throughput T-DNA insertion mutagenesis in rice: a first step towards in silico reverse genetics. Plant J 39: 450-464

Salman H, Abu-Arish A, Oliel S, Loyter A, Klafter J, Granek R, Elbaum M (2005) Nuclear localization signal peptides induce molecular delivery along microtubules. Biophys J 89: 2134-2145

Salomon S, Puchta H (1998) Capture of genomic and T-DNA sequences during double-strand break repair in somatic plant cells. EMBO J 17: 6086-6095

Sangwan RS, Bourgeois Y, Brown S, Vasseur G, Sangwan-Norreel B (1992) Characterization of competent cells and early events of *Agrobacterium*-mediated genetic transformation in *Arabidopsis thaliana*. Planta 188: 439-456

Schafer W, Gorz A, Kahl G (1987) T-DNA integration and expression in a monocot crop plant after induction of Agrobacterium. Nature 327: 529-532

Schmidt-Eisenlohr H, Domke N, Angerer C, Wanner G, Zambryski P, Baron C (1999) Vir proteins stabilize VirB5 and mediate its association with the T pilus of *Agrobacterium tumefaciens*. J Bacteriol 181: 7485-7492

Schrammeijer B, den Dulk-Ras A, Vergunst AC, Jácome JE, Hooykaas PJJ (2003) Analysis of Vir protein translocation from *Agrobacterium tumefaciens* using *Saccharomyces cerevisiae* as a model: evidence for transport of a novel effector protein VirE3. Nucleic Acids Res 31: 860-868

Schrammeijer B, Risseeuw E, Pansegrau W, Regensburg-Tuïnk TJG, Crosby WL, Hooykaas PJJ (2001) Interaction of the virulence protein VirF of *Agrobacterium tumefaciens* with plant homologs of the yeast Skp1 protein. Curr Biol 11: 258-262

Sen P, Pazour GJ, Anderson D, Das A (1989) Cooperative binding of *Agrobacterium tumefaciens* VirE2 protein to single-stranded DNA. J Bacteriol 171: 2573-2580

Shibasaki F, Kondo E, Akagi T, McKeon F (1997) Suppression of signaling through transcription factor NF-AT by interaction between calcineurin and Bcl-2. Nature 386: 728-731

Shibasaki F, Price ER, Milan D, McKeon F (1996) Role of kinases and the phosphatase calcineurin in the nuclear shuttling of transcription factor NF-AT4. Nature 382: 370-373

Silva H, Yoshioka K, Dooner HK, Klessig DF (1999) Characterization of a new *Arabidopsis* mutant exhibiting enhanced disease resistance. Mol Plant-Microbe Interact 12: 1053-1063

Smarrelli J, Watters MT, Diba LH (1986) Response of various cucurbits to infection by plasmid-harboring strains of *Agrobacterium*. Plant Physiol 82: 622-624

Smith HMS, Raikhel NV (1998) Nuclear localization signal receptor importin alpha associates with the cytoskeleton. Plant Cell 10: 1791-1799

Sparrow PAC, Townsend TM, Arthur AE, Dale PJ, Irwin JA (2004) Genetic analysis of *Agrobacterium tumefaciens* susceptibility in *Brassica oleracea*. Theor Appl Genet 108: 644-650

Stomp A-M, Loopstra C, Chilton WS, Sederoff RR, Moore LW (1990) Extended host range of *Agrobacterium tumefaciens* in the genus *Pinus*. Plant Physiol 92: 1226-1232.

Swart S, Logman TJ, Smit G, Lugtenberg BJJ, Kijne JW (1994) Purification and partial characterization of a glycoprotein from pea (*Pisum sativum*) with receptor activity for rhicadhesin, an attachment protein of Rhizobiaceae. Plant Mol Biol 24: 171-183

Szegedi E, Kozma P (1984) Studies on the inheritance of resistance to crown gall disease of grapevine. Vitis 23: 121-126

Tao Y (1998) Isolation and characterization of a cDNA encoding a plant phosphatase implicated in nuclear import of the *Agrobacterium* VirD2 protein. PhD Thesis, Purdue University

Tao Y, Rao PK, Bhattacharjee S, Gelvin SB (2004) Expression of plant protein phosphatase 2C interferes with nuclear import of the *Agrobacterium* T-complex protein VirD2. Proc Natl Acad Sci USA 101: 5164-5169

Thomashow MF, Karlinsey JE, Marks JR, Hurlbert RE (1987) Identification of a new virulence locus in *Agrobacterium tumefaciens* that affects polysaccharide composition and plant cell attachment. J Bacteriol 169: 3209-3216

Tian L, Wang J, Fong MP, Chen M, Cao H, Gelvin SB, Chen ZJ (2003) Genetic control of developmental changes induced by disruption of *Arabidopsis* histone deacetylase 1 (AtHD1) expression. Genetics 165: 399-409

Tingay S, McElroy D, Kalla R, Fieg S, Wang M, Thornton S, Brettell R (1997) *Agrobacterium tumefaciens*-mediated barley transformation. Plant J 11: 1369-1376

Tinland B (1996) The integration of T-DNA into plant genomes. Trends Plant Sci 1: 178-184

Tinland B, Hohn B (1995) Recombination between prokaryotic and eukaryotic DNA: integration of *Agrobacterium tumefaciens* T-DNA into the plant genome. In JK Settoe, ed, Genetic Engineering, Plenum Press, New York pp. 209-229

Tinland B, Koukolikova-Nicola Z, Hall MN, Hohn B (1992) The T-DNA-linked VirD2 protein contains two distinct nuclear localization signals. Proc Natl Acad Sci USA 89: 7442-7446

Toki S (1997) Rapid and efficient *Agrobacterium*-mediated transformation in rice. Plant Mol Biol Rep 15: 16-21

Tzfira T, Citovsky V (2002) Partners-in-infection: host proteins involved in the transformation of plant cells by *Agrobacterium*. Trends Cell Biol 12: 121-129

Tzfira T, Citovsky V (2003) The *Agrobacterium*-plant cell interaction. Taking biology lessons from a bug. Plant Physiol 133: 943-947

Tzfira T, Frankman LR, Vaidya M, Citovsky V (2003) Site-specific integration of *Agrobacterium tumefaciens* T-DNA via double-stranded intermediates. Plant Physiol 133: 1011-1023

Tzfira T, Li J, Lacroix B, Citovsky V (2004a) *Agrobacterium* T-DNA integration: molecules and models. Trends Genet 20: 375-383

Tzfira T, Rhee Y, Chen M-H, Kunik T, Citovsky V (2000) Nucleic acid transport in plant-microbe interactions: the molecules that walk through the walls. Annu Rev Microbiol 54: 187-219

Tzfira T, Vaidya M, Citovsky V (2001) VIP1, an *Arabidopsis* protein that interacts with *Agrobacterium* VirE2, is involved in VirE2 nuclear import and Agrobacterium infectivity. EMBO J 20: 3596-3607

Tzfira T, Vaidya M, Citovsky V (2002) Increasing plant susceptibility to *Agrobacterium* infection by overexpression of the *Arabidopsis* nuclear protein VIP1. Proc Natl Acad Sci USA 99: 10435-10440

Tzfira T, Vaidya M, Citovsky V (2004b) Involvement of targeted proteolysis in plant genetic transformation by *Agrobacterium*. Nature 431: 87-92

Valentine L (2003) *Agrobacterium tumefaciens* and the plant: the David and Goliath of modern genetics. Plant Physiol 133: 948-955

van Attikum H, Bundock P, Hooykaas PJJ (2001) Non-homologous end-joining proteins are required for *Agrobacterium* T-DNA integration. Embo J 20: 6550-6558

van Attikum H, Bundock P, Overmeer RM, Lee LY, Gelvin SB, Hooykaas PJJ (2003) The Arabidopsis AtLIG4 gene is required for the repair of DNA damage, but not for the integration of *Agrobacterium* T-DNA. Nucleic Acids Res 31: 4247-4255

van Attikum H, Hooykaas PJJ (2003) Genetic requirements for the targeted integration of *Agrobacterium* T-DNA in *Saccharomyces cerevisiae*. Nucleic Acids Res 31: 826-832

Vandromme M, Gauthier-Rouviere C, Lamb N, Fernandez A (1996) Regulation of transcription factor localization: Fine-tuning of gene expression. Trends Biochem Sci 21: 59-64 van Wordragen MF, Dons HJM (1992) *Agrobacterium* tumefaciens-mediated transformation of recalcitrant crops. Plant Mol Biol Rep 10: 12-36

Veena, Jiang H, Doerge RW, Gelvin SB (2003) Transfer of T-DNA and Vir proteins to plant cells by *Agrobacterium tumefaciens* induces expression of host genes involved in mediating transformation and suppresses host defense gene expression. Plant J 35: 219-236

Vogel AM, Das A (1992) Mutational analysis of *Agrobacterium tumefaciens* virD2: tyrosine 29 is essential for endonuclease activity. J Bacteriol 174: 303-308

Wagner VT, Matthysse AG (1992) Involvement of vitronectin-like protein in attachment of *Agrobacterium tumefaciens* to carrot suspension culture cells. J Bacteriol 174: 5999-6003

Ward ER, Barnes WM (1988) VirD2 protein of *Agrobacterium tumefaciens* very tightly linked to the 5' end of T-strand DNA. Science 242: 927-930

Wenck AR, Quinn M, Whetten RW, Pullman G, Sederoff R (1999) High-efficiency *Agrobacterium*-mediated transformation of Norway spruce (*Picea abies*) and loblolly pine (*Pinus taeda*). Plant Mol Biol 39: 407-416

Windels P, De Buck S, Van Bockstaele E, De Loose M, Depicker A (2003) T-DNA integration in *Arabidopsis* chromosomes. Presence and origin of filler DNA sequences. Plant Physiol 133: 2061-2068

Ye GN, Stone D, Pang SZ, Creely W, Gonzalez K, Hinchee M (1999) *Arabidopsis* ovule is the target for *Agrobacterium in planta* vacuum infiltration transformation. Plant J 19: 249-257

Yi H, Mysore KS, Gelvin SB (2002) Expression of the *Arabidopsis* histone H2A-1 gene correlates with susceptibility to *Agrobacterium* transformation. Plant J 32: 285-298

Yi H, Sardesai N, Fujinuma T, Chan C-W, Veena, Gelvin SB (2006) Constitutive expression exposes functional redundancy between the *Arabidopsis* histone H2A gene HTA1 and other H2A gene family members. Plant Cell 18: 1575-1589

Young C, Nester EW (1988) Association of the VirD2 protein with the 5' end of T-strands in *Agrobacterium tumefaciens*. J Bacteriol 170: 3367-3374

Zhu Y, Nam J, Carpita NC, Matthysse AG, Gelvin SB (2003a) *Agrobacterium*-mediated root transformation is inhibited by mutation of an *Arabidopsis* cellulose synthase-like gene. Plant Physiol 133: 1000-1010

Zhu Y, Nam J, Humara JM, Mysore KS, Lee LY, Cao H, Valentine L, Li J, Kaiser AD, Kopecky AL, Hwang HH, Bhattacharjee S, Rao PK, Tzfira T, Rajagopal J, Yi H, Veena, Yadav BS, Crane YM, Lin K, Larcher Y, Gelvin MJ, Knue M, Ramos C, Zhao X, Davis SJ, Kim SI, Ranjith-Kumar CT, Choi YJ, Hallan VK, Chattopadhyay S, Sui X, Ziemienowicz A, Matthysse AG, Citovsky V,

Hohn B, Gelvin SB (2003b) Identification of *Arabidopsis rat* mutants. Plant Physiol 132: 494-505

Ziemienowicz A, Görlich D, Lanka E, Hohn B, Rossi L (1999) Import of DNA into mammalian nuclei by proteins originating from a plant pathogenic bacterium. Proc Natl Acad Sci USA 96: 3729-3733

Ziemienowicz A, Merkle T, Schoumacher F, Hohn B, Rossi L (2001) Import of *Agrobacterium* T-DNA into plant nuclei: Two distinct functions of VirD2 and VirE2 proteins. Plant Cell 13: 369-383

Zorreguieta A, Ugalde RA (1986) Formation in *Rhizobium* and *Agrobacterium* spp. of a 235-kilodalton protein intermediate in b-D(1-2) glucan synthesis. J Bacteriol 167: 947-951

Zupan J, Citovsky V, Zambryski PC (1996) *Agrobacterium* VirE2 protein mediates nuclear uptake of single-stranded DNA in plant cells. Proc Natl Acad Sci USA 93: 2392-2397

Zupan J, Muth TR, Draper O, Zambryski PC (2000) The transfer of DNA from *Agrobacterium tumefaciens* into plants: a feast of fundamental insights. Plant J 23: 11-28

Chapter 14

THE ONCOGENES OF *AGROBACTERIUM TUMEFACIENS* AND *AGROBACTERIUM RHIZOGENES*

Monica T. Britton[1], Matthew A. Escobar[2] and Abhaya M. Dandekar[1]

[1]Department of Plant Sciences, University of California, One Shields Ave, Davis, CA 95616, USA; [2]Department of Biological Sciences, California State University San Marcos, 333 S. Twin Oaks Valley Road, San Marcos, CA 92096, USA

Abstract. The common soil bacteria *Agrobacterium tumefaciens* and *Agrobacterium rhizogenes* are unique genetic pathogens capable of fundamentally redirecting plant metabolism in order to generate macroscopic tissue masses (crown galls and hairy roots, respectively) which support the growth of large populations of *Agrobacteria*. Central to pathogenesis is the horizontal transfer of a suite of oncogenes from the tumor-inducing (Ti) plasmids of *A. tumefaciens* and the root-inducing (Ri) plasmids of *A. rhizogenes* into the plant cell genome. These oncogenes alter the synthesis, perception and/or transport of phytohormones *in planta*, leading to the development of the crown gall and hairy root structures from single genetically transformed plant cells. Crown galls and hairy roots become effective sinks that divert plant resources to produce opine compounds that can only be metabolized by the infecting strain of *Agrobacterium*. The basic genetic and biochemical mechanisms underlying *A. tumefaciens* tumorigenesis were initially described over 20 years ago, with the characterization of the *ipt*, *iaaM* and *iaaH* oncogenes. However, the simplistic view of crown gall development as solely a function of ipt-driven cytokinin synthesis and iaaM/iaaH-driven auxin synthesis has recently given way to a more nuanced understanding of the roles of secondary oncogenes in modulating hormone perception and the complex hormone activation cascade in crown galls involving ethylene, abscisic acid and jasmonic acid. The biochemistry and functional significance of specific oncogenes in *A. rhizogenes*-mediated hairy root development is less well understood, but recent work has substantially increased our understanding of the *A. rhizogenes* oncogenes, especially the *rol* genes. Expression of the *rolA*, *B* and *C* oncogenes *in planta* induces a subtle interaction with endogenous plant signal transduction pathways and transcription factors, affecting the local concentrations of several classes of plant hormones. These interactions lead to *de novo* meristem formation in transformed cells, with subsequent differentiation depending on the local hormone balance. This process most often results in the induction of highly branched non-geotropic adventitious roots, the "hairy root" phenotype. Further dissection of the molecular mechanisms underlying *Agrobacterium* pathogenesis should continue to yield broader insights into the understanding of endogenous hormone signaling pathways and tissue differentiation in plants.

1 INTRODUCTION

Following the initial demonstration that crown gall disease was caused by the common soil bacterium *Agrobacterium tumefaciens* (Smith and Townsend, 1907), much of crown gall research was driven by perceived parallels between the development of plant crown galls and animal cancers (reviewed by Braun, 1982). Indeed, there are some striking similarities between crown gall disease and cancer, including (i) the genetic/epigenetic alteration of a cell or group of cells leading to loss of cell cycle control, (ii) subsequent unchecked cell proliferation and the production of a macroscopic, generally undifferentiated tumor, and (iii) diversion/development of vasculature to feed the tumor structure through angiogenesis (in animals) or vascularization (in plants) (Ullrich and Aloni, 2000). The molecular events underlying the development of animal cancers and crown gall disease are now known to be quite different (though see Sauter and Blum, 2003); however, much of the terminology related to crown gall disease has been derived from the parallel to cancer (e.g. oncogenes, tumorigenesis). For the purposes of this review, we will utilize the term oncogenes to refer to the group of genes transferred from *Agrobacterium* to the plant cell which contribute to the development of the crown gall and hairy root structures in plants.

Most of the genes present on the transferred DNA (T-DNA) region of the *Agrobacterium* Ti (tumor inducing) or Ri (root inducing) plasmids have been characterized as either oncogenes or opine-related genes.

Opine-related genes encode proteins responsible for the synthesis and secretion of sugar and amino acid derived opines, which are utilized by the infecting strain of *Agrobacterium* as a carbon and nitrogen source. Upon horizontal transfer of the T-DNA from *Agrobacterium* to a plant cell, expression of the bacterial oncogenes and opine-related genes *in planta* leads to the development of opine-producing crown gall or hairy root structures. Unlike typical prokaryotic genes, oncogenes (and opine-related genes) possess eukaryotic transcriptional regulatory sequences such as TATA boxes, CCAAT boxes and polyadenylation sites, and include protein trafficking sequences for organellar targeting, thus ensuring efficient expression in the plant host (Barker et al., 1983; Klee et al., 1984; Sakakibara et al., 2005). This article will focus specifically on the genetics and biochemistry of the oncogenes of *A. tumefaciens* and *A. rhizogenes*, causal agents of crown gall disease and hairy root disease.

2 THE *A. TUMEFACIENS* ONCOGENES

Several early studies of crown gall disease suggested that *A. tumefaciens*-mediated tumor formation was initiated and maintained by alterations in plant hormone metabolism (reviewed by Binns and Costantino, 1998). Unlike other plant tissues, axenically cultured crown gall tumors were found to proliferate in the absence of exogenously supplied auxins and cytokinins (White and Braun, 1942). In addition, auxin and cytokinin levels were substantially elevated in tumor tissue (Kulescha, 1954; Miller, 1974). In 1977 Chilton et al. demonstrated that *A. tumefaciens* T-DNA is horizontally transferred to plants and integrates into the plant cell genome (Chilton et al., 1977), and later studies demonstrated that a single fragment of the T-DNA is conserved between octopine- and nopaline-type strains of *A. tumefaciens* (Willmitzer et al., 1983). This "common T-DNA" region contains six genes: gene *5*, *iaaM*, *iaaH*, *ipt*, gene *6a* and gene *6b*. As discussed below, nearly all of these common genes were subsequently shown to be oncogenes involved in crown gall tumorigenesis.

2.1 *iaaM*, *iaaH* and auxin synthesis

In a seminal study of the genetic basis of tumorigenesis, Garfinkel and Nester (1980) identified several tumor morphology mutants in a Tn5-mutagenized population of *A. tumefaciens*. Rather than typical undifferentiated, callus-like tumors, *tms* (tumor morphology shooty) mutants produced tumors with emergent differentiated shoots. The *tms* locus was mapped to the T-DNA region of the Ti plasmid (Garfinkel and Nester, 1980; Garfinkel et al., 1981) and tumors from *tms* mutants were found to have substantially reduced auxin levels (Akiyoshi et al., 1983), suggesting the involvement of *tms* in auxin metabolism *in planta*. Sequencing of the *tms* locus revealed two genes (*tms1*/*iaaM* and *tms2*/*iaaH*), both expressed in tumor tissues, whose inactivation results in the *tms* phenotype (Klee et al., 1984). *iaaM* and *iaaH* encode a tryptophan monooxygenase and an indole-3-acetamide hydrolase, respectively, which catalyze the two-step conversion of tryptophan to the auxin indole-3-acetic acid (IAA) (Schröder et al., 1984; Thomashow et al., 1984; Thomashow et al., 1986; van Onckelen et al., 1986) (see *Figure 14-1*).

The activity of the iaaM and iaaH proteins *in planta* leads to the accumulation of free IAA levels in crown gall tumors that are generally more than 10-fold greater than in surrounding tissues, with highest auxin concentrations at the tumor periphery (Weiler and Spanier, 1981; Veselov et al., 2003). In addition, it has recently been suggested that unique flavonoids

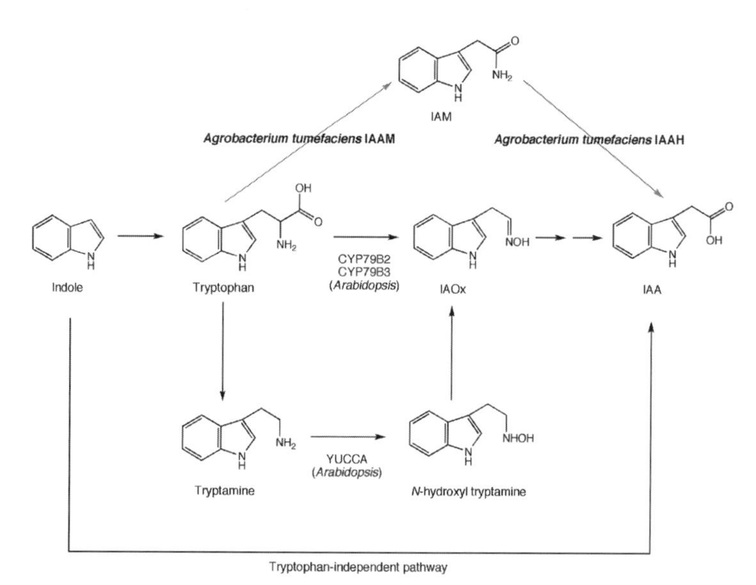

Figure 14-1. A comparison between endogenous plant IAA biosynthesis pathways and *A. tumefaciens* oncogene-mediated IAA biosynthesis pathways. Oncogene-catalyzed reactions are denoted in gray. Abbreviations: IAOx, indole-3-acetaldoxime; IAA, indole-3-acetic acid; IAM indole-3- acetamide. Figure modified from Cohen et al. (2003) and Escobar and Dandekar (2003).

which accumulate in crown gall tumors assist in maintaining high free auxin levels by suppressing basipetal efflux of auxin from tumor cells (Schwalm et al., 2003). Relatively little is known about the transcriptional regulation of the *iaaM* and *iaaH* genes *in planta*, however the homologous *aux1* and *aux2* genes of *A. rhizogenes* appear to be expressed primarily in rapidly dividing cells, such as root meristems, shoot meristems and callus (Gaudin and Jouanin, 1995). As expected, transgenic plants constitutively overexpressing *iaaM* and *iaaH* display global increases in free IAA, with corresponding developmental abnormalities (adventitious rooting, extreme apical dominance, curled leaves) (Klee et al., 1987; Eklöf et al., 1997).

A very recent report by Dunoyer et al. (2006) has reinvigorated discussion of a "secondary" role of *iaaM* and *iaaH* beyond tumorigenesis. This work has shown that T-DNA-encoded genes are targeted by endogenous plant RNA silencing pathways, a defense response previously characterized only in plant-virus interactions. Similar to many viruses, *A. tumefaciens* has developed a mechanism that at least partially counteracts RNA silencing-based plant defenses, allowing high expression of targeted genes such as *iaaM* and *ags* (agropine synthase). In contrast to characterized viral systems, suppression of RNA silencing is probably caused by oncogene-mediated increases in auxin and/or cytokinin levels in transformed cells. In addition, loss-of-function mutants in the RNA interference branch of plant RNA silencing pathways displayed increased susceptibility to transformation (Dunoyer et al., 2006). These results support previous work that has tied oncogenes (e.g., *iaaM/iaaH*) to early events in *A. tumefaciens* pathogenesis. For example, Robinette and Matthysse (1990) found that *A. tumefaciens* can suppress a plant hypersensitive response, and that this suppression is dependent upon *iaaM* and *iaaH*. Likewise, hormone pretreatment of transformation-recalcitrant *Arabidopsis* ecotypes has been shown to drastically increase root transformation efficiency (Chateau et al., 2000). Thus, it appears that the activity of iaaM and iaaH (and potentially other oncogene proteins) not only drives tumor development, but also plays a critical role in circumventing plant-imposed barriers to transformation.

The lack of *iaaM*/*iaaH*-homologous sequences in plant genomes and the presence of clear homologs in the auxin-producing plant pathogens *Pseudomonas savastanoi* pv. *savastanoi* and *Erwinia herbicola* pv. *gypsophilae* suggest that *A. tumefaciens iaaM* and *iaaH* are likely of prokaryotic origin, despite optimization for expression in a eukaryotic host (Yamada et al., 1985; Manulis et al., 1998). Indeed, characterized plant auxin biosynthesis pathways are biochemically distinct from the indole-3-

acetamide pathway mediated by *iaaM* and *iaaH* (*Figure. 14-1*). In *Arabidopsis thaliana*, at least two tryptophan-dependent and one tryptophan-independent pathway operate in IAA synthesis. In the endogenous tryptophan-dependent pathways, indole-3-acetaldoxime (not indole-3-acetamide) appears to be a central intermediate in auxin biosynthesis (Cohen et al., 2003). Thus, the transfer of *iaaM* and *iaaH* from *A. tumefaciens* to the plant cell produces a novel and presumably uncontrolled, auxin biosynthesis pathway.

2.2 *ipt* and cytokinin synthesis

Like *tms*, the *tmr* (tumor morphology rooty) mutant was identified by screening Tn5-mutagenized *A. tumefaciens* (Garfinkel and Nester, 1980). The small, rooty tumors induced by *tmr* strains on tobacco stems and *Kalanchoe* leaves possess very low levels of zeatin-type cytokinins (Akiyoshi et al., 1983). The *tmr* locus lies adjacent to *tms* on *A. tumefaciens* T-DNA and consists of a single protein-coding gene: *tmr/ipt*. The ipt protein is an isopentenyl transferase which catalyzes the condensation of adenosine monophosphate (AMP) and an isoprenoid precursor, the rate limiting step of cytokinin synthesis *in planta* (Åstot et al., 2000). Thus, cytokinin levels in crown gall tumors are often more than 100-fold higher than in surrounding tissues (Weiler and Spanier, 1981) and transgenic plants overexpressing *ipt* display increased cytokinin biosynthesis with corresponding developmental abnormalities (e.g. ectopic shoot development, suppressed apical dominance, delayed leaf senescence) (Estruch et al., 1991b; Eklöf et al., 1997). Results from promoter:reporter gene fusion studies suggest that *ipt* is expressed broadly in all plant organs with highest expression in roots (Neuteboom et al., 1993; Strabala et al., 1993) and that expression of *ipt* is negatively regulated by auxin (Zhang et al., 1996).

Over the past five years, our understanding of cytokinin biosynthesis by endogenous plant pathways and by *ipt* has increased substantially. The isoprenoid substrate of *ipt in planta* was long thought to be dimethylallylpyrophosphate (DMAPP), with the subsequent product, isopentenyladenosine-5'-monophosphate (iPMP), rapidly converted to trans-zeatin by plant-encoded enzymes (Akiyoshi et al., 1984; Barry et al., 1984). However, recent *in vivo* isotope labeling experiments have definitively shown that 1-hydroxy-2-methyl-2-(*E*)-butenyl 4-diphosphate (HMBDP) is the primary isoprenoid substrate of *ipt*, thus generating zeatinriboside-5'-monophosphate (ZMP), which is converted to trans-zeatin (again by plant enzymes) (Åstot et al., 2000; Sakakibara et al., 2005) (*Figure 14-2*).

Though *ipt* lacks a typical chloroplast transit peptide, the protein is plastid-localized in tumor cells, providing access to the HMBDP substrate (Sakakibara et al., 2005). In *Arabidopsis*, a family of putative isopentenyl transferase-encoding genes with weak homology to *ipt* have been identified and denoted *AtIPT1-9* (Kakimoto, 2001; Takei et al., 2001). *Escherichia coli* transformants expressing *AtIPT1* and *AtIPT3-8* secrete cytokinins into culture medium, and the overexpression of *AtIPT1*, *-3*, *-4*, *-5*, *-7* and *-8*, in transgenic plants increases levels of isopentenyladenine-type cytokinins (Takei et al., 2001; Sun et al., 2003; Sakakibara et al., 2005). Most interestingly, AtIPT proteins appear to utilize DMAPP exclusively as their isoprenoid substrate, thereby producing cytokinins via the iPMP-dependent pathway described above (Sun et al., 2003; Sakakibara et al., 2005). Thus, despite their sequence similarities, *A. tumefaciens* ipt and the *Arabidopsis* AtIPTs display divergent substrate specificities, forming parallel pathways for cytokinin production *in planta* (*Figure 14-2*).

A. tumefaciens possesses an additional, *ipt*-homologous, isopentenyl transferase-encoding gene called *tzs* (trans-zeatin synthase), which is located near the *vir* region of the Ti plasmid and displays a prokaryotic

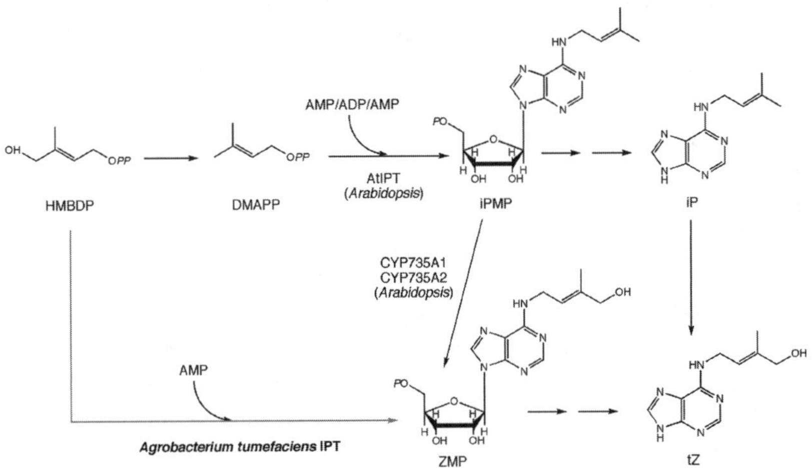

Figure 14-2. A comparison between endogenous plant cytokinin biosynthesis pathways and *A. tumefaciens* oncogene-mediated cytokinin biosynthesis pathways. Oncogene-catalyzed reactions are denoted in gray. Abbreviations: HMBDP, 1-hydroxy-2-methyl-(E)-butenyl 4-diphosphate; DMAPP, dimethylallyl-pyrophosphate; iPMP, isopentenyladenosine-5'-monophosphate; iP, isopentenyladenine; ZMP, zeatinriboside-5'-monophosphate; tZ, trans-zeatin. Figure modified from Åstot et al. (2000) and Sakakibara et al. (2005).

regulatory structure (Akiyoshi et al., 1985). Expression of *tzs*, which is controlled by the VirA/VirG two-compoment regulatory system, allows the bacteria to synthesize zeatin-type cytokinins, which may enhance virulence (Gaudin et al., 1994). Both *ipt* and *tzs* are also highly homologous to isopentenyl transferase genes presumably involved in cytokinin synthesis in other prokaryotes, including *Pseudomonas savastanoi* pv. *savastanoi* and *Xanthomonas oryzae* pv. *oryzae* (Powell and Morris, 1986; Lee et al., 2005).

2.3 Gene *6b*

Transposon insertion into the *tml* (tumor morphology large) locus of octopine strains of *A. tumefaciens* results in a 2-3-fold increase in tumor size on Kalanchoe (Garfinkel et al., 1981). However, inactivation of (octopine) *tml* has little effect on tumorigenesis in tobacco and *tml* has no apparent effect on tumor formation induced by nopaline-type strains of *A. tumefaciens* (Garfinkel et al., 1981; Joos et al., 1983). *tml* was mapped to a 1.25 kb region of the T-DNA encompassing two genes expressed in crown gall tumors: gene *6a* and gene *6b* (Garfinkel et al., 1981; Hooykaas et al., 1988). Gene *6a* was later shown to encode an opine permease with no direct role in tumorigenesis (Messens et al., 1985), but gene *6b* is an oncogene capable of inducing small tumors (independent of the other T-DNA oncogenes) on a limited number of plant hosts (Hooykaas et al., 1988). In addition, several studies have documented various growth abnormalities in transgenic plants overexpressing gene *6b* (e.g. tubular leaves, increased root thickness, ectopic shoot production) (Tinland et al., 1992; Wabiko and Minemura, 1996; Grémillon et al., 2004). In tobacco, the 6B protein is nuclear-localized and interacts with the putative transcription factor NtSIP2, potentially activating transcription of a suite of downstream target genes (Kitakura et al., 2002). In addition, 6B is capable of cell-to-cell movement and can traverse graft unions by entering the vascular system (Grémillon et al., 2004).

Unfortunately, critical questions about the biochemistry and mode of action of 6B remain unanswered. There is some agreement that 6B probably affects the synthesis or perception of hormones *in planta*. Overexpression studies have shown that 6B does not appear to directly increase cytokinin levels in plants (Wabiko and Minemura, 1996). Likewise, the development of *6b*-overexpressing plants is not typical of auxin overproducers and expression of *6b* cannot substitute for auxin in root induction assays (Tinland et al., 1990; Wabiko and Minemura, 1996). Still, 6B may function by altering plant cell sensitivity to cytokinins, by causing small, localized increases in auxin or cytokinin biosynthesis, or by altering the synthesis/perception of entirely different classes of plant hormones (Tinland et al., 1992; Wabiko and Minemura, 1996; Grémillon et al., 2004). Our muddled understanding of the biochemistry and mode of action of 6B is further complicated by substantial tumorigenic and regulatory differences between *6b* alleles from different *A. tumefaciens* strains (Helfer et al., 2002).

2.4 Gene *5*

Inactivation of T-DNA gene *5* generates a subtle phenotype which is apparent only in *iaaM* or *iaaH* mutants: an increase in the total number of shoots generated from the *tms*-type galls (Leemans et al., 1982). Similarly, the gene *5* transcript is abundant in teratomas (green tumors with some shoot and/or leaf development) induced by some nopaline strains of *A. tumefaciens*, but the transcript is almost undetectable in the undifferentiated tumors induced by octopine strains (Joos et al., 1983; Körber et al., 1991) demonstrated that gene *5* encodes an enzyme which catalyzes the synthesis of indole-3-lactate (ILA) from tryptophan. Gene *5*-over-expressing plants display a massive increase in ILA (>1000-fold), a modest decrease in free IAA and an increased tolerance to toxic levels of exogenously supplied IAA (Körber et al., 1991). ILA has almost no detectable auxin-like activity, but appears to compete with IAA in associating with some auxin binding proteins (Körber et al., 1991; Sprunck et al., 1995). Gene *5* is induced by auxin, but this induction is abolished by ILA, suggesting that the gene is regulated by a negative feedback loop (Körber et al., 1991). Thus, it appears that gene *5* can negatively moderate auxin responsiveness, thereby increasing the effective cytokinin:auxin ratio in the tumor and increasing shoot development. However, the lack of a clear phenotype in the gene *5* single mutant suggests that that gene *5* likely plays a non-essential role in tumorigenesis or that redundant activities exist elsewhere in the T-DNA.

2.5 Other *A. tumefaciens* oncogenes

The limited host range *A. tumefaciens* strain AB2/73 is capable of inducing tumors on only a few plant species, including *Lippia canescens*, *Kalanchoe tubiflora* and *Nicotiana glauca* (Otten and Schmidt, 1998).

AB2/73 possesses a very small (~3 kb) and unusual T-DNA region containing only two putative genes: *lso* and *lsn*. The *lsn* gene is highly homologous to nopaline synthase and the *lso* gene displays weak homology to gene *5*. *A. tumefaciens* strains harboring the *lso* gene alone induce small tumors on *Kalanchoe tubiflora* (but not *Nicotiana* spp.), indicating that *lso* does act as an oncogene, at least in some plant species (Otten and Schmidt, 1998). As yet, there are no reports detailing the biochemistry or mode of action of *lso*.

Similarly, the *3'* gene of *A. tumefaciens* strain Ach5 is capable of inducing tumors on *K. tubiflora*, but not other plant species (Otten et al., 1999). Gene *3'* is expressed in Ach5 tumors, but little else is currently known about this gene (Otten et al., 1999).

2.6 Tumorigenesis and hormone interactions

Crown gall tumorigenesis is often simplistically represented as merely the result of auxin and cytokinin overproduction. While it is true that *ipt* and *iaaM*/*iaaH* are the primary drivers of tumor formation, it has recently become apparent that complex hormone activation cascades contribute to crown gall development and morphology. For instance, the ethylene-insensitive tomato mutant *Never ripe* (*Nr*) develops small, smooth, crown gall tumors with an unbroken epidermis, which contrasts with the large, irregular tumors initiated on wild type tomato (Aloni et al., 1998). The high levels of auxins and cytokinins in tomato crown gall tumors are thought to activate ethylene synthesis, as ethylene levels are 50-fold higher in tumor tissues than control tissues (Aloni et al., 1998). The increased ethylene levels cause a decrease in xylem vessel diameter beside and above the tumor and contribute to the size and rough surface texture of the tumor itself, both of which are thought to ensure that the tumor is a major sink for water, supporting its continued growth (Aloni et al., 1998).

Similarly complex hormone interrelationships were recently documented in stem tumors of *Ricinus communis*. Expected increases in auxins and cytokinins were observed in stem tissues two weeks after inoculation with *A. tumefaciens*, but were rapidly followed by large increases in ethylene and abscisic acid (Veselov et al., 2003). These observations likely reflect a hormone signaling cascade in crown gall tumors (and other plant tissues), whereby auxins and/or cytokinins induce ethylene synthesis and ethylene induces the synthesis of abscisic acid (Veselov et al., 2003). Interestingly, levels of jasmonic acid, a plant hormone associated with stress and wounding, also exhibited a specific but transient increase shortly after

A. tumefaciens inoculation. Overall, these results demonstrate that tumorigenesis is influenced by a multitude of plant hormones with *iaaM*/*iaaH* and *ipt* (and auxin and cytokinin levels per se) triggering and acting in concert with multiple endogenous plant hormone biosynthesis pathways.

3 THE *A. RHIZOGENES* ONCOGENES

A. rhizogenes, like *A. tumefaciens*, invokes morphological changes in infected plant tissues and allows growth of transformed tissues *in vitro* in the absence of exogenous plant growth regulators. However, rather than undifferentiated tumors, highly branched, ageotropic roots emerge from sites of *A. rhizogenes* infection. Transformed roots can be regenerated into plants which, in many species, have a characteristic morphology (called the "hairy root" phenotype) that includes stunted growth, shortened internodes, reduced apical dominance, severely wrinkled leaves, atypical flower morphology and reduced fertility (Tepfer, 1984).

A few years after the discovery of the central role of the Ti plasmid in crown gall tumorigenesis, it was determined that *A. rhizogenes*-mediated rhizogenesis was also linked to a plasmid containing a T-DNA region that is transferred into the plant cell genome (Moore et al., 1979; White and Nester, 1980b; White and Nester, 1980a; Chilton et al., 1982). Because the hairy root phenotype closely mimics *A. tumefaciens tmr* mutants, it was initially hypothesized that *A. rhizogenes* possessed a defective Ti plasmid. Insertional mutagenesis demonstrated that the root-inducing (Ri) plasmid pRiA4b possesses a T-DNA region homologous to *tms* (designated T_R-DNA), as well as a second T-DNA region (T_L-DNA). Transposon insertions and deletions in the T_L-DNA as well as in the *aux* (*tms*-homologous) loci were shown to affect or eliminate the hairy root phenotype in *Kalanchoe diagremontiana* (White et al., 1985). Initially, four potential oncogenes were identified in the T_L-DNA and designated *rolA*, *B*, *C* and *D* (root locus).

Expression of the T_R-DNA alone can induce root formation in some plants, but the resulting phenotype is not as strong as when both T_L- and T_R-DNA are introduced together (Vilaine and Casse-Delbart, 1987). This suggests that the T_L-DNA may be required to extend host range in *A. rhizogenes* strains possessing binary T-DNAs (agropine strains) (Porter, 1991; Nilsson and Olsson, 1997). Auxin is also linked to the characteristic ageotropic root phenotype as both transformed and untransformed roots lose their geotropism when exogenous IAA is added (Capone et al., 1989).

In 1986, Slightom and colleagues (Slightom et al., 1986) sequenced the T_L-DNA from two agropine Ri plasmids and found 18 open reading frames (ORFs). The general organization of the T_L-DNA is similar to that found on Ti plasmids, including left and right border sequences and eukaryotic 5′ and 3′ regulatory elements. In addition to *rolA*, *B*, *C* and *D*, ORFs 8, 13 and 14 were also identified as oncogenes. Several of the Ri- and Ti-plasmid oncogenes genes appear to be homologs descended from a single ancestor. These include the Ti genes *tms1/iaaM*, *ons, tml, 5*, *6a* and *6b* as well as the Ri genes *rolB*, $rolB_{TR}$, *rolC*, ORF8, ORF13, ORF14 and ORF18 (Levesque et al., 1988; Otten and Helfer, 2001). It should be noted that cucomopine, mannopine and mikimopine strains of *A. rhizogenes* possess only a single T-DNA which is similar to agropine T_L-DNA, but lacking *rolD*.

In subsequent studies (Cardarelli et al., 1987a; Spena et al., 1987; Capone et al., 1989), various plants were transformed with the four *rol* genes, individually and in combination. These experiments demonstrated that the genes interacted synergistically. However, differences were seen in various plant species, leading Spena to hypothesize that endogenous plant factors may interact with the *rol* gene products, generating different phenotypes in distinctive plants and tissues.

Due to the complexity of the genetic interactions influencing the hairy root phenotype, most subsequent research concentrated on specific genes. Individual genes have been expressed under inducible and constitutive promoters (typically the Cauliflower Mosaic Virus 35S promoter) to observe phenotypes and to measure hormone levels. T-DNA promoters have also been studied using the GUS reporter gene, which has elucidated the pattern of gene expression by tissue type and developmental stage.

3.1 *rolA*

The *rolA* gene is found on all Ri plasmids. However, only the N-terminal half of the protein is highly conserved in studied *A. rhizogenes* strains. Transgenic tobacco expressing *rolA* are short and bushy, with dark green wrinkled leaves and abnormal flowers (Schmulling et al., 1988; Carneiro and Vilaine, 1993). *rolA* is transcribed in phloem cells, with stronger expression in stems and weaker expression in roots and leaves (Sinkar et al., 1988; Carneiro and Vilaine, 1993). The wrinkled leaf phenotype results from inhibition of elongation in leaf parenchyma cells adjacent to the vascular bundles. It has therefore been hypothesized that expression of *rolA* generates a diffusible factor. This hypothesis was supported by reciprocal grafting experiments in which *rolA* rootstocks and scions modified the phenotype of the untransformed plants to which they were grafted (Guivarc'h et al., 1996a).

Expression of the *rolA* gene in tobacco causes dramatic decreases in several classes of hormones, including auxin, cytokinin, gibberellin and abscisic acid. The amount of decrease is highly dependent on developmental stage and tissue type (Dehio et al., 1993). Gibberellic acid (GA) biosynthesis inhibitors also cause similar phenotypes, but treatment with GA does not completely restore *rolA* transformants (Dehio et al., 1993). Although GA_1 and GA_{20} concentrations are reduced, precursors GA_{53} and GA_{19} increase, indicating that a block in the GA 20-oxidase complex may be partially responsible for the phenotype (Moritz and Schmulling, 1998). Despite the lower auxin concentration in *rolA* plants, auxin sensitivity is greatly enhanced, particularly in protoplasts and in young seedlings (Vansuyt et al., 1992; Maurel et al., 1991).

In 1994, Magrelli and colleagues (Magrelli et al., 1994) discovered an intron in the *rolA* 5′ UTR and showed that mutations in the splice site abolish the *rolA* phenotype but do not reduce transcript levels. The intron is a bacterial promoter, with initiation of transcription inside the intron (Pandolfini et al., 2000), allowing the gene to be expressed both in *A. rhizogenes* and in the eukaryotic plant host (where the bacterial promoter is spliced out). In *A. rhizogenes*, *rolA* expression is maximal during stationary growth, suggesting a function in long-term survival of the bacteria in soil, where they are most likely in a nutrient-limited stationary growth phase (Pandolfini et al., 2000).

The exact function of the RolA protein is unknown, but the promoter has sequences similar to known auxin-regulated genes (Carneiro and Vilaine, 1993). Although RolA:GUS fusions localize to the plasma membrane, no transmembrane motifs have been identified, so it is thought to be a non-integral membrane-associated protein (Vilaine et al., 1998). It has also been hypothesized that RolA may interfere with protein degradation pathways, because RolA appears to stabilize GUS activity when the fusion protein is expressed (Barros et al., 2003).

3.2 *rolB*

Early studies indicated that *rolB* was the most important oncogene because, when inactivated, transformation with the remaining oncogenes failed to produce the hairy root phenotype. In many plants, *rolB* alone is sufficient to induce rooting, creating roots that are fast growing, highly branched and non-geotropic (Altamura, 2004). Originally believed to specifically induce roots, *rolB* is now known to stimulate the formation of new meristems that subsequently differentiate into specific organs depending on local hormone concentrations (Altamura, 2004). The *rolB* gene is present in all Ri plasmids, with approximately 60% identity between strains (Meyer et al., 2000).

Expression of *rolB* is limited to phloem parenchyma, rays, pericycle and root, shoot and flower meristems (Altamura et al., 1991). *rolB* transgenics display adventitious root formation and modified shoot morphology (necrotic leaves, altered leaf shape, increased flower size and heterostyly; Schmülling et al., 1989). Experiments with tobacco thin cell layers (TCLs) showed that *rolB* both stimulates adventitious roots and enhances adventitious flowering by affecting the developmental stage at which meristemoids are formed (Altamura et al., 1994). These results strongly support the role of *rolB* in stimulating meristem induction, with local hormone balance determining subsequent differentiation patterns (Altamura et al., 1998). In addition, experiments indicate a close association between auxin and *rolB*. For example, tobacco hairy roots and protoplasts have much higher auxin sensitivity than corresponding untransformed tissues (Shen et al., 1988; Shen et al., 1990; Maurel et al., 1994; Spanò et al., 1998). In carrot, *rolB* alone cannot induce rooting; auxin is required and can be provided by the T_R-DNA *aux* genes (Cardarelli et al., 1987b; Capone et al., 1989).

The *rolB* and *rolC* genes (expressed on opposite strands) are both controlled by a single bi-directional promoter. Experiments using the GUS reporter gene showed that auxin induces *rolB* (increasing expression 20-100 fold) and *rolC* (increasing expression 5 fold) (Maurel et al., 1990). However, the increase in *rolB* transcript does not occur until several hours after auxin is added to the culture (Maurel et al., 1994). In tobacco, the Dof (DNA binding with one finger) domain of the endogenous transcription factor NtBBF1 (*Nicotiana tabacum rolB* domain B factor 1) binds to a *cis*-regulatory element that is required for auxin induction and meristem-specific expression (De Paolis et al., 1996). This binding site is conserved between Ri plasmids with different *rolB* sequences (Handayani et al., 2005). Although the expression patterns of NtBBF1 and *rolB* are very similar and NtBBF1 appears to play a role in regulating *rolB*, NtBBF1 is not sensitive to auxin concentration (Baumann et al., 1999). Another protein, RBF1 (*Rol* Binding Factor 1), binds to a *rolB* promoter domain responsible for expression in certain non-meristem root cells. The concentration of RBF1 does not differ between tobacco transformed with *rolB* and non-transformed plants (Filetici et al., 1997).

The RolB protein encoded by pRiA4 localizes to the plasma membrane and demonstrates tyrosine phosphatase activity, so it has been hypothesized that it alters a kinase/phosphatase cascade in the auxin signal transduction pathway (Filippini et al., 1996). However, an alignment of RolB protein sequences from four different Ri plasmids demonstrated that only pRiA4 RolB contains a conserved CX_5R motif thought to be required for phosphatase activity (Lemcke and Schmulling, 1998b). Therefore, either the other RolB proteins have a different functional motif or only the pRiA4 RolB is a tyrosine phosphatase, which may indicate that this is not the main activity of the protein *in planta*. In contrast to strain A4, the mikimopine strain 1724 posseses a RolB protein which is nuclear localized. pRi1724 RolB has been shown to specifically bind to the tobacco protein Nt14-3-3ωII (Moriuchi et al., 2004). However, no typical nuclear localization signal was found in the sequence of either pRi1724 RolB or Nt14-3-3ωII, so it is possible that both of these proteins interact with another factor for nuclear import.

Despite somewhat conflicting experimental results, an overall hypothesis of *rolB*-induced organogenesis has emerged (Baumann et al., 1999; Meyer et al., 2000; Altamura, 2004). The current hypothesis holds that RolB alters auxin perception (Maurel et al., 1994). Antibodies against an auxin binding protein can block auxin response and a higher concentration of antibodies is needed if *rolB* is expressed, so RolB may either induce the expression of more auxin binding proteins and/or increase their activity (Venis et al., 1992). Auxin and the actions of transcription factor NtBBF1 regulate *rolB* expression. The RolB protein perturbs a signal transduction pathway involved in auxin perception (perhaps through tyrosine-phosphatase activity), increasing auxin sensitivity. This increased sensitivity stimulates the cell to become meristematic. The local level of hormones, the physiological state of the plant and the cell's overall competence determines what type of organ the meristem will differentiate into. It is likely that this process often results in root induction because roots are, in general, the most frequent adventitious organ to be formed.

3.3 *rolB*$_{TR}$ (*rolB* homologue in T$_R$-DNA)

Within the agropine Ri plasmid T_R -DNA, a *rolB* homolog was found and designated *rolB*$_{TR}$. This gene shares 53% nucleotide similarity to *rolB* in the coding sequence, but no similarity in the 5′ or 3′ flanking sequences

(Bouchez and Camilleri, 1990). Tobacco plants expressing 35S:*rolB*$_{TR}$ had slightly wrinkled leaves bent strongly downward, formed shoots at the base of the stem and showed retarded growth. This phenotype is different than that of *rolB* under either its own promoter or the CaMV 35S promoter. An alignment of the protein sequences of *rolB* and *rolB*$_{TR}$ showed two significant differences. First, *rolB*$_{TR}$ lacks a CX_5R motif, and second, *rolB*$_{TR}$ encodes 14 amino acids at the N-terminus which, when deleted, abolish the altered phenotype (Lemcke and Schmulling, 1998b).

3.4 *rolC*

The rolC proteins encoded on the various Ri plasmids share over 65% amino acid identity (Meyer et al., 2000). Plants transformed with *rolC* under its own promoter are short, display reduced apical dominance, have lanceolate leaves and yield early inflorescences with small flowers and poor pollen production (Schmulling et al., 1988). Dwarfing is caused by reduced epidermal cell size in internodes (Oono et al., 1990). In addition, root production is increased compared to untransformed plants, but decreased compared to plants transformed with all the *rol* genes (Palazón et al., 1998). The *rolC* gene is only expressed in phloem companion cells and in root protophloem initial cells (Guivarc'h et al., 1996b). In plants transformed with *rolC* alone, expression is strongest in roots, however when the rest of the T_L-DNA is present, *rolC* is also expressed in leaves (Sugaya et al., 1989; Oono et al., 1990; Leach and Aoyagi, 1991).

Plants expressing 35S:*rolC* are male-sterile and have a more pronounced phenotype including pale green leaves (Schmülling et al., 1988). Seedlings have a higher tolerance to auxin, gibberellins and abscisic acid, but increased sensitivity to cytokinin (Schmülling et al., 1993). The relationship between ethylene and polyamine metabolism is also altered, with a reduction in ethylene production in the flowers and an increase in the accumulation of water-insoluble polyamine conjugates (Martin-Tanguy et al., 1993).

Various experiments have produced conflicting results and hypotheses as to the biochemical activity of RolC. Estruch et al. (1991a) demonstrated that the protein can release free, active cytokinins by cleaving inactive cytokinin glucosides through a β-glycolytic activity. However, this observation was based on *in vitro* activity assays using recombinant RolC. Attempts to measure cytokinin levels *in planta* have produced conflicting results. Although Estruch et al. (1991a) measured increased cytokinin levels, Nilsson et al. (1993), working with a different tobacco cultivar, found that cytokinin levels decreased due to reduced biosynthesis. Faiss et al. (1996), expressing *rolC* under a tetracycline-inducible promoter, found that RolC did not hydrolyze cytokinin glucosides and that the levels of free cytokinins did not change in any plant tissue. It also appears that cytokinin glucosides are sequestered in the vacuole and are thus unavailable to RolC, which is located in the cytoplasm (Faiss et al., 1996; Nilsson and Olsson, 1997). Faiss hypothesized that oligosaccharins may be potential RolC substrates, since they can influence plant development.

It has also been suggested that the effects of RolC could be due to changes in gibberellin levels because a reduction of GA_1 and an increase in GA_{19} concentration was measured in *rolC* transgenics (Nilsson et al., 1993; Schmulling et al., 1993). Treatment with GA_3 restored internode length in 35S-*rolC* plants, but did not restore male fertility (Schmülling et al., 1993). Nilsson et al. (1993) looked at gibberellin levels in different plant tissues and found higher levels in leaves than in internodes. The level of GA_{19} was particularly increased, indicating that RolC might block the conversion of GA_{19} to GA_{20}. GA_1 concentration was reduced, which may explain the reduction in leaf cell size.

Although *rolC* expression is generally limited to phloem companion cells, the gene can be induced in any cell that has been soaked in sucrose (Nilsson et al., 1996). The sucrose-responsive region of the *rolC* promoter overlaps with the sequence that controls expression in phloem cells (identified by Sugaya and Uchimiya, 1992), indicating that the two are linked (Yokoyama et al., 1994). Nilsson and Olsson (1997) hypothesized that sucrose may be a substrate for RolC and that this may assist in the early stages of root initiation, since sucrose promotes cell division. RolC may therefore regulate sugar metabolism and transport by creating a strong sink to promote the unloading of sucrose.

Several endogenous plant nuclear proteins that interact with the *rolC* promoter have been identified (Kanaya et al., 1990; Kanaya et al., 1991; Suzuki et al., 1992; Matsuki and Uchimiya, 1994; Fujii, 1997). Because there is a sequence in the intergenic region between *rolB* and *rolC* that is essential for high promoter activity, some DNA binding proteins may jointly regulate the expression of both genes (Leach and Aoyagi, 1991).

Since *rolC* is induced by sucrose and *rolB* is induced by auxin, these genes will only be expressed in the same cell if there are concentrations of each compound above some threshold levels. The cells in which this naturally occurs are the phloem companion, parenchyma and ray cells, which are normally unresponsive to local auxin concentration. Therefore, sensitivity to auxin must be increased, which could be accomplished by the tyrosine phosphatase activity of RolB. This would activate a signal transduction pathway that ultimately leads to formation of a new meristem. So, RolC may create the sink required to increase sugar production, while RolB increases auxin sensitivity (Nilsson and Olsson, 1997).

3.5 *rolD*

Unlike the other *rol* genes, *rolD* is only found in the T_L-DNA of agropine Ri plasmids. It is also the only *rol* gene that is incapable of inducing root formation on its own (Mauro et al., 1996). The main phenotype of *rolD*-expressing tobacco plants is early and increased flowering, and *reduced* rooting (Mauro et al., 1996). This phenotype is independent of hormones present in culture media (Altamura, 2004). Due to the early transition from vegetative to reproductive growth, some *rolD*-expressing plants develop no vegetative buds. Although flower production is increased, the flowers display heterostyly, which prevents self-fertilization. Manually-selfed plants produce viable seeds (Mauro et al., 1996). However, it should be noted that these experiments were performed using the *rolD* sequence from pRi1855. It has been reported that *rolD* from pRiHRI does not induce flowering (Lemcke and Schmülling, 1998a).

Expression of *rolD* is not tissue specific, but is developmentally regulated. Activity is seen in the elongating and expanding tissues of each organ in adult plants, but never in apical meristems. As the plants age, expression decreases and ceases at senescence (Trovato et al., 1997). Like *rolB*, *rolD* is a late-auxin induced gene, with a lag time of at least four hours. However, while *rolB* promoter activation strengthens with increasing auxin concentration, induction of the *rolD* promoter reaches a maximum at 1 μM IAA and then decreases at higher auxin levels (Mauro et al., 2002). Similar to *rolB*, the *rolD* promoter has a Dof-binding element which is likely involved in auxin induction. Altamura (2004) suggested that *rolD* is involved at a later stage of meristem formation than *rolB* and may be responsible for determining the fate of the meristem, perhaps involving a stress-response.

RolD is a cytosolic protein with a sequence similar to ornithine cyclodeaminase (OCD), a bacterial enzyme which converts ornithine to proline (Trovato et al., 2001). Plants do not have an endogenous OCD protein and produce proline via a different pathway. However, *A. tumefaciens* (but not *A. rhizogenes*) uses an OCD enzyme to catabolize opines derived from arginine (Dessaux et al., 1986; Sans et al., 1988; Schindler et al., 1989).

Thus, the *rolD* phenotype may be related to proline production. Proline is an osmoprotectant related to the stress response in plants. It may also have a role in flowering because high proline levels are found in flowers (Trovato et al., 2001). The *rolD* gene may be responsible for altering the hormone balance in plant tissues and thus inducing flowering. Low levels of auxin are required for flowering, while high auxin concentrations inhibit it (Mauro et al., 1996). This may explain why *rolD* promoter induction decreases as auxin levels rise above a threshold. Trovato (2001) has speculated that proline may increase the biosynthesis of hydroxyproline-rich glycoproteins, cell wall components believed to be involved in the regulation of cell division and extension.

3.6 ORF3n

The T_L-DNA sequence contains several transcribed ORFs other than the *rol* genes (Slightom et al., 1986). Lemcke and Schmülling (1998a) investigated some of these using the *A. rhizogenes* strain HRI. Because ORF3 in pRiHRI is slightly larger than that of pRiA4, it was designated ORF3n. Expression of 35S-ORF3n in transgenic tobacco caused alterations in internode length, leaf morphology and growth. Onset of flowering was delayed and inflorescenses were less dense than in untransformed plants. The tips of upper leaves, sepals and bracts became necrotic.

Shoot formation from ORF3n callus was inhibited on media containing auxin and cytokinin. Plantlets also showed decreased sensitivity to auxin and cytokinin, remaining small and forming less callus than controls. Lemcke and Schmülling (1998a) suggested that ORF3n may act to suppress the dedifferentation of tissues, which may favor the formation of *rol* gene-induced roots from such cells. The ORF3n protein resembles phenolic-modifying enzymes and may be involved in secondary metabolism and/or the transport of hormones.

3.7 ORF8

ORF8 has the longest sequence of any T_L-DNA gene (Slightom et al., 1986), coding for a protein containing 780 amino acids. It is also one of the most conserved genes, with 81% amino acid sequence similarity between pRiA4 and pRi2659 (Ouartsi et al., 2004). Various researchers have expressed full length ORF8 under a CaMV-35S promoter in transgenic tobacco. However, the phenotypes of the resulting plants have differed. Lemcke et al. (2000) reported no morphological changes, but a five-fold increase in IAM production over untransformed tobacco. In contrast, Otten and Helfer (2001) did not observe an increase in IAM concentration in their ORF8 overexpressing plants. Ouartsi et al. (2004) reported altered cotyledon morphology which was attributed to auxin-induced cell expansion and division. Umber et al. (2005) observed significant morphological differences between untransformed and 35S-ORF8 plants, including stunted growth and rough, mottled leaves with thick, fleshy midribs. It is possible that these phenotypic differences may be due to the use of different versions of the CaMV 35S promoter (e.g., tetracycline-inducible, double 35S enhancer, etc.) and/or to the use of different Ri plasmids for the ORF8 coding sequence.

Plant cell tolerance to exogenous auxins and cytokinins is increased by ORF8. Seedlings expressing ORF8 are able to grow on media containing concentrations of auxin that completely inhibit the growth of wild-type and *iaaM* transgenic seedlings (Lemcke and Schmülling, 1998a). Tobacco leaf discs transformed with 35S-ORF8 produce fewer, but thicker, roots and more callus than untransformed leaf discs (Ouartsi et al., 2004). ORF8 expression is dramatically increased by exogenous auxin, but with a lag time of about six hours, indicating that ORF8, like *rolB* and *rolD*, is in the category of late-auxin response genes. However, there is no conserved Dof binding motif in the ORF8 promoter. Instead, the ORF8 promoter has a RAV1 binding sequence that is also found in the promoters of plant genes encoding auxin binding proteins.

The ORF8 N-terminal domain shows some homology to the RolB protein and the C-terminus has a low but significant similarity to the iaaM proteins of *A. tumefaciens* and other plant pathogenic bacteria (Levesque et al., 1988). Lemcke et al. (2000) reported increased IAM production in 35S-ORF8 plants and confirmed the presence of tryptophan monooxygenase activity by expressing the protein in *E. coli*. Otten and Helfer (2001) separately expressed the N-terminal and C-terminal domains of ORF8 and *A. tumefaciens iaaM* in tobacco under the CaMV 35S promoter. Expression of the *A. tumefaciens* iaaM N-terminal domain did not produce any discernable phenotype, while the C-terminal domain was a fully functional iaaM protein. Neither the full-length ORF8 protein, nor either of the domains individually produced IAM. However, when only the N-terminal domain (*Norf8*) was expressed, the transformed plants showed a novel phenotype including chlorotic and necrotic leaves, stunted growth and a large increase in the concentrations of glucose, fructose, sucrose and starch in the leaves (Otten and Helfer, 2001).

Tobacco plants expressing 35S-*Norf8* displayed a "high starch" phenotype resembling those deficient in a sucrose transporter (NtSUT1) that export less sucrose from source leaves than do untransformed plants. Experiments showed that only sugar loading is impaired, because 35S-*Norf8* shoots grafted onto untransformed plants resume growth due to sugars transported from the wild-type leaves (Umber et al., 2002). In contrast, the phenotypically normal 35S-*Corf8* plants (expressing only the C-terminal iaaM-like domain) accumulate less starch than do untransformed plants. Crosses between 35S-*Corf8* and 35S-*Norf8* plants result in heterozygotes with an intermediate phenotype for sucrose and starch accumulation, further demonstrating that the Corf8 domain can reduce carbohydrate accumulation in both *Norf8* and wild-type backgrounds (Umber et al., 2005).

The function of the ORF8 protein has not been determined, but it has been hypothesized that modification of sucrose transport may be related to *rolC* promoter induction by this compound (Ouartsi et al., 2004). It is also possible that high levels of sugars may be beneficial to the bacterium and could contribute to opine synthesis (Otten and Helfer, 2001).

3.8 ORF13

Transformation with the *rol* genes alone is insufficient for hairy root induction in carrot discs. Rhizogenesis is observed only with simultaneous transformation of either the T_R-DNA *aux* genes or ORFs 13 and 14 (Cardarelli et al., 1987a; Capone et al., 1989). In tobacco, when different combinations of *rolB*, *rolC*, ORF13 and ORF14 were used to transform leaf discs, it was found that *rolB* and ORF13 together induced rooting nearly as well as the full length T-DNA (Aoki and Syōno, 1999b). These observations led to the hypothesis that ORF13 may be involved in auxin biosynthesis. However, ORF13, a protein of approximately 200 amino acids that is highly conserved between Ri proteins, has no similarity to auxin biosynthesis genes (Hansen et al., 1997).

The ORF13 promoter is strongly wound-inducible in most tissues. Induction begins five hours after wounding and reaches maximum 17 hours later. This wound-induced expression is seen only in tissues immediately surrounding the wound and is not systemic. The addition of exogenous auxin after wounding increases ORF13 activity, but cytokinin does not (Specq et al., 1994; Hansen et al., 1997). This pattern of wound induction is similar to that of several *A. tumefaciens* T-DNA-encoded genes, including *nos*, *mas* and *6b*. However, there is no similarity between the *A. tumefaciens* and ORF13 promoters. Repeats of an 11 bp motif have been identified in the ORF13 promoter that may be responsible for wound induction (Hansen et al., 1997).

When expressed in tobacco plants under an inducible promoter or the CaMV35S promoter, ORF13 strongly inhibits cell division and elongation in meristems and developing leaves (Lemcke and Schmulling, 1998a). This resulted in slow growth, reduced apical dominance, short internodes, inhibited root elongation and severely wrinkled dark green leaves with increased chlorophyll content. Although flowers are morphologically abnormal, preventing self-fertilization, seed production is normal if manually pollinated (Hansen et al., 1993; Specq et al., 1994; Lemcke and Schmulling, 1998a). Similar to *rolA*, the 35S-ORF13 phenotype is graft transmissible in a non-polar fashion, so ORF13 expression must induce a diffusible substance (Hansen et al., 1993).

ORF13 is the only Ri T-DNA gene that induces cell proliferation. Both carrot discs and tobacco leaf discs inoculated with 35S-ORF13 develop green callus (Hansen et al., 1993; Fründt et al., 1998). Exogenous cytokinin, but not auxin, increases the number of roots produced from 35S-ORF13 tobacco leaf discs, but does not change root induction on untransformed leaf discs, even though there was no difference in endogenous cytokinin levels (Specq et al., 1994; Lemcke and Schmulling, 1998a).

Recent experiments have shown that ORF13 contains a retinoblastoma-binding (Rb) domain. When mutations are introduced into the Rb motif, normal leaf size is restored, but plants still show stunting and reduced apical dominance, so there must be additional functional domains in the protein (Stieger et al., 2004). In mammals, Rb is a tumor suppressor protein that negatively regulates the progression from G1 to S phase in the cell cycle. Rb homologs have been found in maize and tobacco. ORF13 increases the number of mitoses in the shoot apical meristem resulting in cell proliferation, but does not influence meristem structure. This interference in cell cycle regulation results in a premature cessation of organ growth, resulting in smaller leaves. Stieger et al. (2004) hypothesized that ORF13 activates cell division for T-DNA replication and induces dedifferentiation, a prerequisite to cell competence that is required for a new differentiation program. This may improve transformation efficiency by inducing cells to reenter the cell cycle.

3.9 Other *A. rhizogenes* T-DNA genes

Among the additional ORFs in the T_L-DNA, there are two genes which may also contribute to the hairy root phenotype. ORF13a was identified when pRi8196 (mannopine type) was sequenced (Hansen et al., 1991). It is located between ORF13 and ORF14 on the opposite strand. Shorter, homologous ORFs are conserved in the same region of pRiA4 (agropine) and pRi2659 (cucumopine). This gene is transcribed in plants in a tissue specific manner, primarily in leaf vascular tissues (Hansen et al., 1994). Because ORF13a contains motifs common to phorphorylated gene regulatory proteins, Hansen et al. (1994) hypothesized that the protein may interact directly with DNA.

ORF14 is in the same gene family as *rolB*, *rolC*, ORF8 and ORF13 (Levesque et al., 1988) and is flanked by eukaryotic regulatory sequences typical of other expressed T-DNA genes (Slightom et al., 1986). Although overexpression of ORF14 did not produce morphological changes (Lemcke and Schmulling, 1998a), it has been shown to act synergistically with the *rol* genes and ORF13 to improve root induction in carrot and tobacco (Capone et al., 1989; Aoki and Syōno, 1999b). Additional research is needed to understand the actions of these genes *in planta.*

3.10 Plant homologues to Ri genes

Early Southern blots, performed to confirm T-DNA integration, also showed hybridization to genomic DNA of untransformed tobacco plants (White et al., 1982; White et al., 1983). The genomic region of homology was termed cT-DNA (cellular T-DNA) (White et al., 1983). Subsequently, highly conserved homologs of *rolB*, *rolC*, ORF13 and ORF14 were found in several *Nicotiana* species, including *N. glauca* and *N. tabacum* (Furner et al., 1986; Aoki et al., 1994; Meyer et al., 1995). These genes are aligned in the plant genome in the same order as in the Ri T-DNA, but no *rolD* homolog is present (Aoki et al., 1994). The cT-DNA sequences are found in only a subset of *Nicotiana* species (Furner et al., 1986) and are likely present due to infection of an ancestral plant by a bacterium harboring a mikimopine-like Ri plasmid (Suzuki et al., 2002). Analysis of the cT-DNA insertion sites in different species indicates that there were two or three independent transformation events (Suzuki et al., 2002). Transcription of the cT-DNA genes has been shown to occur *in planta* (Meyer et al., 1995; Aoki and Syōno, 1999a; Intrieri and Buiatti, 2001). A detailed review and analysis of these genes has recently been published (Aoki, 2004).

3.11 Ri T-DNA genetic interactions

When the individual Ri genes are viewed together (see *Table 14-1*), common themes do emerge. Three genes (*rolB*, *rolD* and ORF8) are induced by auxin, but do not respond until hours after auxin is introduced, indicating that there is not a direct cause and effect, but the oncogenes instead influence and are influenced by an auxin-induced signal transduction cascade. Previous studies have demonstrated cross-talk between hormones (summarized by Ross and O'Neill, 2001) and have indicated that auxin can influence the gibberellin biosynthesis pathway (O'Neill and Ross, 2002). Both *rolA* and *rolC* expression reduces cell size and both may be involved in blocking GA synthesis. Recently, it has been shown that KNOX (KNOTTED1-like homeobox) transcription factors are involved in the maintenance of cell function in the shoot apical meristem and that this activity is regulated by levels of cytokinin and gibberellins (Jasinski et al., 2005 and reviewed by Hudson, 2005). The KNOX proteins repress the transcription of GA 20-oxidases, leading to reduction of active GA1. Since expression of either *rolA* or *rolC* also leads to an apparent block of the GA 20-oxidase complex, it is possible that one or both of these proteins may influence or interact with the KNOX genes.

It is possible that the products of some of the oncogenes, including RolA, RolC and ORF3n may influence the production of secondary metabolites, including flavonoids. Such compounds, naturally produced by plants, have been shown to regulate auxin reception and transport (Jacobs and Rubery, 1988). It could be hypothesized that, by altering the production of endogenous secondary metabolites, or by catalyzing the production of analogs, a single oncogene could alter the cellular concentration of multiple hormones, as occurs in plants overexpressing *rolA*.

Sucrose also appears to play a significant role in the induction of new meristems. The ORF8 protein modifies the transport of sucrose. This increased sucrose concentration induces the expression of the RolC gene product, which may contribute to sink formation by promoting sucrose unloading. Wounding induces the expression of ORF13, leading to dedifferentiation of cells, promoting their competence and increasing transformation efficiency. This activity may be held in check by the ORF3n protein which appears to suppress dedifferentiation. The RolB gene product stimulates the formation of a new meristem, which will develop into a particular organ depending on the local concentrations of hormones. However,

Table 14-1. Oncogenes of *Agrobacterium Rhizogenes*. This is a summary of the currently hypothesized attributes of the oncogenes of the Ri plasmids (see text for references)

Gene	Protein	Expression	Function
rolA	• Non-integral membrane-associated	• Expressed in both bacteria and plants • May be auxin regulated	• Inhibits cell elongation via diffusible factor • Decreases hormone concentrations • GA biosynthesis block? • Interferes with protein degradation? • Graft transmissible
rolB	• Tyrosine phosphatase • Localizes to plasma membrane but may be transported to nucleus	• Induced by auxin (late) • Inhibited by oligogalacturonides • NtBBF1 and RBF1 transcription factors bind to promoter • Interacts with Nt14-3-3ωII	• Alters auxin perception/sensitivity through perturbation of signal transduction pathway • Stimulates new meristem formation
rolC	• Cleaves cytokinin glucosides or perhaps oligosaccharins • Cytoplasmic	• Induced by sucrose • Plant nuclear proteins interact with promoter	• Reduces cell size • May promote cell division (through sucrose) • GA biosynthesis block? • May regulate sugar metabolism by creating a strong sink to promote unloading of sucrose

Table 14-1(cont). Oncogenes of *Agrobacterium Rhizogenes*. This is a summary of the currently hypothesized attributes of the oncogenes of the Ri plasmids (see text for references)

Gene	Protein	Expression	Function
ORF3n	• Modification of phenolic compounds?	• ?	• Suppression of tissue dedifferentiation?
ORF8	• Tryptophan monooxygenase activity?	• Induced by auxin (late)	• Modifies sucrose transport • N-terminal domain causes sugar/starch accumulation • C-terminal domain reduces sugar/starch accumulation
ORF13	• Retinoblastoma domain	• Induced by wounding	• Perturbs cell cycle to activate cell division • Increases number of mitoses in shoot apical meristem • Induces dedifferentiation (prerequisite to competence) • Graft transmissible

the cellular hormone concentrations are influenced by the activity of the oncogenes as described above. Finally, in cells transformed by agropine plasmids, the RolD protein may provoke a stress response through increased proline production, which may further influence cell division and appears to specifically induce flowering. Proline may also increase the biosynthesis of glycoproteins (rich in hydroxyproline), cell wall structural components believed to be involved in the regulation of cell division (Trovato et al., 2001). Parallel to this, it has been shown that the cytokinin-influenced accumulation of certain cell wall components (dehydrodiconiferyl glucosides) in crown gall tumors may promote cell division (Binns et al., 1987).

4 CONCLUSIONS

Over the past 25 years, the genetics and molecular biology underlying crown gall disease and hairy root disorder have been studied extensively. In the case of *A. tumefaciens*, proliferation of undifferentiated plant cells is triggered by expression of Ti oncogenes that induce the overproduction of auxins and cytokinins. The genes encoded on Ri plasmids, however, influence cellular pathways involved in growth and differentiation in ways that are not yet completely understood. This is complicated by the fact that a great deal of the existing research has utilized tobacco species that already contain endogenous homologs to some Ri genes. These homologs may influence the outcome of such experiments in subtle ways, making more difficult the task of dissecting the molecular and biochemical pathways underlying the morphological changes provoked by Ri oncogenes. As the mechanisms underlying the pathology of crown gall and hairy root are dissected, they can be exploited to develop novel methodologies. As shown in *Figure 14-3*, the technique of RNA interference has been used to silence the *iaaM* and *ipt* oncogenes *in planta*, resulting in crop plants that are functionally resistant to crown gall disease (Escobar et al., 2001; Escobar et al., 2002). In addition, a dexamethasone-inducible *ipt* gene construct has been developed as an efficient selectable marker system for plant transformation (Kunkel et al., 1999). The hairy root phenotype and underlying genes are also being investigated in a number of different systems for plant improvement (see review by Casanova et al., 2005) and have been adapted to produce several high-value compounds, including secondary metabolites, in bioreactors (recently reviewed by Uozumi, 2004).

Figure 14-3. Functional resistance to crown gall disease in oncogene-silenced walnut microshoots. Wild-type (right) and *iaaM/ipt*-silenced (left) walnut shoots were inoculated with virulent *A. tumefaciens* and tumor formation was assayed six weeks post-inoculation. Tumorigenesis is completely suppressed in the oncogene-silenced line (inoculation sites denoted with arrows), while the wild type displays characteristic large, undifferentiated tumor development.

The crown gall and hairy root systems can also be used to elucidate endogenous plant pathways, particularly those involved in hormone utilization and cell division and differentiation. Although the protein coding regions of the *Agrobacterium* oncogenes appear to be of bacterial origin, the eukaryotic promoters and *cis*-regulatory regions allow endogenous plant transcription factors acting in *trans* to induce and regulate expression in different tissues and cell types. Further research into these oncogene pathways will lead to a better understanding of plant development and how it can be manipulated for plant improvement.

5 REFERENCES

Akiyoshi DE, Klee H, Amasino RM, Nester EW, Gordon MP (1984) T-DNA of *Agrobacterium tumefaciens* encodes an enzyme of cytokinin biosynthesis. Proc Natl Acad Sci USA 81: 5994-5998

Akiyoshi DE, Morris RO, Hinz R, Kosuge T, Garfinkel DJ, Gordon MP, Nester EW (1983) Cytokinin/auxin balance in crown gall tumors is regulated by specific loci in the T DNA. Proc Natl Acad Sci USA 80: 407-411

Akiyoshi DE, Reiger DA, Jen G, Gordon MP (1985) Cloning and nucleotide sequence of the *tzs* gene from *Agrobacterium tumefaciens* strain T37. Nucl Acids Res 13: 2773-2788

Aloni R, Wolf A, Feigenbaum P, Avni A, Klee HJ (1998) The never ripe mutant provides evidence that tumor-induced ethylene controls the morphogenesis of *Agrobacterium tumefaciens*-induced crown galls on tomato stems. Plant Physiol 117: 841-849

Altamura MM (2004) *Agrobacterium rhizogenes rolB* and *rolD* genes: regulation and involvement in plant development. Plant Cell Tissue Org Cult 77: 89-101

Altamura MM, Artchilletti T, Capone I, Costantino P (1991) Histological analysis of the expression of *Agrobacterium rhizogenes rolB*-GUS gene fusions in transgenic tobacco. New Phytol 118: 67-78

Altamura MM, Capitani F, Gazza L, Capone I, Costantino P (1994) The plant oncogene *rolB* stimulates the formation of flower and root meristemoids in tobacco thin cell layers. New Phytol 126: 283-293

Altamura MM, D'Angeli S, Capitani F (1998) The protein of *rolB* gene enhances shoot formation in tobacco leaf explants and thin cell layers from plants in different physiological stages. J Exp Bot 49: 1139-1146

Aoki S (2004) Resurrection of an ancestral gene: functional and evolutionary analyses of the Ng*rol* genes transferred from *Agrobacterium* to *Nicotiana*. J Plant Res 117: 329-337

Aoki S, Kawaoka A, Sekine M, Ichikawa T, Fujita T, Shinmyo A, Syōno K (1994) Sequence of the cellular T-DNA in the untransformed genome of *Nicotiana glauca* that is homologous to ORFs 13 and 14 of the Ri plasmid and analysis of its expression in genetic tumours of *N. glauca* x *N. langsdorffii*. Mol Gen Genet 243: 706-710

Aoki S, Syōno K (1999a) Horizontal gene transfer and mutation: Ng*rol* genes in the genome of *Nicotiana glauca*. Proc Natl Acad Sci USA 96: 13229-13234

Aoki S, Syōno K (1999b) Synergistic function of *rolB*, *rolC*, ORF13 and ORF14 of TL-DNA of *Agrobacterium rhizogenes* in hairy root induction in *Nicotiana tabacum*. Plant Cell Physiol 40: 252-256

Astot C, Dolezal K, Nordström A, Wang Q, Kunkel T, Moritz T, Chua N-H, Sandberg G (2000) An alternative cytokinin biosynthesis pathway. Proc Natl Acad Sci USA 97: 14778-14783

Barker R, Idler K, Thompson DV, Kemp J (1983) Nucleotide sequence of the T-DNA region from the *Agrobacterium tumefaciens* octopine Ti plasmid pTi 15955. Plant Mol Biol 2: 335-350

Barros LMG, Curtis RH, Viana AAB, Campos L, Carneiro M (2003) Fused RolA protein enhances beta-glucoronidase activity 50-fold: implication for RolA mechanism of action. Protein Pept Lett 10: 303-311

Barry GF, Rogers SG, Fraley RT, Brand L (1984) Identification of cloned cytokinin biosynthesis gene. Proc Natl Acad Sci USA 81: 4776-4780

Baumann K, De Paolis A, Costantino P, Gualberti G (1999) The DNA binding site of the Dof protein NtBBF1 is essential for tissue-specific and auxin-regulated expression of the *rolB* oncogene in plants. Plant Cell 11: 323-334

Binns AN, Chen RH, Wood HN, Lynn DG (1987) Cell division promoting activity of naturally occurring dehydrodiconiferyl glucosides: Do cell wall components control cell division? Proc Natl Acad Sci USA 84: 980-984

Binns AN, Costantino P (1998) The *Agrobacterium* oncogenes. *In* HP Spaink, A Kondorosi, PJJ Hooykaas, eds, The Rhizobiaceae. Kluwer Academic Publishers, Dordrecht, The Netherlands, pp 251-266

Bouchez D, Camilleri C (1990) Identification of a putative *rolB* gene on the TR-DNA of the *Agrobacterium rhizogenes* A4 Ri plasmid. Plant Mol Biol 14: 617-619

Braun AC (1982) A history of the crown gall problem. *In* G Kahl, J Schell, eds, Molecular Biology of Plant Tumors. Academic Press, New York, pp 155-210

Capone I, Spanò L, Cardarelli M, Bellincampi D, Petit A, Costantino P (1989) Induction and growth properties of carrot roots with different complements of *Agrobacterium rhizogenes* T-DNA. Plant Mol Biol 13: 43-52

Cardarelli M, Mariotti D, Pomponi M, Spanò L, Capone I, Costantino P (1987a) *Agrobacterium rhizogenes* T-DNA genes capable of inducing hairy root phenotype. Mol Gen Genet 209: 475-480

Cardarelli M, Spanò L, Mariotti D, Mauro ML, Van Sluys MA, Costantino P (1987b) The role of auxin in hairy root induction. Mol Gen Genet 208: 457-463

Carneiro M, Vilaine F (1993) Differential expression of the *rolA* plant oncogene and its effect on tobacco development. Plant J 3: 785-792

Casanova E, Trillas MI, Moysset L, Vainstein A (2005) Influence of *rol* genes in floriculture. Biotechnol Adv 23: 3-39

Chateau S, Sangwan RS, Sangwan-Norreel BS (2000) Competence of *Arabidopsis thaliana* genotypes and mutants for *Agrobacterium tumefaciens*-mediated gene transfer: role of phytohormones. J Exp Bot 51: 1961-1968

Chilton M-D, Drummond MH, Merio DJ, Sciaky D, Montoya AL, Gordon MP, Nester EW (1977) Stable incorporation of plasmid DNA into higher plant cells: the molecular basis of crown gall tumorigenesis. Cell 11: 263-271

Chilton M-D, Tepfer DA, Petit A, David C, Casse-Delbart F, Tempé J (1982) *Agrobacterium rhizogenes* inserts T-DNA into the genomes of host plant root cells. Nature 295: 432-434

Cohen JD, Slovin JP, Hendrickson AM (2003) Two genetically discrete pathways convert tryptophan to auxin: more redundancy in auxin biosynthesis. Trends Plant Sci 8: 197-199

De Paolis A, Sabatini S, De Pascalis L, Costantino P, Capone I (1996) A *rolB* regulatory factor belongs to a new class of single zinc finger plant proteins. Plant J 10: 215-223

Dehio C, Grossmann K, Schell J, Schmülling T (1993) Phenotype and hormonal status of transgenic tobacco plants overexpressing the *rolA* gene of *Agrobacterium rhizogenes* T-DNA. Plant Mol Biol 23: 1199-1210

Dessaux Y, Petit A, Tempé J, Demarez M, Legrain C, Wiame JM (1986) Arginine catabolism in *Agrobacterium* strains: role of the Ti plasmid. J Bacteriol 166: 44-50

Dunoyer P, Himber C, Voinnet O (2006) Induction, suppression and requirement of RNA silencing pathways in virulent *Agrobacterium tumefaciens* infections. Nature Genet 38: 258-263

Eklöf S, Åstot C, Blackwell J, Moritz T, Olsson O, Sandberg G (1997) Auxin-cytokinin interactions in wild-type and transgenic tobacco. Plant Cell Physiol 38: 225-235

Escobar MA, Civerolo EL, Summerfelt KR, Dandekar AM (2001) RNAi-mediated oncogene silencing confers resistance to crown gall tumorigenesis. Proc Natl Acad Sci USA 98: 13437-13442

Escobar MA, Dandekar AM (2003) *Agrobacterium tumefaciens* as an agent of disease. Trends Plant Sci 8: 380-386

Escobar MA, Leslie CA, McGranahan GH, Dandekar AM (2002) Silencing crown gall disease in walnut (*Juglans regia* L.). Plant Sci 163: 591-597

Estruch JJ, Chriqui D, Grossmann K, Schell J, Spena A (1991a) The plant oncogene *rolC* is responsible for the release of cytokinins from glucoside conjugates. Embo J 10: 2889-2895

Estruch JJ, Prinsen E, van Onckelen H, Schell J, Spena A (1991b) Viviparous leaves produced by somatic activation of an inactive cytokinin-synthesizing gene. Science 254: 1364-1367

Faiss M, Strnad M, Redig P, Doležal K, Hanuš J, Van Onckelen H, Schmülling T (1996) Chemically induced expression of the *rolC*-encoded B-glucosidase in transgenic tobacco plants and analysis of cytokinin metabolism: *rolC* does not hydrolyze endogenous cytokinin glucosides *in planta*. Plant J 10: 33-46

Filetici P, Moretti F, Camilloni G, Mauro ML (1997) Specific interaction between a *Nicotiana tabacum* nuclear protein and the *Agrobacterium rhizogenes rolB* promoter. J Plant Physiol 151: 159-165

Filippini F, Rossi V, Marin O, Trovato M, Costantino P, Downey PM, Lo Schiavo F, Terzi M (1996) A plant oncogene as a phosphatase. Nature 379: 499-500

Fründt C, Meyer AD, Ichikawa T, Meins F, Jr. (1998) A tobacco homologue of the Ri-plasmid *orf13* gene causes cell proliferation in carrot root discs. Mol Gen Genet 259: 559-568

Fujii N (1997) Pattern of DNA binding of nuclear proteins to the proximal Agrobacterium rhizogenes rolC promoter is altered during somatic embryogenesis of carrot. Gene 201: 55-62

Furner IJ, Huffman GA, Amasino RM, Garfinkel DJ, Gordon MP, Nester EW (1986) An *Agrobacterium* transformation in the evolution of the genus *Nicotiana*. Nature 319: 422-427

Garfinkel DJ, Nester EW (1980) *Agrobacterium tumefaciens* mutants affected in crown gall tumorigenesis and octopine catabolism. J Bacteriol 144: 732-743

Garfinkel DJ, Simpson RB, Ream LW, White FF, Gordon MP, Nester EW (1981) Genetic analysis of crown gall: fine structure map of the T-DNA by site-directed mutagenesis. Cell 27: 143-153

Gaudin V, Jouanin L (1995) Expression of *Agrobacterium rhizogenes* auxin biosynthesis genes in transgenic tobacco plants. Plant Mol Biol 28: 123-136

Gaudin V, Vrain T, Jouanin L (1994) Bacterial genes modifying hormonal balances in plants. Plant Physiol Biochem 32: 11-29

Grémillon L, Helfer A, Clément B, Otten L (2004) New plant growth-modifying properties of the *Agrobacterium* T-*6b* oncogene revealed by the use of a dexamethasone-inducible promoter. Plant J 37: 218-228

Guivarc'h A, Carneiro M, Vilaine F, Pautot V, Chriqui D (1996a) Tissue-specific expression of the *rolA* gene mediates morphological changes in transgenic tobacco. Plant Mol Biol 30: 125-134

Guivarc'h A, Spena A, Noin M, Besnard C, Chriqui D (1996b) The pleiotropic effects induced by the *rolC* gene in transgenic plants are caused by expression restricted to protophloem and companion cells. Transgenic Res 5: 3-11

Handayani NSN, Moriuchi H, Yamakawa M, Yamashita I, Yoshida K, Tanaka N (2005) Characterization of the *rolB* promoter on mikimopine-type pRi1724 T-DNA. Plant Sci 108: 1353-1364

Hansen G, Larribe M, Vaubert D, Tempé J, Biermann B, Montoya A, Chilton M, Brevet J (1991) *Agrobacterium rhizogenes* pRi8196 T-DNA: Mapping and DNA sequence of functions involved in mannopine synthesis and hairy root differentiation. Proc Natl Acad Sci USA 88: 7763-7767

Hansen G, Vaubert D, Clérot D, Brevet J (1997) Wound-inducible and organ-specific expression of ORF13 from *Agrobacterium rhizogenes* 8196 T-DNA in transgenic tobacco plants. Mol Gen Genet 254: 337-343

Hansen G, Vaubert D, Clérot D, Tempé J, Brevet J (1994) A new open reading frame, encoding a putative regulatory protein, in *Agrobacterium rhizogenes* T-DNA. C R Acad Sci Paris, Life Sciences 317: 49-53

Hansen G, Vaubert D, Heron JN, Clérot D, Tempé J, Brevet J (1993) Phenotypic effects of overexpression of *Agrobacterium rhizogenes* T-DNA ORF13 in transgenic tobacco plants are mediated by diffusible factor(s). Plant J 4: 581-585

Helfer A, Pien S, Otten L (2002) Functional diversity and mutational analysis of *Agrobacterium* 6B oncoproteins. Mol Genet Genomics 267: 577-586

Hooykaas PJJ, den Dulk-Ras H, Schilperoort RA (1988) The *Agrobacterium tumefaciens* T-DNA gene *6b* is an *onc* gene. Plant Mol Biol 11: 791-794

Hudson A (2005) Plant meristems: mobile mediators of cell fate. Curr Biol 15: R803-805

Intrieri MC, Buiatti M (2001) The horizontal transfer of *Agrobacterium rhizogenes* genes and the evolution of the genus *Nicotiana*. Mol Phylogenet Evol 20: 100-110

Jacobs M, Rubery PH (1988) Naturally occurring auxin transport regulators. Science 241: 346-349

Jasinski S, Piazza P, Craft J, Hay A, Woolley L, Rieu I, Phillips A, Hedden P, Tsiantis M (2005) KNOX action in *Arabidopsis* is mediated by coordinate regulation of cytokinin and gibberellin activities. Curr Biol 15: 1560-1565

Joos H, Inzé D, Caplan A, Sormann M, Van Montagu M, Schell J (1983) Genetic analysis of T-DNA transcripts in nopaline crown galls. Cell 32: 1057-1067

Kakimoto T (2001) Identification of plant cytokinin biosynthetic enzymes as dimethylallyl diphosphate: ATP/ADP isopentenyltransferases. Plant Cell Physiol 42: 677-685

Kanaya K, Hayakawa K, Uchimiya H (1991) In vitro binding of wheatgerm proteins to the 5'-upstream region of the *rolC* gene of Ri plasmid. Plant Cell Physiol 32: 295-297

Kanaya K, Tabata T, Iwabuchi M, Uchimiya H (1990) Specific binding of nuclear protein from tobacco hairy roots cultured in vitro to a 5'-upstream region of the *rolC* gene of the Ri plasmid. Plant Cell Physiol 31: 941-946

Kitakura S, Fujita T, Ueno Y, Terakura S, Wabiko H, Machida Y (2002) The protein encoded by oncogene *6b* from *Agrobacterium tumefaciens* interacts with a nuclear protein of tobacco. Plant Cell 14: 451-463

Klee H, Montoya A, Horodyski F, Lichtenstein C, Garfinkel D, Fuller S, Flores C, Peschon J, Nester E, Gordon M (1984) Nucleotide sequence of the *tms* genes of the pTiA6NC octopine Ti plasmid: two gene products involved in plant tumorigenesis. Proc Natl Acad Sci USA 81: 1728-1732

Klee HJ, Horsch RB, Hinchee MA, Hein MB, Hoffmann NL (1987) The effects of overproduction of two *Agrobacterium tumefaciens* T-DNA auxin biosynthetic gene products in transgenic petunia plants. Genes Dev 1: 86-96

Körber H, Strizhov N, Staiger D, Feldwisch J, Olsson O, Sandberg G, Palme K, Schell J, Koncz C (1991) T-DNA gene *5* of *Agrobacterium* modulates auxin response by autoregulated synthesis of a growth hormone antagonist in plants. Embo J 10: 3983-3991

Kulescha Z (1954) Croissance et teneur en auxin de divers tissues normaux et tumoraux. L'Annee Biol 3e Ser 30: 319-327

Kunkel T, Niu QW, Chan YS, Chua N-H (1999) Inducible isopentenyl transferase as a high-efficiency marker for plant transformation. Nat Biotechnol 17: 916-919

Leach F, Aoyagi K (1991) Promoter analysis of the highly expressed *rolC* and *rolD* root-inducing genes of *Agrobacterium rhizogenes*: enhancer and tissue-specific DNA determinants are dissociated. Plant Sci 79: 69-76

Lee BM, Park YJ, Park DS, Kang HW, Kim JG, Song ES, Park IC, Yoon UH, Hahn JH, Koo BS, Lee GB, Kim H, Park HS, Yoon KO, Kim JH, Jung CH, Koh NH, Seo JS, Go SJ (2005) The genome sequence of *Xanthomonas oryzae* pathovar *oryzae* KACC10331, the bacterial blight pathogen of rice. Nucleic Acids Res 33: 577-586

Leemans J, Deblaere R, Willmitzer L, De Greeve H, Hernalsteens JP (1982) Genetic identification of functions of TL-DNA transcripts in octopine crown galls. Embo J 1: 147-152

Lemcke K, Prinsen E, van Onckelen H, Schmülling T (2000) The ORF8 gene product of *Agrobacterium rhizogenes* TL-DNA has tryptophan 2-monooxygenase activity. Mol Plant-Microbe Interact 13: 787-790

Lemcke K, Schmülling T (1998a) Gain of function assays identify non-*rol* genes from *Agrobacterium rhizogenes* TL-DNA that alter plant morphogenesis or hormone sensitivity. Plant J 15: 423-433

Lemcke K, Schmülling T (1998b) A putative *rolB* gene homologue of the *Agrobacterium rhizogenes* TR-DNA has different morphogenetic activity in tobacco than *rolB*. Plant Mol Biol 36: 803-808

Levesque H, Delepelaire P, Rouzé P, Slightom J, Tepfer D (1988) Common evolutionary origin of the central portions of the Ri TL-DNA of *Agrobacterium rhizogenes* and the Ti T-DNAs of *Agrobacterium tumefaciens*. Plant Mol Biol 11: 731-744

Magrelli A, Langenkemper K, Dehio C, Schell J, Spena A (1994) Splicing of the *rolA* transcript of *Agrobacterium rhizogenes* in *Arabidopsis*. Science 266: 1986-1988

Manulis S, Haviv-Chesner A, Brandl MT, Lindow SE, Barash I (1998) Differential involvement of indole-3-acetic acid biosynthetic pathways in pathogenicity and epiphytic fitness of *Erwinia herbicola* pv. *gypsophilae*. Mol Plant-Microbe Interact 11: 634-642

Martin-Tanguy J, Corbineau F, Burtin D, Ben-Hayyim G, Tepfer D (1993) Genetic transformation with a derivative of *rolC* from *Agrobacterium rhizogenes* and treatment with a-aminoisobutyric acid produce similar phenotypes and reduce ethylene production and the accumulation of water-insoluble polyamine-hydroxycinnamic acid conjugates in tobacco flowers. Plant Sci 93: 63-76

Matsuki R, Uchimiya H (1994) A 43-kDa nuclear tobacco protein interacts with a specific single-stranded DNA sequence from the 5'-upstream region of the *Agrobacterium rhizogenes rolC* gene. Gene 141: 201-205

Maurel C, Brevet J, Barbier-Brygoo H, Guern J, Tempé J (1990) Auxin regulates the promoter of the root-inducing *rolB* gene of *Agrobacterium rhizogenes* in transgenic tobacco. Mol Gen Genet 223: 58-64

Maurel C. Barbier-Brygoo H, Spena A, Tempé J. Guern, J. (1991) Single *rol* genes from the *Agrobacterium rhizogenes* T_L-DNA alter some of the cellular responses to auxin in *Nicotiana tabacum*. Plant Physiol 97: 212-216

Maurel C, Leblanc N, Barbier-Brygoo H, Perrot-Rechenmann C, Bouvier-Durand M, Guern J (1994) Alterations of auxin perception in *rolB*-transformed tobacco protoplasts. Time course of *rolB* mRNA expression and increase in auxin sensitivity reveal multiple control by auxin. Plant Physiol 105: 1209-1215

Mauro ML, De Lorenzo G, Costantino P, Bellincampi D (2002) Oligogalacturonides inhibit the induction of late but not of early auxin-responsive genes in tobacco. Planta 215: 494-501

Mauro ML, Trovato M, Paolis AD, Gallelli A, Costantino P, Altamura MM (1996) The plant oncogene *rolD* stimulates flowering in transgenic tobacco plants. Dev Biol 180: 693-700

Messens E, Lenaerts A, van Montagu M, Hedges RW (1985) Genetic basis for opine secretion from crown gall tumor cells. Mol Gen Genet 199: 344-348

Meyer AD, Ichikawa T, Meins F, Jr. (1995) Horizontal gene transfer: regulated expression of a tobacco homologue of the *Agrobacterium rhizogenes rolC* gene. Mol Gen Genet 249: 265-273

Meyer AD, Tempé J, Costantino P (2000) Hairy root: a molecular overview. Functional analysis of *Agrobacterium rhizogenes* T-DNA genes. *In* G Stacy, NT Keen, eds, Plant Microbe Interactions. APS Press, St. Paul, MN, pp 93-139

Miller CO (1974) Ribosyl-trans-zeatin, a major cytokinin produced by crown gall tumor tissue. Proc Natl Acad Sci USA 71: 334-338

Moore L, Warren G, Strobel G (1979) Involvement of a plasmid in the hairy root disease of plants caused by *Agrobacterium rhizogenes*. Plasmid 2: 617-626

Moritz T, Schmülling T (1998) The gibberellin content of *rolA* transgenic tobacco plants is specifically altered. J Plant Physiol 153: 774-776

Moriuchi H, Okamoto C, Nishihama R, Yamashita I, Machida Y, Tanaka N (2004) Nuclear localization and interaction of RolB with plant 14-3-3 proteins correlates with induction of adventitious roots by the oncogene *rolB*. Plant J 38: 260-275

Neuteboom STC, Hulleman E, Schilperoort RA, Hoge HC (1993) *In planta* analysis of the *Agrobacterium tumefaciens* T-*cyt* gene promoter: identification of an upstream region essential for promoter activity in leaf, stem and root cells of transgenic tobacco. Plant Mol Biol 22: 923-929

Nilsson O, Little CHA, Sandberg G, Olsson O (1996) Expression of two heterologous promoters, *Agrobacterium rhizogenes rolC* and cauliflower mosaic virus 35S, in the stem of transgenic hybrid aspen plants during the annual cycle of growth and dormancy. Plant Mol Biol 31: 887-895

Nilsson O, Moritz T, Imbault N, Sandberg G, Olsson O (1993) Hormonal characterization of transgenic tobacco plants expressing the *rolC* gene of *Agrobacterium rhizogenes* TL-DNA. Plant Physiol 102: 363-371

Nilsson O, Olsson O (1997) Getting to the root: The role of the *Agrobacterium rhizogenes rol* genes in the formation of hairy roots. Physiol Plant 100: 463-473

O'Neill DP, Ross JJ (2002) Auxin regulation of the gibberelllin pathway in pea. Plant Physiol 130: 1974-1982

Oono Y, Kanaya K, Uchimiya H (1990) Early flowering in transgenic tobacco plants possessing the *rolC* gene of *Agrobacterium rhizogenes* Ri plasmid. Jpn J Genet 68: 7-16

Otten L, Helfer A (2001) Biological activity of the *rolB*-like 5' end of the A4-*orf8* gene from the *Agrobacterium rhizogenes* TL-DNA. Mol Plant-Microbe Interact 14: 405-411

Otten L, Salomone J-Y, Helfer A, Schmidt J, Hammann P, De Ruffray P (1999) Sequence and functional analysis of the left-hand part of the T-region from the nopaline-type Ti plasmid, pTiC58. Plant Mol Biol 41: 765-776

Otten L, Schmidt J (1998) A T-DNA from the *Agrobacterium tumefaciens* limited-host-range strain AB2/73 contains a single oncogene. Mol Plant-Microbe Interact 11: 335-342

Ouartsi A, Clérot D, Meyer AD, Dessaux Y, Brevet J, Bonfill M (2004) The T-DNA ORF8 of the cucumopine-type *Agrobacterium rhizogenes* Ri plasmid is involved in auxin response in transgenic tobacco. Plant Sci 166: 5577-5567

Palazón J, Cusidó RM, Roig C, Piñol MT (1998) Expression of the *rolC* gene and nicotine production in transgenic roots and their regenerated plants. Plant Cell Rep 17: 384-390

Pandolfini T, Storlazzi A, Calabria E, Defez R, Spena A (2000) The spliceosomal intron of the *rolA* gene of *Agrobacterium rhizogenes* is a prokaryotic promoter. Mol Microbiol 35: 1326-1334

Robinette D, Matthysse AG (1990) Inhibition by *Agrobacterium tumefaciens* and *Pseduomonas savastanoi* of development of the hypersensitive response elicited by *Pseudomonas syringae* pv. phaseolicola. J Bact 172: 5742-5749

Porter JR (1991) Host range and implications of plant infection by *Agrobacterium rhizogenes*. Crit Rev Plant Sci 10: 387-421

Powell GK, Morris RO (1986) Nucleotide sequence and expression of a *Pseudomonas savastanoi* cytokinin biosynthetic gene: homology with *Agrobacterium tumefaciens tmr* and *tzs* loci. Nucl Acids Res 14: 2555-2565

Ross J, O'Neill J (2001) New interactions between classical plant hormones. Trends Plant Sci 6: 2-4

Sakakibara H, Kasahara H, Ueda N, Kojima M, Takei K, Hishiyama S, Asami T, Okada K, Kamiya Y, Yamaya T, Yamaguchi S (2005) *Agrobacterium tumefaciens* increases cytokinin production in plastids by modifying the biosynthetic pathway in the host plant. Proc Natl Acad Sci USA 102: 9972-9977

Sans N, Schindler U, Schröder J (1988) Ornithine cyclodeaminase from Ti plasmid C58: DNA sequence, enzyme properties and regulation of activity by arginine. Eur J Biochem 173: 123-130

Sauter C, Blum S (2003) Regression of lung lesions in Hodgkin's disease by antibiotics: case report and hypothesis on the etiology of Hodgkin's disease. Am J Clin Oncol 26: 92-94

Schindler U, Sans N, Schröder J (1989) Ornithine cyclodeaminase from octopine Ti plasmid Ach5: identification, DNA sequence, enzyme properties, and comparison with gene and enzyme from nopaline Ti plasmid C58. J Bacteriol 171: 847-854

Schmülling T, Fladung M, Grossmann K, Schell J (1993) Hormonal content and sensitivity of transgenic tobacco and potato plants expressing single *rol* genes of *Agrobacterium rhizogenes* T-DNA. Plant J 3: 371-382

Schmülling T, Schell J, Spena A (1988) Single genes from *Agrobacterium rhizogenes* influence plant development. Embo J 7: 2621-2629

Schmülling T, Schell J, Spena A (1989) Promoters of the *rolA*, *B*, and *C* genes of *Agrobacterium rhizogenes* are differentially regulated in transgenic plants. Plant Cell 1: 665-670

Schröder G, Waffenschmidt S, Weiler EW, Schröder J (1984) The T-region of Ti plasmids codes for an enzyme synthesizing indole-3-acetic acid. Eur J Biochem 138: 387-391

Schwalm K, Aloni R, Langhans M, Heller W, Stich S, Ullrich CI (2003) Flavonoid-related regulation of auxin accumulation in *Agrobacterium tumefaciens*-induced plant tumors. Planta 218: 163-178

Shen WH, Davioud E, David C, Barbier-Brygoo H, Tempé J, Guern J (1990) High sensitivity to auxin is a common feature of hairy root. Plant Physiol 94: 554-560

Shen WH, Petit A, Guern J, Tempé J (1988) Hairy roots are more sensitive to auxin than normal roots. Proc Natl Acad Sci USA 85: 3417-3421

Sinkar VP, Pythoud F, White FF, Nester EW, Gordon MP (1988) *rolA* locus of the Ri plasmid directs developmental abnormalities in transgenic tobacco plants. Genes Dev 2: 688-697

Slightom JL, Durand-Tardif M, Jouanin L, Tepfer D (1986) Nucleotide sequence analysis of TL-DNA of *Agrobacterium rhizogenes* agropine type plasmid. Identification of open reading frames. J Biol Chem 261: 108-121

Smith EF, Townsend CO (1907) A plant tumor of bacterial origin. Science 25: 671-673

Spanò L, Mariotti D, Cardarelli M, Branca C, Costantino P (1998) Morphogenesis and auxin sensitivity of transgenic tobacco with different complements of Ri T-DNA. Plant Physiol 87: 479-483

Specq A, Hansen G, Vaubert D, Clérot D, Heron JN, Tempé J, Brevet J (1994) Studies on hairy root T-DNA: regulation and properties of ORF13 from *Agrobacterium rhizogenes* 8196. *In* Plant Pathogenic Bacteria, Versailles (France), pp 465-468

Spena A, Schmülling T, Koncz C, Schell JS (1987) Independent and synergistic activity of *rol A*, *B* and *C* loci in stimulating abnormal growth in plants. Embo J 6: 3891-3899

Sprunck S, Jacobsen HJ, Reinard T (1995) Indole-3-lactic acid is a weak auxin analogue but not an anti auxin. J Plant Growth Regul 14: 191-197

Stieger PA, Meyer AD, Kathmann P, Fründt C, Niederhauser I, Barone M, Kuhlemeier C (2004) The *orf13* T-DNA gene of *Agrobacterium rhizogenes* confers meristematic competence to differentiated cells. Plant Physiol 135: 1798-1808

Strabala TJ, Crowell DN, Amasino RM (1993) Levels and location of expression of the *Agrobacterium tumefaciens* pTiA6 *ipt* gene promoter in transgenic tobacco. Plant Mol Biol 21: 1011-1021

Sugaya S, Hayakawa K, Handa T, Uchimiya H (1989) Cell-specific expression of the *rolC* gene of the TL-DNA of Ri plasmid in transgenic tobacco plants. Plant Cell Physiol 305: 649-653

Sugaya S, Uchimiya H (1992) Deletion analysis of the 5'-upstream region of the *Agrobacterium rhizogenes* Ri plasmid *rolC* gene required for tissue-specific expression. Plant Physiol 99: 464-467

Sun J, Niu Q-W, Tarkowski P, Zheng B, Tarkowska D, Sandberg G, Chua N-H, Zuo J (2003) The Arabidopsis AtIPT8/PGA22 gene encodes an isopentenyl transferase that is involved in de novo cytokinin biosynthesis. Plant Physiol 131: 167-176

Suzuki A, Kato A, Uchimiya H (1992) Single-stranded DNA of 5'-upstream region of the *rolC* gene interacts with nuclear proteins of carrot cell cultures. Biochem Biophys Res Commun 188: 727-733

Suzuki K, Yamashita I, Tanaka N (2002) Tobacco plants were transformed by *Agrobacterium rhizogenes* infection during their evolution. Plant J 32: 775-787

Takei K, Sakakibara H, Sugiyama T (2001) Identification of genes encoding adenylate isopentenyltransferase, a cytokinin biosynthesis enzyme, in *Arabidopsis thaliana*. J Biol Chem 276: 26405-26410

Tepfer D (1984) Transformation of several species of higher plants by *Agrobacterium rhizogenes*: sexual transmission of the transformed genotype and phenotype. Cell 37: 959-967

Thomashow LS, Reeves S, Thomashow MF (1984) Crown gall oncogenesis: evidence that a T-DNA gene from the *Agrobacterium* Ti plasmid pTiA6 encodes an enzyme that catalyzes synthesis of indoleacetic acid. Proc Natl Acad Sci USA 81: 5071-5075

Thomashow MF, Hughly S, Buchholz WG, Thomashow LS (1986) Molecular basis for the auxin-independent phenotype of crown gall tumor tissue. Science 231: 616-618

Tinland B, Fournier P, Heckel T, Otten L (1992) Expression of a chimaeric heat-shock-inducible *Agrobacterium 6b* oncogene in *Nicotiana rustica*. Plant Mol Biol 18: 921-930

Tinland B, Rohfritsch O, Michler P, Otten L (1990) *Agrobacterium tumefaciens* T-DNA gene *6b* stimulates *rol*-induced root formation, permits growth at high auxin concentrations and increases root size. Mol Gen Genet 223: 1-10

Trovato M, Maras B, Linhares F, Costantino P (2001) The plant oncogene *rolD* encodes a functional ornithine cyclodeaminase. Proc Natl Acad Sci USA 98: 13449-13453

Trovato M, Mauro ML, Costantino P, Altamura MM (1997) The *rolD* gene from *Agrobacterium rhizogenes* is developmentally regulated in transgenic tobacco. Protoplasma 197: 111-120

Ullrich CI, Aloni R (2000) Vascularization is a general requirement for growth of plant and animal tumours. J Exp Bot 51: 1951-1960

Umber M, Clément B, Otten L (2005) The T-DNA oncogene A4-*orf8* from *Agrobacterium rhizogenes* A4 induces abnormal growth in tobacco. Mol Plant-Microbe Interact 18: 205-211

Umber M, Voll L, Weber A, Michler P, Otten L (2002) The *rolB*-like part of the *Agrobacterium rhizogenes orf8* gene inhibits sucrose export in tobacco. Mol Plant-Microbe Interact 15: 956-962

Uozumi N (2004) Large-scale production of hairy root. Adv Biochem Eng Biotechnol 91: 75-103

van Onckelen H, Prinsen E, Inzé D, Rüdelsheim P, Van Lijsebettens M, Follin A, Schell J, Van Montagu M, De Greef J (1986) *Agrobacterium* T-DNA gene *1* codes for tryptophan 2-monooxygenase activity in tobacco crown gall cells. FEBS Lett 198: 357-360

Vansuyt G, Vilaine F, Tepfer M, Rossigno M (1992) *rolA* modulates the sensitivity to auxin of the proton translocation catalyzed by the plasma membrane H+-ATPase in transformed tobacco. FEBS Lett 298: 89-92

Venis MA, Napier RM, Barbier-Brygoo H, Maurel C, Perrot-Rechenmann C, Guern J (1992) Antibodies to a peptide from the maize auxin-binding protein have auxin agonist activity. Proc Natl Acad Sci USA 89: 7208-7212

Veselov D, Langhans M, Hartung W, Aloni R, Feussner I, Götz C, Veselova S, Schlomski S, Dickler C, Bächmann K, Ullrich CI (2003) Development of *Agrobacterium tumefaciens* C58-induced plant tumors and impact on host shoots are controlled by a cascade of jasmonic acid, auxin, cytokinin, ethylene and abscisic acid. Planta 216: 512-522

Vilaine F, Casse-Delbart F (1987) Independent induction of transformed roots by the TL and TR regions of the Ri plasmid of agropine type *Agrobacterium rhizogenes*. Mol Gen Genet 206: 17-23

Vilaine F, Rembur J, Chriqui D, Tepfer M (1998) Modified development in transgenic tobacco plants expressing a rolA::GUS translational fusion and subcellular localization of the fusion protein. Mol Plant-Microbe Interact 11: 855-859

Wabiko H, Minemura M (1996) Exogenous phytohormone-independent growth and regeneration of tobacco plants transgenic for the *6b* gene of *Agrobacterium tumefaciens* AKE10. Plant Physiol 112: 939-951

Weiler EW, Spanier K (1981) Phytohormones in the formation of crown gall tumors. Planta 153: 326-337

White FF, Garfinkel DJ, Huffman GA, Gordon MP, Nester EW (1983) Sequence homologous to *Agrobacterium rhizogenes* T-DNA in the genomes of uninfected plants. Nature 301: 348-350

White FF, Ghidossi G, Gordon MP, Nester EW (1982) Tumor induction by *Agrobacterium rhizogenes* involves the transfer of plasmid DNA to the plant genome. Proc Natl Acad Sci USA 79: 3193-3197

White FF, Nester EW (1980a) Hairy root: plasmid encodes virulence traits in *Agrobacterium rhizogenes*. J Bacteriol 141: 1134-1141

White FF, Nester EW (1980b) Relationship of plasmids responsible for hairy root and crown gall tumorigenicity. J Bacteriol 144: 710-720

White FF, Taylor BH, Huffman GA, Gordon MP, Nester EW (1985) Molecular and genetic analysis of the transferred DNA regions of the root-inducing plasmid of *Agrobacterium rhizogenes*. J Bacteriol 164: 33-44

White PR, Braun AC (1942) A cancerous neoplasm of plants. Autonomous bacteria-free crown gall tissue. Cancer Res 2: 597-617

Willmitzer L, Dhaese P, Schreier PH, Schmalenbach W, Van Montagu M, Schell J (1983) Size, location and polarity of T-DNA-encoded transcripts in nopaline crown gall tumors; common transcripts in octopine and nopaline tumors. Cell 32: 1045-1056

Yamada T, Palm CJ, Brooks B, Kosuge T (1985) Nucleotide sequences of the *Pseudomonas savastanoi* indoleacetic acid genes show homology with *Agrobacterium tumefaciens* T-DNA. Proc Natl Acad Sci USA 82: 6522-6526

Yokoyama R, Hirose T, Fujii N, Aspuria ET, Kato A, Uchimiya H (1994) The *rolC* promoter of *Agrobacterium rhizogenes* Ri plasmid is activated by sucrose in transgenic tobacco plants. Mol Gen Genet 244: 15-22

Zhang XD, Letham DS, Zhang R, Higgins TJV (1996) Expression of the isopentenyl transferase gene is regulated by auxin in transgenic tobacco tissues. Transgenic Res 5: 57-65

Chapter 15

BIOLOGY OF CROWN GALL TUMORS

Roni Aloni[1] and Cornelia I. Ullrich[2]

[1]Department of Plant Sciences, Tel Aviv University, Tel Aviv 69978, Israel; [2]Institute of Botany, Darmstadt University of Technology, Schnittspahnstr. 3, 64287 Darmstadt, Germany

Abstract. Specific adaptive mechanisms for water and nutrient acquisition and the suppression of shoot and root differentiation characterize crown gall tumor development. Strong vascularization like in animal and human tumors is the most prominent and important feature of tumor proliferation. Vascular bundles consisting of phloem and xylem are from the onset of tumor initiation functionally connected to the host bundle. At the host/tumor interface the vessel number is considerably increased and interrupted by multiseriate rays. These altered structures enhance water flow into the tumor parenchyma and, together with the disruption of epidermis and cuticle, substantially support tumor transpiration. Expression of the T-DNA-encoded genes for abundant auxin and cytokinin biosynthesis trigger a cascade of further phytohormones, which are essential for tumor development as well. Auxin accumulation is particularly enhanced by the expression of the T-DNA-located gene *6b* for phenylpropanoids, hence for flavonoid biosynthesis. Spatio-temporal distribution patterns of the bioactive free and conjugated auxin and cytokinins, ethylene and abscisic acid match well the sites of highest chalcone synthase (*CHS*) expression and hence flavonoid concentration. Flavonoids accumulate at the sites of strongest free auxin accumulation and prevent basipetal auxin efflux, thus maintaining high auxin and cytokinin concentrations for induction and development of the vascular system. The considerable auxin- and cytokinin-enhanced ethylene emission is causally related with the development of the enlarged xylem in the tumor/host interface and the aerenchyma, which is important for aerobic energy metabolism; ethylene finally induces the accumulation of abscisic acid (ABA) in the tumor and host leaves. ABA in turn leads to diminished shoot water loss by enhancing closure of host leaf stomata, so that a stronger water supply to the tumor is guaranteed. In addition, ABA accumulation in the tumor periphery enhances accumulation of osmoprotectants such as sucrose and proline, to prevent tumor desiccation. Tumors accumulate high solute concentrations. The expression of root-specific K^+-influx channels (*AKT1* and *AtKC1*) is up-regulated while genes of anion transporters at the plasma membrane are down-regulated; therefore, an important role is attributed to phloem transport for xylem-derived nutrient import into the tumor parenchyma. The phloem sieve element/companion cell complex is well coupled to the tumor parenchyma by numerous plasmodesmata. Spatio-temporal analysis of the activity of sucrose degrading enzymes and of sugar accumulation confirm symplastic metabolite phloem unloading. In conclusion, predominantly auxin and cytokinin-induced ethylene have a key role for successful tumor establishment by tumor vascularization and, together with cuticular disruption, by redirecting of water flow and symplastic phloem unloading of carbohydrate, amino acid and anion import.

1 INTRODUCTION

Crown gall disease is the most fascinating of all known phytobacterioses due to its unique mechanism of transfer of the bacterial T-DNA into the higher plant genome, to its use in modern gene technology, to the perfect biological control by *Agrobacterium radiobacter* strains K 84 and 1026 and, last not least, to the phytohormone-controlled tumor structure specialized for just nutrient and water redirection from the host plant. The detection of particular tissue structures in plant tumors, namely the continuous vascular bundles, made them a unique model system to study phytohormone-controlled vascular bundle development, including membrane pumps, channels, specific carriers within vascular bundles and hence xylem and phloem long distance transport mechanisms related to tumor adaptations of water relations and carbohydrate metabolism.

Some benign human tumors are related to bacteria such as *Bartonella* (*Rochalimaea*) *henselae* which belongs to the same α-2 subgroup of proteobacteria as *Agrobacterium tumefaciens*. Therefore, the relationship between bacteria and certain plant, animal, and human malignant tumors, termed "crown gall hypothesis", is suggested as challenge for research if further tumors are incited by bacteria, such as Hodgkin's disease (Sauter and Blum, 2003; Dehio, 2005). It is expected that comparison of plant, animal and human tumor development may provide further insights into general principles of cancer pathogenesis and eventually leads to new strategies for tumor prevention or therapy.

2 CROWN GALL VASCULARIZATION

For a long time crown galls were regarded to be masses of cells, containing unorganized patterns of tracheary elements, vessels which are not at all or only scarcely connected with the host vascular system (Bopp and Leppla, 1964; Gordon, 1982; Weiler and Schröder, 1987; Tarbah and Goodman, 1988; Sachs, 1991; Schell et al., 1994; Agrios, 2004). Since the tumors reach diameters of up to 30 cm special structural and physiological pathways for efficient nutrient and water supply extending up to the rapidly growing periphery were suspected (Malsy et al., 1992; Aloni et al., 1995). Until then only vessels (tracheary elements) were found. Due to their prominent lignified structures they are much easier to recognize than sieve elements. Furthermore, because of the complicated three-dimensional organization and tree-like architecture of the tumor bundles, in thin tissue sections the tracheary elements appear as if they were idioblasts without connection to the host bundle (Agrios, 2004), or only scarcely connected to the main host bundle (Kupila-Ahvenniemi and Therman, 1968; Beiderbeck, 1977).

Re-investigation of the tumor structure by preparing a few mm thick sections, either after staining with aniline blue and viewing under UV-light (Malsy et al., 1992) or after clearing with lactic acid and staining with lacmoid (Aloni et al., 1995) made the three-dimensional structure of both, vessels and sieve elements, simultaneously visible throughout the tumors. This sophisticated structure became apparent as soon as 2 to 3 d after infection with *Agrobacterium tumefaciens* C58 or A281. Both vessels and sieve tubes differentiate with the growing tumor in all studied host plant species, *i.e., Ricinus, Vitis, Trifolium, Kalanchoë, Lycopersicon, Cucurbita and Arabidopsis*. Apparently, strong vascularization is a precondition of crown gall tumor proliferation in plants just as neovascularization is essential for development of animal and human tumors (Folkman, 1971; Gimbrone et al., 1974). Moreover, the induction of a neoangiogenic process by a bacterium is not restricted to *Agrobacterium* and higher plants. In bacillary angiomatosis the human pathogenic bacterium *Bartonella henselae* was found to cause vasculoproliferative disorders and to induce the secretion of vasculoproliferative cytokines from infected host cells, similar to vascular endothelial growth factors (VEGF), a major factor in tumor angiogenesis (Kempf et al., 2002; Dehio, 2004).

While leaf tumors are characterized by a dense net of single-stranded vessels and sieve elements (Malsy et al., 1992), stem tumors develop concentric bundles with inner xylem surrounded by phloem (*Figure 15-1* and

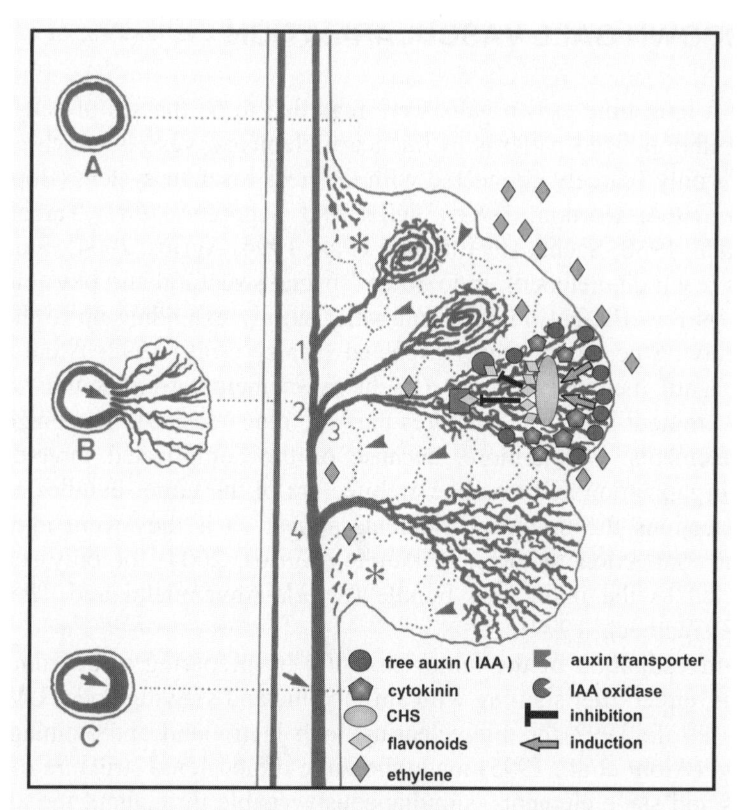

Figure 15-1. Longitudinal diagram with vascularization pattern (thick lines) and phloem anastomoses (dotted lines and arrowheads) in crown gall tumor and host stem. The three diagrams of cross sections illustrate the symmetric structure of a healthy vascular stem system above the tumor (A), the pathological host xylem (arrow) with multiseriate rays in a median position (B), and the asymmetrically enhanced vascular differentiation below the tumor (C). 1 and 2 mark the connecting sites of globular vascular bundles to the host vascular system, while 3 and 4 mark the base of tree-like branched bundles. Asterisks mark regenerative phloem fibres restricted to the upper and lower basal regions of the tumor. Color symbols indicate how T-DNA-elevated IAA concentration leads to enhanced chalcone synthase (*CHS*) expression, which intensifies the synthesis of flavonoid aglycones. These inhibit IAA oxidase. Flavonoids and IAA plus cytokinin (CK) –triggered ethylene impair the basipetal IAA transport (within the same cell as flavonoid synthesis) to regulate sufficiently high free-auxin concentration for tumor vascularization and continuous proliferation. (Follows Aloni et al., 1995, Figure. 1 and Schwalm et al., 2003)

Aloni et al., 1995). Bundles extend close to the tumor surface and are interconnected by a dense net of phloem anastomoses (*Figure 15-1*, *Figure 15-2* A and C and Aloni et al., 1995; Ullrich and Aloni, 2000). The vascular bundles are either of tree-like or of circular structure, the latter most probably induced by pathological circular flow of auxin (Aloni et al.,

1995). Amazingly, fibres are almost absent from the tumors, or are produced only at the border between the tumor and the host (Aloni et al., 1995), probably due to low gibberellin concentrations in relation to the high auxin and cytokinin concentrations; giberellins are known to be a limiting factor for fibre differentiation (Aloni, 1979). Tumor-induced ethylene also inhibits fibre formation; in the ethylene insensitive *Never ripe* tomato mutant therefore fibres are produced also inside crown galls (Aloni et al., 1998).

Usually tumors do not differentiate shoots and roots, indicating high inhibitory and mutually interacting concentrations of auxin and cytokinin. *Kalanchoë daigremontiana* and tobacco leaf and stem tumors however tend to produce teratomes and adventitious roots, depending on the position of tumor induction, *i.e.*, the variable balance of auxin and cytokinin in the upper part of the stem and in leaves (Hooykaas et al., 1982). When tumors on *Trifolium* runners get into contact with humid soil numerous roots proliferate (Schwalm et al., 2003).

Voluminous organs like crown galls are supposed to be endangered by anaerobiosis and hence to posses prevailing glycolytic metabolism, as known from animal and human tumors (Warburg, 1930; Aisenberg, 1961; Pedersen, 1978). To encounter this, plant tumors develop a dense net of vessels with oxygen-transporting water flow. In addition, in the periphery where less bundles are found in the non-transformed former host cortex tissue, spectacular aerenchymas develop (*Figure 15-2*a), initiated by the strong auxin-induced ethylene production (Wächter et al., 1999; Pavlovkin et al., 2002). Ethylene is known to promote cell enlargement of cells and to dissolving middle lamellae thus inducing large lacunae, which provide better aeration for the otherwise compact tissue. Less vascularized and compact inner tumor parts and also the phloem are probably predominantly energized by glycolysis since several genes of the glycolytic pathway such as pyruvate decarboxylase and alcohol dehydrogenase were found to be upregulated (Deeken et al., 2005). However, expression of these genes has not yet been attributed to defined locations or analyzed for their time dependence during tumor development

As a general feature, the structures of the host stem xylem and phloem in the host/tumor interface and below are substantially different. The stem diameter considerably increases adjacent and below galls due to vigorous xylem enhancement; vessel diameters are much smaller, rays remain unlignified and become multiseriate (*Figure 15-2d* and Aloni et al., 1995; Aloni et al., 1998; Wächter et al., 1999). Generally, tumorized plants are much smaller, the shoot remains at about 25% of the size of non-infected control plants (Wächter et al., 2003).

3 ONCOGENE-INDUCED PHYTOHORMONE CASCADE

Tumors were believed to be chimerical structures of transformed and non-transformed cells, containing only 10-25% transformed cells, which overproduce auxin and cytokinins and were assumed to be homogeneously distributed among the majority of non-transformed cells (Sacristan and Melchers, 1977; Ooms et al., 1982; Van Slogteren et al., 1983). Taking into account the enucleate vascular structural peculiarities of the tumors, recent re-investigation based on analysis of the structure (living parenchyma *vs.* dead tracheary elements), mRNA and DNA marker analyses and on expression of *GUS* fused to the bacterial T-DNA, revealed a transformation rate of up to 100% (*Figure 15-2*b and Rezmer et al., 1999). Correspondingly, in these transformed tissues mitochondrial activity was remarkably high as revealed by staining with 3-(4,5-dimethylthiazolyl)-2,5-diphenyl-tetrazolium bromide (MTT) (Rezmer et al., 1999; Wächter et al., 1999; Pavlovkin et al., 2002; Schwalm et al., 2003). Regarding this high transformation rate the elevated phytohormone concentrations found in plant tumors become conceivable.

The transfer and expression of the T-DNA-encoded oncogenes, *iaaM* and *iaaH*, for auxin biosynthesis, from tryptophan via indole-acetamide, as well as cytokinin synthesis via isopentenyltransferase (*ipt*), are well documented for crown gall tumors (Weiler and Schröder, 1987; Zambryski et al., 1989).

3.1 Auxin

Free auxin concentrations in plant tissues are usually low, about 20-500 pmol g^{-1} fw in young stems of tobacco and *Ricinus* (Sitbon et al., 1991; Veselov et al., 2003). In contrast, auxin accumulation in crown galls can be abnormally high with up to 500-fold increase over that found in control tissues (Weiler and Spanier, 1981; Kado, 1984). Since auxin is known to be involved in vascular bundle development (Aloni and Zimmermann, 1983; Aloni, 2004), the elevated auxin concentration is a major factor promoting tumor vascularization.

By immunolocalization with monoclonal and polyclonal IAA antibodies, strong labeling can be detected over the entire tumor, and most pronounced around the differentiating vascular bundles in xylem parenchyma cells (Veselov et al., 2003). Cytoplasmic free IAA, bound to cell proteins by fixation of the tissue with N-(3-dimethylethylaminopropyl)-N'-ethylcarbodiimide-HCl (EDC), and IAA conjugates in storage pools as detected with this method, reveal a gradient with the highest label in the tumor periphery (*Figure 15-2e* and Schwalm et al., 2003). In sieve elements, the IAA label is restricted to the small cytoplasmic layer, whereas the companion cells with their high cytoplasm content show strong overall label (Veselov et al., 2003). In the pathological ray cells of the host stem-tumor interface IAA is scarcely labeled.

In contrast to IAA immunolocalization, enhanced expression of the IAA-responsive promoter of the soybean gene *GH3*, fused to the *GUS* gene (β-glucuronidase), in tumors of *Trifolium repens* reflects the concentration of bioactive free auxin. The latter proved to be highest in the tumor periphery where the vascular cells differentiate (Schwalm et al., 2003), corresponding to the considerable *iaaM* and indole-3-glycerate-phosphate-synthase (*IGS*) expression in contrast to non-infected hypocotyls (Schwalm et al., 2003). Only a low constitutive expression of nitrilase (*NIT*), the key enzyme of the tryptophan-dependent auxin biosynthesis pathway, was apparent in both tumor and non-infected hypocotyls of *Arabidopsis* (Schwalm et al., 2003). Tumors also rapidly proliferate in the rosette of *DR5::GUS* transformants of *Arabidopsis thaliana* (Col background); they contain high concentrations of free auxin, indicated by strong *DR5::GUS* expression in the tumor lobes (*Figure 15-2*g and Schwalm et al., 2003).

The concentration of free auxin in *Ricinus* stem tumors, as detected by GC-ECD, was rapidly at a maximum 2 weeks after tumor initiation and was up to 13-fold higher than in the control tissue. The concentration then declined to a minimum after 4-5 weeks and showed a second peak at 7 weeks. The transient decrease can be related to disruption of epidermis and cuticle at that developmental state, causing increased aeration and possibly oxidation of free auxin (*Figure 15-3a* and Veselov et al., 2003).

Free auxin concentration in *Ricinus* control roots is about 40% of that in the control stem tissue. Root auxin concentration clearly increases in plants with a developing stem tumor, reaching more than twice those in the controls, and this during the time period of more than 3 weeks after tumor initiation. When auxin starts to be retained in the tumors at 2 weeks *post inoculationem* (pi), the concentration in the roots transiently decreases. Between 3 and 4 weeks, auxin concentration drops even below that of the control roots (*Figure 15-3b* and Veselov et al., 2003), indicating an efficient regulation of auxin accumulation by specific crown gall physiology.

3.1.1 Regulation of auxin accumulation

The conspicuously increased diameter of the xylem below stem tumors was interpreted to be due to initial strong basipetal auxin flow out of the tumors and auxin accumulation by auxin-induced ethylene production (Aloni et al., 1995). Obviously, crown galls develop a mechanism to retain and accumulate auxin at concentrations far beyond those of healthy host tissues, where the accumulation of auxin is regulated by its basipetal export or inactivation by conjugation or oxidation (Normanly et al., 1995; Palme and Gälweiler, 1999; Bartel et al., 2001). Basipetal cellular auxin efflux, namely the PIN protein function, is specifically inhibited by the herbicide 1-N-naphthylphthalamic acid (NPA) (Geldner et al., 2001). As natural regulators endogenous flavonoids were found to displace NPA from its binding site on membranes, to block polar auxin efflux, to stimulate auxin accumulation and to inhibit or activate auxin oxidase (Stenlid, 1976). Likewise, in tumors of *Trifolium repens*, flavonoids detected by the diphenylboric acid 2-aminoethyl ester (DPBA) reagent, accumulate predominantly in the lobes of highest auxin concentrations (Schwalm et al., 2003). These flavonoids were identified by RP-HPLC as a tumor-specific flavone aglycone, namely 7,4'-dihydroxyflavone (DHF), which is absent in the host stem. Further tumor-specific aglycones were detected and identified as the flavonoid formononetin and the pterocarpan medicarpin (Schwalm et al., 2003), which is biosynthetically derived from formononetin (Heller and Forkmann, 1993). Two additional isoflavones, presumably 2'-hydroxyformononetin and vestitone, two intermediates of medicarpin biosynthesis, and a pterocarpan, presumably 4-methoxymedicarpin, were detected. DHF is known to support auxin accumulation, whereas formononetin stimulates auxin breakdown by peroxidase in rhizobia-induced nodules (Mathesius, 2001). Thus a well-balanced but high auxin concentration seems to be maintained in crown galls.

Flavonoids in tumors were analyzed in detail by their fluorescence spectra in clover tissue sections upon treatment with DPBA, using the 2PLSM in the two-photon mode (Schwalm et al., 2003). In addition, flavonoid fluorescence was compared with the expression of chalcone synthase (*CHS*), using *A. tumefaciens*-inoculated transformants of *Trifolium repens* containing the *GUS*-fused auxin-responsive promoter (*GH3*) or chalcone synthase (*CHS2*) genes. The most prominent DHF fluorescence was localized at the sites of strongest *GH3::GUS* and auxin-inducible *CHS2::GUS* expression in the tumor that was differentially modulated by auxin in the vascular tissue. *CHS* mRNA expression changes, as revealed

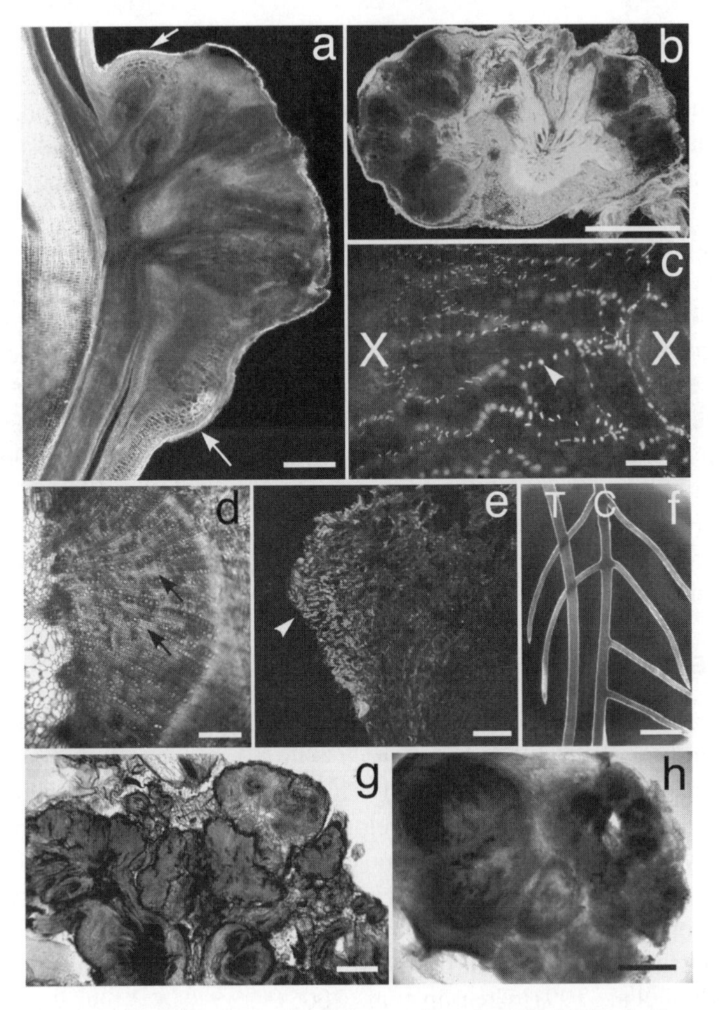

Figure 15-2. Vascularization, transformation, auxin and cytokinin distribution pattern in crown gall tumors. (**a**) Vascularized tumor regions rapidly proliferate (dark blue), distinct aerenchyma development in non-transformed tissue in *Ricinus* (arrows); lacmoid staining. (**b**) Highly transformed 7-week-old *Arabidopsis* rosette tumor, revealed by GUS expression (blue) of the wild-type T-DNA (A281 p35S gus int). (**c**) Tumor bundles (X) are interconnected by a dense net of phloem anastomoses (arrowhead) in *Ricinus*; aniline blue staining. (**d**) 5-week-old *Ricinus* tumor with unlignified multiseriate rays (arrows) and increased number of small vessels (blue); toluidine blue staining. (**e**) Immunolocalization of total auxin in an 8-week-old *Trifolium* tumor, revealing an auxin gradient with the intensity decreasing from the periphery (arrowhead) inwards; red autofluorescence identifies vessels. (**f**) Lack of lateral roots in tumorized (T) in comparison with non-infected control *Ricinus* (C) grown under identical conditions. (**g**) Auxin-induced *DR5::GUS* and (**h**) Cytokinin-induced *ARR5::GUS* expression (blue) in proliferating 5-week-old *Arabidopsis* tumors. Bars = 75 μm (e), 250 μm (c), 500 μm (d,g,h), 2 mm (a,b,f). (From (Aloni et al., 1995; Wächter et al., 1999; Mistrik et al., 2000; Ullrich and Aloni, 2000; Schwalm et al., 2003).

by Northern blotting and RT-PCR, corresponded well to the auxin concentration profile in tumors and roots of tumorized plants (*Figure 15-3*a,b and h). Application of DHF to *Trifolium* stems that were apically pretreated with α-naphthaleneacetic acid (NAA) inhibits *GH3::GUS* expression in a fashion similar to NPA. As a decisive result, tumor, root and shoot growth is poor in inoculated *tt4*(85) flavonoid-deficient *CHS* mutants of *Arabidopsis*. It is concluded that CHS-dependent flavonoid aglycones are possibly endogenous regulators of the basipetal auxin flux, thereby leading to free-auxin accumulation in *A. tumefaciens*-induced tumors. This, in turn, triggers vigorous proliferation and vascularization of the tumor tissue and suppresses their further differentiation.

These data on the essential role of flavonoids in tumor proliferation are obviously further supported by recent findings that the T-DNA-located gene *6b* regulates the phenylpropane expression and flavonoid synthesis in tobacco (Gàlis et al., 2004; Kakiuchi et al., 2005). Hence not only the phytohormone genes of the T-DNA are required for tumor development but also the not yet well understood function of gene *6b*.

3.1.2 Enhancement of tumor induction by host plant auxin

The presence of high auxin concentrations in wounds enhances the stable integration of the T-DNA in *Arabidopsis* (Chateau et al., 2000) and in *Vitis vinifera* (Creasap et al., 2005). This explains why in grapevine cuttings inoculated wounds below developing and auxin-producing leaf buds produce vigorously growing tumors in contrast to wounds opposite the buds. Accordingly, naturally occurring crown galls develop on grapevine only in spring, when the dormancy callose plugs in the phloem are degraded during the incipient basipetal auxin flow (Aloni and Peterson, 1991; Aloni et al., 1991; Creasap et al., 2005).

3.2 Cytokinins

Accumulation of cytokinins, even up to 1600times higher concentrations, was found in plant tumors (Kado, 1984). In comparison with auxin distribution, a more distinct pattern of cytokinin localization (tZ and trans-ZR) was revealed by immunofluorescence with monoclonal and polyclonal antibodies (Veselov et al., 2003). The strongest fluorescence was detected within the concentric tumor bundles in the parenchyma cells of developing phloem and xylem. Sieve elements were only weakly labeled in contrast to companion cells. Thus CKs ensure vascular bundle differentiation and consequently permanent tumor growth. No or little CKs were found in the tumor periphery. Within the pathological xylem, predominantly parenchyma cells were labeled, in contrast to the multiseriate rays.

In *ARR5::GUS* transformants of *Arabidopsis* high free bioactive CK concentrations were visualized throughout the developing tumors (*Figure 15-2*h).

Following the kinetics of CK concentrations in *Ricinus* stem tumors in parallel with IAA accumulation, similarly the concentration of zeatin riboside (tZR) rapidly increased, reaching a maximum of 140-175 pmol g^{-1} fresh wt at 2 weeks pi and remaining so until at least 7 weeks pi, with at least a 10-fold increase over the control (*Figure 15-3*d and Veselov et al., 2003). The amount of free zeatin (tZ) clearly peaked at 4 weeks pi with an eightfold increase (*Figure 15-3*c). The large cytokinin nucleotide pool steadily increased during tumor development and attained 16times that of the controls.

Tumorized plants develop only poor roots with particularly few lateral roots (*Figure 15-2*f and Mistrik et al., 2000). It is conceivable that CKs are released from the tumors and are basipetally transported within the phloem and, together with ethylene, inhibit root growth as both phytohormones do in healthy roots as well (Aloni et al., 2005).

3.3 Ethylene

Not only IAA but also CKs were found to induce and enhance ethylene biosynthesis and to modulate each other's response by cross-talk between their signaling pathways (Vogel et al., 1998; Hall et al., 1999). The structural peculiarities such as vessels of decreasing diameter and the multiplication of rays that remain unlignified suggested that ethylene plays a crucial role in tumor development (Aloni et al., 1995). Indeed, ethylene emission by the tumors was shown to be up to 140 times that by control stems of tomato and *Ricinus* (Aloni et al., 1998; Wächter et al., 1999), as determined with different methods, *i.e.*, process gas chromatography and photo-acoustic laser spectroscopy. Ethylene emission upon IAA and CK accumulation started later and was maximum as late as 5 weeks pi (*Figure 15-3*e). Inhibitors of ethylene synthesis or perception, either AVG or ethylene insensitive *Never ripe* mutants, suppress crown gall growth and epinastic responses of neighboring leaves; vascularization is also completely inhibited and hence tumor growth and cuticular rupture (Aloni et al., 1998; Wächter et al., 2003).

During ethylene biosynthesis oxygen plays a decisive role. Whereas ACC synthase activity is stimulated by oxygen deficiency in some plant species, ACC-oxidase needs oxygen as substrate. Exposing tumors to nitrogen gassing (*i.e.* anaerobiosis) completely suppressed and increasing oxygen concentrations enhanced the ethylene emission, indicating that tumor tissue is not under oxygen deficiency like animal or human tumors, hence glycolytic activities will be normal. This is in accordance with MTT-staining, resulting from the reduction by cellular NADH, revealed high reducing activity of the tumor and indicated elevated mitochondrial respiration. Indeed, the respiratory, N_2-inhibitable, CO_2 emission (J'_{CO2}) and O_2 uptake rates considerably increased with tumor growth, beginning 2 weeks pi (Marx and Ullrich-Eberius, 1988; Wächter et al., 2003). Good aeration is supported by the peripheral aerenchyma of the tumor and in the former cortex tissue. It is characterized by cell enlargement, lysis and large intercellular spaces (Wächter et al., 1999). In contrast to water-logged plants, tumors emit more ethylene in re-watered plants after mild water deficiency (Wächter et al., 1999). Also the 1-aminocyclopropane-1-carboxylic acid (ACC) content was up to 75 times higher in tumors (Wächter et al., 1999), with an early maximum 2 weeks pi. ACC content in control stems and in stems above and below tumors was very low and similar. This means that ACC is not exported from the tumors. Ethylene is known to induce leaf epinasty. Correspondingly, downward bending of cotyledons and leaves are typical for tumorized plants (Aloni et al., 1998; Wächter et al., 1999; Veselov et al., 2003).

In ethylene-insensitive *Never ripe* tomato mutants infected with *A. tumefaciens*, tumor growth is largely prevented (Aloni et al., 1998; Aloni and Ullrich, 2002), in spite of integration and expression of the T-DNA-located oncogenes (*iaaH, iaaM, ipt*). Accordingly, ethylene is a crucial determinant of crown gall development.

3.4 Abscisic acid

The signal transduction chains of ethylene and ABA are partly overlapping and interfering. From application of various inhibitors of ethylene or ABA biosynthesis and from the use of ethylene-insensitive or ABA-deficient tomato mutants, it was concluded that auxin- and cytokinin-induced ethylene triggers the synthesis of abscisic acid, most likely by enhancing the 9'-cis-neoxanthin dioxygenase activity and not by inhibiting the ABA catabolism (Hansen and Grossmann, 2000).

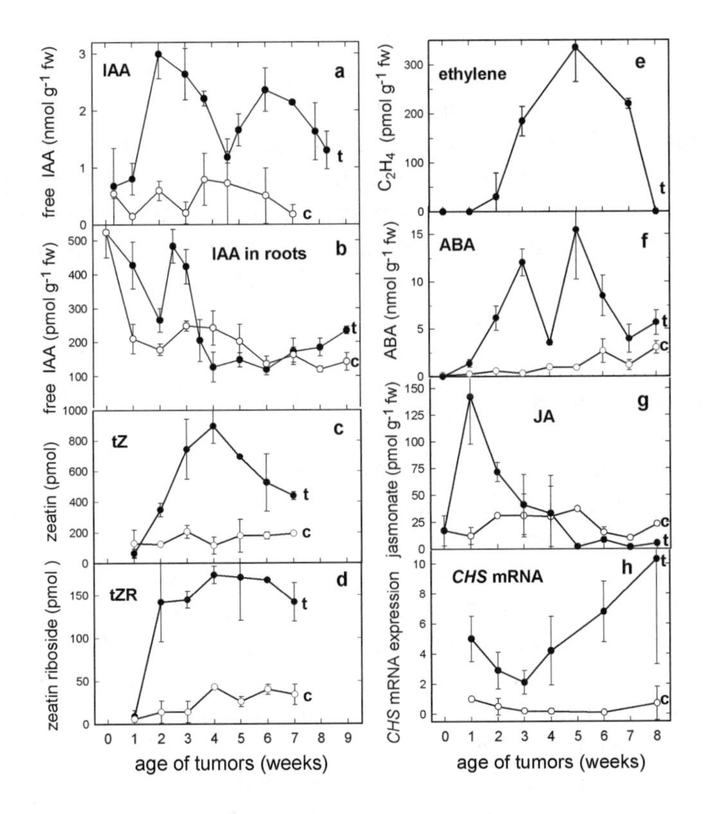

Figure15-3. Comparison of the time course of phytohormone accumulation in *Ricinus* stem tumors with chalcone synthase (*CHS*) expression in *Trifolium* stem tumors. (a) Free auxin accumulation in tumors (t) and (b) in the 2-cm apical section of roots of tumorized and control (c) plants. (c) trans-zeatin (tZ) and (d) trans-zeatin riboside (tZR) accumulation (pmol g-1 fw). (e) Ethylene emission from tumors. (f) Abscisic acid (ABA) concentration. (g) Methyl-jasmonate (JA) outburst in tumors; wounded and unwounded but non-infected tissues had the same concentration (c). (h) *CHS* mRNA expression, displayed as relative transcription rate from Northern blots of *CHS2::GUS* transformants. (From Wächter et al., 1999; Schwalm et al., 2003; Veselov et al., 2003).

ABA distribution, as detected in 3-week-old tumors by immunofluorescence with monoclonal antibodies, showed an interesting pattern throughout the tumors (Wächter et al., 2003). Distinct fluorescence was detected around the vascular bundles and in the very peripheral cell layers. At cellular resolution, the ABA-specific fluorescence was localized in parenchyma cells adjacent to vessels.

Measurements of the ABA concentration in leaves above epinastic cotyledons of tumorized plants revealed a fourfold increase over that in control leaves (Veselov et al., 2003), which was effective in decreasing leaf transpiration (Schurr et al., 1996; Wächter et al., 2003). While ABA accumulation in leaves is usually correlated with wilting symptoms in plants suffering from severe drought, leaves of plants stressed by tumor- or ethrel-released ethylene remained highly turgescent (Veselov et al., 2003).

The ABA concentration in the tumors increased simultaneously with the increase in ethylene emission and was maximum at 5 weeks pi, which is 36-fold higher than in the control (*Figure 15-3*f and Veselov et al., 2003). ABA accumulates also in the phloem sap, mainly below and slightly above the tumor (Mistrik et al., 2000).

3.5 Jasmonic acid

Jasmonic acid (JA) is regarded as a wound and stress hormone, which induces plant responses like those to ABA (Feussner and Wasternack, 2002). However, in plant tumors the kinetics of JA accumulation is rather different from that of ABA (Veselov et al., 2003). As early as 1 week pi its accumulation is maximum but then rapidly decreases, after 4 weeks even below the concentration in the control stem (*Figure 15-3*g). The initial rapid increase is not due to a mere wound effect, since concentrations in both unwounded healthy and wounded but non-inoculated tissue were the same. The different kinetics suggests an early role for JA in tumor induction, perhaps an additional mechanism that rapidly increases the free auxin concentration by inducing the expression of IAA-amino acid hydrolase JR3 (Rojo et al., 1998). The presence of sufficiently high auxin and cytokinin concentrations enhanced stable T-DNA integration in various *Arabidopsis* genotypes (Chateau et al., 2000). An early pathological but continuing JA-inducing wound effect may be caused by bacterial activity with cell wall degrading enzymes such as pectinase, ligninase, xylanase and cellulase as encoded in the bacterial chromosome (Wood et al., 2001).

3.6 Interactive reactions of JA, IAA, CK, ethylene and ABA

Phytohormone signaling pathways and their physiological effects permanently interfere with each other during the continuous proliferation of crown gall tumors. Jasmonic acid probably enhances the accumulation of host-derived auxin in the inoculated wound, thus supporting the transfer of the T-DNA from agrobacteria into the host cell. Upon stable integration and expression of the prokaryotic phytohormone genes in the plant genome, T-DNA-dependent auxin and cytokinin synthesis entails a vigorous production of ethylene, which in turn, together with auxin-enhanced flavonoid synthesis, supports free auxin accumulation in the tumor by preventing its basipetal efflux. Only such high auxin and cytokinin concentrations enable the continuous differentiation of functional phloem and xylem in the tumors, while they simultaneously suppress root and shoot differentiation out of the tumor parenchyma. Subsequently, ethylene induces the accumulation of abscisic acid, not only in the tumor periphery for activating osmoprotectants but also in the host leaves where it re-directs the nutrient bearing water and sieve tube flow via the auxin and ethylene-dependent enlarged xylem and phloem into the tumor by inducing host leaf stomata closure.

4 ENHANCEMENT OF WATER AND SOLUTE TRANSPORT

Prerequisite of water and nutrient transport out of the host vessels into those of the tumor is a tight connection of both. Revising old assumptions of lacking connections of the host bundle with the tumor vessels, the transport of negatively-charged dyes, amido black, acid fuchsin and the fluorescent pyrenetrisulfonate clearly visualized a continuous and efficient water flow through the vessels from the host stem into the tumor, up to its surface (Schurr et al., 1996).

4.1 Water transport

Vascular differentiation and epidermal and cuticlar disruption are further requirements for efficient nutrient-bearing water flow into the tumors for tumor growth. Water flow dramatically increases during the first 3 weeks of tumor development. However, tumor water loss contributes little change to the water flow to host shoots. The water vapour conductance (g_{H2O}) of tumors rises rapidly within the first week after infection, with the strongest increase between the 2nd and 3rd week pi (Wächter et al., 2003) up to the 27fold of the control stem. The highly irregular tumor surface area increases almost linearly with tumor growth. g_{H2O} thus appears to be related to the tumor surface area, combined with the timing of rupture and break-up of epidermis and cuticle during the first 3 weeks of tumor growth. g_{H2O} of leaves of non-infected plants is about 10 times that of tumors during the light period. During the dark period, g_{H2O} of the leaves decreases to a value which is only half of that of the tumors, *i.e.* tumor g_{H2O} is twice that of the leaves. The net transpiration rate (J_{H2O}) of the total transpiring surface of a 3-week-old tumor (10 cm^2) at a host plant with 5 leaves (400 cm^2) revealed that in the light J_{H2O} is about 1500 times greater than that of the tumor, which means the latter would thus be negligible. However, the transpiration rate of leaves of tumor-infected plants is typically only 10% of that of non-infected plants (Veselov et al., 2003), due to the induction of ABA production by tumor-emitted ethylene (Mistrik et al., 2000; Veselov et al., 2003), because ethylene causes closure of stomata by inducing ABA accumulation. The remaining transpiration in leaves of tumorized plants is still 150 times that of the tumor in light. But this phytohormone signaling from the tumor to the host shoot has to be regarded as the crucial step in the water regime of the host plant with substantial consequences for nutrient partitioning between the host shoot and the tumor. The leaves closest to the tumor have a spectacular 10-fold lower g_{H2O} than leaves of non-tumorized plants, but it clearly increases with increasing distance from the tumor. Similarly, leaves at the shoot apex of non-tumorized plants, when supplied with 0.01 to 0.1% ethrel in the nutrient solution, develop typical epinastic symptoms, similar to leaves of ethylene-emitting tumorized plants. The ABA concentration of ethrel-treated leaves increased by 15-fold and g_{H2O} was about 40 times smaller. ABA concentration in the host leaves of tumorized plants was almost 4 times that of control plant leaves, which is in line with the assumption of ethylene enhancing ABA synthesis (Veselov et al., 2003).

4.2 Regulation of inorganic nutrient accumulation

Such an increased water flow carries considerable nutrient amounts into the tumors, in particular K^+, phosphate and sulfate, either directly from the vessels within the tumor (K^+) or via xylem/phloem exchange in the host shoot ($H_2PO_4^-$ and SO_4^{2-}) (Klein, 1952; Wood and Braun, 1965; Marx and Ullrich-Eberius, 1988; Mistrik et al., 2000; Deeken et al., 2003; Wächter et al., 2003; Deeken et al., 2005). In *Kalanchoë* leaf tumors accumulation is for K^+ the 5-fold, for Cl^- the 2-fold, for $H_2PO_4^-$ the 8-fold (Marx and Ullrich-Eberius, 1988) on average, in *Arabidopsis* tumors: for $H_2PO_4^-$ the 23-fold and for SO_4^{2-} the 12-fold (Deeken et al., 2005). In contrast, the NH_4^+ concentration is only half of that of the control and NO_3^- concentration is even considerably (8 times) lower in *Ricinus* tumors than in control stems (Mistrik et al., 2000). In accordance there is no nitrate reducing activity detectable in tumors in contrast to the control stem or leaves (0, 0.7, 5.4 μmol g^{-1} fw h^{-1}, respectively). NO_3^- and $H_2PO_4^-$ uptake in roots of tumorized *Ricinus* plants is less than 50% of the control plants, on a g FW basis, together with the poorly developed lateral root system of tumorized plants (*Figure 15-2*f), resulting in severely diminished total anion availability for the host plant.

To know how the inorganic nutrients efficiently accumulate in the tumor sink tissue, plasma membrane electropotentials (E_m) were determined (Pavlovkin et al., 2002). Xylem and phloem parenchyma cells and stem/tumor-located rays hyperpolarized to about -170 mV (controls were -120 to -140 mV), indicating high plasma membrane proton pump activities, as suggested earlier by Ramaiah and Mookerjee (Ramaiah and Mookerjee, 1982). In fact, two p-type H^+-ATPase genes are upregulated in *Arabidopsis* tumors (Deeken et al., 2005). Cell K^+ concentrations largely matched the respective E_m (Pavlovkin et al., 2002). The patterns of individual cell electropotentials in excised tissue sections were supplemented by whole organ voltage measurements. These trans-tumor electropotentials confirm the findings of respiration-dependent and phytohormone-stimulated high plasma membrane proton pump activity in intact tumors, mainly in the xylem and phloem parenchyma cells (Pavlovkin et al., 2002). Xylem parenchyma (XP) cells in roots and shoots are known to be highly active in ion release into the xylem vessels or in absorbing ions out of the vessels in an energy-dependent process (Läuchli et al., 1971; De Boer and Prins, 1985). The existence of back-to-back electrogenic H^+ pump activity across the boundary membrane of the organ surface and across the xylem parenchyma symplast/xylem interface has been demonstrated (Okamoto et al., 1978), providing strong evidence for the operation of XP pumps. These assumed PM H^+-ATPases in the XP plasma membrane were confirmed by cytochemical localization (Winter-Sluiter et al., 1977). Furthermore, it was shown that the activity (release of Pi from PPi) and protein level (by Western blot analysis) of the tonoplast V-PPase and V-ATPase remains high in tumor tissue, whereas activity and protein of the PPase decreases in the host stem during growth (Fischer-Schliebs et al., 1998). PPase is suggested to have special functions in young developing and growing tissues by utilizing pyrophosphate produced in particularly active metabolism and by pumping of K^+ for vacuolization (Suzuki and Kasamo, 1993).

Recently, voltage-dependent XP-located K^+ channels, outward directed (SKOR), inward directed (KIRC and AKT1) and non-specific channels have been identified. Some channels responded to water stress and ABA decreased the outward directed K^+ current. Auxin hyperpolarized XP and enhanced K^+ absorption from the vessels (De Boer and Prins, 1985;

Roberts and Tester, 1995; Gaymard et al., 1998). Accordingly, in *Arabidopsis* peduncle (flower stalk) tumors the expression and activity (patch-clamp measurements) of the root-specific Shaker-like K^+ channel *AKT1* and its modulating subunit *AtKC1* are upregulated and enhanced. However, the shoot-specific and IAA and ABA-activated K^+ channels *AKT2/3* and *GORK* are suppressed. The XP-specific SKOR channels, which secrete K^+ into the vessels, are even completely absent from the tumor tissue (Deeken et al., 2003). AKT1 are hyperpolarization-activated K^+ uptake channels, AKT2/3 voltage-independent and GORK activated channels upon depolarization and mediate K^+ efflux. Since *Arabidopsis* mutants that lack a functional AKT1 channel only poorly support tumor growth, functional AKT1 K^+ channels seem to be essential for tumor growth, but not for tumor induction. This channel profile is similar to that of heterotrophically growing cells. However, since the IAA and ABA-transcriptionally-inducible channels *GORK, AKT2/3, KAT1* and *KAT2* are down-regulated in tumors, yet unknown tumor factors other than IAA and ABA must up-regulate *AKT1* and *AtKC1* transcription and overrule the ABA control of *GORK* and *AKT2/3* expression (Deeken et al., 2003).

4.3 Phloem transport and symplastic unloading

Besides inorganic ions also high concentrations of organic nutrients, such as amino acids and sugars, accumulate in crown galls, whereas concentrations of organic acid, in particular malate, are considerably lower in the tumors (Neish and Hibbert, 1943-44; Brucker and Schmidt, 1959; Marx and Ullrich-Eberius, 1988; Malsy et al., 1992; Pradel et al., 1996; Mistrik et al., 2000; Wächter et al., 2003; Deeken et al., 2005). In *Kalanchoë* leaf tumors, concentration of sucrose was 14times, fructose 40times, glucose 25times, and total amino acids up to 10times higher than in the control leaf tissue. In *Ricinus* stem tumors, sucrose was maximally increased to 20times and amino acids up to 40times the control values (Mistrik et al., 2000; proline up to 40times, Wächter et al., 2003). Since solute concentrations change with time and in local distribution (periphery *vs.* center) during tumor development, it is very important for interpretation in relation to tumor metabolism to consider the spatio-temporal concentration changes of solutes and of the enzymes involved as well. The concentration of opines, the T-DNA-encoded and specific *A. tumefaciens* substrates, namely nopaline in C58-induced *Kalanchoë* leaf tumors, was as high as that of proline; both were about 25 to 50% of the concentration of glutamic and aspartic acid, respectively, which were present at the highest concentrations of all amino acids in the tumors (Malsy et al., 1992).

Since the photosynthetic activity of leaf crown galls is only about 25% of that of control leaves (Marx and Ullrich-Eberius, 1988), organic nutrients have to be imported via the phloem. Upon the finding of acid cell wall invertase activity, which was regarded to indicate apoplastic phloem unloading (Weil and Rausch, 1990), the mode of phloem unloading in tumors was explored by several methods, to know whether it is symplastic or apoplastic unloading: A structural indication for a well-coupled system of sieve element/companion cells (SE/CC) to tumor parenchyma cells is the detection of primary pit fields, indicating plasmodesmata between these cells, by callose fluorescence. Fluorescent dyes such as Lucifer Yellow, iontophoretically injected into sieve elements and fluorescein, applied to orthostichy leaves above stem tumors, move from the SE/CC into the tumor parenchyma (Pradel et al., 1996). In addition, using non-destructive imaging techniques, such as the movement of potato virus X (PVX) expressing the green fluorescent protein (GFP) as a fusion to the viral coat protein (Pradel et al., 1999), both carboxyfluorescein and PVX are symplastically unloaded into tumor tissue and subsequently show extensive cell-to-cell movement within the parenchyma tissue of the tumor, in contrast to non-infected stem tissue. Not only the phloem, but also the ray parenchyma cells are functional in lateral transmission of both solutes and virus across the stem. These results were recently confirmed for the mechanism of phloem unloading in the syncytium of nematode-induced root giant cells (Hoth et al., 2005).

The tumor-specific symplastic phloem unloading as predominant mechanism of nutrient import into tumor parenchyma may explain the down-regulation of gene expression of specific plasma membrane-located uptake carriers, such as transporters of NO_3^-, $H_2PO_4^-$, SO_4^{2-}, sucrose, glucose and amino acids, as found in a genome-wide approach (Deeken et al., 2005). The fact that, e.g., the tumor NO_3^- concentration is very low and nitrate reductase activity is undetectable, together with the down-regulation of the NO_3^-, H_2PO_4 and SO_4^{2-}-tumor transporters, suggests that these anions are first transported into the host shoot, where NO_3^- is reduced and converted into amino acids, which are then loaded into the host phloem like the leaf-derived sucrose, but also $H_2PO_4^-$ and SO_4^{2-}; all these solutes are finally symplastically unloaded into the tumor parenchyma. Hence, not only the tumor xylem is well coupled with the host vessels, but also the sieve elements are tightly, functionally and effectively connected to the phloem of the host and are symplastically unloaded into the tumor parenchyma tissue.

5 KINETICS AND FUNCTION OF THE SUGAR-CLEAVING ENZYMES SUCROSE SYNTHASE, ACID CELL WALL AND VACUOLAR INVERTASE

Tumor water loss upon disruption of its surface is followed by accumulation of the osmoprotectants, sucrose and proline, in the tumor periphery, shifting hexose to sucrose in favor of sugar signals for maturation and desiccation tolerance (Wächter et al., 2003). Concurrent activities and sites of action for enzymes of sucrose metabolism changes are the following: Vacuolar invertase predominates during initial import of sucrose into the symplastic continuum, corresponding to hexose concentrations in expanding tumors. Later, sucrose synthase and cell wall invertase activities rise in the tumor periphery to modulate both sucrose accumulation and decreasing turgor for import by metabolization. Sites of abscisic acid immunolocalization correlate with both central vacuolar invertase and peripheral cell wall invertase. Roles in vascular bundles are indicated by sucrose synthase immunolocalization in xylem parenchyma for inorganic nutrient uptake and in the phloem, where the resolution allowed sucrose synthase identification in sieve elements and companion cells. The time course of acid cell wall invertase activity is clearly independent of initial phloem unloading, which confirms the existence of symplastic phloem unloading, excluding a major function of acid cell wall invertase in primary phloem unloading of sucrose. The osmoprotectant proline accumulates almost simultaneously with the activity of the acid cell wall invertase (Wächter et al., 2003).

6 CONCLUSIONS

The available data indicate auxin (*iaaH* and *iaaM*) and cytokinin (*ipt*) - induced key roles for ethylene-dependent vascularization and cuticular disruption in the re-direction of water flow with up-regulated genes for K^+-influx channels and with symplastic phloem unloading of carbohydrates, amino acids and metabolizable anions for successful tumor establishment. The gradually auxin-induced and T-DNA gene *6b* related flavonoid-dependent auxin retention is regarded as additional major factor for the unusual auxin accumulation in crown gall tumors, as summarized in *Figure 15-1*.

7 ACKNOWLEDGEMENTS

Thanks are due to the Deutsche Forschungsgemeinschaft (DFG) for continuous support of our research on crown gall tumors (to C.I.U.) and to Dr. Rosalia Deeken, Univ Würzburg, Germany, for supplying unpublished results.

8 REFERENCES

Agrios GN (2004) Bacterial Galls. Plant Pathology. 5th ed. Elsevier Academic Press, Amsterdam

Aisenberg AC (1961) The glycolysis and respiration of tumors. Academic Press, New York

Aloni A, Aloni E, Langhans M, Ullrich CI (2005) Cytokinin-dependent root apical dominance, regulation of root vascular differentiation, root gravitropism and the control of lateral root initiation. Ann Bot

Aloni R (1979) Role of auxin and gibberellin in differentiation of primary phloem fibres. Plant Physiol 63: 609-614

Aloni R (2004) The induction of vascular tissue by auxin. *In* PJ Davies, ed, Plant Hormones: Biosynthesis, Signal Transduction, Action! Kluwer, Dordrecht, pp 471-492

Aloni R, Peterson C (1991) Seasonal changes in callose levels and fluorescein translocation in the phloem of *Vitis vinifera* L. IAWA Bull 12: 223-234

Aloni R, Pradel KS, Ullrich CI (1995) The three-dimensional structure of vascular tissues in *Agrobacterium tumefaciens*-induced crown galls and in the host stem of *Ricinus communis* L. Planta 196: 597-605

Aloni R, Raviv A, Peterson C (1991) The role of auxin in the renewal of dormancy callose and resumption in the phloem activity in *Vitis vinifera* L. Can J Bot 69: 1825-1832

Aloni R, Ullrich CI (2002) Tumor-induced ethylene controls crown gall morphogenesis. *In* Plant Physiology Online Essay 221 http://wwwplantphysnet,

Aloni R, Wolf A, Feigenbaum P, Avni A, Klee HJ (1998) The never ripe mutant provides evidence that tumor-induced ethylene controls the morphogenesis of *Agrobacterium tumefaciens*-induced crown galls on tomato stems. Plant Physiol 117: 841-849

Aloni R, Zimmermann MH (1983) The control of vessel size and density along the plant axis. Differentiation 24: 203-208

Bartel B, LeClere S, Magidin M, Zolman BK (2001) Inputs to the active indole-3-acetic acid pool: *de novo* synthesis, conjugate hydrolysis and indole-3-butyric acid β-oxidation. J Plant Growth Regul 20: 198-216

Beiderbeck R (1977) Pflanzliche Tumoren. Ulmer Verlag, Stuttgart

Bopp M, Leppla E (1964) Ein Vergleich der Histogenese der Wurzelhalsgallen an Blättern und Sprossachsen von *Kalanchoe daigremontiana*. Planta 61: 36-55

Brucker W, Schmidt WAK (1959) Zum Zuckerstoffwechsel des Kallus- und crown-gall-Gewebes von *Datura* und *Daucus*. Ber Dtsch Bot Ges 72: 321-332

Chateau S, Sangwan RS, Sangwan-Norreel BS (2000) Competence of *Arabidopsis thaliana* genotypes and mutants for *Agrobacterium tumefaciens*-mediated gene transfer: role of phytohormones. J Exp Bot 51: 1961-1968

Creasap JE, Reid CL, Goffinet MC, Aloni R, Ullrich C, Burr TJ (2005) Effect of wound position, auxin and *Agrobacterium vitis* strain F2/5 on wound healing and crown gall in grapevine. Phytopathology 95: 362-367

De Boer AH, Prins HBA (1985) Xylem perfusion of tap root segments of *Plantago maritima*: the physiological significance of electrogenic xylem pumps. Plant Cell Environ 8: 587-594

Deeken R, Engelmann J, Efetova M, Müller T, Kaiser W, Palme K, Schartl M, Dandekar T, Hedrich R (2005) An integrated view of gene expression and solute profiles in *Arabidopsis* tumour cells: a genome-wide approach.

Deeken R, Ivashikina N, Czirjak T, Philippar K, Becker D, Ache P, Hedrich R (2003) Tumour development in *Arabidopsis thaliana* involves the Shaker-like K^+ channels AKT1 and AKT2/3. Plant J 34: 778-787

Dehio C (2004) Molecular and cellular basis of *Bartonella* pathogenesis. Annu Rev Microbiol 58: 365-390

Dehio C (2005) *Bartonella*-host-cell interactions and vascular tumour formation. Nat Rev Microbiol 3: 621-631

Feussner I, Wasternack C (2002) The lipoxygenase pathway. Annu Rev Plant Biol 53: 275-297

Fischer-Schliebs E, Ratajczak R, Weber P, Tavakoli N, Ullrich CI, Lüttge U (1998) Concordant time-dependent patterns of activities and enzyme protein amounts of V-PPase and V-ATPase in induced (flowering and CAM or tumour) and non-induced plant tissues. Bot Acta 111: 130-136

Folkman J (1971) Tumor angiogenesis: therapeutic implications. New England J Med 285: 1182-1186

Gàlis I, Kakiuchi Y, Šimek P, Wabiko H (2004) *Agrobacterium tumefaciens* AK-*6b* gene modulates phenolic compound metabolism in tobacco. Phytochemistry 65: 169-179

Gaymard F, Pilot G, Lacombe B, Bouchez D, Bruneau D, Boucherez J, Michaux-Ferrière N, Thibaud J-B, Sentenac H (1998) Identification and distribution of a Shaker-like outward channel involved in K+ release into the xylem sap. Cell 94: 647-655

Geldner N, Friml J, Stierhof YD, Jürgens G, Palme K (2001) Auxin transport inhibitors block PIN1 cycling and vesicle trafficking. Nature 413: 425-428

Gimbrone MAJ, Cotran RS, Leapman SB, Folkman J (1974) Tumor growth and neovascularization: an experimental model using the rabbit cornea. J Natl Cancer Inst 52: 413-427

Gordon MP (1982) Reversal of crown gall tumors. *In* G Kahl, JS Schell, eds, Molecular biology of plant tumors. Academic Press, New York, pp 415-426

Hall AE, Chen QG, Findell JL, Schaller GE, Bleecker AB (1999) The relationship between ethylene binding and dominant insensitivity conferred by mutant forms of the ETR1 ethylene receptor. Plant Physiol 121: 291-299

Hansen H, Grossmann K (2000) Auxin-induced ethylene triggers abscisic acid biosynthesis and growth inhibition. Plant Physiol 124: 1437-1448

Heller W, Forkmann G (1993) Biosynthesis of flavonoids. *In* JB Harborne, ed, The Flavonoids: Advances in Research Since 1986. Chapman and Hall, London, pp 499-535

Hooykaas PJJ, Ooms G, Schilperoort RA (1982) Tumors induced by different strains of *Agrobacterium tumefaciens*. *In* G Kahl, JS Schell, eds, Molecular biology of plant tumors. Academic Press, New York, pp 373-390

Hoth A, Schneidereit A, Lauterbach C, Scholz-Starke J, Sauer N (2005) Nematode infection triggers the *de novo* formation of unloading phloem that allows macromolecular trafficking of green fluorescent protein into syncytia. Plant Physiol 138: 383-392

Kado CI (1984) Phytohormone-mediated tumorigenesis by plant pathogenic bacteria. *In* DPS Verma, T Hohn, eds, Genes Involved in Microbe-Plant Interactions. Springer-Verlag, Wien, pp 311-336

Kakiuchi Y, Gàlis I, Tamogami S, Wabiko H (2005) Reduction of polar auxin transport in tobacco by the tumorigenic *Agrobacterium tumefaciens* AK-*6b* gene. Planta

Kempf VA, Hitziger N, Riess T, Autenrieth IB (2002) Do plant and human pathogens have a common pathogenicity strategy? Trends Microbiol 10: 269-275

Klein RM (1952) Nitrogen and phosphorous fractions, respiration, and structure of normal and crown-gall tissue of tomato. Plant Physiol 27: 335-354

Kupila-Ahvenniemi S, Therman E (1968) Morphogenesis of crown gall. Adv in Morphogen 7: 45-78

Läuchli A, Spurr AR, Epstein E (1971) Lateral transport of ions into the xylem of corn roots. Plant Physiol 48: 118-124

Malsy S, Van Bel AJE, Kluge M, Hartung W, Ullrich CI (1992) Induction of crown galls by *Agrobacterium tumefaciens* (strain C 58) reverses assimilate translocation and accumulation in *Kalanchoe daigremontiana*. Plant Cell Environ 15: 519-529

Marx S, Ullrich-Eberius CI (1988) Solute accumulation and electrical membrane potential in *Agrobacterium tumefaciens*-induced crown galls in *Kalanchoe daigremontiana*. Plant Sci 57: 27-36

Mathesius U (2001) Flavonoids induced in cells undergoing nodule organogenesis in white clover are regulators of auxin breakdown by peroxidase. J Exp Bot 52: 419-426

Mistrik I, Pavlovkin J, Wächter R, Pradel KS, Schwalm K, Hartung W, Mathesius U, Stöhr C, Ullrich CI (2000) Impact of *Agrobacterium tumefaciens*-induced stem tumours on NO3- uptake in *Ricinus communis*. Plant Soil 226: 87-98

Neish AC, Hibbert H (1943-44) Studies on plant tumors. II. Carbohydrate metabolism of normal and tumor tissue of beet root. Arch Biochem 3: 141-157

Normanly J, Slovin JP, Cohen JD (1995) Rethinking auxin biosynthesis and metabolism. Plant Physiol 107: 323-329

Okamoto H, Ichino K, Katou K (1978) Radial electrogenic activity in the stem of *Vigna unguiculata*: involvement of spatially separate pumps. Plant Cell Environ 1: 279-284

Ooms G, Bakker A, Molendijk L, Wullems GJ, Gordon MP, Nester EW, Schilperoort RA (1982) T-DNA organization in homogeneous and heterogeneous octopine-type crown gall tissues of *Nicotiana tabacum*. Cell 30: 589-597

Palme K, Gälweiler L (1999) PIN-pointing the molecular basis of auxin transport. Curr Opin Plant Biol 2: 375-381

Pavlovkin J, Okamoto H, Wächter R, Läuchli A, Ullrich CI (2002) Evidence for high activity of xylem parenchyma and ray cells in the interface of host stem and *Agrobacterium tumefaciens*-induced tumours of *Ricinus communis*. J Exp Bot 53: 1143-1154

Pedersen PL (1978) Tumor mitochondria and the bioenergetics of cancer cells. Prog Exp Tumor Res 22: 190-274

Pradel KS, Rezmer C, Krausgrill S, Rausch T, Ullrich CI (1996) Evidence for symplastic phloem unloading with a concomitant high level of acid cell-wall invertase in *Agrobacterium tumefaciens*-induced plant tumors. Bot Acta 109: 397-404

Pradel KS, Ullrich CI, Santa Cruz S, Oparka KJ (1999) Symplastic continuity in *Agrobacterium tumefaciens*-induced tumours. J Exp Bot 50: 183-192

Ramaiah KVA, Mookerjee A (1982) A comparative study of membrane related phenomena in normal and crown-gall tissues of red beet (*Beta vulgaris* L.). Experientia 38: 1324-1325

Rezmer C, Schlichting R, Wächter R, Ullrich CI (1999) Identification and localization of transformed cells in *Agrobacterium tumefaciens*-induced plant tumors. Planta 209: 399-405

Roberts SK, Tester M (1995) Inward and outward K^+-selective currents in the plasma membrane of protoplasts from maize root cortex and stele. Plant J 8: 811-825

Rojo E, Titarenko E, Leon J, Berger S, Vancanneyt G, Sanchez-Serrano JJ (1998) Reversible protein phosphorylation regulates jasmonic acid-dependent and -independent wound signal transduction pathways in *Arabidopsis thaliana.* Plant J 13: 153-165

Sachs T (1991) Callus and tumor development. *In* T Sachs, ed, Pattern formation in plant tissues. Cambridge University Press, Cambridge, pp 38-55

Sacristan MD, Melchers G (1977) Regeneration of plants from 'habituated' and '*Agrobacterium*-transformed' single-cell clones of tobacco. Mol Gen Genet 152: 111-117

Sauter C, Blum S (2003) Regression of lung lesions in Hodgkin's disease by antibiotics: case report and hypothesis on the etiology of Hodgkin's disease. Am J Clin Oncol 26: 92-94

Schell J, Koncz C, Spena A, Palme K, Walden R (1994) The role of phytohormones in plant growth and development. *In* Proc V Int Bot Congr, Tokyo/Yokohama, Japan, pp 38-48

Schurr U, Schuberth B, Aloni R, Pradel KS, Schmundt D, Jähne B, Ullrich CI (1996) Structural and functional evidence for xylem-mediated water transport and high transpiration in *Agrobacterium tumefaciens*-induced tumors of *Ricinus communis*. Bot Acta 109: 405-411

Schwalm K, Aloni R, Langhans M, Heller W, Stich S, Ullrich CI (2003) Flavonoid-related regulation of auxin accumulation in *Agrobacterium tumefaciens*-induced plant tumors. Planta 218: 163-178

Sitbon F, Sundberg B, Olsson O, Sandberg G (1991) Free and conjugated indoleacetic acid (IAA) contents in transgenic tobacco plants expressing the *iaaM* and *iaaH* IAA biosynthesis gene from *Agrobacterium tumefaciens*. Plant Physiol 95: 480-485

Stenlid G (1976) Effects of flavonoids on the polar transport of auxins. Physiol Plant 38: 262-266

Suzuki K, Kasamo K (1993) Effects of aging on the ATP- and pyrophosphate-dependent pumping of protons across the tonoplast isolated from pumpkin cotyledons. Plant Cell Physiol 34: 613-619

Tarbah FA, Goodman RN (1988) Anatomy of tumor development in grape stem tissue inoculated with *Agrobacterium tumefaciens* biovar 3, strain AG 63. Mol Plant-Microbe Interact 32: 455-466

Ullrich CI, Aloni R (2000) Vascularization is a general requirement for growth of plant and animal tumours. J Exp Bot 51: 1951-1960

Van Slogteren GMS, Hoge JHC, Hooykaas PJ, Schilperoort RA (1983) Clonal analysis of heterogeneous crown gall tumor tissues in wild-type and shooter mutant strains of *Agrobacterium tumefaciens*-expression of T-DNA genes. Plant Mol Biol 2: 321-333

Veselov D, Langhans M, Hartung W, Aloni R, Feussner I, Götz C, Veselova S, Schlomski S, Dickler C, Bächmann K, Ullrich CI (2003) Development of *Agrobacterium tumefaciens* C58-induced plant tumors and impact on host shoots are controlled by a cascade of jasmonic acid, auxin, cytokinin, ethylene and abscisic acid. Planta 216: 512-522

Vogel JP, Woeste KE, Theologis A, Kieber JJ (1998) Recessive and dominant mutations in the ethylene biosynthetic gene *ACS5* of *Arabidopsis* confer cytokinin insensitivity and ethylene overproduction, respectively. Proc Natl Acad Sci USA 95: 4766-4771

Wächter R, Fischer K, Gäbler R, Kühnemann F, Urban W, Bögemann GM, Voesenek LACJ, Blom CWPM, Ullrich CI (1999) Ethylene production and ACC-accumulation in *Agrobacterium tumefaciens*-induced plant tumours and their impact on tumour and host stem structure and function. Plant Cell Environ 22: 1263-1273

Wächter R, Langhans M, Aloni R, Götz G, Weilmünster A, Koops A, Temguia L, Mistrik I, Pavlovkin J, Rascher U, Schwalm K, Koch KE, Ullrich CI (2003) Vascularization, high-volume solution flow, and localized roles for enzymes of sucrose metabolism during tumorigenesis by *Agrobacterium tumefaciens*. Plant Physiol 133: 1024-1037

Warburg O (1930) The Metabolism of Tumors. Arnold Constable, London

Weil M, Rausch T (1990) Cell wall invertase in tobacco crown gall cells: enzyme properties and regulation by auxin. Plant Physiol 94: 1575-1581

Weiler EW, Schröder J (1987) Hormone genes and crown gall disease. Trends Biochem Sci 12: 271-275

Weiler EW, Spanier K (1981) Phytohormones in the formation of crown gall tumors. Planta 153: 326-337

Winter-Sluiter E, Läuchli A, Kramer D (1977) Cytochemical localization of K^+-stimulated adenosine triphosphatase activity in xylem parenchyma cells of barley roots. Plant Physiol 60: 923-927

Wood DW, Setubal JC, Kaul R, Monks DE, Kitajima JP, Okura VK, Zhou Y, Chen L, Wood GE, Almeida Jr. NF, Woo L, Chen Y, Paulsen IT, Eisen JA, Karp PD, Bovee Sr. D, Chapman P, Clendenning J, Deatherage G, Gillet W, Grant C, Kutyavin T, Levy R, Li MJ, McClelland E, Palmieri P, Raymond C, Rouse R, Saenphimmachak C, Wu Z, Romero P, Gordon D, Zhang S, Yoo H,

Tao Y, Biddle P, Jung M, Krespan W, Perry M, Gordon-Kamm B, Liao L, Kim S, Hendrick C, Zhao ZY, Dolan M, Chumley F, Tingey SV, Tomb JF, Gordon MP, Olson MV, Nester EW (2001) The genome of the natural genetic engineer *Agrobacterium tumefaciens* C58. Science 294: 2317-2323

Wood HN, Braun AC (1965) Studies on the net uptake of solutes by normal and crown-gall tumor cells. Proc Natl Acad Sci USA 54: 1532-1538

Zambryski PC, Tempé J, Schell J (1989) Transfer and function of T-DNA genes from *Agrobacterium* Ti and Ri plasmids in plants. Cell 56: 193-201

Chapter 16

THE CELL-CELL COMMUNICATION SYSTEM OF *AGROBACTERIUM TUMEFACIENS*

Catharine E. White and Stephen C. Winans

Department of Microbiology, Cornell University, Ithaca, NY 14853, USA

Abstract. The Ti plasmids of *Agrobacterium tumefaciens* carry almost all of the genes required for the formation of crown gall tumors and for the utilization of opines that are produced by these tumors. These plasmids also encode a cell-cell signalling (quorum sensing) system that is homologous to the LuxR-LuxI system of *Vibrio fischeri*. The LuxI orthologue TraI synthesizes a specific *N*-acylhomoserine lactone (AHL). This AHL is a diffusible signalling molecule and, when it accumulates to a sufficiently high concentration, it interacts with the LuxR-type transcription activator TraR. The *traR* gene is induced by particular opines, causing quorum sensing in this bacterium to occur only in the presence of these compounds. TraR activates genes required for conjugal transfer and vegetative replication of the Ti plasmid. In this chapter, we discuss the quorum sensing system of *A. tumefaciens* from a molecular perspective, and speculate on the possible roles this system may have in virulence and plant colonization.

1 INTRODUCTION

Over the past 20 years, it has become evident that many types of bacteria are far more social than had been previously thought. Populations of bacterial cells appear to use a variety of types of chemical signalling to coordinate diverse activities, especially those activities requiring large numbers of bacterial cells. For example, many types of pathogenic bacteria communicate during colonization of plant or animal hosts, and thereby coordinate their attack. It is thought that single cells are more susceptible to host defences than a population of bacteria that coordinately express genes involved in virulence. For example, species of *Erwinia* that cause soft rot on plants do not produce lytic enzymes in their plant hosts until a threshold population density is reached (Perombelon, 2002; Smadja et al., 2004). In biofilm formation, communication between neighbouring cells must occur for the complex structures that are associated with biofilms to form (Parsek and Greenberg, 2005). Swarming motility is another example of behaviour where cell-cell communication is a critical component (Eberl et al., 1996; Givskov et al., 1998).

The importance of cell-cell communication in bacteria is also reflected by the fact that these systems have evolved numerous times. Many gram positive bacteria use short peptides as intercellular signalling molecules (Dunny and Leonard, 1997; Lyon and Novick, 2004). The plant pathogen *Ralstonia solanacearum* uses palmitic acid methyl ester (PAME) to signal (Brumbley et al., 1993; Flavier et al., 1997), while the actinomycetes use γ-butyrolactones (Chater and Horinouchi, 2003). Many different types of bacteria produce and respond to a furanosyl borate diester (designated AI-2), first described as an intercellular signal in *Vibrio harveyi* and often thought of as a universal bacterial "esperanto" (Chen et al., 2002; Xavier and Bassler, 2003).

Many proteobacteria use small molecules called *N*-acylhomoserine lactones (AHLs) for communication. The paradigm for AHL mediated signalling is the system found in the marine animal symbiont *Vibrio fischeri*. This system includes an AHL synthase (LuxI) and an intracellular receptor of the signal, a transcriptional regulator called LuxR (Engebrecht et al., 1983; Engebrecht and Silverman, 1984, 1987). The AHL is synthesized at some basal rate and can freely diffuse across the cellular membrane (Kaplan and Greenberg, 1985). When it accumulates to some threshold level, it binds to the LuxR protein, which then forms dimers and activates transcription of target genes that direct bioluminescence (Choi and Greenberg, 1991, 1992a, 1992b; Stevens et al., 1994). *Vibrio fischeri* colonizes the light organs of certain species of squid and fish to high population densities. The signal molecule therefore accumulates in the light organ, and active LuxR-AHL complexes activate genes responsible for the production of bioluminescence.

The term quorum sensing is often used for these systems to reflect the importance of population density for the accumulation of the signal molecules, which must reach some threshold level before a coordinated response occurs (Fuqua et al., 2001). However, it should also be pointed out that a diffusion barrier must also be present for the system to work, so that the signal can accumulate in the growing population (Redfield, 2002).

LuxR-LuxI type systems have been described in a number of plant pathogens, including *Pseudomonas syringae*, *P. aeruginosa*, *Erwinia carotovora*, and *Ralstonia solanacearum* (reviewed in von Bodman et al., 2003), and also in *Agrobacterium tumefaciens*. In *A. tumefaciens*, the *traR* and *traI* genes are highly conserved, and are associated with Ti plasmids or related catabolic plasmids (Farrand, 1998). Similar systems have also been identified on the symbiosis plasmids in the related nitrogen-fixing plant symbionts, such as *Rhizobium* sp. (Wisniewski-Dye and Downie, 2002; Gonzalez and Marketon, 2003). Thus this system in *A. tumefaciens*, which has been intensively studied, is an important model for quorum sensing in this agriculturally important family. Most studies of this system have focussed on the molecular biology of TraI, TraR, and the other proteins involved in regulating the activity of TraR. Our knowledge of the importance and activity of the TraR-TraI system in pathogenesis and establishment of *A. tumefaciens* on infected plants is much more limited. In this review, we will focus on the current state of our knowledge of this system from a molecular perspective. In the last section, we will speculate on the adaptive significance of this system.

2 A MODEL OF QUORUM SENSING IN *A. TUMEFACIENS*

Almost all of the genes that are involved in the quorum sensing system in *A. tumefaciens* are located on the Ti plasmids, including *traI* and *traR* themselves. TraI synthesizes specifically *N*-3-oxooctanoyl-L-homoserine lactone (OOHL)(Fuqua and Winans, 1994; More et al., 1996). When this signal accumulates to a threshold level (in the nanomolar range), it binds to its intracellular target the LuxR-type protein TraR (Piper et al., 1993; Zhang et al., 1993; Fuqua and Winans, 1994; Hwang et al., 1994). Active TraR-OOHL complexes can then form dimers which bind with high specificity to DNA sequences called *tra* boxes at target promoters on the Ti plasmid (Zhu and Winans, 1999; Qin et al., 2000; Zhu and Winans, 2001). A number of proteins negatively regulate the activity of TraR, including TraM, TrlR (only on octopine-type Ti plasmids), and the products of the *attKLM* operon (located on the At or 'cryptic' plasmid). It should be noted that four additional *luxR* homologs are present in the genome sequence of C58 (Goodner et al., 2001; Wood et al., 2001), though no additional *luxI* homologs are apparent. It is not known whether the products of these genes are functional as AHL receptors. A model of the TraR-TraI system in octopine-type Ti plasmids is presented in *Figure 16-1*, and discussed in the following sections.

2.1 Regulation of *tra* gene expression

Most studies of quorum sensing in *A. tumefaciens* have focussed on either the octopine-type Ti plasmids (such as pTiR10) or the nopaline-type Ti plasmids such as pTiC58. However, almost all of the genes involved in quorum sensing (*traR*, *traI*, and *traM*) and the genes regulated by TraR-OOHL complexes (the conjugation and replication genes of the Ti plasmid) are highly conserved. A composite gene map and sequence is available for the octopine-type plasmids R10, A6, B6, Ach5, and 15966, and for the nopaline-type pTiC58 from the genome sequence of strain C58 (Zhu et al., 2000; Goodner et al., 2001; Wood et al., 2001; also see Chapter 4 for more details).

On both types of plasmids, expression of the *traR* gene is positively regulated by the presence of specific opines: octopine for octopine-type and agrocinopines A and B for nopaline-type Ti plasmids (Genetello et al., 1977; Kerr et al., 1977). On octopine-type Ti plasmids, *traR* is the last gene of the *occ* operon, which is divergently transcribed from *occR* (Fuqua and Winans, 1996b). In the presence of octopine, OccR activates expression of the *occ* operon, which directs octopine uptake and utilization (*Figure 16-1*). On nopaline-type Ti plasmids, *traR* is the fourth gene in the five gene *arc* operon, which is not related to the *occ* operon described above (Beck von Bodman et al., 1992). The *arc* operon, required for agrocinopine A and B utilization, is divergently transcribed from the *acc* operon, the first gene of which is *accR*. In the presence of agrocinopine A or B, repression of both the *arc* and *acc* promoters by AccR is relieved, resulting in gene expression (Beck von Bodman et al., 1992). For chrysopine-type Ti plasmids, the conjugal opines are agrocinopines C and D, and *traR* is thought to be negatively regulated by AccR (Oger and Farrand, 2001). Agrocinopines C and D are also required for *traR* expression on pTiBo542, although in this case the transcription regulator has not been identified

(Ellis et al., 1982). Finally, pAtK84b of the non-pathogenic *A. radiobacter* has two copies of *traR* (Oger and Farrand, 2002). One of these is thought to be regulated by NocR in response to nopaline, and the other is repressed by AccR.

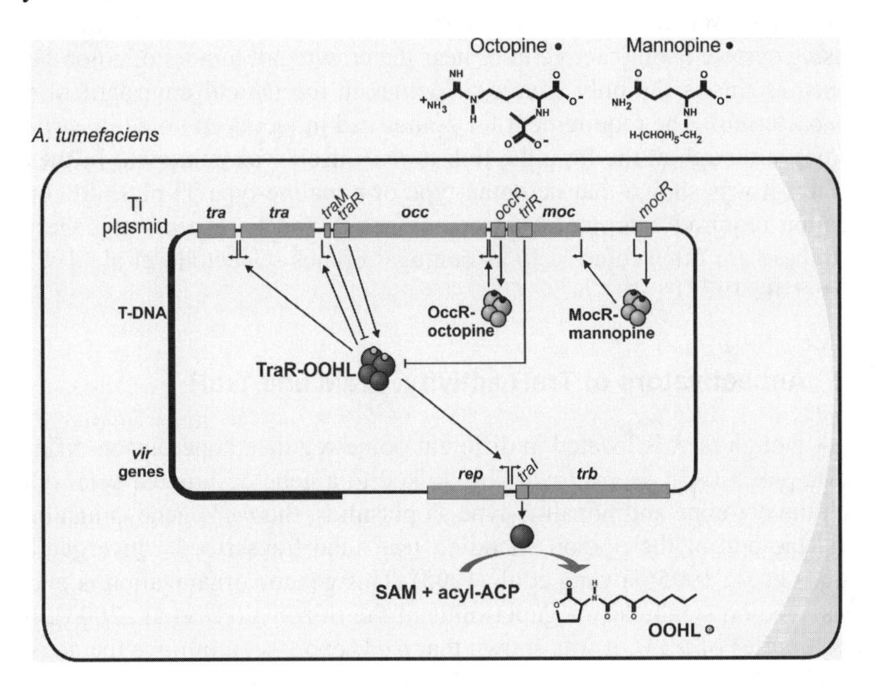

Figure16-1. A model of quorum sensing in *A. tumefaciens*. Transcription of *tra*R is activated only in response to octopine via OccR. Binding of OOHL (produced by TraI) induces TraR dimerization, and TraR-OOHL complexes then activate transcription of the two *tra* operons, *tra*M, and the *tra*I-*trb* and *rep* operons (promoters are indicated with bent arrows). Activity of TraR-OOHL complexes can be directly inhibited by the products of *tra*M and *trl*R. The approximate locations of the T-DNA and *vir* genes are shown for reference. S-adenosyl methionine (SAM) and 3-oxooctanoyl-ACP (acyl-ACP) are the precursors for OOHL synthesis (More et al., 1996)

A comparison of *traR* regulation on the various Ti plasmids described above reveals that the control of *traR* gene expression by opines must have evolved independently a number of times. The genes in the operons which contain *traR* on each plasmid are not related, and they are regulated by at least two different mechanisms in response to different opines. For example, OccR is a transcriptional activator and a member of the LysR family (Wang et al., 1992; Fuqua and Winans, 1996b). In contrast, AccR resembles the Lac repressor of *E. coli* (Beck von Bodman et al., 1992). In this case, the inducing agrocinopines are expected to bind to AccR resulting in a dramatic decrease in affinity for its binding sites at target promoters, although this has not yet been demonstrated experimentally.

As described above, expression of the *traR* genes requires opines (Fuqua and Winans, 1996b; Piper et al., 1999). Therefore, the quorum sensing system is only active in or near the crown gall tumors of colonized plants, as this is the only source of opines in the natural environment of this bacterium. The requirement for opines had in fact been noted in earlier studies, although at the time the link to *traR* itself was unknown. In these studies, it was shown that octopine-type or nopaline-type Ti plasmid conjugation required octopine or agrocinopines A and B respectively, therefore these are often referred to as conjugal opines (Genetello et al., 1977; Kerr et al., 1977).

2.2 Antiactivators of TraR activity: TraM and TraR

Although *traR* is located in different opine-regulated operons on different Ti plasmids, it seems invariably linked to a gene designated *traM*. On both the octopine and nopaline-type Ti plasmids, the *traM* gene is just beyond the end of the operon encoding *traR*, and transcribed convergently (Fuqua et al., 1995; Hwang et al., 1995). This genetic organization is even conserved on symbiosis megaplasmids of the rhizobia (He et al., 2003). In early studies of *traM*, it was shown that *traM* expression inhibits the activity of TraR. Over-expression of *traM* inhibits TraR activity, while a *traM* null mutation results in an increase in TraR activity on both the nopaline and octopine-type Ti plasmids (Fuqua et al., 1995; Hwang et al., 1995). Further studies have shown that TraM directly disrupts TraR activity through protein-protein interactions (Hwang et al., 1999), although the exact mechanism of inhibition is not entirely clear.

Mutational studies of TraM and TraR have demonstrated that the C-terminal region of TraM and the DNA-binding region of TraR are required for the formation of TraR-TraM complexes and inhibition of TraR activity (Hwang et al., 1999). This suggests that perhaps TraM directly inhibits TraR activity through sequestering the DNA-binding surface of TraR such that it cannot bind to *tra* boxes. However, TraM can both block binding of TraR to *tra* box DNA and disrupt TraR-DNA complexes that have already formed (Luo et al., 2000). Therefore, if the surfaces of TraR that are contacted by TraM and DNA overlap, then TraM must be able to out-compete DNA in binding to TraR. The affinities of TraR for TraM and TraR for DNA are both in the nanomolar range (Zhu and Winans, 1999; Swiderska et al., 2001).

Two crystal structures of TraM have been published, and in both cases the protein was crystallized in the absence of TraR (Chen et al., 2004; Vannini et al., 2004). TraM is a small protein (approximately 11 kDa) consisting of two anti-parallel α helices per monomer (*Figure 16-2*). In both structures, TraM crystallized as dimers, with specific and extensive associations along the length of each subunit. The dimerization interface is predominantly hydrophobic, and a large surface area is buried between the two subunits, relative to the overall size of the dimer. The regions of the protein involved in dimerization had also been predicted in a mutagenesis study (Qin et al., 2004b). Although these structures of TraM are useful in generating a model of how this protein may inhibit TraR activity, the mechanism is not clear. One of the structural studies suggests that a complex forms consisting of two monomers of TraM and two dimers of TraR, supported by gel filtration chromatography experiments with both proteins (Vannini et al., 2004). This bulky complex was thought to block both TraR dimers from contacting DNA. In the second structural study, gel filtration experiments with TraR and TraM were also performed, but suggested that one monomer of TraR binds to one or two monomers of TraM (Chen et al., 2004). In the same study, it was shown that when monomers of a TraM dimer are covalently linked, they are able to bind to but not inactivate TraR. Furthermore, residues of TraM that are important for initial binding to TraR are different than those required for inactivation, suggesting that

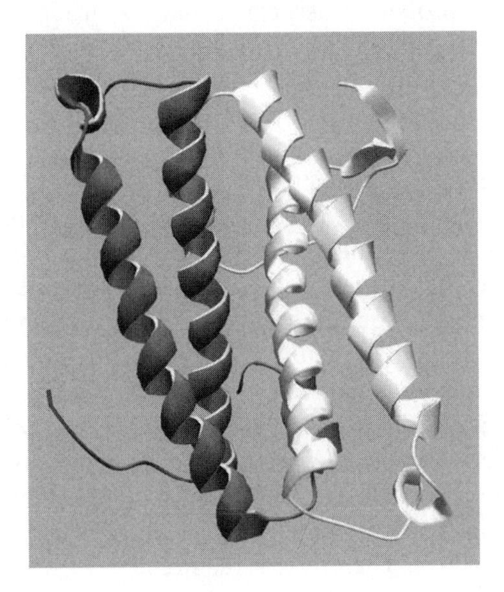

Figure 16-2. Ribbon model of a TraM dimer. One monomer is white, the other gray. Note the extensive dimerization interface (Chen et al., 2004; Vannini et al., 2004).

these are two separate steps in the interaction (Swiderska et al., 2001). These data led to the model that TraR and TraM dimers bind to each other, and then the homodimers dissociate to form the anti-activation complex (Chen et al., 2004). Both DNA-binding domains of a TraR dimer are required for *tra* box recognition, therefore a disruption of TraR dimerization would result in a disruption of DNA binding (Chai et al., 2001).

The role that TraM plays in quorum sensing is not entirely clear, although it is thought to have a conserved function as discussed above. In both nopaline and octopine-type Ti plasmids, the *traM* gene is activated by TraR-OOHL complexes (*Figure 16-1*) (Fuqua et al., 1995; Hwang et al., 1995). This suggests that a negative feedback loop forms to attenuate the activity of TraR. Indeed, in the *traM* null mutants, TraR is both more active and responsive to lower concentrations of OOHL than in the wild type strain (Fuqua et al., 1995; Hwang et al., 1995). Recently, it has also been demonstrated that *traM* is activated in response to a plant-released phenolic (acetosyringone) which is a *vir* gene inducer (Cho and Winans, 2005). This overlap in regulation between the virulence and quorum sensing systems may be a mechanism to avoid concurrent expression of the Type IV secretion system for T-DNA transfer (induced by acetosyringone), and the Type IV secretion system for Ti plasmid conjugation (induced by TraR and OOHL).

Another protein that inhibits TraR activity through direct interactions is TrlR, although this protein is associated only with the octopine-type Ti plasmids (Oger et al., 1998). The *trlR* gene is almost identical to *traR*, except for one frame-shift mutation that prevents translation of the C-terminal DNA-binding domain, resulting in a truncated protein lacking a DNA-binding domain. If the frame-shift is corrected by site-directed mutagenesis, the result is a functional copy of TraR (Zhu and Winans, 1998). In a study using purified proteins, it was shown that TrlR inhibits TraR activity directly by forming inactive heterodimers (Chai et al., 2001).

The *trlR* gene is the fifth gene in the six-gene *mot* operon, most of the other genes of which are required for uptake of the opine mannopine (Oger et al., 1998). Mannopine is required for the expression of this operon, probably through relieving repression by MocR, which is related to the LacI repressor of *E. coli* (*Figure 16-1*) (Oger et al., 1998; Zhu and Winans, 1998). A null mutation of the *trlR* gene results in an increase in TraR activity, while overexpression of *trlR* from a multi-copy plasmid causes a significant decrease in TraR activity (Oger et al., 1998; Zhu and Winans, 1998; Chai et al., 2001). Therefore, a requirement of mannopine for *trlR* gene expression fully explains the negative effect of this opine on TraR activity. The fact that *trlR* is nearly identical to *traR*, and that both lie in opine-regulated operons suggests that *trlR* arose from a gene duplication event (Oger et al., 1998). As *trlR* is present on all octopine-type Ti plasmids studied to date (and always in the *mot* operon), it is not a laboratory artifact (Zhu and Winans, 1998). It is intriguing that one opine, octopine, up-regulates TraR activity while another, mannopine, results in a decrease in activity. As it was also found that favoured catabolites, such as succinate, glutamine, and tryptone block *trlR* expression, it has been suggested that TrlR functions to attenuate the energetically expensive process of conjugation when nutrients are limited (Chai et al., 2001).

2.3 Regulation of TraR activity through OOHL turnover

AHLs are intrinsically unstable at alkaline pH, as they are susceptible to hydrolysis of the homoserine lactone ring. A report published only four years ago identified a lactonase of *Bacillus cereus*, called AiiA, which is capable of enzymatically inactivating AHLs by the same mechanism (Dong et al., 2001). This lactonase therefore converts *N*-acylhomoserine lactones into their corresponding *N*-acyl homoserine. Identification of the AiiA homologue AttM on the pAtC58 plasmid of strain C58 (Goodner et al., 2001; Wood et al., 2001), prompted interest in the possibility of OOHL turnover in *A. tumefaciens* as yet another mechanism for regulating TraR activity. AHL utilization as an energy and nitrogen source had been reported for *Variovorax paradoxus* and later *Pseudomonas aeruginosa* (Leadbetter and Greenberg, 2000; Huang et al., 2003). Three groups have subsequently demonstrated that AttM does have activity as a lactonase against OOHL (Zhang et al., 2002a; Carlier et al., 2003; Y. Chai and S. C. Winans, unpublished data).

Although the reports of OOHL depletion by AttM speculated that the *attM* gene evolved by selection for this substrate, it now appears more likely that AttM evolved by selection for catabolism of γ-butyrolactone (GBL), a compound similar to AHLs but lacking an acyl chain and amine (Carlier et al., 2004). This conclusion was drawn when it was discovered that GBL induces transcription of the *attM* gene. This gene lies within the *attKLM* operon, the other two genes of which encode dehydrogenases, and all three protein products are thought to act in a single pathway. This operon is divergently transcribed from *attJ*, which encodes a repressor that resembles IclR. In an *attJ* null mutant, OOHL does not accumulate to appreciable levels, suggesting that AttJ represses *attKLM* expression (Zhang et al., 2002a). Three different groups tested different AHLs, including OOHL, for activation of *attKLM*, expecting that these compounds might relieve repression by AttJ through direct interactions (Zhang et al., 2002a;

Carlier et al., 2004; Y. Chai and S. C. Winans, unpublished data). However, none of the AHLs tested had an effect on *attKLM* expression, leading to the conclusion that AHLs are not inducers of AttJ and perhaps not the natural substrates of this pathway. One group reported that *attKLM* in strain A6 is activated only at high population densities and in response to nitrogen or carbon starvation via the ppGpp stress response (Zhang et al., 2002a; Zhang et al., 2004). However, population density and nutrient limitations were not identified as important for *attKLM* induction in strain C58 (Carlier et al., 2004; Y. Chai and S. C. Winans, unpublished data). Perhaps these differences in regulation are due to differences between the two strains.

In one study, a number of additional compounds with lactone rings, other than AHLs, were tested for their effects on *attKLM* expression (Carlier et al., 2004). Among these compounds, only GBL caused induction of the operon. A more recent study has demonstrated *in vitro* that this effect occurred through direct interactions between GBL and AttJ (Y. Chai and Winans, unpublished data). GBL binding to AttJ results in a dramatic decrease in the affinity of AttJ for *attKLM* promoter DNA.

These results suggest that GBL is the natural substrate for AttKLM, and in fact the wild type strain C58 can grow on GBL as a sole carbon source, while growth on OOHL alone is not supported (Carlier et al., 2004). In the same study, it was predicted that the intermediates of GBL

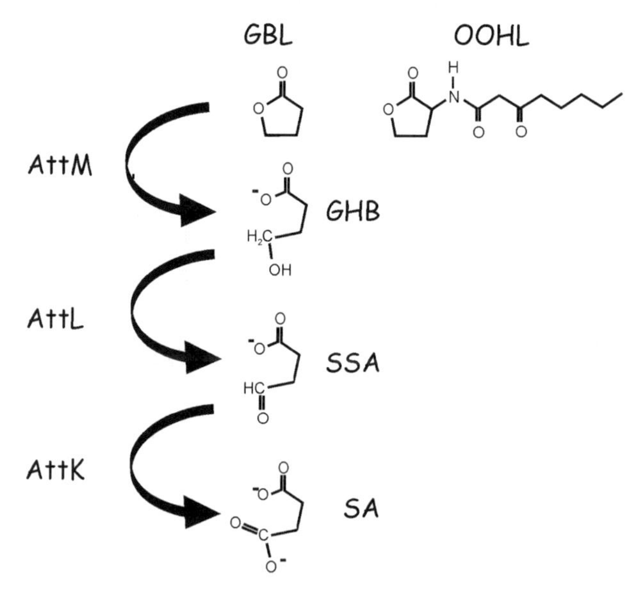

Figure 16-3. GBL degradation pathway via AttKLM. At least the first step of this pathway can lead to ring opening of AHLs.

degradation via AttKLM are γ-hydroxybutyrate (GHB) from activity of AttM, and succinate semialdehyde (SSA) from activity of AttL (*Figure 16-3*). SSA was predicted to be converted to succinic acid (SA) via AttK. Strain C58 was able to grow on either GBL or GHB as the sole carbon source, but an *attKLM* null mutant could not (Carlier et al., 2004). Not only GBL but also the intermediates of the pathway, GHB and SSA, are inducers of *attKLM* expression. Furthermore, when C58 was cultured in medium supplemented with *attKLM* inducers (GBL, GHB, or SSA), OOHL did not accumulate (Carlier et al., 2004).

It is not yet clear what the role of the *attKLM* operon is in quorum sensing in nature. It is possible that TraR and AttKLM are seldom active simultaneously in *A. tumefaciens* in its natural environment, and therefore the overlap in substrate specificity might not be significant. AttKLM may be active when the bacteria are in soil (rather then on host plants), in which case *A. tumefaciens* would be able to degrade AHL signals of neighbouring bacteria. Alternatively, AttKLM may play some role in attenuating the activity of TraR in the presence of opines on infected plants. GBL has been reported to accumulate in the tissue of at least some plants, and the AttK and AttM proteins have been shown, in a proteomics study, to accumulate in *A. tumefaciens* when exposed to tomato roots (Lee and Shibamoto, 2000; Rosen et al., 2003). Further studies to assess the role of AttKLM in possible 'quorum-quenching' during growth on infected plants, or for survival in soil should be quite interesting.

3 STRUCTURE AND FUNCTION STUDIES OF TRAR

A number of studies have been published that focus on the biochemistry of the TraR protein, and in 2002, two crystal structures of TraR complexed with OOHL and bound to DNA also became available. As TraR is a central component of the quorum sensing system, and interacts with OOHL, DNA, the transcription machinery, and two antiactivators that act through different mechanisms, there is clearly interest in understanding the function of this protein at a biochemical and structural level.

In initial attempts to overexpress TraR in either *E. coli* or *A. tumefaciens*, it was noticed that the protein would not accumulate in soluble fractions, but rather formed insoluble inclusion bodies. This observation led one group to suggest that apo-TraR may associate with the cytoplasmic membranes, and is released only upon binding to OOHL (Qin et al., 2000). The TraR-OOHL complexes could then dimerize and activate transcription at target promoters. Another group found that when OOHL was added to cultures during overexpression, the abundance of TraR increased dramatically (Zhu and Winans, 2001). OOHL could increase the abundance of TraR only if it was added during overexpression, not after. Pulse-chase experiments in the same study revealed that when only mildly expressed, the apo-protein does not accumulate in the cell at all, but is rapidly degraded by proteases. In fact, proteolysis of TraR in the absence of OOHL occurs so rapidly that the ligand must exert its effect on TraR solubility as soon as the protein is synthesized (Zhu and Winans, 1999). Accumulation of apo-TraR could not be detected, while in the presence of OOHL, TraR was stable in *A. tumefaciens* for more than 30 minutes (Zhu and Winans, 1999). As OOHL stabilizes TraR against proteolysis, it must trigger some conformational change in the protein. The observation that this effect is most likely to occur during translation led to the suggestion that OOHL serves as a scaffold for TraR folding.

A number of other proteins, in both eukaryotes and bacteria, have been identified that require their cognate ligands for protein folding and stability against proteolysis (Dyson and Wright, 2002). In most cases, these proteins are involved in highly time-dependent processes (such as regulation of transcription). It has been suggested that this requirement of inducing ligand for protein stability and accumulation results in a highly controlled and 'fast-response' switch from inactive to active forms of the protein (Shoemaker et al., 2000). It is also possible that incorporation of ligand into the protein folding process results in an increase in specificity of binding (Dyson and Wright, 2002). When TraR is expressed at native levels in *A. tumefaciens*, it binds to OOHL with extremely high specificity (Zhu et al., 1998). Furthermore, the affinity of TraR for OOHL is also extremely high: the ligand can be removed only by extensive dialysis in the presence of mild detergents (Zhu and Winans, 2001).

The observation that OOHL is required for accumulation of soluble TraR allowed successful over-expression and purification of stable TraR-OOHL complexes. In gel filtration chromatography experiments, purified complexes were shown to exist in solution as dimers (Qin et al., 2000; Zhu and Winans, 2001). TraR-OOHL dimers bind to the 18 bp *tra* boxes at TraR-dependent promoters with high specificity, and only TraR-OOHL complexes, promoter DNA, and RNA polymerase are required for transcription activation *in vitro* (Zhu and Winans, 1999). This strongly suggested that TraR contacts RNA polymerase directly to activate transcription, as discussed below in Section 4.3.

Following these studies, complexes containing TraR, OOHL, and operator DNA were crystallized and the structures solved by two different groups (Vannini et al., 2002; Zhang et al., 2002b). To date, these are the only structures available for any LuxR-type proteins. TraR crystallized as dimers, as had been expected, and each monomer has two domains: an N-terminal domain (NTD) that binds to OOHL, and a C-terminal domain (CTD) that binds to the DNA (*Figure 16-4*). Each subunit bound one molecule of OOHL, confirming an earlier study (Zhu and Winans, 1999).

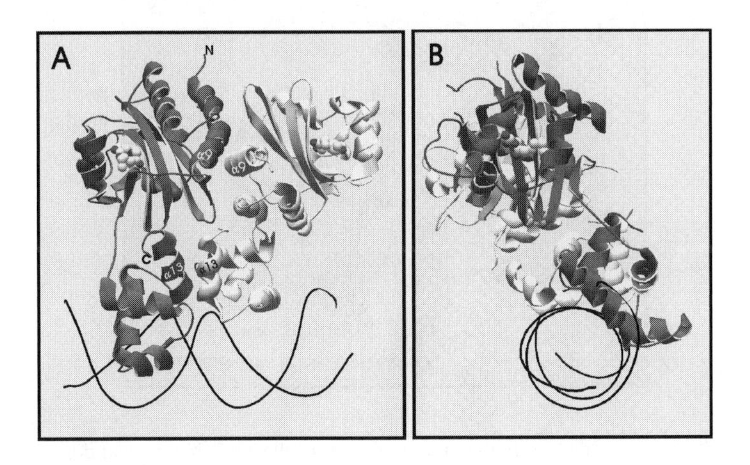

Figure 16-4. Ribbon model of a TraR-OOHL dimer bound to DNA. One monomer is light gray, the other is dark gray, and the backbone of the DNA is shown in black. The OOHL, bound in the N-terminal domain of each monomer, is in space-fill and CPK colors. **a**. The N-terminus (N) and C-terminus (C) of the left monomer are shown for reference, and α helix 9 (α9) and 13 (α13) of each dimer are also marked. These helices are involved in dimerization of the N-terminal domains and the C-terminal domains respectively. **b**. View of the same model, but along the long axis of the DNA to highlight the overall asymmetry of the structure (Vannini et al., 2002; Zhang et al., 2002b).

Each dimer has two dimerization interfaces, the most extensive of which is between α helix 9 of the NTD of each monomer (*Figure 16-4*) (Vannini et al., 2002; Zhang et al., 2002b). Mutational studies confirmed that a number of residues buried in the hydrophobic interface between these two helices are critical to dimerization (Luo et al., 2003). A less extensive dimerization interface is present between the last helix of each CTD of the dimer. The two dimerization interfaces described above are distinct (i.e. they are not continuous). In addition, within each monomer, the NTD is connected to the CTD by a 12-residue, unstructured linker (Vannini et al., 2002; Zhang et al., 2002b). This results in a high degree of flexibility of the N-terminal domains of the dimer relative to the C-terminal domains. In both crystal structures, the CTDs have a two-fold axis of symmetry, and the NTDs also have a two-fold axis of symmetry, but the

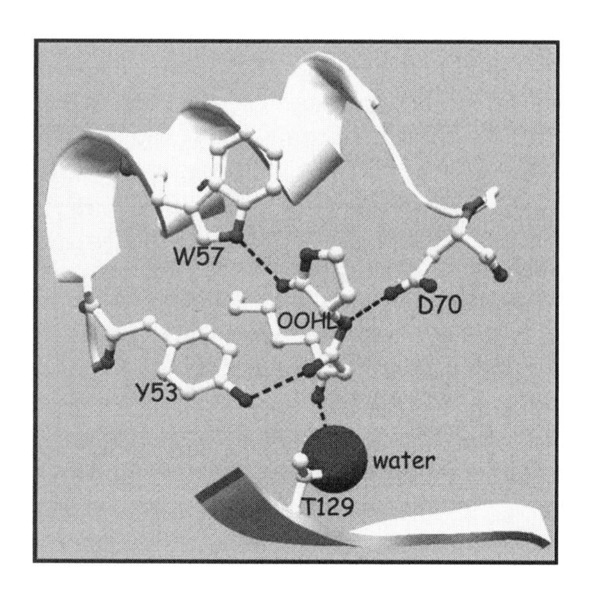

Figure16-5. Model of OOHL in the binding pocket of the TraR N-terminal domain. A portion of an α helix and β sheet are shown for reference. The OOHL and the four residues of TraR that have hydrogen bonds (dashed lines) with OOHL are shown in CPK colors. A water molecule mediates the hydrogen bond between T129 and the 3-oxo group of OOHL (Vannini et al., 2002; Zhang et al., 2002b).

overall structure of the dimer bound to the DNA is highly asymmetric (*Figure 16-4*b).

The N-terminal domain of each monomer is an α-β-α structure (Vannini et al., 2002; Zhang et al., 2002b). The bound OOHL is buried within the core of the domain between the β sheet and a layer of α helix. In fact, the ligand is completely engulfed in the protein and protected from bulk solvent, further supporting the model that OOHL is involved in protein folding. For OOHL to gain access to the ligand binding pocket, major structural rearrangements must occur in the protein. Contacts between the ligand and the residues in the binding pocket are extensive, and include many hydrophobic and van der Waals packing interactions between the 8-carbon tail and non-polar residue side-chains. In addition, all of the polar groups of OOHL participate in hydrogen bonds with nearby residues, satisfying partial charges such that they do not disrupt the integrity of the domain core (*Figure 16-5*). The importance of these polar contacts to ligand binding and therefore protein stabilization was confirmed in two mutational studies (Luo et al., 2003; Chai and Winans, 2004).

The intimate and extensive contacts between OOHL and residues in the binding pocket not only support the model that OOHL is involved in protein folding, but also clarify why the specificity of binding is so high. For example, many AHLs involved in quorum sensing in other bacteria do not have the 3-oxo group on the fatty acid tail. Binding of C8-acyl homoserine lactone (C8-HSL) instead of 3-oxo-C8-HSL to TraR would presumably not satisfy the partial charge of T129, which otherwise interacts with the 3-oxo group, resulting in a disruption of the stability of the domain core. It is interesting to note that the hydrogen bond between T129 and the 3-oxo group is in fact mediated by a water molecule, also bound in the core of the domain (Vannini et al., 2002; Zhang et al., 2002b). In an attempt to alter the specificity of binding from 3-oxo-C8-HSL to C8-HSL, single site substitutions of this residue were constructed, to increase hydrophobicity at that position and exclude the water molecule from the binding pocket (Chai and Winans, 2004). A number of substitution mutants were identified that were stabilized by C8-HSL, however these had broadened rather than altered specificity. Another variable feature of AHLs is the length of the fatty acid tail. In the same study described above, substitutions of residues in the binding pocket that pack around the 8 carbon tail of OOHL were made to increase hydrophobic bulk at the end of the pocket, in the hopes of altering specificity for 3-oxo-C6-HSL. Again, a handful of broadened specificity mutants were identified, although the substitutions also decreased the stability of the protein (with either AHL). These studies demonstrate that the intimate and multiple associations between TraR and OOHL are indeed important for both specificity of binding and protein stability. By coupling these two roles, the ligand specificity is enhanced and not easily altered.

There is considerable interest in understanding how OOHL stabilizes TraR against proteolysis, as the requirement of ligand for TraR stability is expected to be a critical point in the regulation of TraR activity. In a recent study, an attempt was made to generate random mutations of the N-terminal domain (in the full-length protein) that have constitutive activity (Chai and Winans, 2005a). A constitutive mutant was identified that was active *in vivo*. However, the mutation arose from an N-terminal fusion of another protein (an aminoglycoside *N*-acetyltransferase) encoded on the same plasmid rather than a point mutation within the NTD of TraR itself. To determine if this result was due to a general effect of an N-terminal fusion or specific to the aminoglycoside *N*-acetyltransferase, a library of random N-terminal fusions to TraR was constructed and screened for constitutive activity. Five additional fusion proteins were identified that were active *in vivo* in the absence of OOHL. These results suggest that the

N-terminal fusions were sequestering the unfolded or partially unfolded NTD of apo-TraR such that it is less accessible to proteases.

The C-terminal DNA binding domain of TraR is a four-helix bundle containing a highly conserved helix-turn-helix (HTH) DNA binding motif as the central two helices (Vannini et al., 2002; Zhang et al., 2002b). The structure of the domain itself is quite conserved, and places the LuxR family in the larger NarL-FixJ super-family based on both protein sequence and structural similarity (Fuqua and Greenberg, 2002). All members of this super-family have very similar DNA binding domains, and in fact the high-resolution structures of this conserved domain in two other members of this family, NarL and GerE of *Bacillus subtilus*, both superimpose very well with the TraR CTD, with surprisingly little deviation between all three (Ducros et al., 2001; Maris et al., 2002). In both crystal structures of TraR-OOHL-DNA complexes, the binding site is the consensus *tra* box, an 18 bp sequence with perfect dyad symmetry and the strongest native binding site for TraR (Vannini et al., 2002; Zhang et al., 2002b). In the structures, the *tra* box has a smooth 30 degree bend toward the sides of the protein, although the DNA retains a B-form conformation. The recognition helix (the second helix of the HTH motif) of each monomer is bound in the major groove of each half-site. The only sequence-specific interactions between the dimer and its binding site occur between three residues of the recognition helix and four bases in each half-site. However, the interface between TraR and the DNA is quite extensive, and many polar and hydrophobic contacts between residues of TraR and the DNA backbone contribute to affinity of binding (Vannini et al., 2002; Zhang et al., 2002b; C. E. White and S. C. Winans, unpublished data).

4 TRAR IN TRANSCRIPTION ACTIVATION

4.1 Activation of the Ti plasmid conjugation genes

The Ti plasmid conjugation genes are arranged in three operons, all of which are dependent on TraR-OOHL for transcription (Piper et al., 1993; Fuqua and Winans, 1994; Fuqua and Winans, 1996a; Li et al., 1999). Almost all of the genes (*tra* and *trb*) and their organization are highly conserved on all Ti plasmids that have been studied to date (Farrand et al., 1996; Li et al., 1998; Zhu et al., 2000). However, the *tra* and *trb* genes appear to have diverse origins (Farrand et al., 1996; Zhu et al., 2000). For example, all of the *trb* genes are most similar to IncP-type plasmid transfer systems, while one of the *tra* genes (*traA*) is not at all similar to any of the IncP-type *tra* genes.

The *tra* genes of the *traAFBH* and *traCDG* operons are expected to be important for conjugal DNA processing, based on similarity to other conjugal transfer systems (Zhu et al., 2000). The *traA*, *traF*, and *traG* genes were shown to be critical for conjugal transfer in pTiC58 (Farrand et al., 1996). Although *traB* was not critical, the intact gene did enhance transfer frequency. Recently, it has been shown that *traC* and *traD* are also required for maximal efficiency of plasmid transfer (H. Cho and Winans, unpublished data). The *traH* gene is not required for conjugal transfer and its function is unknown (Farrand et al., 1996; Zhu et al., 2000).

The *traAFBH* and *traCDG* operons are divergently transcribed from one *tra* box, called *tra* box I (also the consensus and strongest *tra* box) (Fuqua and Winans, 1996a). At both of the promoters, the *tra* box overlaps the -35 element of the RNA polymerase (RNAP) binding site (*Figure 16-6*). This structure resembles the class II-type promoters, first described for CRP (Busby and Ebright, 1999). At this type of promoter, the activator (in this case TraR) is in a position to make multiple contacts with RNAP. These contacts activate transcription by recruiting polymerase to the promoter and can also be involved in later steps in transcription initiation. In Section 4.3 below, we discuss interactions between TraR and RNAP at TraR-dependent promoters.

The origin of transfer of the conjugation system (*oriT*) is also located in the intergenic region between the two *tra* operons, specifically between *tra* box I and *traA* (Zhu et al., 2000; H. Cho and S. C. Winans, unpublished data). It is predicted that the TraA protein is a conjugal relaxase that binds to the *oriT* and covalently binds to one DNA strand (Farrand et al., 1996; Zhu et al., 2000). The TraC and TraD proteins are thought to be accessory proteins that may form a complex with TraA and enhance its function at *oriT* (Farrand et al., 1996; H. Cho and S. C. Winans, unpublished data).

Most genes of the *traI-trb* operon are thought to be involved in the Type IV secretion system for transfer of Ti plasmid DNA from conjugal donors to recipient cells (Fuqua and Winans, 1996a). The two exceptions are *traI* and *trbK*. As described above, *traI* encodes the OOHL synthase, while *trbK* may be involved in entry exclusion (Li et al., 1999). Activation of *traI* expression via TraR-OOHL represents a positive feedback loop, as has been described for a number of other LuxR-LuxI type systems (Whitehead et al., 2001). The *traI-trb* operon is activated from *tra* box II and also has a class II type promoter as described above for the *tra* operons (*Figure 16-6*) (Fuqua and Winans, 1996a). A strict requirement of TraR and OOHL for expression of this operon has been demonstrated for both the nopaline and octopine-type Ti plasmids (Fuqua and Winans, 1996a; Li et al., 1999).

4.2 Activation of the Ti plasmid vegetative replication genes

The vegetative replication genes, and their promoter architecture, are also thought to be highly conserved on the Ti plasmids (Li and Farrand, 2000; Zhu et al., 2000). The replication genes are all in one operon, *repABC*, which is closely related to *rep* systems of other large plasmids in members of the Rhizobiaceae (Li and Farrand, 2000). RepA and RepB are thought to be involved in plasmid partitioning (Williams and Thomas, 1992; Moller-Jensen et al., 2000). RepC is strictly required for vegetative replication (Li and Farrand, 2000; Pappas and Winans, 2003a, 2003b). An additional small gene, *repD*, has recently been identified, and lies in the intergenic region between *repA* and *repB* (Chai and Winans, 2005b). Although this gene is translated, the protein product has no known function,

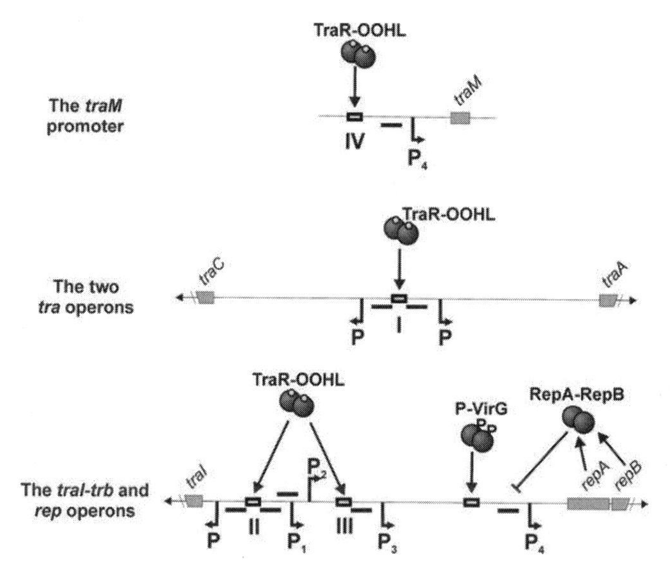

Figure 16-6. Models for activation by TraR-OOHL complexes at TraR-dependent promoters. The *tra* boxes are marked with open squares and correspond to *tra* boxes I, II, III, and IV as described in the text. Thick bars represent the -35 and -10 elements (combined) of each promoter, and bent arrows indicate transcription start sites. The four promoters of the *rep* operon are labelled (P1 to P4). Activation by phospho-VirG and repression by RepA-RepB complexes are also shown for P4.

and a null *repD* mutant does not have an observable effect on replication. In addition, a gene encoding a small non-translated antisense RNA, *repE*, lies in the intergenic region between *repB* and *repC* (Chai and Winans, 2005c). This RNA attenuates the expression of *repC*, most likely by forming a complex with the *rep* transcript (near the *repC* gene) to either terminate transcription or block translation.

The *repABC* operon is divergently transcribed from *traI-trb*, but the promoter architecture for *rep* is quite complex, and has been well characterized in the octopine-type plasmids (*Figure 16-6*). In the intergenic spacer between *traI-trb*, there are two *tra* boxes (*tra* boxes II and III), both of which are involved in *repABC* activation, and a total of four *rep* promoters (Fuqua and Winans, 1996a; Pappas and Winans, 2003a, 2003b). Three of these promoters are TraR-dependent. Two of these (*repAP1* and *repAP3*) are class II type, and are activated from *tra* box II and *tra* box III respectively (Pappas and Winans, 2003a). At class II promoters, the activator binding site overlaps the -35 element of the promoter as described above (Busby and Ebright, 1999). The third TraR-dependent promoter (*repAP2*) is also activated from *tra* box II, but resembles a class I type promoter (Pappas and Winans, 2003a). At these types of promoters, the activator binds farther upstream but on the same face of the DNA as RNAP (Busby and Ebright, 1999).

Activation of the *rep* operon by these three TraR-dependent promoters collectively enhances Ti plasmid copy number 8 fold (Pappas and Winans, 2003a). TraR also stimulates copy number of the nopaline-type Ti plasmid (Li and Farrand, 2000). The intergenic spacer between *traI-trb* and *rep*, with its two *tra* boxes, is strongly conserved. Furthermore, the genetic linkage of these two operons is also conserved in some plasmids of *Rhizobium* sp. (Li and Farrand, 2000; Zhu et al., 2000). Therefore, it is plausible that activation of other *rep* operons by TraR will be a conserved feature of Ti plasmids.

The fourth *rep* promoter (*repAP4*) is not TraR-dependent (Pappas and Winans, 2003b). This promoter is critical to plasmid maintenance in the absence of inducing concentrations of OOHL and conjugal opines. The *repAP4* promoter is negatively auto-regulated by RepA and RepB. Both of these proteins interact to form a repression loop, with RepA bound to specific sites near P4, and RepB bound to target DNA sites between *repA* and *repB* (Pappas and Winans, 2003b; Chai and Winans, 2005b). Recently, it was observed in expression studies using microarrays that the plant-released phenolic acetosyringone (AS) also induces *repABC* transcription (Cho and Winans, 2005). AS is perceived by the membrane-bound protein

VirA, which then phosphorylates VirG, and phospho-VirG can then activate transcription of the Ti plasmid virulence genes (see chapter 6). A VirG binding site (*vir* box) was identified at *repAP4*, and specific binding of phospho-VirG to this site was demonstrated *in vitro* (Cho and Winans, 2005). Although *repAP4* is activated in response to AS, this promoter is not dependent on phospho-VirG for expression. Activation at *repAP4* via phospho-VirG in response to AS increases the Ti plasmid copy number approximately 4 fold. Increased Ti plasmid copy number would likely increase the expression of all Vir proteins, as well as the production of T-strands that are transferred to plant cells.

4.3 TraR-OOHL interactions with RNA polymerase

As described above, TraR activates expression of the *tra*, *trb*, and *rep* genes of the Ti plasmids. The *traM* gene is also activated by TraR in response to OOHL (Fuqua et al., 1995; Hwang et al., 1995). A *tra* box (*tra* box IV) has been identified that is centred approximately 60 nucleotides from the transcription start site of the *traM* promoter on the octopine-type Ti plasmids (White and Winans, 2005 and C. Fuqua, personal communication). Therefore, this is expected to be a class I type promoter, as described above for *repAP2*.

Direct contacts between TraR and RNAP are predicted to occur, as activation *in vitro* requires only TraR-OOHL complexes, RNAP, and promoter DNA (Zhu and Winans, 1999). As mentioned above, activators at class II type promoters are in a position to make multiple contacts with RNAP. At class I type promoters, as the activator and RNAP binding sites do not overlap, the only region of RNAP that the activator can contact is the C-terminal domain of the α subunit (αCTD), which is connected to the rest of RNAP by a flexible linker (Busby and Ebright, 1999). Therefore, it is predicted that TraR can contact at least the αCTD to recruit RNAP to TraR-dependent promoters.

A mutagenesis study has identified a putative RNAP contact site, or activating region (AR1), on the C-terminal domain of TraR (White and Winans, 2005). This region consists of at least six surface-exposed residues that are critical for activation, but not for DNA binding, and therefore are expected to contact RNAP. In the same study, it was shown that AR1 is critical in activation at both class I and class II type TraR-dependent promoters, suggesting that this region contacts specifically the αCTD of RNAP. Two residues of LuxR that overlap with the TraR AR1 had been previously identified as being critical for activation (Egland and Greenberg, 2001).

In another study, two residues of the N-terminal domain were also identified as being critical for activation but not DNA binding (Luo and Farrand, 1999). Mutations of these residues disrupted interactions with the αCTD of RNAP in surface plasmon resonance studies (Qin et al., 2004a). It is unlikely that TraR would have two separate αCTD contact sites on its surface. However, due to the flexibility of the NTDs of TraR relative to the CTDs (*Figure 16-4*b), it is possible that AR1 on the CTD and the two additional residues of the NTD could approach each other closely enough on one side of the protein to form a single activating patch. This could be important for activation at the divergent TraR-dependent promoters described above, however, this remains to be determined. In addition, as TraR activates transcription at both class I and class II type promoters, it is likely that there is at least one more RNAP contact site on the surface of the protein.

5 QUORUM SENSING IN TUMOURS AND INFECTED PLANTS

In this chapter, we have discussed the regulation of *traR* gene expression by specific opines (depending on the type of Ti plasmid), and activation of genes required for both Ti plasmid conjugal transfer and vegetative replication by TraR-OOHL complexes. As both the *traR* and *traI* genes are located on the Ti plasmid, all conjugal donors in a population both release and detect OOHL. The result is that conjugal donors detect signals from other donors rather than from recipients that do not carry a Ti plasmid (Zhu et al., 2000). In addition, sequence examination suggests that the Ti plasmids may not have a potent entry exclusion system (Farrand, 1998). Therefore, it is quite likely that donors conjugate with other donors, perhaps as a mechanism to increase overall Ti plasmid copy number in the population. In the presence of opines, an increase in Ti plasmid copy number (by both replication and conjugation) in a population of *A. tumefaciens* may well be adaptive, as it would increase the expression of opine catabolic genes (Li and Farrand, 2000).

The quorum sensing system may confer an advantage for epiphytic or endophytic colonization of infected plants by non-tumour associated bacteria. Opines diffuse quite readily through plant tissue, and can be detected in plant roots even when the site of synthesis is much closer to the terminal bud (Savka et al., 1996). Therefore, any *A. tumefaciens* strain (including the non-pathogenic *A. radiobacter*) that carry plasmids with opine utilization genes should have a growth advantage on a plant producing the corresponding opine(s). This growth advantage has been demonstrated in competition studies of opine and non-opine utilizing strains on root surfaces of transgenic opine-producing plants (Oger et al., 1997; Savka and Farrand, 1997). In the presence of opines, the quorum sensing system can be induced if OOHL accumulates, and may increase the efficiency of opine-utilization through gene dosage effects, as discussed above. This would allow populations of *A. tumefaciens* to out-compete their non-opine utilizing neighbours. An increase in Ti plasmid copy number (again, resulting in an increase in virulence gene copy numbers) in an endophytic population may also be important for inducing secondary tumours on infected plants. Several reports on the recovery of *A. tumefaciens* from the vasculature (most likely xylem) of infected plants have been published (Suit and Eardley, 1935; Tarbah and Goodman, 1987; Bouzar et al., 1995; Marti et al., 1999). Movement of *A. tumefaciens* within plant tissue is expected to be critical for the formation of secondary tumours.

Although we often think of quorum sensing as a mechanism for measuring population densities, barriers to AHL diffusion may be just as important for AHL accumulation (Redfield, 2002). To the authors' knowledge, only one group has reported studies of AHL diffusion in plant tissue (Fray et al., 1999; Newton and Fray, 2004). In those studies, the diffusion of two different AHLs (C6-HSL and 3-oxo-C12-HSL) from their source in leaf chloroplasts of transgenic plants was compared. The more soluble C6-HSL (with a shorter fatty acid tail) was detected at the surface of leaves, while the less soluble 3-oxo-C12-HSL did not diffuse efficiently from the chloroplasts. Diffusion of 3-oxo-C8-HSL (OOHL) in plant tissue has not been reported. Diffusion of these molecules in plants is also likely to be affected by the type of tissue. For example, OOHL is not likely to accumulate in vascular cells such as xylem.

6 ACKNOWLEGEMENTS

We would like to thank the members of our laboratory for helpful discussions and critical review of the manuscript. Research in the authors' laboratory is supported by the National Institutes of Health (Grant GM41892).

7 REFERENCES

Beck von Bodman S, Hayman GT, Farrand SK (1992) Opine catabolism and conjugal transfer of the nopaline Ti plasmid pTiC58 are coordinately regulated by a single repressor. Proc Natl Acad Sci USA 89: 643-647

Bouzar H, Chilton WS, Nesme X, Dessaux Y, Vaudequin V, Petit A, Jones JB, Hodge NC (1995) A new *Agrobacterium* strain isolated from aerial tumors on *Ficus benjamina* L. Appl Environ Microbiol 61: 65-73

Brumbley SM, Carney BF, Denny TP (1993) Phenotype conversion in *Pseudomonas solanacearum* due to spontaneous inactivation of *PhcA*, a putative *LysR* transcriptional regulator. J Bacteriol 175: 5477-5487

Busby S, Ebright RH (1999) Transcription activation by catabolite activator protein (CAP). J Mol Biol 293: 199-213

Carlier A, Chevrot R, Dessaux Y, Faure D (2004) The assimilation of gamma-butyrolactone in *Agrobacterium tumefaciens* C58 interferes with the accumulation of the N-acyl-homoserine lactone signal. Mol Plant-Microbe Interact 17: 951-957

Carlier A, Uroz S, Smadja B, Fray R, Latour X, Dessaux Y, Faure D (2003) The Ti plasmid of *Agrobacterium tumefaciens* harbors an *attM*-paralogous gene, *aiiB*, also encoding N-Acyl homoserine lactonase activity. Appl Environ Microbiol 69: 4989-4993

Chai Y, Winans SC (2004) Site-directed mutagenesis of a LuxR-type quorum-sensing transcription factor: alteration of autoinducer specificity. Mol Microbiol 51: 765-776

Chai Y, Winans SC (2005a) Amino-terminal protein fusions to the TraR quorum-sensing transcription factor enhance protein stability and autoinducer-independent activity. J Bacteriol 187: 1219-1226

Chai Y, Winans SC (2005b) RepB protein of an *Agrobacterium tumefaciens* Ti plasmid binds to two adjacent sites between *repA* and *repB* for plasmid partitioning and autorepression. Mol Microbiol 56: 1574-1585

Chai Y, Winans SC (2005c) A small antisense RNA downregulates expression of an essential replicase protein of an *Agrobacterium tumefaciens* Ti plasmid. Mol Microbiol 56: 1574-1585

Chai Y, Zhu J, Winans SC (2001) TrlR, a defective TraR-like protein of *Agrobacterium tumefaciens*, blocks TraR function *in vitro* by forming inactive TrlR:TraR dimers. Mol Microbiol 40: 414-421

Chater KF, Horinouchi S (2003) Signalling early developmental events in two highly diverged *Streptomyces* species. Mol Microbiol 48: 9-15

Chen G, Malenkos JW, Cha MR, Fuqua C, Chen L (2004) Quorum-sensing antiactivator TraM forms a dimer that dissociates to inhibit TraR. Mol Microbiol 52: 1641-1651

Chen X, Schauder S, Potier N, Van Dorsselaer A, Pelczer I, Bassler BL, Hughson FM (2002) Structural identification of a bacterial quorum-sensing signal containing boron. Nature 415: 545-549

Cho H, Winans SC (2005) VirA and VirG activate the Ti plasmid *repABC* operon, elevating plasmid copy number in response to wound-released chemical signals. Proc Natl Acad Sci USA 102: 14843-14848

Choi SH, Greenberg EP (1991) The C-terminal region of the *Vibrio fischeri* LuxR protein contains an inducer-independent *lux* gene activating domain. Proc Natl Acad Sci USA 88: 11115-11119

Choi SH, Greenberg EP (1992a) Genetic dissection of DNA binding and luminescence gene activation by the *Vibrio fischeri* LuxR protein. J Bacteriol 174: 4064-4069

Choi SH, Greenberg EP (1992b) Genetic evidence for multimerization of LuxR, the transcriptional activator of *Vibrio fischeri* luminescence. Mol Mar Biol Biotechnol 1: 408-413

Dong YH, Wang LH, Xu JL, Zhang HB, Zhang XF, Zhang LH (2001) Quenching quorum-sensing-dependent bacterial infection by an N-acyl homoserine lactonase. Nature 411: 813-817

Ducros VM, Lewis RJ, Verma CS, Dodson EJ, Leonard G, Turkenburg JP, Murshudov GN, Wilkinson AJ, Brannigan JA (2001) Crystal structure of GerE, the ultimate transcriptional regulator of spore formation in *Bacillus subtilis*. J Mol Biol 306: 759-771

Dunny GM, Leonard BA (1997) Cell-cell communication in gram-positive bacteria. Annu Rev Microbiol 51: 527-564

Dyson HJ, Wright PE (2002) Coupling of folding and binding for unstructured proteins. Curr Opin Struct Biol 12: 54-60

Eberl L, Winson MK, Sternberg C, Stewart GS, Christiansen G, Chhabra SR, Bycroft B, Williams P, Molin S, Givskov M (1996) Involvement of N-acyl-L-hormoserine lactone autoinducers in controlling the multicellular behaviour of *Serratia liquefaciens*. Mol Microbiol 20: 127-136

Egland KA, Greenberg EP (2001) Quorum sensing in *Vibrio fischeri*: analysis of the LuxR DNA binding region by alanine-scanning mutagenesis. J Bacteriol 183: 382-386

Ellis JG, Kerr A, Petit A, Tempé J (1982) Conjugal transfer of nopaline and agropine Ti-plasmids - the role of agrocinopines. Mol Gen Genet 186: 269-273

Engebrecht J, Nealson K, Silverman M (1983) Bacterial bioluminescence: isolation and genetic analysis of functions from *Vibrio fischeri*. Cell 32: 773-781

Engebrecht J, Silverman M (1984) Identification of genes and gene products necessary for bacterial bioluminescence. Proc Natl Acad Sci USA 81: 4154-4158

Engebrecht J, Silverman M (1987) Nucleotide sequence of the regulatory locus controlling expression of bacterial genes for bioluminescence. Nucleic Acids Res 15: 10455-10467

Farrand SK (1998) Conjugal plasmids and their transfer. *In* HP Spaink, A Kondorosi, PJJ Hooykaas, eds, The *Rhizobiaceae*: molecular biology of model plant-associated bacteria. Kluwer Academic Publishers, Dordrecht, The Netherlands, pp 199-233

Farrand SK, Hwang I, Cook DM (1996) The tra region of the nopaline-type Ti plasmid is a chimera with elements related to the transfer systems of RSF1010, RP4, and F. J Bacteriol 178: 4233-4247

Flavier AB, Clough SJ, Schell MA, Denny TP (1997) Identification of 3-hydroxypalmitic acid methyl ester as a novel autoregulator controlling virulence in *Ralstonia solanacearum*. Mol Microbiol 26: 251-259

Fray RG, Throup JP, Daykin M, Wallace A, Williams P, Stewart GS, Grierson D (1999) Plants genetically modified to produce N-acylhomoserine lactones communicate with bacteria. Nat Biotechnol 17: 1017-1020

Fuqua C, Burbea M, Winans SC (1995) Activity of the *Agrobacterium* Ti plasmid conjugal transfer regulator TraR is inhibited by the product of the traM gene. J Bacteriol 177: 1367-1373

Fuqua C, Greenberg EP (2002) Listening in on bacteria: acyl-homoserine lactone signalling. Nat Rev Mol Cell Biol 3: 685-695

Fuqua C, Parsek MR, Greenberg EP (2001) Regulation of gene expression by cell-to-cell communication: acyl-homoserine lactone quorum sensing. Annu Rev Genet 35: 439-468

Fuqua C, Winans SC (1996a) Conserved *cis*-acting promoter elements are required for density-dependent transcription of *Agrobacterium tumefaciens* conjugal transfer genes. J Bacteriol 178: 435-440

Fuqua C, Winans SC (1996b) Localization of OccR-activated and TraR-activated promoters that express two ABC-type permeases and the traR gene of Ti plasmid pTiR10. Mol Microbiol 20: 1199-1210

Fuqua WC, Winans SC (1994) A LuxR-LuxI type regulatory system activates *Agrobacterium* Ti plasmid conjugal transfer in the presence of a plant tumor metabolite. J Bacteriol 176: 2796-2806

Genetello C, Van Larebeke N, Holsters M, De Picker A, Van Montagu M, Schell J (1977) Ti plasmids of *Agrobacterium* as conjugative plasmids. Nature 265: 561-563

Givskov M, Ostling J, Eberl L, Lindum PW, Christensen AB, Christiansen G, Molin S, Kjelleberg S (1998) Two separate regulatory systems participate in control of swarming motility of *Serratia liquefaciens* MG1. J Bacteriol 180: 742-745

Gonzalez JE, Marketon MM (2003) Quorum sensing in nitrogen-fixing rhizobia. Microbiol Mol Biol Rev 67: 574-592

Goodner B, Hinkle G, Gattung S, Miller N, Blanchard M, Qurollo B, Goldman BS, Cao Y, Askenazi M, Halling H, Mullin L, Houmiel K, Gordon J, Vaudin M, Iartchouk O, Epp A, Liu F, Wollam C, Allinger M, Doughty D, Scott C, Lappas C, Markelz B, Flanagan C, Crowell C, Gurson J, Lomo C, Sear C, Strub G, Cielo C, Slater S (2001) Genome sequence of the plant pathogen and biotechnology agent *Agrobacterium tumefaciens* C58. Science 294: 2323-2328

He X, Chang W, Pierce DL, Seib LO, Wagner J, Fuqua C (2003) Quorum sensing in *Rhizobium* sp. strain NGR234 regulates conjugal transfer (tra) gene expression and influences growth rate. J Bacteriol 185: 809-822

Huang JJ, Han JI, Zhang LH, Leadbetter JR (2003) Utilization of acyl-homoserine lactone quorum signals for growth by a soil pseudomonad and *Pseudomonas aeruginosa* PAO1. Appl Environ Microbiol 69: 5941-5949

Hwang I, Cook DM, Farrand SK (1995) A new regulatory element modulates homoserine lactone-mediated autoinduction of Ti plasmid conjugal transfer. J Bacteriol 177: 449-458

Hwang I, Li PL, Zhang L, Piper KR, Cook DM, Tate ME, Farrand SK (1994) TraI, a LuxI homologue, is responsible for production of conjugation factor, the Ti plasmid N-acylhomoserine lactone autoinducer. Proc Natl Acad Sci USA 91: 4639-4643

Hwang I, Smyth AJ, Luo ZQ, Farrand SK (1999) Modulating quorum sensing by antiactivation: TraM interacts with TraR to inhibit activation of Ti plasmid conjugal transfer genes. Mol Microbiol 34: 282-294

Kaplan HB, Greenberg EP (1985) Diffusion of autoinducer is involved in regulation of the *Vibrio fischeri* luminescence system. J Bacteriol 163: 1210-1214

Kerr A, Manigault P, Tempé J (1977) Transfer of virulence *in vivo* and *in vitro* in *Agrobacterium*. Nature 265: 560-561

Leadbetter JR, Greenberg EP (2000) Metabolism of acyl-homoserine lactone quorum-sensing signals by *Variovorax paradoxus*. J Bacteriol 182: 6921-6926

Lee KG, Shibamoto T (2000) Antioxidant properties of aroma compounds isolated from soybeans and mung beans. J Agric Food Chem 48: 4290-4293

Li PL, Everhart DM, Farrand SK (1998) Genetic and sequence analysis of the pTiC58 trb locus, encoding a mating-pair formation system related to members of the type IV secretion family. J Bacteriol 180: 6164-6172

Li PL, Farrand SK (2000) The replicator of the nopaline-type Ti plasmid pTiC58 is a member of the *repABC* family and is influenced by the TraR-dependent quorum-sensing regulatory system. J Bacteriol 182: 179-188

Li PL, Hwang I, Miyagi H, True H, Farrand SK (1999) Essential components of the Ti plasmid *trb* system, a type IV macromolecular transporter. J Bacteriol 181: 5033-5041

Luo ZQ, Farrand SK (1999) Signal-dependent DNA binding and functional domains of the quorum-sensing activator TraR as identified by repressor activity. Proc Natl Acad Sci USA 96: 9009-9014

Luo ZQ, Qin Y, Farrand SK (2000) The antiactivator TraM interferes with the autoinducer-dependent binding of TraR to DNA by interacting with the C-terminal region of the quorum-sensing activator. J Biol Chem 275: 7713-7722

Luo ZQ, Smyth AJ, Gao P, Qin Y, Farrand SK (2003) Mutational analysis of TraR. Correlating function with molecular structure of a quorum-sensing transcriptional activator. J Biol Chem 278: 13173-13182

Lyon GJ, Novick RP (2004) Peptide signaling in *Staphylococcus aureus* and other Gram-positive bacteria. Peptides 25: 1389-1403

Maris AE, Sawaya MR, Kaczor-Grzeskowiak M, Jarvis MR, Bearson SM, Kopka ML, Schröder I, Gunsalus RP, Dickerson RE (2002) Dimerization allows DNA target site recognition by the NarL response regulator. Nat Struct Biol 9: 771-778

Marti R, Cubero J, Daza A, Piquer J, Salcedo CI, Morente C, Lopez MM (1999) Evidence of migration and endophytic presence of *Agrobacterium tumefaciens* in rose plants. Eur J Plant Pathol 105: 39-50

Moller-Jensen J, Jensen RB, Gerdes K (2000) Plasmid and chromosome segregation in prokaryotes. Trends Microbiol 8: 313-320

More MI, Finger LD, Stryker JL, Fuqua C, Eberhard A, Winans SC (1996) Enzymatic synthesis of a quorum-sensing autoinducer through use of defined substrates. Science 272: 1655-1658

Newton JA, Fray RG (2004) Integration of environmental and host-derived signals with quorum sensing during plant-microbe interactions. Cell Microbiol 6: 213-224

Oger P, Farrand SK (2001) Co-evolution of the agrocinopine opines and the agrocinopine-mediated control of TraR, the quorum-sensing activator of the Ti plasmid conjugation system. Mol Microbiol 41: 1173-1185

Oger P, Farrand SK (2002) Two opines control conjugal transfer of an *Agrobacterium* plasmid by regulating expression of separate copies of the quorum-sensing activator gene traR. J Bacteriol 184: 1121-1131

Oger P, Kim KS, Sackett RL, Piper KR, Farrand SK (1998) Octopine-type Ti plasmids code for a mannopine-inducible dominant-negative allele of traR, the quorum-sensing activator that regulates Ti plasmid conjugal transfer. Mol Microbiol 27: 277-288

Oger P, Petit A, Dessaux Y (1997) Genetically engineered plants producing opines alter their biological environment. Nat Biotechnol 15: 369-372

Pappas KM, Winans SC (2003a) A LuxR-type regulator from *Agrobacterium tumefaciens* elevates Ti plasmid copy number by activating transcription of plasmid replication genes. Mol Microbiol 48: 1059-1073

Pappas KM, Winans SC (2003b) The RepA and RepB autorepressors and TraR play opposing roles in the regulation of a Ti plasmid repABC operon. Mol Microbiol 49: 441-455

Parsek MR, Greenberg EP (2005) Sociomicrobiology: the connections between quorum sensing and biofilms. Trends Microbiol 13: 27-33

Perombelon MCM (2002) Potato diseases caused by soft rot erwinias: an overview of pathogenesis. Plant Pathol 51: 1-12

Piper KR, Beck von Bodman S, Farrand SK (1993) Conjugation factor of *Agrobacterium tumefaciens* regulates Ti plasmid transfer by autoinduction. Nature 362: 448-450

Piper KR, Beck Von Bodman S, Hwang I, Farrand SK (1999) Hierarchical gene regulatory systems arising from fortuitous gene associations: controlling quorum sensing by the opine regulon in *Agrobacterium*. Mol Microbiol 32: 1077-1089

Qin Y, Luo ZQ, Farrand SK (2004a) Domains formed within the N-terminal region of the quorum-sensing activator TraR are required for transcriptional activation and direct interaction with RpoA from *Agrobacterium*. J Biol Chem 279: 40844-40851

Qin Y, Luo ZQ, Smyth AJ, Gao P, Beck von Bodman S, Farrand SK (2000) Quorum-sensing signal binding results in dimerization of TraR and its release from membranes into the cytoplasm. Embo J 19: 5212-5221

Qin Y, Smyth AJ, Su S, Farrand SK (2004b) Dimerization properties of TraM, the antiactivator that modulates TraR-mediated quorum-dependent expression of the Ti plasmid tra genes. Mol Microbiol 53: 1471-1485

Redfield RJ (2002) Is quorum sensing a side effect of diffusion sensing? Trends Microbiol 10: 365-370

Rosen R, Matthysse AG, Becher D, Biran D, Yura T, Hecker M, Ron EZ (2003) Proteome analysis of plant-induced proteins of *Agrobacterium tumefaciens*. FEMS Microbiol Ecol 44: 355-360

Savka MA, Black RC, Binns AN, Farrand SK (1996) Translocation and exudation of tumor metabolites in crown galled plants. Mol Plant-Microbe Interact 9: 310-313

Savka MA, Farrand SK (1997) Modification of rhizobacterial populations by engineering bacterium utilization of a novel plant-produced resource. Nat Biotechnol 15: 363-368

Shoemaker BA, Portman JJ, Wolynes PG (2000) Speeding molecular recognition by using the folding funnel: the fly-casting mechanism. Proc Natl Acad Sci USA 97: 8868-8873

Smadja B, Latour X, Faure D, Chevalier S, Dessaux Y, Orange N (2004) Involvement of N-acylhomoserine lactones throughout plant infection by *Erwinia carotovora* subsp. atroseptica (*Pectobacterium atrosepticum*). Mol Plant-Microbe Interact 17: 1269-1278

Stevens AM, Dolan KM, Greenberg EP (1994) Synergistic binding of the *Vibrio fischeri* LuxR transcriptional activator domain and RNA polymerase to the *lux* promoter region. Proc Natl Acad Sci USA 91: 12619-12623

Suit RF, Eardley EA (1935) Secondary tumor formation on herbaceous hosts induced by *Pseudomonas tumefaciens*. Scientific Agriculture 15: 345-357

Swiderska A, Berndtson AK, Cha MR, Li L, Beaudoin GM, 3rd, Zhu J, Fuqua C (2001) Inhibition of the *Agrobacterium tumefaciens* TraR quorum-sensing regulator. Interactions with the TraM anti-activator. J Biol Chem 276: 49449-49458

Tarbah FA, Goodman RN (1987) Systemic spread of *Agrobacterium tumefaciens* biovar 3 in the vascular system of grapes. Phytopathology 77: 915-920

Vannini A, Volpari C, Di Marco S (2004) Crystal structure of the quorum-sensing protein TraM and its interaction with the transcriptional regulator TraR. J Biol Chem 279: 24291-24296

Vannini A, Volpari C, Gargioli C, Muraglia E, Cortese R, De Francesco R, Neddermann P, Marco SD (2002) The crystal structure of the quorum sensing protein TraR bound to its autoinducer and target DNA. Embo J 21: 4393-4401

von Bodman SB, Bauer WD, Coplin DL (2003) Quorum sensing in plant-pathogenic bacteria. Annu Rev Phytopathol 41: 455-482

Wang L, Helmann JD, Winans SC (1992) The *A. tumefaciens* transcriptional activator OccR causes a bend at a target promoter, which is partially relaxed by a plant tumor metabolite. Cell 69: 659-667

White CE, Winans SC (2005) Identification of amino acid residues of the *Agrobacterium tumefaciens* quorum-sensing regulator TraR that are critical for positive control of transcription. Mol Microbiol 55: 1473-1486

Whitehead NA, Barnard AM, Slater H, Simpson NJ, Salmond GP (2001) Quorum-sensing in Gram-negative bacteria. FEMS Microbiol Rev 25: 365-404

Williams DR, Thomas CM (1992) Active partitioning of bacterial plasmids. J Gen Microbiol 138: 1-16

Wisniewski-Dye F, Downie JA (2002) Quorum-sensing in *Rhizobium*. Antonie Van Leeuwenhoek 81: 397-407

Wood DW, Setubal JC, Kaul R, Monks DE, Kitajima JP, Okura VK, Zhou Y, Chen L, Wood GE, Almeida Jr. NF, Woo L, Chen Y, Paulsen IT, Eisen JA, Karp PD, Bovee Sr. D, Chapman P, Clendenning J, Deatherage G, Gillet W, Grant C, Kutyavin T, Levy R, Li MJ, McClelland E, Palmieri P, Raymond C, Rouse R, Saenphimmachak C, Wu Z, Romero P, Gordon D, Zhang S, Yoo H, Tao Y, Biddle P, Jung M, Krespan W, Perry M, Gordon-Kamm B, Liao L, Kim S, Hendrick C, Zhao ZY, Dolan M, Chumley F, Tingey SV, Tomb JF, Gordon MP, Olson MV, Nester EW (2001) The genome of the natural genetic engineer *Agrobacterium tumefaciens* C58. Science 294: 2317-2323

Xavier KB, Bassler BL (2003) LuxS quorum sensing: more than just a numbers game. Curr Opin Microbiol 6: 191-197

Zhang HB, Wang C, Zhang LH (2004) The quormone degradation system of *Agrobacterium tumefaciens* is regulated by starvation signal and stress alarmone (p)ppGpp. Mol Microbiol 52: 1389-1401

Zhang HB, Wang LH, Zhang LH (2002a) Genetic control of quorum-sensing signal turnover in *Agrobacterium tumefaciens*. Proc Natl Acad Sci USA 99: 4638-4643

Zhang L, Murphy PJ, Kerr A, Tate ME (1993) *Agrobacterium* conjugation and gene regulation by N-acyl-L-homoserine lactones. Nature 362: 446-448

Zhang RG, Pappas T, Brace JL, Miller PC, Oulmassov T, Molyneaux JM, Anderson JC, Bashkin JK, Winans SC, Joachimiak A (2002b) Structure of a bacterial quorum-sensing transcription factor complexed with pheromone and DNA. Nature 417: 971-974

Zhu J, Beaber JW, More MI, Fuqua C, Eberhard A, Winans SC (1998) Analogs of the autoinducer 3-oxooctanoyl-homoserine lactone strongly inhibit activity of the TraR protein of *Agrobacterium tumefaciens*. J Bacteriol 180: 5398-5405

Zhu J, Oger PM, Schrammeijer B, Hooykaas PJ, Farrand SK, Winans SC (2000) The bases of crown gall tumorigenesis. J Bacteriol 182: 3885-3895

Zhu J, Winans SC (1998) Activity of the quorum-sensing regulator TraR of *Agrobacterium tumefaciens* is inhibited by a truncated, dominant defective TraR-like protein. Mol Microbiol 27: 289-297

Zhu J, Winans SC (1999) Autoinducer binding by the quorum-sensing regulator TraR increases affinity for target promoters *in vitro* and decreases TraR turnover rates in whole cells. Proc Natl Acad Sci USA 96: 4832-4837

Zhu J, Winans SC (2001) The quorum-sensing transcriptional regulator TraR requires its cognate signaling ligand for protein folding, protease resistance, and dimerization. Proc Natl Acad Sci USA 98: 1507-1512

Chapter 17

HORIZONTAL GENE TRANSFER

Nobukazu Tanaka

Center for Gene Science, Hiroshima University, Kagamiyama 1-4-2, Higashi-Hiroshima, Hiroshima 739-8527, Japan

Abstract. A quarter-century ago, a sequence homologous to the Ri plasmid (pRi) T-DNA of *Agrobacterium rhizogenes* was detected in the genome of untransformed tree tobacco, *Nicotiana glauca*, and was named "cellular T-DNA" (cT-DNA). The origin of the homologous sequences in tobacco remained unknown for a long period, but at present, it has been clearly demonstrated that the cT-DNA is the pRi T-DNA inserted by ancient infection with mikimopine-type *A. rhizogenes*. The cT-DNA of *N. glauca* is composed of an imperfect inverted repeat and it contains homologues of some pRi T-DNA genes involved in adventitious root formation and opine synthesis, which are called Ng*rolB*, Ng*rol*C, NgORF13, NgORF14, and Ng*mis*. In spite of the footprint of ancient insertion of pRi T-DNA, these homologues are still expressed not only in genetic tumors of F1 hybrids of *N. glauca* x *N. langsdorffii* but also in some organs of *N. glauca*, although at a low level. The cT-DNA is also found in some other species of the genus *Nicotiana*, with mikimopine-type cT-DNA contained in at least three, *N. tomentosa*, *N. tomentosiformis*, and *N. tabacum*. Therefore, there is a possibility that multiple infection events occurred independently in several ancestors of *Nicotiana*. Furthermore, some plant species in different families also contain cT-DNA-like sequences, although the details are still unknown. Tumors are spontaneously generated on some plants in the absence of tumorigenic microorganisms. Hybrid plants of *Nicotiana* species also form genetic tumors, but the mechanism of this tumorigenesis is still unknown. One of the parents of the hybrid usually contains cT-DNA, implying that it is the causal factor of tumorigenesis, although the causal association between the cT-DNA and tumorigenesis remains unsolved. Since pRi-transgenic plants exhibit a peculiar phenotype, the so-called "hairy root syndrome", which shows advantageous traits in some cases, ancient pRi-transformed plants might also have predominated in competition with parental plants or survived under a harsh climate. Therefore, the insertion events of T-DNA into the genome of plants might have influenced their evolution, resulting in the creation of new plant species.

1 INTRODUCTION

Infectious *Agrobacterium* species such as *A. tumefaciens*, *A. rhizogenes*, and *A. rubi* invade dicotyledonous plants and gymnosperms and form new growths on them by a process of natural transformation, followed by integration of the T-DNA region of their large plasmid and finally by expression of oncogenes on the T-DNA that govern development of the growths. The transformed cells are triggered to produce opine, which serves as a nutrient and bacterial conjugation signal for the *Agrobacterium* parasite. This phenomenon of inducing synthesis of a compound useful for the parasite is called "genetic colonization", one of the typical examples of a host-parasite relationship. In what era did genetic colonization start? Have colonized plants ever received benefits from such colonization? And, if the host-parasite relationship has collapsed for some reason, what happened to the formerly colonized plants and their progeny that contain only the T-DNA? Although we don't know the nascent era of "genetic colonization", we can find the footprint of early horizontal gene transfer from *Agrobacterium* to plants in some present-day plants.

2 FOOTPRINT OF HORIZONTAL GENE TRANSFER FROM *AGROBACTERIUM* TO TOBACCO PLANTS

2.1 Cellular T-DNA (cT-DNA) in wild plants of tree tobacco, *Nicotiana glauca*

In early investigations of transformation by the tumor-inducing plasmid (pTi) or root-inducing plasmid (pRi) of *Agrobacterium*, most researchers must have believed that there was no significant homology to the T-DNA in the plant genome. Contrary to expectations, however, some plant genomes seemed to contain sequences homologous to their T-DNAs. The most conspicuous example was in the genome of tree tobacco, *Nicotiana glauca*. White et al. (1982) tried to detect pRiA4b sequence in transgenic *Nicotiana glauca* by infection with *A. rhizogenes* strain A4. By DNA gel blot analysis using pRiA4b as a probe, they detected a portion of pRiA4b in the *A. rhizogenes*-infected axenic tissue. Surprisingly, a unique DNA sequence, highly homologous to a portion of pRi, was found in the genomic DNA of uninfected *N. glauca*. Their further analysis showed evidence that the homologous sequences in *N. glauca* are indeed within the T-DNA of pRiA4, referred to as the cellular T-DNA (cT-DNA, White et al., 1983). For a certain time after this discovery, the cT-DNA in the genome of *N. glauca* was believed to be an endogenous oncogene or an origin of pRi T-DNA, a so-called "protooncogene". However, the cT-DNA of *N. glauca* could have been a recent integration of pRi T-DNA during cultivation in the laboratory or on the farm. To exclude this possibility, Furner et al. (1986) obtained five varieties of *N. glauca*, three collected from wild-type lines in Peru, Bolivia and Paraguay, where native vegetations are distributed (Goodspeed, 1954), and their genomic DNA was analyzed by gel blot analysis. Although the signal patterns differed slightly, cT-DNA was found in all the varieties. Therefore, the endogenous T-DNA sequence must have preceded the speciation of *N. glauca*. To extend the confirmation of this T-DNA homologous sequence as being endogenous, the same group attempted to isolate this homologous region directly from the genome of *N. glauca* to determine its nucleotide sequence (*Figure 17-1*). The determined sequence comprised an imperfect inverted repeat, called the left arm and the right arm. The left arm, containing *rolB* and *rolC* homologues, is longer than the right arm, containing only *rolC* homologues. The coding region of the *rolB* homologue (Ng*rolB*) is shorter than that of *rolB* in pRiA4 (Ri*rolB*) because of a premature stop codon, and the coding region of *rolC* homologue in the right arm (Ng*rolC*R) is interrupted by a 1 bp deletion only 100 bp into the gene and a later 32 bp substitution that generates a stop codon in the original reading frame; the *rolC* homologue in the left arm (Ng*rolC*L) is full length. However, the sequences of Ng*rolB* and Ng*rolC*R beyond the early stop codons, though shorter, are strongly homologous to each of the corresponding *rol* genes in pRi T-DNA, suggesting later mutation. The large imperfect inverted repeat of the cT-DNA found in *N. glauca* is similar to the integration pattern of T-DNA (Kwok et al., 1985), although no insertion of filler DNA is involved (De Buck et al., 1999). These results suggest the ancient transfer of T-DNA from *A. rhizogenes* to *N. glauca* and its subsequent passage in evolutionary time, although which type of pRi was involved in this transfer is unclear because the unique opine synthase gene is absent. Determination of the nucleotide sequence of cT-DNA was further extended, resulting in the identification of open reading frame (ORF) 13 and ORF14 homologues (called NgORF13L and NgORF14L, respectively, Aoki et al., 1994) as shown in *Figure 17-1.* They also found a portion of an unknown ORF outside of NgORF14, sharing no homology with either *rolD* or other oncogenes of pTi and pRi T-DNA, or with plant genomic DNA sequences available at that time. Since the determined nucleotide sequence also did not contain the core of the canonical T-DNA border sequence (Journin et al., 1989), the complete resolution of the origin of the cT-DNA was seemingly deadlocked.

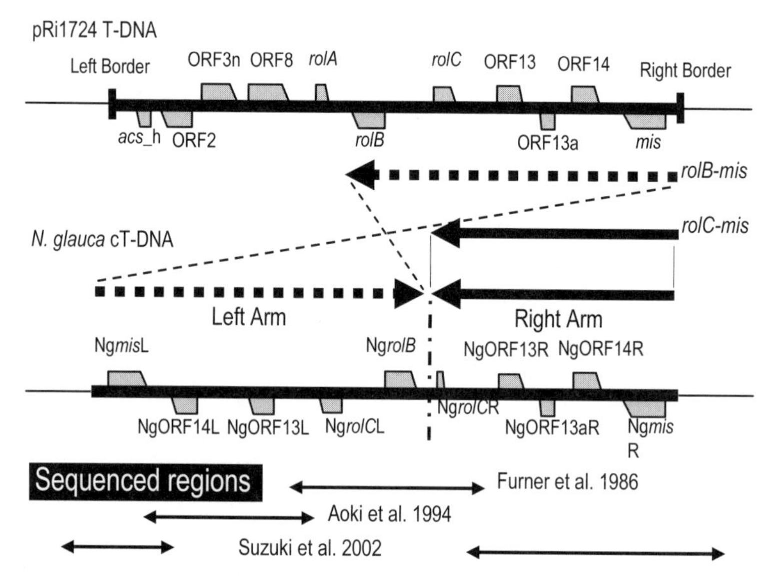

Figure 17-1. Structure of cT-DNA in *Nicotiana glauca* genome. (**top**) The region of the pRi1724 T-DNA and its flanking region are illustrated. A line and a broken line with an arrowhead indicate the component regions of cT-DNA. (**middle**) The region of cT-DNA and its flanking region is indicated. The lines and the broken line with an arrow head indicate the imperfect inverted repeat. (**bottom**) Lines with arrowheads at both ends indicate the regions sequenced by each of three groups. Cited from Suzuki et al. (2002) with kind permission from Blackwell Publishing Ltd.

The homology of the unknown ORF was found in an unexpected T-DNA gene. Suzuki et al. (2001) successfully identified a new ORF of pRi1724 T-DNA in Japanese *A. rhizogenes* strain MAFF301724 as a mikimopine synthase gene (*mis*). A portion of *mis* showed strong homology to the unknown ORF in cT-DNA. Therefore, they determined the nucleotide sequence of the remaining cT-DNA region (*Figure 17-1*) and found a couple of complete *mis* homologues in each arm, called Ng*mis*L and Ng*mis*R (Suzuki et al., 2002). Moreover, on the outer sides of these homologues, the homology between the pRi1724 T-DNA and *N. glauca* cT-DNA completely disappeared just at the right border position (Suzuki et al., 2002). Since such regions lacking homology to pRi1724 T-DNA were present in the genome of most other cT-DNA-less *Nicotiana* species, they must be of tobacco genome origin (Suzuki et al., 2002). These results indicated that the cT-DNA of *N. glauca* does not contain protooncogenes but is rather a footprint of *A. rhizogenes* infection. Finally, Suzuki et al. (2002) considered that the complete cT-DNA region is composed of the left arm of 7968 bp and the right arm of 5778 bp, originating from the T-DNA of mikimopine-type pRi similar to the present-day pRi1724. In the left arm of the cT-DNA region, ORFs called Ng*rolB*, Ng*rolC*L, NgORF13L, NgORF14L, and Ng*mis*L and in the right arm of the cT-DNA region, Ng*rolC*R with a truncated C-terminus, NgORF13R, NgORF13aR, NgORF14R, and Ng*mis*R corresponding to the ORFs on pRi1724 are present (*Figure 17-1* and *Table 17-1*). The very high nucleotide sequence homology between the left and right arms is about 96.9%, which indicates that each of the arms originated from the same T-DNA. The GC content in the cT-DNA is 44.7%, whereas that outside the left and right arms is 35.7 and 39.3%, indicating that cT-DNA is heterogeneous to the genomic DNA of *N. glauca*. The unanswered question is why the origin of cT-DNA in *N. glauca* is of the mikimopine-type pRi.

The original birthplace of the plants in the genus *Nicotiana* is unknown; however, beginning with Goodspeed (1954), most researchers have believed that the genus originated in South America. On the other hand, strains of *A. rhizogenes* containing mikimopine-type pRi have yet to be isolated outside Japan. One possibility is that mikimopine-type *A. rhizogenes* may also have originated in South America, then been transferred to Japan via imported plants.

2.2 cT-DNA is present in quite a few species of the genus *Nicotiana*

The genus *Nicotiana* contains approximately 70 species, divided into three subgenera, *Rustica*, *Tabacum*, and *Petunioides* (Goodspeed, 1954; Japan Tobacco Inc., 1990; Aoki and Ito, 2000, 2001). Is the cT-DNA found in other species in the genus *Nicotiana* besides *N. glauca*? Furner et al. (1986) reported that at least 4 species, *N. otophora*, *N. tomentosi-*

formis, *N. tomentosa*, and *N. benavidesii* share homology with pRi T-DNA and *N. tabacum* also shows significant homology, though the large region, corresponding to the *rolB* and *rolC* loci, is absent. However, some polymerase chain reaction (PCR) products have been obtained from the genome of wild-type *N. tabacum* that share homologies with *rolB*, *rolC*, ORF13 and their intergenic regions in pRi T-DNA (Frundt et al., 1998). The *rolC* homologue (*trolC*) and two ORF13 homologues (*torf13-1* and *torf13-2*) were analyzed in particular detail (Meyer et al., 1995; Frundt et al., 1998). Intrieri and Buiatti (2001) showed that cT-DNA was detected in at least 15 species (*N. glauca*, *N. cordifolia*, *N. benavidesii*, *N. tomentosiformis*, *N. otophora*, *N. setchelli*, *N. tabacum*, *N. arentsii*, *N. acuminata*, *N. miersi*, *N. bigelovii*, *N. debneyi*, *N. gossei*, *N. suaveolens*, and *N. exigua*) belonging to 7 subgenera of the genus *Nicotiana*, including the previously

Table 17-1. Oncogenes on pRi T-DNA and introgressed genes in the genome of *Nicotiana* plants. Numeric values indicate the number of nucleotide residues (bp). +, present; -, absent or undetected. (Furner et al., 1986; Slightom et al., 1986; Hansen et al., 1991; Aoki et al., 1994; Serino et al., 1994; Tanaka et al., 1994; Meyer et al., 1995; Frundt et al., 1998; Intrieri and Buiatti, 2001; Moriguchi et al., 2001; Suzuki et al., 2001; Suzuki et al., 2002)

	rolB	*rolC*	ORF13	ORF13a	ORF14	*mis*
pRi1724	837	540	594	270	564	984
pRi2659	837	540	594	270	564	-
pRiA4b	777	540	600	-	555	-
pRi8196	762	534	591	327	552	-
				-	558	
N. glauca (-) cT-DNA left arm	633	540	585	312	558	975
N. glauca (-) cT-DNA right arm	-	131	591	-	+	984
N. tabacum (-)	+	543	594	-	+	+
N. tomentosa	+	+	+	-	+	+
N. tomemtosiformis	+	+	+	-	+	+
N. otophora	+	+	+	-	+	-
N. benavidesii	+	+	+	-	+	-
N. cordifolia	+	+	+	-	+	-
N. setcheli	-	+	-	-	-	-
N. arentsii	-	+	-	-	-	-
N. acuminata	-	+	-	-	-	-
N. miersi (-)	+	-	-	-	-	-
N. bigelovii (-)	+	-	-	-	-	-
N. debneyi (-)	-	+	-	-	-	-
N. gossei	-	+	-	-	-	-
N. suaveolens (-)	-	+	-	-	-	-
N. exigua	-	+	-	-	-	-

reported species. The species that share homology with pRi T-DNA are listed in *Table 17-1*.

2.3 Phylogenetic analysis of cT-DNA genes and their evolution

At least 15 species in 7 sections in all three subgenera of the genus *Nicotiana* seem to contain cT-DNA in their genomes (Furner et al., 1986; Intrieri and Buiatti, 2001). Although other species in the genus *Nicotiana* may also contain cT-DNA, no further reports have yet appeared. This means that not every *Nicotiana* species contains cT-DNA, suggesting that it might have been deleted during their long evolutionary process or might have suffered no integration event. The examination of whether the cT-DNA exists in their genome, the homology levels between each of the *Nicotiana* species, and homologies with the present-day pRi T-DNA open the way for elucidating the evolution of the genus *Nicotiana*. To compare the nucleotide sequences of *rolB*, *rolC*, ORF13, and ORF14, Intrieri and Buiatti (2001) performed a phylogenetic analysis. They concluded that the *Nicotiana rol* genes seem to follow *Nicotiana* species evolution, being clustered into two groups. One includes *N. glauca* and *N. cordifolia* (the section *Paniculatae* in the subgenus *Rustica*), and the other includes the species in the subgenus *Tabacum*. If the pRi T-DNA were introduced only once into the ancient tobacco genome, such phylogenetic results must be shown. However, *rolC* in *N. debneyi* (the section *Suaveolens* in the subgenus *Petunioides*) showed high homology to Ng*rolC*, although the two species are removed from one another in tobacco evolution (Goodspeed, 1954). These results suggest that some independent infection by *A. rhizogenes* occurred in ancient *Nicotiana* plants. The origin of cT-DNA in *N. glauca* has been clarified as an integration of the mikimopine-type pRi T-DNA (Suzuki et al., 2002). Is the origin of cT-DNA in the other species of *Nicotiana* the same pRi T-DNA? If not, which type (or types) of pRi T-DNA is its origin? The nucleotide sequences of T-DNA of four different opine-type pRi have been determined so far (Slightom et al., 1986; Bouchez and Tourneur, 1991; Camilleri and Jouanin, 1991; Hansen et al., 1991; Serino et al., 1994; Tanaka et al., 1994; Moriguchi et al., 2001; Suzuki et al., 2001). To identify the closest pRi-T-DNA, a phylogenetic analysis among each of the *rol* genes on cT-DNA and pRiA4, pRi8196, and pRi1724 T-DNA was performed (Intrieri and Buiatti, 2001). However, the present-day pRi *rol* genes clustered with each other, giving no hint for the understanding of which pRi(s) is a possible origin of the cT-DNA in each *Nicotiana* species. Suzuki et al. (2002) also attempted to identify the origin of cT-DNA with multiple alignment and phylogenetic analysis using *rol* homologues in T-DNA of four pRis and in cT-DNA of *N. glauca*. However, they have been unable to provide confirmation of the origin of cT-DNA, showing that the divergence patterns of the present-day *A. rhizogenes* strains seem to follow a completely different logic.

On the other hand, the origin of cT-DNA must be determined by opine typing of pRi T-DNA. Suzuki et al. (2002) performed DNA gel blot analysis using the *mis* homologue, the Ng*mis*R gene, to detect homologous sequences in 12 *Nicotiana* species in 3 subgenera. Sequence homologous to Ng*mis*R was found in the genomes of *N. glauca*, *N. tomentosa*, *N. tomentosiformis*, and *N. tabacum*. However, the signal pattern in *N. glauca* differed from those in the species belonging to the subgenus *Tabacum*. Furthermore, in the subgenus *Tabacum*, signal patterns of three restriction enzymes found in *N. tomentosa* were different from those in *N. tomentosiformis* and *N. tabacum*, which looked identical. These results indicate that the cT-DNA in *N. tomentosa* differs from that of the two other species. *N. tabacum* is an amphidiploid species that is a hybrid of *N. tomentosiformis* and *N. sylvestlis* (Goodspeed, 1954), and since the signal patterns of three restriction enzymes of *N. tabacum* were the same as those of *N. tomentosiformis* and were not found in the genome of *N. sylvestris*, the *mis* homologue in *N. tabacum* originated in *N. tomentosiformis*.

To determine whether the integration site is the same in each species containing the cT-DNA, Suzuki et al. (2002) further performed DNA gel blot analysis using a fragment outside the left arm and a fragment outside the right arm as probes, because they must be of tobacco genome origin. When 11 *Nicotiana* species other than *N. glauca* were examined, sequences homologous to both regions outside of the cT-DNA of *N. glauca* were found in all of their genomes, as a single subgenus-specific restriction fragment for diploid species or as a double one for amphidiploid species corresponding to those of the parent species. Surprisingly, the sequences homologous to the outside of the cT-DNA in *N. glauca* were not contiguous to the cT-DNA in the genome of *N. tomentosa*, *N. tomentosiformis*, or *N. tabacum*; that is, the location of cT-DNA in the genome of *N. glauca* differs from that of these three species. Suzuki et al. (2002) suggested that at least four independent pRi T-DNA insertion events occurred in the genus *Nicotiana*, namely in *N. glauca*, *N. tomentosa*, *N. tomentosiformis*, and *N. otophora* (*Figure 17-2*). The former three insertions occurred by mikimopine-type pRi, although these insertion events must have been independent, whereas the latter was by a different opine-type pRi. Intrieri and Buiatti (2001) and Aoki (2004) also proposed the hypothesis of single and multiple ancient infections of *A. rhizogenes*. Phylogenetic analyses using such opine typing is applicable for determining divergence in the genus *Nicotiana.*

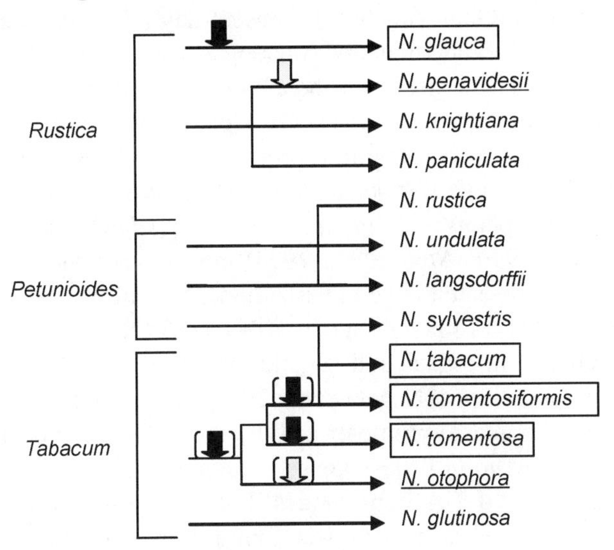

Figure 17-2. Scheme of the divergence o part of the genus *Nicotiana.* The tree shows only the relative divergence but not the evolutionary distance. Black and white arrows indicate the deduced insertion events by mikimopine-type or unknown opine-type pRi T-DNAs. The species containing mikimopine-type pRi T-DNA are boxed and those containing unknown opine-type pRi T-DNA are underlined. Cited from Suzuki et al. (2002) with kind permission from Blackwell Publishing Ltd.

Exactly when each infection event occurred is still unknown; however, Suzuki et al. (2002) estimates that the insertion of T-DNA into *N. tomentosiformis* occurred before the formation of *N. tabacum*, which was less than 6 million years ago (Okamuro and Goldberg, 1985). As mentioned later, only a few studies have identified cT-DNA in other plant genera, suggesting that the genus *Nicotiana* is particularly susceptible to insertion of *A. rhizogenes* pRi T-DNA.

2.4 Expression of the oncogenes on the cT-DNA

The genes on the cT-DNA in the genomes of *N. glauca* and other *Nicotiana* species must be the footprint of ancient infection by an *A. rhizogenes*-like bacterium harboring pRi. pRi-transgenic plants exhibit a specific, typical phenotype, the so-called "hairy root syndrome", with dwarfing caused by shortened internode distances, loss of apical dominance, alterations in organs, increases in root mass and decreasing rates of fertilization (Tepfer, 1984). However, plants in *Nicotiana* species containing cT-DNA display no such phenotype. Have these genes been preserved to express and maintain their function in present-day *Nicotiana* plants or have they completely lost expression, and function as pseudogenes? (Sinkar et al., 1988) suggested that the genes on thc cT-DNA must be silent because the transcripts from the genes of cT-DNA were undetectable in *N. glauca* (Taylor et al., 1985). However, transcripts of Ng*rolB*, Ng*rolC*, NgORF13 and NgORF14 were initially detected in genetic tumors (mentioned again later) on F1 hybrid plants of *N. glauca* and *N. langsdorffii* but at a low level (Ichikawa et al., 1990; Aoki et al., 1994), and transcripts of Ng*rolB* and Ng*rolC* were also detected in stem tissue and callus of *N. glauca* (Aoki and Syono, 1999b). The promoter regions of Ng*rolB*, Ng*rolC* and NgORF13 were also analyzed in organs and genetic tumors of F1 hybrid plants of *N. glauca* and *N. langsdorffii* (Nagata et al., 1995, 1996; Udagawa et al., 2004). A GUS (β-glucuronidase)-promoter activity assay for these cT-DNA genes revealed that expression of these genes is regulated similarly to that of the corresponding pRi T-DNA homologues, but with less than half the promoter activity. Ng*rolB* expression was found in meristematic zones, while Ng*rolC* expression was predominantly in vascular systems of various organs (Nagata et al., 1995, 1996). NgORF13 expression was detected in tumors and in vascular bundles and parenchymatous tissue in normal-type F1 transgenic plants, and moreover, was induced by wounding, similarly to RiORF13 expression (Udagawa et al., 2004). Ng*rolB* expression was promoted by auxin, similarly to Ri*rolB*, probably due to the auxin-responsive *cis*-element ACTTTA, which is bound by the trans-factor NtBBF1 (Baumann et al., 1999) in the Ng*rolB* as well as the Ri*rolB* promoter (Handayani et al., 2005). Transcripts from Ng*rolB*, Ng*rolC*, NgORF13, and NgORF14 in leaves and callus of *N. glauca* and *N. langsdorffii* were also investigated by reverse transcriptase-polymerase chain reaction (RT-PCR), and every gene was expressed in callus of *N. glauca*, but only Ng*rolC* was expressed in leaves of *N.glauca*, whereas no gene expressions were detected in *N. langsdorffii* (Intrieri and Buiatti, 2001). Transcripts from Ng*mis*L and Ng*mis*R were detected by RT-PCR because their accumulation was very low. These *mis* homologues could be amplified by RT-PCR using primers specific for each of the *mis* homologues, indicating that these *mis* homologues are also not pseudogenes (Suzuki et al., 2002). Interestingly, the expression levels of Ng*mis*L and Ng*mis*R in each organ correlated inversely (Suzuki et al., 2002).

In the case of *N. tabacum*, the expression of *trolC* and *torf13* genes under some conditions was investigated by RNA gel blot analysis in more detail (Meyer et al., 1995; Frundt et al., 1998). The *trolC* transcript accumulated in shoot tip and in upper and middle leaves. In suspension cultured cells, high *trolC* expression was observed in the exponential growth phase. The effect of auxin and cytokinin on *trolC* expression was investigated using callus tissues derived from leaves of *N. tabacum*; cytokinin treatment appeared to induce *trolC* expression and was effective even in the presence of auxin. When the expression of *torf13* was also investigated, *torf13* transcript accumulated to high levels in shoot tip, upper leaf and middle leaf of mature plants, which was similar to the expression pattern of *trolC*, although weaker signals were also detected in stem, lower leaf and root. The expression of *torf13* was further determined in different flower parts. Strong signals were found in petal and sepal, whereas no signals were in stigma, ovary or stamen. In experiments of phytohormone effects using leaf discs of *N. tabacum*, *torf13* transcripts declined by day 13 with auxin and more slowly with cytokinin, whereas the combination of auxin and cytokinin accelerated the decline by day 8. Therefore, the expression of *trolC* and *torf13* is regulated not only developmentally but also by phytohormones (Meyer et al., 1995; Frundt et al., 1998).

In conclusion, the oncogenes on cT-DNA are not pseudogenes and they are apparently expressed at low levels in present-day tobacco plants.

2.5 Function of the oncogenes on the cT-DNA

The genes that are a remnant of ancient *A. rhizogenes* infection have been expressed through a long period. However, no appearance of hairy roots has been observed on *N. glauca* plants. This may be due to the low expression level of Ng*rol* genes, or may result from a decrease or loss of their functions. Out of the oncogenes in pRi T-DNA, *rolB* gene expression seems to be more important for hairy root induction, because introduction of the *rolB* gene alone can induce adventitious root formation. However, the Ng*rolB* gene alone or in combination with Ng*rolC*, NgORF13, and NgORF14 did not generate adventitious roots (Aoki and Syono, 1999b, 1999a). Based on comparison with the amino acid sequences of RiRolB proteins, NgRolB might be a C-terminus-truncated mutated RolB protein, because the glutamine and glutamate residues at the 212 and 242 positions seem to have been replaced by termination codons caused by two point mutations, and the sequence shows no adventitious root induction capability (Aoki and Syono, 1999b). Moriuchi et al. (2004) reported that RolB protein seems to induce adventitious roots by association with a tobacco 14-3-3 protein. To interact with 14-3-3, most of the amino acid residues of RiRolB seem to be required. A RiRolB deletion mutant, which has the same C-terminal truncation as the NgRolB protein, showed no interaction with 14-3-3, resulting in no capability of adventitious root induction (Moriuchi et al., 2004). Therefore, impaired interaction with 14-3-3 may be responsible for blocking the capability for adventitious root induction by the NgRolB protein. However, Aoki and Syono (1999b) did a very exciting trial to infer the sequence of the ancient Ng*rolB* gene just after T-DNA insertion into *N. glauca*. Base substitution at these two positions by site-directed mutagenesis led to production of an NgRolB protein with the predicted full-length form and the capability for adventitious root induction. Transgenic plants overexpressing such resurrected Ng*rolB* exhibited morphogenetic abnormalities. This means that the ancient Ng*rolB* might have a function similar to the present-day Ri*rolB*. On the other hand, without any changes, NgORF13, and NgORF14 have retained their function, which supplements the rooting capability of *rolB* in unknown way for a long period (Aoki and Syono, 1999a). Ng*rolC* seems also to retain its function because the morphogenetic abnormalities of the transgenic plants are similar to those of Ri*rolC* ones (Aoki and Syono, 1999b), although the fundamental function of *rolC* remains unknown. The overexpression of *torf13-1* in carrot cells showed cell proliferation similar to that induced by ORF13 of pRi8196 (Frundt et al., 1998).

In *N. glauca* plants, mikimopine is not detected. To determine whether the *mis* homologue, Ng*mis*, catalyzes mikimopine synthesis, the protein encoded by the full-length Ng*mis*R homologue was produced in *E. coli*. The purified protein is able to synthesize mikimopine under the same conditions used for analysis of Mis activity (Suzuki et al., 2001; Suzuki et al., 2002). Therefore, Ng*mis*R must have retained mikimopine synthesis activity with no crucial change since T-DNA insertion.

3 Other CT-DNAs

3.1 cT-DNAs outside of the genus *Nicotiana*

Is cT-DNA found only in the genus *Nicotiana*? The answer is no; the existence of cT-DNAs in plants outside of the genus *Nicotiana* has been reported by several groups. By DNA gel blot analysis, the genomic DNA of normal carrot (*Daucus carota*) showed homology to pRi1855 T-DNA (Spanò et al., 1982). The genomes of field bindweed (*Convolvulus arvensis*) and carpet bugleweed (*Ajuga reptans*) also seem to contain sequences homologous to the T-DNAs of pRi8196 and pRi1724, respectively (Tepfer, 1982; Tanaka et al., 1998). Apples have sequences homologous not only to TL-DNA but also to TR-DNA of pRiA4 (Lambert and Tepfer, 1992). In some plants in the Brassicaceae, the presence of sequences homologous to *rol* genes and their transcripts was suggested (K. Syono, Japan Woman's University, personal communication). By more careful experiments, the spectrum of the plants possessing cT-DNA is likely to spread further.

3.2 Presence of cT-DNA originating from pTi T-DNA

Compared to reports about the presence of sequences homologous to pRi T-DNA, those homologous to pTi T-DNA are very few. There are two reports in which sequences homologous to pTi in the normal genomic DNA of *N. tabacum* were mentioned a quarter century ago. Thomashow et al. (1980) reported that genomic DNA of *N. tabacum* var. Xanthi had a sequence homologous to a fragment containing a portion of TL-DNA and TR-DNA of pTi-$B_6$806, although the hybridization signal was weak. Yang and Simpson (1981) also reported homology between pTi-T37 and genomic DNA of *N. tabacum* var. Xanthi and var. Havana. However, the sequence of pTi-T37 that was homologous to the genomic DNA of *N. tabacum* seemed to be outside of the T-DNA. Unfortunately, further observations have never been pursued. Kung (1989) made reference to a DNA gel blot analysis to detect homologous sequences in the genomes of thirteen species and nine genetic tumorous hybrids of *Nicotiana* using pTiB6S3 as a probe. However, no sequence homology between pTi and any of the *Nicotiana* genomes was mentioned.

4 GENETIC TUMORS

4.1 Genetic tumors on interspecific hybrids in the genus *Nicotiana*

In interspecific hybrids of some plant genera, tumors occur in the absence of any detectable external causal agents, such as bacteria, fungi, or plant tumor viruses. Such neoplastic growths are called genetic tumors. Genetic tumors are generated not only on interspecific hybrids (the genus *Lycopersicon*, *Datura*, *Bryophyllum*, *Gossypium*, *Lilium*, and so on) but also on inbreds (the genus *Melilotus*, *Thea*, *Sorghum*, *Pharbitis*, *Picea*, etc., see review by Ahuja, 1998). In particular, genetic tumors on interspecific hybrids of the genus *Nicotiana* are the most famous and have been well-studied (*Figure 17-3*). Genetic tumors on interspecific tumor-prone hybrids in the genus *Nicotiana* were first reported by Kostoff (1930). There are approximately 70 species in the genus *Nicotiana* and about 300 different interspecific hybrids have been reported (Smith, 1968). Among the hybrids, about 30 combinations produce tumors with good reproducibility. Tumors develop in reciprocal crosses, suggesting that nuclear genes from both species are involved in tumor formation in the hybrid combination. Each partner in the combination used for tumor formation falls into one of two groups, so-called "plus" and "minus"; the former mainly consists of species in the section *Alatae* with 9 or 10 chromosome pairs, while the latter consists of a variety of species belonging to diverse sections of the genus *Nicotiana*, generally with 12 chromosome pairs (Näf, 1958) as shown in *Figure 17-4*. With a few exceptions, the crosses between these two groups produce tumor-prone hybrids, whereas the crosses within either group never produce them (*Figure 17-4*). In an effort to detect association between tumor formation and a particular chromosome, tumorous F1 progeny between the amphidiploid *N. debneyi-tabacum* in the minus group and *N. longiflora* in the plus group (*N. debneyi-tabacum* x *N. longiflora*; Ahuja, 1962) was repeatedly back-crossed to *N. debneyi-tabacum*, finally yielding hybrid derivatives carrying a single chromosome derived from *N. longiflora* in an *N. debneyi-tabacum* genetic background (Ahuja, 1966). By a similar approach, hybrid derivatives containing a single chromosome of *N. longiflora* in an *N. tabacum* genetic background (Ahuja, 1962) and inverse hybrid derivatives carrying a single chromosome of *N. glauca* (minus) in an otherwise *N. langsdorffii* (plus) background (Smith, 1988) were obtained. These hybrid derivatives with a particular chromosome or its fragment developed tumors, although, in the cases of a single chromosome of a minus group plant on the genetic background of a plus group plant, tumors on the hybrid derivatives were smaller than those on the F1 hybrid plants (Ahuja, 1962, 1966; Smith, 1988). From these results, the two complementary genetic systems in the hybrid are likely to be quite different in terms of their contribution to tumor formation. Therefore, Ahuja (1968) proposed that the species belonging to the plus group have a gene or a locus designated as an initiator (*I*) for initiation of tumorigenesis and the species belonging to the minus group have genes or loci (*ee*) for enhancement and expression of tumors, and that the presence of both *I* and *ee* elements and their interaction in a hybrid plant must lead to initiate tumorigenesis and develop tumors. Tumorigenesis is usually initiated during and following the flowering phase, but is also triggered by various stresses such as wounding, chemical treatment, and irradiation.

Figure 17-3. Genetic tumors on the stem of aged plant of *N. langsdorffii* × *N. glauca* hybrid (Courtesy of Dr. K. Syono).

4.2 Are cT-DNA genes related to genetic tumor formation?

Interestingly, it appears that most species belonging to the minus group harbor cT-DNA (*Table 17-1*). This appearance reminds us that the genes in the cT-DNA may be involved in the formation of genetic tumors on the hybrid *Nicotiana* plants. However, since there are no reports showing a relationship between the *ee* genes and cT-DNA genes so far, this attractive hypothesis has not been validated yet. As already stated, in genetic tumors on F1 hybrids of *N. glauca* x *N. langsdorffii*, Ng*rolB*, Ng*rolC*, NgORF13 and NgORF14 genes are transcribed (Ichikawa et al., 1990; Aoki et al., 1994). On the other hand, these genes are expressed in a regulated pattern in some organs of normal-type F1 hybrid plants, similarly to their counterparts in pRi T-DNA (Nagata et al., 1995, 1996; Udagawa et al., 2004). Once tumorigenesis is initiated by aging or stress, these genes are transcribed in the developing outgrowth in a regulated order and pattern. This means that strong expression of Ng*rol* genes is related to tumor formation on an F1 hybrid. However, it has not been determined whether the formation of tumors is caused by the expression of Ng*rol* genes. The Ng*rolB* gene encodes the C-terminal truncated version of full-length RiRolB, showing no rooting function under the control of its own promoter and no morphological abnormality due to overexpression (Aoki and Syono, 1999b). With or without the Ng*rolB* gene, the other Ng*rol* genes, Ng*rolC*, NgORF13 and NgORF14, under the control of their own promoters do not show tumorigenesis either alone or in combination (Aoki and Syono, 1999a). Moreover, the stem tissue of *N. glauca* usually accumulates transcripts of the four Ng*rol* genes. Similarly, the *trolC* and *torf13* transcripts also accumulate in some organs of *N. tabacum* (Meyer et al., 1995; Frundt et al., 1998). These observations suggest that the high expression of these Ng*rol* genes might be unrelated to the initiation of tumor formation. However, Ng*rolC*, NgORF13 or *torf13* overexpression cause hairy root syndrome-like morphological alterations on transgenic tobacco plants (Aoki and Syono, 1999b, 1999a), and cell proliferation on carrot disks (Fründt et al., 1998). Therefore, cT-DNA genes may be responsible for the enhancement of genetic tumor development.

To grow and differentiate *in vitro*, cells from normal plants generally require essential phytohormones such as auxin and cytokinin. However, genetic tumor cells from some combinations of *Nicotiana* species can grow autonomously without hormones, similar to the crown gall tumors induced by *A. tumefaciens* (Ahuja, 1998). This suggests that these phytohormones must contribute to genetic tumor formation. The role of auxin in genetic tumor formation has been disputed, because opinions differ as to whether or not a positive correlation between the onset of genetic tumor formation and the indoleacetic acid (IAA) level in the tumor-prone tissues exists (Bayer, 1967; Ames and Mistretta, 1975; Ichikawa et al., 1989; Fujita et al., 1991). On the other hand, a higher cytokinin level is likely to be related to tumorigenesis in tumor-prone hybrid tissues (Feng et al., 1990; Nandi et al., 1990a), although it is unstable, so its level likely changes during the course of culture (Nandi et al., 1990b). Using *A. tumefaciens* containing insertion mutants of the *tms* (*iaaM* and *iaaH*) and *tmr* (*ipt*) loci of pTi, effects of overproduction of endogenous auxin and cytokinin were investigated on tumorigenesis of *N. glauca* and *N. langsdorffii*, their tumorous and non-tumorous (mutant) hybrids, and some other *Nicotiana* species (Nachmias et al., 1987). Although the mutants with a non-tumorous phenotype were complemented by introduction of the wild-type *ipt* gene, the results suggested that there are different genetic controls for initial cell proliferation and for continued autotrophic growth (Nachmias et al., 1987). On the other hand, Feng et al. (1990) reported that the tumorigenesis of X-ray-induced non-tumorous mutants of *N. glauca* × *N. langsdorffii* was restored by introduction of *ipt* and by addition of cytokinin. Therefore, cytokinin is likely to be involved in genetic tumor formation. Of course, as the tumorigenesis must involve not only cytokinin but also auxin, an aberrant control of these phytohormone biosynthetic genes during genetic tumor formation is presumed. Although we may expect that the phytohormone biosynthetic gene such as *iaaM*, *iaaH* and *ipt* originated from pTi T-DNA, these genes have not yet been identified in the genome of *Nicotiana* plants, as described above. After all, the relationship between cT-DNA and genetic tumor formation remains to be solved.

5 ADVANTAGE OF CT-DNA AND CREATION OF NEW SPECIES

In some species, shoots are directly regenerated from hairy roots with comparative ease in the natural environment, resulting in the production of whole plants exhibiting the hairy root syndrome (Tepfer, 1984). In contrast, although sometimes showing a shooty teratoma, crown gall tumors with high production of auxin and cytokinin are recalcitrant to regeneration of normal-appearing whole plants. This may be one of the reasons why the remnants of pRi T-DNA are predominantly found in the genome of wild plants. When pRi-transgenic plants with a severe hairy root phenotype are compared to parental plants, most people must wonder if the genetic background is identical with the exception of the T-DNA genes. Among the properties of the hairy root phenotype, an increase in root mass must be most effective to survive under adverse circumstances. For example, *Ajuga reptans* plants transformed by pRi displayed a 1.5-fold increase in root mass compared to that of normal plants (*Figure 17-5*), although the whole plant mass was similar (Tanaka and Matsumoto, 1993). Indeed, increase in root mass seems to be advantageous for tolerance of a dry environment (H. Kamada, University of Tsukuba, personal communication). Therefore, ancient pRi-transformed plants with an increase in root mass might have shown drought tolerance, surviving in a suddenly dry environment. An alteration of flower organs with reduced fertility in pRi-transgenic plants is disadvantageous for their reproduction. However, on *rolABC*-transgenic plants in some species, an increased number of fertile flowers is exhibited (Casanova et al., 2005). Early flowering or a switch

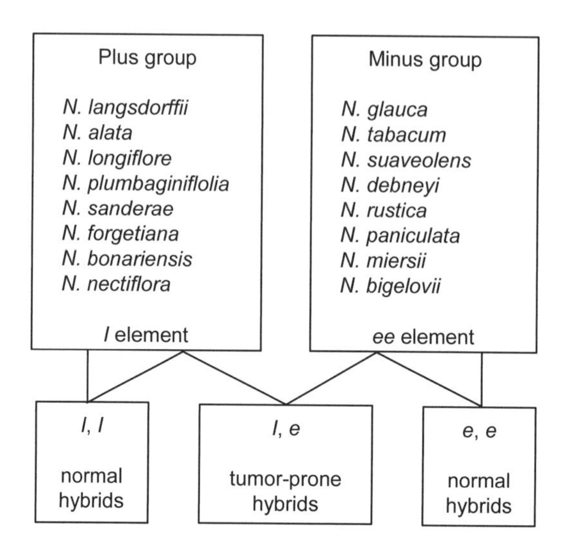

Figure 17-4. Combinations of tumor-prone interspecific hybrid (Näf, 1958). *I* and *ee* indicate hypothetical tumor-inducing elements contained in each group (Ahuja, 1968). Cited from Fujita (1994) with kind permission from Japanese Society for Plant Cell and Molecular Biology, with some modifications.

from biennial to annual flowering without vernalization can also occur on pRi-or *rolC*-transgenic plants (Limami et al., 1998; Casanova et al., 2005). These flowering properties areadvantageous to reproduce such transgenic plants over the untransformed parental plants. On the other hand, opines are strain-specific metabolites and each opine can only be catabolized by its corresponding strain, and therefore utilized as carbon and nitrogen sources for invading bacteria (Dessaux et al., 1993). However, Oger et al. (1997, 2000) reported that genetically engineered plants producing opines alter their biological environment, in particular, soil microflora such as root-associated bacterial populations. The *mis* homologues with intact enzyme activity are also transcribed in *N. glauca* plants, albeit in low amounts (Suzuki et al., 2002). The ancient mikimopine-producing tobacco plants might also have changed their soil environment. Since no invading bacteria containing *mis* homologues usually live in present-day *N. glauca* and *N. tabacum* plants, the role of mikimopine synthesized at such low levels is interesting. After all, it is likely that transformation by T-DNA insertion might have a strong influence on divergence of a new species if such a species had adapted to the environment better than the previous one.

It is hypothesized that horizontal gene transfer such as T-DNA insertion caused divergence of a new species from its former evolutionary process. Although Broothaerts et al. (2005) reported that some bacteria outside *Agrobacterium* also showed potential for horizontal gene transfer to plants, *Agrobacterium* is most likely to have influenced plant evolution (Gelvin, 2005).

Figure17-5. General views of untransformed and pRi-transgenic plants of *Ajuga reptans* cultured for 1 month in a green house (Tanaka and Matsumoto, 1993).

In spite of their increase in root mass, transgenic plants expressing *rolB* at high levels frequently display disadvantages such as stunting, necrosis and wrinkling of leaves, and reduced flower induction or development due to its cytotoxity (Röder et al., 1994). However, although the expression of Ng*rolB*, Ng*rolC*, NgORF13, and NgORF14 continue in present-day *N. glauca* plants following T-DNA insertion, no such phenotype seems to be exhibited, most likely due to the truncated Ng*rolB* sequence (Ichikawa et al., 1990; Aoki et al., 1994; Aoki and Syono, 1999b). This natural modification of the horizontally transferred DNA sequences during their evolution might also influence the new divergence process. We would refer to such natural events in evaluating the potential consequences of the dissemination of transgenic plants into the environment.

6 ACKNOWLEDGEMENTS

I wish to thank Dr. K. Syono (Japan Woman's University), Dr. S. Aoki, (University of Tokyo, Japan), Dr. H. Kamada (University of Tsukuba, Japan), Dr. H. Wabiko (Akita Prefectural University, Japan), Dr. T. Fujita (Hokkaido University, Japan) and Dr. D. Tepfer (INRA, France) for helpful discussion and providing pictures or materials.

7 REFERENCES

Ahuja MA (1962) A cytogenetic study of heritable tumors in *Nicotiana* species hybrids. Genetics 47: 865-880

Ahuja MA (1966) Genetic tumors and introgression in *Nicotiana*. Genetics 54: 316

Ahuja MA (1968) An hypothesis and evidence concerning the genetic components controlling tumor formation in *Nicotiana*. Mol Gen Genet 103: 176-184

Ahuja MA (1998) Genetic tumors in Nicotiana and other plants. Q Rev Biol 73: 439-462

Ames IH, Mistretta PW (1975) Auxin: its role in genetic tumor induction. Plant Physiol 56: 744-746

Aoki S (2004) Resurrection of an ancestral gene: functional and evolutionary analyses of the Ng*rol* genes transferred from *Agrobacterium* to *Nicotiana*. J Plant Res 117: 329-337

Aoki S, Ito M (2000) Molecular phylogeny of *Nicotiana* (Solanaceae) based on the nucleotide sequence of the *matK* gene. Plant Bio 2: 316-324

Aoki S, Ito M (2001) Consideration of the classification system and phytogeography of the *Nicotiana* (Solanaceae) by the analysis of molecular phylogeny. Pro Jap Soc Pl Tax 16: 49-67

Aoki S, Kawaoka A, Sekine M, Ichikawa T, Fujita T, Shinmyo A, Syono K (1994) Sequence of the cellular T-DNA in the untransformed genome of *Nicotiana glauca* that is homologous to ORFs 13 and 14 of the Ri plasmid and analysis of its expression in genetic tumours of *N. glauca* × *N. langsdorffii*. Mol Gen Genet 243: 706-710

Aoki S, Syono K (1999a) Function of Ng*rol* genes in the evolution of *Nicotiana glauca*: Conservation of the function of NgORF13 and NgORF14 after ancient infection by an *Agrobacterium rhizogenes*-like ancestor. Plant Cell Physiol 40: 222-230

Aoki S, Syono K (1999b) Horizontal gene transfer and mutation: Ng*rol* genes in the genome of *Nicotiana glauca*. Proc Natl Acad Sci USA 96: 13229-13234

Baumann K, De Paolis A, Costantino P, Gualberti G (1999) The DNA binding site of the Dof protein NtBBF1 is essential for tissue-specific and auxin-regulated expression of the *rolB* oncogene in plants. Plant Cell 11: 323-334

Bayer MH (1967) Thin-layer chromatography of auxin and inhibitors in *Nicotiana glauca, N. langsdorffii* and three of their tumor-forming hybrids. Planta 72: 329-337

Bouchez D, Tourneur J (1991) Organization of the agropine synthesis region of the T-DNA of the Ri plasmid from *Agrobacterium rhizogenes*. Plasmid 25: 27-39

Broothaerts W, Mitchell HJ, Weir B, Kaines S, Smith LM, Yang W, Mayer JE, Roa-Rodriguez C, Jefferson RA (2005) Gene transfer to plants by diverse species of bacteria. Nature 433: 629-633

Camilleri C, Jouanin L (1991) The TR-DNA region carrying the auxin synthesis genes of the *Agrobacterium rhizogenes* agropine-type plasmid pRiA4: nucleotide sequence analysis and introduction into tobacco plants. Mol Plant-Microbe Interact 4: 155-162

Casanova E, Trillas MI, Moysset L, Vainstein A (2005) Influence of *rol* genes in floriculture. Biotechnol Adv 23: 3-39

De Buck S, Jacobs A, Van Montagu M, Depicker A (1999) The DNA sequences of T-DNA junctions suggest that complex T-DNA loci are formed by a recombination process resembling T-DNA integration. Plant J 20: 295-304

Dessaux Y, Petit A, Tempé J (1993) Chemistry and biochemistry of opines, chemical mediators of parasitism. Phytochemistry 34: 31-38

Feng XH, Dube SK, Bottino PJ, Kung SD (1990) Restoration of shooty morphology of a nontumorous mutant of *Nicotiana glauca* × *N. langsdorffii* by cytokinin and the isopentenyltransferase gene. Plant Mol Biol 15: 407-420

Fründt C, Meyer AD, Ichikawa T, Meins F, Jr. (1998) A tobacco homologue of the Ri-plasmid *orf13* gene causes cell proliferation in carrot root discs. Mol Gen Genet 259: 559-568

Fujita T (1994) Screening of genes related to tumor formation in tobacco genetic tumors. Plant Tiss Cult Lett 11: 171-177

Fujita T, Ichikawa T, Syono K (1991) Changes in morphology, levels of endogenous IAA and protein composition in relation to the development of tobacco genetic tumor induced in the dark. Plant Cell Physiol 32: 169-177

Furner IJ, A. HG, Amasino RM, Garfinkel DJ, Gordon MP, Nester EW (1986) An *Agrobacterium* transformation in the evolution of the genus *Nicotiana*. Nature 319: 422-427

Gelvin SB (2005) Agricultural biotechnology: gene exchange by design. Nature 433: 583-584

Goodspeed TH (1954) The Genus Nicotiana, Chronica Botanica. Waltham, Massachusetts

Handayani NSN, Moriuchi H, Yamakawa M, Yamashita I, Yoshida K, Tanaka N (2005) Characterization of the *rolB* promoter on mikimopine-type pRi1724 T-DNA. Plant Sci 108: 1353-1364

Hansen G, Larribe M, Vaubert D, Tempé J, Biermann BJ, Montoya AL, Chilton M-D, Brevet J (1991) *Agrobacterium rhizogenes* pRi8196 T-DNA: mapping and DNA sequence of functions involved in mannopine synthesis and hairy root differentiation. Proc Natl Acad Sci USA 88: 7763-7767

Ichikawa T, Kobayashi M, Nakagawa S, Sakurai A, Syono K (1989) Morphological observations and qualitative and quantitative studies of auxins after induction of tobacco genetic tumor. Plant Cell Physiol 30: 57-63

Ichikawa T, Ozeki Y, Syono K (1990) Evidence for the expression of the *rol* genes of *Nicotiana glauca* in genetic tumors of *N. glauca* X *N. langsdorffii.* Mol Gen Genet 220: 177-180

Japan Tobacco Inc. (1990) Illustrated Book of The Genus *Nicotiana.* Seibundo Shinkosha Publishing Co., Tokyo, Japan

Intrieri MC, Buiatti M (2001) The horizontal transfer of *Agrobacterium rhizogenes* genes and the evolution of the genus *Nicotiana.* Mol Phylogenet Evol 20: 100-110

Journin L, Bouchezs D, Drong RF, Tepfer D, Slightom JL (1989) Analysis of TR-DNA/plant junctions in the genome of *Convolvulus arvensis* clone transformed by *Agrobacterium rhizogenes* strain A4. Plant Mol Biol 12: 72-85

Kostoff D (1930) Tumors and other malformations on certain *Nicotiana* hybrids. Zentralbl Bakteriol Parasitenkd Infectionskr Hyg Abt 1: Org 81: 244-260

Kung SD (1989) Genetic tumors in *Nicotiana.* Bot Bull Acad Sinica 30: 231-240

Kwok WW, Nester EW, Gordon MP (1985) Unusual plasmid DNA organization in an octopine crown gall tumor. Nucleic Acids Res 13: 459-471

Limami MA, Sun LY, Douat C, Helgeson J, Tepfer D (1998) Natural genetic transformation by *Agrobacterium rhizogenes* . Annual flowering in two biennials, belgian endive and carrot. Plant Physiol 118: 543-550

Meyer AD, Ichikawa T, Meins F, Jr. (1995) Horizontal gene transfer: regulated expression of a tobacco homologue of the *Agrobacterium rhizogenes rolC* gene. Mol Gen Genet 249: 265-273

Moriguchi K, Maeda Y, Satou M, Hardayani NS, Kataoka M, Tanaka N, Yoshida K (2001) The complete nucleotide sequence of a plant root-inducing (Ri) plasmid indicates its chimeric structure and evolutionary relationship between tumor-inducing (Ti) and symbiotic (Sym) plasmids in *Rhizobiaceae*. J Mol Biol 307: 771-784

Moriuchi H, Okamoto C, Nishihama R, Yamashita I, Machida Y, Tanaka N (2004) Nuclear localization and interaction of RolB with plant 14-3-3 proteins correlates with induction of adventitious roots by the oncogene *rolB*. Plant J 38: 260-275

Nachmias B, Ugolini S, Ricci MD, Pellegrini MG, Bogani P, Bettini P, Inzé D, Buiatti M (1987) Tumor formation and morphogenesis on different *Nicotiana* sp. and hybrids induced by *Agrobacterium tumefaciens* T-DNA mutants. Dev Genet 8: 61-71

Näf U (1958) Studies on tumor formation in *Nicotiana hybrids*. I. The classification of the parents into two etiologically significant groups. Growth 22: 167-180

Nagata N, Kosono S, Sekine M, Shinmyo A, Syono K (1995) The regulatory functions of the *rolB* and *rolC* genes of *Agrobacterium rhizogenes* are conserved in the homologous genes (Ng*rol*) of *Nicotiana glauca* in tobacco genetic tumors. Plant Cell Physiol 36: 1003-1012

Nagata N, Kosono S, Sekine M, Shinmyo A, Syono K (1996) Different expression patterns of the promoter of the Ng*rolB* and Ng*rolC* genes during the development of tobacco genetic tumors. Plant Cell Physiol 37: 489-493

Nandi SK, de Klerk GJM, Parker CW, Palni LSM (1990a) Endogenous cytokinin levels and metabolism of zeatin riboside in genetic tumour tissues and non-tumourous tissues of tobacco. Physiol Plant 78: 197-204

Nandi SK, Palni LSM, Parker CW (1990b) Dynamic of endogenous cytokinin during the growth cycle of a hormone-autotrophic genetic tumor line of tobacco. Plant Physiol 94: 1084-1089

Oger P, Mansouri H, Dessaux Y (2000) Effect of crop rotation and soil cover on alteration of the soil microflora generated by the culture of transgenic plants producing opines. Mol Ecol 9: 881-890

Oger P, Petit A, Dessaux Y (1997) Genetically engineered plants producing opines alter their biological environment. Nat Biotechnol 15: 369-372

Okamuro JK, Goldberg RB (1985) Tobacco single-copy DNA is highly homologous to sequences present in the genomes of its diploid progenies. Mol Gen Genet 198: 290-298

Röder FT, Schmulling T, Gatz C (1994) Efficiency of the tetracycline-dependent gene expression system: complete suppression and efficient induction of the *rolB* phenotype in transgenic plants. Mol Gen Genet 243: 32-38

Serino G, Clerot D, Brevet J, Costantino P, Cardarelli M (1994) *rol* genes of *Agrobacterium rhizogenes* cucumopine strain: sequence, effects and pattern of expression. Plant Mol Biol 26: 415-422

Sinkar VP, White FF, Furner IJ, Abrahamsen M, Pythoud F, Gordon MP (1988) Reversion of aberrant plants transformed with *Agrobacterium rhizogenes* is associated with the transcriptional inactivation of the TL-DNA genes. Plant Physiol 86: 584-590

Slightom JL, Durand-Tardif M, Jouanin L, Tepfer D (1986) Nucleotide sequence analysis of TL-DNA of *Agrobacterium rhizogenes* agropine type plasmid. Identification of open reading frames. J Biol Chem 261: 108-121

Smith HH (1968) Recent cytogenetic studies in the genus *Nicotiana*. Adv Genet 14: 1-54

Smith HH (1988) The inheritance of genetic tumors in *Nicotiana hybrids*. J Heredity 79: 277-283

Spanò L, Pomponi M, Costantino P, van Slogteren GMS, Tempé J (1982) Identification of T-DNA in the root-inducting plasmid of the agropine type *Agrobacterium rhizogenes* 1855. Plant Mol Biol 1: 291-300

Suzuki K, Tanaka N, Kamada H, Yamashita I (2001) Mikimopine synthase (*mis*) gene on pRi1724. Gene 263: 49-58

Suzuki K, Yamashita I, Tanaka N (2002) Tobacco plants were transformed by *Agrobacterium rhizogenes* infection during their evolution. Plant J 32: 775-787

Tanaka N, Ikeda T, Oka A (1994) Nucleotide sequence of the *rol* region of the mikimopine-type root-inducing plasmid pRi1724. Biosci Biotechnol Biochem 58: 548-551

Tanaka N, Matsumoto T (1993) Regenerants from *Ajuga* hairy roots with high productivity of 20-hydroxyecdysone. Plant Cell Rep 13: 87-90

Tanaka N, Yamakawa M, Yamashita I (1998) Characterization of transcription of genes involved in hairy root induction on pRi1724 core-T-DNA in two *Ajuga reptans* hairy root lines. Plant Sci 137: 95-105

Taylor BH, White FF, Nester EW, Gordon MP (1985) Transcription of *Agrobacterium rhizogenes* A4 T-DNA. Mol Gen Genet 201: 546-553

Tepfer D (1982) La transfromation génétique de plants supérieures par *Agrobacterium rhizogenes*. *In* 2ed Colloque sur les Recherches Fruitéres - Bordeaux, pp 47-59

Tepfer D (1984) Transformation of several species of higher plants by *Agrobacterium rhizogenes*: sexual transmission of the transformed genotype and phenotype. Cell 37: 959-967

Thomashow MF, Nutter R, Montoya AL, Gordon MP, Nester EW (1980) Integration and organization of Ti plasmid sequences in crown gall tumors. Cell 19: 729-739

Udagawa M, Aoki S, Syono K (2004) Expression analysis of the NgORF13 promoter during the development of tobacco genetic tumors. Plant Cell Physiol 45: 1023-1031

White FF, Garfinkel DJ, Huffman GA, Gordon MP, Nester EW (1983) Sequence homologous to *Agrobacterium rhizogenes* T-DNA in the genomes of uninfected plants. Nature 301: 348-350

White FF, Ghidossi G, Gordon MP, Nester EW (1982) Tumor induction by *Agrobacterium rhizogenes* involves the transfer of plasmid DNA to the plant genome. Proc Natl Acad Sci USA 79: 3193-3197

Yang F, Simpson RB (1981) Revertant seedling from crown gall tumors retain a portion of the bacterial Ti plasmid sequences. Proc Natl Acad Sci USA 78: 4151-4155

Chapter 18

AGROBACTERIUM-MEDIATED TRANSFORMATION OF NON-PLANT ORGANISMS

Jalal Soltani, G. Paul H. van Heusden and Paul J.J. Hooykaas

Department of Molecular and Developmental Genetics, Institute of Biology, Leiden University, Wassenaarseweg 64, 2333 AL Leiden, The Netherlands

Abstract. During the last decade it became clear that the ability of *Agrobacterium* to transform host organisms is not restricted to plants, but that numerous other organisms are transformable by *Agrobacterium* under laboratory conditions. It has been shown that *Agrobacterium*-mediated transformation is possible for at least 80 different non-plant species. Most of these organisms are fungi including yeasts, but also mammalian cells and algae can be transformed. *Agrobacterium*-mediated transformation is not restricted to eukaryotes as *Agrobacterium* is also able to act on the gram positive bacterium *Streptomyces lividans*. In general, the procedures for the transformation of different organisms are similar, but each organism has its own conditions for optimal transformation efficiency. Nowadays *Agrobacterium*-mediated transformation is the method of choice for the transformation of various fungi as transformation efficiencies are much higher than with other methods and the transformation protocols are relatively facile. *Agrobacterium* can transfer not only DNA but also proteins to the host organisms through its type four secretion system. This protein transfer has been shown for both plants and the yeast *Saccharomyces cerevisiae*. A major issue in the transformation of eukaryotic cells is the integration of the foreign DNA at random positions in the genome rather than at specific locations. The ability of *Agrobacterium* to transform the yeast *S. cerevisae* offers the possibility to use the many experimental tools available for this organism to fully unravel the mechanisms involved in the *Agrobacterium*-mediated transformation process. This is especially relevant as in contrast to most other organisms *S. cerevisiae* has a very efficient system for targeted integration of DNA fragments via homologous recombination. Knowledge of this system has already led to an increased frequency of targeted integration in the yeast *Kluyveromyes lactis*, in the filamentous fungus *Neurospora crassa* and the plant *Arabidopsis thaliana*. The ability of *Agrobacterium* to transfer T-DNA to a wide variety of eukaryotic and some prokaryotic organisms may have important implications for evolution. Future research has to show whether *Agrobacterium*-mediated transformation contributed to horizontal gene transfer between microorganisms in the rhizosphere.

1 INTRODUCTION

Agrobacterium tumefaciens is a plant pathogen, which causes crown gall by genetic transformation in more than 600 dicotyledonous plant species belonging to 90 families (De Cleene and De Ley, 1976). Although tumors are not formed on monocots, *Agrobacterium* can transform such plant species including the cereals as well (Hooykaas-Van Slogteren et al., 1984; Ishida et al., 1996). As discussed in previous chapters, its capability to transform plants is widely used in plant biotechnology and plant research. This is based on the presence in *Agrobacterium* of a large Ti plasmid, which contains a set of genes (the virulence genes) that can mobilize a segment of the Ti plasmid called the T-DNA, which is surrounded by a direct repeat (border repeat) of 24 bp, to plant cells. The *vir*-genes are activated in the presence of plant-specific phenolic compounds such as acetosyringone. Through the VirA chemoreceptor this leads to the activation of the transcription regulator VirG by phosphorylation, which then mediates transcription of the other *vir*-genes. Besides *vir*A and *vir*G the *vir*B operon (with 11 genes) and the *vir*D operon are essential for transformation. The *vir*B operon encodes the Type IV Secretion System (T4SS) which delivers the T-DNA and a number of virulence proteins into the plant cells and the *vir*D operon encodes proteins involved in the production of the single stranded DNA copy of the T-DNA that is transferred to the plant cell. Other *vir*-genes have an accessory role in transformation. Especially the *vir*E2 gene is important, as it encodes a single stranded DNA binding protein that coats the T-strand in the plant cell and thus protects it against nucleases. In its absence plant transformation occurs with a 1000-10.000 fold lower efficiency. During the last decade it became clear that the ability of *Agrobacterium* to transform host organisms is not restricted to plants, but that numerous other eukaryotic and even prokaryotic organisms are transformable by *Agrobacterium* under laboratory conditions. Since the pioneering work on the yeast *Saccharomyces cerevisiae* (Bundock et al., 1995), *Agrobacterium*-mediated transformation of other yeasts and many fungi has been shown (reviewed in Hooykaas et al., 2005; Michielse et al., 2005). In addition, *Agrobacterium*-mediated transformation of algae (Cheney et al., 2001; Kumar et al., 2004), mammalian cells (Kunik et al., 2001) and the gram positive bacterium *Streptomyces lividans* (Kelly and Kado, 2002) has been reported.

At the moment genomes from many organisms have been sequenced or are being sequenced. For effective functional genomics of these organisms and for their application in biotechnology highly efficient and facile genetic transformation protocols are needed. In this respect, *Agrobacterium*-mediated transformation is becoming a very effective tool. The more traditional transformation methods such as particle bombardment and the use of polyethylene glycol treated cells or protoplasts, have several drawbacks including the low transformation efficiency, the difficulty to control the copy number, the loss of molecular integrity of the DNA, the need to generate protoplasts and the limits on the size of the DNA (van den Eede et al., 2004). On the other hand, for *Agrobacterium*-mediated transformation it is possible to use different kinds of intact host cells such as conidia, mycelia, sexual spores and fruiting body tissues from fungi without the need to make protoplasts (Michielse et al., 2005). Furthermore, it is possible to transfer relatively large segments of DNA, up to 150 kb with no or only little rearrangements (Hamilton et al., 1996) and the transformation frequencies are higher than with traditional transformation methods (de Groot et al., 1998; Bundock et al., 1999; Amey et al., 2002; Campoy et al., 2003; Fitzgerald et al., 2003; Meyer et al., 2003; Idnurm et al., 2004; Michielse et al., 2004a; Rodriguez-Tovar et al., 2005). Also some species such as *Agaricus bisporous* and *Calonecteria morganii* that could not be transformed by traditional methods turned out to be transformable by *Agrobacterium* (de Groot et al., 1998; Malonek and Meinhardt, 2001). The potential of *Agrobacterium* to generate transformants having only a single integrated copy of the transgene in the genome, has been shown not only for plants but also for yeasts, fungi and mammalian cells (de Groot et al., 1998; Bundock et al., 1999; Kunik et al., 2001; Bundock et al., 2002). This property makes *Agrobacterium* potentially a very powerful tool for insertial mutagenesis, gene tagging and gene targeting (Michielse et al., 2005).

Indeed, *Agrobacterium* has been used as a tool for insertional mutagenesis in plants, such as *Arabidopsis thaliana*, *Nicotiana* species, and *Oryza sativa* (Koncz et al., 1989; Koncz et al., 1992; Krysan et al., 1999; Jeon et al., 2000) and more recently in different genera of fungi such as the model eukaryote *S. cerevisiae* (Bundock et al., 2002), the symbiotic fungus *Hebeloma cylindrosporum* (Combier et al., 2003), the biocontrol agents *Beauveria bassiana* (Leclerque et al., 2004) and *Coniothyrium minitans* (Rogers et al., 2004), the phytopathogens *Ascochyta rabiei* (Mogensen et al., 2006) and *Magnaprthe grisea* (Li et al., 2003), and the human pathogens *Aspergillus fumigatus* (Sugui et al., 2005) and *Cryptococcus neoformans* (Idnurm et al., 2004).

The ability of *Agrobacterium* to genetically modify the yeast *S. cerevisiae* offers the possibility to use the many experimental tools available for this organism to study the transformation process in detail. A major issue in the transformation of eukaryotic cells is the integration of the foreign DNA at random positions in the genome rather than at specific locations. In contrast to most eukaryotic organisms, *S. cerevisiae* very efficiently integrates the foreign DNA by homologous recombination, allowing targeted integration at specific chromosomal locations. Comparison of the integration processes occurring in *S. cerevisae* with those occurring in other eukaryotes may unravel the factors required for targeted integration of the foreign DNA.

In this chapter we will review the *Agrobacterium*-mediated transformation of non-plant species. Several aspects of the *Agrobacterium*-mediated transformation of non-plant organisms especially fungi have been discussed in recent reviews (Hooykaas, 2005; Michielse et al., 2005; Lacroix et al., 2006). Here, we focus on the transformation of both fungi and other non-plant organisms.

2 NON-PLANT ORGANISMS TRANSFORMED BY *AGROBACTERIUM*

Since the observation that *Agrobacterium* is not only capable of transforming plant cells, but also cells from the yeast *S. cerevisae*, more than 55 genera of different non-plant organisms have been successfully transformed by *Agrobacterium* (*Table 18-1*). Most of these organisms are fungi, but also algae and mammalian cells have been transformed. *Agrobacterium*-mediated transformation is not restricted to eukaryotes as *Agrobacterium* is also able to transform the gram positive bacterium *Streptomyces lividans*.

Table 18-1. Non-plant organisms transformed by *A. tumefaciens* (Modified from Michielse et al., 2005)

Species	Reference(s)
Prokaryotes	
Gram-positive bacteria	
Streptomyces lividans	(Kelly and Kado, 2002)
Eukaryotes	
Algae	
Chlamydomonas reinhardtii	(Kumar et al., 2004)
Porphyra yezoensis	(Cheney et al., 2001)
Fungi	
Oomycetes	
Pythium ultimum var. *sporangiiferum*	(Vijn and Govers, 2003)
Phytophthora infestans	(Vijn and Govers, 2003)
Phytophthora palmivora	(Vijn and Govers, 2003)
Zygomycetes	
Blakeslea trispora	(Michielse et al., 2005)
Mucor circinelloides	(Nyilasi et al., 2003)
Mucor miehei	(Monfort et al., 2003)
Rhizopus oryzae	(Michielse et al., 2004c)
Ascomycetes	
Ascochyta rabiei	(White and Chen, 2005; Mogensen et al., 2006)
Aspergillus awamori	(de Groot et al., 1998; Gouka et al., 1999; Michielse et al., 2004a)
Aspergillus fumigatus	(Sugui et al., 2005)
Aspergillus giganteus	(Meyer et al., 2003)
Aspergillus niger	(de Groot et al., 1998)
Beauveria bassiana	(dos Reis et al., 2004; Fang et al., 2004; Leclerque et al., 2004)
Blastomyces dermatiditis	(Brandhorst et al., 2002 ; Sullivan et al., 2002)
Botrytis cinerea	(Rolland et al., 2003)
Calonectria morganii	(Malonek and Meinhardt, 2001)
Candida albicans	(Michielse et al., 2005)
Candida glabrata	(Michielse et al., 2005)
Candida tropicalis	(Michielse et al., 2005)
Ceratocystis resinifera	(Loppnau et al., 2004)
Chaetomium globosum	(Gao and Yang, 2005)
Claviceps pururea	(Michielse et al., 2005)
Coccidiodes immitis	(Abuodeh et al., 2000)
Coccidiodes posadasii	(Kellner et al., 2005)

Species	Reference(s)
Colletotrichum.destructivum	(O'Connell et al., 2004)
Colletotrichum gloeosporioides	(de Groot et al., 1998)
Colletotrichum graminicola	(Flowers and Vaillancourt, 2005)
Colletotrichum lagenarium	(Tsuji et al., 2003)
Colletotrichum trifolii	(Takahara et al., 2004)
Coniothyrium minitans	(Rogers et al., 2004; Li et al., 2005)
Cryphonectria parasitica	(Park and Kim, 2004)
Fusarium circinatum	(Covert et al., 2001)
Fusarium culmorum	(Michielse et al., 2005)
Fusarium oxysporum	(Mullins et al., 2001; Takken et al., 2004; Khang et al., 2005)
Fusarium venenatum	(de Groot et al., 1998)
Glarea lozoyensis	(Zhang et al., 2003)
Heterobasidion annosum	(Samils et al., 2006)
Helminthosporium turcicum	(Degefu and Hanif, 2003)
Histoplasma capsulatum	(Sullivan et al., 2002)
Kluyveromyces lactis	(Bundock et al., 1999; Kooistra et al., 2004)
Leptosphaeria biglobosa	(Eckert et al., 2005)
Leptosphaeria maculans	(Gardiner et al., 2004; Gardiner and Howlett, 2004; Eckert et al., 2005)
Magnaporthe grisea	(Rho et al., 2001; Li et al., 2003; Khang et al., 2005)
Metarhizium anisopliae var. acridum	(Michielse et al., 2005)
Monascus purpureus	(Campoy et al., 2003)
Monilinia fructicola	(Dai et al., 2003)
Mycosphaerella fijiensis	(Michielse et al., 2005)
Mycosphaerella graminicola	(Zwiers and de Waard, 2001)
Neurospora crassa	(de Groot et al., 1998)
Oculimacula acuformis	(Eckert et al., 2005)
Oculimacula yallundae	(Eckert et al., 2005)
Ophiostoma floccosum	(Michielse et al., 2005)
Ophiostoma piceae	(Tanguay and Breuil, 2003)
Ophiostoma piliferum	(Hoffman and Breuil, 2004)
Paecilomyces fumosoroseus	(Michielse et al., 2005)
Paracoccidioides brasiliensis	(Leal et al., 2004)
Penicillium chrysogenum	(Sun et al., 2002)
Saccharomyces cerevisiae	(Bundock et al., 1995; Piers et al., 1996; Risseeuw et al., 1996; Bundock et al., 2002)
Trichoderma asperellum	(Michielse et al., 2005)
Trichoderma atroviride	(Zeilinger, 2004)
Trichoderma harzianum	(Michielse et al., 2005)
Trichoderma longibrachiatum	(Michielse et al., 2005)

Species	Reference(s)
Trichoderma reesei	(de Groot et al., 1998)
Tuber borchii	(Grimaldi et al., 2005)
Venturia inaequalis	(Fitzgerald et al., 2003; Fitzgerald et al., 2004)
Verticillium dahliae	(Dobinson et al., 2004)
Verticillium fungicola	(Amey et al., 2002; Amey et al., 2003)
Basidiomycetes	
Agaricus bisporus	(de Groot et al., 1998; Chen et al., 2000; Mikosch et al., 2001)
Cryptococcus gattii	(McClelland et al., 2005)
Cryptococcus neoformans	(Idnurm et al., 2004 ; McClelland et al., 2005)
Hebeloma cylindrosporum	(Pardo et al., 2002; Combier et al., 2003)
Hypholoma sublateritium	(Godio et al., 2004)
Laccaria bicolor	(Kemppainen et al., 2006)
Omphalotus olearius	(Michielse et al., 2005)
Paxillus involutus	(Pardo et al., 2002)
Phaffia rhodozyma	(Michielse et al., 2005)
Phanerocheate chrysosporium	(Sharma et al., 2006)
Pisolithus tinctorius	(Rodriguez-Tovar et al., 2005)
Pisolithus microcarpus	(Pardo et al., 2005)
Suillus bovines	(Hanif et al., 2002; Pardo et al., 2002)
Mammalian cells	
Human cells	
Clonal pheochromocytoma PC12 neuronal cells	(Kunik et al., 2001)
Embryonic kidney 293 cells	(Kunik et al., 2001)
HeLa R19 cells	(Kunik et al., 2001)

Foreign DNA can be engineered to allow stable integration in extranuclear DNA, such as plastids and mitochondrial DNA. For instance, chloroplast DNA has successfully been engineered for resistance to herbicides, insects, disease and drought, and for the production of biopharmaceuticals (reviewed by Daniell et al., 2005). Most of the methods used for transformation of extranuclear DNA are based on polyethylene glycol treated protoplasts. More than a decade ago, two studies on *Agrobacterium*-mediated transformation of plastid DNA have been published (De Block et al., 1985; Venkateswarlu and Nazar, 1991), but these works have not been reproduced.

3 EXPERIMENTAL ASPECTS OF *AGROBACTERIUM*-MEDIATED TRANSFORMATION OF NON-PLANT ORGANISMS

Many different protocols for the transformation of non-plant organisms have been developed. In general, the procedures for the transformation of the different organisms are similar. For example, the binary system is standard for use in both plants and non-plant organisms. Most transformations of non-plant organisms are performed by co-cultivation of *Agrobacterium* and recipient cells on a solid support. On the other hand, each organism requires its own optimal conditions to obtain maximal transformation frequencies. An optimized protocol for the *Agrobacterium*-mediated transformation of the yeast *S. cerevisiae* has recently been published (Hooykaas et al., 2006) which is the basic protocol for other fungi as well.

3.1 *Agrobacterium* strains

Various *Agrobacterium* strains have been used for the transformation of non-plant organisms, e.g. LBA4404, EHA105, and LBA1100. Systematic comparisons of different strains in relation to transformation frequencies have not been published, making it difficult to say which strain is best to use. The use of *Agrobacterium* strains derived from the supervirulent A281 strain which has a high level of vir gene expression, resulted in higher transformation frequencies in *S. cerevisiae*, *Monascus purpureus*, *Phytophthora infestans* and *Cryphonectria parasitica* (Piers et al., 1996; Campoy et al., 2003; Vijn and Govers, 2003; Park and Kim, 2004). The introduction of a ternary plasmid carrying the virG mutant gene coding for the constitutively active Vir-GN54D protein into *Agrobacterium* strain LBA1100 resulted in a considerable improvement in the transformation efficiency of *P. infestans* (Vijn and Govers, 2003).

3.2 Requirement of acetosyringone

During plant transformation the T-DNA transfer machinery of *Agrobacterium* is induced by phenolic compounds such as acetosyringone originating from the wounded plant cells. Also for transformation of non-plant organisms the virulence system has to be induced and in most transformation protocols the addition of acetosyringone to the induction medium is required. On the other hand, it has been reported that acetosyringone was not necessary for transformation of the alga *Chlamydomonas reinhardtii* (Kumar et al., 2004).

Addition of acetosyringone not only to the induction medium, but also to the *Agrobacterium* pre-culture medium, improved transformation frequencies of the fungi *Beauveria bassiana*, *Fusarium oxysporum*, and *Magnaporthe grisea* (Mullins et al., 2001; Rho et al., 2001; Leclerque et al., 2004). Furthermore, omission of acetosyringone from the pre-culture medium delayed the formation of transformants. In contrast, addition of acetosyringone to the pre-culture medium did not affect transformation frequencies of the fungi *Hebeloma cylindrosporum* and *Colletotrichum trifolii* (Combier et al., 2003; Takahara et al., 2004). Moreover, *Agrobacterium*-mediated transformation of mammalian cells was possible without the addition of acetosyringone to the pre-culture of *Agrobacterium* (Kunik et al., 2001).

3.3 Effect of co-cultivation conditions

The conditions under which *Agrobacterium* is co-cultivated with the recipient organism have a major influence on transformation frequencies. Transformation efficiency is influenced by the ratio between *Agrobacterium* and recipient, the length of the co-cultivation period, temperature, pH, and the choice of filters. Increasing the amount of *Agrobacterium* cells relative to the recipient cells in the co-cultivation mixture may lead to an increase in the transformation frequency. However, addition of too many *Agrobacterium* cells can result in a decrease in transformation efficiency (Meyer et al., 2003). Several studies have shown that each organism has an optimal combination of co-cultivation period and temperature to obtain a maximum number of transformants (Mullins et al., 2001; Rho et al., 2001; Combier et al., 2003; Meyer et al., 2003; Rolland et al., 2003; Gardiner and Howlett, 2004; Michielse et al., 2004a). In most transformation protocols optimal transformation is achieved at room temperature. An interesting aspect of *Agrobacterium*-mediated transformation of mammalian cells is that it occurred at 37°C after pre-growth of *Agrobacterium* at 28°C (Kunik et al., 2001). The effect of pH during co-cultivation on the transformation frequency was tested in *Agrobacterium*-mediated transformation of *C. trifolii* and *Colletotrichum lagenarium* (Tsuji et al., 2003; Takahara et al., 2004). It was found that the optimal pH, leading to the highest transformation frequency, is between 5.0 and 5.3. Also for transformation of the yeast *S. cerevisiae* a small deviation from the optimal pH of 5.3 already resulted in a considerably lower transformation frequency (J. Soltani, unpublished observation). The optimal pH also depends on the *Agrobacterium* strain used, as the pH requirements for optimal *vir*-gene induction are slightly different for the different *Agrobacterium* strains (Turk et al., 1991). For efficient *Agrobacterium*-mediated transformation cells are co-cultivated on a solid support such as nitrocellulose filters, Hybond, filter paper, cellophane sheets, and polyvinylidene difluoride.

3.4 Markers used for *Agrobacterium*-mediated transformation

The vectors used for *Agrobacterium*-mediated transformation of non-plant organisms have similar requirements as the vectors used for plant transformations. The DNA sequences located inside the T-DNA borders will be transferred to the recipient cells and for selection of transformants a suitable selection marker is required. Frequently used markers in both plants and non-plant systems are different antibiotic resistance genes from bacterial plasmids. Also an herbicide resistance gene (*bar*) has been used as a selection marker in fungi (Fang et al., 2004). It is important that these markers are controlled by a promoter active in the host organism. For transformation of the yeast *S. cerevisiae* both auxotrophic markers such as *URA3* and *TRP1* as well as dominant resistance markers such as the G418 resistance marker *KAN-MX* are being used (Bundock et al., 1995; Piers et al., 1996; van Attikum, 2003). Uracil auxotrophy markers have also been used during *Agrobacterium*-mediated transformation of filamentous fungi (Gouka et al., 1999; Sullivan et al., 2002; Michielse et al., 2005).

4 ROLE OF VIRULENCE PROTEINS IN THE *AGROBACTERIUM*-MEDIATED TRANSFORMATION OF NON-PLANT ORGANISMS

4.1 Chromosomally-encoded virulence proteins

Chromosomally-encoded virulence proteins (Chv proteins) are necessary for T-DNA transfer to plants. However, their role in transformation of non-plant organisms is not well-established. For *Agrobacterium*-mediated transformation of the yeast S. *cerevisiae* the chromosomal virulence operons *chvA*, *chvB*, and *exoC* which are required for bacterial attachment to plant cells are not required (Piers et al., 1996). On the other hand,

Agrobacterium-mediated transformation of mammalian cells depends on the presence of the *chvA* and *chvB* genes (Kunik et al., 2001). Reversely, it was reported that inactivation of one of the chromosomal genes involved in the biosynthesis of cellulose fibrils increases the frequency of transformation of *Aspergillus awamori* (Michielse et al., 2005).

4.2 Ti-plasmid encoded virulence proteins

The role of the *Agrobacterium* Ti-plasmid encoded virulence proteins (Vir proteins) in plant transformation is well studied. Some information is also available on the role of the Vir proteins in the transformation of non-plant organisms. The *virA*, *virB*, *virD* and *virG* genes are essential not only for plant transformation, but also for transformation of non-plant organism such as the yeast *S. cerevisiae* (Bundock et al., 1995) and the fungus *A. awamori* (Michielse et al., 2004b). Although inactivation of *virE2* almost eliminates the ability of *Agrobacterium* to transform plants, transformation of *S. cerevisiae* by such mutant is only 10-fold reduced (Bundock et al., 1995) and transformation of *A. awamori* only less than 2-fold (Michielse et al., 2004a). Nevertheless, as in plants *A. awamori* transformants had left-border truncations (Michielse et al., 2004b), indicating that VirE2 in fungi as in plants helps to protect the T-strand against nucleases. The deletion of *virC2* reduced the transformation efficiency of *A. awamori* about 13-fold. Transformants in this case were characterized by the presence of complex T-DNA structures containing multicopy and truncated T-DNAs and vector backbone sequences (Michielse et al., 2004b). This suggests that VirC2 plays a role in correct T-DNA border processing and is required for single-copy T-DNA integration.

5 TARGETED INTEGRATION OF T-DNA

Agrobacterium-mediated transformation of the yeast *S. cerevisiae* can result in random insertion of the T-DNA into the yeast genome by non-homologous end joining as is the common mechanism for T-DNA integration into the plant chromosome (*Figure 18-1*a) (Bundock and Hooykaas, 1996). However, when DNA sequences homologous to those of the yeast genome are present, the DNA fragment will mostly integrate into the genome by homologous recombination (*Figure 18-1b*). This will result in integration of the T-DNA at a predetermined location of the chromosome. Integration of the T-DNA via homologous recombination occurs approximately 50-100-fold more efficient than via non-homologous recombination in the yeast (Bundock et al., 1995). When the T-DNA contains a yeast replicator such as an autonomously replicating sequence (ARS) or the replicator of the 2μ plasmid, the T-DNA will be maintained in the yeast cell as a replicative plasmid (Bundock et al., 1995; Piers et al., 1996), after circularization of the T-DNA (*Figure 18-1*C).

In *S. cerevisiae* the integration of T-DNA by homologous recombination is very efficient. However, for most other organisms T-DNA insertion mainly occurs via non-homologous recombination, even when the T-DNA fragment has extensive sequence homology to the host chromosome. For the application of *Agrobacterium*-mediated transformation in functional genomics or biotechnology it is of great importance to improve the efficiency of integration via homologous recombination over non-homologous recombination. By using the yeast *S. cerevisiae* as a model, it was found that the proteins mediating T-DNA integration are the proteins involved in double strand break (DSB) repair of the genomic DNA (van Attikum et al., 2001; van Attikum and Hooykaas, 2003). Indeed, T-DNA integrates at preformed DSBs in both plant (Salomon and Puchta, 1998) and yeast chromosomes (van Attikum, 2003) (Reviewed in Tzfira et al., 2004). Several studies in *S. cerevisiae* have shed light on the mechanisms involved in DNA break repair, which may occur by homologous recombination or by non-homologous end joining (reviewed by Pâques and Haber, 1999; Jackson, 2002; Symington, 2002; West, 2003; Dudasova et al., 2004; Krogh and Symington, 2004; Daley et al., 2005). These mechanisms are summarized in *Figure 18-2*. Homologous recombination is initiated by a chromosomal double strand break (DSB) followed by the nucleolytic resection of the ends of the double stands break in the 5' to 3' direction. In yeast, and most likely also in mammals, this reaction relies on the Mre11 nuclease activity in a multiprotein complex consisting of Rad50, Mre11, and Xrs2 (*Figure 18-2*A). The Rad52 protein is able to bind to the ends of double strand breaks. This has led to the proposal that this Rad52 binding channels to repair by homologous recombination instead of to non-homologous end joining. For T-DNA integration at double strand breaks by homologous recombination Rad52 is essential (van Attikum and Hooykaas, 2003). The Rad51 nucleofilament interacts with undamaged DNA molecules and searches for a region with extensive homology. This process is influenced by other proteins such as replication protein A (Rpa), Rad52 and Rad54. When homology is found for example on the T-DNA, Rad51 catalyzes a strand invasion reaction in which the 3' protruding end of the damaged molecule invades the undamaged DNA molecule. The 3' end of the damaged DNA

(a) Non-homologous end joining

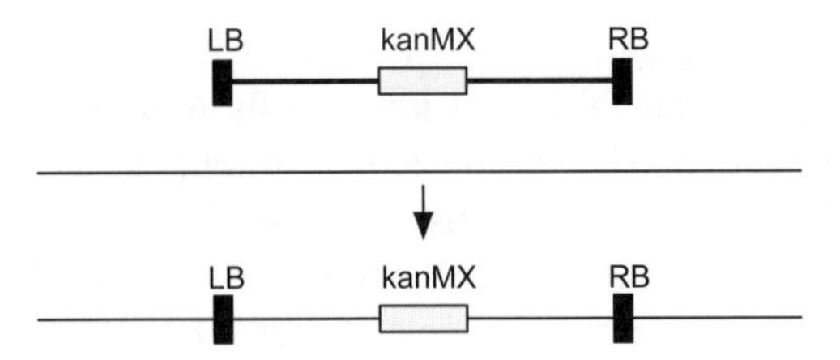

(b) Homologous recombination

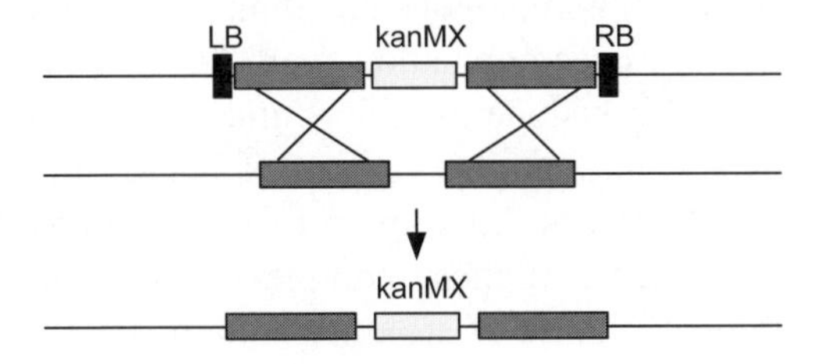

(c) Circularization into plasmids

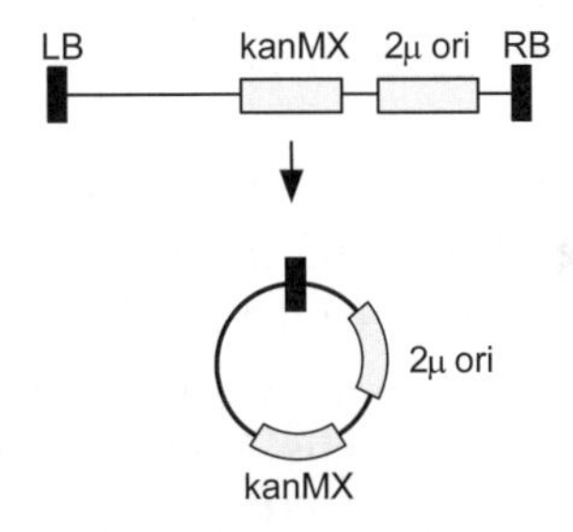

Figure 18-1. Mechanisms involved in targeted and non-targeted integration of T-DNA and circularization into plasmids. (A). *Agrobacterium*-mediated transformation of the yeast *S. cerevisiae* can result in random insertion of the T-DNA into the yeast genome by non-homologous end joining as is the common mechanism for T-DNA integration into the plant chromosome. (B) When DNA sequences homologous to those of the yeast genome are present, the DNA fragment will mostly integrate into the genome by homologous re-combination. (C) When the T-DNA contains a yeast replicator such as an autonomously replicating sequence (ARS) or the replicator of the 2μ plasmid, the T-DNA will be maintained in the yeast cell as a replicative plasmid after circularization of the T-DNA. (LB, left border; RB, right border; 2μ ori, origin of replication from the yeast 2μ plasmid)

molecule is then extended by a DNA polymerase that copies information from the undamaged DNA molecule and the ends are ligated by a DNA ligase, resulting in the integration of the T-DNA into the chromosome.

Non-homologous end joining is initiated by a double strand break, followed by binding of the Ku70 and Ku80 proteins to the ends (*Figure18-2b*). With the help of other proteins the break is then sealed restoring the original sequence or with small deletions. Heterologous DNA sequences, including those of transposable elements and the *Agrobacterium* T-DNA can be captured during this process and be integrated at the break site. In *S. cerevisiae* at least six genes are required for efficient integration of T-DNA via non-homologous end joining: *YKU70*, *LIG4*, *RAD50*, *MRE11*, *XRS2* and *SIR4* (van Attikum et al., 2001). Interestingly, *RAD50*, *MRE11* and *XRS2* are also involved in (meiotic) homologous recombination, but not in T-DNA integration by homologous recombination (van Attikum, 2003). Recently, it has been shown that also histone modifiers and ATP-dependent chromatin-remodeling complexes are recruited to sites of DNA damage (reviewed by Peterson and Côté, 2004; van Attikum and Gasser, 2005). Some plant homologs of the components of these complexes have already been found which show positive effects on or are differentially expressed during *Agrobacterium*-mediated transformation (Veena et al., 2003; Zhu et al., 2003; Loyter et al., 2005). For T-DNA integration at double strand breaks by non-homologous end joining Yku70 is essential (van Attikum, 2003). Thus, in *S. cerevisiae* the Rad52 and Yku70 proteins play a critical role in determining whether the T-DNA is integrated via homologous recombination or via non-homologous end joining (van Attikum, 2003).

Microhomology between the T-DNA and the chromosomal DNA plays a role in the initial steps of the non-homologous end joining process (Tinland, 1996; Wurtele et al., 2003; Daley et al., 2005). As a result truncations or deletions of the ends of the T-DNA may occur. This has been observed not only in plants but also in the yeast *S. cerevisiae*, in filamentous fungi and in mammalian cells (Bundock and Hooykaas, 1996; de Groot et al., 1998; Kunik et al., 2001; Mullins et al., 2001; van Attikum, 2003; Leclerque et al., 2004).

It is expected that in the absence of the non-homologous end joining proteins Ku70 or Ku80 integration by homologous recombination will become relatively more frequent. Indeed, after mutation of the *YKU70* or*YKU80* genes no integration by non-homologous end joining was seen in *S. cerevisiae* (van Attikum and Hooykaas, 2003). Recently, it was shown that in the yeast *Kluyveromyces lactis* disruption of *YKU80* led to a large

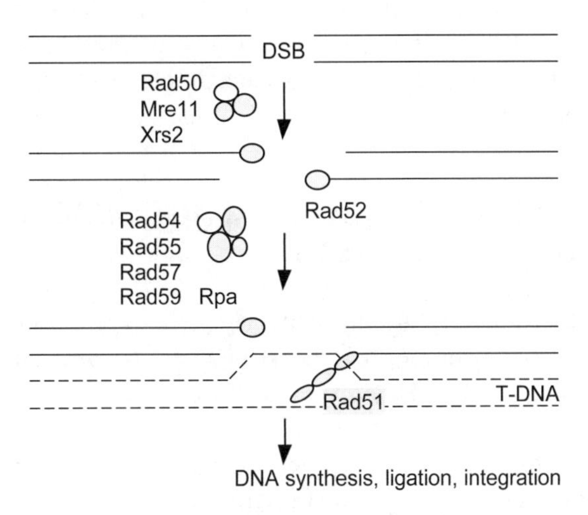

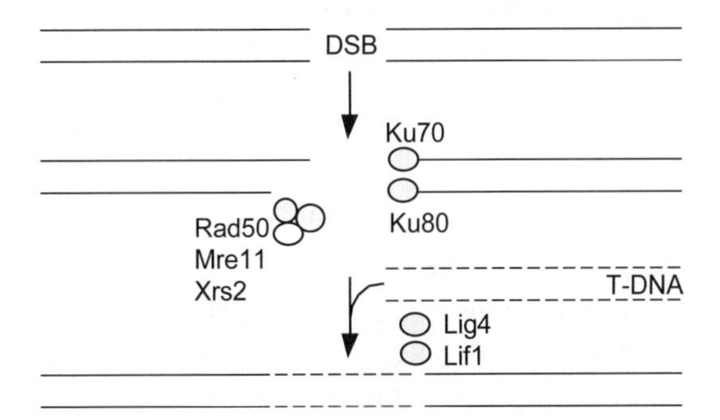

Figure18-2. Overview of the mechanisms involved in T-DNA integration via homologous recombination (A) and non-homologous end joining (B) in S. cerevisiae. DSB, double strands break. For further description see text. Adapted from Jackson (2002) and van Attikum (2003).

increase in integration of T-DNA by homologous recombination up to 97% of transformants analyzed (Kooistra et al., 2004). Also, in the filamentous fungus *Neurospora crassa* disruption of *YKU70* or *YKU80* vastly increased the integration of exogenous DNA into the genome by homologous recombination at a frequency of 100% (Ninomiya et al., 2004). These results support the great potential of manipulating the recombination machinery in optimizing targeted integration of T-DNA.

Furthermore, as in plants in *S. cerevisiae* T-DNA can also be integrated via gap repair (Risseeuw et al., 1996). This integration event supports the model for the integration of T-DNA as a single strand in cooperation with VirD2 protein (reviewed in Tzfira et al., 2004).

6 PROTEIN TRANSFER FROM *AGROBACTERIUM* TO NON-PLANT HOSTS

During the transformation process *Agrobacterium* transfers not only T-DNA but also a number of its virulence proteins into the host cell (Vergunst et al., 2000). This protein transfer is not restricted to plant cells, but it has been shown that the VirE2, VirE3 and VirF proteins can be transferred into cells from the yeast *Saccharomyces cerevisiae* as well (Schrammeijer et al., 2003). To study protein transfer from *Agrobacterium* to yeast, the Cre recombinase reporter assay for translocation has been used. Protein fusions between Cre and Vir proteins were expressed in *Agrobacterium*. Transfer of the Cre-Vir fusion proteins from *Agrobacterium* to yeast can be monitored by a selectable excision event resulting from site-specific recombination mediated by Cre on a *lox*-flanked transgene in yeast. This assay illustrates the potential of *Agrobacterium* to introduce genome modifying enzymes into eukaryotic cells. As the signal for transport by the type four secretion system (TFSS) lies in the 30 C-terminal amino acids of transferred proteins, coupling of this transport signal to the C-terminus of heterologous proteins may allow their mobilization from *Agrobacterium* to eukaryotic target cells (Vergunst et al., 2005). This property of *Agrobacterium* is promising for its application in protein therapy of both plant and non-plant cells.

7 PROSPECTS

Agrobacterium-mediated transformation has become a widely used tool for transformation of different types of eukaryotic cells. Especially for the transformation of various fungi, it has great advantages over other transformation methods. The efficiencies are much higher and the transformation protocols are relatively facile. It has also been shown that multiple copies of the T-DNA can be integrated by homologous recombination at a predetermined position of the genome of *A. awamori* allowing a high level of expression of the introduced gene of interest (Gouka et al., 1999). T-DNA integration at random positions in the genome of most eukaryotic organisms makes *Agrobacterium* a very useful tool for random mutagenesis and random gene tagging. *Agrobacterium* can transfer not only DNA but also proteins to the host organisms through its type four secretion system (TFSS). This protein transfer has been shown to occur independently of DNA transfer to both plants and the yeast *S. cerevisiae*. Most likely, protein transfer occurs during the transformation of all host cells, irrespective of their origin. Because of this property, *Agrobacterium* has a great potential for use in protein therapies.

A major issue in the transformation of eukaryotic cells is the integration of the foreign DNA at random positions in the genome rather than at specific locations. In contrast to most other organisms, *S. cerevisiae* has a very efficient system for targeted integration of DNA fragments via homologous recombination, but will integrate the DNA at random positions if homology with the genome is lacking. By the use of the well developed genetics of *S. cerevisiae* it was possible to identify key factors that control DNA integration by homologous recombination and non-homologous end-joining, respectively. As these key proteins are strongly conserved in other eukaryotes (from fungi to plants and animals) the knowledge obtained from yeast may be directly applicable in these other organisms to improve the frequency of targeted integration. Indeed, by disruption of *YKU70* or *YKU80* in the yeast *K. lactis* (Kooistra et al., 2004) and in the filamentous fungus *Neurospora crassa* (Ninomiya et al., 2004) the relative efficiency of targeted integration increased. Recently, it has been shown that expression in the plant *Arabidopsis thaliana* of the *S. cerevisae RAD54* gene which is involved in chromatin remodeling, resulted in an increase of the integration of T-DNA by homologous recombination by two orders of magnitude (Shaked et al., 2005).

The ability of *Agrobacterium* to transfer T-DNA to a wide variety of eukaryotic and some prokaryotic organisms may have important consequences for evolution. In the rhizosphere where *vir* inducers are readily available and numerous microorganisms are living in close proximity, it is very likely that *Agrobacterium*-mediated transformation of non-plant organisms is occurring. This process may contribute to horizontal gene transfer. Future research has to show whether such horizontal gene transfer has contributed to evolution.

8 ACKNOWLEDGEMENTS

We would like to thank the Ministry of Science, Research and Technology of Iran for the scholarship to J.S. We would like to thank Peter Hock for preparing the figures. The work in our laboratory was supported by grants from the organizations for Applied Research (STW) and Earth and Life Sciences (ALW) with financial aid from the Netherlands Organization of Scientific Research (NWO).

9 REFERENCES

Abuodeh RO, Orbach MJ, Mandel MA, Das A, Galgiani JN (2000) Genetic transformation of *Coccidioides immitis* facilitated by *Agrobacterium tumefaciens*. J Infect Dis 181: 2106-2110

Amey RC, Athey-Pollard A, Burns C, Mills PR, Bailey A, Foster GD (2002) PEG-mediated and *Agrobacterium*-mediated transformation in the mycopathogen *Verticillium fungicola*. Mycol Res 106: 4-11

Amey RC, Mills PR, Bailey A, Foster GD (2003) Investigating the role of a *Verticillium fungicola* beta-1,6-glucanase during infection of *Agaricus bisporus* using targeted gene disruption. Fungal Genet Biol 39: 264-275

Brandhorst TT, Rooney PJ, Sullivan TD, Klein B (2002) Molecular genetic analysis of *Blastomyces dermatitidis* reveals new insights about pathogenic mechanisms. Int J Med Microbiol 292: 363-371

Bundock P, den Dulk-Ras A, Beijersbergen A, Hooykaas PJJ (1995) Trans-kingdom T-DNA transfer from *Agrobacterium tumefaciens* to *Saccharomyces cerevisiae*. EMBO J 14: 3206-3214

Bundock P, Hooykaas PJJ (1996) Integration of *Agrobacterium tumefaciens* T-DNA in the *Saccharomyces cerevisiae* genome by illegitimate recombination. Proc Natl Acad Sci USA 93: 15272-15275

Bundock P, Mroczek K, Winkler AA, Steensma HY, Hooykaas PJ (1999) T-DNA from *Agrobacterium tumefaciens* as an efficient tool for gene targeting in *Kluyveromyces lactis*. Mol Gen Genet 261: 115-121

Bundock P, van Attikum H, den Dulk-Ras A, Hooykaas PJ (2002) Insertional mutagenesis in yeasts using T-DNA from *Agrobacterium tumefaciens*. Yeast 19: 529-536

Campoy S, Perez F, Martin JF, Gutierrez S, Liras P (2003) Stable transformants of the azaphilone pigment-producing *Monascus purpureus* obtained by protoplast transformation and *Agrobacterium*-mediated DNA transfer. Curr Genet 43: 447-452

Chen X, Stone M, Schlagnhaufer C, Romaine CP (2000) A fruiting body tissue method for efficient *Agrobacterium*-mediated transformation of *Agaricus bisporus*. Appl Environ Microbiol 66: 4510-4513

Cheney D, Metz B, Stiller J (2001) *Agrobacterium*-mediated genetic transformation in the macroscopic marine red alga *Porphyra yezoensis*. J Phycol Suppl 37: 11

Combier JP, Melayah D, Raffier C, Gay G, Marmeisse R (2003) *Agrobacterium tumefaciens*-mediated transformation as a tool for insertional mutagenesis in the symbiotic ectomycorrhizal fungus *Hebeloma cylindrosporum*. FEMS Microbiol Lett 220: 141-148

Covert SF, Kapoor P, Lee M, Briley A, Nairn CJ (2001) *Agrobacterium*-mediated transformation of *Fusarium circinatum*. Mycol Res 105: 259-264

Dai Q, Sun Z, Schnabel G (2003) Development of spontaneous hygromycin B resistance in *Monilinia fructicola* and Its impact on growth rate, morphology, susceptibility to demethylation inhibitor fungicides, and sporulation. Phytopathol 93: 1354-1359

Daley JM, Palmbos PL, Wu D, Wilson TE (2005) Nonhomologous end joining in yeast. Annu Rev Genet 39: 431-451

Daniell H, Kumar S, Dufourmantel N (2005) Breakthrough in chloroplast genetic engineering of agronomically important crops. Trends Biotechnol 23: 238-245

De Block M, Schell J, Van Montagu M (1985) Chloroplast transformation by *Agrobacterium tumefaciens*. EMBO J 4: 1367-1372

De Cleene M, De Ley J (1976) The host range of crown gall. Bot Rev 42: 389-466

de Groot MJ, Bundock P, Hooykaas PJJ, Beijersbergen AG (1998) *Agrobacterium tumefaciens*-mediated transformation of filamentous fungi. Nat Biotechnol 16: 839-842

Degefu Y, Hanif M (2003) *Agrobacterium tumefaciens*-mediated transformation of *Helminthosporium turcicum*, the maize leaf-blight fungus. Arch Microbiol 180: 279-284

Dobinson KF, Grant SJ, Kang S (2004) Cloning and targeted disruption, via *Agrobacterium tumefaciens*-mediated transformation, of a trypsin protease gene from the vascular wilt fungus *Verticillium dahliae*. Curr Genet 45: 104-110

dos Reis MC, Pelegrinelli Fungaro MH, Delgado Duarte RT, Furlaneto L, Furlaneto MC (2004) *Agrobacterium tumefaciens*-mediated genetic transformation of the entomopathogenic fungus *Beauveria bassiana*. J Microbiol Methods 58: 197-202

Dudasova Z, Dudas A, Chovanec M (2004) Non-homologous end-joining factors of *Saccharomyces cerevisiae*. FEMS Microbiol Rev 28: 581-601

Eckert M, Maguire K, Urban M, Foster S, Fitt B, Lucas J, Hammond-Kosack K (2005) *Agrobacterium tumefaciens*-mediated transformation of *Leptosphaeria* spp. and *Oculimacula* spp. with the reef coral gene DsRed and the jellyfish gene *gfp*. FEMS Microbiol Lett 253: 67-74

Fang W, Zhang Y, Yang X, Zheng X, Duan H, Li Y, Pei Y (2004) *Agrobacterium tumefaciens*-mediated transformation of *Beauveria bassiana* using an herbicide resistance gene as a selection marker. J Invertebr Pathol 85: 18-24

Fitzgerald A, Van Kan JA, Plummer KM (2004) Simultaneous silencing of multiple genes in the apple scab fungus, *Venturia inaequalis*, by expression of RNA with chimeric inverted repeats. Fungal Genet Biol 41: 963-971

Fitzgerald AM, Mudge AM, Gleave AP, Plummer KM (2003) *Agrobacterium* and PEG-mediated transformation of the phytopathogen *Venturia inaequalis*. Mycol Res 107: 803-810

Flowers JL, Vaillancourt LJ (2005) Parameters affecting the efficiency of *Agrobacterium tumefaciens*-mediated transformation of *Colletotrichum graminicola*. Curr Genet 48: 380-388

Gao XX, Yang Q (2005) *Agrobacterium tumefaciens*-mediated transformation of *Chaetomium globosum* and its T-DNA insertional mutagenesis. Wei Sheng Wu Xue Bao 45: 129-131

Gardiner DM, Cozijnsen AJ, Wilson LM, Pedras MS, Howlett BJ (2004) The sirodesmin biosynthetic gene cluster of the plant pathogenic fungus *Leptosphaeria maculans*. Mol Microbiol 53: 1307-1318

Gardiner DM, Howlett BJ (2004) Negative selection using thymidine kinase increases the efficiency of recovery of transformants with targeted genes in the filamentous fungus *Leptosphaeria maculans*. Curr Genet 45: 249-255

Godio RP, Fouces R, Gudina EJ, Martin JF (2004) *Agrobacterium tumefaciens*-mediated transformation of the antitumor clavaric acid-producing basidiomycete *Hypholoma sublateritium*. Curr Genet 46: 287–294

Gouka RJ, Gerk C, Hooykaas PJ, Bundock P, Musters W, Verrips CT, de Groot MJ (1999) Transformation of *Aspergillus awamori* by *Agrobacterium tumefaciens*-mediated homologous recombination. Nat Biotechnol 17: 598-601

Grimaldi B, de Raaf MA, Filetici P, Ottonello S, Ballario P (2005) *Agrobacterium*-mediated gene transfer and enhanced green fluorescent protein visualization in the mycorrhizal ascomycete *Tuber borchii*: a first step towards truffle genetics. Curr Genet 48: 69-74

Hamilton CM, Frary A, Lewis C, Tanksley SD (1996) Stable transfer of intact high molecular weight DNA into plant chromosomes. Proc Natl Acad Sci USA 93: 9975-9979

Hanif M, Pardo AG, Gorfer M, Raudaskoski M (2002) T-DNA transfer and integration in the ectomycorrhizal fungus *Suillus bovinus* using hygromycin B as a selectable marker. Curr Genet 41: 183-188

Hoffman B, Breuil C (2004) Disruption of the subtilase gene, albin1, in *Ophiostoma piliferum*. Appl Environ Microbiol 70: 3898-3903

Hooykaas PJ (2005) Transformation mediated by *Agrobacterium tumefaciens*. *In* JS Tkacz, L Lange, eds, Advances in Fungal Biotechnology for Industry, Agriculture and Medicine. Kluwer Acad./Plenum Publ., New York, pp 41-65

Hooykaas PJ, Dulk-Ras A, Bundock P, Soltani J, van Attikum H, van Heusden GPH (2006) *Agrobacterium*-mediated transformation of the yeast. *In* K Wang, ed, *Agrobacterium* Protocols. Humana Press, pp 465-473

Hooykaas-Van Slogteren GMS, Hooykaas PJJ, Schilperoort RA (1984) Expression of Ti plasmid genes in monocotyledonous plants infected with *Agrobacterium tumefaciens*. Nature 311: 763-764

Idnurm A, Reedy JL, Nussbaum JC, Heitman J (2004) *Cryptococcus neoformans* virulence gene discovery through insertional mutagenesis. Eukaryot Cell 3: 420-429

Ishida Y, Saito H, Ohta S, Hiei Y, Komari T, Kumashiro T (1996) High efficiency transformation of maize (*Zea mays* L.) mediated by *Agrobacterium tumefaciens*. Nat Biotechnol 14: 745-750

Jackson SP (2002) Sensing and repairing DNA double-strand breaks. Carcinogenesis 23: 687-696

Jeon JS, Lee S, Jung KH, Jun SH, Jeong DH, Lee J, Kim C, Jang S, Yang K, Nam J, An K, Han MJ, Sung RJ, Choi HS, Yu JH, Choi JH, Cho SY, Cha SS, Kim SI, An G (2000) T-DNA insertional mutagenesis for functional genomics in rice. Plant J 22: 561-570

Kellner EM, Orsborn KI, Siegel EM, Mandel MA, Orbach MJ, Galgiani JN (2005) *Coccidioides posadasii* contains a single 1,3-{beta}-glucan synthase gene that appears to be essential for growth. Eukaryot Cell 4: 111-120

Kelly BA, Kado CI (2002) *Agrobacterium*-mediated T-DNA transfer and integration into the chromosome of *Streptomyces lividans*. Mol Plant Pathol 3: 125-134

Kemppainen M, Circosta A, Tagu D, Martin F, Pardo AG (2006) *Agrobacterium*-mediated transformation of the ectomycorrhizal symbiont *Laccaria bicolor* S238N. Mycorrhiza 16: 19-22

Khang CH, Park SY, Lee YH, Kang S (2005) A dual selection based, targeted gene replacement tool for *Magnaporthe grisea* and *Fusarium oxysporum*. Fungal Genet Biol 42: 483-492

Koncz C, Martini N, Mayerhofer R, Koncz-Kalman Z, Korber H, Redei GP, Schell J (1989) High-frequency T-DNA-mediated gene tagging in plants. Proc Natl Acad Sci USA 86: 8467-8471

Koncz C, Nemeth K, Redei GP, Schell J (1992) T-DNA insertional mutagenesis in Arabidopsis. Plant Mol Biol 20: 963-976

Kooistra R, Hooykaas PJ, Steensma HY (2004) Efficient gene targeting in *Kluyveromyces lactis*. Yeast 21: 781-792

Krogh BO, Symington LS (2004) Recombination proteins in yeast. Annu Rev Genet 38: 233-271

Krysan PJ, Young JC, Sussman MR (1999) T-DNA as an insertional mutagen in Arabidopsis. Plant Cell 11: 2283-2290

Kumar SV, Misquitta RW, Reddy VS, Rao BJ, Rajam MV (2004) Genetic transformation of the green alga *Chlamydomonas reinhardtii* by *Agrobacterium tumefaciens*. Plant Sci 166: 731-738

Kunik T, Tzfira T, Kapulnik Y, Gafni Y, Dingwall C, Citovsky V (2001) Genetic transformation of HeLa cells by *Agrobacterium*. Proc Natl Acad Sci USA 98: 1871-1876

Lacroix B, Tzfira T, Vainstein A, Citovsky V (2006) A case of promiscuity: *Agrobacterium*'s endless hunt for new partners. Trends Genet 22: 29-37

Leal CV, Montes BA, Mesa AC, Rua AL, Corredor M, Restrepo A, McEwen JG (2004) *Agrobacterium tumefaciens*-mediated transformation of *Paracoccidioides brasiliensis*. Med Mycol 42: 391-395

Leclerque A, Wan H, Abschutz A, Chen S, Mitina GV, Zimmermann G, Schairer HU (2004) *Agrobacterium*-mediated insertional mutagenesis (AIM) of the entomopathogenic fungus *Beauveria bassiana*. Curr Genet 45: 111-119.

Li HY, Pan CY, Chen H, Zhao CJ, Lu GD, Wang ZH (2003) Optimization of T-DNA insertional mutagenesis and analysis of mutants of *Magnaporthe grisea*. Sheng Wu Gong Cheng Xue Bao 19: 419-423

Li M, Gong X, Zheng J, Jiang D, Fu Y, Hou M (2005) Transformation of *Coniothyrium minitans*, a parasite of *Sclerotinia sclerotiorum*, with *Agrobacterium tumefaciens*. FEMS Microbiol Lett 243: 323-329

Loppnau P, Tanguay P, Breuil C (2004) Isolation and disruption of the melanin pathway polyketide synthase gene of the softwood deep stain fungus *Ceratocystis resinifera*. Fungal Genet Biol 41: 33-41

Loyter A, Rosenbluh J, Zakai N, Li J, Kozlovsky SV, Tzfira T, Citovsky V (2005) The plant VirE2 interacting protein 1. A molecular link between the *Agrobacterium* T-Complex and the host cell chromatin? Plant Physiol 138: 1318-1321

Malonek S, Meinhardt F (2001) *Agrobacterium tumefaciens*-mediated genetic transformation of the phytopathogenic ascomycete *Calonectria morganii*. Curr Genet 40: 152-155

McClelland CM, Chang YC, Kwon-Chung KJ (2005) High frequency transformation of *Cryptococcus neoformans* and *Cryptococcus gattii* by *Agrobacterium tumefaciens*. Fungal Genet Biol 42: 904-913

Meyer V, Mueller D, Strowig T, Stahl U (2003) Comparison of different transformation methods for *Aspergillus giganteus*. Curr Genet 43: 371-377

Michielse CB, Hooykaas PJ, van den Hondel CA, Ram AF (2005) *Agrobacterium*-mediated transformation as a tool for functional genomics in fungi. Curr Genet 48: 1-17

Michielse CB, Ram AF, Hooykaas PJ, Hondel CA (2004a) Role of bacterial virulence proteins in *Agrobacterium*-mediated transformation of *Aspergillus awamori*. Fungal Genet Biol 41: 571-578

Michielse CB, Ram AF, Hooykaas PJ, van den Hondel CA (2004b) *Agrobacterium*-mediated transformation of *Aspergillus awamori* in the absence of full-length VirD2, VirC2, or VirE2 leads to insertion of aberrant T-DNA structures. J Bacteriol 186: 2038-2045

Michielse CB, Salim K, Ragas P, Ram AF, Kudla B, Jarry B, Punt PJ, van den Hondel CA (2004c) Development of a system for integrative and stable transformation of the zygomycete *Rhizopus oryzae* by *Agrobacterium*-mediated DNA transfer. Mol Genet Genomics 271: 499-510

Mikosch TS, Lavrijssen B, Sonnenberg AS, van Griensven LJ (2001) Transformation of the cultivated mushroom *Agaricus bisporus* (Lange) using T-DNA from *Agrobacterium tumefaciens*. Curr Genet 39: 35-39

Mogensen EG, Challen MP, Strange RN (2006) Reduction in solanapyrone phytotoxin production by *Ascochyta rabiei* transformed with *Agrobacterium tumefaciens*. FEMS Microbiol Lett 255: 255-61

Monfort A, Cordero L, Maicas S, Polaina J (2003) Transformation of *Mucor miehei* results in plasmid deletion and phenotypic instability. FEMS Microbiol Lett 224: 101-106

Mullins ED, Chen X, Romaine P, Raina R, Geiser DM, Kang S (2001) *Agrobacterium*-mediated transformation of *Fusarium oxysporum*: an efficient tool for insertional mutagenesis and gene transfer. Phytopathol 91: 173-180

Ninomiya Y, Suzuki K, Ishii C, Inoue H (2004) Highly efficient gene replacements in *Neurospora* strains deficient for nonhomologous end-joining. Proc Natl Acad Sci USA 101: 12248-12253

Nyilasi I, Acs K, Lukacs G, Papp T, Kasza Z, Vagvolgyi C (2003) *Agrobacterium tumefaciens*-mediated transformation of *Mucor circinelloides*. *In* First FEMS Congress Eur Microbiol, Ljubljana, pp 13-16.

O'Connell R, Herbert C, Sreenivasaprasad S, Khatib M, Esquerre-Tugaye MT, Dumas B (2004) A novel *Arabidopsis-Colletotrichum pathosystem* for the molecular dissection of plant-fungal interactions. Mol Plant-Microbe Interact 17: 272-282

Pâques F, Haber JE (1999) Multiple pathways of recombination induced by double-strand breaks in *Saccharomyces cerevisiae*. Microbiol Mol Biol Rev 63: 349-404

Pardo AG, Hanif M, Raudaskoski M, Gorfer M (2002) Genetic transformation of ectomycorrhizal fungi mediated by *Agrobacterium tumefaciens*. Mycol Res 106: 132-137

Pardo AG, Kemppainen M, Valdemoros D, Duplessis S, Martin F, Tagu D (2005) T-DNA transfer from *Agrobacterium tumefaciens* to the ectomycorrhizal fungus *Pisolithus microcarpus*. Rev Argent Microbiol 37: 69-72

Park SM, Kim DK (2004) Transformation of a filamentous fungus *Cryphonectria parasitica* using *Agrobacterium tumefaciens*. Biotechnol Bioprocess Eng 9: 217-222

Peterson CL, Côté J (2004) Cellular machineries for chromosomal DNA repair. Genes Dev 18: 602-616

Piers KL, Heath JD, Liang X, Stephens KM, Nester EW (1996) *Agrobacterium tumefaciens*-mediated transformation of yeast. Proc Natl Acad Sci USA 93: 1613-1618

Rho HS, Kang S, Lee YH (2001) *Agrobacterium tumefaciens*-mediated transformation of plant pathogenic fungus, *Magnaporthe grisea*. Mol Cells 12: 407-411

Risseeuw E, Franke-van Dijk ME, Hooykaas PJ (1996) Integration of an insertion-type transferred DNA vector from *Agrobacterium tumefaciens* into the *Saccharomyces cerevisiae* genome by gap repair. Mol Cell Biol 16: 5924-5932

Rodriguez-Tovar AV, Ruiz-Medrano R, Herrera-Martinez A, Barrera-Figueroa BE, Hidalgo-Lara ME, Reyes-Marquez BE, Cabrera-Ponce JL, Valdes M, Xoconostle-Cazares B (2005) Stable genetic transformation of the ectomycorrhizal fungus *Pisolithus tinctorius*. J Microbiol Methods 63: 45-54

Rogers CW, Challen MP, Green JR, Whipps JM (2004) Use of REMI and *Agrobacterium*-mediated transformation to identify pathogenicity mutants of the biocontrol fungus, *Coniothyrium minitans*. FEMS Microbiol Lett 241: 207-214

Rolland S, Jobic C, Fevre M, Bruel C (2003) *Agrobacterium*-mediated transformation of *Botrytis cinerea*, simple purification of monokaryotic transformants and rapid conidia-based identification of the transfer-DNA host genomic DNA flanking sequences. Curr Genet 44: 164-171

Salomon S, Puchta H (1998) Capture of genomic and T-DNA sequences during double-strand break repair in somatic plant cells. EMBO J. 17: 6086-6095

Samils N, Elfstrand M, Lindner Czederpiltz DL, Fahleson J, Olson A, Dixelius C, Stenlid J (2006) Development of a rapid and simple *Agrobacterium tumefaciens*-mediated transformation system for the fungal pathogen *Heterobasidion annosum*. FEMS Microbiol Lett 255: 82-88

Schrammeijer B, Dulk-Ras Ad A, Vergunst AC, Jurado Jacome E, Hooykaas PJ (2003) Analysis of Vir protein translocation from *Agrobacterium tumefaciens* using *Saccharomyces cerevisiae* as a model: evidence for transport of a novel effector protein VirE3. Nucleic Acids Res 31: 860-868

Shaked H, Melamed-Bessudo C, Levy AA (2005) High frequency gene targeting in *Arabidopsis* plants expressing the yeast RAD54 gene. Proc Natl Acad Sci USA 102: 12265-12269

Sharma KK, Gupta S, Kuhad RC (2006) *Agrobacterium* mediated delivery of marker genes to *Phanerochaete chrysosporium* mycelial pellets: a model transformation system for white-rot fungi. Biotechnol Appl Biochem (in press)

Sugui JA, Chang YC, Kwon-Chung KJ (2005) *Agrobacterium tumefaciens*-mediated transformation of *Aspergillus fumigatus*: an efficient tool for insertional mutagenesis and targeted gene disruption. Appl Environ Microbiol 71: 1798-1802

Sullivan TD, Rooney PJ, Klein BS (2002) *Agrobacterium tumefaciens* integrates transfer DNA into single chromosomal sites of dimorphic fungi and yields homokaryotic progeny from multinucleate yeast. Eukaryot Cell 1: 895-905

Sun CB, Kong QL, Xu WS (2002) Efficient transformation of *Penicillium chrysogenum* mediated by *Agrobacterium tumefaciens* LBA4404 for cloning of *Vitreoscilla* hemoglobin gene. Electronic J Biotech 5: 21-28

Symington LS (2002) Role of RAD52 epistasis group genes in homologous recombination and double-strand break repair. Microbiol Mol Biol Rev 66: 630-670

Takahara H, Tsuji G, Kubo Y, Yamamoto M, Toyoda K, Inagaki Y, Ichinose Y, Shiraishi T (2004) *Agrobacterium tumefaciens*-mediated transformation as a tool for random mutagenesis of *Colletotrichum trifolii*. J Gen Plant Pathol 70: 93-96

Takken FL, Van Wijk R, Michielse CB, Houterman PM, Ram AF, Cornelissen BJ (2004) A one-step method to convert vectors into binary vectors suited for *Agrobacterium*-mediated transformation. Curr Genet 45: 242-248

Tanguay P, Breuil C (2003) Transforming the sapstaining fungus *Ophiostoma piceae* with *Agrobacterium tumefaciens*. Can J Microbiol 49: 301-304

Tinland B (1996) The integration of T-DNA into plant genomes. Trends Plant Sci 1: 178-184

Tsuji G, Fujii S, Fujihara N, Hirose C, Tsuge S, Shiraishi T, Kubo Y (2003) *Agrobacterium tumefaciens*-mediated transformation for random insertional mutagenesis in *Colletotrichum lagenarium*. J Gen Plant Pathol 69: 230-239

Turk SC, Melchers LS, den Dulk-Ras H, Regensburg-Tuink AJ, Hooykaas PJ (1991) Environmental conditions differentially affect *vir* gene induction in different *Agrobacterium* strains. Role of the VirA sensor protein. Plant Mol Biol 16: 1051-1059

Tzfira T, Li J, Lacroix B, Citovsky V (2004) *Agrobacterium* T-DNA integration: molecules and models. Trends Genet 20: 375-383

van Attikum H (2003) Genetic requirements for the integration of *Agrobacterium* T-DNA in the eukaryotic genome. PhD. Leiden University, Leiden, The Nethrlands

van Attikum H, Bundock P, Hooykaas PJJ (2001) Non-homologous end-joining proteins are required for *Agrobacterium* T-DNA integration. EMBO J 20: 6550-6558

van Attikum H, Gasser SM (2005) The histone code at DNA breaks: a guide to repair? Nat Rev Mol Cell Biol 6: 757-765

van Attikum H, Hooykaas PJJ (2003) Genetic requirements for the targeted integration of *Agrobacterium* T-DNA in *Saccharomyces cerevisiae*. Nucleic Acids Res 31: 826-832

van den Eede G, Aarts H, Buhk HJ, Corthier G, Flint HJ, Hammes W, Jacobsen B, Midtvedt T, van der Vossen J, von Wright A, Wackernagel W, Wilcks A (2004) The relevance of gene transfer to the safety of food and feed derived from genetically modified (GM) plants. Food Chem Toxicol 42: 1127-1156

Veena, Jiang H, Doerge RW, Gelvin SB (2003) Transfer of T-DNA and Vir proteins to plant cells by *Agrobacterium tumefaciens* induces expression of host genes involved in mediating transformation and suppresses host defense gene expression. Plant J 35: 219-236

Venkateswarlu K, Nazar RN (1991) Evidence for T-DNA mediated gene targeting to tobacco chloroplasts. Biotechnology (NY) 9: 1103-1105

Vergunst AC, Schrammeijer B, den Dulk-Ras A, de Vlaam CMT, Regensburg-Tuink TJ, Hooykaas PJJ (2000) VirB/D4-dependent protein translocation from *Agrobacterium* into plant cells. Science 290: 979-982

Vergunst AC, van Lier MC, den Dulk-Ras A, Grosse Stüve TA, Ouwehand A, Hooykaas PJ (2005) Positive charge is an important feature of the C-terminal transport signal of the VirB/D4-translocated proteins of *Agrobacterium*. Proc Natl Acad Sci USA 102: 832-837

Vijn I, Govers F (2003) *Agrobacterium tumefaciens* mediated transformation of the oomycete plant pathogen *Phytophthora infestans*. Mol Plant Pathol 4: 459-467

West SC (2003) Molecular views of recombination proteins and their control. Nat Rev Mol Cell Biol 4: 435-445

White D, Chen W (2005) Genetic transformation of *Ascochyta rabiei* using *Agrobacterium*-mediated transformation. Curr Genet 21: 1-9

Wurtele H, Little KC, Chartrand P (2003) Illegitimate DNA integration in mammalian cells. Gene Ther 10: 1791-1799

Zeilinger S (2004) Gene disruption in *Trichoderma atroviride* via *Agrobacterium*-mediated transformation. Curr. Genet. 45: 45: 54–60

Zhang A, Lu P, Dahl-Roshak AM, Paress PS, Kennedy S, Tkacz JS, An Z (2003) Efficient disruption of a polyketide synthase gene (*pks1*) required for melanin synthesis through *Agrobacterium*-mediated transformation of *Glarea lozoyensis*. Mol Genet Genomics 268: 645-655

Zhu Y, Nam J, Humara JM, Mysore K, Lee LY, Cao H, Valentine L, Li J, Kaiser A, Kopecky A, Hwang HH, Bhattacharjee S, Rao P, Tzfira T, Rajagopal J, Yi HC, Yadav VBS, Crane Y, Lin K, Larcher Y, Gelvin M, Knue M, Zhao X, Davis S, Kim SI, Kumar CTR, Choi YJ, Hallan V, Chattopadhyay S, Sui X, Ziemienowitz A, Matthysse AG, Citovsky V, Hohn B, Gelvin SB (2003) Identification of *Arabidopsis rat* mutants. Plant Physiol 132: 494-505

Zwiers LH, de Waard MA (2001) Efficient *Agrobacterium tumefaciens*-mediated gene disruption in the phytopathogen *Mycosphaerella graminicola*. Curr Genet 39: 388-393

Chapter 19

THE BIOETHICS AND BIOSAFETY OF GENE TRANSFER

Kathrine H. Madsen[1] and Peter Sandøe[2]

Danish Centre for Bioethics and Risk Assessment (CeBRA) [1]Danish Institute of Agricultural Sciences. Research Centre Flakkebjerg, DK-4200 Slagelse, Denmark; [2]Royal Veterinary and Agricultural University, Rolighedsvej 25 DK-1958 Frederiksberg C, Denmark

Abstract. From the early stages of genetic engineering legal frameworks were set up to ensure the safe development of this technology. These regulatory frameworks focus primarily on risks to human health and the environment, and the concepts of substantial equivalence and familiarity seem to be the two universally adopted principles on which risk assessments are based. Despite this focus on risk prevention, genetically modified (GM) crops have given rise to controversies over the last 10-15 years. It is argued that one reason for this is that the early regulatory frameworks did not adequately address the concerns that seem to underlie public resistance to GM crops. Some of these concerns are about risks which lie beyond the issues addressed by the authorities who approve GM crops. Awareness of these concerns has led to a tightening of the regulatory requirements in the European Union where, among other things, indirect and long-term environmental effects are now included. Other major socio-economic concerns — e.g. lack of demonstrated usefulness to society, and the consumer's right to choose non-GM food products — have been debated. This debate has led to regulations designed to permit the co-existence of GM growers and non-GM growers in several EU Member States. The discussion about GM crops therefore relates both to risks to human health and the environment and a wider range of concerns such as usefulness, risks to society and a number of other ethical concerns.

1 INTRODUCTION

The use of gene transfer to create genetically modified (GM) crops has given rise to serious controversy. This controversy has not arisen merely from conflicts of interest. It also reflects differences in the way in which the technology, its risks, and its potential benefits, are perceived.

Scientists, plants breeders and many farmers view GM crops as an opportunity to overcome some of the limits of conventional plant breeding. For them, genetic modification simply furthers the development of crops which deliver high yields and are less vulnerable to a variety of problems currently facing conventional crops. Gene transfer is just one more technology with which to improve and develop plant production. Previous developments in high-yielding, intensive agriculture were perceived fundamentally positive and on this background gene transfer, similarly, is merely a step in the direction of enhancing and developing agriculture.

For many consumers and citizens, on the other hand, GM-food has been viewed as something that is not only dangerous, but also ethically highly problematic. For these stakeholders developments in pesticides, artificial fertilizers and other technologies underlying modern intensive agriculture threaten both human health and nature. Organic production and associated developments are welcome. Public discussion of GM crops has offered an occasion to express their concerns about GM crops, and in this discussion the new crops have virtually become a symbol of what they do not like about intensive agriculture.

Politicians in developed countries often have an ambivalent attitude to the new crops. On the one hand, they are keen to benefit from the opportunities offered by modern technology. On the other, they are sensitive to the worries and concerns of their citizens. From the early stages of the development of gene technology a legal framework has been set up which, at one and the same time, should allow the technology to develop and ensure that this development is safe. The key issue in the regulation has been biosafety. In order to be allowed to release GM crops and use them for food production it is necessary to demonstrate that they do not give rise to significant risks to human health and the environment.

However, it has turned out that regulation focusing solely on biosafety does not really satisfy those who are worried. Hence controversy has escalated as a rising number of crops with genetically modified traits have undergone assessment by the authorities. This is where bioethics enters the picture. The aim of bioethics is to articulate and critically discuss the values and moral principles at play in discussions of the development and use of modern biotechnology.

This chapter aims to clarify the concerns underlying the debate about the development and use of GM crops. To begin with (Section 1.1), we describe how governments on both sides of the Atlantic, in their differing ways, have tried to regulate GM crops to ensure that human health and the environment are protected. Here we also examine the underlying framework of risk analysis. Secondly (Section 2), we argue that, even when the risks alone are considered, there may be different views about which risks are relevant to consider, and about how much knowledge is needed to decide that a given level of risk is acceptable. Finally (Section 3), we shall argue that the controversy over GM crops raises some issues that are not covered within the framework of risk analysis relating to human health and the environment.

1.1 Responding to biosafety concerns: regulation

During the late 1980s and early 1990s the regulatory frameworks on both sides of the Atlantic had to deal with concerns about biosafety issues arising from the production and release of genetically modified organisms. These regulations focused on potential risks to human health and environmental safety.

Two key factors affected the development of the regulatory frameworks for GM crops: pressure from agricultural products industry to obtain certainty that their products could be marketed provided regulatory requirements were fulfilled; and public concerns about the safety. Later on, international obligations — e.g. Cartagena Biosafety Protocol — added to the need for regulation (Jaffe, 2004). Different approaches were used to regulate genetically modified organisms: in the USA, Canada and Argentina the regulation focused primarily on the product, whereas in the EU it is process-based, meaning that all organisms produced by genetic engineering must be approved by the regulatory system prior to release (Nap et al., 2003).

1.1.1 Product-based regulation

In Canada the product-based approach focused on the novel traits or attributes introduced into the plant, and on all plants or products with new characteristics not previously used in Canada irrespective of how these organisms were developed (Nap et al., 2003).

In USA health and safety laws written prior to the advent of modern biotechnology are used in connection with genetically engineered products. USDA-APHIS (United States Department of Agriculture – Animal and Plant Health Inspection Service) regulates organisms and products that are known or suspected to be plant pests or to pose a plant pest risk, including those that have been altered or produced through genetic engineering. The EPA (Environmental Protection Agency) has jurisdiction over planting and food and feed uses of pesticides engineered into plants. The FDA (Food and Drug Administration) has jurisdiction over food and feed uses of all foods derived from plants. Depending on its intended use, a product may or may not be reviewed by all three regulatory agencies. For example, a food crop developed using genetic engineering to produce a pesticide in its own tissue may be reviewed by all three regulatory agencies. A common example of this type of product is the maize into which scientists have inserted a gene isolated from the soil bacterium, *Bacillus thuringiensis* (Bt). The Bt gene encodes a pesticide and when this gene is inserted into the plant, the plant can produce the Bt-pesticidal substance (FGUSA, 2006). Not all transgenic crops are subject to mandatory risk assessment. If the transformation technology includes use of *Agrobacterium*, which is known to cause crown gall in a wide range of broadleaved plants, then the USDA's regulation of plant pests will be applied. However, if a gene gun is used to transform a non-food crop, and the genetic construct is not a pesticide protein, the crop in question could fall between the jurisdictions of the three authorities (Jaffe, 2004).

1.1.2 Process-based regulation

The EU regulatory framework is a form of process-based regulation. Here any organism that falls under the definition of a GMO is prohibited unless it is approved by the authorities. Process-based regulation and product-based regulation are, however, similar in a number of ways. For example, in both kinds of regulation risk assessment is performed case-by-case and based on the same framework for the risk assessment procedures.

EU regulation of GMOs has been revised several times since the first directive *'on the deliberate release into the environment of genetically modified organisms'* (90/220/EEC) came into force at the beginning of the 1990s. The directive was revised a few years ago in order to incorporate provisions meeting concerns about long-term and indirect effects. The new regulatory framework, laid down in the revised directive '*on the deliberate release into the environment of genetically modified organisms and repealing Council Directive 90/220/EEC*' (2001/18/EC), is currently the legal basis of environmental risk assessment for GMOs. This directive stresses that the risk assessment must "identify and evaluate potential adverse effects of the GMO, either direct or indirect, immediate or delayed, on human health and the environment, which the deliberate release or the placing on the market of GMOs may have". The principal difference between the previous and current directive (i.e. 2001/18/EC) is that it is now possible to include indirect and long-term adverse effects in the risk assessment.

Recently, GMOs which are intended for food or feed uses have been required to be approved according to standards for food and feed safety through the two EU regulations: one '*on food and feed*' (regulation (EC) 1829/2003) and the other '*concerning the traceability and labelling of genetically modified organisms and the traceability of food and feed products produced from genetically modified organisms and amending Directive 2001/18/EC*' (regulation (EC) 1830/2003). A GMO that is not intended for food or feed use may still be approved within the framework of Directive 2001/18/EC. However, no food or feed uses can be approved within the framework of this directive alone.

As stated in (EC) 1830/2003, "genetically modified food and feed should only be authorised for placing on the Community market after a scientific evaluation of the highest possible standard, to be undertaken under the responsibility of the European Food Safety Authority (EFSA), of any risks which they present for human and animal health and, as the case may be, for the environment". This means that risk assessments of food and feed safety are no longer automatically performed at the national level, but will now be centralized at EFSA.

In recently published guidance notes, the EFSA introduces the comparative approach based on the assumption "that traditionally cultivated crops have gained a history of safe use for the normal consumer or animal and the environment. These crops can serve a baseline for the environmental and food/feed safety. To this end the concept of familiarity and substantial equivalence were developed". The guidance notes indicate that "the safety assessment of GMOs consists of two steps, i.e. a comparative analysis to identify differences, followed by an assessment of the environmental and food/feed safety or nutritional impact of the identified differences including both intended and unintended differences".

Two regulations (1829/1830) furthermore state that foods produced from genetically engineered products must be labelled even when the GM material cannot be measured in the final food product. This means, for example, that sugar will be labelled when there is no trace of DNA left in it, but products from farm animals fed with GM feed will, as previously, not be labelled. The threshold for labelling the measurable incorporation of GM material will be a content of 0.9% such material in any ingredient contained in the food product.

At an international level, process-based regulation is apparent in the Cartagena Protocol (http://www.biodiv.org/biosafety), which seeks to protect biological diversity from the potential risks posed by living GMOs. The primary objective of this protocol is to establish an advance, informed agreement procedure for ensuring that countries are provided with the information necessary to make informed decisions before agreeing to the import of such organisms. The protocol requires parties to make decisions on the import of GMOs for intentional introduction into the environment in accordance with scientifically sound risk assessments. The general principles include, among others, the following concepts: risk assessment should be carried out in a scientifically sound and transparent manner; lack of scientific knowledge or scientific consensus should not necessarily be interpreted as indicating a particular level of risk, an absence of risk, or an acceptable risk; risks should be considered in the context of risks posed by the non-modified recipients or parental organisms; and risks should be assessed on a case-by-base basis. The protocol came into force in 2003 and is currently ratified or signed by 129 states — but not, however, by USA (CPBiodiv, 2006).

1.2 Risk analysis

In the regulatory apparatus described above GMOs are being assessed within the framework of risk analysis. The main idea of this framework is that decisions about whether or not to accept the release of GMOs, or about whether or not to allow the GMOs to enter the human food chain, should be based on scientific risk assessment. These risk assessments should be carried out by scientists who are not themselves involved in the decisions about whether or not the GMOs satisfy the requirements of the law.

It is internationally accepted that risk assessment of GMOs should focus on two particular areas, human health and environmental hazard. This universal view emerged from the early discussions of this technology. At a conference in Asilomar, California, in 1976, it was concluded that gene technology offers options for almost unlimited combinations of the genetic material that biology has produced, with possibly far-ranging consequences. This led to systems for biological containment to prevent harm to people plants or animals (Anonymous, 1985). In the USA, the EPA entered into the picture in 1985 as a result of concerns from leading environmentalists (Lund, 1986). However, the general development in society, in response to other technologies, also made it natural to focus on human health and the environment. There is a long tradition of using risk assessment to ensure food safety dating back to 1930s (ACS, 1998), whereas concerns about risks to the environment are of more recent origin, starting in the beginning of the 1970s when national environmental protection agencies and similarly titled authorities were established on both sides of the Atlantic as a direct response to widespread industrial pollution of, and other consequences for, the natural environment and human health (USEPA, 2006). In the EU, common regulation regarding environmental protection was first discussed in 1972 (Anonymous, 2003). The topics of health and the environment were therefore an obvious choice when regulatory procedures were developed for the products of gene technology.

Two concepts appear to have been incorporated into most regulatory frameworks governing GMOs: that of *substantial equivalence*, which is used to assess risk to human health; and that of *familiarity*, which is used in environmental risk assessment.

1.2.1 Food safety risk assessment

The concept of substantial equivalence embodies the idea that existing organisms used as food, or as a source of food, can be used as the basis for comparison when assessing the safety of GMOs to human health (OECD, 1993a). The EU guidance notes specify that the internationally known concept of substantial equivalence should be used to identify differences between GM crop-derived foods or feed and their non-GM counterparts unless no appropriate comparator can be identified, in which case a comprehensive safety and nutritional assessment should be carried out. The food and feed safety assessment should take account of the following issues: potential toxicity and allergenicity, compositional and nutritional characteristics, the influence of processing on the properties of the food or feed, the potential for changes in dietary intake, and potential long-term nutritional impact (EFSA, 2004a).

To date, food/feed safety assessments have not found any substantial differences in composition, nor in the production, of substances that are of concern to human health. In 2002 the Royal Society (2002) in the UK concluded that there seems to be no evidence that the (then) current GM crops were more likely to be harmful to human health or cause allergic reactions than conventional crops. The Society further believed that the health risks associated with use of viral DNA sequences were negligible, and that the consumption of genes introduced into GM plants posed no significant risk to human health — although it also recommended that the regulation of food safety for novel foods should be tightened.

1.2.2 Environmental risk assessment

According to the OECD (1993b) the concept of familiarity enters the environmental risk assessment of GMOs in the following way: a risk assessment should be "based on the characteristics of the organism, the introduced trait, the environment into which the organism is introduced, the interactions between these, and the intended application. Knowledge of and experience with any or all of these provides familiarity which plays an important role". As an example of the practical application of the familiarity principle, the recently revised EU directive on the deliberate release into the environment of genetically modified organisms (EU Directive 2001/18/EC) states, in Annex II, that: "information from releases of similar organisms and organisms with similar traits and their interaction with similar environments can assist in the environmental risk assessment"; and it is indicated that an assessment of both the direct and indirect effect, as well as the immediate and delayed effects, must be included. These effects include: "disease to animals or plants, including toxic and, where appropriate, allergenic effects; effects on the dynamics of populations of species in the receiving environment, and on the genetic diversity of each of these populations altered susceptibility to pathogens facilitating the dissemination of infectious diseases and/or creating new reservoirs or vectors; compromising prophylactic or therapeutic medical, veterinary, or plant protection treatments; and, effects on biogeochemistry (biogeochemical cycles), particularly carbon and nitrogen recycling through changes in soil decomposition of organic material". Adverse effects may occur directly or indirectly through mechanisms which may include: the spread of the GMO(s) in the environment; the transfer of the inserted genetic material to other organisms, or to the same organism, whether genetically modified or not; phenotypic and genetic instability; interactions with other organisms; changes in management, including, where applicable, in agricultural practices. Furthermore, companies may now be required to monitor environmental effects, and an ethical committee may evaluate ethical issues of a general nature.

Risks to the environment are often difficult to assess, since they range from risks relating to the preservation of genetic diversity and the composition of the natural landscape, on the one hand, to agricultural concerns, such as the risk of GM crops causing uncontrollable weeds to develop, on the other (Madsen and Sandøe, 2005).

One problem which has received a lot of attention (see Table 1) is that of gene flow through pollen, where out-crossing GM plants fertilize non-GM crops or wild relatives, thus unintentionally transferring the GM gene to the progeny of other varieties or wild relatives. Ellstrand et al. (1999) concluded that 12 out of the world's 13 most important crops were able to hybridize with wild relatives, so including this mechanism in a risk assessment of GM crops in respect of gene spread is in general an extremely important part of the assessment of environmental risks.

Despite the fact that GM crops are now being assessed for their risks both to the environment and human health, the controversy has not come to an end. There is no simple explanation for this situation. Part of the cause may be that, paradoxically, the very fact that the crops are being assessed seems to indicate that there is something to be worried about. Another cause may be that some of those who are worried are concerned about risks other than those that are being assessed. This is what is going to be looked at in the next section.

2 WHICH RISKS ARE RELEVANT?

Of course, it is a far from simple matter to achieve a unanimous decision, among the various stakeholders, about *what* risks are relevant when it comes to GM crops. And since risk analysis will always be based on some kind of decision about what to look for, and how far and deep to look, there is ample room for disagreements about the extent to which GM crops pose a risk. To make this point we will start by arguing that risk assessment always takes place within a more or less well-defined risk window.

2.1 The risk window

Jensen et al. (2003) provide the following description of scientific risk assessment: such assessment, they say, is based on scientific and technical data, but these data must "fit into a normative framework that is not of scientific nature. This normative framework stems from the decision problem of whether or not a given application for releasing and marketing a particular GMO should be approved. The framing of this decision problem, and the further framing of the questions that the risk assessment is required to answer, depend on a number of value judgements concerning the criteria for approval and, consequently, the risks it is considered relevant to assess. Hence, an environmental risk assessment views the world through a 'risk window' that only makes visible that which has been predefined as relevant risks; and the particular size and structure of the 'risk window' depends on value judgements as to what is considered to be an adverse effect within what is considered the relevant horizon of time and space".

In other words, the scientific risk assessment of a GMO is not a 'mechanistic process', but rather a process dependent on the context (when, where) and the personnel (who) performing the evaluation. In the following we shall try to support this statement by, first, offering a crude analysis of the risks that have been judged relevant by scientists publishing on the risks associated with GM crops; and, secondly, showing that the risk window goes hand in hand with the regulatory requirements. Finally, we shall show that, even within the scientific risk window, there are discrepancies among the experts when it comes to the interpretation of available data.

2.1.1 What risks associated with GM crops have scientists judged relevant?

To obtain a crude indication of the kinds of adverse effect associated with GM crops that scientists have concentrated upon we performed a literature survey using the database Web-of-Science. This database covers approximately 8,500 research journals. We searched for genetically modified plants and risks, models or experiments. This gave a total of 2044 hits, which were then searched in order to determine the number of publications addressing each key issue per year (*Table 19-1*).

These results are summarized graphically in *Figure 19-1*. It can be seen, for example, that vertical gene flow and herbicide resistance both attract a fair amount of scientific attention, whereas soil micro-organisms and fitness of insects living on GM crops seem to have been considered less.

Note that the human health (food safety) issue is a relatively minor concern among the issues identified in the literature search. However, according to the latest Eurobarometer survey, concerns about GM food, in particular, may indicate that Europeans are more concerned about food safety than the environmental impact of agri-food biotechnologies (Gaskell et al., 2003).

2.1.2 The risk window has changed with new regulation

As mentioned previously, the risk window defined in the EU regulation has expanded with the move from the former to the current directive. One of the most plausible explanations for this expansion is that herbicide-resistant (HR) crops are the most abundant modified crop, currently covering 82% of the total area with GM crops (James, 2005).

Table 19-1. Key issues in scientific risk assessment of GM crops. The search included different keywords, which were used to summarize the annual quantity of scientific publications addressing on each key issue

Key issue	Keywords
Gene flow (vertical)	Vertical gene flow, transgene escape, outcrossing, out-crossing, out crossing, cross-pollination, cross pollination, pollen dispersal, gene introgression, transfer of genes, spontaneous hybridization, hybrid, vertical gene transfer
Gene flow (horizontal)	Horizontal gene flow
Non-target effects	Beneficial insect, non target, non-target, non-target, honey bee, tritrophic or multitropic) interaction, insect predator, food web
Fitness GM plants	Competitive ability, fitness, fecundity, not (insects or Lepidoptera)
Fitness insects	Fitness and insect or Lepidoptera
Agricultural management	Co-existence, coexistence, seed purity, integrated pest management
Herbicide resistance	Herbicide (or glyphosate or glufosinate) resistance (or tolerance)
Pest resistance	Insect resistance, pathogen resistance
Soil living microorganisms	Microbial community, bacterial community
Human health	Allergy, allergenic, human health, health risk, food safety, vaccine, immunotoxic
Socioeconomic	Intellectual property right, plant breeders' right, patenting, labelling, regulatory requirements, regulation, precautionary principle
Public	Attitudes, ethic, public concern

For these crops the major environmental risk seems to be connected with herbicide use. In particular, there is a worry that tolerant or resistant weeds and crop volunteers will develop, and that this will lead to environmentally unacceptable increases in herbicide use when farmers increase doses, or mix herbicides having a different mode of action, in order to control weeds. The herbicides used on HR crops are often believed to be less environmentally problematic than those used on similar conventional varieties. However, they are often highly effective in controlling weeds, and thus may leave fields with lower weed numbers than their conventional counterparts. Some people believe this to be an environmental issue in itself, because it may reduce the habitat available to other organisms (Madsen and Sandøe, 2005).

2.1.3 Scientists sometimes have different values – the MON 863 maize example

For a long time now, scientists have agreed that the current GM crops present no risk to human health. In 2004, however, some scientists had doubts about a particular new type of Bt-maize called MON 863 which had been developed to resist attacks on its roots by larvae of the corn root worm. In one of the toxicological tests conducted on this crop, a 90-day feeding study involving rats, the rats reacted differently from the control rats receiving normal feed.

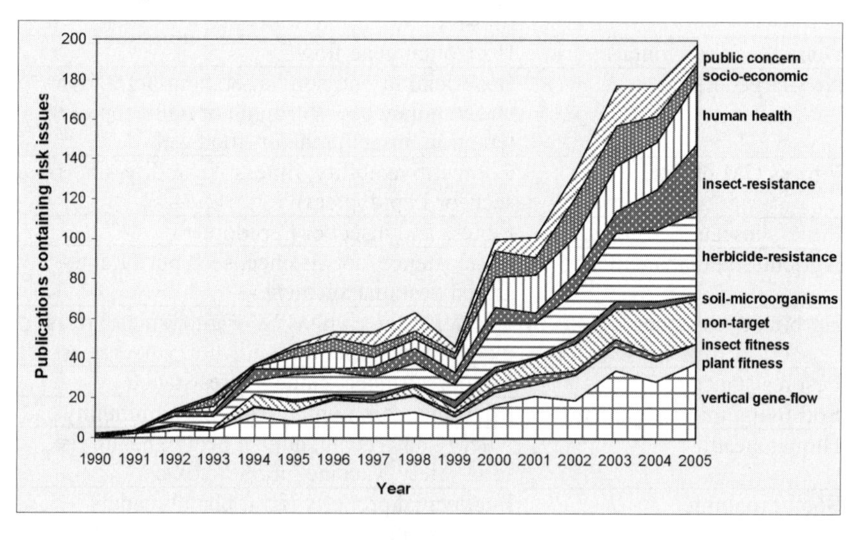

Figure 19-1. Number of references per year incorporating specific risk keywords relating to GM crops.

According to researchers running the study, these differences were not significant, nor of the type to cause concern. The scientific panel on GMOs in EFSA came to a similar conclusion, and EFSA recommended MON 863 for approval by EU politicians (EFSA, 2004b). In the course of the approval process, this recommendation was sent to the national authorities of Member States, and at this point a French scientist expressed doubts about whether it would be safe to approve the maize on the basis of these data. Later the NGO Greenpeace asked to see the report with the data; they were denied access as the report was confidential. Greenpeace did, however, obtain permission for a scientist who was critical of GMOs to read and comment on the report. This scientist concluded that the data could not be interpreted to show that the MON 863 would be a food hazard; but that nor could it be concluded that it was safe either, and that therefore additional experiments were needed (Greenpeace, 2005).

This case and the resulting controversy raise several questions about the risk assessment. First, how can an expert panel unanimously agree that the data did not give rise to a genuine concern when several scientists beyond the panel were to become concerned? A plausible explanation is that panel members had similar features from the beginning in order to be appointed for this job. Second, the members may influence each other by collectively drawing conclusions about the scientific data put forward. Third, the scientist disagreed about the quantity of experimental data needed to make an informed decision; and fourth, this case raised questions about transparency of the process and information on which the decision was made; in particular, the critics here were not allowed full access to the report in question.

3 CONCERNS BEYOND RISK ASSESSMENT

Section 1 described the scientific-technical frameworks set up by various authorities to protect human health and the environment. However, it is a second aim of regulation, in general, to meet public concerns about uses of technology, and thus to ensure that the public will trust that the authorities have the technological developments under control. However, both large population surveys within the EU and focus group interviews in Denmark make it clear that past regulatory approaches have not properly dealt with people's worries about GMOs (Lassen et al., 2002; Gaskell et al., 2003).

The 2002 Eurobarometer survey showed that, in general, and after a decade of decline, optimism about biotechnology had increased to levels last seen in the early 1990s. For GM crops and food specifically, support seems to have stabilised across Europe between 1999 and 2002. However, in 2002 the majority of Europeans still did not support GM foods. Such foods were not perceived to be useful and were felt to present risks to society (Gaskell et al., 2003). This suggests that lack of usefulness is one of the main concerns for many Europeans. In the following section, therefore, we try to unravel the underlying arguments about usefulness.

3.1 Usefulness

A GM crop can be beneficial, or have a positive impact, in at least two distinct ways: by being profitable for the producer, or by fulfilling important societal needs (Madsen et al., 2003). Global figures show that 90 million ha were sown with GM crops in 2005, with approximately 38% of these in the developing world. In view of this, it can hardly be denied that GM crops benefit the farmers growing them in developed and developing countries around the world (James, 2005). It has also been estimated that if these crops were grown in the EU, there would be significant yield increases, savings for growers, and pesticide use reductions (Gianessi et al., 2003).

However, when members of the general public insist that GM crops must be useful, they typically seem to have the second definition of usefulness in mind: that GM crops must fulfil important societal needs. A GM crop can fulfil such needs in several ways: 1) by giving us more healthy food, 2) by mitigating the environmental impact of agriculture, 3) by producing raw materials which at present require costly industrial processing, or 4) by improving the situation in developing countries and feeding a rising world population. HR crops have been developed chiefly for agronomic benefits, and therefore the usefulness of these crops has not been obvious to the general public in Europe (Lassen et al., 2002). The latest Eurobarometer survey asked respondents if they would buy GM foods offering particular benefits. The most persuasive reason for buying a GM food product was that it contained reduced pesticide residue (approximately 40% tended to agree). This was followed by environmental benefit. Less than 25% respondents would buy GM food just because it was cheaper. The report comments that there may well be a difference between a person's response as a citizen and as a consumer — if these crops were actually on the market, more people would in practice probably buy the GM product because it had a lower price (assuming it did). However, in connection with every one of the benefits set out, the majority of respondents said that they would not buy GM food (Gaskell et al., 2003).

3.2 Other socioeconomic issues

In the developing world, GM crops may present a direct socioeconomic dilemma if they are introduced without prior public acceptance in importing countries such as those in the EU. This was realised in 2002 when Zambia and Zimbabwe rejected maize with GM content as food aid. Zambia's Vice President, Enoch Kavindele, explained to UN aid workers that their decision to reject some of these foods was made in response to fears that they would lose the European market if they started growing GM foods (Meron, 2002). In Zimbabwe the government ended up grinding the maize grains, thus ensuring that farmers could not use the maize seed (Biotik, 2002). In Zambia, however, it seems that local people later broke into the stores and stole the GM maize (Wendo, 2003).

A more diffuse socioeconomic issue which has often been discussed in developed countries concerns seed companies and the agrochemical industry. Large businesses like these, which are regarded as having an invidious association with (or as actually being) monopolies, are often perceived as the major driving force in the development of HR crops. Many people resent developments towards monopolisation. Instead they wish to protect the smaller plant-breeding companies, and to secure influence over development at community level (Madsen and Sandøe, 2001). Tied in with this attitude is concern about the 'patenting of life', as it is often put. A patent gives its holder exclusive controlling rights over an innovation for a substantial number of years, while society gets access to the information in the patent for further research. Society may benefit from investments being made within the area of biotechnology, but many people are alarmed at the idea that private companies will have exclusive rights over the utilisation of nature (Madsen and Sandøe, 2005).

3.3 The consumer's right to choose – co-existence

Another issue overlooked within the framework of risk analysis is the effect that the cultivation of GM crops may have on consumer choice.

Even with strict regulations governing the cultivation and segregation of GM and non-GM crops, trace-levels of GM material (via gene flow, the contamination of seed lots and so on) cannot be avoided for some products; and some people perceive this as a violation of freedom of choice.

There is a specific issue here about impacts on *organic* production, since gene-flow from GM crops may undermine the claim of an organic producer to be GM-free. This issue has given rise to strong reactions from organic producers and consumer organisations.

Responding to this problem, the EU commission recommended in 2003 that Member States issue guidelines on the development of national strategies and best practice to ensure the co-existence of genetically modified crops with conventional and organic farming. 'Co-existence' here refers to the ability of farmers to make a practical choice between conventional, organic and GM-crop production — a choice meeting the legal requirements for labelling and/or purity standards. The conditions under which European farmers work are extremely diverse. For this reason the Commission, expressed a preference for an approach that would leave it up to Member States to develop and implement management measures for co-existence. The role of the Commission would include gathering and coordinating relevant information based on on-going studies at community and national level, and offering advice and issuing guidelines which may assist Member States in establishing best practice for co-existence (Commission recommendation of 23 July 2003, 2003/556/EC).

So far, only Germany, Denmark, Italy and five regions of Austria have laws regulating GMO cultivation. The main Dutch farming organisations have reached a voluntary agreement; another eight countries are drafting legislation, and in this process Spain, Luxembourg, Portugal, Poland and the Czech Republic are most advanced (Smith, 2005).

The Danish regulation, which was the first to be introduced, stipulates certain kinds of crop cultivation and management practice. Thus growers of GM crops must follow rules on the distance between fields grown with GM crops and neighbouring conventional or organic fields; neighbours must be informed if they have fields within a certain distance, depending on the GM crop, in speech; farmers must attend a course in the cultivation and management of GM crops; and information about the whereabouts of these fields must be available to the public (Danish law about growing of genetically modified crops, LOV nr 436 af 09/06/2004).

3.4 Other moral concerns

The issues presented centre on the consequences or impact of agricultural uses of gene technology. However, it is clear that some people worry that the technology as such is unnatural.

In the 1999 Eurobarometer survey, the following two statements were presented: "Even if GM food has advantages, it is fundamentally unnatural", and "GM foods threatens the natural order of things." In response, 45% and 38%, respectively, strongly agreed with these statements, and 27% and 31% somewhat agreed (INRA, 2000). This indicates that perceived naturalness is an important factor in the public's assessment of GM foods — and thus also in their assessment of GM plants. During a series of Danish focus group interviews, issues connected with nature and naturalness were spontaneously taken up within all of the groups, again suggesting that concerns about the violation of nature play an important role in the discussions about genetic engineering.

To some people, the terms 'nature' and 'naturalness' appear to be connected with serious moral concern about departures from what is natural. This refers to the perception that nature itself embodies a guiding principle, or incorporates an inherent order of things, that reaches beyond the influence of mankind (Lassen et al., 2002). Midgley (2000) has suggested that this perception may be grounded in our traditional understanding of nature. In this understanding, each species is represented as having been carefully optimized to fit its ecological niche through the process of natural selection. In myths, moreover, transgressions across the boundaries of species have lead to monsters. From this perspective, gene technology violates the sanctity of species, and this admittedly imprecise concept of sanctity is fundamental in our current understanding of the world. The 'natural order' argument thus refers to a *moral* critique reaching beyond the scientific evaluation of the risks associated with genetic modification.

However, for those making use of the "nature as safety mechanism" argument, the notion of unnaturalness is clearly a proxy for criticism of, or doubt about, the effectiveness of existing risk assessment procedures. They are concerned about potential risks to the environment arising from the combination of hereditary material moving across natural boundaries and the limits of scientific foresight of long-term consequences (Madsen et al., 2002). These people appear to link concern about GM crops being unnatural to risk issues.

Moral questions can be difficult to discuss and reach consensus on in society. In part this is because it is difficult to find common ground on which to base the discussion. It is also because people balance such concerns differently. Nevertheless, frameworks to achieve clarity and address value questions have been formulated, and in the following section we present one such approach.

3.4.1 Ethical criteria

In 2002 proposals about the overall argumentative framework within which the various concerns could be balanced against each other were made in a Danish government report on ethics and genetic engineering. This report was based on a debate book from the BioTIK group, i.e. a group of experts from natural science, sociology and philosophy brought together by the Minister of Economic and Business Affairs. The framework consists of a list of four 'ethical criteria'. These criteria may be interpreted in two ways: either as necessary conditions to be fulfilled (criteria proper), or as a set of factors to be considered in the risk assessment and decision process before a final decision is made. The proposed criteria are: (A) the technology should be employed for the economic and, most important, qualitative benefit of humans, society and nature; (b) respect must be shown for the autonomy, dignity, integrity and vulnerability of living beings; (c) the burdens and benefits associated with the technology must be distributed fairly; and (d) decisions to use the technology must be taken with openness and respect for the individual human being's right to self-determination (Biotik-Gruppen, 1999).

Neither this framework nor any other suggested approach to the handling of value questions — e.g. Mepham (1996), Carr and Levidow (2000), Madsen et al. (2002) — have yet been put into use, although ethical questions are gradually appearing within the regulatory framework. Thus, for example, one EU directive, 2001/18/EC, states that the EU commission must report annually on any ethical issues rose by GMOs and may recommend amendments to the directive 2001/18/EC. However, a recommendation from an ethical committee cannot stop, delay or change the procedure for approval of a GMO and will, therefore, have a limited impact on the approval process of any specific GMO.

4 CONCLUSIONS

The aim of this chapter has been to review the concerns to which GM crops give rise. It has been argued that the framework of risk analysis on which most current legislation is based assumes certain values, and in particular the desire to protect human health and the environment. In a way, these values are uncontroversial, which is supported by the fact that the protection of human health and environment appear to be universally adopted in regulatory apparatus in many different parts of the world, although of course different countries have adopted different approaches to GMO regulation. However, in practice the framework seems to rely on two assumptions that are far from uncontroversial. The first is that the risk assessments that underlie the current regulation of GM crops are sufficient. The second is that risks to human health and the environment are the only significant concerns to which GM crops give rise. (Obviously the second of these assumptions tends to fortify the first.) In this chapter we have argued that there is some genuine discussion to be had about these two claims. The debate about GM crops is a dialogue not only about the risks defined by the regulations, but also, ideally, about a broader range of concerns about usefulness, wider risks to society and distinctively ethical concerns.

5 ACKNOWLEDGEMENTS

The authors wish to thank Danida, Denmark for financial support.

6 REFERENCES

[ACS] American Chemical Society (1998) Understanding risk analysis. A short guide for health, safety, and environmental policy making. Internet edition

Anonymous (1985) Indenrigsministeriets gensplejsningsudvalg. Genteknologi og sikkerhed. Betænkning nr. 1043, København 1985

Anonymous (2003) European Parliament Fact Sheets. http://www.europarl.eu.int/factsheets/4_9_1_en.htm.

Biotik (2002) Zimbabwe modtager alligevel GMO nødhjælp fra USA. Summary after *The Washington Post* 10.08.2002. http://www.biotik.dk/nyheder/alle/uge_200233/15828/ (accessed 7 May 2006).

Biotik-Gruppen (1999) De genteknologiske valg. Et debatoplæg udarbejdet af BioTIK-gruppen. Erhvervsministeriet. Statens Information, København K, Denmark, (in Danish).

Carr S, Levidow L (2000) Exploring the links between science, risk, uncertainty, and ethics in regulatory controversies about genetically modified crops. J Agri Enviro Ethics 12: 29-39

[CPBiodiv] Cartagena Protocol on Biosafety (2006) Cartagena Protocol on Biosafety. http://www.biodiv.org/biosafety/default.asp (accessed 16 February 2006).

EFSA (2004a) Guidance document of the Scientific Panel on Genetically Modified Organisms for the risk assessment of genetically modified plants and derived food and feed. the EFSA Journal 99: 1-94

EFSA (2004b) Statement of the GMO Panel on an evaluation of the 13-week rat feeding study on MON 863 maize, submitted by the German authorities to the European Commission, http://www.efsa.eu.int/science/gmo/statements/666_en.html (accessed May 2006)

Ellstrand NC, Prentice HC, Hancock JF (1999) Gene flow and introgression from domesticated plants into their wild relatives. Annu Rev Ecol Syst 30: 539-563

[FGUSA] Federal Government of the United States of America (2006) Federal Government of the United States of America [FGUSA] United States regulatory agencies Unified Biotechnology Website. http://usbiotechreg.nbii.gov/ (accessed 16 February 2006)

Gaskell G, Allum N, Stares S (2003) Europeans and biotechnology in 2002. Eurobarometer 58,0 (2nd Edition: March 21st 2003), A report to the EC Directorate General for Research from the project 'Life Sciences in European Society' QLG7CT-1999-00286. (http://europa.eu.int/comm/public_opinion/archives/ebs/ebs_177_en.pdf, accessed 28 March 2006)

Gianessi L, Sankula S, Reigner N (2003) Plant Biotechnology: Potential Impact for improving pest management in European agriculture. A summary of nine case studies, December 2003. The National Center for Food and Agricultural Policy, Washington, DC

Greenpeace (2005) June 2005 – Bacground briefing: MON863. Monsanto's GM corn: Unfit for rats, ufit for humans. http://eu.greenpeace.org/downloads/gmo/Mon863June05.pdf (accesssed May 2006)

INRA (2000) Eurobarometer 52.1: The Europeans and biotechnology. March 15, 2000. (http://europa.eu.int/comm/public_opinion/archives/eb_special.htm, accessed 8 March 04)

Jaffe G (2004) Regulating transgenic crops: a comparative analysis of different regulatory processes. Transgenic Res 13: 5-19

James C (2005) Preview, Global status of commercialized biotech/GM crops: 2005. ISAAA Briefs 34. ISAAA, Ithaca, NY

Jensen KK, Gamborg C, Madsen KH, Jorgensen RB, von Krauss MK, Folker AP, Sandoe P (2003) Making the EU "risk window" transparent: the normative foundations of the environmental risk assessment of GMOs. Environ Biosafety Res 2: 161-171

Lassen J, Madsen KH, Sandøe P (2002) Ethics and genetic engineering – lessons to be learned from GM foods. Bioprocess Biosystems Engineering 24: 263-271

Lund E (1986) Gensplejsning. Naryana Press, Denmark, Hekla

Madsen KH, Holm PB, Lassen J, Sandøe P (2002) Ranking genetically modified plants according to familiarity. J Agri Environ Ethics 15: 267-278

Madsen KH, Sandoe P (2005) Ethical reflections on herbicide-resistant crops. Pest Manag Sci 61: 318-325

Madsen KH, Sandøe P (2001) Herbicide resistant sugar beets – What is the problem? J Agri Enviro Ethics 14: 161-168

Madsen KH, Sandøe P, Lassen J (2003) Genetically modified crops: A US farmer's versus an EU citizen's point of view. Acta Agriculturae Scandinavica Sect B, Soil and Plant Science Supplementum I: 60-67

Mepham B (1996) Ethical analysis of food biotechnologies: an evaluative framework. *In* B Mepham, ed, Food Ethics: Professional Ethics. Routledge, New York, pp 101-119

Meron TM (2002) Africa Bites the Bullet on Genetically Modified Food Aid. Worldpress Org. http://www.worldpress.org/Africa/737.cfm#down (accessed 12 Dec. 05)

Midgley M (2000) Biotechnology and monstrosity: Why we should pay attention to the 'Yuk factor', Hastings Center Report 30 (5), pp 7-15

Nap JPJ, Metz PL, Escaler M, Conner AJ (2003) The release of genetically modified crops into the environment. Part I. Overview of current status and regulations. Plant J 33: 1-18

OECD (1993a) Safety evaluation of foods derived by modern biotechnology, concepts and principles. *In*. Organisation for Economic Co-operation and Development, Paris, France

OECD (1993b) Safety considerations for Biotechnology: Scale-up of crop plants. Organisation for Economic Co-operation and Development, Paris, France.

Smith J (2005) Lawmaking on Genetic (GMO) Food is Minefield For EU. GENET archive. http://www.gene.ch/genet/2005/Mar/msg00013.html (assessed 12 Dec. 05)

Royal Society (2002) Genetically modified plants for food use and human health - an update. Policy document 4/02, February 2002, 20 p (2002) (www.royalsoc.as.uk)

USEPA US Environmental Protection Agency (2006) US Environmental Protection Agency [USEPA]. http://www.epa.gov/earthday/history.htm (accessed 16 February 2006)

Wendo C (2003) Uganda tries to learn from Zambia's GM food controversy. The Lancet 361, February 8, 500

Chapter 20

AGROBACTERIUM-MEDIATED GENE TRANSFER: A LAWYER'S PERSPECTIVE

Carol Nottenburg[1] and Carolina Roa Rodríguez[2]

[1]Cougar Patent Law, Seattle, WA, USA; [2]Australian National University, Canberra, ACT, Australia

Abstract. Whether or not you agree with patent protection for *Agrobacterium*-mediated transformation technology or for other basic platform technologies, "the times, they are a-changing". In the United States, patents are awarded for many types of biotechnology inventions, including nucleic acid sequences, bacterium containing a vector construct, transgenic plants and methods of making transgenic plants. Both companies and non-profit institutes are affected by such patents. Here, some of the impacts of patents are discussed followed by a mini-primer on key points about patents and patent documents. In the final section, we present a patent landscape of *Agrobacterium*-mediated transformation of plants and discuss a number of key patents impacting research and development.

1 INTRODUCTION -WHY SHOULD A SCIENTIST CARE ABOUT A LAWYER'S VIEW OF *AGROBACTERIUM*?

The legal system is often viewed by the public as inaccessible and incomprehensible; in addition, scientists may feel as if patents and other legalities are being foisted on them. Readers of this book may well just want to get on with their research and wish that patents would just "go away"; some may even resent or ignore patents for a variety of reasons. While we empathize, the reality is that patents are not only here to stay, but are increasing in number (*Figure 20-1*) and importance for basic science research and researchers – for better or for worse.

As a scientist in the 1980s and early 1990s, one of us - Nottenburg - knew precious little about patents. Because her appointment was in a non-profit research institution, patents were not on any list of "things to worry about". When confronted with a material transfer agreement from a company for obtaining a reagent, she read it but with a fair bit of uncertainty and lack of understanding. Luckily, her institution had an Office of Technology Transfer (OTT) that could help, especially in negotiating away a clause that endangered independence in controlling the project's outcomes.

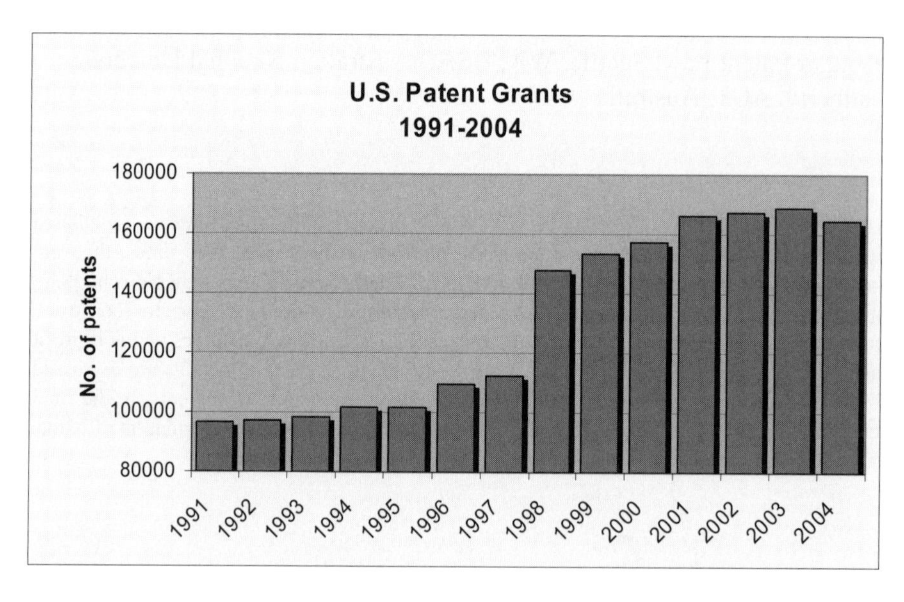

Figure 20-1. U.S. Patent Grants by Year

Even more luckily, a colleague told her about OTT because the administration hadn't even alerted the faculty of its existence, let alone its services.

Since then, a number of substantial and widespread changes has taken place, notably increased patenting and commercialization of biotechnologies and a bigger presence of OTTs in universities and research institutes. Whereas, for most of Nottenburg's scientific career the world of patents and other intellectual properties (IP) in biotechnology belonged primarily to the realm of private sector, now public and private non-profit institutions are increasingly involved in this legal world. For example, during the five-year period from 1981-1985, a mere 0.59 percent (1,887) of United States patents were granted to inventors who assigned to an entity whose name contained "University", subsequent five-year periods saw the percentages (and number) of patents steadily climb to 2.15 percent (13,940) for the latest period from 1996-2000 (Nottenburg et al., 2002).

What hasn't substantially changed though is a scientist's knowledge and understanding of patents and related law areas. Knowledge and understanding are not to be confused with general awareness of patents or the use of legal "buzzwords", which certainly has substantially increased.

> "You can know the name of a bird in all the languages of the world, but when you're finished, you'll know absolutely nothing whatever about the bird... So let's look at the bird and see what its doing – that's what counts. I learned very early the difference between knowing the name of something and knowing something."
>
> – Richard Feynman

Our advice to scientists is to learn enough about patents to be able to use the system to your advantage. This doesn't mean that you have to become one of "them", but by participating in the patent system you become empowered to use patents as a tool to help accomplish your goals.

So why should scientists know something of patents? Because more emphasis is being placed on commercial realization of scientific research, because patents are a type of scientific literature, and because infringement can be costly.

1.1 Commercialization of research results

The emphasis on commercial realization of scientific research is readily apparent in both for-profit and non-profit settings. The terms "for-profit" and "non-profit" are used to indicate the main function of the referred-to entity. While "for-profit" institutions are almost always private companies, "non-profit" institutions include both public entities (e.g., state universities) and private entities (e.g., research institutions). In many fields, including agriculture and health sciences, patents play a key role in R&D and product development. Since the Bayh-Dole Act was enacted in 1980, non-profit organizations have increasingly pursued patents and commercial partners for exploiting their patents. Specifically, the Bayh-Dole Act ceded federal interest in ownership and licensing of inventions to institutions receiving federal funding. Analyses have documented that Bayh-Dole was a major contributor to the substantial growth of technology industries, especially in biotechnologies (Mireles 2004; Press and Washburn 2000; Wolf and Zilberman 2001). According to the Association of University Technology Managers (AUTM), from 1980 to 2004, more than 4500 companies total (with 462 companies in 2004 alone) have spun out from research in U.S. universities, hospitals and research institutes (AUTM 2005). Clearly, the Bayh-Dole Act has irrevocably altered the face of universities and moved them into a closer relationship with the corporate world. And part and parcel of this closer relationship are patents. The AUTM Surveys report that since 1993, institutions participating in the surveys have received a total of about 34,500 U.S. patents. In 2004 alone, 183 institutions filed over 10,500 new patent applications, a 16.7 percent increase in filings over 2003 (AUTM 2005).

In addition, some granting agencies now require the grant recipient to provide an analysis of the patents that may affect commercialization of the research subject matter. For example, several of the Research and Development Corporations in Australia (non-profit corporations funded under a Federal Act) request such assurances. For the agricultural R&D Corporations in Australia part of their mandate is to invest in research for the greatest benefit to their stakeholders: the farmers and the Commonwealth. As an exporter of agriculture products, Australian industries must be sensitive not only to domestic intellectual property but also to that of other countries. In Nottenburg's experience at CAMBIA (an Australian non-profit research institute focused on agricultural biotechnology and patent resources), grant applications from two different R&D Corporations required disclosure of any patents that could affect the ability of research results to be commercialized. As well, for certain of its grants, the Gates Foundation also requires due diligence – an investigation into patents that might limit the ability to practice the expected research outcomes.

A third avenue of pressure driving non-profits to patenting is the possibility of financial returns from research back to the organization. Pressure is driven at least in part because of the on-going issues regarding federal funding for science – a mere 2% increase for National Institutes of Health (NIH) from FY 04 to FY 06 (NIH Office of Budget) and no change for National Science Foundation (Meeks 2005, NSF 2005) from FY 04 to FY 06. A continuing climate of funding uncertainty may motivate an increasing emphasis of non-profit organizations to seek financial returns from research, including reimbursement for legal fees (AUTM 2005).

Non-profit institutions that obtain patents typically realize a return on this investment by selling (or licensing) its patent rights to others, such as existing commercial entities, new spin-off companies, or other research institutions. Some highly publicized patents that have been licensed widely have generated a very substantial income for the host institution. For example, Stanford University received over $254 million from licensing the now-expired Boyer-Cohen patent claiming basic recombinant DNA technology; this single patent generated over half of its total licensing income (Feldman et al., 2005).

Furthermore, federal funding for research is shifting to applied research. In a 10-year period from 1993 to 2003, funding for basic research in life sciences (including agriculture) increased 2.4-fold while funding for applied research increased 3.5-fold (NSF SRS 2004). Applied research is necessarily more commercially oriented, which goes hand-in-hand with filing patent applications.

For commercial entities, patents and determining freedom to operate is not a new story – it's a necessary part of a business plan. Large companies are usually patent-savvy, while small to medium sized businesses may lack the necessary legal skills and often don't avail themselves of publicly available patent information (EPO 2003). Most telling is that larger companies were more likely to have access to patent databases or services. Yet, more than 80% of all companies consider the information in patents to be important or very important. Thus, not only scientists, but also science and business administrators in both public and private sectors, will be well-served to have at least some practical understanding of patents.

1.2 Advantages for scientific research

Scientific research can benefit from patents, apart from providing protection for inventions. An often overlooked benefit is that patent documents comprise a substantial body of scientific literature. Companies especially do not always publish in traditional scientific journals. For example, a search of United States granted patents since 1976 for those owned by

"Monsanto" and containing the term "*Agrobacterium*" yielded 208 results, while the same search terms applied to PubMed yielded only 17 results. Furthermore, patent documents contain very detailed descriptions of experimental methods and materials, whereas, journal articles may be very skimpy on the details. Rather than trying to deduce for example the structure of a particular vector construct, look for the corresponding patent document, and with luck, details of experiments and materials will be there. (but, *cf* Dam 1999, arguing that patent disclosure is often insufficient to practice the invention, and Lichtman et al., 2000).

Unfortunately, recognition and comprehension of the importance of patents do not always coincide; many factors discourage use of patent information, including difficulty of access, amount of time involved, difficulty in reading documents, and inability to fully understand the information.

1.3 The myth of the "experimental use exception"

The right to use a patented invention for research is a concern in both non-profit and commercial settings. Many, if not most, university scientists assume that patent law does not apply to their basic research. Several factors seem to reinforce their perception, including academic culture, which typically is removed from legal concerns, and lack of guidance or even basic information and education from host institutions. Indeed, review of the patent or intellectual property policy documents of a number of United States universities revealed that while they all provided information about invention disclosures, patenting process, revenue sharing mechanisms and similar subjects, none of them discussed infringement of other's patents (or copyrights). Thus, academic researchers are often shocked to discover that, except for some very limited statutory exemptions that rarely apply to them, no general research exemption exists in the United States for using other people's patented technologies. In contrast to patents, the U.S. Plant Variety Protection Act (PVPA, 7 U.S.C. §2321 et seq.) provides for a research exemption. Under the Act, a protected variety may be used and reproduced in plant breeding or other bona fide research. The UPOV Convention (an international agreement that governs plant varieties), has a similar exemption.

In developed countries that have implemented an "experimental use" exemption for patents, such as Germany and Great Britain, the exempt use is for experimenting *on* the invention. Exceptions given for *using* the invention as intended in non-commercial settings are more likely to be found in developing countries (Commission on Intellectual Property Rights 2002). The difference between these types of experimental use are illustrated by two scenarios: "*experimental use of the invention*" – performing PCR (polymerase chain reaction) to verify presence or absence of a specific DNA sequence; and "*experimenting on the invention*" – testing different additives in PCR reactions to find one that markedly improves amplification.

The U.S. Congress has the authority to legislate a general research use exemption, but so far has only enacted a few very narrow exemptions in the medical field. In 1984, the Drug Price and Patent Term Restoration Act allowed drug companies, particularly those that market generic drugs, to proceed with pre-market approval testing of competing drugs or veterinary biological products during the life of the relevant patent. The use of patented herbicides to test new herbicide-tolerant cultivars, however, would not fall within this exemption. Without this exemption, pharmaceutical companies enjoyed an extension of exclusivity after patent expiration while generic manufacturers collected the data necessary for FDA approval. Implementation of the Act was meant to ensure that consumers received the advantages of generic drug prices. In addition, 35 U.S.C. § 287 (c)(1) grants an exemption to medical practitioners performing a medical or surgical procedure that would otherwise be an infringement. The exemption only applies to methods of treating human patients and does not apply to medical instruments or their use.

Except for the few legislated exceptions, the United States has a very proscriptive research use exception, so proscriptive that for all practical purposes, there is no exception for basic research in any kind of setting (e.g., university, non-profit research institute, commercial entity). In the course of the development of patent law in the United States, courts have repeatedly refused to find a general research or experimental use exemption, even in infringement actions against the United States Government, where there is a clear absence of a profit motive for using the patented inventions (*Pitcairn v. United States*).

In this landscape, the researcher at a university or other non-profit organization who uses a patented method or composition research is infringing, even if used without any overt profit motive. This point was hammered home in *Madey v. Duke University*, in which use of a patented laser device for research, academic, or experimental purposes at a non-profit university was held as an infringing use. The decision followed a line of earlier cases in which using an invention for furthering legitimate business purposes is infringing conduct. With respect to a university, its "legitimate business objectives include educating and enlightening students and faculty participating in the projects" and serve "to increase the status of the institution and lure lucrative research grants, students and faculty." Thus, sounded the death knell of an experimental use defense – until such time, if ever, that Congress enacts an exemption.

Do not panic, however. There appears to be a *de facto* exemption (an exemption based on reality rather than based on law) in the United States. The number of patent suits filed in United States District Courts against non-profit organizations is extremely few, so few that Congress does not believe that universities suffer a high or actual risk. In 1990, the House Committee on the Judiciary, which has jurisdiction over patent matters, recommended a broad research exemption (House of Rep. 1990), but in opposing the exemption, one Representative questioned the need for the exemption, challenging universities to come forward to show how the existing patent law was harming them (*ibid*). We assume that the evidence simply was not there because the exemption was never passed. Moreover, in the United States, the 11th Amendment of the Constitution protects State institutions from being sued in federal courts unless they consent to the suit or implicitly waive their immunity. Although Congress has attempted several times to make a law that removes State immunity from patent infringement actions, none of the Acts has passed muster in the U.S. Supreme Court (*Florida Prepaid 1999*). Although eventually Congress is likely to succeed in passing legislation that will abrogate States' rights and withstand the scrutiny of the Supreme Court.

Even in the absence of a research exemption, non-profit organizations likely have only a very minor risk of patent infringement exposure. It would be poor public relations for a patentee company to sue a non-profit organization for infringement, and it is likely that a jury would sympathize with the defendant. In addition, the type of remedy imposed is unlikely to be severe from the institute's point of view. In *Roche Products v. Bolar Pharmaceutical Co.*, a key experimental use exemption case, the patent owner urged that the data generated during the infringing activity be confiscated and destroyed. The Court however, expressed a preference for monetary damages and admonished that injunctions are an equitable remedy and by no means a mandatory remedy. Although difficult to predict with certainty, damages owed by a non-profit infringer would likely be limited, possibly to the cost of a license, as use of the technology within a non-profit organization would not generally cause a company to lose profits. Thus, weighed against the significant expenses of litigation, a corporation is unlikely to pursue such a suit except for very significant matters.

Furthermore, patentee corporations stand to gain some advantages by having researchers do some of their research and widely adopt technologies that the corporation can then license. For example, CAMBIA owns rights to β-glucuronidase (GUS), which was widely used by researchers in non-profit organizations who ultimately moved to corporations and continued using GUS. While CAMBIA grants non-commercial research in non-profit settings a cost-free license, fees are charged for using GUS in commercial research.

Thus, although there is no research exemption for non-profit institutions, it is unlikely that infringement suits will be filed against universities and research institutes in cases where the nature of the research is clearly non-commercial.

1.4 Freedom-to-commercialize and anti-commons problems

Generally, the main concern voiced by scientists and other inventors is whether their great idea is patentable. For someone eating, breathing, and dreaming about the big breakthrough, patenting the invention assumes prime importance. Sad to say, almost always the most important issue is not whether the idea is patentable, but can you practice your own invention?

It often surprises people that someone else's patent may be more important than their own. That importance follows from the nature of patent rights, which are a grant to *exclude* others from making, using, selling, offering for sale or importing the patented invention. Or when the invention is a method, the rights additionally allow the patent holder to exclude others importing at least the product obtained directly by that process for the purposes of using, selling, and offering for sale. (Patent rights are set forth in 35 U.S.C. §271 and article 28 of TRIPs (Agreement on Trade-related Aspects of Intellectual Property Rights), which requires WTO member countries to provide essentially these same rights.) Note that none of these rights actually grants the patent holder a right to practice her own invention.

How is that possible? A trivial example illustrates this conundrum.

Imagine that Company A owns a patent with this claim: *A pencil comprising No. 2 lead.* The pencil that is made and sold is shown to the left in

Figure 20-2. Pencils

Figure 20-2. Some time after this patent issues, another patent issues to Company B with the claim: *A pencil comprising (i) No. 2 lead and (ii) an eraser attached to one end.* An example of such a pencil is shown on the right in *Figure 20-1.* Company B cannot make, sell, etc. its pencil without permission from Company A because the claim in Company A's patent encompasses pencils with or without erasers. (This statement is true when a claim uses the term "comprising", which means that the elements listed in the claim are the minimum elements required. Additional, unnamed elements (e.g., eraser) are covered under the patent claim.) The reverse however, is not true: Company A does not need permission from Company B as long as A's pencils do not have both No. 2 lead and an eraser. Therefore, A's patent dominates B's patent. The result is that Company B has no freedom-to-commercialize without permission from Company A.

The thicket of patents that capture various aspects of a technology and that are required to practice the technology without infringing is sometimes referred to as "anticommons" (Heller and Eisenberg 1998). Under an anticommons scenario, resources (i.e., technologies) are prone to underuse because the technology is vested in the hands of multiple owners, each of whom have the right to exclude others at the same time that none has consolidated a right to use. This theory is by no means proven. A lively debate centers on whether or not there is an anticommons and, if so, its impact on research and development (Epstein and Kuhlik, 2004; Lichtman 2006; Mireles 2004; Stern and Murray 2005). In any case, the debate is about overall patenting trends and does not look at or define the patent thicket and extent of anticommons for a particular technology. Delineating the patent thicket for a particular technology requires exhaustive patent searching and claim analysis of the key patents. In the next section, we present a primer on patents to lay a foundation for the patent landscape analysis.

2 SOME BASICS ABOUT PATENTS

Remember that the patent owner's right is exclusionary: she may exclude others from making, using, selling, offering to sell, and importing the patented invention and importing a product made by a process that is patented in the importing country. To determine if someone is infringing a patent, the allegedly infringing product is compared to the claims.

2.1 Claims define the "metes and bounds" of protection

The *claims* are the most important part of a patent. Not the title, not the text, not the examples, and not the drawings. It is the claims that define the boundaries of the patent owner's rights. Don't fall into the trap of concluding that the title or the abstract or the general description found in the text of the patent indicates what is patented. For example, United States Patent No. 6,074,877 is titled "Process for transforming monocotyledonous plants". From the title, you might conclude that these patent owners have protected a method for transforming all monocot plants. The claims, however, refer only to transformation of cereal plants, and furthermore an embryogenic callus must be wounded first or treated with an enzyme that degrades cell walls prior to transferring DNA into the cells with *Agrobacterium*. A bit different than what the title implied.

In order to avoid infringement, the meaning of claims must be determined. While the purpose of claims is to clearly demarcate the extent of the patent owner's rights, the meaning and scope of the claimed invention are not always clear from just reading the claim. Proper claim interpretation is achieved by reading the claims in the context of the specification and in the context of the "prosecution history" (the back and forth negotiations of the claim language between the patent applicant and the patent office). In this case above, for example, claim 1 recites:

> A process for the stable integration of a DNA, comprising a gene that is functional in a cell of a cereal plant, wherein said DNA is integrated into the nuclear genome of said cereal plant, said process comprising the steps of:
>
> (a) providing a compact embryogenic callus of said cereal plant;
>
> (b) wounding said compact embryogenic callus or treating said compact embryogenic callus with a cell wall degrading enzyme for a period of time so as not to cause a complete disruption of tissues,

and transferring said DNA into the nuclear genome of a cell in said compact embryogenic callus by means of *Agrobacterium*-mediated transformation to generate a transformed cell; and

(c) regenerating a transformed cereal plant from said transformed cell.

Upon first reading this claim most people will think that they understand its meaning. But read the claim again, or several times again, and ask yourself if you are sure what the inventors mean by the terms: "cereal plants", "wounding" "embryogenic callus", and "enzyme that degrades cell walls". Is there only a single meaning for these terms? Has the inventor provided her own definitions? How do you uncover the true meaning of these terms? First, read the text of the patent, especially looking for a section titled "Definitions" or for phrases like "as used herein" or "cereal'plant refers to". Oftentimes, the inventor will directly define a term. If no explicit definition exists, then try deducing a meaning from how the term is used when the invention describes the invention. Clues might also be found in the prosecution history posted online by the U.S. Patent and Trademark Office for patents and pending, published patent applications. Uncovering the meaning may take a bit of detective work. Don't despair; it can be difficult (and time-consuming) to accomplish.

Claim basics. Claims are written in a way peculiar to patents. A claim is always written as a single sentence, composed of two parts – the preamble and the body – with a transition word or phrase between them.

- The preamble is an introductory statement that names the thing that is to be claimed. For example, "A method for making a genetically modified plant."
- The body of a claim defines the elements or steps of the named thing or method.

The transition words or phrases commonly used are "comprising," "consisting of" and "consisting essentially of" and have very distinct meanings ("Consisting essentially of" is rarely, if ever, used in biotechnology and is not discussed here). "Comprising" (also "having") means that the claim encompasses all the elements listed and, moreover, can include additional, unnamed elements. For example, if a claim recites elements (A and B), an individual that uses elements (A and B) or elements (A, B, and C) is infringing, whereas using the single elements (A) or (B) is not infringing. The same rationale applies to the pencils example above, where A = No. 2 lead and B = an eraser.

In contrast, the transition"consisting of" has more limited scope. "Consisting of" means that the device (or method) has the recited elements (or steps) and no more. For example, if a claim recites (A and B) and the individual uses only (A), or (A and C), or even (A, B, and C), the claim is not infringed. (A′ and B) also escapes the claim, where A′ is a modified version of A. With respect to the pencils, C could be a rubber finger grip and A′ could be No. 3 lead.

Claims also come in two flavors: independent and dependent. An independent claim stands alone. It includes all the necessary limitations and can be read without reference to any other claim. A dependent claim refers back to another claim and includes all the limitations of the referred-to claim. Thus, when analyzing claims in a patent for infringement or claim scope, only the independent claims need to be considered.

2.2 A patent application is not a patent

Are claims in a patent *application* important? The answer is both yes and no. Yes, because the claims indicate the invention intended for protection and may indicate the scope of protection that is desired. No, because claims in a patent application have no force. A patent application is NOT the same as a patent. Claims in a published patent application have not been examined by a national patent office and may not be representative of a scope that will ultimately be granted.

During the application process, the patent text is published 18 months after the earliest filing. The publications print the claims as filed. Sometimes the claims are written much broadly than is actually patentable. As the application is examined by a patent office and claim language negotiated, the claims may shrink in scope. In contrast, the specification of a granted patent is nearly always identical as filed; typos and obvious errors may be fixed, material can be deleted, but new matter is not allowed to be added to the text after filing.

Because the claims in an application are what the applicant hopes for and not what she will necessarily receive, it is important to know whether you are looking at a granted patent or a patent application.

2.3 Parts of a patent document

A patent document has three main sections:

1. a cover page which presents bibliographic information,

2. a specification, which describes the invention, and
3. claims, which define the scope of activity from which the patentee has the right to exclude others unless they sign licensing agreements.

The cover page presents mainly bibliographic information. The information provides notice mainly of historical facts and identifying elements, such as application filing date and serial number. None of it, including the abstract, has legal import for interpreting the patent. Nevertheless, in a departure from previous decisions, the Federal Circuit in *Hill-Rom* used the abstract for aid in interpreting the claims.

The specification is also called the disclosure. It contains a description of the invention that must satisfy certain writing requirements. The layout and content of a specification may vary somewhat from country to country.

The specification has particular value as an aid to interpreting the scope of the claims. Thus, a patent specification is drafted both to satisfy the written requirements for patentability as well as to define claim scope. With this in mind, we will examine each major section of the specification and analyze what purpose is being accomplished. For a more detailed guide to "How to Read a Patent" the reader can referee to the tutorial area of the CAMBIA web site (www.patentlens.net) or on the Cougar Patent Law web site (www.cougarlaw.com).

"Background of the Invention" is typically drafted for a jury audience or a patent examiner. Selected art in the field is discussed to emphasize differences with the current invention, and to point out the need for the current invention. "Summary of the Invention", which is distinct from the abstract, is meant to discuss the invention (i.e., the claims) rather than the disclosure as a whole. Often, the summary will discuss advantages of the invention or how it solves the problems existing in the art.

"Detailed Description of the Invention" is the meatiest section of a patent. Its purpose is to adequately and accurately describe the invention. There are generally two sections: (1) a general explanation of the invention and how to practice it; and (2) specific examples of how to practice the invention. Many new readers find the purposes of these two sections confounding and assume that the examples set forth how the invention will be practiced. Rather, examples are meant only "to illustrate, but in no way to limit, the scope of claimed invention."

In the first section the invention is described in its broadest sense, to show that the inventors have a broad view of the scope of the elements. Preferred embodiments of the invention are often described. Such embodiments are generally more limited versions of the broadest concept and are provided for support of the claims. Definitions of key terms are often provided and are extremely important in interpreting the scope of the claims.

Although a patent application does not require examples, in practice, they can often assist in showing patentability (e.g., enablement). The examples may or may not actually have been performed by the inventors. "Working" examples present completed undertakings. "Prophetic" examples are hypothetical undertakings and are always written in the present or future tense. Typically, the examples demonstrate practice of one or more specific embodiments of the invention.

And now armed with the basics on patents and tools to read them, let us enter the patent world of *Agrobacterium* as a vehicle for plant transformation.

3 *AGROBACTERIUM*-MEDIATED TRANSFORMATION AND PATENT LAW

An analysis of the patent landscape of *Agrobacterium*-mediated transformation begins with determining the scope of the subject matter. Patent documents pertaining to the subject are then obtained by searching on-line databases, such as CAMBIA's Patent Lens (www.patentlens.net), Esp@cenet at the European Patent Office (ep.espacenet.com), patents at U.S. Patent and Trademark Office (www.uspto.gov/patft/index/html), PatBase (www.patbase.com), and Delphion (www.delphion.com). An exhaustive patent search is conducted, typically based on a combination of keywords, scientists' names, organizations, and patent office classification codes. If the number of documents obtained is quite large, a rapid screening method may be used to reduce the pile to a manageable number of key documents. In our rapid screening method, the document text is skimmed, especially the background and the summary of the invention sections and the claims read.

A broad claim , which encompass a relatively large part of the field, is the criterion for identifying a key patent. Especially key patents may dominate others (like the pencil example above). The relationship of patents found can be visualized either as a pyramid, with the broadest patent claims at the apex and each row (tier) moving down the pyramid to the base contains patents with successively narrower claims, or as Venn diagrams, with the broadest claims residing in the outermost circle. For a patent landscape review generally only the top 1-3 tiers are analyzed.

Patent analysis focuses on interpreting the meaning of the claims because patent rights are delimited by the scope of the claims. Claim scope analysis requires an understanding of the science and the invention described in the patent text as well as the legal rules that define claim interpretation. Briefly, claims are interpreted by their plain language, by the definitions or context used in the patent text, and by statements made to the patent office during the back and forth negotiations called prosecution (see also discussion in section 2.1 above). Other resources, including dictionaries and experts, can be consulted as needed. For the patent landscape analyses of the sort presented here, the meaning and scope of claims are established only from looking within the "four corners" of the patent. Due to the volume of key document s, time and resource constraints, and the purpose of the landscape paper, we did not consult the prosecution history or other sources.

For the analysis of patent landscape surrounding *Agrobacterium*-mediated transformation of plants, the subject matter of the technology for inclusion was decided on before any patent search. The following analysis is based on Roa-Rodríguez and Nottenburg (2003). *Figure 20-3* illustrates various aspects of transformation as a patent lawyer might draw it out.

In this simplistic scheme, gene vector(s) that contain a *vir* region and a T-DNA with a gene of interest on the same or separate vectors are constructed, *Agrobacterium* containing the vectors are prepared, plant cells or tissues are incubated with the *Agrobacterium*, and transgenic plants are grown. Underneath the main pathway, it is helpful to list alternatives for each component or method step. It is these possibilities that form the contours of the subject matter.

Looking at the diagram, let's examine some pieces that might be patentable (assuming that the requirements of patentability are met):

- Vectors for transforming plants
- Genes of the vectors
- Transgene
- Methods for making vectors
- Methods for making *Agrobacterium* with engineered vector
- *Agrobacterium* containing engineered vectors
- Improved *Agrobacterium* strains for transformation
- Methods of preparing plant tissue for transformation
- Methods of transforming specific plants
- Transformed plants and plant cells.

Agrobacterium itself is not patentable under patent law. Only inventions are patentable; a naturally-occurring bacterium is not considered an invention even under the most magnanimous patent laws. In addition to that, some countries do not allow patents for plants or native sequence genes or others, but for purposes of this chapter, we consider transgenic plants and gene sequences patentable – especially as initial patenting of *Agrobacterium*-mediated transformation was primarily done in U.S., which considers all of these subject matters patentable.

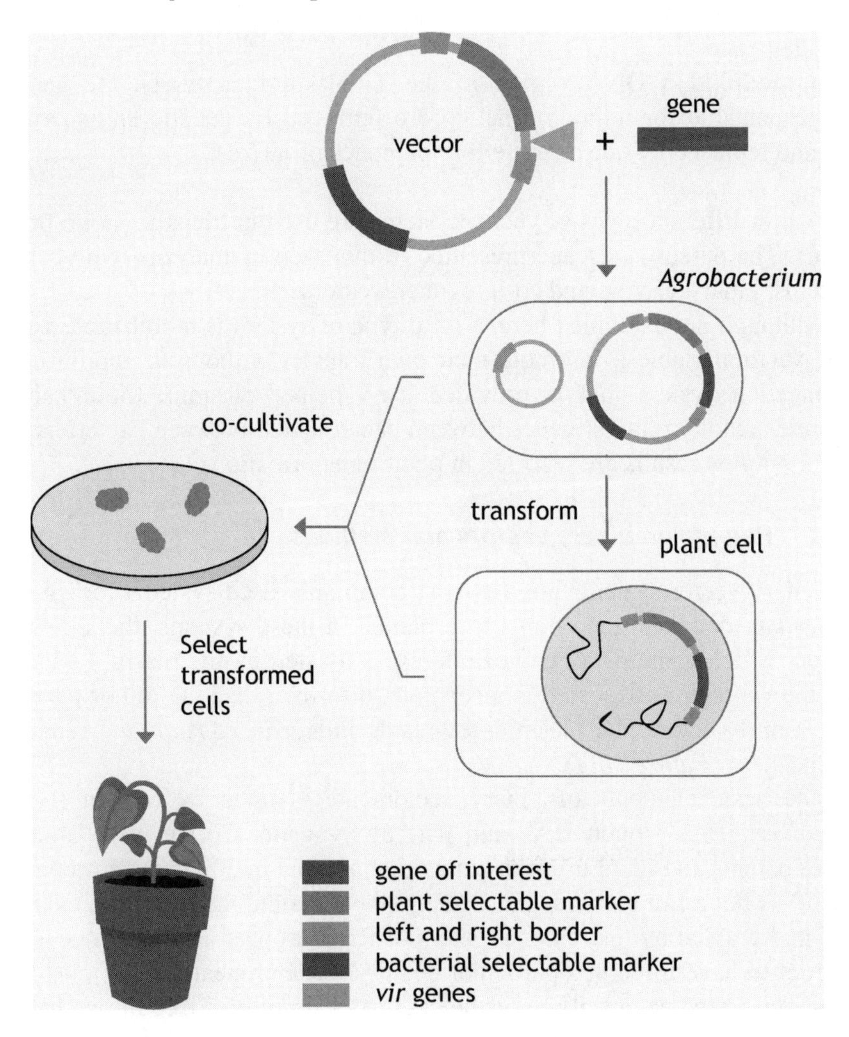

Figure 20-3. A lawyer's view of the transformation process

3.1 Vectors for transformation

The basic elements of vectors designed for *Agrobacterium*-mediated transformation are derived from the native Ti-plasmid. The necessary elements are:

- **T-DNA border sequences**, or at least the right border, which initiates the integration of the T-DNA region into the plant genome;
- ***vir* genes**, which are required for transfer of the T-DNA region to the plant, and
- **a modified T-DNA region** of the Ti plasmid, in which the genes responsible for tumor formation are removed by genetic engineering and replaced by one or more foreign genes of interest.

A few different types of vector systems are used in transformation protocols. The patent landscape presented in this section analyzes two vector systems: binary vectors and co-integrated vectors.

Although not examined here, a third type of system is mobilizable vectors, vectors unable to promote their own transfer without an appropriate conjugation system that is provided by a helper plasmid. Mobilizable plasmids readily transfer genes between bacteria, and between bacteria and fungi, but few results are reported in plant transformation.

3.1.1 Patents on binary vectors and methods

Binary vector systems are the most commonly used systems for *Agrobacterium*-mediated gene transfer to plants. In these systems, the T-DNA region, which contains a gene of interest is located in one plasmid vector and the *vir* region is located in a separate, disarmed (lacking tumor genes) Ti plasmid vector. The plasmids co-reside in *Agrobacterium* and remain independent (*Figure 20-4*).

The basic elements of binary vectors are claimed by Mogen (now called Syngenta Mogen B.V. and part of Syngenta Co.) in two United States patents and one European patent that expired in 2004. (A "European patent" is not a Europe-wide patent but rather a patent that has been examined and granted by the regional patent office, European Patent Office and in order to have force in a particular country, the European patent must be registered in the patent office of the country. Currently, 31 countries have signed on to the European Patent Convention. Alternatively, a patent application may be examined and granted by individual patent offices in European countries.). *Table 20-1* presents an overview of these three patents.

Table 20-1. Binary vector patents

Patent No:	US 4,940,838	US 5,464,763	EP 120 516 B1
Filing date:	23 Feb 1984	23 Dec 1993	21 Feb 1984
Issue date:	10 July 1990	07 Nov 1995	23 Oct 1991
Expiration date	10 July 2007	07 Nov 2012	21 Feb 2004

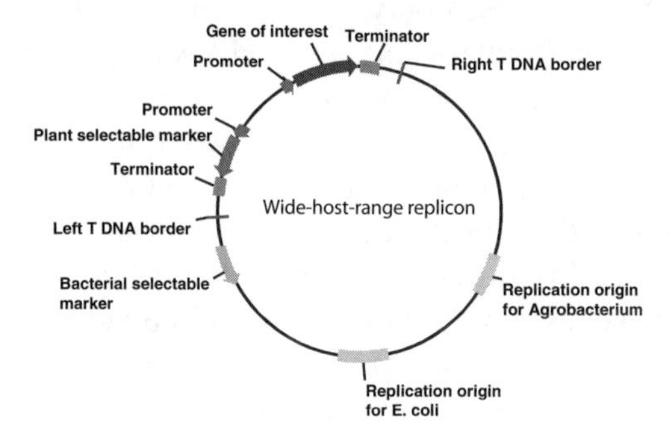

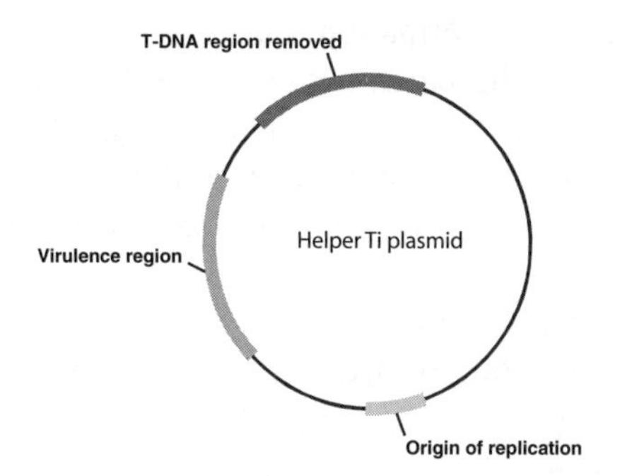

Figure 20-4. Binary vector system

Broadly speaking, these patents describe methods for transforming dicotyledonous plants with *Agrobacterium* that contains binary vectors. More specifically, in claim 2 of US 4,940,838, the invention is:

Agrobacterium containing two plasmids, in which

1. one of the plasmids has foreign DNA residing in a T-region but lacks a *vir* gene region, and
2. the other of which has a *vir* gene region but lacks the T-region.

Other claims in this patent recite using this *Agrobacterium* for transforming plants and plant cells. But unlike claim 2, method claims in the patent are limited specifically to transformation of dicot plants and cells. This difference begets an interesting conundrum: if you make and use the

Agrobacterium of claim 2 to transform monocot cells, are you infringing this patent?

There is little doubt that *using* the claimed *Agrobacterium* to transform monocots is not infringing. Making the claimed *Agrobacterium*, whether intended or not for transformation of monocots, appears to be literally infringing. (This analysis is based only upon the published specification and claims and as such is preliminary and informal.) An argument can be made however, that the claimed *Agrobacterium,* which is constructed for transforming monocots, is not infringing: in 1984 when the patent application was filed transformation of monocots was not routine and the inventors could not have known that the claimed *Agrobacterium* would successfully transform monocots. This situation is analogous to US 5,561,236, owned by Plant Genetic Systems (now a part of Bayer Crop Science), first filed in 1987, which claims "plants" transformed with an herbicide resistance gene. The term "plants" was interpreted by the Federal Circuit court in *Plant Genetic Systems N.V. v. DeKalb Genetics Corp.* to mean only "dicot plants" because monocot transformation was not routine at the time nor was it adequately taught in the patent. Based on this court decision, it is likely that the *Agrobacterium* of US 4,940,838 is limited to uses in dicots (The Federal Circuit is the only appeals court in the United States that reviews patent cases. Its decisions can only be reviewed by the U.S. Supreme Court).

The claims in US 5,464,763 additionally limit the binary plasmid vectors by requiring that only foreign DNA be contained between the 23 bp borders of the T-region. Furthermore, the two plasmids lack a region of homology, which insures that the two plasmids do not recombine. The now-expired European patent issued with claims equivalent to US 5,464,763.

Other patents are directed to variations of the basic binary vector system. Variations include: *Agrobacterium* with multiple copies of T-DNA and *vir* integrated in the bacterial chromosome (US 5,149,645 owned by Leiden University and Schilperoort, but method claims are limited to transforming plants in either the Liliaceae or Amaryllidaceae families); *Agrobacterium* with more than one binary vector (US 6,265,638 B1, EP 1 117 816 B1, CA 2,344,700 C, and AU 764100 B2 owned by Pioneer Hi-Bred). In other variations, Syngenta Mogen B.V patents claim vectors that can integrate into plant genomes by homologous recombination (US 5,501,967 and EP 436 007 B1). In this scheme, the vector has a region homologous to a part of a target locus in the plant, permitting insertion of a gene of interest or a specific mutation in a particular locus of a plant genome.

Other binary vectors have been devised to suit different needs of plant transformation. For example, some desirable features of binary plasmids include different origins of replication, a large maximum size of the insert a binary plasmid can carry, and the size of the plasmid. These features are claimed in a number of patents (see e.g., US Patent 6,165,780 filed by The National Institute of Agrobiological Resources (Japan) in which the binary plasmid has origins of replication that maintain either a low or high copy number).

In summary, the key patents on basic binary vectors, those that claim the basic elements, were granted to Mogen in the United States (two patents) and in Europe (one patent, which has expired). The claims in these patents encompass essentially any two vector system located in the same *Agrobacterium* strain having (i) a T-region in one vector, and (ii) a *vir* region in another vector.

While it is difficult to form a conclusion that will apply to every reader, overall these patents likely encompass many transformation protocols in common use, with one major exception. The claims are limited to methods of transformation of dicotyledonous (dicots) plants only; use for transformation of monocotyledonous plants is not covered. Keeping this limitation in mind, users of this binary vector system in the United States should consider these patents when crafting a commercial research strategy.

3.1.2 Patents on co-integrated vectors

A co-integrated plasmid vector is the product of homologous recombination between a small plasmid of bacterial origin and an *Agrobacterium* Ti plasmid. In general, the *Agrobacterium* Ti plasmid lacks tumor-causing genes ("disarmed" Ti plasmid) and the small vector plasmid is engineered to carry a gene of interest between a right and a left T-DNA border of the T-DNA region. Recombination takes place through a single crossover event in a homologous region present in both plasmids.

Although co-integrated vectors have become less popular in recent years due to some difficulties encountered in engineering them, they are still used to a certain extent when modified, for example, to allow site-specific recombination of the plasmids within *Agrobacterium*.

A co-integrated plasmid or hybrid Ti plasmid contains at least (i) a *gene of interest*, located between the left and right T-DNA border sequences, and (ii) a *vir* region, which allows the transfer of the gene of interest located between the two border sequences into the plant genome. Two plasmids are required for the assembly of a co-integrated plasmid. They are (i) a *vector molecule containing a gene of interest* to be transferred into a plant and a homologous region, and (ii) a *Ti plasmid* containing the *vir* region and a homologous region but does not contain tumor-inducing genes.

Max-Planck Society and Monsanto Company have both been granted patents on basic co-integrated vectors. Patents awarded in Europe, Australia, Japan and Russia have all expired. Notably, no patent has been granted in the United States. Most likely any application filed in the U.S. has either been abandoned or is in interference, which is a procedure to determine who was the first inventor in time. Unfortunately, if a patent does issue in the U.S. it will likely have a 17-year term from the *date of issue*.

Be aware that although the patents have expired on the most basic features of co-integrated vectors, other patents on improvements or modifications of co-integrated vector systems may still be in force (see, e.g., US 5,635,381).

3.2 Tissue types for transformation

The efficiency of T-DNA transfer via *Agrobacterium* to a plant varies considerably, not only among plant species and cultivars, but also among tissues. Various protocols for *Agrobacterium*-mediated transformation of plants use leaves, shoot apices, roots, hypocotyls, cotyledons, seeds and calli derived from various parts of a plant. In other methods, the transformed tissue is not removed from the plant but left in its natural environment, thus the transformation takes place *in planta*.

Patents directed specifically to different tissues to be transformed are relatively few, but the scope of their protection is rather broad. Some of the patents referred to in this section are considered key patents for widely used technologies by the research community. With the exception of Japan Tobacco's patents directed to callus and immature embryo transformation of a monocotyledonous plant, claims in these patents are not restricted to the type or species of plant to be transformed.

3.2.1 Callus transformation

Japan Tobacco was granted two patents, one in the United States and the other in Australia (now expired). The United States patent claims a method for transforming a monocot-derived tissue with *Agrobacterium*. Moreover, the tissue must be dedifferentiated or undergo dedifferentiation by culture for at least 7 days.

US 5,591,616   issued 7 Jan 1997   expires 7 Jan 2014

In an allowed United States patent application, Japan Tobacco claims transformation of a dedifferentiated monocot tissue cultured from 1 to 6 days. As filed, the tissue is not limited to any particular plant tissue; it just has to be immature.

US 2002/0178463 when issued, will expire 07 Jan 2014

3.2.2 Immature embryo transformation

Protected by Japan Tobacco in Australia and in Europe, the patents claim transformation of the scutellum of an immature embryo of a monocot plant with *Agrobacterium.* The transformation process takes place before the tissue has differentiated into a callus. To date, no related patent has issued in the United States.

AU 687863 B	issued 5 Mar 1998	expires 1 Sep 2014
EP 672 752 B1	issued 26 May 2004	expires 1 Sep 2014

3.2.3 In planta transformation

Three different entities have patents in this area. Cotton Inc., Rhobio and Paradigm Genetics claim transformation of a plant tissue *in situ* with *Agrobacterium*. Cotton Inc. claims the injection of *Agrobacterium* into floral or meristematic tissue, whereas the other entities' patents do not limit the type of tissue (*Table 20-2*).

Table 20-2. Patents on *in planta* transformation

Patent No.	Issue Date	Expiry Date	Patent Owner
US 5,994,624	30 Nov 1999	20 Oct 2017	Cotton Inc
AU 752717 B2	26 Sep 2002	19 Oct 2018	Cotton Inc
US 6,353,155 B1	5 Mar 2002	30 Jun 2020	Paradigm Genetics
EP 1 171 621	7 Dec 2005	19 April 2020	Rhobio
AU 775949 B2	19 Aug 2004	19 April 2020	Rhobio

3.2.4 Floral transformation

Floral transformation is basically an *in planta* method that is very popular for transformation of *Arabidopsis thaliana* (Brassicaceae), one of the best known model plants in genomic studies, and is also suitable for the transformation of monocotyledonous plants. A U.S. patent assigned to Rhône-Poulenc Agro is described further in section 3.3.1 below. The U.S. patents granted to Cotton Inc. and Paradigm Genetics for *in planta* transformation (discussed above) also include claims for transformation of floral or meristematic tissue. Unlike the Cotton Inc. patent where *Agrobacterium* is injected into the floral tissue, in the Paradigm Genetics patent, the floral tissue is immersed into a diluted suspension of *Agrobacterium* cells (*Table 20-3*).

Table 20-3. Patents on floral transformation

Patent No.	Issue Date	Expiry Date	Patent Owner
US 5,994,624	30 Nov 1999	20 Oct 2017	Cotton Inc
AU 752717 B2	26 Sep 2002	19 Oct 2018	Cotton Inc
US 6,353,155 B1	5 Mar 2002	30 Jun 2020	Paradigm Genetics
US 6,037,522	14 Mar 2000	22 June 2018	Rhône-Poulenc Agro

3.2.5 Seed transformation

Two groups have pending patents or patent applications on transformation of plants using seed as target tissue: (1) the Agricultural Biotechnology Research Center (ABRC) of Shanxi (China) in China, the U.S., and Canada, and (2) Scigen Harvest Co. (Korea) in Korea. As filed, the U.S. and Canadian applications of ABRC generally claim applying *Agrobacterium* to germinating seed, without further treatment of the seed. The scope of a related Chinese patent is unknown, as it is only available in Chinese language. The Scigen method uses needle-wounded seed as target tissue in combination with *Agrobacterium tumefaciens*.

Table 20-4. Patents on seed transformation

Patent No.	Issue Date*	Expiry Date*	Patent Owner
WO 20/66599 A2	29 Aug 2002	not applicable†	Scigen Harvest Co
US 2002/0184663 A1	5 Dec2002	19 Feb 2022	Agricultural Biotechnology
AU 762964 B2	10 Jul 2003	14 Oct 2019	Protein Research Trust
EP 1 121 452 A1	8 Aug 2001	14 Oct 2019	Protein Research Trust

*Publication date and estimated expiry date (if the application is granted) are given for patent applications.
†WO patent applications do not have an expiry date because they can only become patents if converted to national applications.

In addition, the Protein Research Trust of South Africa has an granted Australian patent and a European patent application directed to transforming seed with a mixture of *Agrobacterium* and a wetting or surfactant agent that enhances or facilitates the penetration of the bacterium into the explant (*Table 20-4*).

3.2.6 Pollen transformation

A patent related to this topic was granted to the United States Department of Agriculture (USDA) in the U.S., Europe and Australia. In this invention, *Agrobacterium* containing a foreign gene is applied to pollen, allowing the pollen to take up the bacteria and germinate. The transformed pollen fertilizes a second plant to obtain transgenic seed, which is then germinated to obtain a transgenic plant (*Table 20-5*).

Table 20-5. Patents on pollen transformation

Patent No.	Issue Date	Expiry Date	Patent Owner
US 5,929,300	27 Jul 1999	15 July 2017	US Dept. of Agriculture
EP 996 328 B1	5 Mar 2003	14 July 2018	US Dept. of Agriculture
AU 733080 B	3 May 2001	14 July 2018	US Dept. of Agriculture

3.2.7 Shoot apex transformation

Transformation of an excised shoot apical tissue by inoculating the tissue with *A. tumefaciens* is disclosed by Texas A & M University in a granted United States patent (US 5,164,310; issued 17 Nov 1992; expires 17 Nov 2009). Applications filed in Europe and in Australia have been abandoned.

Additionally, in a U.S. application filed in 2003, Texas A&M University disclosed a transformation method consisting in direct inoculation of *Agrobacterium* into a shoot apex still attached to a plant seedling. The application was abandoned in 2005.

3.2.8 Summary

Transformation of pollen with *Agrobacterium* is fairly broadly protected in the United States and in Australia. Similarly, shoot apex transformation is protected in the United States, except that the bacterium used in this case is specifically *A. tumefaciens.* Thus, use of other species of *Agrobacterium* to transform apical shoots from any plant may fall outside of the scope of the claimed invention.

There appears to be more room to avoid infringing patents on *in planta* and callus transformation. A United States patent owned by Cotton Inc. particularly claims: use of a needleless device to inject *Agrobacterium* into floral or meristematic tissue. Thus, use of a different device or other tissues may bring the method outside the scope of the claimed invention. In the case of Paradigm Genetics' U.S. patent, the suspensions of *Agrobacterium* cells are of specified density. If one uses aqueous solutions that do not conform to the specified dilutions and density, the method may fall outside the terms of the claimed invention.

With respect to callus transformation claimed by Japan Tobacco, at least in the United States, the monocot tissue must be at least seven days old. Sometime in 2006 however, a new Japan Tobacco patent will issue that claims use of any immature monocot tissue with a one-day minimum culture period. With the grant of this patent, Japan Tobacco's protection is expanded in the U.S. for essentially all *Agrobacterium* transformations of immature tissue of a monocot.

Seed transformation with *Agrobacterium* seems to be an area still to be explored. Not only are dominant patents absent in the major jurisdictions, but no patents appear to be either filed or granted in the USA to date.

In addition, there are inventions where the plant tissue or cell type employed in the *Agrobacterium*-mediated transformation protocol is not defined. This is the case of a pending U.S. application filed by Monsanto in 2003 (US 2003/0204875 A1). One limitation in the disclosed invention is the use of a non-specified agent to inhibit the growth of *Agrobacterium* in the co-culture medium with the plant cells or tissue. The addition of such agent is said to facilitate the generation of transformed plants with low number of inserted copies. If granted as filed, avoidance of the patent may be possible by forgoing a growth inhibiting agent.

3.3 Patents on transformation of monocots

Monocots (monocotyledonous) comprise one of the large divisions of angiosperm plants (flowering plants with seeds protected within a vessel). They are herbaceous plants with parallel veined leaves and have an embryo with a single cotyledon, as opposed to dicot plants (dicotyledonous), which have an embryo with two cotyledons.

Most of the important staple crops of the world, the so-called cereals, such as wheat, barley, rice, maize, sorghum, oats, rye and millet, are monocots. Other monocots include food crops such as onion, garlic, ginger, banana, plantain, yam and asparagus.

Agrobacterium-mediated transformation of commercially important monocots was first attained in rice and maize in the mid 1990s. Following these achievements, other monocot crops were successfully transformed and refinements of techniques led to improved regeneration of transformed monocot tissue.

3.3.1 General methods for transforming monocots

Japan Tobacco (in Japan), Rhône-Poulenc Agro (in France), and the National Institute of Agrobiological Resources (in Japan) own patents or pending patent applications directed to methods for *Agrobacterium*-mediated transformation of any monocot with a gene of interest (*Table 20-6*). The main differences among the patent claims lie in:

- the type of plant tissue or explant used for the transformation process, and
- the use of additional treatments, such as vacuum infiltration or adding phenolic compounds, to facilitate the transformation process.

Table 20-6. U.S., European and Australian patents for transforming monocots

Patent No.	Issue Date*	Expiry Date*	Patent Owner
US 5,591,616	7 Jan 1997	7 Jan 2014	Japan Tobacco
AU 667939 B	18 Apr 1996	6 Jul 2013	Japan Tobacco
EP 0 604 662 A1	6 Jul 1994	6 Jul 2013	Japan Tobacco
AU 687863 B	5 Mar 1998	1 Sep 2014	Japan Tobacco
EP 0 672 752 B1	26 May 2004	1 Sep 2014	Japan Tobacco
US 6,037,522	14 Mar 2000	23 Jun 2018	Rhône-Poulenc
EP 1 198 985 A1	24 Apr 2002	22 Jul 2019	Nat'l Inst. of Agrobiological Resources
AU 775233 B2	22 Jul 2004	22 Jul 2019	Nat'l Inst. of Agrobiological Resources

*Publication date and estimated expiry date (if the application is granted) are given for pending patent applications

Japan Tobacco is typically considered to have the broadest patent in this area. In two different sets of patents, it claims the transformation of a monocot callus during a dedifferentiation process and the transformation of the scutellum of an immature embryo prior to dedifferentiation. Thus, these patents cover transformation of monocot tissues that are widely and commonly used.

Rhône-Poulenc Agro (now Bayer Crop Science) claims the transformation of a monocot inflorescence via *Agrobacterium*. The inflorescence can be dissected and then transformed. Alternatively, callus formation is induced from an inflorescence in culture, and the derived callus is then transformed with *Agrobacterium* and regenerated into a plant. The invention is thus limited to transformation of a monocot inflorescence. Transformation of other monocot tissues are not claimed.

The National Institute of Agrobiological Resources (Japan) has a PCT application and an Australian granted patent that discloses a method for transforming a monocot by treatment of intact seed with *Agrobacterium* containing a recombinant gene of interest. In the Australian patent, the seed to be transformed is a germinated seed pre-cultured in a medium with 2,4 D for four or five days.

3.3.2 Gramineae and cereals

Gramineae is one of the largest families of monocot plants. Mostly herbaceous, grass-like plants, this family includes several important staple crops (cereals) such as wheat, rice, maize, sorghum, barley, oats, and millet. It also encompasses plants such as bamboos, palms, and foraging grasses (e.g., turf grass, king grass (*Pennisetum purpureum*), and *Brachiaria*). Therefore, patents addressing the Gramineae family embrace cereals, but patents directed to cereals *do not embrace* all Gramineae.

Gramineae transformation. The United States and Australian patents granted to the University of Toledo and the United States patent granted to Goldman and Graves belong to the same patent family (*Table 20-7*). They all claim a method for transforming seedlings of a Gramineae with a *vir*$^+$ *Agrobacterium*. Claims of both United States patents limit the inoculation of the bacterium to a particular area in the seedling, where cells divide rapidly and wounding takes place prior the inoculation.

Remarkably, the United States patent granted to Goldman and Graves (US 6,020,539) – and licensed by the University of Toledo (*Table 20-7*) – also contains broad claims to the transformation of Gramineae with *Agrobacterium*. One particular claim (claim 22) encompasses any Gramineae, constituting one of the broadest claims recently issued in the area of plant transformation technologies. United States patents claiming *Agrobacterium* transformation of any tissue of a Gramineae might be dominated by this patent. The grant of this patent has wreaked havoc in the scientific community and multiple parties with interest in *Agrobacterium*-mediated transformation of Gramineae.

Cereal transformation. Plant Genetic Systems (now part of Bayer Crop Science) has granted United States and European patents disclosing the transformation of any cereal with *Agrobacterium* (*Table 20-7*). The most limiting elements in the claims are the wounding of a cereal embryogenic callus and the enzymatic disruption of a tissue cell wall before transformation. The claims of the European patent do not recite enzymatic degradation but do additionally encompass different transformation methods besides *Agrobacterium*.

Table 20-7. Patents directed to transformation of Gramineae and cereals

Patent No.	Issue Date	Expiry Date	Patent Owner
Gramineae			
US 5,187,073	16 Feb 1993	16 Feb 2010	University of Toledo
AU 606874 B2	21 Feb 1991	30 Jan 2003	University of Toledo
US 6,020,539	1 Feb 2000	1 Feb 2017	University of Toledo
US 2002/0002711 A1	3 Jan 2002*	Abandoned	University of Toledo
Cereals			
US 6,074,877	13 June 2000	23 June 2013	Plant Genetic Systems
EP 0 955 371 B1	22 Feb 2006	21 Nov 2011	Plant Genetic Systems

*Publication date

3.4 Patents on transformation of dicots

Dicotyledonous plants (dicots) are the second major group of plants within the *Angiospermae* division (flowering plants with seeds protected in vessels). The mature leaves have veins in a net-like pattern, and the flowers have four or five parts. Apart from cereals and grasses that belong to the monocot group, most of the fruits, vegetables, spices, roots and tubers, which constitute a very important part of our daily diet, are dicots. In addition, all legumes, beverages (e.g., coffee and cocoa), and a great variety of flowers, oil seeds, fibers, and woody plants belong to the dicot group.

3.4.1 General transformation methods

There are fewer patents on general methods for *Agrobacterium*-mediated transformation of dicots than for transformation of monocots. A broad patent directed to transformation of dicots using an *Agrobacterium* strain lacking functional tumor genes was granted a few years ago to Washington University (US 6,051,757; issued 18 Apr 2000; expires 18 Apr 2017)

Although issued in 2000 in the United States, this patent has an initial priority date of 1983. Thus, the prosecution process took approximately 17 years until the patent was finally granted. The patent appears to be one of the broadest in scope granted in the area of *Agrobacterium* transformation. Moreover, the patent rights under this patent may overlap with the rights already granted in previous patents related to transformation of dicots with *Agrobacterium*.

One of the distinctive features in the patent claims is that the cytokinin function in the Ti plasmid is knocked out in order to obtain a non-tumorigenic (disarmed) *Agrobacterium* strain. Disarmed strains lacking functional tumorigenic genes are typically used in protocols of *Agrobacterium*-mediated transformation. The present patent thus may constitute a blow for a widely used and standard procedure carried to transform dicot plants.

Other patent applications in this area may still be in interference proceedings (procedures that determine who is the first inventor-in-time) at the U.S. Patent and Trademark Office. While many press releases from Monsanto Co., Syngenta, Bayer CropSciences, and Max Planck Society, allude to the settlement of interference proceedings, details are notably lacking.

In 2003, Monsanto was granted a U.S. patent that claims the transformation of a dicot cell or tissue with *Agrobacterium* (US 6,603,061 B1; issued 3 Aug 2003; expires 29 July 2019). The distinctive element in the claims of this patent is the use of an antibiotic that inhibits or suppresses the growth of *Agrobacterium* during the inoculation phase. According to the applicants, this procedure promotes the generation of transformed plants with low copy inserts and improves the transformation efficiency. The possible effects of this patent in the U.S. depend on the extent of the practice of using a growth inhibiting agent in the inoculation medium.

Most of the other major patents in this area claim transformation of dicots in conjunction with the use of co-integrated or binary vectors, the vectors being the main subject matter of the claimed inventions. These patents have been reviewed above in sections on "Binary vectors" and "Co-integrated vectors".

3.4.2 *Transformation of cotton*

Of the patents directed to transformation of various dicot plants, the patents directed to transformation of cotton present an interesting study. Five different entities have patents and applications in this area.

Table 20-8. Patents owned by Agracetus (Monsanto)

Patent No.	Issue Date	Expiry Date	Comments
US 5,004,863	2 Apr 1991	2 Apr 2008	Re-examined twice: in 1992 and 2000 Currently in interference
US 5,159,135	27 Oct 1992	27 Oct 2009	Re-examined once: in 2000 Currently in interference
EP 270 355 B1	16 Mar 1994	2 Dec 2007	Originally opposed by Monsanto, but opposition withdrawn

The most contentious of the patents belong to Agracetus (now owned by Monsanto), in particular, two patents in the United States and one in Europe directed to transformation of immature cotton plants with *A. tumefaciens* (*Table 20-8*).

In the first patent, US 5,004,863, the claims are directed to a method of transforming cotton plants where:

- hypocotyl cotton tissue is used as the target of transformation;
- non-oncogenic *Agrobacterium* contains a Ti plasmid having a T-DNA region containing both a foreign gene and a selectable resistance gene;
- formation of somatic embryo is induced in the transformed tissue; and
- cotton plants are regenerated.

The claims of the second patent, US 5,159,135, are directed to transformed cotton seeds and plants. In its broadest claims, the cotton seed and cotton plants contain a gene of interest whose product – a protein or a negative strand RNA – confers a detectable trait.

The contentious nature of these patents is reflected by requests for reexaminations in the U.S. and opposition in Europe. Nothing changed however as a result of the challenges: the claims were upheld in two reexaminations, and the opposition in Europe brought by Monsanto was dropped when Monsanto bought Agracetus. Currently, the U.S. patents are in an interference proceeding to determine who was the first inventor-in-time of the claimed invention. The other party (or parties) involved in the interference are not known.

Calgene (also owned by Monsanto) has a United States patent, as well as European and Australian patents, claiming transformation of cotton (*Table 20-9*). In both Agracetus' and Calgene's inventions hypocotyl cotton tissue is transformed with *Agrobacterium.* In Calgene's patent, the hypocotyl tissue is cut from a seedling that has been grown in the dark and the transformed tissue is further cultured on callus initiation media that contains a selective agent and is plant hormone-free.

Another strong player in the area of genetically engineered cotton is Mycogen, an affiliate of Dow AgroSciences LLC (*Table 20-9*). It has a large portfolio of patents and applications related to cotton transformation and regeneration, mostly to transgenic herbicide-resistant cotton. Of the several patents granted in the U.S., one of them (US 6,620,990 B1) claims methods to obtain transformed cotton plants from embryogenic callus transformed with *Agrobacterium*. The invention also includes additional tissues for transformation such as cotyledons and zygotic embryos. Mycogen's non-U.S. patents additionally claim other tissues for transformation (i.e., hypocotyl), and specific media elements and conditions (i.e., time length, light and temperature) for the incubation with *Agrobacterium* and recovery of transformed explants.

In contrast to those discussed above, Cotton Inc. and The Institute of Molecular Agrobiology disclose the use of meristematic cells of apical shoot tips of cotton, and cotton petiole and root callus, respectively, as tissues to be transformed with either *Agrobacterium* (Cotton Inc.) or *A. tumefaciens* (Institute of Molecular Agrobiology).

Besides the mentioned U.S. patent application, Cotton Inc. has patents granted in Europe and in Australia with broader claims. These claims recite transformation of meristematic apical shoot tips of *any* plant in addition to cotton, with *any* DNA-based transforming agent. Thus, the patents are not limited to the use of *Agrobacterium* as agent for transformation.

Table 20-9. Patents on cotton transformation

Patent No.	Issue Date*	Expiry Date*	Patent Owner
US 5,846,797	8 Dec 1998	4 Oct 2015	Calgene (Monsanto)
EP 0 910 239 B1	5 Dec 2001	4 Oct 2016	Calgene (Monsanto)
AU 727910 B2	4 Jan 2001	4 Oct 2016	Calgene (Monsanto)
US 6,620,990 B1	16 Sep 2003	14 Sep 2017†	Mycogen
EP 0 344 302 B1	31 Mar 1999	2005 lapsed	Mycogen
EP 0 899 341 A2	3 Mar 1999	16 Nov 2008	Mycogen
AU 632038 B2	17 Dec 1992	16 Nov 2008	Mycogen
AU 668915 B2	23 May 1996	16 Mar 2013	Mycogen
AU 708250 B2	29 Jul 1999	23 Aug 2016	Mycogen
US 2003/208795 A1	6 Nov 2003	19 Feb 2018	Cotton Inc.
EP 1 056 334 B1	8 Sep 2004	18 Feb 2019	Cotton Inc.
AU 747514 B2	16 May 2002	18Feb 2019	Cotton Inc.
EP 1 159 436 A1	5 Dec 2001	10 Mar 2019	Institute of Molecular Agrobiology
EP 1 194 579 A1	10 Apr 2002	11 Jun 2019	Institute of Molecular Agrobiology
AU 777365 B2	14 Oct 2004	10 Mar 2019	Institute of Molecular Agrobiology
AU 782198 B2	7 Jul 2005	11 Jun 2019	Institute of Molecular Agrobiology
US 2004/009601 A1	15 Jan 2004	15 Jul 2022	University of California
US 6,483,013	19 Nov 1999	19 May 2019	Bayer BioScience
EP 1 183 377 A1	6 Mar 2002	18 May 2020	Bayer BioScience
AU 772686 B2	6 May 2004	18 May 2020	Bayer BioScience

*Publication date and estimated expiry date (if the patent application is granted) are given for pending patent applications.

†U.S. patent subject to a terminal disclaimer. Thus, it is possible that the expiry date is earlier than the date provided here.

The University of California has filed a U.S patent application describing the transformation of *Gossypium* clusters of cells or callus with *Agrobacterium*. This patent application has the most recent priority date (15 Jul 2002) among the cotton transformation patents.

Finally, patents and applications by Bayer BioScience (*Table 20-9*), which now owns Aventis CropScience, disclose the use of cotton embryogenic callus as target tissue for transformation with *Agrobacterium*. The addition of a plant phenolic compound prior or during the transformation of the cotton tissue for *vir* gene induction constitutes a disclosed improvement of cotton transformation methodology.

Given the thicket of proprietary technology in the area of cotton transformation, those who would like to enter this field and those who already are in it should exercise caution when deciding on the tools and methods to use for R&D on cotton transformation and generation of transgenic plants. An opinion should be sought as to whether and which licenses would need to be obtained.

3.5 *Agrobacterium* and Rhizobiaceae

A review of the key patents and claims presented above reveals an interesting fact: all these inventions are claimed in combination with *Agrobacterium* or *Agrobacterium tumefaciens*. What exactly is *Agrobacterium*? Taxonomy of bacteria is an evolving field. As more biochemical and genetic data of species become available, criteria for classifying organisms changes. Since the time that most all of the above patents were filed, *Agrobacterium tumefaciens* has been reclassified as *Rhizobium radiobacter* (Young et al., 2001). While *Agrobacterium* have long been recognized as belonging to the Rhizobiaceae group, the International Nomenclature Commission believes that *Agrobacterium* is an artificial genus and so have recommended a major overhaul of the genus. This belief is by no means universal. Considerable debate has ensued as to whether or not *Agrobacterium* is an artificial species (Farrand et al., 2003). One rationale for keeping *Agrobacterium* as a separate genus is its pathogenicity (Farrand et al., 2003).

As shown in this chapter and in the technology landscape "*Agrobacterium*-mediated transformation of plants" at the CAMBIA web site (Roa-Rodríguez and Nottenburg, 2003), the key patents lie mostly in the hands of large, international agriculture biotechnology companies. These companies not only actively sue each other for infringement but also cross-license their technologies. For example, in February 2004, Syngenta

International AG and Monsanto Co. announced an agreement to cross-license proprietary *Agrobacterium*-mediated transformation technology (Jones 2005). The same month, Bayer CropScience, the Max-Planck Society and Monsanto Co. announced an agreement that ended approximately 12 years of interference. Under the agreement, the entities entered into a cross-licensing situation (Monsanto 2005) (Curiously, no U.S. patent seems to have issued after this settlement, and there is a worrisome aspect that it will have a 17-year patent term from the *date of issue*). Absent though are press releases that "Small Agri-biotech Co." has licensed transformation technology from the multi-national companies.

Rather than submit to the patent stranglehold on transformation technologies, CAMBIA in Canberra, Australia, pursued an invention to "design around" the patents. (In the interest of full disclosure, at the time of the invention, one of us (Carol Nottenburg) was in-house patent counsel at CAMBIA and remains its patent counsel.) To this end, several strains in the Rhizobiaceae family, including *Rhizobium* spp., *Sinorhizobium meliloti,* and *Mesorhizobium loti*, were made competent for gene transfer into plants by acquisition of both a disarmed Ti plasmid and a suitable binary vector. (Broothaerts et al., 2005) These non-*Agrobacterium* strains produced transgenic rice, tobacco and *Arabidopsis* plants. Patent applications for this technology (called TransBacter™) were initially filed in the United States (US 2005/289667 and US 2005/289672). Moreover, licenses for the technology are available.

4 CONCLUSIONS

Intellectual property is often not very well understood by the research community, especially those in the public sector. All too often rumors and misstatements about patents are passed along from researcher to researcher. This is an unfortunate situation; however, it is understandable as scientists are not generally familiar with reading and understanding patents.

Because of increasing importance and emphasis on patents, in the non-profit sector as well as the for-profit sector, scientists are well-advised to become familiar with basics of intellectual property, especially patents. This chapter was prepared to assist and inform researchers using or interested in *Agrobacterium*-mediated transformation of plants.

The information in this chapter is not exhaustive, but consists of selected topics in patent law and selected areas of transformation. To satisfy the myriad questions and issues raised by the research or the interests of each person who reads this chapter requires considerably more space than was allotted. Instead, this paper should open the door into the patent world and furnish platform knowledge from which additional self-directed investigation can be performed.

The patent landscape section touches on portions of the *Agrobacterium*-mediated transformation process that are most widely used. The landscape of only a single crop - cotton - is evaluated. As shown, the private sector holds many of the key patent positions. Although the basic technologies are not widely licensed by companies, many of the key patents have expired or will expire in the next few years. As demonstrated with *Rhizobium*-mediated transformation, creative thinking to come up with alternatives for the patents still in force can be empowering and fruitful.

5 ACKNOWLEDGEMENTS

The authors thank CAMBIA and, in particular, the members of the IP Resource Group – Doug Ashton, Nick dos Remedios, Greg Quinn, Chris Pratt, and Richard A Jefferson – for their support of the landscape paper "*Agrobacterium*-mediated transformation of plants", which formed the basis for part of this chapter.

6 REFERENCES

AUTM, Association of University Technology Managers (2005) AUTM U.S. licensing survey: FY 2004. http://www.autm.org/surveys

Broothaerts W, Mitchell HJ, Weir B, Kalnes S, Smith, LMA, Yang W, Mayer J, Roa-Rodríguez C, Jefferson RA (2005) Gene transfer to plants by diverse species of bacteria. Nature 433: 629-633

CIPR, Commission on Intellectual Property Rights (2002) Implementation of the TRIPS agreement by developing countries. Study paper 7 of the report of the commission on intellectual property rights. http://www.iprcommission.org

Dam K (1999) "Intellectual Property and the Academic Enterprise", Chicago Economics Working Paper No. 68 (http://www.ssrn.com, abstract id=1665420)

Epstein RA, Kuhlik B (2004) Is there a Biomedical Anticommons? Regulation 27: 54-58, Available at SSRN:http://ssrn.com/abstract=568401

EPO, European Patent Office (2003) Usage profiles of patent information among current and potential users. http://www.european-patent-office.org/news/info/survey2003/index.php

Farrand SK, vanBerkum PB, Oger P (2003) *Agrobacterium* is a definable genus of the family *Rhizobiaceae*. Int J Syst Evol Microbiol 53: 1681-1687

Feldman M, Colaianni A, Liu K (2005) Commercializing Cohen-Boyer 1980-1997. DRUID Working Paper No. 05-21, http://www.druid.dk

Florida Prepaid Postsecondary Education Expense Board v. College Savings Bank et al. 119 S.Ct. 2199 (1999)

Heller MA, Eisenberg RS (1998) Can Patents Deter Innovation? The Anticommons in Biomedical Research. Science 280 (5364): 698-701

Hill-Rom Co., Inc. v. Kinetic Concepts, Inc., 209 F.3d 1337 (Fed. Cir. 2000)

H.R. Rep. No., 960, 101st Cong., 2d Sess, 1990

Jones PBC (2005) Patent challenges to agbiotech technologies in 2004. ISB News Report February 2005

Lichtman DG (2006) Patent Holdouts and the Standard-Setting Process. U Chicago Law and Economics, Olin Working Paper No. 292 Available at SSRN: http://ssrn.com/abstract=902646

Lichtman D, Baker S, Kraus K (2000) "Strategic disclosure in the patent system", U Chicago Law & Economics, Olin Working Paper No. 107, (http://www.ssrn.com, abstract id=243414)

Madey v. Duke University, 307 F.3d 1351, (Fed. Cir. 2002)

Meeks RL (2005) President's FY 2006 budget requests level R&D funding. NSF 05-322 October 2005

Mireles MS (2004) An examination of patents, licensing, research tools, and the tragedy of the anticommons in biotechnology innovation. U of Mich J of Law Reform 38: 141-235

Monsanto (2005) Bayer CropScience, Max-Planck Society, Monsanto Company resolve *Agrobacterium* patent dispute. Press release 4 February 2005, http://www.monsanto.com

National Institutes of Health, Office of Budget, http://officeofbudget.od.nih.gov

NSF (2005) National Science Foundation, Division of Science Resources Statistics http://www.nsf.gov/statistics.

NSF SRS (2004) National Science Foundation, Division of Science Resources Statistics, *Federal R&D Funding by Budget Function: Fiscal Years 2003-05*, NSF 05-303, Project Officer, Ronald L. Meeks (Arlington, VA 2004).

Nottenburg C, Pardey PG, Wright BD (2002) Accessing other peoples' technology for non-profit research. Aust J of Agric and Res Econ 48: 389-416

Pitcairn v. United States, 547 F.2d 1106 (Ct. Cl. 1976), *cert. denied,* 434 U.S. 1051 (1978)

Plant Genetic Systems N.V. v. DeKalb Genetics Corp. 315 F.3d 1335 (Fed. Cir. 2003)

Press E, Washburn J (2000) The kept university. The Atl Monthly 285: 39-54

Roa-Rodríguez C, Nottenburg C (2003) *Agrobacterium*-mediated transformation of plants, CAMBIA Patent Lens (http://www.patentlens.net)

Roche Products v. Bolar Pharmaceutical Co., 733 F.2d 858 (Fed. Cir. 1984), *cert. denied,* 469 U.S. 856 (1984)

Stern S, Murray FE (2005) Do Formal Intellectual Property Rights Hinder the Free Flow of Scientific Knowledge? An Empirical Test of the Anti-Commons Hypothesis. NBER Working Paper No. W11465. Available at SSRN:http: //sssrn.com/abstract=755701

Wolf S, Zilberman D (2001) *Institutional Innovation* in Agriculture. Natural Res Mgmt and Policy 19: 1-394 Kluwer Academic Publishers

Young JM, Keykendall LD, Martinez-Romero E, Kerr A, Sawada H (2001) A revision of Rhizobium Frank 1889, with an amended description of the genus, and the inclusion of all species of *Agrobacterium* Conn 1942 and *Allorhizobium undicola* de Lajudie et al. 1998 as new combinations: *Rhizobium radiobacter, R. rhizogenes, R. rubi, R. undicola and R. vitis*. Int J Syst Evol Microbiol 51: 89-103

Index

O

P

Q

R

T

U

V

Printed In The United States Of America